# Lecture Notes in Computer Science 6848

Commenced Publication in 1973
Founding and Former Series Editors:
Gerhard Goos, Juris Hartmanis, and Jan van Leeuwen

Axel Polleres   Claudia d'Amato
Marcelo Arenas   Siegfried Handschuh
Paula Kroner   Sascha Ossowski
Peter Patel-Schneider (Eds.)

# Reasoning Web

## Semantic Technologies
## for the Web of Data

7th International Summer School 2011
Galway, Ireland, August 23-27, 2011
Tutorial Lectures

 Springer

Volume Editors

Axel Polleres
DERI, National University of Ireland, Galway, Ireland/Siemens AG, Austria
E-mail: axel@polleres.net

Claudia d'Amato
University of Bari, Computer Science Department, Bari, Italy
E-mail: claudia.damato@di.uniba.it

Marcelo Arenas
Pontificia Universidad Católica de Chile, Santiago de Chile, Chile
E-mail: marenas@ing.puc.cl

Siegfried Handschuh
DERI, National University of Ireland, Galway, Ireland
E-mail: siegfried.handschuh@deri.org

Paula Kroner
Skytec AG, Oberhaching, Germany
E-mail: paula.kroner@skytecag.com

Sascha Ossowski
Universidad Rey Juan Carlos, Móstoles (Madrid), Spain
E-mail: sascha.ossowski@urjc.es

Peter Patel-Schneider
Bell Labs Research, Alcatel-Lucent, Murray Hill, NJ, USA
E-mail: pfps@research.bell-labs.com

ISSN 0302-9743        e-ISSN 1611-3349
ISBN 978-3-642-23031-8        ISBN 978-3-642-23032-5 (eBook)
DOI 10.1007/978-3-642-23032-5
Springer Heidelberg Dordrecht London New York

Library of Congress Control Number: 2011933830

CR Subject Classification (1998): H.4, H.3, I.2, H.5, C.2, D.2

LNCS Sublibrary: SL 3 – Information Systems and Application, incl. Internet/Web
and HCI

Typesetting: Camera-ready by author, data conversion by Scientific Publishing Services, Chennai, India
Printed on acid-free paper
Springer is part of Springer Science+Business Media (www.springer.com)

# Preface

This volume contains the lecture notes of the 7th Reasoning Web Summer School 2011 held during August 23-27, 2011 in Galway, Ireland.

The Reasoning Web Summer School has become a well-established event in the area of applications of reasoning techniques on the Web both targeting scientific discourse of established researchers and attracting young researchers to this emerging field. After the previous successful editions in Malta (2005), Lisbon (2006), Dresden (2007 and 2010), Venice (2008), and Bressanone-Brixen (2009), the 2011 edition moved to the west of Ireland, hosted by the Digital Enterprise Research Institute (DERI) at the National University of Ireland, Galway. By co-locating this year's summer school with the 5th International Conference on Web Reasoning and Rule Systems (RR2011)[1] we hope to have further promoted interaction between researchers, practitioners and students.

The 2011 school programme focused around the central topic of applications of reasoning for the emerging "Web of Data," with 12 exciting lectures. Along with the lecture slides which will be made available on the summer school's Website[2] the chapters in the present book provide educational material and references for further reading. The excellent overview articles provided by the lecturers did not only serve as accompanying material for the students of the summer school itself: we are happy to present a volume that also provides the general reader an entry point to various topics related to reasoning over Web data.

The first four chapters are devoted to foundational topics, providing introductory material to the Resource Description Framework (RDF) and Linked Data principles (Chap. 1), Description Logics as a foundation of the Web Ontology Language (OWL)(Chap. 2), the query language SPARQL and its usage together with OWL (Chap. 3), as well as database foundations relevant to efficient and scalable RDF processing (Chap. 4).

Based on these foundations, Chap. 5 presents approaches for scalable OWL reasoning over Linked Data, whereafter the following two chapters introduce rules and logic programming techniques relevant for Web reasoning (Chap. 6) and particularly the combination of rule-based reasoning with OWL (Chap. 7).

Chapter 8 takes a closer look at models for the Web of data. Chapter 9 discusses the important issue of trust management methodologies for the Web. The last two chapters continue on non-standard reasoning methods for the Semantic Web: Chap. 10 discusses the application of inductive reasoning methods for the Semantic Web which are also applied in software analysis, whereas Chap. 11

---

[1] Proceedings of this event are available in a separate volume also published in Springer's LNCS series.

[2] http://reasoningweb.org/2011/

focuses on an approach that combines logical and probabilistic reasoning for Web data integration.

The school also had an additional lecture on constraint programming and combinatorial optimisation.

We want to thank all the lecturers and authors of the present volume— without your effort and enthusiasm this school would not have been possible. We are further grateful to the members of the Programme Committee and the sub-reviewers who provided feedback to help the authors improve their articles and tailor them to the audience of the school. Last, but not least, we thank the local organisation team and all sponsors who supported this event financially: The European COST Action IC0801 "Agreement Technologies," the European FP7 project LOD2, the Artificial Intelligence Journal (AIJ), as well as our industry sponsors Alcatel-Lucent, IOS Press, Siemens AG, Skytec AG, Storm Technology, and the Office of Naval Research Global (ONRG).

June 2011

Axel Polleres
Claudia D'Amato
Marcelo Arenas
Siegfried Handschuh
Paula Kroner
Sascha Ossowski
Peter Patel-Schneider

# Organisation

## Programme Committee

| | |
|---|---|
| Marcelo Arenas | Pontificia Universidad Católica de Chile, Chile |
| Claudia D'Amato | University of Bari, Italy |
| Siegfried Handschuh | DERI, National University of Ireland, Galway |
| Paula Kroner | Skytec AG, Oberaching, Germany |
| Sascha Ossowski | University Rey Juan Carlos, Spain |
| Peter Patel-Schneider | Bell Labs Research, Alcatel-Lucent, USA |
| Axel Polleres (chair) | DERI, National University of Ireland, Galway / Siemens AG, Austria |

## Local Organisation

| | |
|---|---|
| Alessandra Mileo | DERI, National University of Ireland, Galway |
| Maria Smyth | DERI, National University of Ireland, Galway |

## Additional Reviewers

Stefan Bischof
Renaud Delbru
Alberto Fernandez Gil
Ramón Hermoso
Nuno Lopes
Alessandra Mileo
Vit Novacek
Alexandre Passant
Jodi Schneider

# Sponsors

## Platinum Sponsors

*The Artificial Intelligence Journal*

COST Action IC0801 "Agreement-Technologies"

The Office of Naval Research Global (ONRG)

## Gold Sponsors

Siemens AG Österreich

Storm Technology

## Other Sponsors

- The EU FP7 Integrated Project "LOD2"
- The Marie Curie IRSES Action "Net$^2$"
- Alcatel-Lucent
- IOS Press
- Skytec AG
- The Digital Enterprise Research Institute (DERI)

# Table of Contents

# Introduction to Linked Data
# and Its Lifecycle on the Web

Sören Auer, Jens Lehmann, and Axel-Cyrille Ngonga Ngomo

AKSW, Institut für Informatik, Universität Leipzig, Pf 100920, 04009 Leipzig
lastname@informatik.uni-leipzig.de
http://aksw.org

**Abstract.** With Linked Data, a very pragmatic approach towards achieving the vision of the Semantic Web has recently gained much traction. The term Linked Data refers to a set of best practices for publishing and interlinking structured data on the Web. While many standards, methods and technologies developed within by the Semantic Web community are applicable for Linked Data, there are also a number of specific characteristics of Linked Data, which have to be considered. In this article we introduce the main concepts of Linked Data. We present an overview of the Linked Data lifecycle and discuss individual approaches as well as the state-of-the-art with regard to extraction, authoring, linking, enrichment as well as evolution of Linked Data. We conclude the chapter with a discussion of issues, limitations and further research and development challenges of Linked Data.

## 1 Introduction

One of the biggest challenges in the area of intelligent information management is the exploitation of the Web as a platform for data and information integration as well as for search and querying. Just as we publish unstructured textual information on the Web as HTML pages and search such information by using keyword-based search engines, we will soon be able to easily publish structured information, reliably interlink this information with other data published on the Web and search the resulting data space by using more expressive querying beyond simple keyword searches.

The Linked Data paradigm has evolved as a powerful enabler for the transition of the current document-oriented Web into a Web of interlinked Data and, ultimately, into the Semantic Web. The term Linked Data here refers to a set of best practices for publishing and connecting structured data on the Web. These best practices have been adopted by an increasing number of data providers over the past three years, leading to the creation of a global data space that contains many billions of assertions - the Web of Linked Data (cf. Figure 1).

In this chapter we give an overview of recent development in the area of Linked Data management. The different stages in the linked data life-cycle are depicted in Figure 2.

Information represented in unstructured form or adhering to other structured or semi-structured representation formalisms must be mapped to the RDF data model (*Extraction*). Once there is a critical mass of RDF data, mechanisms have to be in place to store, index and query this RDF data efficiently (*Storage & Querying*). Users must have the opportunity to create new structured information or to correct and extend existing

A. Polleres et al. (Eds.): Reasoning Web 2011, LNCS 6848, pp. 1–75, 2011.

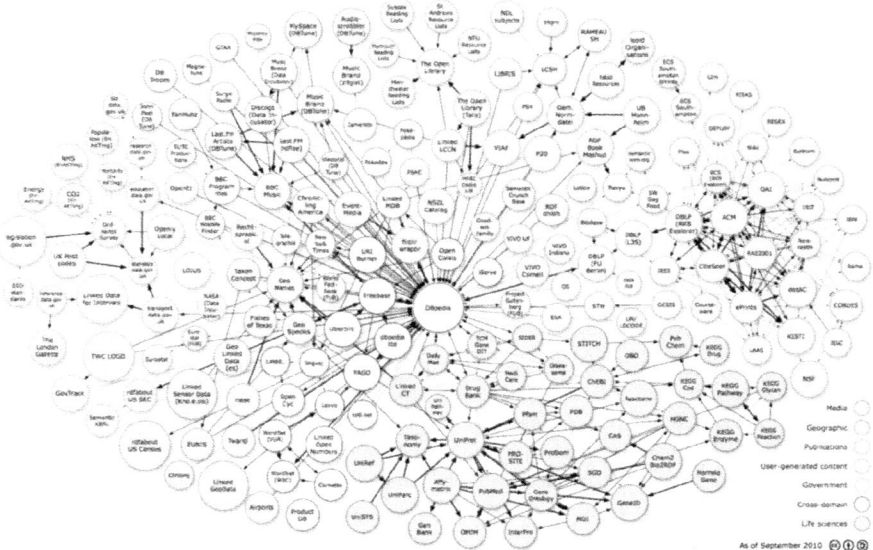

**Fig. 1.** Overview of some of the main Linked Data knowledge bases and their interlinks available on the Web. (This overview is published regularly at `http://lod-cloud.net` and generated from the Linked Data packages described at the dataset metadata repository `ckan.net`.).

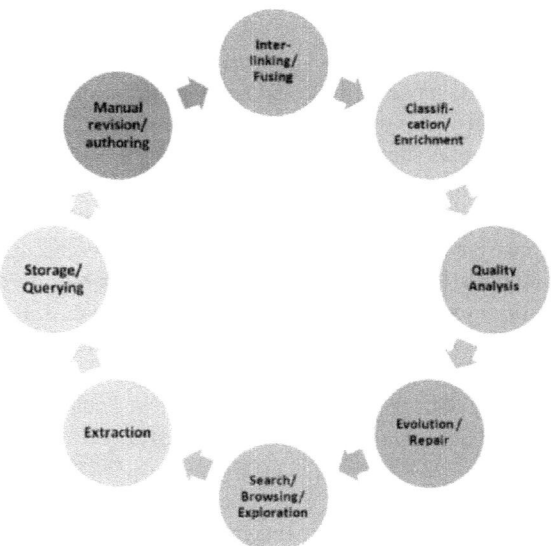

**Fig. 2.** The Linked Data life-cycle

ones (*Authoring*). If different data publishers provide information about the same or related entities, links between those different information assets have to be established

(*Linking*). Since Linked Data primarily comprises instance data we observe a lack of classification, structure and schema information. This deficiency can be tackled by approaches for enriching data with higher-level structures in order to be able to aggregate and query the data more efficiently (*Enrichment*). As with the Document Web, the Data Web contains a variety of information of different quality. Hence, it is important to devise strategies for assessing the quality of data published on the Data Web (*Quality Analysis*). Once problems are detected, strategies for repairing these problems and supporting the evolution of Linked Data are required (*Evolution & Repair*). Last but not least, users have to be empowered to browse, search and explore the structure information available on the Data Web in a fast and user friendly manner (*Search, Browsing & Exploration*).

These different stages of the linked data life-cycle do not exist in isolation or are passed in a strict sequence, but mutually fertilize themselves. Examples include the following:

- The detection of mappings on the schema level, will directly affect instance level matching and vice versa.
- Ontology schema mismatches between knowledge bases can be compensated for by learning which concepts of one are equivalent to which concepts of the other knowledge base.
- Feedback and input from end users can be taken as training input (i.e. as positive or negative examples) for machine learning techniques in order to perform inductive reasoning on larger knowledge bases, whose results can again be assessed by end users for iterative refinement.
- Semantically-enriched knowledge bases improve the detection of inconsistencies and modelling problems, which in turn results in benefits for interlinking, fusion, and classification.
- The querying performance of the RDF data management directly affects all other components and the nature of queries issued by the components affects the RDF data management.

As a result of such interdependence, we envision the Web of Linked Data to realize an improvement cycle for knowledge bases, in which an improvement of a knowledge base with regard to one aspect (e.g. a new alignment with another interlinking hub) triggers a number of possible further improvements (e.g. additional instance matches).

The use of Linked Data offers a number of significant benefits:

- *Uniformity*. All datasets published as Linked Data share a uniform data model, the RDF statement data model. With this data model all information is represented in facts expressed as triples consisting of a subject, predicate and object. The elements used in subject, predicate or object positions are mainly globally unique IRI/URI entity identifiers. At the object position also literals, i.e. typed data values can be used.
- *De-referencability*. URIs are not just used for identifying entities, but since they can be used in the same way as URLs they also enable locating and retrieving resources describing and representing these entities on the Web.

- *Coherence.* When an RDF triple contains URIs from different namespaces in subject and object position, this triple basically establishes a link between the entity identified by the subject (and described in the source dataset using namspace A) with the entity identified by the object (described in the target dataset using namespace B). Through the typed RDF links, data items are effectively interlinked.
- *Integrability.* Since all Linked Data source share the RDF data model, which is based on a single mechanism for representing information, it is very easy to attain a syntactic and simple semantic integration of different Linked Data sets. A higher level semantic integration can be achieved by employing schema and instance matching techniques and expressing found matches again as alignments of RDF vocabularies and ontologies in terms of additional triple facts.
- *Timeliness.* Publishing and updating Linked Data is relatively simple thus facilitating a timely availability. In addition, once a Linked Data source is updated it is straightforward to access and use the updated data source, since time consuming and error prune extraction, transformation and loading is not required.

The development of research approaches, standards, technology and tools for supporting the Linked Data lifecycle data is currently one of the main challenges. Developing adequate and pragmatic solutions to these problems can have a substantial impact on science, economy, culture and society in general. The publishing, integration and aggregation of statistical and economic data, for example, can help to obtain a more precise and timely picture of the state of our economy. In the domain of health care and life sciences making sense of the wealth of structured information already available on the Web can help to improve medical information systems and thus make health care more adequate and efficient. For the media and news industry, using structured background information from the Data Web for enriching and repurposing the quality content can facilitate the creation of new publishing products and services. Linked Data technologies can help to increase the flexibility, adaptability and efficiency of information management in organizations, be it companies, governments and public administrations or online communities. For end-users and society in general, the Data Web will help to obtain and integrate required information more efficiently and thus successfully manage the transition towards a knowledge-based economy and an information society.

*Intended audience.* This chapter is part of the lecture notes of the ReasoningWeb Summer School 2011. As such it is primarily intended for postgraduate (PhD or MSc) students, postdocs, and other young researchers investigating aspects related to the Data Web. However, the chapter might also be beneficial to senior researchers wishing to learn about Linked Data issues related to their own fields of research. Most parts of this chapter should be self-contained. However, we committed a detailed description of

**Table 1.** Juxtaposition of the concepts Linked Data, Linked Open Data and Open Data

| Representation \ degree of openness | Possibly closed | Open (cf. opendefinition.org) |
|---|---|---|
| *Structured data model* (i.e. XML, CSV, SQL etc.) | **Data** | **Open Data** |
| *RDF data model* (published as Linked Data) | **Linked Data** (LD) | **Linked Open Data** (LOD) |

SPQRQL, since SPARQL is already tackled by the lecture *Using SPARQL with RDFS and OWL entailment* by Birte Glimm later in this book.

*Structure of this chapter.* This chapter aims to explain the foundations of Linked Data and introducing the different aspects of the Linked Data lifecycle by highlighting a particular approach and providing references to related work and further reading. We start by briefly explaining the principles underlying the Linked Data paradigm in Section 2. The first aspect of the Linked Data lifecycle is the extraction of information from unstructured, semi-structured and structured sources and their representation according to the RDF data model (Section 3). We omit the storage and querying aspect, since this is already well covered by the *Using SPARQL with RDFS and OWL entailment* chapter of this book. We present the user friendly authoring and manual revision aspect of Linked Data with the example of Semantic Wikis in Section 4. The interlinking aspect is tackled in Section 5 and gives an overview on the LIMES framework. We describe how the instance data published and commonly found on the Data Web can be enriched with higher level structures in Section 6. We present an approach for the pattern-based evolution of Linked Data knowledge-bases in Section 7. Due to space limitations we omit a detailed discussion of the quality analysis as well as search, browsing and exploration aspects of the Linked Data lifecycle in this chapter.

## 2   The Linked Data Paradigm

In this section we introduce the basic principles of Linked Data. The section is partially based on the Section 2 from [48]. The term Linked Data refers to a set of best practices for publishing and interlinking structured data on the Web. These best practices were introduced by Tim Berners-Lee in his Web architecture note Linked Data[1] and have become known as the Linked Data principles. These principles are:

- Use URIs as names for things.
- Use HTTP URIs so that people can look up those names.
- When someone looks up a URI, provide useful information, using the standards (RDF, SPARQL).
- Include links to other URIs, so that they can discover more things.

The basic idea of Linked Data is to apply the general architecture of the World Wide Web [56] to the task of sharing structured data on global scale. The document Web is built on the idea of setting hyperlinks between Web documents that may reside on different Web servers. It is built on a small set of simple standards: Uniform Resource Identifiers (URIs) and their extension Internationalized Resource Identifiers (IRIs) as globally unique identification mechanism [19], the Hypertext Transfer Protocol (HTTP) as universal access mechanism [39], and the Hypertext Markup Language (HTML) as a widely used content format [52]. Linked Data builds directly on Web architecture and applies this architecture to the task of sharing data on global scale.

---

[1] http://www.w3.org/DesignIssues/LinkedData.html

## 2.1   Resource Identification with IRIs

To publish data on the Web, the data items in a domain of interest must first be identified. These are the things whose properties and relationships will be described in the
data, and may include Web documents as well as real-world entities and abstract concepts. As Linked Data builds directly on Web architecture [56], the Web architecture
term resource is used to refer to these *things of interest*, which are in turn identified by
HTTP URIs. Linked Data uses only HTTP URIs, avoiding other URI schemes such as
URNs [82] and DOIs[2]. HTTP URIs make good names for two reasons:

1. They provide a simple way to create globally unique names in a decentralized fashion, as every owner of a domain name or delegate of the domain name owner may
   create new URI references.
2. They serve not just as a name but also as a means of accessing information describing the identified entity.

## 2.2   De-referencability

Any HTTP URI should be de-referencable, meaning that HTTP clients can look up the
URI using the HTTP protocol and retrieve a description of the resource that is identified
by the URI. This applies to URIs that are used to identify classic HTML documents,
as well as URIs that are used in the Linked Data context to identify real-world objects
and abstract concepts. Descriptions of resources are embodied in the form of Web documents. Descriptions that are intended to be read by humans are often represented as
HTML. Descriptions that are intended for consumption by machines are represented
as RDF data. Where URIs identify real-world objects, it is essential to not confuse the
objects themselves with the Web documents that describe them. It is therefore common
practice to use different URIs to identify the real-world object and the document that
describes it, in order to be unambiguous. This practice allows separate statements to be
made about an object and about a document that describes that object. For example, the
creation year of a painting may be rather different to the creation year of an article about
this painting. Being able to distinguish the two through use of different URIs is critical
to the consistency of the Web of Data.

  The Web is intended to be an information space that may be used by humans as
well as by machines. Both should be able to retrieve representations of resources in a
form that meets their needs, such as HTML for humans and RDF for machines. This
can be achieved using an HTTP mechanism called content negotiation [39]. The basic
idea of content negotiation is that HTTP clients send HTTP headers with each request
to indicate what kinds of documents they prefer. Servers can inspect these headers and
select an appropriate response. If the headers indicate that the client prefers HTML then
the server will respond by sending an HTML document If the client prefers RDF, then
the server will send the client an RDF document.

  There are two different strategies to make URIs that identify real-world objects de-
referencable [105]. Both strategies ensure that objects and the documents that describe
them are not confused and that humans as well as machines can retrieve appropriate
representations.

---

[2] http://www.doi.org/hb.html

*303 URIs.* Real-world objects can not be transmitted over the wire using the HTTP protocol. Thus, it is also not possible to directly de-reference URIs that identify real-world objects. Therefore, in the 303 URI strategy, instead of sending the object itself over the network, the server responds to the client with the HTTP response code 303 See Other and the URI of a Web document which describes the real-world object. This is called a *303 redirect.* In a second step, the client de-references this new URI and gets a Web document describing the real-world object.

*Hash URIs.* A widespread criticism of the 303 URI strategy is that it requires two HTTP requests to retrieve a single description of a real-world object. One option for avoiding these two requests is provided by the hash URI strategy. The hash URI strategy builds on the characteristic that URIs may contain a special part that is separated from the base part of the URI by a hash symbol (#). This special part is called the fragment identifier. When a client wants to retrieve a hash URI the HTTP protocol requires the fragment part to be stripped off before requesting the URI from the server. This means a URI that includes a hash cannot be retrieved directly, and therefore does not necessarily identify a Web document. This enables such URIs to be used to identify real-world objects and abstract concepts, without creating ambiguity [105].

Both approaches have their advantages and disadvantages. Section 4.4. of the W3C Interest Group Note Cool URIs for the Semantic Web compares the two approaches [105]: Hash URIs have the advantage of reducing the number of necessary HTTP round-trips, which in turn reduces access latency. The downside of the hash URI approach is that the descriptions of all resources that share the same non-fragment URI part are always returned to the client together, irrespective of whether the client is interested in only one URI or all. If these descriptions consist of a large number of triples, the hash URI approach can lead to large amounts of data being unnecessarily transmitted to the client. 303 URIs, on the other hand, are very flexible because the redirection target can be configured separately for each resource. There could be one describing document for each resource, or one large document for all of them, or any combination in between. It is also possible to change the policy later on.

## 2.3   RDF Data Model

The RDF data model [1] represents information as sets of statements, which can be visualized as node-and-arc-labeled directed graphs. The data model is designed for the integrated representation of information that originates from multiple sources, is heterogeneously structured, and is represented using different schemata. RDF aims at being employed as a *lingua franca*, capable of moderating between other data models that are used on the Web.

In RDF, information is represented in statements, called RDF triples. The three parts of each triple are called its subject, predicate, and object. A triple mimics the basic structure of a simple sentence, such as for example:

```
Burkhard Jung     is the mayor of     Leipzig
   (subject)          (predicate)      (object)
```

The following is the formal definition of RDF triples as it can be found in the W3C RDF standard [1].

**Definition 1 (RDF Triple).** *Assume there are pairwise disjoint infinite sets $I$, $B$, and $L$ (IRIs, blank nodes, and RDF literals, respectively). A triple $(v_1, v_2, v_3) \in (I \cup B) \times I \times (I \cup B \cup L)$ is called an RDF triple. In this tuple, $v_1$ is the subject, $v_2$ the predicate and $v_3$ the object. We denote the union $I \cup B \cup L$ as by $T$ called RDF terms.*

The main idea is to use IRIs as identifiers for entities in the subject, predicate and object positions in a triple. Data values can be represented in the object position as literals. Furthermore, the RDF data model also allows in subject and object positions the use of identifiers for unnamed entities (called blank nodes), which are not globally unique and can thus only be referenced locally. However, the use of blank nodes is discouraged in the Linked Data context as we discuss below. Our example fact sentence about Leipzig's mayor would now look as follows:

```
<http://leipzig.de/id>
              <http://example.org/p/hasMayor>
                              <http://Burkhard-Jung.de/id> .
    (subject)          (predicate)              (object)
```

This example shows, that IRIs used within a triple can originate from different namespaces thus effectively facilitating the mixing and mashing of different RDF vocabularies and entities from different Linked Data knowledge bases. A triple having identifiers from different knowledge bases at subject and object position can be also viewed as an typed link between the entities identified by subject and object. The predicate then identifies the type of link. If we combine different triples we obtain an RDF graph.

**Definition 2 (RDF Graph).** *A finite set of RDF triples is called RDF graph. The RDF graph itself represents an resource, which is located at a certain location on the Web and thus has an associated IRI, the graph IRI.*

An example of an RDF graph is depicted in Figure 3. Each unique subject or object contained in the graph is visualized as a node (i.e. oval for resources and rectangle for literals). Predicates are visualized as labeled arcs connecting the respective nodes. There are a number of synonyms being used for RDF graphs, all meaning the essentially the same but stressing different aspects of an RDF graph, such as *RDF document* (file perspective), *knowledge base* (collection of facts), *vocabulary* (shared terminology), *ontology* (shared logical conceptualization).

*Problematic RDF features in the Linked Data Context.* Besides the features mentioned above, the RDF Recommendation [1] also specifies some other features. In order to make it easier for clients to consume data only the subset of the RDF data model described above should be used. In particular, the following features are problematic when publishing RDF as Linked Data:

 – *RDF reification* should be avoided if possible, as reified statements are rather cumbersome to query with the SPARQL query language. In many cases using reification to publish metadata about individual RDF statements can be avoided by attaching the respective metadata to the RDF document containing the relevant triples.

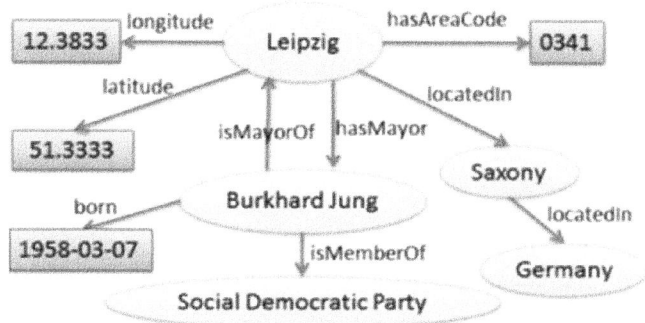

**Fig. 3.** Example RDF graph containing 9 triples describing the city of Leipzig and its mayor

- *RDF collections* and *RDF containers* are also problematic if the data needs to be queried with SPARQL. Therefore, in cases where the relative ordering of items in a set is not significant, the use of multiple triples with the same predicate is recommended.
- The scope of *blank nodes* is limited to the document in which they appear, meaning it is not possible to create links to them from external documents. In addition, it is more difficult to merge data from different sources when blank nodes are used, as there is no URI to serve as a common key. Therefore, all resources in a data set should be named using IRI references.

## 2.4  RDF Serializations

The initial official W3C RDF standard [1] comprised a serialization of the RDF data model in XML called *RDF/XML*. Its rationale was to integrate RDF with the existing XML standard, so it could be used smoothly in conjunction with the existing XML technology landscape. Unfortunately, RDF/XML turned out to be rather difficult to understand for the majority of potential users, since it requires to be familiar with two data models (i.e. the tree-oriented XML data model as well as the statement oriented RDF datamodel) and interactions between them, since RDF statements are represented in XML. As a consequence, with N-Triples, Turtle and N3 a family of alternative text-based RDF serializations was developed, whose members have the same origin, but balance different between readability for humans and machines. Later in 2009, RDFa (RDF Annotations, [2]) was standardized by the W3C in order to simplify the integration of HTML and RDF and to allow the joint representation of structured and unstructured content within a single source HTML document. Another RDF serialization, which is particularly beneficial in the context of JavaScript web applications and mashups is the serialization of RDF in JSON. In the sequel we present each of these RDF serializations in some more detail. Figure 5 presents an example serialized in the most popular serializations.

*N-Triples.*  This serialization format was developed specifically for RDF graphs. The goal was to create a serialization format which is very simple. N-Triples are easy to parse and generate by software. They are a subset of *Notation 3* and *Turtle* but lack, for

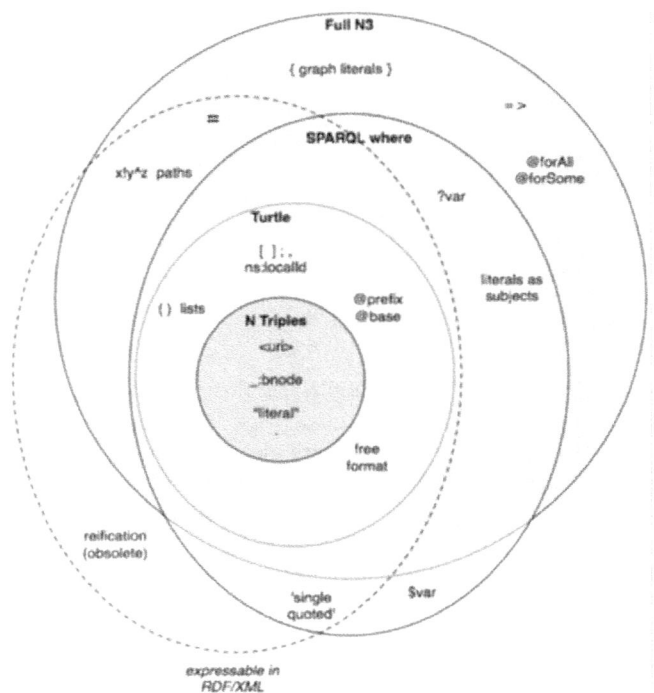

**Fig. 4.** Various textual RDF serializations as subsets of N3 (from [18])

example, shortcuts such as CURIEs. This makes them less readable and more difficult to create manually. Another disadvantage is that N-triples use only the 7-bit US-ASCII character encoding instead of UTF-8.

*Turtle.* Turtle (Terse RDF Triple Language) is a subset of, and compatible with, Notation 3 and a superset of the minimal N-Triples format (cf. Figure 4). The goal was to use the essential parts of Notation 3 for the serialization of RDF models and omit everything else. Turtle became part of the SPARQL query language for expressing graph patterns. Turtle, just like Notation 3, is human-readable, and can handle the "%" character in URIs (required for encoding special characters) as well as IRIs due to its UTF-8 encoding.

*Notation 3.* N3 (Notation 3) was devised by Tim Berners-Lee and developed for the purpose of serializing RDF. The main aim was to create a very human-readable serialization. Hence, an RDF model serialized in N3 is much more compact than the same model in RDF/XML but still allows a great deal of expressiveness even going beyond the RDF data model in some aspects. Since, the encoding for N3 files is UTF-8 the use of IRIs does not pose a problem.

*RDF/XML.* The RDF/XML syntax [80] is standardized by the W3C and is widely used to publish Linked Data on the Web. However, the syntax is also viewed as difficult for humans to read and write, and therefore consideration should be given to using

**N-Triples**

```
1  <http://dbpedia.org/resource/Leipzig> <http://dbpedia.org/property/hasMayor>
2      <http://dbpedia.org/resource/Burkhard_Jung> .
3  <http://dbpedia.org/resource/Leipzig> <http://www.w3.org/2000/01/rdf-schema#label>
4      "Leipzig"@de .
5  <http://dbpedia.org/resource/Leipzig> <http://www.w3.org/2003/01/geo/wgs84_pos#lat>
6      "51.333332"^^<http://www.w3.org/2001/XMLSchema#float> .
```

**Turtle**

```
1  @prefix rdf: <http://www.w3.org/1999/02/22-rdf-syntax-ns#> .
2  @prefix rdfs="http://www.w3.org/2000/01/rdf-schema#> .
3  @prefix dbp="http://dbpedia.org/resource/> .
4  @prefix dbpp="http://dbpedia.org/property/> .
5  @prefix geo="http://www.w3.org/2003/01/geo/wgs84_pos#> .
6
7  dbp:Leipzig  dbpp:hasMayor  dbp:Burkhard_Jung ,
8               rdfs:label      "Leipzig"@de ,
9               geo:lat         "51.333332"^^xsd:float ,
```

**RDF/XML**

```
1  <?xml version="1.0"?>
2  <rdf:RDF xmlns:rdf="http://www.w3.org/1999/02/22-rdf-syntax-ns#"
3           xmlns:rdfs="http://www.w3.org/2000/01/rdf-schema#"
4           xmlns:dbpp="http://dbpedia.org/property/"
5           xmlns:geo="http://www.w3.org/2003/01/geo/wgs84_pos#">
6    <rdf:Description rdf:about="http://dbpedia.org/resource/Leipzig">
7      <property:hasMayor rdf:resource="http://dbpedia.org/resource/Burkhard_Jung" />
8      <rdfs:label xml:lang="de">Leipzig</rdfs:label>
9      <geo:lat rdf:datatype="http://www.w3.org/2001/XMLSchema#float">51.3333</geo:lat>
10   </rdf:Description>
11  </rdf:RDF>
```

**RDFa**

```
1  <?xml version="1.0" encoding="UTF-8"?>
2  <!DOCTYPE html PUBLIC "-//W3C//DTD XHTML+RDFa 1.0//EN"
3      "http://www.w3.org/MarkUp/DTD/xhtml-rdfa-1.dtd">
4  <html version="XHTML+RDFa 1.0" xml:lang="en" xmlns="http://www.w3.org/1999/xhtml"
5        xmlns:rdf="http://www.w3.org/1999/02/22-rdf-syntax-ns#"
6        xmlns:rdfs="http://www.w3.org/2000/01/rdf-schema#"
7        xmlns:dbpp="http://dbpedia.org/property/"
8        xmlns:geo="http://www.w3.org/2003/01/geo/wgs84_pos#">
9    <head><title>Leipzig</title></head>
10   <body about="http://dbpedia.org/resource/Leipzig">
11     <h1 property="rdfs:label" xml:lang="de">Leipzig</h1>
12     <p>Leipzig is a city in Germany. Leipzig's mayor is
13      <a href="Burkhard_Jung" rel="dbpp:hasMayor">Burkhard Jung</a>. It is located
14      at latitude <span property="geo:lat" datatype="xsd:float">51.3333</span>.</p>
15   </body>
16  </html>
```

**RDF/JSON**

```
1  {
2    "http://dbpedia.org/resource/Leipzig" : {
3        "http://dbpedia.org/property/hasMayor":
4          [ { "type":"uri",  "value":"http://dbpedia.org/resource/Burkhard_Jung" } ],
5        "http://www.w3.org/2000/01/rdf-schema#label":
6          [ { "type":"literal", "value":"Leipzig", "lang":"en" } ] ,
7        "http://www.w3.org/2003/01/geo/wgs84_pos#lat":
8          [ { "type":"literal", "value":"51.3333",
9              "datatype":"http://www.w3.org/2001/XMLSchema#float" } ]
10     }
11  }
```

**Fig. 5.** Different RDF serializations of three triples from Figure 3

other serializations in data management and curation workflows that involve human intervention, and to the provision of alternative serializations for consumers who may wish to eyeball the data. The MIME type that should be used for RDF/XML within HTTP content negotiation is `application/rdf+xml`.

*RDFa.* RDF in Attributes (RDFa, [2]) was developed for embedding RDF into XHTML pages. Since it is an extension to the XML based XHTML, UTF-8 and UTF-16 are used for encoding. The "%" character for URIs in triples can be used because RDFa tags are not used for a part of a RDF statement. Thus IRIs are usable, too. Because RDFa is embedded in XHTML, the overhead is higher compared to other serialization technologies and also reduces the readability.

*RDF/JSON.* JavaScript Object Notation (JSON) was developed for easy data interchange between applications. JSON, although carrying JavaScript in its name and being a subset of JavaScript, meanwhile became a language independent format which can be used for exchanging all kinds of data structures and is widely supported in different programming languages. Compared to XML, RDF/JSON requires less overhead with regard to parsing and serializing. There is a non-standardized specification[3] for RDF serialization in JSON. Text in JSON and, thus, also RDF resource identifiers are encoded in Unicode and hence can contain IRIs.

# 3   Extraction

Information represented in unstructured form or adhering to a different structured representation formalism must be mapped to the RDF data model in order to be used within the Linked Data life-cycle. In this section we give an overview on some relevant approaches for extracting RDF from unstructured and structured sources.

## 3.1   From Unstructured Sources

The extraction of structured information from unstructured data sources (especially text) has been a central pillar of *natural language processing* (NLP) and *Information Extraction* (IE) for several decades. With respect to the extraction of RDF data from unstructured data, three sub-disciplines of NLP play a central role: *Named Entity Recognition* (NER) for the extraction of entity labels from text, *Keyword/Keyphrase Extraction* (KE) for the recognition of central topics and *Relationship Extraction* (RE, also called relation mining) for mining the properties which link the entities and keywords described in the data source. A noticeable additional task during the migration of these techniques to Linked Data is the extraction of suitable IRIs for the discovered entities and relations, a requirement that was not needed before. In this section, we give a short overview of approaches that implement the required NLP functionality. Then we present a framework that applies machine learning to boost the quality of the RDF extraction from unstructured data by merging the results of NLP tools.

---

[3] `http://n2.talis.com/wiki/RDF_JSON_Specification`

**Named Entity Recognition.** The goal of NER is to discover instances of a predefined classes of entities (e.g., persons, locations, organizations) in text. NER tools and frameworks implement a broad spectrum of approaches, which can be subdivided into three main categories: dictionary-based, rule-based, and machine-learning approaches. The first systems for NER implemented dictionary-based approaches, which relied on a list of NEs and tried to identify these in text [120,5]. Following work that showed that these approaches did not perform well for NER tasks such as recognizing proper names [104], rule-based approaches were introduced. These approaches rely on hand-crafted rules [27,112] to recognize NEs. Most rule-based approaches combine dictionary and rule-based algorithms to extend the list of known entities. Nowadays, handcrafted rules for recognizing NEs are usually implemented when no training examples are available for the domain or language to process [84].

When training examples are available, the methods of choice are borrowed from supervised machine learning. Approaches such as Hidden Markov Models [127], Maximum Entropy Models [30] and Conditional Random Fields [40] have been applied to the NER task. Due to scarcity of large training corpora as necessitated by machine learning approaches, semi-supervised [93,83] and unsupervised machine learning approaches [85,36] have also been used for extracting NER from text. [83] gives an exhaustive overview of approaches for NER.

**Keyphrase Extraction.** Keyphrases/Keywords are multi-word units (MWUs) which capture the main topics of a document. The automatic detection of such MWUs has been an important task of NLP for decades but due to the very ambiguous definition of what an appropriate keyword should be, current approaches to the extraction of keyphrases still display low F-scores [59]. From the point of view of the Semantic Web, the extraction of keyphrases is a very similar task to that of finding tags for a given document. Several categories of approaches have been adapted to enable KE, of which some originate from research areas such as summarization and information retrieval (IR). Still, according to [58], the majority of the approaches to KE implement combinations of statistical, rule-based or heuristic methods [43,87] on mostly document [79], keyphrase [115] or term cohesion features [92]. [59] gives a overview of current tools for KE.

**Relation Extraction.** The extraction of relations from unstructured data builds upon work for NER and KE to determine the entities between which relations might exist. Most tools for RE rely on pattern-based approaches. Some early work on pattern extraction relied on supervised machine learning [46]. Yet, such approaches demanded large amount of training data, making them difficult to adapt to new relations. The subsequent generation of approaches to RE aimed at bootstrapping patterns based on a small number of input patterns and instances. For example, [25] presents the Dual Iterative Pattern Relation Expansion (DIPRE) and applies it to the detection of relations between authors and titles of books. This approach relies on a small set of seed patterns to maximize the precision of the patterns for a given relation while minimizing their error rate of the same patterns. Snowball [3] extends DIPRE by a new approach to the generation of seed tuples. Newer approaches aim to either collect redundancy information from the whole Web [91] or Wikipedia [121,126] in an unsupervised manner or to use linguistic analysis [47,86] to harvest generic patterns for relations.

**URI Disambiguation.** One important problem for the integration of NER tools for Linked Data is the retrieval of IRIs for the entities to be manipulated. In most cases, the URIs can be extracted from generic knowledge bases such as DBpedia by comparing the label found in the input data with the `rdfs:label` or `dc:title` of the entities found in the knowledge base. Furthermore, information such as the type of NEs can be used to filter the retrieved IRIs via a comparison of the `rdfs:label` of the `rdf:type` of the URIs with the name of class of the NEs. Still in many cases (e.g., Leipzig, Paris), several entities might bear the same label.

## 3.2   The FOX Framework

Several frameworks have been developed to implement the functionality above for the Data Web including OpenCalais[4] and Alchemy[5]. Yet, these tools rely mostly on one approach to perform the different tasks at hand. In this section, we present the FOX (Federated knOwledge eXtraction) framework[6], which makes use of the diversity of the algorithms available for NER, KE and RE to generate high-quality RDF.

The architecture of FOX consists of *three main layers* as shown in Figure 6. The *machine learning* layer implements interfaces for accommodating ensemble learning techniques such as simple veto algorithms but also neural networks. It consists of *two main modules*. The *training module* allows to load training data so as to enable FOX to learn the best combination of tools and categories for achieving superior recall and precision on the input training data. Depending on the training algorithm used, the user can choose to tune the system for either precision or recall. When using neural networks for example, the user can decide to apply a higher threshold for the output neurons, thus improving the precision but potentially limiting the recall. The *prediction module* allows to run FOX by loading the result of a training session and processing the input data according to the tool-category combination learned during the training phase. Note that the same learning approach can by applied to NER, KE, RE and URI lookup as they call all be modelled as classification tasks.

The second layer of FOX is the *controller*, which coordinates the access to the modules that carry out the language processing. The controller is aware of each of the modules in its backend and carries out the initialisation of these modules once FOX is started. Furthermore, it collects the results from the backend modules and invokes the results of a training instance to merge the results of these tools.

The final layer of FOX is the *tool layer*, wherein all NLP tools and services integrated in FOX can be found. It is important to notice that the tools per se are not trained during the learning phase of FOX. Rather, we learn of the models already loaded in the tools to allow for the best prediction of named entities in a given domain.

The ensemble learning implemented by FOX was evaluated in the task of NER by integrating three NER tools (Stanford NER, Illinois NER and a commercial tool) and shown to lead to an improvement of more than 13% in F-Score (see Figure 7) when combining three tools, therewith even outperforming commercial systems.

---

[4] http://www.opencalais.com
[5] http://www.alchemyapi.com
[6] http://aksw.org/projects/fox

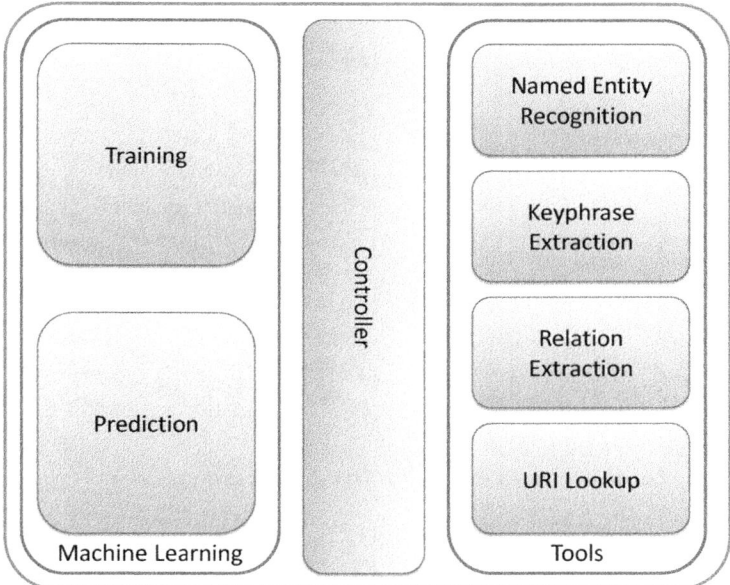

**Fig. 6.** FOX Architecture

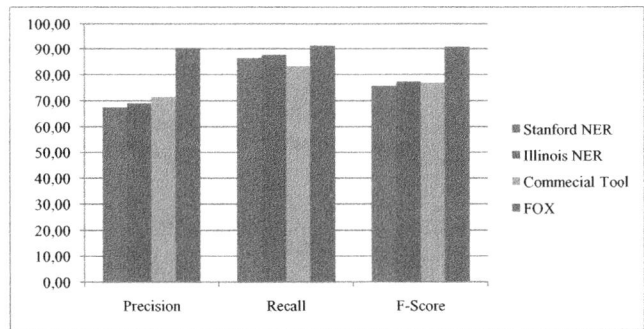

**Fig. 7.** Comparison of precision, recall and F-score of the best runs of FOX and its components on NER

### 3.3   From Structured Sources

Structured knowledge, e.g. relational databases and XML, is the backbone of many (web) applications. Extracting or converting this knowledge to RDF is a long-standing research goal in the Semantic Web community. A conversion to RDF allows to integrate the data with other sources and perform queries over it. In this lecture, we focus on the conversion of relational databases to RDF (see Figure 8). In the first part, we summarize material from a recent relational database to RDF (RDB2RDF) project report. After that, we describe the mapping language R2RML, which is a language for expressing database to RDF conversion mappings.

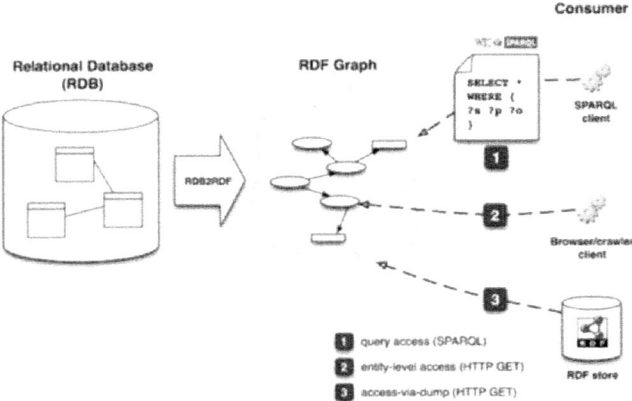

**Fig. 8.** Illustration of RDB to RDF conversion
Source: http://www.w3.org/2001/sw/rdb2rdf/use-cases/

| Approach | Automation (a) | Domain or database semantics-driven (b) | Access paradigm (c) | Mapping language (d) | Domain reliance (e) |
|---|---|---|---|---|---|
| Dartgrid [17] | Manual | Domain | SPARQL | Visual Tool | dependent |
| Hu et al. [11] | Auto | Both | ETL | intern | dependent |
| Tirmizi et al. [16] | Auto | DB | ETL | FOL | general |
| Li et al. [12] | Semi | DB | ETL | n/a | general |
| DB2OWL [10] | Semi | DB | SPARQL | R2O | general/dependent |
| RDBToOnto [6] | Semi | DB+M | ETL | Visual Tool | general |
| Sahoo et al. [15] | Manual | Domain | ETL | XSLT | dependent |
| R2O [13] | Manual | DB+M | SPARQL | R2O | dependent |
| D2RQ [4] | Auto | DB+M | LD, SPARQL | D2RQ | general |
| Virtuoso RDF View [5, 9] | Semi | DB+M | SPARQL | own | general |
| Triplify | Manual | Domain | LD | SQL | general |

Table 4: An integrated overview of mapping approaches. Criteria for classification were merged, some removed, fields were completed, when missing. DB+M means that the semi-automatic approach can later be customized manually

**Fig. 9.** Table comparing relevant approaches from [7]

**Triplify and RDB2RDF Survey report.** The table displayed in Figure 9 is taken from the Triplify WWW paper [7]. The survey report [103] furthermore contained a chart(see Figure 10) showing the reference framework for classifying the approaches and an extensive table classifying the approaches (see Figure 11).

The following criteria can be extracted:

*Automation Degree.* Degree of mapping creation automation.
**Values:** Manual, Automatic, Semi-Automatic.

*Domain or Database Semantics Driven.* Some approaches are tailored to model a domain, sometimes with the help of existing ontologies, while others attempt to extract domain information primarily from the given database schema with few other resources used (domain or database semantics-driven). The latter often results in a table-to-class, column-to-predicate mapping.Some approaches also use a (semi) automatic approach based on the database, but allow manual customization to model domain semantics.

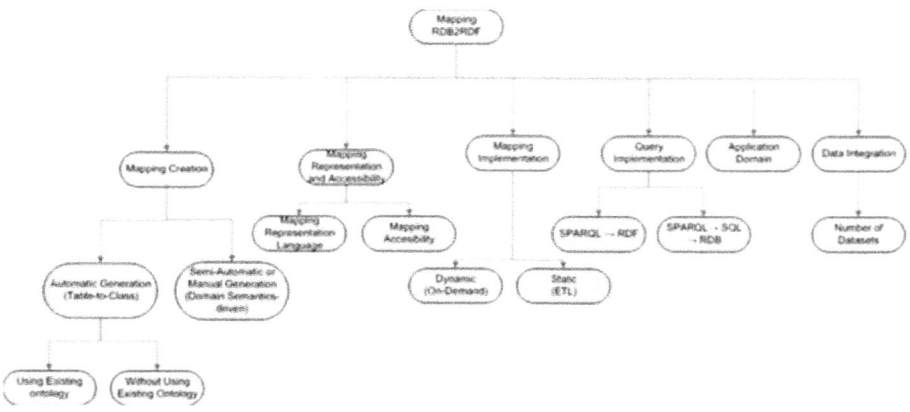

**Fig. 10.** Reference framework by [103]

**Values:** Domain, DB (database), DB+M (database and later manual customisation), Both (Domain and DB).

*Access Paradigm.* Resulting access paradigm (ETL [extract transform load], Linked Data, SPARQL access). Note that the access paradigm also determines whether the resulting RDF model updates automatically. ETL means a one time conversion, while Linked Data and SPARQL always process queries versus the original database.
**Values:** SPARQL, ETL, LD.

*Mapping Language.* The used mapping language as an important factor for reusability and initial learning cost.
**Values:** Visual Tool, intern (internal self-designed language), FOL, n/a (no information available), R2O, XSLT, D2RQ, proprietary, SQL.

*Domain reliance.* Domain reliance (general or domain-dependent): requiring a pre-defined ontology is a clear indicator of domain dependency.
**Values:** Dependent, General.

*Type.* Although not used in the table the paper discusses four different classes:
**Values:** Alignment, Database Mining, Integration, Languages/Servers.

**R2RML - RDB to RDF Mapping Language.** The R2RML working draft[7] specifies an RDF notation for mapping relational tables, views or queries into RDF. The primary area of applicability of this is extracting RDF from relational databases, but in special cases R2RML could lend itself to on-the-fly translation of SPARQL into SQL or to converting RDF data to a relational form. The latter application is not the primary intended use of R2RML but may be desirable for importing linked data into relational stores. This is possible if the constituent mappings and underlying SQL objects constitute updateable views in the SQL sense.

---

[7] http://www.w3.org/TR/r2rml/

| PROJECTS | MAPPING CREATION | MAPPING REPRESENTATION AND ACCESSIBILITY | | MAPPING IMPLEMENTATION | QUERY IMPLEMENTATION | APPLICATION DOMAIN | DATA INTEGRATION | |
|---|---|---|---|---|---|---|---|---|
| | Automatic (Table-to-Class) or Manual/Semi-Automatic (Domain Semantics-driven) | Representation Language | Mapping Access | Static (ETL) or Dynamic | SPARQL → RDF or SPARQL→SQL→RDB | | Yes/No | Number of Datasets |
| 1 Hu et al, 2007 | Automatic (with use of existing ontology) | First Order Logic formulae or Horn Clauses | Files | None Specified | None Specified | Generic | Enables (through contextual mappings) | Potentially Multiple |
| 2 Kashyap et al, 2007 | Manual/Semi-Automatic (Domain Semantics-driven) | Mediator Framework Classes | Mapping mediator | Dynamic | SPARQL→SQL→RDB | Life Sciences | Enables | Potentially Multiple |
| 3 DB2OWL (Cullot et al, 2007) | Automatic (Table to Class) | R2O language | R2O mapping document | | SPARQL→SQL→RDB | Generic | Enables | Potentially Multiple |
| 4 Tirmizi et al 2008 | Automatic (Table to Class, SQL-DDL to RDF) | First Order Logic | None specified | Static | None Specified | Generic | No | None |
| 5 SOAM (Li et al, 2005) | Automatic (Table to Class) with user input | Logic Rules | Implemented as part of system | Static | Potentially SPARQL (on generated populated ontology) | Generic (Case Study Economics) | No | None |
| 6 Sahoo et al, 2008 | Manual/Semi-Automatic (Domain Semantics-driven) | XPath expressions | XSLT document | Static | SPARQL | Life Sciences | Yes | Test included five (Gene, Biological Pathway) |
| 7 Byrne, 2008 | Manual/Semi-Automatic (Domain Semantics-driven) | SKOS vocabulary | RDF document | Static | SPARQL | Cultural Heritage | No | None |
| 8 Green et al, 2008 | Manual/Semi-Automatic (Domain Semantics-driven) | D2RQ language | D2RQ mapping file | Dynamic | SPARQL→SQL→RDB | Ordnance Survey | Yes | Multiple |
| 9 Virtuoso RDF View (Blakeley, 2007) | Both (user-specified) | SPASQL-based Meta Schema Language | Quad Storage | Both | Both | Generic | Enables | Potentially Multiple |
| 10 D2RQ (Bizer et al, 2007) | Both (user-specified) | D2RQ language | D2RQ mapping file | Both | Both | Generic | Enables | Potentially Multiple |
| 11 R2O (Barrasa et al, 2006) | Both (user-specified) | R2O language | R2O mapping document | Both | Both | Generic | Enables | Potentially Multiple |
| 12 Dartgrid (Wu et al, 2006) | Automatic (Table to Class) | XML File | Visualized Mapping tool | Dynamic | SPARQL→SQL→RDB (Provide search and query interface) | Life Science (Traditional Chinese Medicine, TCM) | Yes | Test included databases for herb, compound formulas, disease, drug, TCM treatment |
| 13 RDBtoOnto (Cerbah, 2008) | Automatic (Table to Class, allows user intervention) | Constraint rules | Not explicitly stored | Static | Potentially SPARQL (on generated populated ontology) | Generic | No | None |
| 14 Asio Tools | Automatic (Table to | OWL Full based language | File based | Both | SPARQL→SQL→RDB | Generic | Enables | Potentially Multiple |

**Fig. 11.** Comparison of approaches from [103]

Data integration is often mentioned as a motivating use case for the adoption of RDF. This integration will very often be between relational databases which have logical entities in common, each with its local schema and identifiers.Thus, we expect to see relational to RDF mapping use cases involving the possibility of a triple coming from multiple sources. This does not present any problem if RDF is being extracted but does lead to complications if SPARQL queries are mapped into SQL. In specific, one will end up with potentially very long queries consisting of joins of unions. Most of the joins between terms of the unions will often be provably empty and can thus be optimized away. This capability however requires the mapping language to be able to express metadata about mappings, i.e. that IRIs coming from one place are always disjoint from IRIs coming from another place. Without such metadata optimizing SPARQL to SQL translation is not possible, which will significantly limit the possibility of querying collections of SQL databases through a SPARQL end point without ETL-ing the mapped RDF into an RDF store.

RDF is emerging as a format for interoperable data publishing. This does not entail that RDF were preferable as a data warehousing model. Besides, for large warehouses, RDF is far from cost competitive with relational technology, even though LOD2 expects to narrow this gap. Thus it follows that on the fly mapping of SPARQL to SQL will be important. Regardless of the relative cost or performance of relational or RDF technology, it is not a feasible proposition to convert relational warehouses to RDF in general, rather existing investments must be protected and reused. Due to these reasons, R2RML will have to evolve in the direction of facilitating querying of federated relational resources.

## 4    Authoring with Semantic Wikis

Semantic Wikis are an extension to conventional, text-based Wikis. While in conventional Wikis pages are stored as blocks of text using a special Wiki markup for structuring the display of the text and adding links to other pages, semantic Wikis aim at adding rich structure to the information itself. To this end, two initially orthogonal approaches have been used: a) extending the markup language to allow semantic annotations and links with meaning or b) building the Wiki software directly with structured information in mind. Nowadays, both approaches have somewhat converged, for instance Semantic MediaWiki [61] also provides forms for entering structured data (see Figure 12). Characteristics of both approaches are summarized in Table 2 for the two prototypical representatives of both approaches, i.e. Semantic MediaWiki and OntoWiki.

*Extending Wikis with Semantic Markup.* The benefit of a Wiki system comes from the amount of interlinking between Wiki pages. Those links clearly state a relationship

**Table 2.** Conceptual differences between Semantic MediaWiki and OntoWiki

|                   | Semantic MediaWiki | OntoWiki  |
|-------------------|--------------------|-----------|
| *Managed entities* | Articles           | Resources |
| *Editing*          | Wiki markup        | Forms     |
| *Atomic element*   | Text blob          | Statement |

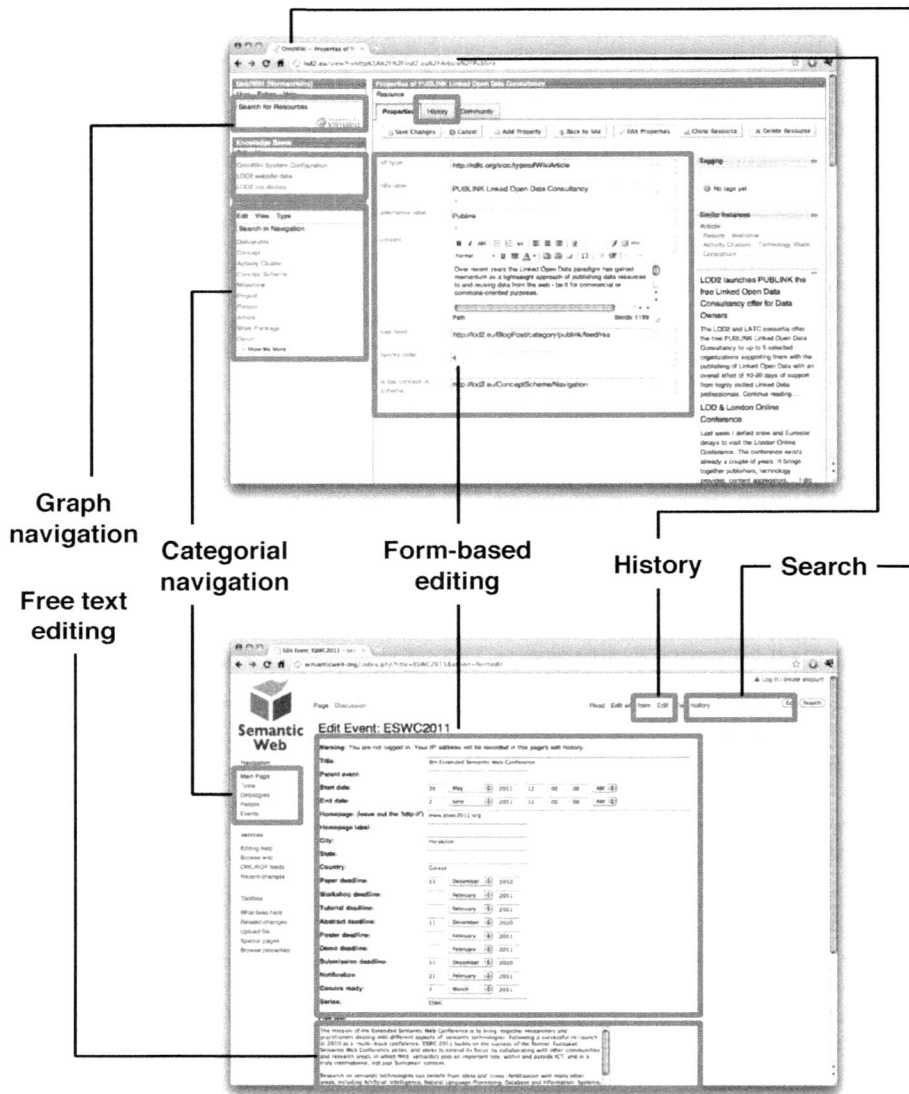

Graph navigation

Free text editing

Categorial navigation

Form-based editing

History

Search

**Fig. 12.** Comparison of Semantic MediaWiki and OntoWiki GUI building blocks

between the linked-to and the linking page. However, in conventional Wiki systems this relationship cannot be made explicit. Semantic Wiki systems therefore add a means to specify typed relations by extending the Wiki markup with semantic (i.e. typed) links. Once in place, those links form a knowledge base underlying the Wiki which can be used to improve search, browsing or automatically generated lists and category pages. Examples of approaches for extending Wikis with semantic markup can be found in [61,106,12,90,110]. They represent a straightforward combination of existing Wiki

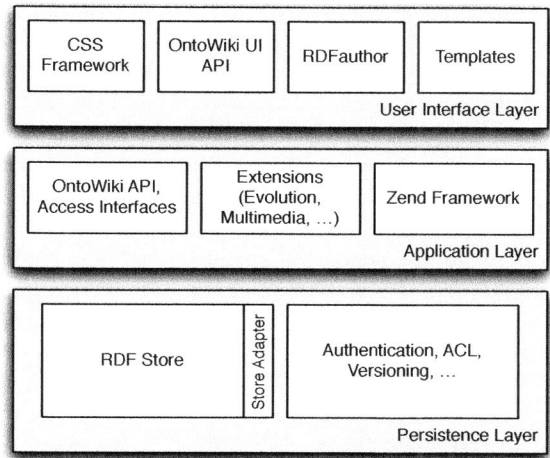

**Fig. 13.** Overview of OntoWiki's architecture with extension API and Zend web framework (modified according to [49])

systems and the Semantic Web knowledge representation paradigms. Yet, we see the following obstacles:

***Usability*:** The main advantage of Wiki systems is their unbeatable usability. Adding more and more syntactic possibilities counteracts ease of use for editors.

***Redundancy*:** To allow the answering of real-time queries to the knowledge base, statements have to be additionally kept in a triple store. This introduces a redundancy, which complicates the implementation.

***Evolution*:** As a result of storing information in both Wiki texts and triple store, supporting evolution of knowledge is difficult.

*Wikis for Editing Structured Data.* In contrast to text-based systems, Wikis for structured data – also called Data Wikis – are built on a structured model of the data being edited. The Wiki software can be used to add instances according to the schema or (in some systems) edit the schema itself. One of those systems is OntoWiki[8] [9] which bases its data model on RDF. This way, both schema and instance data are represented using the same low-level model (i.e. statements) and can therefore be handled identically by the Wiki.

### 4.1    OntoWiki - A Semantic Data Wiki

OntoWiki started as an RDF-based data wiki with emphasis on collaboration but has meanwhile evolved into a comprehensive framework for developing Semantic Web applications [49]. This involved not only the development of a sophisticated extension interface allowing for a wide range of customizations but also the addition of several access and consumption interfaces allowing OntoWiki installations to play both a provider and a consumer role in the emerging Web of Data.

---

[8] Available at: http://ontowiki.net

OntoWiki is inspired by classical Wiki systems, its design, however, (as mentioned above) is independent and complementary to conventional Wiki technologies. In contrast to other semantic Wiki approaches, in OntoWiki text editing and knowledge engineering (i. e. working with structured knowledge bases) are not mixed. Instead, OntoWiki directly applies the Wiki paradigm of "making it easy to correct mistakes, rather than making it hard to make them" [70] to collaborative management of structured knowledge. This paradigm is achieved by interpreting knowledge bases as *information maps* where every node is represented visually and interlinked to related resources. Furthermore, it is possible to enhance the knowledge schema gradually as well as the related instance data agreeing on it. As a result, the following requirements and corresponding features characterize OntoWiki:

*Intuitive display and editing* of instance data should be provided in generic ways, yet enabling domain-specific presentation of knowledge.

*Semantic views* allow the generation of different views and aggregations of the knowledge base.

*Versioning and evolution* provides the opportunity to track, review and roll-back changes selectively.

*Semantic search* facilitates easy-to-use full-text searches on all literal data, search results can be filtered and sorted (using semantic relations).

*Community support* enables discussions about small information chunks. Users are encouraged to vote about distinct facts or prospective changes.

*Online statistics* interactively measures the popularity of content and activity of users.

*Semantic syndication* supports the distribution of information and their integration into desktop applications.

OntoWiki enables the easy creation of highly structured content by distributed communities. The following points summarize some limitations and weaknesses of OntoWiki and thus characterize the application domain:

*Environment*: OntoWiki is a Web application and presumes all collaborators to work in a Web environment, possibly distributed.

*Usage Scenario*: OntoWiki focuses on knowledge engineering projects where a single, precise usage scenario is either initially (yet) unknown or not (easily) definable.

*Reasoning*: Application of reasoning services was (initially) not the primary focus.

## 4.2   Generic and Domain-Specific Views

OntoWiki can be used as a tool for presenting, authoring and managing knowledge bases adhering to the RDF data model. As such, it provides generic methods and views, independent of the domain concerned. Two generic views included in OntoWiki are the resource view and the list view. While the former is generally used for displaying all known information about a resource, the latter can present a set of resources, typically instances of a certain concept. That concept must not necessarily be explicitly defined as rdfs:Class or owl:Class in the knowledge base. Via its faceted browsing, OntoWiki allows the construction of complex concept definitions, with a pre-defined class as a starting point by means of property value restrictions. These two views are sufficient for browsing and editing all information contained in a knowledge base in a generic way. For domain-specific use cases, OntoWiki provides an easy-to-use extension interface that enables the integration of custom components. By providing such a custom view, it is even possible to hide completely the fact that an RDF knowledge base is worked on. This permits OntoWiki to be used as a data-entry frontend for users with a less profound knowledge of Semantic Web technologies.

### 4.3  Workflow

With the use of RDFS [24] and OWL [94] as ontology languages, resource definition is divisible into different layers: a terminology box for conceptual information (i.e. classes and properties) and an assertion box for entities using the concepts defined (i.e. instances). There are characteristics of RDF which, for end users, are not easy to comprehend (e.g. *classes* can be defined as *instances* of `owl:Class`). OntoWiki's user interface, therefore, provides elements for these two layers, simultaneously increasing usability and improving a user's comprehension for the structure of the data. After starting and logging in into OntoWiki with registered user credentials, it is possible to select one of the existing ontologies. The user is then presented with general information about the ontology (i.e. all statements expressed about the knowledge base as a resource) and a list of defined classes, as part of the conceptual layer.

After starting and logging in into OntoWiki with registered user credentials, it is possible to select one of the existing knowledge bases. The user is then presented with general information about the ontology (i.e. all statements expressed about the knowledge base as a resource) and a list of defined classes, as part of the conceptual layer. By selecting one of these classes, the user obtains a list of the class' instances. OntoWiki applies basic `rdfs:subClassOf` reasoning automatically. After selecting an instance from the list – or alternatively creating a new one – it is possible to manage (i.e. insert, edit and update) information in the details view.OntoWiki focuses primarily on the assertion layer, but also provides ways to manage resources on the conceptual layer. By enabling the visualization of schema elements, called *System Classes* in the OntoWiki nomenclature, conceptional resources can be managed in a similar fashion as instance data.

### 4.4  Authoring

Semantic content in OntoWiki is represented as resource descriptions. Following the RDF data model representing one of the foundations of the Semantic Web vision, resource descriptions are represented (at the lowest level) in the form of *statements*. Each of these statements (or triples) consist of a *subject* which identifies a resource as well as a *predicate* and an *object* which together represent data about said resource in a fashion reminiscent of key-value pairs. By means of RDFa [2], these statements are retained in the HTML view (i.e. user interface) part and are thus accessible to client-side techniques like JavaScript.

Authoring of such content is based on said client-side representation by employing the RDFauthor approach [114]: views are declared in terms of the model language (RDF) which allows the underlying model be restored. Based on this model, a user interface can be generated with the model being providing all the domain knowledge

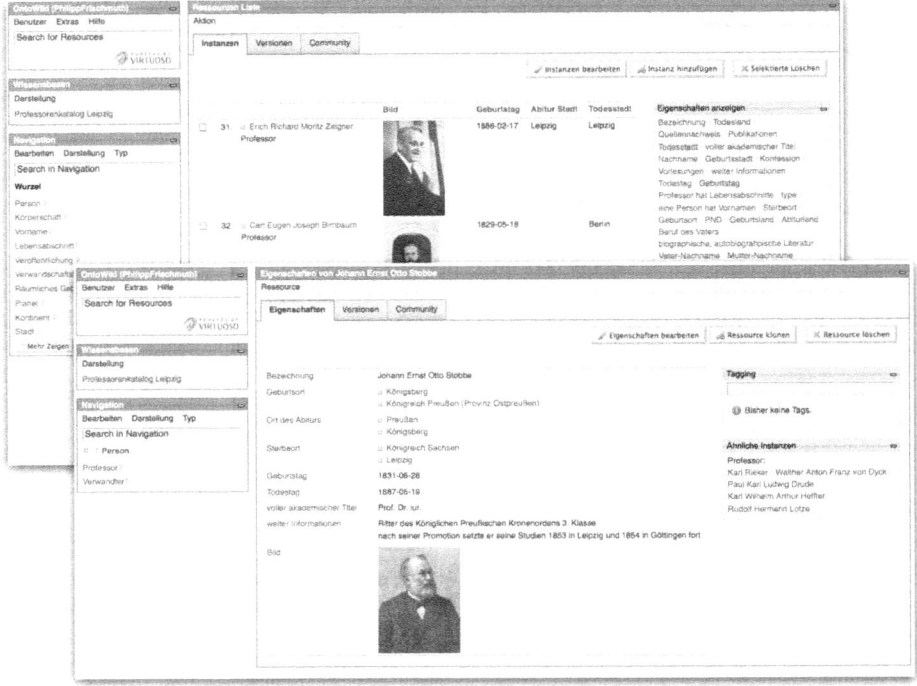

**Fig. 14.** OntoWiki views: (background) A tabular list view, which contains a filtered list of resources highlighting some specific properties of those resources and (foreground) a resource view which allows to tag and comment a specific resource as well as editing all property values

required to do so. The RDFauthor system provides an extensible set of authoring widgets specialized for certain editing tasks. RDFauthor was also extended by adding capabilities for automatically translating literal object values between different languages. Since the semantic context is known to the system, these translation functionality can be bound to arbitrary characteristics of the data (e. g. to a certain property or a missing language).

*Versioning & Evolution.* As outlined in the wiki principles, keeping track of all changes is an important task in order to encourage user participation. OntoWiki applies this concept to RDF-based knowledge engineering in that all changes are tracked on the statement level [10]. These low-level changes can be grouped to reflect application- and domain-specific tasks involving modifications to several statements as a single versioned item. Provenance information as well as other metadata (such as time, user or context) of a particular changeset can be attached to each individual changeset. All changes on the knowledge base can be easily reviewed and rolled-back if needed. The loosely typed data model of RDF encourages continuous evolution and refinement of knowledge bases. With *EvoPat*, OntoWiki supports this in a declarative, pattern-based manner (cf. section 7).

## 4.5   Access Interfaces

In addition to human-targeted graphical user interfaces, OntoWiki supports a number of machine-accessible data interfaces. These are based on established Semantic Web standards like SPARQL or accepted best practices like publication and consumption of Linked Data.

*SPARQL Endpoint.*  The SPARQL recommendation not only defines a query language for RDF but also a protocol for sending queries to and receiving results from remote endpoints[9]. OntoWiki implements this specification, allowing all resources managed in an OntoWiki be queried over the Web. In fact, the aforementioned RDFauthor authoring interface makes use of SPARQL to query for additional schema-related information, treating OntoWiki as a remote endpoint in that case.

*Linked Data.*  Each OntoWiki installation can be part of the emerging Linked Data Web. According to the Linked Data publication principles (cf. section 2), OntoWiki makes all resources accessible by its IRI (provided, the resource's IRI is in the same namespace as the OntoWiki instance). Furthermore, for each resource used in OntoWiki additional triples can be fetches if the resource is de-referenceable.

*Semantic Pingback.*  Pingback is an established notification system that gained wide popularity in the blogsphere. With Semantic Pingback [113], OntoWiki adapts this idea to Linked Data providing a *notification mechanism* for resource usage. If a Pingback-enabled resource is mentioned (i. e. linked to) by another party, its pingback server is notified of the usage. Provided, the Semantic Pingback extension is enabled all resources used in OntoWiki are pinged automatically and all resources defined in OntoWiki are Pingback-enabled.

## 4.6   Exploration Interfaces

For exploring semantic content, OntoWiki provides several exploration interfaces that range from generic views over search interfaces to sophisticated querying capabilities for more RDF-knowledgable users. The subsequent paragraphs give an overview of each of them.

*Knowledge base as an information map.*  The compromise between, on the one hand, providing a generic user interface for arbitrary RDF knowledge bases and, on the other hand, aiming at being as intuitive as possible is tackled by regarding knowledge bases as *information maps*. Each node at the information map, i. e. RDF resource, is represented as a Web accessible page and interlinked to related digital resources. These Web pages representing nodes in the information map are divided into three parts: a left sidebar, a main content section and a right sidebar. The left sidebar offers the selection of content to display in the main content section. Selection opportunities include the set of available knowledge bases, a hierarchical browser and a full-text search.

---

[9] http://www.w3.org/TR/rdf-sparql-protocol/

*Full-text search.* The full-text search makes use of special indexes (mapped to proprietary extensions to the SPARQL syntax) if the underlying knowledge store provides this feature, else, plain SPARQL string matching is used. In both cases, the resulting SPARQL query is stored as an object which can later be modified (e. g. have its filter clauses refined). Thus, full-text search is seamlessly integrated with faceted browsing (see below).

*Content specific browsing interfaces.* For domain-specific use cases, OntoWiki provides an easy-to-use extension interface that enables the integration of custom components. By providing such a custom view, it is even possible to hide completely the fact that an RDF knowledge base is worked on. This permits OntoWiki to be used as a data-entry frontend for users with a less profound knowledge of Semantic Web technologies.

*Faceted-browsing.* Via its faceted browsing, OntoWiki allows the construction of complex concept definitions, with a pre-defined class as a starting point by means of property value restrictions. These two views are sufficient for browsing and editing all information contained in a knowledge base in a generic way.

*Query-builder.* OntoWiki serves as a SPARQL endpoint, however, it quickly turned out that formulating SPARQL queries is too tedious for end users. In order to simplify the creation of queries, we developed the *Visual Query Builder*[10] (VQB) as an OntoWiki extension, which is implemented in JavaScript and communicates with the triple store using the SPARQL language and protocol. VQB allows to visually create queries to the stored knowledge base and supports domain experts with an intuitive visual representation of query and data. Developed queries can be stored and added via drag-and-drop to the current query. This enables the reuse of existing queries as building blocks for more complex ones.

### 4.7   Applications

**Catalogous Professorum.** The World Wide Web, as an ubiquitous medium for publication and exchange, already significantly influenced the way historians work: the online availability of catalogs and bibliographies allows to efficiently search for content relevant for a certain investigation; the increasing digitization of works from historical archives and libraries, in addition, enables historians to directly access historical sources remotely. The capabilities of the Web as a medium for collaboration, however, are only starting to be explored. Many, historical questions can only be answered by combining information from different sources, from different researchers and organizations. Also, after original sources are analyzed, the derived information is often much richer, than can be captured by simple keyword indexing. These factors pave the way for the successful application of knowledge engineering techniques in historical research communities.

In [99] we report about the application of an adaptive, semantics-based knowledge engineering approach using OntoWiki for the development of a prosopographical knowledge base. In prosopographical research, historians analyze common characteristics of historical groups by studying statistically relevant quantities of individual biographies. Untraceable periods of biographies can be determined on the basis of such

---

[10] http://aksw.org/Projects/OntoWiki/Extension/VQB

accomplished analyses in combination with statistically examinations as well as patterns of relationships between individuals and their activities.

In our case, researchers from the historical seminar at Universität Leipzig aimed at creating a prosopographical knowledge base about the life and work of professors in the 600 years history of Universität Leipzig ranging from the year 1409 till 2009 - the *Catalogus Professorum Lipsiensis* (CPL). In order to enable historians to collect, structure and publish this prosopographical knowledge an ontological knowledge model was developed and incrementally refined over a period of three years. The community of historians working on the project was enabled to add information to the knowledge base using an adapted version of OntoWiki. For the general public, a simplified user interface[11] is dynamically generated based on the content of the knowledge base. For access and exploration of the knowledge base by other historians a number of access interfaces was developed and deployed, such as a graphical SPARQL query builder, a relationship finder and plain RDF and Linked Data interfaces. As a result, a group of 10 historians supported by a much larger group of volunteers and external contributors collected information about 1,300 professors, 10,000 associated periods of life, 400 institutions and many more related entities.

The benefits of the developed knowledge engineering platform for historians are twofold: Firstly, the collaboration between the participating historians has significantly improved: The ontological structuring helped to quickly establish a common understanding of the domain. Collaborators within the project, peers in the historic community as well as the general public were enabled to directly observe the progress, thus facilitating peer-review, feedback and giving direct benefits to the contributors. Secondly, the ontological representation of the knowledge facilitated original historical investigations, such as historical social network analysis, professor appointment analysis (e.g. with regard to the influence of cousin-hood or political influence) or the relation between religion and university. The use of the developed model and knowledge engineering techniques is easily transferable to other prosopographical research projects and with adaptations to the ontology model to other historical research in general. In the long term, the use of collaborative knowledge engineering in historian research communities can facilitate the transition from largely individual-driven research (where one historian investigates a certain research question solitarily) to more community-oriented research (where many participants contribute pieces of information in order to enlighten a larger research question). Also, this will improve the reusability of the results of historic research, since knowledge represented in structured ways can be used for previously not anticipated research questions.

**OntoWiki Mobile.** As comparatively powerful mobile computing devices are becoming more common, mobile web applications have started gaining in popularity. An important feature of these applications is their ability to provide *offline functionality* with local updates for later synchronization with a web server. The key problem here is the reconciliation, i. e. the problem of potentially *conflicting updates* from *disconnected clients*. Another problem current mobile application developers face is the plethora of mobile application development platforms as well as the incompatibilities between

---

[11] Available at: http://www.uni-leipzig.de/unigeschichte/professorenkatalog/

them. *Android* (Google), *iOS* (Apple), *Blackberry OS* (RIM), *WebOS* (HP/Palm), *Symbian* (Nokia) are popular and currently widely deployed platforms, with many more proprietary ones being available as well. As a consequence of this fragmentation, realizing a special purpose application, which works with many or all of these platforms is extremely time consuming and inefficient due to the large amount of duplicate work required.

The W3C addressed this problem, by enriching HTML in its 5th revision with access interfaces to local storage (beyond simple cookies) as well as a number of devices and sensors commonly found on mobile devices (e. g. GPS, camera, compass etc.). We argue, that in combination with semantic technologies these features can be used to realize a *general purpose*, mobile collaboration platform, which can support the long tail of mobile special interest applications, for which the development of individual tools would not be (economically) feasible.

In [34] we present the *OntoWiki Mobile* approach realizing a mobile semantic collaboration platform based on the OntoWiki. It comprises specifically adopted user interfaces for browsing, faceted navigation as well as authoring of knowledge bases. It allows users to collect instance data and refine the structured knowledge bases on-the-go. OntoWiki Mobile is implemented as an *HTML5 web application*, thus being completely mobile device platform independent. In order to allow offline use in cases with restricted network coverage (or in order to avoid roaming charges) it uses the novel HTML5 local storage feature for replicating parts of the knowledge base on the mobile device. Hence, a crucial part of OntoWiki Mobile is the advanced conflict resolution for RDF stores. The approach is based on a combination of the EvoPat [100] method for data evolution and ontology refactoring along with a versioning system inspired by distributed version control systems like Git. OntoWiki Mobile is a generic, application domain agnostic tool, which can be utilized in a wide range of very different usage scenarios ranging from instance acquisition to browsing of semantic data on the go. Typical OntoWiki Mobile usage scenarios are settings where users need to author and access semantically structured information on the go or in settings where users are away from regular power supply and restricted to light-weight equipment (e. g. scientific expeditions).

**Semantics-based Requirements Engineering.** Semantic interoperability, linked data, and a shared conceptual foundation become increasingly important prerequisites in software development projects that are characterized by spatial dispersion, large numbers of stakeholders, and heterogeneous development tools. The SoftWiki OntoWiki extension [74] focuses specifically on semantic collaboration with respect to requirements engineering. Potentially very large and spatially distributed groups of stakeholders, including developers, experts, managers, and average users, shall be enabled to collect, semantically enrich, classify, and aggregate software requirements. OntoWiki is used to support collaboration as well as interlinking and exchange of requirements data. To ensure a shared conceptual foundation and semantic interoperability, we developed the SoftWiki Ontology for Requirements Engineering (SWORE) that defines core concepts of requirement engineering and the way they are interrelated. For instance, the ontology defines frequent relation types to describe requirements interdependencies such as details, conflicts, related to, depends on, etc. The flexible SWORE design allows for easy extension. Moreover, the requirements can be linked to external resources, such as publicly available domain knowledge or company-specific policies. The whole process is called semantification of requirements. It is envisioned as an evolutionary process: The requirements are successively linked to each other and to further concepts in a collaborative way, jointly by all stakeholders. Whenever a requirement is formulated, reformulated, analyzed, or exchanged, it might be semantically enriched by the respective participant.

# 5   Automatic Linking

The fourth Linked Data Principle, i.e., "Include links to other URIs, so that they can discover more things" (cf. section 2) is the most important Linked Data principle as it enables the paradigm change from data silos to interoperable data distributed across the Web. Furthermore, it plays a key role in important tasks such as cross-ontology question answering [20,75], large-scale inferences [116,81] and data integration [76,17]. Yet, while the number of triples in Linked Data sources increases steadily and has surpassed 26 billions[12], links between knowledge bases still constitute less than 5% of these triples. The goal of linking is to tackle this sparseness so as to transform the Web into a platform for data and information integration as well as for search and querying.

## 5.1   Instance Matching

*Linking* can be generally defined as *connecting things that are somehow related*. In the context of Linked Data, linking is especially concerned with establishing typed links between entities (i.e., classes, properties or instances) contained in knowledge bases. Over the last years, several link discovery frameworks have been developed to address the lack of typed links between the different knowledge bases on the Linked Data web. Overall, two main types of link discovery frameworks can be differentiated. The first category implements *ontology matching* techniques and aims to establish links between the ontologies underlying two data sources. The second and more prominent category of approaches, dubbed *instance matching*, aims to discover links between instances contained in two data sources. It is important to notice that while ontology and instance matching are similar to schema matching [96,95] and record linkage [124,33,22] respectively (as known in the research area of databases), linking on the Web of Data is a more generic and thus more complex task, as it is not limited to finding equivalent entities in two knowledge bases. Rather, it aims at finding semantically related entities and establishing typed links between them, most of these links being imbued with formal properties (e.g., transitivity, symmetry, etc.) that can be used by reasoners and other application to infer novel knowledge. In this section, we will focus on the discovery of links between instances. An overview of ontology matching techniques is given in [37].

Formally, instance matching can be defined as follows:

**Definition 3 (Instance Matching).** *Given two sets $S$ (source) and $T$ (target) of instances, a semantic similarity measure $\sigma : S \times T \to [0, 1]$ and a threshold $\theta \in [0, 1]$, the goal of instance matching task is to compute the set $M = \{(s, t), \sigma(s, t) \geq \theta\}$.*

---

[12] http://lod-cloud.net/

```
 1  <?xml version="1.0" encoding="UTF-8"?>
 2  <!DOCTYPE LIMES SYSTEM "limes.dtd">
 3  <LIMES>
 4    <PREFIX>
 5      <NAMESPACE>http://www.w3.org/1999/02/22-rdf-syntax-ns#</NAMESPACE>
 6      <LABEL>rdf</LABEL>
 7    </PREFIX>
 8    <PREFIX>
 9      <NAMESPACE>http://www.w3.org/2002/07/owl#</NAMESPACE>
10      <LABEL>owl</LABEL>
11    </PREFIX>
12    <PREFIX>
13      <NAMESPACE>http://data.linkedct.org/resource/linkedct/</NAMESPACE>
14      <LABEL>linkedct</LABEL>
15    </PREFIX>
16    <PREFIX>
17      <NAMESPACE>http://bio2rdf.org/ns/mesh#</NAMESPACE>
18      <LABEL>meshr</LABEL>
19    </PREFIX>
20    <SOURCE>        <ID> linkedct </ID>
21                   <ENDPOINT> http://data.linkedct.org/sparql </ENDPOINT>
22                   <VAR> ?x </VAR>
23                   <PAGESIZE> 5000 </PAGESIZE>
24                   <RESTRICTION> ?x rdf:type linkedct:condition </RESTRICTION>
25                   <PROPERTY> linkedct:condition_name </PROPERTY>
    </SOURCE>
26    <TARGET>       <ID> mesh </ID>
27                   <ENDPOINT> http://mesh.bio2rdf.org/sparql </ENDPOINT>
28                   <VAR> ?y </VAR>
29                   <PAGESIZE> 5000 </PAGESIZE>
30                   <RESTRICTION> ?y rdf:type meshr:Concept </RESTRICTION>
31                   <PROPERTY> dc:title </PROPERTY>
    </TARGET>
32    <METRIC>       levenshtein(x.linkedct:condition_name, y.dc:title) </METRIC>
33    <EXEMPLARS>    70 </EXEMPLARS>
34    <ACCEPTANCE>
35        <THRESHOLD> 0.9 </THRESHOLD> <RELATION> owl:sameAs </RELATION>
36        <FILE>diseases\_accepted.nt</FILE> </ACCEPTANCE>
37    <REVIEW>
38        <THRESHOLD> 0.8 </THRESHOLD> <RELATION> owl:sameAs </RELATION>
39        <FILE>diseases\_review.nt</FILE> </REVIEW>
40  </LIMES>
```

**Fig. 15.** Example of a link specification for the LIMES framework

In general, the specification for a matching is described by using a *link specification* (sometimes called *linkage decision rule* [55]). Figure 15 shows an example of such a link specification for the instance matching framework LIMES[13]. This specification links diseases in LinkedCT[14] with diseases in Bio2RDF[15] via the owl:sameAs property when the Levenshtein similarity [71] (i.e., the similarity derived from the edit distance) of their labels is greater or equal to 0.9. Instances whose labels bear a similarity between 0.8 and 0.9 are returned for manual examination.

## 5.2   Challenges

Two key challenges arise when trying to discover links between two sets of instances: the computational complexity of the matching task *per se* and the selection of an appropriate link specification.

---

[13] http://limes.sf.net
[14] http://linkedct.org
[15] http://bio2rdf.org

The first challenge is intrinsically related to the link discovery process. The time complexity of a matching task can be measured by the number of comparisons necessary to complete this task. When comparing a source knowledge base $S$ with a target knowledge base $T$, the completion of a matching task requires a-priori $O(|S||T|)$ comparisons, an impractical proposition as soon as the source and target knowledge bases become large. For example, discovering duplicate cities in DBpedia [6] alone would necessitate approximately $0.15 \times 10^9$ similarity computations. Hence, the provision of time-efficient approaches for the reduction of the time complexity of link discovery is a key requirement to instance linking frameworks for Linked Data.

The second challenge of the link discovery process lies in the selection of an appropriate link specification. The configuration of link discovery frameworks is usually carried out manually, in most cases simply by guessing. Yet, the choice of a suitable link specification measure is central for the discovery of satisfactory links. The large number of properties of instances and the large spectrum of measures available in literature underline the complexity of choosing the right specification manually[16]. Supporting the user during the process of finding the appropriate similarity measure and the right properties for each mapping task is a problem that still needs to be addressed by the Linked Data community. Methods such as supervised and active learning can be used to guide the user in need of mapping to a suitable linking configuration for his matching task. Yet, these methods could not be used so far because of the time complexity of link discovery, thus even further amplifying the need for time-efficient methods. In the following, we give a short overview of existing frameworks for Link Discovery on the Web of Data. Subsequently, we present a time-efficient framework for link discovery in more detail.

## 5.3  Approaches to Instance Matching

Current frameworks for link discovery can be subdivided into two main categories: *domain-specific* and *universal* frameworks. Domain-specific link discovery frameworks aim at discovering links between knowledge bases from a particular domain. One of the first domain-specific approaches to carry out instance linking for Linked Data was implemented in the *RKBExplorer*[17] [45] with the aim of discovering links between entities from the domain of academics. Due to the lack of data available as Linked Data, the RKBExplorer had to extract RDF from heterogeneous data source so as to populate its knowledge bases with instances according to the AKT ontology[18]. Especially, instances of persons, publications and institutions were retrieved from several major metadata websites such as ACM and DBLP. The linking was implemented by the so-called Consistent Reference Service (CRS) which linked equivalent entities by comparing properties including their type and label. So far, the CRS is limited to linking objects in the knowledge bases underlying the RKBExplorer and cannot be used for other tasks without further implementation.

Another domain-specific tool is GNAT [97], which was developed for the music domain. It implements several instance matching algorithms of which the most

---

[16] The SimMetrics project (`http://simmetrics.sf.net`) provides an overview of strings similarity measures.

[17] `http://www.rkbexplorer.com`

[18] `http://www.aktors.org/publications/ontology/`

sophisticated, the online graph matching algorithm (OGMA), applies a similarity propagation approach to discover equivalent resources. The basic approach implemented by OGMA starts with a single resource $s \in S$. Then, it retrieves candidate matching resources $t \in T$ by comparing properties such as `foaf:name` for artists and `dc:title` for albums. If $\sigma(s, t) \geq \theta$, then the algorithm terminates. In case a disambiguation is needed, the resourced related to $s$ and $t$ in their respective knowledge bases are compared and their similarity value is cumulated to recompute $\sigma(s, t)$. This process is iterated until a mapping resource for $s$ is found in $T$ or no resource matches.

Universal link discovery frameworks are designed to carry out mapping tasks independently from the domain of the source and target knowledge bases. For example, RDF-AI [107], a framework for the integration of RDF data sets, implements a five-step approach that comprises the preprocessing, matching, fusion, interlinking and post-processing of data sets. RDF-AI contains a series of modules that allow for computing instances matches by comparing their properties. Especially, it contains translation modules that allow to process the information contained in data sources before mapping. By these means, it can boost the precision of the mapping process. These modules can be configured by means of XML-files. RDF-AI does not comprise means for querying distributed data sets via SPARQL[19]. In addition, it suffers from not being time-optimized. Thus, mapping by using this tool can be very time-consuming.

A time-optimized approach to link discovery is implemented by the SILK framework [119]. SILK implements several approaches to minimize the time necessary for mapping instances from knowledge bases. First, it allows to specify restrictions on the instances to be loaded from each knowledge base. Furthermore, it uses rough index pre-matching to reach a quasi-linear time-complexity. In addition, it allows to specify blocking strategies to reduce the runtime of the system. It can be configured by using the SILK-Link Specification Language, which is based on XML. The drawback of the pre-matching approach and of most blocking approaches is that their recall is not guaranteed to be 1. Thus, some links can be lost during the instance-matching process.

It is important to notice that the task of discovering links between knowledge bases is related with record linkage [124,33] and de-duplication [22]. The database community has produced a vast amount of literature on efficient algorithms for solving these problems. Different blocking techniques such as standard blocking, sorted-neighborhood, bigram indexing, canopy clustering and adaptive blocking [16,21,60] have been developed to address the problem of the quadratic time complexity of brute force comparison methods. The idea is to filter out obvious non-matches efficiently before executing the more detailed and time-consuming comparisons. In the following, we present a state-of-the-art framework that implements lossless instance matching based on a similar idea in detail.

## 5.4 LIMES

LIMES (Link Discovery Framework for metric spaces) is an instance-matching framework that implements time-efficient approaches for the discovery of links between Linked Data sources. It addresses the scalability problem of link discovery by utilizing the *triangle inequality* in metric spaces to compute pessimistic estimates of instance

---

[19] http://www.w3.org/TR/rdf-sparql-query/

similarities. Based on these approximations, LIMES can filter out a large number of instance pairs that cannot suffice the matching condition set by the user. The real similarities of the remaining instances pairs are then computed and the matching instances are returned.

**Mathematical Framework.** In the remainder of this section, we use the following notations:

1. $A$ is an affine space,
2. $m, m_1, m_2, m_3$ symbolize metrics on $A$,
3. $x, y$ and $z$ represent points from $A$ and
4. $\alpha, \beta, \gamma$ and $\delta$ are scalars, i.e., elements of $\mathbb{R}$.

**Definition 4 (Metric space).** *A metric space is a pair $(A, m)$ such that $A$ is an affine space and $m : A \times A \to \mathbb{R}$ is a function such that for all $x, y$ and $z \in A$*

1. $m(x, y) \geq 0$ $(M_1)$ *(non-negativity),*
2. $m(x, y) = 0 \Leftrightarrow x = y$ $(M_2)$ *(identity of indiscernibles),*
3. $m(x, y) = m(y, x)$ $(M_3)$ *(symmetry) and*
4. $m(x, z) \leq m(x, y) + m(y, z)$ $(M_4)$ *(triangle inequality).*

Note that the definition of a matching based on a similarity function $\sigma$ can be rewritten for metrics $m$ as follows:

**Definition 5 (Instance Matching in Metric Spaces).** *Given two sets $S$ (source) and $T$ (target) of instances, a metric $m : S \times T \to [0, \infty[$ and a threshold $\theta \in [0, \infty[$, the goal of instance matching task is to compute the set $M = \{(s, t) | m(s, t) \leq \theta\}$.*

Example of metrics on strings include the Levenshtein distance and the block distance. However, some popular measures such as JaroWinkler [123] do not satisfy the triangle inequality and are consequently not metrics.

The rationale behind the LIMES framework is to make use of the boundary conditions entailed by the triangle inequality (TI) to reduce the number of comparisons (and thus the time complexity) necessary to complete a matching task. Given a metric space $(A, m)$ and three points $x, y$ and $z$ in A, the TI entails that

$$m(x, y) \leq m(x, z) + m(z, y). \tag{1}$$

Without restriction of generality, the TI also entails that

$$m(x, z) \leq m(x, y) + m(y, z), \tag{2}$$

thus leading to the following boundary conditions in metric spaces:

$$m(x, y) - m(y, z) \leq m(x, z) \leq m(x, y) + m(y, z). \tag{3}$$

Inequality 3 has two major implications. The first is that the distance from a point $x$ to any point $z$ in a metric space can be approximated given the distance from $x$ to a reference point $y$ and the distance from the reference point $y$ to $z$. Such a reference point

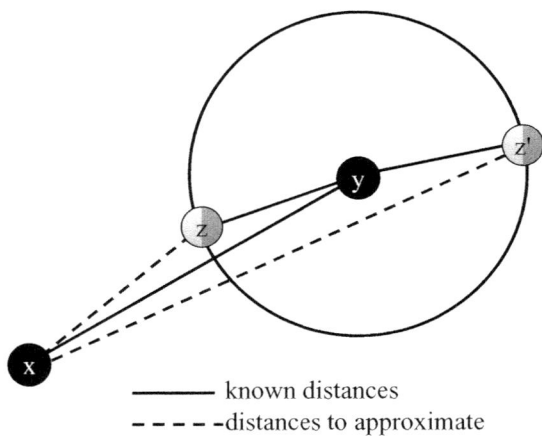

———— known distances
– – – –distances to approximate

**Fig. 16.** Approximation of distances via exemplars. The lower bound of the distance from $x$ to $z$ can be approximated by $m(x, y) - m(y, z)$.

is called an *exemplar* following [44]. The role of an exemplar is to be used as a sample of a portion of the metric space $A$. Given an input point $x$, knowing the distance from $x$ to an exemplar $y$ allows to compute lower and upper bounds of the distance from $x$ to any other point $z$ at a known distance from $y$. An example of such an approximation is shown in Figure 16. In this figure, all the points on the circle are subject to the same distance approximation. The distance from $x$ to $z$ is close to the lower bound of inequality 3, while the distance from $x$ to $z'$ is close to the upper bound of the same inequality.

The second implication of inequality 3 is that the distance from $x$ to $z$ can only be smaller than $\theta$ if the lower bound of the approximation of the distance from $x$ to $z$ via *any exemplar* $y$ is also smaller than $\theta$. Thus, if the lower bound of the approximation of the distance $m(x, z)$ is larger than $\theta$, then $m(x, z)$ itself must be larger than $\theta$. Formally,

$$m(x, y) - m(y, z) > \theta \Rightarrow m(x, z) > \theta. \tag{4}$$

Supposing that all distances from instances $t \in T$ to exemplars are known, reducing the number of comparisons simply consists of using inequality 4 to compute an approximation of the distance from all $s \in S$ to all $t \in T$ and computing the real distance only for the $(s, t)$ pairs for which the first term of inequality 4 does not hold. This is the core of the approach implemented by LIMES.

**Computation of Exemplars.** The core idea underlying the computation of exemplars in LIMES is to select a set of exemplars in the metric space underlying the matching task in such a way that they are distributed uniformly in the metric space. One way to achieve this goal is by ensuring that the exemplars display a high dissimilarity. The approach used by LIMES to generate exemplars with this characteristic is shown in Algorithm 1.

Let $n$ be the desired number of exemplars and $E$ the set of all exemplars. In step 1 and 2, LIMES initializes $E$ by picking a random point $e_1$ in the metric space $(T, m)$ and

---

**Algorithm 1.** Computation of Exemplars

---

**Require:** Number of exemplars $n$
**Require:** Target knowledge base $T$
  1. Pick random point $e_1 \in T$
  2. Set $E = E \cup \{e_1\}$;
  3. Compute the distance from $e_1$ to all $t \in T$
**while** $|E| < n$ **do**
    4. Get a random point $e'$ such that $e' \in argmax_t \sum_{t \in T} \sum_{e \in E} m(t, e)$
    5. $E = E \cup \{e'\}$;
    6. Compute the distance from $e'$ to all $t \in T$
**end while**
  7. Map each point in $t \in T$ to one of the exemplars $e \in E$ such that $m(t, e)$ is minimal
**return** $E$

---

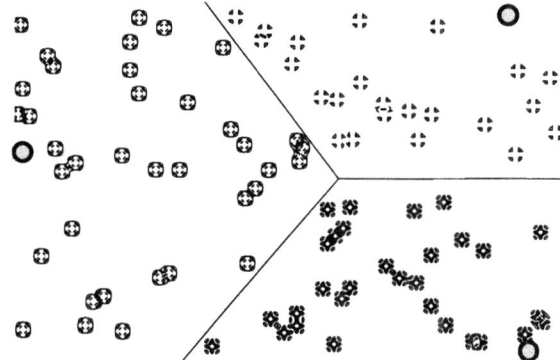

**Fig. 17.** Mapping of points to three exemplars in a metric space. The exemplars are displayed as gray disks.

setting $E = \{e_1\}$. Then, it computes the similarity from the exemplar $e_1$ to every other point in $T$ (step 3). As long as the size of $E$ has not reached $n$, LIMES repeats steps 4 to 6: In step 4, a point $e' \in T$ such that the sum of the distances from $e'$ to the exemplars $e \in E$ is maximal (there can be many of these points) is chosen randomly. This point is chosen as new exemplar and consequently added to $E$ (step 5). Then, the distance from $e'$ to all other points in $T$ is computed (step 6). Once $E$ has reached the size $n$, LIMES terminates the iteration. Finally, each point is mapped to the exemplar to which it is most similar (step 7) and the exemplar computation terminates (step 8). This algorithm has a constant time complexity of $O(|E||T|)$.

An example of the results of the exemplar computation algorithm ($|E| = 3$) is shown in Figure 17. The initial exemplar was the leftmost exemplar in the figure.

**Matching Based on Exemplars.** The instances associated with an exemplar $e \in E$ in step 7 of Algorithm 1 are stored in a list $L_e$ sorted in descending order with respect to the distance to $e$. Let $\lambda_1^e...\lambda_m^e$ be the elements of the list $L_e$. The goal of matching an instance $s$ from a source knowledge base to a target knowledge base w.r.t. a metric $m$ is

---

**Algorithm 2.** LIMES' Matching algorithm

---

**Require:** Set of exemplars $E$
**Require:** Instance $s \in S$
**Require:** Metric m
**Require:** threshold $\theta$

  1. $M = \emptyset$
  **for** $e \in |E|$ **do**
    **if** $m(s, e) \leq \theta$ **then**
      2. $M = M \cup \{e\}$
      **for** $i = 1...|L_e|$ **do**
        **if** $(m(s, e) - m(e, \lambda_i^e)) \leq \theta$ **then**
          **if** $m(s, \lambda_i^e) \leq \theta$ **then**
            3. $M = M \cup \{(s, \lambda_i^e)\}$
          **end if**
        **else**
          break
        **end if**
      **end for**
    **end if**
  **end for**
  **return** $M$

---

to find all instances $t$ of the target knowledge source such that $m(s, t) \leq \theta$, where $\theta$ is a given threshold. LIMES achieves this goal by using the matching algorithm based on exemplars shown in Algorithm 2.

LIMES only carries out a comparison when the approximation of the distance is less than the threshold. Moreover, it terminates the similarity computation for an exemplar $e$ as soon as the first $\lambda^e$ is found such that the lower bound of the distance is larger than $\theta$. This break can be carried out because the list $L_e$ is sorted, i.e., if $m(s, e) - m(e, \lambda_i^e) > \theta$ holds for the $i^{th}$ element of $L_e$, then the same inequality holds for all $\lambda_j^e \in L_e$ with $j > i$. In the worst case, LIMES' matching algorithm has the time complexity $O(|S||T|)$, leading to a total worst time complexity of $O((|E| + |S|)|T|)$, which is larger than that of brute force approaches. However, as the results displayed in Figure 18 show, a correct parameterization of LIMES leads to significantly smaller numbers of comparisons and runtimes.

**Implementation.** The LIMES framework consists of seven main modules (see Figures 19 and 20) of which each can be extended to accommodate new or improved functionality[20].

LIMES carries out matching processes as follows: First, the *controller* calls the *I/O-module*, which reads the configuration file and extracts all the information necessary to carry out the comparison of instances, including the URL of the SPARQL endpoints of the knowledge bases, the restrictions on the instances to map (e.g., their type), the expression of the metric and the threshold to be used. An example of such a configuration file is given in Figure 15. Note that the LIMES configuration takes similarity

---

[20] The framework is open source and can be found at `http://limes.sf.net`

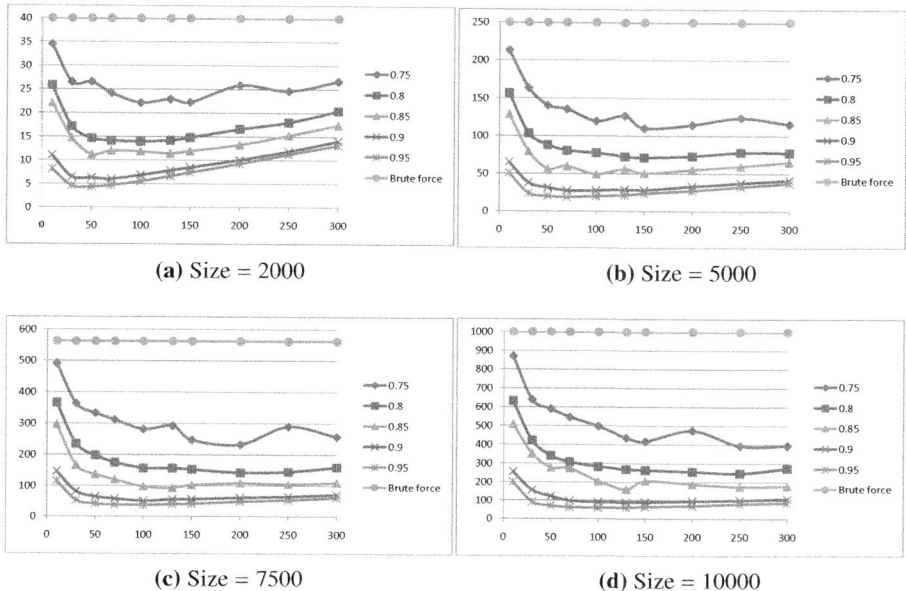

**Fig. 18.** Comparisons required by LIMES for different numbers of exemplars on knowledge bases of different sizes. The x-axis shows the number of exemplars, the y-axis the number of comparisons in multiples of $10^5$.

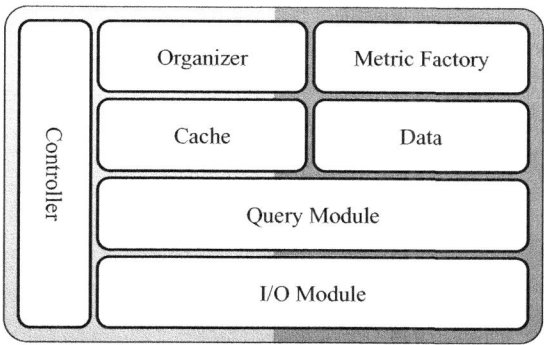

**Fig. 19.** Architecture of LIMES

(and not distance) thresholds as input for the users' convenience. Given that the configuration file is valid w.r.t. the LIMES Specification Language (LSL)[21], the *query module* is called. This module uses the configuration for the target and source knowledge bases to retrieve instances and properties from the SPARQL-endpoints of the source and target knowledge bases that adhere to the restrictions specified in the configuration file.

---

[21] The DTD for the LSL can be found in the LIMES distribution along with a user manual at: http://limes.sf.net

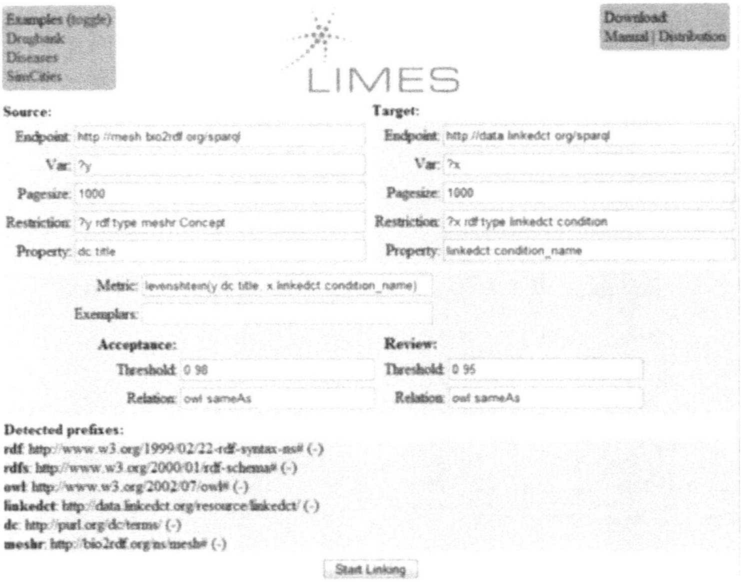

**Fig. 20.** The LIMES Web dashboard

LIMES can also retrieve data for linking from other data sources including CSV files describing tabular data or vector data. The query module writes its output into a *cache* which can either be a file (for large number of instances) or main memory. For the user's convenience, the cache module serializes the data retrieved from remote sources to a file to ensure swift subsequently matching processes with other data sources. Once all instances have been stored in the cache, the controller calls the *organizer module*. This module carries out two tasks: first, it computes the exemplars of the source knowledge base. Then, it uses the exemplars to compute the matchings from the source to the target knowledge base. Finally, the *I/O-module* is called to serialize the results, i.e. to write the results in a user-defined output stream according to a user-specified format, e.g., in an N-Triples file.

## 5.5  Conclusion

We presented and discussed semi-automatic linking instance matching approaches for Linked Data and the challenges they face. In addition, we gave an overview of several state-of-the-art approaches for instance matching for Linked Data. Finally we presented a time-efficient approach for instance matching . Time-efficient frameworks that address the performance challenge also provide the backbone necessary to address other substantial challenges of link discovery on the Web of Data. The main quality criteria of matching tasks are performance, precision/recall and ease-of-use. These quality criteria are not unconnected. Rather, they influence each other strongly. In particular, performance can be traded for precision/recall and substantially increase the ease-of-use for users with faster matching algorithms by using supervised matching. On the other hand, increasing precision/recall with faster matching algorithms can be carried out by involving the user into an interactive feedback loop, where he/she feeds hints on the quality of matching results back into the system. These hints can be used subsequently for automatically adjusting thresholds and/or matcher configurations. Such semi-automatic adjustments of thresholds and matcher configurations can in turn dramatically increase the ease-of-use for the users of a matching framework, since they are released from the burden of creating a comprehensive a-priori configuration.

# 6 Enrichment

The term *enrichment* in this chapter refers to the (semi-)automatic extension of a knowledge base schema. It describes the process of increasing the expressiveness and semantic richness of a knowledge base. Usually, this is achieved by adding or refining terminological axioms.

Enrichment methods can typically be applied in a *grass-roots* approach to knowledge base creation. In such an approach, the whole ontological structure is not created upfront, but evolves with the data in a knowledge base. Ideally, this enables a more agile development of knowledge bases. In particular, in the context of the Web of Linked Data such an approach appears to be an interesting alternative to more traditional ontology engineering methods. Amongst others, Tim Berners-Lee advocates to get "raw data now"[22] and worry about the more complex issues later.

Knowledge base enrichment can be seen as a sub-discipline of ontology learning. Ontology learning is more general in that it can rely on external sources, e.g. written text, to create an ontology. The term knowledge base enrichment is typically used when already existing data in the knowledge base is analysed to improve its schema.

Enrichment methods span several research areas like knowledge representation and reasoning, machine learning, statistics, natural language processing, formal concept analysis and game playing. Considering the variety of methods, we structure this section as follows: First, we give an overview of different types of enrichment and list some typical methods and give pointers to references, which allow the reader to obtain more information on a topic. In the second part, we describe a specific software – the ORE tool – in more detail.

## 6.1 State of the Art and Types of Enrichment

Ontology enrichment usually involves applying heuristics or machine learning techniques to find axioms, which can be added to an existing ontology. Naturally, different techniques have been applied depending on the specific type of axiom.

One of the most complex tasks in ontology enrichment is to find *definitions* of classes. This is strongly related to Inductive Logic Programming (ILP) [88] and more specifically supervised learning in description logics. Research in those fields has many applications apart from being applied to enrich ontologies. For instance, it is used in the life sciences to detect whether drugs are likely to be efficient for particular diseases. Work on learning in description logics goes back to e.g. [28,29], which used

---

[22] http://www.ted.com/talks/tim_berners_lee_on_the_next_web.html

so-called *least common subsumers* to solve the learning problem (a modified variant of the problem defined in this article). Later, [15] invented a refinement operator for $\mathcal{ALER}$ and proposed to solve the problem by using a top-down approach. [35,53,54] combine both techniques and implement them in the YINYANG tool. However, those algorithms tend to produce very long and hard-to-understand class expressions. The algorithms implemented in DL-Learner [67,68,62,69] overcome this problem and investigate the learning problem and the use of top down refinement in detail. DL-FOIL [38] is a similar approach, which is based on a mixture of upward and downward refinement of class expressions. They use alternative measures in their evaluation, which take the open world assumption into account, which was not done in ILP previously. Most recently, [66] implements appropriate heuristics and adaptations for learning definitions in ontologies. The focus in this work is efficiency and practical application of learning methods. The article presents plugins for two ontology editors (Protégé and OntoWiki) as well stochastic methods, which improve previous methods by an order of magnitude. For this reason, we will analyse it in more detail in the next subsection. The algorithms presented in the article can also learn *super class axioms*.

A different approach to learning the definition of a named class is to compute the so called *most specific concept* (msc) for all instances of the class. The most specific concept of an individual is the most specific class expression, such that the individual is instance of the expression. One can then compute the *least common subsumer* (lcs) [14] of those expressions to obtain a description of the named class. However, in expressive description logics, an msc does not need to exist and the lcs is simply the disjunction of all expressions. For light-weight logics, such as $\mathcal{EL}$, the approach appears to be promising.

Other approaches, e.g. [72] focus on learning in hybrid knowledge bases combining ontologies and *rules*. Ontology evolution [73] has been discussed in this context. Usually, hybrid approaches are a generalisation of concept learning methods, which enable powerful rules at the cost of efficiency (because of the larger search space). Similar as in knowledge representation, the tradeoff between expressiveness of the target language and efficiency of learning algorithms is a critical choice in symbolic machine learning.

Another enrichment task is *knowlege base completion*. The goal of such a task is to make the knowledge base complete in a particular well-defined sense. For instance, a goal could be to ensure that all subclass relationships between named classes can be inferred. The line of work starting in [102] and further pursued in e.g. [13] investigates the use of *formal concept analysis* for completing knowledge bases. It is promising, although it may not be able to handle noise as well as a machine learning technique. A Protégé plugin [108] is available. [117] proposes to improve knowledge bases through relational exploration and implemented it in the *RELExO framework*[23]. It focuses on simple relationships and the knowledge engineer is asked a series of questions. The knowledge engineer either must positively answer the question or provide a counterexample.

[118] focuses on learning *disjointness* between classes in an ontology to allow for more powerful reasoning and consistency checking. To achieve this, it can use the ontology itself, but also texts, e.g. Wikipedia articles corresponding to a concept. The article

---

[23] http://code.google.com/p/relexo/

**Table 3.** Work in ontology enrichment grouped by type or aim of learned structures

| Type/Aim | References |
|---|---|
| Taxonomies | [125] |
| Definitions | often done via ILP approaches such as [67,68,69,66,38,35,53,54,15], genetic approaches [62] have also been used |
| Super Class Axioms | [66] |
| Rules in Ontologies | [72,73] |
| Disjointness | [118] |
| Properties of properties | usually via heuristics |
| Alignment | challenges: [109], recent survey: [26] |
| Completion | formal concept analysis and relational exploration [13,117,108] |

includes an extensive study, which shows that proper modelling disjointness is actually a difficult task, which can be simplified via this ontology enrichment method.

Another type of ontology enrichment is schema mapping. This task has been widely studied and will not be discussed in depth within this chapter. Instead, we refer to [26] for a survey on ontology mapping.

There are further more light-weight ontology enrichment methods. For instance, *taxonomies* can be learned from simple tag structures via heuristics. Similarly, "properties of properties" can be derived via simple statistical analysis. This includes the detection whether a particular property might be symmetric, function, reflexive, inverse functional etc. Similarly, domains and ranges of properties can be determined from existing data. Enriching the schema with domain and range axioms allows to find cases, where properties are misused via OWL reasoning.

In the following subsection, we describe an enrichment approach for learning definitions and super class axioms in more detail. The algorithm was recently developed by the first authors and is described in full detail in [66].

## 6.2   Class Expression Learning in DL-Learner

The Semantic Web has recently seen a rise in the availability and usage of knowledge bases, as can be observed within the Linking Open Data Initiative, the TONES and Protégé ontology repositories, or the Watson search engine. Despite this growth, there is still a lack of knowledge bases that consist of sophisticated schema information and instance data adhering to this schema. Several knowledge bases, e.g. in the life sciences, only consist of schema information, while others are, to a large extent, a collection of facts without a clear structure, e.g. information extracted from data bases or texts. The combination of sophisticated schema and instance data allows powerful reasoning, consistency checking, and improved querying possibilities. We argue that being able to learn OWL class expressions[24] is a step towards achieving this goal.

*Example 1.* As an example, consider a knowledge base containing a class Capital and instances of this class, e.g. London, Paris, Washington, Canberra etc. A machine

---

[24] http://www.w3.org/TR/owl2-syntax/#Class_Expressions

learning algorithm could, then, suggest that the class `Capital` may be equivalent to one
of the following OWL class expressions in Manchester OWL syntax[25]:

> `City and isCapitalOf at least one GeopoliticalRegion`
> `City and isCapitalOf at least one Country`

Both suggestions could be plausible: The first one is more general and includes cities
that are capitals of states, whereas the latter one is stricter and limits the instances to
capitals of countries. A knowledge engineer can decide which one is more appropriate,
i.e. a semi-automatic approach is used, and the machine learning algorithm should guide
her by pointing out which one fits the existing instances better. Assuming the knowledge
engineer decides for the latter, an algorithm can show her whether there are instances
of the class `Capital` which are neither instances of `City` nor related via the property
`isCapitalOf` to an instance of `Country`.[26] The knowledge engineer can then continue
to look at those instances and assign them to a different class as well as provide more
complete information; thus improving the quality of the knowledge base. After adding
the definition of `Capital`, an OWL reasoner can compute further instances of the class
which have not been explicitly assigned before.

Using machine learning for the generation of suggestions instead of entering them
manually has the advantage that 1.) the given suggestions fit the instance data,
i.e. schema and instances are developed in concordance, and 2.) the entrance barrier
for knowledge engineers is significantly lower, since understanding an OWL class expression is easier than analysing the structure of the knowledge base and creating a
class expression manually. Disadvantages of the approach are the dependency on the
availability of instance data in the knowledge base and requirements on the quality of
the ontology, i.e. modelling errors in the ontology can reduce the quality of results.

Overall, we describe the following in this chapter:

- extension of an existing learning algorithm for learning class expressions to the
  ontology engineering scenario,
- presentation and evaluation of different heuristics,
- showcase how the enhanced ontology engineering process can be supported with
  plugins for Protégé and OntoWiki,
- evaluation of the presented algorithm with several real ontologies from various domains.

The adapted algorithm for solving the learning problems, which occur in the ontology
engineering process, is called *CELOE (Class Expression Learning for Ontology Engineering)*. It was implemented within the open-source framework DL-Learner.[27] DL-
Learner [63,64] leverages a modular architecture, which allows to define different types
of components: knowledge sources (e.g. OWL files), reasoners (e.g. DIG[28] or OWL
API based), learning problems, and learning algorithms. In this overview, we focus on
the latter two component types, i.e. we define the class expression learning problem in
ontology engineering and provide an algorithm for solving it.

---

[25] For details on Manchester OWL syntax (e.g. used in Protégé, OntoWiki) see [51].

[26] This is not an inconsistency under the standard OWL open world assumption, but rather a hint
towards a potential modelling error.

[27] http://dl-learner.org

[28] http://dl.kr.org/dig/

**Learning Problem.** The process of learning in logics, i.e. trying to find high-level explanations for given data, is also called *inductive reasoning* as opposed to *inference* or *deductive reasoning*. The main difference is that in deductive reasoning it is formally shown whether a statement follows from a knowledge base, whereas in inductive learning new statements are invented. Learning problems, which are similar to the one we will analyse, have been investigated in *Inductive Logic Programming* [88] and, in fact, the method presented here can be used to solve a variety of machine learning tasks apart from ontology engineering.

In the ontology learning problem we consider, we want to learn a formal description of a class $A$, which has (inferred or asserted) instances in the considered ontology. In the case that $A$ is already described by a class expression $C$ via axioms of the form $A \sqsubseteq C$ or $A \equiv C$, those can be either refined, i.e. specialised/generalised, or relearned from scratch by the learning algorithm. To define the class learning problem, we need the notion of a *retrieval* reasoner operation $R_{\mathcal{K}}(C)$. $R_{\mathcal{K}}(C)$ returns the set of all instances of $C$ in a knowledge base $\mathcal{K}$. If $\mathcal{K}$ is clear from the context, the subscript can be omitted.

**Definition 6 (class learning problem).** *Let an existing named class $A$ in a knowledge base $\mathcal{K}$ be given. The class learning problem is to find an expression $C$ such that $R_{\mathcal{K}}(C) = R_{\mathcal{K}}(A)$.*

Clearly, the learned expression $C$ is a description of (the instances of) $A$. Such an expression is a candidate for adding an axiom of the form $A \equiv C$ or $A \sqsubseteq C$ to the knowledge base $\mathcal{K}$. If a solution of the learning problem exists, then the used base learning algorithm (as presented in the following subsection) is complete, i.e. guaranteed to find a correct solution if one exists in the target language and there are no time and memory constraints (see [68,69] for the proof). In most cases, we will not find a solution to the learning problem, but rather an approximation. This is natural, since a knowledge base may contain false class assignments or some objects in the knowledge base are described at different levels of detail. For instance, in Example 1, the city "Apia" might be typed as "Capital" in a knowledge base, but not related to the country "Samoa". However, if most of the other cities are related to countries via a role isCapitalOf, then the learning algorithm may still suggest City and isCapitalOf at least one Country since this describes the majority of capitals in the knowledge base well. If the knowledge engineer agrees with such a definition, then a tool can assist him in completing missing information about some capitals.

According to Occam's razor [23] simple solutions of the learning problem are to be preferred over more complex ones, because they are more readable. This is even more important in the ontology engineering context, where it is essential to suggest simple expressions to the knowledge engineer. We measure simplicity as the *length* of an expression, which is defined in a straightforward way, namely as the sum of the numbers of concept, role, quantifier, and connective symbols occurring in the expression. The algorithm is biased towards shorter expressions. Also note that, for simplicity the definition of the learning problem itself does enforce coverage, but not prediction, i.e. correct classification of objects which are added to the knowledge base in the future. Concepts with high coverage and poor prediction are said to *overfit* the data. However, due to the strong bias towards short expressions this problem occurs empirically rarely in description logics [69].

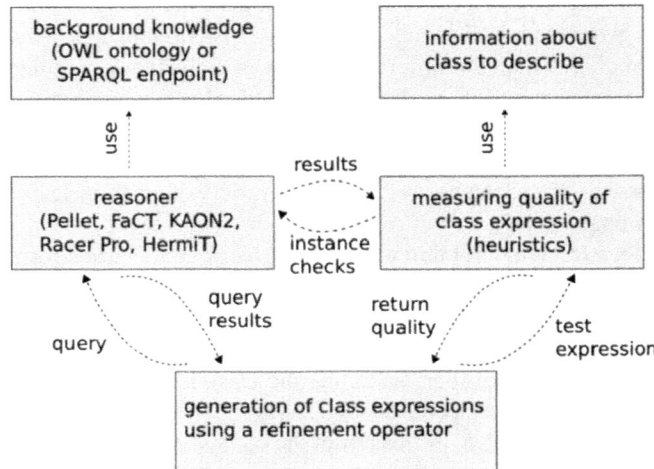

**Fig. 21.** Outline of the general learning approach in CELOE: One part of the algorithm is the generation of promising class expressions taking the available background knowledge into account. Another part is a heuristic measure of how close an expression is to being a solution of the learning problem. Figure adapted from [50].

**Base Learning Algorithm.** Figure 21 gives a brief overview of the *CELOE* algorithm, which follows the common "generate and test" approach in ILP. This means that learning is seen as a search process and several class expressions are generated and tested against a background knowledge base. Each of those class expressions is evaluated using a heuristic, which is described in the next section. A challenging part of a learning algorithm is to decide which expressions to test. In particular, such a decision should take the computed heuristic values and the structure of the background knowledge into account. For CELOE, we use the approach described in [68,69] as base, where this problem has already been analysed, implemented, and evaluated in depth. It is based on the idea of *refinement operators*:

**Definition 7 (refinement operator).** *A* quasi-ordering *is a reflexive and transitive relation. In a quasi-ordered space* $(S, \leq)$ *a* downward (upward) refinement operator $\rho$ *is a mapping from $S$ to $2^S$, such that for any $C \in S$ we have that $C' \in \rho(C)$ implies $C' \leq C$ $(C \leq C')$. $C'$ is called a* specialisation (generalisation) *of $C$.*

Refinement operators can be used for searching in the space of expressions. As ordering we can use subsumption. (Note that the subsumption relation $\sqsubseteq$ is a quasi-ordering.) If an expression $C$ subsumes an expression $D$ ($D \sqsubseteq C$), then $C$ will cover all examples which are covered by $D$. This makes subsumption a suitable order for searching in expressions as it allows to prune parts of the search space without losing possible solutions.

The approach we used is a top-down algorithm based on refinement operators as illustrated in Figure 22 (more detailed schemata can be found in the slides[29] of the

---

[29] http://reasoningweb.org/2010/teaching-material/lehmann.pdf

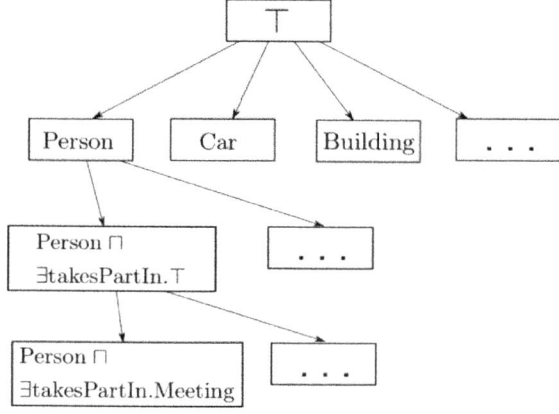

**Fig. 22.** Illustration of a search tree in a top down refinement approach

ontology learning lecture of Reasoning Web 2010 [65]). This means that the first class expression, which will be tested is the most general expression ($\top$), which is then mapped to a set of more specific expressions by means of a downward refinement operator. Naturally, the refinement operator can be applied to the obtained expressions again, thereby spanning a *search tree*. The search tree can be pruned when an expression does not cover sufficiently many instances of the class $A$ we want to describe. One example for a path in a search tree spanned up by a downward refinement operator is the following ($\leadsto$ denotes a refinement step):

$$\top \leadsto \texttt{Person} \leadsto \texttt{Person} \sqcap \texttt{takesPartinIn.} \top$$

$$\leadsto \texttt{Person} \sqcap \texttt{takesPartIn.Meeting}$$

The heart of such a learning strategy is to define a suitable refinement operator and an appropriate search heuristics for deciding which nodes in the search tree should be expanded. The refinement operator in the considered algorithm is defined in [69]. It is based on earlier work in [68] which in turn is built on the theoretical foundations of [67]. It has been shown to be the best achievable operator with respect to a set of properties (not further described here), which are used to assess the performance of refinement operators. The learning algorithm supports conjunction, disjunction, negation, existential and universal quantifiers, cardinality restrictions, hasValue restrictions as well as boolean and double datatypes.

### 6.3    Finding a Suitable Heuristic

A heuristic measures how well a given class expression fits a learning problem and is used to guide the search in a learning process. To define a suitable heuristic, we first need to address the question of how to measure the accuracy of a class expression. We introduce several heuristics, which can be used for *CELOE* and later evaluate them.

We cannot simply use supervised learning from examples directly, since we do not have positive and negative examples available. We can try to tackle this problem by using the existing instances of the class as positive examples and the remaining instances as negative examples. This is illustrated in Figure 23, where $\mathcal{K}$ stands for the knowledge base and $A$ for the class to describe. We can then measure accuracy as the number of correctly classified examples divided by the number of all examples. This can be computed as follows for a class expression $C$ and is known as *predictive accuracy* in Machine Learning:

$$predacc(C) = 1 - \frac{|R(A) \setminus R(C)| + |R(C) \setminus R(A)|}{n} \quad n = |Ind(\mathcal{K})|$$

Here, $Ind(\mathcal{K})$ stands for the set of individuals occurring in the knowledge base. $R(A) \setminus R(C)$ are the false negatives whereas $R(C) \setminus R(A)$ are false positives. $n$ is the number of all examples, which is equal to the number of individuals in the knowledge base in this case. Apart from learning definitions, we also want to be able to learn super class axioms ($A \sqsubseteq C$). Naturally, in this scenario $R(C)$ should be a superset of $R(A)$. However, we still do want $R(C)$ to be as small as possible, otherwise $\top$ would always be a solution. To reflect this in our accuracy computation, we penalise false negatives more than false positives by a factor of $t$ ($t > 1$) and map the result to the interval $[0, 1]$:

$$predacc(C, t) = 1 - 2 \cdot \frac{t \cdot |R(A) \setminus R(C)| + |R(C) \setminus R(A)|}{(t + 1) \cdot n} \quad n = |Ind(\mathcal{K})|$$

While being straightforward, the outlined approach of casting class learning into a standard learning problem with positive and negative examples has the disadvantage that the number of negative examples will usually be much higher than the number of positive examples. As shown in Table 4, this may lead to overly optimistic estimates. More importantly, this accuracy measure has the drawback of having a dependency on the number of instances in the knowledge base.

Therefore, we investigated further heuristics, which overcome this problem, in particular by transferring common heuristics from information retrieval to the class learning problem:

1. *F-Measure:* $F_\beta$-Measure is based on *precision* and *recall* weighted by $\beta$. They can be computed for the class learning problem without having negative examples. Instead, we perform a retrieval for the expression $C$, which we want to evaluate. We can then define precision as the percentage of instances of $C$, which are also instances of $A$ and recall as percentage of instances of $A$, which are also instances of $C$. This is visualised in Figure 23. F-Measure is defined as harmonic mean of precision and recall. For learning super classes, we use $F_3$ measure by default, which gives recall a higher weight than precision.

2. *A-Measure:* We denote the arithmetic mean of precision and recall as A-Measure. Super class learning is achieved by assigning a higher weight to recall. Using the arithmetic mean of precision and recall is uncommon in Machine Learning, since it results in too optimistic estimates. However, we found that it is useful in super class learning, where $F_n$ is often too pessimistic even for higher $n$.

3. *Generalised F-Measure:* Generalised F-Measure has been published in [31] and extends the idea of F-measure by taking the three valued nature of classification in OWL/DLs into account: An individual can either belong to a class, the negation of a class or none of both cases can be proven. This differs from common binary

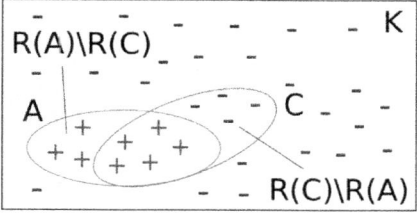 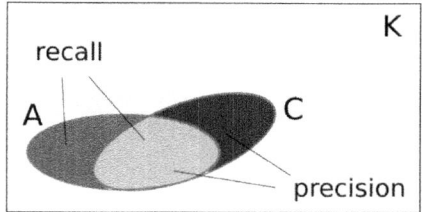

**Fig. 23.** Visualisation of different accuracy measurement approaches. $\mathcal{K}$ is the knowledge base, $A$ the class to describe and $C$ a class expression to be tested. Left side: Standard supervised approach based on using positive (instances of $A$) and negative (remaining instances) examples. Here, the accuracy of $C$ depends on the number of individuals in the knowledge base. Right side: Evaluation based on two criteria: recall (*Which fraction of $R(A)$ is in $R(C)$?*) and precision (*Which fraction of $R(C)$ is in $R(A)$?*).

classification tasks and, therefore, appropriate measures have been introduced (see [31] for details). Adaption for super class learning can be done in a similar fashion as for F-Measure itself.

4. *Jaccard Distance:* Since $R(A)$ and $R(C)$ are sets, we can use the well-known Jaccard coefficient to measure the similarity between both sets.

We argue that those four measures are more appropriate than predictive accuracy when applying standard learning algorithms to the ontology engineering use case. Table 4 provides some example calculations, which allow the reader to compare the different heuristics.

*Efficient Heuristic Computation.* Several optimisations for computing the heuristics are described in [66]. In particular, adapted approximate reasoning and stochastic approximations are discussed. Those improvements have shown to lead to order of magnitude gains in efficiency for many ontologies. We refrain from describing those methods in this chapter.

**The Protégé Plugin.** After implementing and testing the described learning algorithm, we integrated it into *Protégé* and *OntoWiki*. Together with the Protégé developers, we extended the Protégé 4 plugin mechanism to be able to seamlessly integrate the DL-Learner plugin as an additional method to create class expressions. This means that the knowledge engineer can use the algorithm exactly where it is needed without any additional configuration steps. The plugin has also become part of the official Protégé 4 repository, i.e. it can be directly installed from within Protégé.

A screenshot of the plugin is shown in Figure 24. To use the plugin, the knowledge engineer is only required to press a button, which then starts a new thread in the background. This thread executes the learning algorithm. The used algorithm is an *anytime algorithm*, i.e. at each point in time we can always see the currently best suggestions. The GUI updates the suggestion list each second until the maximum runtime – 10 seconds by default – is reached. This means that the perceived runtime, i.e. the time after which only minor updates occur in the suggestion list, is often only one or two seconds for small ontologies. For each suggestion, the plugin displays its accuracy.

**Table 4.** Example accuracies for selected cases (eq = equivalence class axiom, sc = super class axiom). The images on the left represent an imaginary knowledge base $\mathcal{K}$ with 1000 individuals, where we want to describe the class $A$ by using expression $C$. It is apparent that using predictive accuracy leads to impractical accuracies, e.g. in the first row $C$ cannot possibly be a good description of $A$, but we still get 80% accuracy, since all the negative examples outside of $A$ and $C$ are correctly classified.

| illustration | pred. acc. | | F-Measure | | A-Measure | | Jaccard |
|---|---|---|---|---|---|---|---|
| | eq | sc | eq | sc | eq | sc | |
| K:1000 C:100 A:100 | 80% | 67% | 0% | 0% | 0% | 0% | 0% |
| K:1000 C:200 A:100 | 90% | 92% | 67% | 73% | 75% | 88% | 50% |
| K:1000 C:400 A:100 | 70% | 75% | 40% | 48% | 63% | 82% | 25% |
| K:1000 10 C:100 10 A:100 | 98% | 97% | 90% | 90% | 90% | 90% | 82% |
| K:1000 A:100 C:50 | 95% | 88% | 67% | 61% | 75% | 63% | 50% |

When clicking on a suggestion, it is visualized by displaying two circles: One stands for the instances of the class to describe and another circle for the instances of the suggested class expression. Ideally, both circles overlap completely, but in practice this will often not be the case. Clicking on the plus symbol in each circle shows its list of individuals. Those individuals are also presented as points in the circles and moving the mouse over such a point shows information about the respective individual. Red points show potential problems detected by the plugin. Please note that we use closed world reasoning to detect those problems. For instance, in our initial example, a capital which is not related via the property isCapitalOf to an instance of Country is marked red. If there is not only a potential problem, but adding the expression would render the ontology inconsistent, the suggestion is marked red and a warning message is displayed. Accepting such a suggestion can still be a good choice, because the problem often lies elsewhere in the knowledge base, but was not obvious before, since the ontology was not sufficiently expressive for reasoners to detect it. This is illustrated by a screencast available from the plugin homepage,[30] where the ontology becomes inconsistent after

---

[30] http://dl-learner.org/wiki/ProtegePlugin

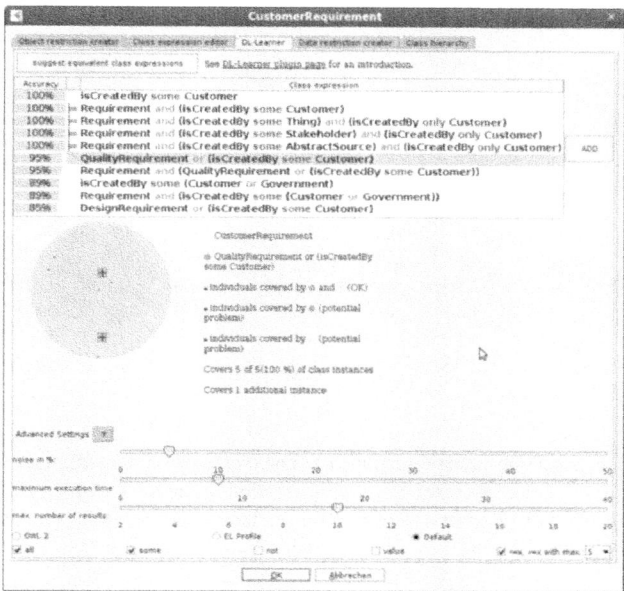

**Fig. 24.** A screenshot of the DL-Learner Protégé plugin. It is integrated as additional tab to create class expressions in Protégé. The user is only required to press the "suggest equivalent class expressions" button and within a few seconds they will be displayed ordered by accuracy. If desired, the knowledge engineer can visualize the instances of the expression to detect potential problems. At the bottom, optional expert configuration settings can be adopted.

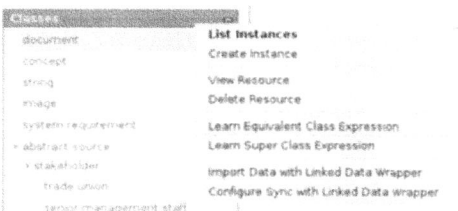

**Fig. 25.** The DL-Learner plugin can be invoked from the context menu of a class in OntoWiki

adding the axiom, and the real source of the problem is fixed afterwards. Being able to make such suggestions can be seen as a strength of the plugin.

The plugin allows the knowledge engineer to change expert settings. Those settings include the maximum suggestion search time, the number of results returned and settings related to the desired target language, e.g. the knowledge engineer can choose to stay within the OWL 2 EL profile or enable/disable certain class expression constructors. The learning algorithm is designed to be able to handle noisy data and the visualisation of the suggestions will reveal false class assignments so that they can be fixed afterwards.

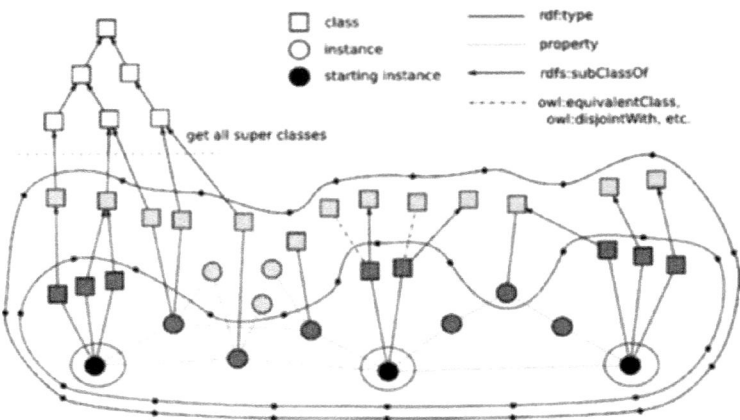

**Fig. 26.** Extraction with three starting instances. The circles represent different recursion depths. The circles around the starting instances signify recursion depth 0. The larger inner circle represents the fragment with recursion depth 1 and the largest outer circle with recursion depth 2. Figure taken from [50].

**The OntoWiki Plugin.** Analogous to Protégé, we created a similar plugin for OntoWiki (cf. section 4). OntoWiki is a lightweight ontology editor, which allows distributed and collaborative editing of knowledge bases. It focuses on wiki-like, simple and intuitive authoring of semantic content, e.g. through inline editing of RDF content, and provides different views on instance data.

Recently, a fine-grained plugin mechanism and extensions architecture was added to OntoWiki. The DL-Learner plugin is technically realised by implementing an OntoWiki component, which contains the core functionality, and a module, which implements the UI embedding. The DL-Learner plugin can be invoked from several places in OntoWiki, for instance through the context menu of classes as shown in Figure 25.

The plugin accesses DL-Learner functionality through its WSDL-based web service interface. Jar files containing all necessary libraries are provided by the plugin. If a user invokes the plugin, it scans whether the web service is online at its default address. If not, it is started automatically.

A major technical difference compared to the Protégé plugin is that the knowledge base is accessed via SPARQL, since OntoWiki is a SPARQL-based web application. In Protégé, the current state of the knowledge base is stored in memory in a Java object. As a result, we cannot easily apply a reasoner on an OntoWiki knowledge base. To overcome this problem, we use the DL-Learner fragment selection mechanism described in [50]. Starting from a set of instances, the mechanism extracts a relevant fragment from the underlying knowledge base up to some specified recursion depth. Figure 26 provides an overview of the fragment selection process. The fragment has the property that learning results on it are similar to those on the complete knowledge base. For a detailed description we refer the reader to the full article.

The fragment selection is only performed for medium to large-sized knowledge bases. Small knowledge bases are retrieved completely and loaded into the reasoner. While the fragment selection can cause a delay of several seconds before the learning

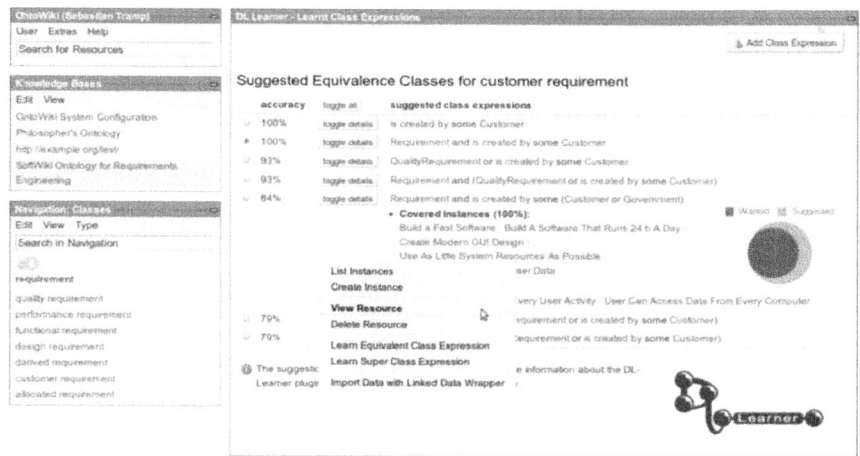

**Fig. 27.** Screenshot of the result table of the DL-Learner plugin in OntoWiki

algorithm starts, it also offers flexibility and scalability. For instance, we can learn class expressions in large knowledge bases such as DBpedia in OntoWiki.[31]

Figure 27 shows a screenshot of the OntoWiki plugin applied to the SWORE [98] ontology. Suggestions for learning the class "customer requirement" are shown in Manchester OWL Syntax. Similar to the Protégé plugin, the user is presented a table of suggestions along with their accuracy value. Additional details about the instances of "customer requirement", covered by a suggested class expressions and additionally contained instances can be viewed via a toggle button. The modular design of OntoWiki allows rich user interaction: Each resource, e.g. a class, property, or individual, can be viewed and subsequently modified directly from the result table as shown for "design requirement" in the screenshot. For instance, a knowledge engineer could decide to import additional information available as Linked Data and run the CELOE algorithm again to see whether different suggestions are provided with additional background knowledge.

**Evaluation.** To evaluate the suggestions made by our learning algorithm, we tested it on a variety of real-world ontologies of different sizes and domains. Please note that we intentionally do not perform an evaluation of the machine learning technique as such on existing benchmarks, since we build on the base algorithm already evaluated in detail in [69]. It was shown that this algorithm is superior to other supervised learning algorithms for OWL and at least competitive with the state of the art in ILP. Instead, we focus on its use within the ontology engineering scenario. The goals of the evaluation are to 1. determine the influence of reasoning and heuristics on suggestions, 2. to evaluate whether the method is sufficiently efficient to work on large real-world ontologies.

To perform the evaluation, we wrote a dedicated plugin for the Protégé ontology editor. This allows the evaluators to browse the ontology while deciding whether the

---

[31] OntoWiki is undergoing an extensive development, aiming to support handling such large knowledge bases. A release supporting this is expected for the first half of 2012.

**Table 5.** Statistics about test ontologies

| Ontology | #logical axioms | #classes | #object properties | #data properties | #individuals | DL expressivity |
|---|---|---|---|---|---|---|
| SC Ontology[32] | 20081 | 28 | 8 | 5 | 3542 | $\mathcal{AL}(\mathcal{D})$ |
| Adhesome[33] | 12043 | 40 | 33 | 37 | 2032 | $\mathcal{ALCHN}(\mathcal{D})$ |
| GeoSkills[34] | 14966 | 613 | 23 | 21 | 2620 | $\mathcal{ALCHOIN}(\mathcal{D})$ |
| Eukariotic[35] | 38 | 11 | 1 | 0 | 11 | $\mathcal{ALCON}$ |
| Breast Cancer[36] | 878 | 196 | 22 | 3 | 113 | $\mathcal{ALCROF}(\mathcal{D})$ |
| Economy[37] | 1625 | 339 | 45 | 8 | 482 | $\mathcal{ALCH}(\mathcal{D})$ |
| Resist[38] | 239 | 349 | 134 | 38 | 75 | $\mathcal{ALUF}(\mathcal{D})$ |
| Finance[39] | 16014 | 323 | 247 | 74 | 2466 | $\mathcal{ALCROIQ}(\mathcal{D})$ |
| Earthrealm[40] | 931 | 2364 | 215 | 36 | 171 | $\mathcal{ALCHO}(\mathcal{D})$ |

suggestions made are reasonable. The plugin works as follows: First, all classes with at least 5 inferred instances are determined. For each such class, we run CELOE with different settings to generate suggestions for definitions. Specifically, we tested two reasoners and five different heuristics. The two reasoners are standard Pellet and Pellet combined with approximate reasoning (not described in detail here). The five heuristics are those described in Section 6.3. For each configuration of CELOE, we generate at most 10 suggestions exceeding a heuristic threshold of 90%. Overall, this means that there can be at most 2 * 5 * 10 = 100 suggestions per class – usually less, because different settings of CELOE will still result in similar suggestions. This list is shuffled and presented to the evaluators. For each suggestion, the evaluators can choose between 6 options (see Table 6):

1 the suggestion improves the ontology (improvement)
2 the suggestion is no improvement and should not be included (not acceptable) and
3 adding the suggestion would be a modelling error (error)

In the case of existing definitions for class $A$, we removed them prior to learning. In this case, the evaluator could choose between three further options:

4 the learned definition is equal to the previous one and both are good (equal +)
5 the learned definition is equal to the previous one and both are bad (equal -) and
6 the learned definition is inferior to the previous one (inferior).

---

[32] http://www.mindswap.org/ontologies/SC.owl
[33] http://www.sbcny.org/datasets/adhesome.owl
[34] http://i2geo.net/ontologies/current/GeoSkills.owl
[35] http://www.co-ode.org/ontologies/eukariotic/2005/06/01/eukariotic.owl
[36] http://acl.icnet.uk/%7Emw/MDM0.73.owl
[37] http://reliant.teknowledge.com/DAML/Economy.owl
[38] http://www.ecs.soton.ac.uk/~aoj04r/resist.owl
[39] http://www.fadyart.com/Finance.owl
[40] http://sweet.jpl.nasa.gov/1.1/earthrealm.owl

**Table 6.** Options chosen by evaluators aggregated by class. FIC stands for the fast instance checker, which is an approximate reasoning procedure.

| reasoner/heuristic | improvement | equal quality (+) | equal quality (-) | inferior | not acceptable | error | missed improvements in % | selected position on suggestion list (incl. std. deviation) | avg. accuracy of selected suggestion in % |
|---|---|---|---|---|---|---|---|---|---|
| Pellet/F-Measure | 16.70 | 0.44 | 0.66 | 0.00 | 64.66 | 17.54 | 14.95 | 2.82 ± 2.93 | 96.91 |
| Pellet/Gen. F-Measure | 15.24 | 0.44 | 0.66 | 0.11 | 66.60 | 16.95 | 16.30 | 2.78 ± 3.01 | 92.76 |
| Pellet/A-Measure | 16.70 | 0.44 | 0.66 | 0.00 | 64.66 | 17.54 | 14.95 | 2.84 ± 2.93 | 98.59 |
| Pellet/pred. acc. | 16.59 | 0.44 | 0.66 | 0.00 | 64.83 | 17.48 | 15.22 | 2.69 ± 2.82 | 98.05 |
| Pellet/Jaccard | 16.81 | 0.44 | 0.66 | 0.00 | 64.66 | 17.43 | 14.67 | 2.80 ± 2.91 | 95.26 |
| Pellet FIC/F-Measure | 36.30 | 0.55 | 0.55 | 0.11 | 52.62 | 9.87 | 1.90 | 2.25 ± 2.74 | 95.01 |
| Pellet FIC/Gen. F-M. | 33.41 | 0.44 | 0.66 | 0.00 | 53.41 | 12.09 | 7.07 | 1.77 ± 2.69 | 89.42 |
| Pellet FIC/A-Measure | 36.19 | 0.55 | 0.55 | 0.00 | 52.84 | 9.87 | 1.63 | 2.21 ± 2.71 | 98.65 |
| Pellet FIC/pred. acc. | 32.99 | 0.55 | 0.55 | 0.11 | 55.58 | 10.22 | 4.35 | 2.17 ± 2.55 | 98.92 |
| Pellet FIC/Jaccard | 36.30 | 0.55 | 0.55 | 0.11 | 52.62 | 9.87 | 1.90 | 2.25 ± 2.74 | 94.07 |

We used the default settings of CELOE, e.g. a maximum execution time of 10 seconds for the algorithm. The knowledge engineers were five experienced members of our research group, who made themselves familiar with the domain of the test ontologies. Each researcher worked independently and had to make 998 decisions for 92 classes between one of the options. The time required to make those decisions was approximately 40 working hours per researcher. The raw agreement value of all evaluators is 0.535 (see e.g. [4] for details) with 4 out of 5 evaluators in strong pairwise agreement (90%). The evaluation machine was a notebook with a 2 GHz CPU and 3 GB RAM.

Table 6 shows the evaluation results. All ontologies were taken from the Protégé OWL[41] and TONES[42] repositories. We randomly selected 5 ontologies comprising instance data from these two repositories, specifically the Earthrealm, Finance, Resist, Economy and Breast Cancer ontologies (see Table 5).

The results in Table 6 show which options were selected by the evaluators. It clearly indicates that the usage of approximate reasoning is sensible. The results are, however, more difficult to interpret with regard to the different employed heuristics. Using predictive accuracy did not yield good results and, surprisingly, generalised F-Measure also had a lower percentage of cases where option 1 was selected. The other three heuristics generated very similar results. One reason is that those heuristics are all based on precision and recall, but in addition the low quality of some of the randomly selected test ontologies posed a problem. In cases of too many very severe modelling errors, e.g. conjunctions and disjunctions mixed up in an ontology or inappropriate domain and

---

[41] http://protegewiki.stanford.edu/index.php/Protege_Ontology_Library
[42] http://owl.cs.manchester.ac.uk/repository/

range restrictions, the quality of suggestions decreases for each of the heuristics. This is the main reason why the results for the different heuristics are very close. Particularly, generalised F-Measure can show its strengths mainly for properly designed ontologies. For instance, column 2 of Table 6 shows that it missed 7% of possible improvements. This means that for 7% of all classes, one of the other four heuristics was able to find an appropriate definition, which was not suggested when employing generalised F-Measure. The last column in this table shows that the average value of generalised F-Measure is quite low. As explained previously, it distinguishes between cases when an individual is instance of the observed class expression, its negation, or none of both. In many cases, the reasoner could not detect that an individual is instance of the negation of a class expression, because of the absence of disjointness axioms and negation in the knowledge base, which explains the low average values of generalised F-Measure. Column 4 of Table 6 shows that many selected expressions are amongst the top 5 (out of 10) in the suggestion list, i.e. providing 10 suggestions appears to be a reasonable choice.

In general, the improvement rate is only at about 35% according to Table 6 whereas it usually exceeded 50% in preliminary experiments with other real-world ontologies with fewer or less severe modelling errors. Since CELOE is based on OWL reasoning, it is clear that schema modelling errors will have an impact on the quality of suggestions. As a consequence, we believe that the CELOE algorithm should be combined with ontology debugging techniques. We have obtained first positive results in this direction and plan to pursue it in future work. However, the evaluation also showed that CELOE does still work in ontologies, which probably were never verified by an OWL reasoner.

*Summary.* We presented the CELOE learning method specifically designed for extending OWL ontologies. Five heuristics were implemented and analysed in conjunction with CELOE along with several performance improvements. A method for approximating heuristic values has been introduced, which is useful beyond the ontology engineering scenario to solve the challenge of dealing with a large number of examples in ILP [122]. Furthermore, we biased the algorithm towards short solutions and implemented optimisations to increase readability of the suggestions made. The resulting algorithm was implemented in the open source DL-Learner framework. We argue that CELOE is the first ILP based algorithm, which turns the idea of learning class expressions for extending ontologies into practice. CELOE is integrated into two plugins for the ontology editors Protégé and OntoWiki and can be invoked using just a few mouse clicks.

## 7   Pattern-Based Evolution

Facilitating the smooth evolution of knowledge bases on the Data Web is still a major challenge. The importance of addressing this challenge is amplified by the shift towards employing agile knowledge engineering methodologies (such as Semantic Wikis, cf. section 4), which particularly stress the evolutionary aspect of the knowledge engineering process.

As an example, how the evolution on the Web of Data can be facilitated, we present in this section the *EvoPat* approach [100], which is heavily inspired by software refactoring. In software engineering, refactoring techniques are applied to improve software quality, to accommodate new requirements or to represent domain changes. The term refactoring refers to the process of making persistent and incremental changes to a system's internal structure without changing its observable behavior, yet improving the quality of its design and/or implementation [42]. Refactoring is based on two key concepts: *code smells* and *refactorings*. Code smells are an informal but still useful characterization of patterns of bad source code. Examples of code smells are "too long method" and "duplicate code". Refactorings are piecemeal transformations of source code which keep the semantics while removing (totally or partly) a code smell. For example, the "extract method" refactoring extracts a section of a "long method" into a new method and replaces it by a call to the new method, thus making the original method shorter (and clearer).

Compared to software source code refactoring, where refactorings must be performed manually or with limited programmatic support, the situation in knowledge base evolution on the Data Web is more advantageous. On the Data Web we have the unified RDF data model, which is the basis for both, data and ontologies. With *EvoPat* we exploit the RDF data model by devising a pattern-based approach for the data evolution and ontology refactoring of RDF knowledge bases. The approach is based on the definition of *basic evolution patterns*, which are represented declaratively and can capture atomic evolution and refactoring operations on the data and schema levels. In essence, a basic evolution pattern consists of two main components: 1) a *SPARQL SELECT query template* for selecting objects, which will be changed and 2) a *SPARQL/Update query template*, which is executed for every returned result of the SELECT query. In order to accommodate more advanced and domain-specific data evolution and refactoring strategies, we define a compound evolution pattern as a linear combination of several simple ones.

To obtain a comprehensive catalog of evolution patterns, we performed a survey of possible evolution patterns with a combinatorial analysis of all possible before/after combinations. Starting with the basic constituents of a knowledge base (i. e. graphs, properties and classes), we consider all possible combinations of the elements potentially being affected by an evolution pattern and the prospective result after application of the evolution pattern. This analysis led to a comprehensive library of 24 basic and compound evolution patterns. The catalog is not meant to be exhaustive but covers the most common knowledge base evolution scenarios as confirmed by a series of interviews with domain experts and knowledge engineers. The EvoPat approach was implemented as an extension for OntoWiki (cf. section 4).

This section is structured as follows: First, we describe the evolution pattern concepts and survey possible evolution patterns. Then, we showcase our implementation and present our work in the light of related approaches.

## 7.1  Concepts

The EvoPat approach is based on the rationale of working as closely as possible with the RDF data model and the common ontology construction elements, i. e. classes, instances as well as datatype and object properties. With EvoPat we also aim at delegating bulk of the work during evolution processing to the underlying triple store. Hence, for the definition of evolution patterns we employ a combination of different SPARQL query templates. In order to ensure modularity and facilitate reusability of evolution patterns our definition of evolution patterns is twofold: *basic evolution patterns* accommodate atomic ontology evolution and data migration operations, while *compound evolution patterns* represent sequences of either basic or other compound evolution patterns in order to capture more complex and domain specific evolution scenarios. The application of a particular evolution pattern to a concrete knowledge base is performed with the help of the EvoPat *pattern execution algorithm*. In order to optimally assist a knowledge engineer we also define the concept of a *bad smell* in a knowledge base. We describe these individual EvoPat components in more detail in the remainder of this paper.

**Evolution Pattern.** Figure 28 describes the general composition of EvoPat evolution patterns. Bad smells (depicted in the lower left of Figure 28 have a number of basic or compound evolution patterns associated, which are triggered once a bad smell is traced. Basic and compound evolution patterns can be annotated with descriptive attributes, such as a label for the pattern, a textual description and other metadata such as the author of the pattern the creation date, revision etc.

*Basic Evolution Pattern (BP).* A basic evolution pattern consists of two main components: 1. a SPARQL SELECT query template for selecting objects, which will be changed and 2. a SPARQL/Update query template, which is executed for every returned

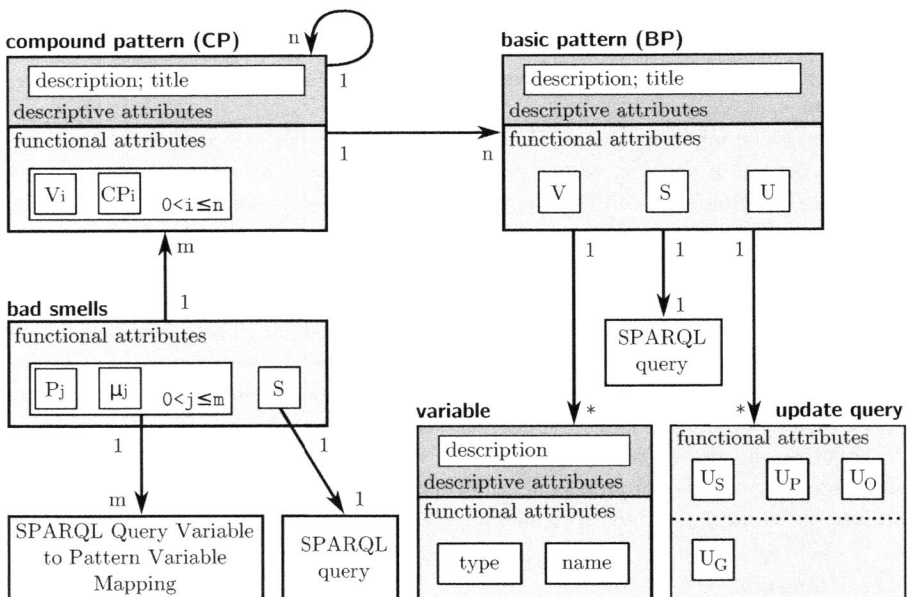

**Fig. 28.** Pattern composition with descriptive attributes, functional attributes and cardinality restrictions

result of the SELECT query. In addition, the placeholders contained in both query templates are typed in order to facilitate the classification and choreography of different evolution patterns. Please note, that in the following we will use the term variable for placeholders contained in SPARQL query templates. These should not be confused with variables contained in SPARQL graph patterns, which, however, do not play any particular role in this article. The following definition describes basic evolution patterns formally:

**Definition 8 (Basic Evolution Pattern).** *A basic evolution pattern is a tuple $(V, S, U)$, where $V$ is a set of typed variables, $S$ is a SPARQL query template with placeholders for the variables from $V$, and $U$ is a SPARQL/Update query template with placeholders referring to a result set which is generated by the SPARQL query template $S$.*

**Listing 1.1.** Basic Evolution Pattern example: moving axioms from one property to another

```
V: dtProp type: PROPERTY
   objProp type: PROPERTY
   p type: TEMP
   o type: TEMP
S: SELECT DISTINCT * WHERE {
     %dtProp% %p% %o% .
     FILTER (
       !sameTerm(%p%, rdfs:range) &&
       !sameTerm(%p%, rdf:type)
     )
   }
U: INSERT: %objProp% %p% %o% .
   DELETE: %dtProp% %p% %o% .
```

Listing 1.1 shows a basic evolution pattern, which moves axioms from one property to another. Lines 1-4 define the typed variables used in the pattern. Lines 5-11 contain the SELECT query template, while lines 12-13 contain the SPARQL/Update query template to be executed for each result of the SELECT query.

*Query preprocessor.* In order to give a SPARQL query for previously unknown entities (since they are selected by the pattern SPARQL query), we introduce an extension to SPARQL that defines two additional types of variables and preprocessor functions:

- *Pattern variables* are enclosed in % characters and will be replaced with the corresponding entity. *Input variables* are defined by the user applying the pattern (e. g. on which entity the pattern is to operate.). *Temp variables* are variables to which query results from the pattern SPARQL query are bound. They can be used in the SPARQL/Update query of the same pattern to describe triple updates. In Listing 1.1, line 12 the variable %objProp% is used to bind the newly created object property.
- *Preprocessor functions* are a means of performing certain actions with the entities bound to a variable. If e. g. the user wants URIs of a certain format or change the datatype of a created literal value, those functions can be used. We provide a number of pre-defined functions for the most common use cases.

*Compound Evolution Pattern (CP).* Basic evolution patterns alone are not sufficient to cover arbitrary evolution scenarios. Especially on higher abstraction levels of represented domain knowledge, it is feasible to represent ontology changes on the same level of abstraction. To this end, we define compound evolution patterns, consisting of several evolution patterns that are subsequently applied to a knowledge base.

**Definition 9 (Compound Evolution Pattern).** *Let* $0 < i \leq n$, $P_i$ *be (basic or compound) patterns and* $V_i$ *the corresponding sets of unbound variables in* $P_i$. *A sequence* $CP := (V_i, P_i)$ *of patterns is called a* compound pattern *(CP)*.

An example of a compound pattern for transforming a datatype property into an object property (including instance transformation) is given in listing 1.2. It consists of the following four basic sub patterns: moving property axioms, deleting datatype property, transforming instance data and creating object property.

**Listing 1.2.** Compound Evolution Pattern example: transforming a datatype into an object property while maintaining instance consistency

```
languagelanguage
// Sub pattern 1: (move axioms from dtProp to objProp)
V: dtProp type: PROPERTY
   objProp type: PROPERTY
   p type: TEMP
   o type: TEMP
S: SELECT DISTINCT * WHERE {
       %dtProp% %p% %o% .
       FILTER (
          !sameTerm(%p%,rdfs:range) &&
          !sameTerm(%p%,rdf:type)
       )
   }
U: INSERT: %objProp% %p% %o% .
   DELETE: %dtProp% %p% %o% .

// Sub pattern 2: (delete dtProp)
V: dtProp type: PROPERTY
   p type: TEMP
   o type: TEMP
S: SELECT DISTINCT * WHERE {
       %dtProp% %p% %o% .
   }
U: DELETE: %dtProp% %p% %o% .

// Sub pattern 3: (transform instance data)
V: dtProp type: PROPERTY
   inst type: TEMP
   o type: TEMP
   objProp: PROPERTY
S: SELECT DISTINCT * WHERE {
       %inst% %dtProp% %o% .
   }
U: INSERT:
   %inst% %objProp% getTempUri(getNamespace(%objProp%),%o%).
   getTempUri(getNamespace(%objProp%),%o%) rdfs:label %o%.
   DELETE: %inst% %dtProp% %o%

// Sub pattern 4: (create property)
V: objProp type: PROPERTY
S:
U: INSERT: %objProp% rdf:type owl:ObjectProperty .
```

**Evolution Pattern Processing.** Algorithm 3 outlines the evolution pattern processing. The algorithm uses an evolution pattern $P$, a graph $G$ and a set of variable bindings $B$ as input. Depending on the type of pattern (basic or compound) the following steps are performed.

---

**Algorithm 3.** Pattern execution sequence

---

**Require:** Pattern $P$
**Require:** RDF graph $G$
**Require:** Variable bindings $B$
  **if** $F$ is Basic Pattern **then**
    substitute variables in SPARQL Query according to $B$
    execute preprocessor functions in $P$
    $QR :=$ SPARQL query result of $P$ on $G$
    **for all** update patterns of $P$ as $UP$ **do**
      **if** $UP$ has graph **then**
        active graph $AG$ = graph of $UP$
      **else**
        active graph $AG$ = default graph $G$
      **end if**
      substitute variables in $UP$ according to $B$
      generate changes $CS$ of $UP$ on $AG$ with $QR$
      apply changes $CS$ to $AG$
    **end for**
  **else**
    **for all** basic patterns in compound pattern $P$ as $SP$ **do** //maintain correct order
      execute Base Pattern $SP$ //see above
    **end for**
  **end if**

---

*Basic pattern.* If $P$ is a basic pattern, the variables in the query are substituted with respect to their binding in $B$. Each of the update patterns contained in $P$ is processed as follows:

1. If the update pattern sets an explicit graph, the active graph is set to that graph, else it is set to the default graph.
2. The variables in the update pattern are substituted according to $B$
3. Changes are determined by executing the SPARQL query in $P$ on $G$.
4. The changes are then applied to the active graph.

*Compound pattern.* Compound patterns are resolved to basic patterns. For each of the basic patterns the above steps are performed. The output of the algorithm is a set of changes on the respective graphs.

**Bad Smells.** In order to assist knowledge engineers and domain experts as much as possible with the evolution of a knowledge base we also provide a formal definition for a bad smell in a certain knowledge base. In essence, a bad smell is represented via a SPARQL SELECT query, which detects a suspicious structure in a knowledge base. In most scenarios, there will be one (or multiple) evolution patterns addressing exactly the issue raised by a certain bad smell. Hence, we allow to assign one (or multiple) evolution patterns to the bad smell for resolving that issue. In order to further automatize the resolving of bad smells each evolution pattern can be assigned with a mapping from the bad smells result set to the variables used in the evolution patterns.

**Definition 10 (Bad smell).** *A bad smell is a tuple* $(S, (P_i, \mu_i))$, *where S is a SPARQL query and* $(P_i, \mu_i)$ *is a list of possible evolution patterns* $P_i$ *for resolving the bad smell with an associated mapping* $\mu_i$, *which maps results of S to the variables in* $P_i$.

**Listing 1.3.** Bad smell example: selecting statements for which the datatype of the object doesn't match the `rdfs:range` of the property

```
SELECT ?s ?p ?o
WHERE {
  ?s ?p ?o .
  ?p a owl:DatatypeProperty .
  ?p rdfs:range ?range .
  FILTER (DATATYPE(?o) != ?range)
}
```

An example of a bad smell is given in listing 1.3. It selects all statements whose object is a literal with a datatype that does not match the `rdfs:range` of the property of that statement. The result set from the bad smell query can be directly applied as input to a pattern that typecasts literal values to the correct datatype.

In certain cases a knowledge base evolution can be even performed completely automatically. This is the case if and only if both of the following conditions are met.

– The bad smell can only be resolved by exactly one evolution pattern and
– the mapping to the evolution pattern's variables is complete in the sense that all variables will be assigned values from the bad smell's query result set.

**Serialization in RDF.** To facilitate the exchange and reuse of previously defined evolution patterns we developed an RDF serialization, i.e. an RDF vocabulary for representing evolution patterns[43]. Together with an updated log publishing (such as e.g. proposed in [8]) on the Linked Data Web this facilitates the creation of an evolution ecosystem, where generic and domain specific evolution patterns are shared and reused and data cleansing and migration strategies can be also performed in network of linked knowledge bases.

### 7.2  Pattern Survey and Classification

In order to obtain a comprehensive catalog of evolution patterns we pursued a threefold strategy: (1) we performed a comprehensive literature review, (2) we looked at all combinatorial combinations of before/after states and (3) we conducted a number of interviews with knowledge engineers and domain experts, which were involved in medium-scale knowledge base construction projects and retrospectively reviewed the evolution of these knowledge bases.

*Literature review.* Most work concerned with ontology evolution patterns identifies a number of useful patterns but gives only an informal description which cannot be used for implementing an evolution software system. In [89], evolution patterns that work on the ontology level are identified. A classification of evolution patterns in four levels of abstraction is presented in [57]. The levels identified by the authors helped us in

---

[43] The vocabulary for representing evolution patterns is available at:
http://ns.aksw.org/Evolution/

**Table 7.** Combinatorially possible before/after evolution states. C, P, G stand for class, property, graph respectively. The '+' indicates that multiple entities of the same type participate in the evolution pattern. Impossible combinations are blackened out.

|      | ∅  | C+ | P+ | G+ | PC | PG | CG |
|------|----|----|----|----|----|----|----|
| ∅    | ok | ok | ok | ok |    |    |    |
| C+   | ok | ok | ok |    |    |    | ok |
| P+   | ok | ok | ok |    | ok | ok |    |
| G+   | ok |    |    | ok |    |    |    |
| PC   |    |    | ok |    |    |    |    |
| PG   |    |    | ok |    |    |    |    |
| CG   |    | ok |    |    |    |    |    |

our classification system. In the interviews we conducted, the need for representational changes was identified. Thus, we added another layer that deals with syntactic changes to resources (i.e. renaming a URI). The authors of [32] present a number of patterns with formally defined participants and execution steps. We extended the approach, providing a pattern behavior in the form of SPARQL/Update queries that can directly be built into Semantic Web applications.

*Combinatorial analysis.* In order to ensure, that we achieved a comprehensive coverage of all possible evaluation patterns we followed a combinatorial analysis. We considered all possible combinations of ontology construction elements (i.e. classes, properties and (sub-)graphs) which are potentially affected by the application of a basic evolution pattern and the possible combinations of remaining elements after the pattern has been applied. All possible combinations are displayed in Table 7. For each of the potentially possible combinations we performed an analysis whether evolution patterns actually exist in practice. The results of this analysis are also summarized in Table 8. Combinations where possible patterns can be represented as combinations of basic evolution patterns are marked with a white background. Those combinations were no basic evolution patterns exist are blackened out.

*Interviews and retrospective coverage checks.* In order to ground our findings from the literature review and combinatorial analysis, we had an in-depth look at several medium- to large-scale knowledge base construction projects. These included in particular the Vakantieland e-tourism knowledge base for the Netherlands [78], the Leipzig Professors Catalog [11] and the development of an ontology for the energy sector, which was performed by our industry partner Business Intelligence GmbH. We also retrospectively reviewed the evolution of these knowledge bases and analyzed to what extend the previously defined evolution patterns would cover the evolution steps found.

## 7.3   Implementation

The EvoPat approach was implemented as an extension to OntoWiki (cf. section 4). The general architecture of the EvoPat extension is depicted in Figure 29. It consists of four distinct components. Core of the EvoPat implementation is the *pattern engine*, which

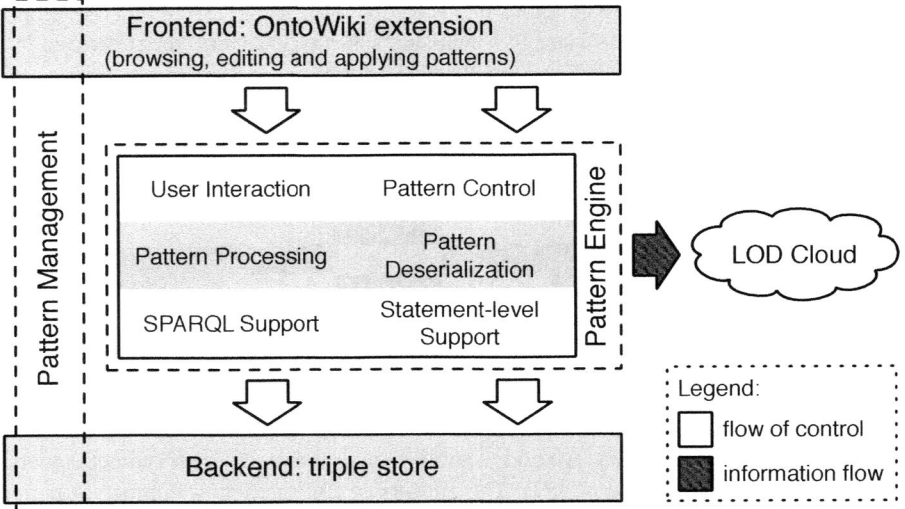

**Fig. 29.** System architecture with internal functional units and provided services. Patterns are exposed as Linked Data.

in particular handles processing, storing, versioning and exposing evolution patterns as Linked Data on the data web. It interacts via SPARQL with a triple store representing the EvoPat *backend*. The EvoPat *frontend* facilitates the user friendly browsing/selection, configuration and application of evolution patterns. The *pattern management component* as a logical component spans several architectural layers. It implements the required APIs as needed by both the user interface and the backend for managing patterns.

Different versions of ontologies resulting from applying evolution patterns can be managed through OntoWiki's versioning component. Similar to database transactions, the changes on the statement level that result from applying a certain evolution pattern can be grouped and versioned as a single change.

Figure 30 showcases the EvoPat user interface with the pattern editor and the pattern execution. The pattern editor allows to create basic and compound evolution patterns. A user-friendly form is generated, where the descriptive attributes, the variables used in the pattern and the respective SPARQL SELECT and UPDATE queries can be filled in. For pattern execution (as shown in the upper left part of Figure 30), the EvoPat implementation generates a form based on the variables definition of the evolution pattern at hand. Employing the typing of the variable a type ahead search simplifies the selection of concrete values for the variables.

*Scalability evaluation.* One of the main goals of developing EvoPat was to push as much of the evolution pattern processing down to the underlying triple store. In order to evaluate whether EvoPat lives up to this promise we evaluated the processing of selected evolution patterns with different knowledge base sizes. The results of the evaluation are summarized in Table 9. We used the *Catalogus Professorum Lipiensis* knowledge base and simply created three different versions of it in different sizes, by simply copying

**Table 8.** Overview of valid evolution patterns on four levels of abstraction

| | | *Ontology level (OWL)* |
|---|---|---|
| Before | After | Description |
| ∅ | ∅ | Trivial empty pattern (no actions taken) |
| ∅ | C+, P+ or G+ | Creating class, property or graph |
| C+, P+ or G+ | ∅ | Deleting class, property or graph |
| C+ | C+ | Subclassing, union, merging, splitting classes |
| P+ | P+ | Property axioms (functional, symmetric, domain, range, etc.) |
| G+ | G+ | Graph merging and splitting, graph annotation |
| C+ | P+ | Remodeling from class membership to distinct property value |
| P+ | C+ | Remodeling from distinct property value to class membership |
| C+ | CG | Class extraction from named graph |
| CG | C+ | Merging classes into graph |
| P | PC | Converting datatype to object property |
| PC | P | Converting object to datatype property (incl. axioms) |
| | | *Instance and data level (RDFS)* |
| Input | Output | Description |
| $I^*$ | $I$ | Instances merging |
| $I^*, C^*$ | $I^*$ | Instances reclassification |
| $I^*, P. O$ | $I^*$ | Adding data to instances |
| $I^*, P. P^*$ | $I^*$ | Generating data from existing instances data |
| $I^*, L^*$ | $I^*$ | Converting literal property values to resources |
| $I^*, R^*$ | $I^*$ | Converting resources to literal property values |
| $I^*(, P^*, O^*)$ | $I^*$ | Moving data (predicates and objects) from one instance to another |
| | | *Entity level (RDF)* |
| Input | Function | Description |
| Literal, datatype | Setting datatype on literal | Datatype added, changed or removed |
| Literal, language | Setting language on literal | Literal language added, changed or removed |
| RegExp search/ replace | regexp replace | Performs a regular expression search and replace on literal value |
| | | *Syntactic/representational level (RDF/XML, N3, etc.)* |
| Input | Function | Description |
| URI. namespace | Set URI prefix | Changes prefixes for a resource |
| URI. local name | Set local name | Changes local name of a resource |

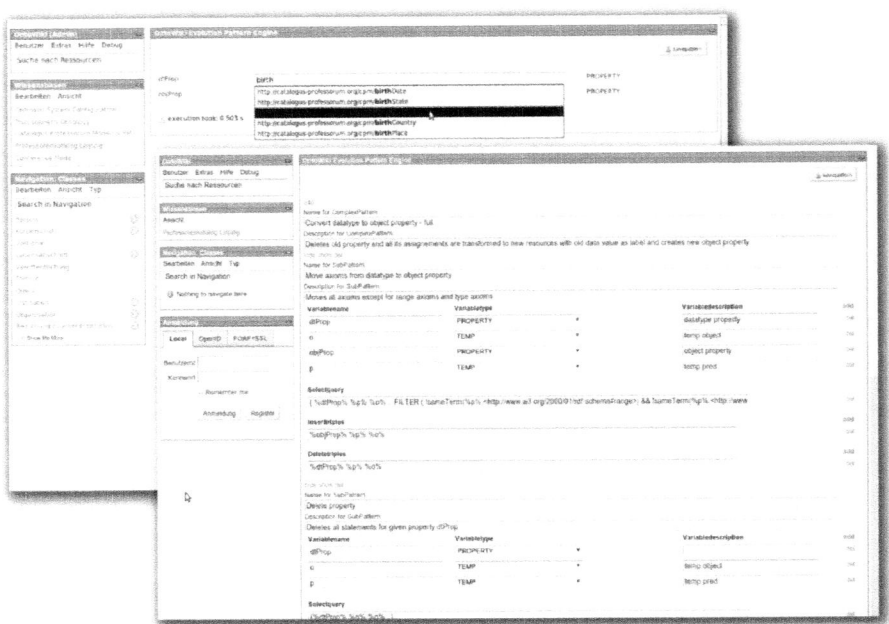

**Fig. 30.** EvoPat user interface showing pattern editor (right) and pattern execution view (left)

the data. The results of the performance evaluation show, that the evolution pattern processing grows linearly with the knowledge base size. As a consequence, EvoPat can be used with arbitrarily large knowledge bases, since the performance of the evolution pattern processing primarily depends on the speed of the underlying triple store.

### 7.4   Other Evolution Approaches

Ontology evolution has constantly been under research during the past two decades. In recent years a ramp-up is evident due to Semantic Web research activity, thus providing a more user-centric view on ontology evolution.

A comprehensive overview on the field of ontology change is given in [41]. The authors conduct an extensive literature review, extracting and defining common vocabulary as a base for discussion. They define ontology evolution as a "response to a change in the domain or conceptualization". The term ontology evolution, as used in this paper, covers what Flouris et al. refer to as ontology translation and by which they mean changes in the syntactical representation of the ontology (e. g. changing the URI of a resource).

The most closely related approach for formally specifying modular evolution patterns in a declarative manner is a categorization of pattern-based change operators in [57]. The paper defines four levels of abstraction of an ontology (element, element context, domain-specific and generic abstract level) to whose elements the said operators

**Table 9.** Scalability evaluation with two compound patterns on Catalogus Professorum Lipsiensis. The benchmarks were performed in three different sizes of the original knowledge base: original size (150K triples), 3 × the size (450K triples), 5 × the size (750K triples). Figures are quoted for two patterns each KB size.

|  | **Pattern exec.** [s] | **Affect. rsrc.** [pcs] | **Throughput** [$\frac{pcs}{s}$] |
|---|---|---|---|
| *KB size: 1 × 150K triples* | | | |
| Datatype to Object Property | 8.593 | 1300 | 151.3 |
| Class merging | 5.949 | 1500 | 252.1 |
| *KB size: 3 × 150K triples* | | | |
| Datatype to Object Property | 24.813 | 3900 | 157.2 |
| Class merging | 17.753 | 4500 | 253.4 |
| *KB size: 5 × 150K triples* | | | |
| Datatype to Object Property | 39.822 | 6500 | 163.2 |
| Class merging | 30.603 | 7500 | 245.1 |

can be applied. Taking into account the Data Web infrastructure, our approach defines an additional level on the representation layer.

Stojanovic et al. in [111] define three requirements for ontology evolution: 1) ensuring consistency, 2) allowing the user supervision of evolution and 3) advice for continuous ontology refinement. In addition, the authors identify six phases of ontology evolution, namely 1) capturing, 2) representation, 3) semantics of change, 4) implementation, 5) propagation and 6) validation of changes. The KAON API[44], implementing the approach, also introduced by the authors. Furthermore, they identify the need for representing changes on different levels of granularity. To cope with different methods of applying changes to an ontology, they introduce basic evolution strategies, which define the steps of a complex evolution process. For a given change request there are usually more than on applicable strategy, resulting in different ontologies. Seen in a broader sense, these basic evolution strategies can be combined into so called advanced evolution strategies, of which they introduce four. EvoPat's compound patterns are similar in nature to Stojanovic's basic evolution strategies, but differ in the inclusion of explicit declarative semantics by means of SPARQL/Update queries.

An interesting approach to ontology evolution with particular respect to consistency management is given by Djedidi and Aufaure [32]. A process model, a pattern and a versioning layer are proposed. If applying a change pattern results in a match to an inconsistency pattern, an alternative pattern is automatically applied by the proposed system. Furthermore, a quality assessment step is integrated into the process. The system can thus alleviate the need for user interaction by applying quality-improving patterns in an automated fashion.

Noy and Klein determine in [89] to what extent ontology evolution resembles schema evolution, which has been extensively researched in the database community. By arguing that different versions of an ontology have to be kept in parallel, they conclude that the traditional distinction between schema evolution and schema versioning is not

---

[44] http://kaon.semanticweb.org/developers

applicable to ontology evolution and ontology versioning. Even though, EvoPat distinguishes between versioning and evolution, both subsystems are closely related and cannot be used exclusively. All evolutionary changes are automatically versioned and can be reverted at any time.

A declarative update language for RDF graphs, named RUL is defined in [77]. RUL is based on the RDF query language RQL and the RDF view language RVL and ensures consistency on the RDF and RDFS levels. It, therefore, contains *primitive*, *set-oriented* and *complex* updates as compositions of primitive or complex ones. Primitive RUL updates are similar in expressiveness to SPARQL 1.1 updates. Complex updates are expressed by means of fine-grained updates on class and property instance level. Our basic evolution patterns with variable placeholders are similar to the set-oriented RUL updates (i. e. repeating the same query for several bindings). Additionally, we, however, define a functional extension that allows for arbitrarily replacing entities in a preprocessor-like manner.

Finally, applying the software engineering concept of *code smell* [42] to ontologies has been inspired by the work of Rosenfeld et al. [101]. They use *bad smells* in a Semantic Wiki context for triggering refactoring operations.

## 7.5  Conclusion

We introduced an approach to pattern-based evolution of RDF knowledge bases. By considering the complete stack of Semantic Web knowledge representation techniques including its syntactic infrastructure as opposed to just the ontology layer, our approach fulfills additional requirements identified for example in user interviews (cf. section 7.2). We provide a concrete implementation that leverages the plug-in architecture of OntoWiki, our semantic collaboration platform and framework. Thus, our implementation can make use of existing functionality of the OntoWiki framework like versioning of RDF knowledge bases. Compared to existing approaches for knowledge base evolution, the declarative, pattern-based EvoPat approach has a number of advantages:

- EvoPat is a *unified method*, which works for both data evolution and ontology refactoring.
- The modularized, *declarative* definition of evolution patterns is relatively simple compared to an imperative description of evolution. It allows domain experts and knowledge engineers to amend the ontology structure and modify data with just a few clicks.
- Combined with our RDF representation of evolution patterns and their exposure on the Linked Data Web, EvoPat facilitates the development of an *evolution pattern ecosystem*, where patterns can be shared and reused on the Data Web.
- The declarative definition of bad smells and corresponding evolution patterns promotes the (semi-)automatic *improvement of information quality*.

On the limitations side, EvoPat, currently, only ensures consistency through the definition of consistency-preserving patterns by the knowledge engineer. User-defined patterns can, however, lead to inconsistent knowledge bases. An approach that ensures consistency by proposing only those patterns whose application will not result in an inconsistent ontology, would thus be desirable. A straightforward (but admittedly not very scalable) solution to this problem is to combine EvoPat with a reasoner and test the application of a pattern employing the reasoner before its actual application in order to ensure correctness.

As opposed to bad smells, which indicate modeling problems, a promising approach is also to share and reuse modeling best practices. A problem which must be solved in this regard, is the formalization and elicitation of a user's modeling requirements. A related idea for future work is the consumption of Linked Data. Our current implementation publishes evolution patterns on the Data Web but makes no use of gathering further information about resources. Doing so, could deliver hints for the applicability of specific patterns.

In a number of application projects we learned, that a key factor for the success of a knowledge engineering project is the efficient co-design of knowledge-bases and knowledge-based applications. Through the declarative definition of evolution with EvoPat it becomes possible to (semi-)automatize this co-design, since a knowledge base refactoring can trigger code refactoring and vice versa.

## 8   Outlook and Future Challenges

Although the different approaches for aspects of the Linked Data life-cycle as presented in this chapter are already working together, much more effort must be done to further integrate them in ways that they mutually fertilize themselves. The discovery of new links or the authoring of new resource descriptions, for example, should automatically trigger the enrichment of the linked knowledge bases. The enrichment in turn can trigger the application of inconsistency detection and repair techniques. The browsing and exploration paths followed by end-users can be taken into account for machine learning techniques to refine the knowledge bases etc. etc. Ultimately, when the different aspects of Linked Data management are fully integrated we envision the Web of Data becoming a washing machine for knowledge. A progress in one particular aspect will automatically trigger improvements in many other ones as well. In the following we outline some research challenges and promising research directions regarding some of the Linked Data management aspects.

*Extraction.*   One promising research direction with regard to the extraction from unstructured sources is the development of standardized, LOD enabled integration interfaces between existing NLP tools. An open question is whether and how efficient bidirectional synchronization between extraction source and target knowledge base can be established. With regard to the extraction from structured sources (e.g. relational, XML) we need a declarative syntax and semantics for data model transformations. Some orthogonal challenges include the use of LOD as background knowledge and the representation and tracing of provenance information.

*Authoring.*   Current Semantic Wikis still suffer from a lack of scalability. Hence, an important research and development target are large-scale Semantic Wikis, which include functionality for access control and provenance. In order to further flexibilize and simplify the authoring an adaptive choreography of editing widgets based on underlying data structures is needed. Also, the joint authoring of unstructured and structured

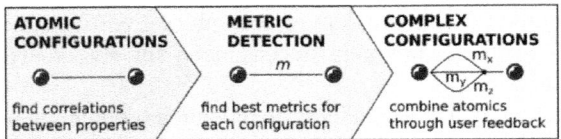

**Fig. 31.** Three steps to the supervised discovery of linking configurations

sources (i.e. HTML/RDFa authoring) and better support for the integrated semantic annotation of other modalities such as images, audio, video is of paramount importance.

*Towards Zero-Configuration Linking.* While link discovery frameworks such as LIMES make it easier to compute links and link candidates for large instance sets, they do not address the second challenge of instance linking, i.e., the discovery of appropriate linking configurations for Linked Data. First steps in this direction follow the simple yet efficient approach for discovering such configurations shown in Figure 31.

The first step consists of detecting *atomic configurations* automatically. We call a configuration atomic when the mapping condition of this configuration links exactly one property of the source with one property of the target instances. The detection of an atomic configuration is carried out by applying a simple metric to all frequent properties of the source and target instances. For each property pair, the average best matches are computed for a limited number of source instances to ensure acceptable runtimes. Only those metric pairs which generate average best match values beyond that of random noise are considered in the subsequent steps. This step can be very time-demanding as mappings have to be carried out for each possible combination of property pairs. Thus, efficient mapping approaches such as LIMES are central for reducing the waiting time of the user.

The second step consists of detecting the *best metric for each atomic configuration*. For this purpose, the similarity of the matching instances detected by using atomic configurations is computed anew by using other metrics. The metric that yields the highest average best match score is then considered the best metric for matching these properties. In most cases, atomic configurations are sufficient to achieve high precision [119]. Still, complex configurations that combine several properties are needed for some difficult tasks.

The third step aims to compute such *complex configurations* by using active learning. Here, the user is presented with the best matching pairs from each of the current configurations. These results are sorted according to the average of the similarity of matching instances pairs. The user can then choose sets of pairs that yield promising results. For $n$ chosen configurations $c_1, ..., c_n$, a set of configurations of the form $\sum_{i=1...n} \alpha_i c_i$ are created, whereby the coefficients $\alpha_i$ can either be set by the user or generated automatically. The resulting configurations are then used as next input for the third step until the user selects a configuration as being optimal. Finally, the results of the chosen configuration on all instances from the source data set are returned.

**Acknowledgments.** We would like to thank our colleagues from the AKSW research group in Leipzig as well as the LOD2 and LATC project consortia, without whom writing this chapter would not have been possible. In particular, we would like to thank Christian Bizer and Tom Heath, whose Chaper 2 of the book 'Linked Data – Evolving the Web into a Global Data Space' [48] served as a blueprint for Section 2; Sebastian Hellmann, Jörg Unbehauen for their contributions to Section 3, Sebastian Tramp, Michael Martin, Norman Heino, Phillip Frischmuth and Thomas Riechert for their contributions to the development of OntoWiki as described in Section 4. We are also very extremely grateful to the recommendations provided by the anonymous reviewers from the Reasoning Web 20011 Programme Committee for this chapter. This work was supported by a grant from the European Union's 7th Framework Programme provided for the projects LOD2 (GA no. 257943), LATC (GA no. 256975) and the Eureka project SCMS.

# References

1. Resource description framework (RDF): Concepts and abstract syntax. Technical report, W3C (February 2004)
2. Adida, B., Birbeck, M., McCarron, S., Pemberton, S.: RDFa in XHTML: Syntax and processing – a collection of attributes and processing rules for extending XHTML to support RDF. W3C Recommendation (October 2008), http://www.w3.org/TR/rdfa-syntax/
3. Agichtein, E., Gravano, L.: Snowball: Extracting relations from large plain-text collections. In: ACM DL, pp. 85–94 (2000)
4. Agresti, A.: An Introduction to Categorical Data Analysis, 2nd edn. Wiley Interscience, Hoboken (1997)
5. Amsler, R.: Research towards the development of a lexical knowledge base for natural language processing. SIGIR Forum 23, 1–2 (1989)
6. Auer, S., Bizer, C., Kobilarov, G., Lehmann, J., Cyganiak, R., Ives, Z.G.: DBpedia: A nucleus for a web of open data. In: Aberer, K., Choi, K.-S., Noy, N., Allemang, D., Lee, K.-I., Nixon, L.J.B., Golbeck, J., Mika, P., Maynard, D., Mizoguchi, R., Schreiber, G., Cudré-Mauroux, P. (eds.) ASWC 2007 and ISWC 2007. LNCS, vol. 4825, pp. 722–735. Springer, Heidelberg (2007)
7. Auer, S., Dietzold, S., Lehmann, J., Hellmann, S., Aumueller, D.: Triplify: Light-weight linked data publication from relational databases. In: Quemada, J., León, G., Maarek, Y.S., Nejdl, W. (eds.) Proceedings of the 18th International Conference on World Wide Web, WWW 2009, Madrid, Spain, April 20-24, 2009, pp. 621–630. ACM Press, New York (2009)
8. Auer, S., Dietzold, S., Lehmann, J., Hellmann, S., Aumueller, D.: Triplify: light-weight linked data publication from relational databases. In: Quemada, J., León, G., Maarek, Y.S., Nejdl, W. (eds.) Proceedings of the 18th International Conference on World Wide Web, WWW 2009, Madrid, Spain, April 20-24, pp. 621–630. ACM Press, New York (2009)
9. Auer, S., Dietzold, S., Riechert, T.: OntoWiki – A tool for social, semantic collaboration. In: Cruz, I., Decker, S., Allemang, D., Preist, C., Schwabe, D., Mika, P., Uschold, M., Aroyo, L.M. (eds.) ISWC 2006. LNCS, vol. 4273, pp. 736–749. Springer, Heidelberg (2006)
10. Auer, S., Herre, H.: A versioning and evolution framework for RDF knowledge bases. In: Virbitskaite, I., Voronkov, A. (eds.) PSI 2006. LNCS, vol. 4378, pp. 55–69. Springer, Heidelberg (2007)

11. Augustin, C., Kuchta, B., Morgenstern, U., Riechert, T.: Datenbank und website catalogus professorum lipsiensis. ein sozialstatistisches analyseinstrumentarium und seine repräsentation im netz. In: Schattkowsky, M., Metasch, F. (eds.) Biografische Lexika im Internet. Bausteine, vol. 14, TUDPress, Verlag der Wissenschaften GmbH, Dresden (2009)
12. Aumüller, D.: Semantic Authoring and Retrieval within a Wiki (WikSAR). In: ESWC (May 2005), http://wiksar.sf.net
13. Baader, F., Ganter, B., Sattler, U., Sertkaya, B.: Completing description logic knowledge bases using formal concept analysis. In: IJCAI 2007. AAAI Press, Menlo Park (2007)
14. Baader, F., Sertkaya, B., Turhan, A.-Y.: Computing the least common subsumer w.r.t. a background terminology. J. Applied Logic 5(3), 392–420 (2007)
15. Badea, L., Nienhuys-Cheng, S.-H.: A refinement operator for description logics. In: Cussens, J., Frisch, A.M. (eds.) ILP 2000. LNCS (LNAI), vol. 1866, pp. 40–59. Springer, Heidelberg (2000)
16. Baxter, R., Christen, P., Churches, T.: A comparison of fast blocking methods for record linkage. In: KDD 2003 Workshop on Data Cleaning, Record Linkage, and Object Consolidation (2003)
17. Ben-David, D., Domany, T., Tarem, A.: Enterprise data classification using semantic web technologies. In: Patel-Schneider, P.F., Pan, Y., Hitzler, P., Mika, P., Zhang, L., Pan, J.Z., Horrocks, I., Glimm, B. (eds.) ISWC 2010, Part II. LNCS, vol. 6497, pp. 66–81. Springer, Heidelberg (2010)
18. Berners-Lee, T.: Notation 3 (1998), http://www.w3.org/DesignIssues/Notation3.html
19. Berners-Lee, T., Fielding, R.T., Masinter, L.: Uniform resource identifiers (URI): Generic syntax. Internet RFC 2396 (August 1998)
20. Bhagdev, R., Chapman, S., Ciravegna, F., Lanfranchi, V., Petrelli, D.: Hybrid search: Effectively combining keywords and semantic searches. In: Bechhofer, S., Hauswirth, M., Hoffmann, J., Koubarakis, M. (eds.) ESWC 2008. LNCS, vol. 5021, pp. 554–568. Springer, Heidelberg (2008), http://dx.doi.org/10.1007/978-3-540-68234-9_41, doi:10.1007/978-3-540-68234-9_41
21. Bilenko, M., Kamath, B., Mooney, R.J.: Adaptive blocking: Learning to scale up record linkage. In: ICDM 2006, pp. 87–96. IEEE, Los Alamitos (2006)
22. Bleiholder, J., Naumann, F.: Data fusion. ACM Comput. Surv. 41(1), 1–41 (2008)
23. Blumer, A., Ehrenfeucht, A., Haussler, D., Warmuth, M.K.: Occam's razor. In: Readings in Machine Learning, pp. 201–204. Morgan Kaufmann, San Francisco (1990)
24. Brickley, D., Guha, R.V.: RDF Vocabulary Description Language 1.0: RDF Schema. In: W3C Recommendation, W3C (February 2004), http://www.w3.org/TR/2004/REC-rdf-schema-20040210/
25. Brin, S.: Extracting patterns and relations from the world wide web. In: Atzeni, P., Mendelzon, A.O., Mecca, G. (eds.) WebDB 1998. LNCS, vol. 1590, pp. 172–183. Springer, Heidelberg (1999)
26. Choi, N., Song, I.-Y., Han, H.: A survey on ontology mapping. SIGMOD Record 35(3), 34–41 (2006)
27. Coates-Stephens, S.: The analysis and acquisition of proper names for the understanding of free text. Computers and the Humanities 26, 441–456 (1992) 10.1007/BF00136985
28. Cohen, W.W., Borgida, A., Hirsh, H.: Computing least common subsumers in description logics. In: AAAI 1992, pp. 754–760 (1992)
29. Cohen, W.W., Hirsh, H.: Learning the CLASSIC description logic: Theoretical and experimental results. In: KR 1994, pp. 121–133. Morgan Kaufmann, San Francisco (1994)
30. Curran, J.R., Clark, S.: Language independent ner using a maximum entropy tagger. In: Proceedings of the seventh conference on Natural language learning at HLT-NAACL 2003, vol. 4, pp. 164–167. Association for Computational Linguistics, Morristown (2003)

31. d'Amato, C., Fanizzi, N., Esposito, F.: A note on the evaluation of inductive concept clas-sification procedures. In: Gangemi, A., Keizer, J., Presutti, V., Stoermer, H. (eds.) SWAP 2008. CEUR Workshop Proceedings, vol. 426, CEUR-WS.org (2008)

32. Djedidi, R., Aufaure, M.-A.: ONTO-EVO⁴L an Ontology Evolution Approach Guided by Pattern Modeling and Quality Evaluation. In: Link, S., Prade, H. (eds.) FoIKS 2010. LNCS, vol. 5956, pp. 286–305. Springer, Heidelberg (2010)

33. Elmagarmid, A.K., Ipeirotis, P.G., Verykios, V.S.: Duplicate record detection: A survey. IEEE Transactions on Knowledge and Data Engineering 19, 1–16 (2007)

34. Ermilov, T., Heino, N., Tramp, S., Auer, S.: OntoWiki Mobile - Knowledge Management in your Pocket. In: Antoniou, G., Grobelnik, M., Simperl, E., Parsia, B., Plexousakis, D., De Leenheer, P., Pan, J. (eds.) ESWC 201. LNCS, vol. 6644, Springer, Heidelberg (2011)

35. Esposito, F., Fanizzi, N., Iannone, L., Palmisano, I., Semeraro, G.: Knowledge-intensive in-duction of terminologies from metadata. In: McIlraith, S.A., Plexousakis, D., van Harmelen, F. (eds.) ISWC 2004. LNCS, vol. 3298, pp. 441–455. Springer, Heidelberg (2004)

36. Etzioni, O., Cafarella, M., Downey, D., Popescu, A.-M., Shaked, T., Soderland, S., Weld, D.S., Yates, A.: Unsupervised named-entity extraction from the web: an experimental study. Artif. Intell. 165, 91–134 (2005)

37. Euzenat, J., Shvaiko, P.: Ontology matching. Springer, Heidelberg (2007)

38. Fanizzi, N., d'Amato, C., Esposito, F.: DL-FOIL Concept Learning in Description Log-ics. In: Železný, F., Lavrač, N. (eds.) ILP 2008. LNCS (LNAI), vol. 5194, pp. 107–121. Springer, Heidelberg (2008)

39. Fielding, R., Gettys, J., Mogul, J., Frystyk, H., Masinter, L., Leach, P., Berners-Lee, T.: Hypertext transfer protocol – http/1.1 (rfc 2616). Request For Comments (1999), http://www.ietf.org/rfc/rfc2616.txt (accessed July 7, 2006)

40. Finkel, J.R., Grenager, T., Manning, C.: Incorporating non-local information into informa-tion extraction systems by gibbs sampling. In: Proceedings of the 43rd Annual Meeting on Association for Computational Linguistics, ACL 2005, pp. 363–370. Association for Computational Linguistics, Morristown (2005)

41. Flouris, G., Manakanatas, D., Kondylakis, H., Plexousakis, D., Antoniou, G.: Ontology change: classification and survey. Knowledge Eng. Review 23(2), 117–152 (2008)

42. Fowler, M.: Refactoring: Improving the Design of Existing Code. Addison-Wesley, Reading (1999)

43. Frank, E., Paynter, G.W., Witten, I.H., Gutwin, C., Nevill-Manning, C.G.: Domain-specific keyphrase extraction. In: Proceedings of the Sixteenth International Joint Conference on Artificial Intelligence, IJCAI 1999, pp. 668–673. Morgan Kaufmann, San Francisco (1999)

44. Frey, B.J., Dueck, D.: Clustering by passing messages between data points. Science 315, 972–976 (2007)

45. Glaser, H., Millard, I.C., Sung, W.-K., Lee, S., Kim, P., You, B.-J.: Research on linked data and co-reference resolution. Technical report, University of Southampton (2009)

46. Grishman, R., Yangarber, R.: Nyu: Description of the Proteus/Pet system as used for MUC-7 ST. In: MUC-7, Morgan Kaufmann, San Francisco (1998)

47. Harabagiu, S., Bejan, C.A., Morarescu, P.: Shallow semantics for relation extraction. In: IJCAI, pp. 1061–1066 (2005)

48. Heath, T., Bizer, C.: Linked Data - Evolving the Web into a Global Data Space. Synthesis Lectures on the Semantic Web:Theory and Technology, vol. 1. Morgan & Claypool, San Francisco (2011)

49. Heino, N., Dietzold, S., Martin, M., Auer, S.: Developing semantic web applications with the ontowiki framework. In: Pellegrini, T., Auer, S., Tochtermann, K., Schaffert, S. (eds.) Networked Knowledge - Networked Media. SCI, vol. 221, pp. 61–77. Springer, Heidelberg (2009)

50. Hellmann, S., Lehmann, J., Auer, S.: Learning of OWL class descriptions on very large knowledge bases. Int. J. Semantic Web Inf. Syst. 5(2), 25–48 (2009)
51. Horridge, M., Patel-Schneider, P.F.: Manchester syntax for OWL 1.1. In: OWLED 2008 (2008)
52. HTML 5: A vocabulary and associated APIs for HTML and XHTML. W3C Working Draft (August 2009), http://www.w3.org/TR/2009/WD-html5-20090825/
53. Iannone, L., Palmisano, I.: An Algorithm Based on Counterfactuals for Concept Learning in the Semantic Web. In: Ali, M., Esposito, F. (eds.) IEA/AIE 2005. LNCS (LNAI), vol. 3533, pp. 370–379. Springer, Heidelberg (2005)
54. Iannone, L., Palmisano, I., Fanizzi, N.: An algorithm based on counterfactuals for concept learning in the semantic web. Applied Intelligence 26(2), 139–159 (2007)
55. Inan, A., Kantarcioglu, M., Bertino, E., Scannapieco, M.: A hybrid approach to private record linkage. In: ICDE, pp. 496–505 (2008)
56. Jacobs, I., Walsh, N.: Architecture of the world wide web. In: World Wide Web Consortium, Recommendation REC-webarch-20041215, vol. 1 (December 2004)
57. Javed, M., Abgaz, Y.M., Pahl, C.: A Pattern-Based Framework of Change Operators for Ontology Evolution. In: Meersman, R., Herrero, P., Dillon, T. (eds.) OTM 2009 Workshops. LNCS, vol. 5872, pp. 544–553. Springer, Heidelberg (2009)
58. Kim, S.N., Kan, M.-Y.: Re-examining automatic keyphrase extraction approaches in scientific articles. In: Proceedings of the Workshop on Multiword Expressions: Identification, Interpretation, Disambiguation and Applications, MWE 2009, pp. 9–16. Association for Computational Linguistics, Stroudsburg (2009)
59. Kim, S.N., Medelyan, O., Kan, M.-Y., Baldwin, T.: Semeval-2010 task 5: Automatic keyphrase extraction from scientific articles. In: Proceedings of the 5th International Workshop on Semantic Evaluation, SemEval 2010, pp. 21–26. Association for Computational Linguistics, Stroudsburg (2010)
60. Köpcke, H., Thor, A., Rahm, E.: Comparative evaluation of entity resolution approaches with fever. In: Proc. VLDB Endow., vol. 2(2), pp. 1574–1577 (2009)
61. Krzsch, M., Vrandecic, D., Vökel, M., Haller, H., Studer, R.: Semantic wikipedia. Journal of Web Semantics 5, 251–261 (2007)
62. Lehmann, J.: Hybrid Learning of Ontology Classes. In: Perner, P. (ed.) MLDM 2007. LNCS (LNAI), vol. 4571, pp. 883–898. Springer, Heidelberg (2007)
63. Lehmann, J.: DL-Learner: learning concepts in description logics. Journal of Machine Learning Research (JMLR) 10, 2639–2642 (2009)
64. Lehmann, J.: Learning OWL Class Expressions. PhD thesis, University of Leipzig, PhD in Computer Science (2010)
65. Lehmann, J.: Ontology learning. In: Proceedings of Reasoning Web Summer School (2010)
66. Lehmann, J., Auer, S., Bühmann, L., Tramp, S.: Class expression learning for ontology engineering. Journal of Web Semantics 9, 71–81 (2011)
67. Lehmann, J., Hitzler, P.: Foundations of Refinement Operators for Description Logics. In: Blockeel, H., Ramon, J., Shavlik, J., Tadepalli, P. (eds.) ILP 2007. LNCS (LNAI), vol. 4894, pp. 161–174. Springer, Heidelberg (2008)
68. Lehmann, J., Hitzler, P.: A refinement operator based learning algorithm for the $\mathcal{ALC}$ description logic. In: Blockeel, H., Ramon, J., Shavlik, J., Tadepalli, P. (eds.) ILP 2007. LNCS (LNAI), vol. 4894, pp. 147–160. Springer, Heidelberg (2008)
69. Lehmann, J., Hitzler, P.: Concept learning in description logics using refinement operators. Machine Learning Journal 78(1-2), 203–250 (2010)
70. Leuf, B., Cunningham, W.: The Wiki Way: Collaboration and Sharing on the Internet. Professional. Addison-Wesley Professional, Reading (2001)
71. Levenshtein, V.I.: Binary codes capable of correcting deletions, insertions, and reversals. Technical Report 8 (1966)

72. Lisi, F.A.: Building rules on top of ontologies for the semantic web with inductive logic programming. Theory and Practice of Logic Programming 8(3), 271–300 (2008)
73. Lisi, F.A., Esposito, F.: Learning SHIQ+log rules for ontology evolution. In: SWAP 2008. CEUR Workshop Proceedings, vol. 426, CEUR-WS.org (2008)
74. Lohmann, S., Heim, P., Auer, S., Dietzold, S., Riechert, T.: Semantifying requirements engineering the softwiki approach. In: Proceedings of the 4th International Conference on Semantic Technologies (I-SEMANTICS 2008), J.UCS, pp. 182–185 (2008)
75. Lopez, V., Uren, V., Sabou, M.R., Motta, E.: Cross ontology query answering on the semantic web: an initial evaluation. In: K-CAP 2009, pp. 17–24. ACM, New York (2009)
76. Ma, L., Sun, X., Cao, F., Wang, C., Wang, X., Kanellos, N., Wolfson, D., Pan, Y.: Semantic enhancement for enterprise data management. In: Bernstein, A., Karger, D.R., Heath, T., Feigenbaum, L., Maynard, D., Motta, E., Thirunarayan, K. (eds.) ISWC 2009. LNCS, vol. 5823, pp. 876–892. Springer, Heidelberg (2009)
77. Magiridou, M., Sahtouris, S., Christophides, V., Koubarakis, M.: RUL: A declarative update language for RDF. In: Gil, Y., Motta, E., Benjamins, V.R., Musen, M.A. (eds.) ISWC 2005. LNCS, vol. 3729, pp. 506–521. Springer, Heidelberg (2005)
78. Martin, M.: Exploring the netherlands on a semantic path. In: Auer, S., Bizer, C., Müller, C., Zhdanova, A. (eds.) Proceedings of the 1st Conference on Social Semantic Web, Leipzig, Germany. LNI, vol. 113, p. 179. Bonner Köllen Verlag (2007) ISSN 1617-5468
79. Matsuo, Y., Ishizuka, M.: Keyword Extraction From A Single Document Using Word Co-Occurrence Statistical Information. International Journal on Artificial Intelligence Tools 13(1), 157–169 (2004)
80. McBride, B., Beckett, D.: Rdf/xml syntax specification. W3C Recommendation (February 2004)
81. McCusker, J., McGuinness, D.: Towards identity in linked data. In: Proceedings of OWL Experiences and Directions Seventh Annual Workshop (2010)
82. Moats, R.: Urn syntax. Internet RFC 2141 (May 1997)
83. Nadeau, D.: Semi-Supervised Named Entity Recognition: Learning to Recognize 100 Entity Types with Little Supervision. PhD thesis, University of Ottawa (2007)
84. Nadeau, D., Sekine, S.: A survey of named entity recognition and classification. Linguisticae Investigationes 30(1), 3–26 (2007)
85. Nadeau, D., Turney, P., Matwin, S.: Unsupervised named-entity recognition: Generating gazetteers and resolving ambiguity, pp. 266–277 (2006)
86. Nguyen, D.P.T., Matsuo, Y., Ishizuka, M.: Relation extraction from wikipedia using subtree mining. In: AAAI, pp. 1414–1420 (2007)
87. Nguyen, T., Kan, M.-Y.: Keyphrase Extraction, pp. 317–326. Scientific Publications, Singapore (2007)
88. Nienhuys-Cheng, S.-H., de Wolf, R. (eds.): Foundations of Inductive Logic Programming. LNCS, vol. 1228. Springer, Heidelberg (1997)
89. Noy, N.F., Klein, M.C.A.: Ontology Evolution: Not the Same as Schema Evolution. Knowl. Inf. Syst. 6(4), 428–440 (2004)
90. Oren, E.: SemperWiki: A Semantic Personal Wiki. In: Decker, S., Park, J., Quan, D., Sauermann, L. (eds.) Proc. of Semantic Desktop Workshop at the ISWC, Galway, Ireland, November 6, vol. 175 (2005)
91. Pantel, P., Pennacchiotti, M.: Espresso: Leveraging generic patterns for automatically harvesting semantic relations. In: ACL, pp. 113–120. ACL Press (2006)
92. Park, Y., Byrd, R.J., Boguraev, B.K.: Automatic glossary extraction: beyond terminology identification. In: Proceedings of the 19th International Conference on Computational Linguistics, COLING 2002, vol. 1, pp. 1–7. Association for Computational Linguistics, USA (2002)

93. Pasca, M., Lin, D., Bigham, J., Lifchits, A., Jain, A.: Organizing and searching the world wide web of facts - step one: the one-million fact extraction challenge. In: Proceedings of the 21st National Conference on Artificial Intelligence, vol. 2, pp. 1400–1405. AAAI Press, Menlo Park (2006)

94. Patel-Schneider, P.F., Hayes, P., Horrocks, I.: OWL Web Ontology Language - Semantics and Abstract Syntax. W3c:rec, W3C (February 10, 2004),
http://www.w3.org/TR/owl-semantics/

95. Rahm, E.: Schema Matching and Mapping. Springer, Heidelberg (2011)

96. Rahm, E., Bernstein, P.A.: A survey of approaches to automatic schema matching. The VLDB Journal 10, 334–350 (2001)

97. Raimond, Y., Sutton, C., Sandler, M.: Automatic interlinking of music datasets on the semantic web. In: 1st Workshop about Linked Data on the Web (2008)

98. Riechert, T., Lauenroth, K., Lehmann, J., Auer, S.: Towards semantic based requirements engineering. In: Proceedings of the 7th International Conference on Knowledge Management, I-KNOW (2007)

99. Riechert, T., Morgenstern, U., Auer, S., Tramp, S., Martin, M.: Knowledge engineering for historians on the example of the *catalogus professorum lipsiensis*. In: Patel-Schneider, P.F., Pan, Y., Hitzler, P., Mika, P., Zhang, L., Pan, J.Z., Horrocks, I., Glimm, B. (eds.) ISWC 2010, Part II. LNCS, vol. 6497, pp. 225–240. Springer, Heidelberg (2010)

100. Rieß, C., Heino, N., Tramp, S., Auer, S.: EvoPat – pattern-based evolution and refactoring of RDF knowledge bases. In: Patel-Schneider, P.F., Pan, Y., Hitzler, P., Mika, P., Zhang, L., Pan, J.Z., Horrocks, I., Glimm, B. (eds.) ISWC 2010, Part I. LNCS, vol. 6496, pp. 647–662. Springer, Heidelberg (2010)

101. Rosenfeld, M., Fernández, A., Díaz, A.: Semantic Wiki Refactoring. A strategy to assist Semantic Wiki evolution. In: Proceedings of the Fifth Workshop on Semantic Wikis (SemWiki 2010), co-located with 7th European Semantic Web Conference, ESWC 2010 (2010)

102. Rudolph, S.: Exploring relational structures via FLE. In: Wolff, K.E., Pfeiffer, H.D., Delugach, H.S. (eds.) ICCS 2004. LNCS (LNAI), vol. 3127, pp. 196–212. Springer, Heidelberg (2004)

103. Sahoo, S.S., Halb, W., Hellmann, S., Idehen, K., Thibodeau Jr, T., Auer, S., Sequeda, J., Ezzat, A.: A survey of current approaches for mapping of relational databases to rdf (January 2009)

104. Sampson, G.: How fully does a machine-usable dictionary cover english text. Literary and Linguistic Computing 4(1) (1989)

105. Sauermann, L., Cyganiak, R.: Cool uris for the semantic web. W3C Interest Group Note (December 2008)

106. Schaffert, S.: Ikewiki: A semantic wiki for collaborative knowledge management. In: Proceedings of the 1st International Workshop on Semantic Technologies in Collaborative Applications (STICA), (2006)

107. Scharffe, F., Liu, Y., Zhou, C.: Rdf-ai: an architecture for rdf datasets matching, fusion and interlink. In: Proc. IJCAI 2009 IR-KR Workshop (2009)

108. Sertkaya, B.: OntocomP system description. In: Grau, B.C., Horrocks, I., Motik, B., Sattler, U. (eds.) Proceedings of the 22nd International Workshop on Description Logics (DL 2009, Oxford, UK, July 27-30. CEUR Workshop Proceedings, vol. 477, CEUR-WS.org (2009)

109. Shvaiko, P., Euzenat Ten, J.: challenges for ontology matching. Technical report (August 1, 2008)

110. Souzis, A.: Building a Semantic Wiki. IEEE Intelligent Systems 20(5), 87–91 (2005)

111. Stojanovic, L., Maedche, A., Motik, B., Stojanovic, N.: User-Driven Ontology Evolution Management. In: Gómez-Pérez, A., Benjamins, V.R. (eds.) EKAW 2002. LNCS (LNAI), vol. 2473, p. 285. Springer, Heidelberg (2002)

112. Thielen, C.: An approach to proper name tagging for german. In: Proceedings of the EACL 1995 SIGDAT Workshop (1995)
113. Tramp, S., Frischmuth, P., Ermilov, T., Auer, S.: Weaving a social data web with semantic pingback. In: Cimiano, P., Pinto, H.S. (eds.) EKAW 2010. LNCS, vol. 6317, pp. 135–149. Springer, Heidelberg (2010)
114. Tramp, S., Heino, N., Auer, S., Frischmuth, P.: RDFauthor: Employing rDFa for collaborative knowledge engineering. In: Cimiano, P., Pinto, H.S. (eds.) EKAW 2010. LNCS, vol. 6317, pp. 90–104. Springer, Heidelberg (2010)
115. Turney, P.D.: Coherent keyphrase extraction via web mining. In: Proceedings of the 18th International Joint Conference on Artificial Intelligence, pp. 434–439. Morgan Kaufmann, San Francisco (2003)
116. Urbani, J., Kotoulas, S., Maassen, J., van Harmelen, F., Bal, H.: OWL reasoning with webPIE: Calculating the closure of 100 billion triples. In: Aroyo, L., Antoniou, G., Hyvönen, E., ten Teije, A., Stuckenschmidt, H., Cabral, L., Tudorache, T. (eds.) ESWC 2010. LNCS, vol. 6088, pp. 213–227. Springer, Heidelberg (2010)
117. Völker, J., Rudolph, S.: Fostering web intelligence by semi-automatic OWL ontology refinement. In: Web Intelligence, pp. 454–460. IEEE, Los Alamitos (2008)
118. Völker, J., Vrandečić, D., Sure, Y., Hotho, A.: Learning Disjointness. In: Franconi, E., Kifer, M., May, W. (eds.) ESWC 2007. LNCS, vol. 4519, pp. 175–189. Springer, Heidelberg (2007)
119. Volz, J., Bizer, C., Gaedke, M., Kobilarov, G.: Discovering and Maintaining Links on the Web of Data. In: Bernstein, A., Karger, D.R., Heath, T., Feigenbaum, L., Maynard, D., Motta, E., Thirunarayan, K. (eds.) ISWC 2009. LNCS, vol. 5823, pp. 650–665. Springer, Heidelberg (2009)
120. Walker, D., Amsler, R.: The use of machine-readable dictionaries in sublanguage analysis. Analysing Language in Restricted Domains (1986)
121. Wang, G., Yu, Y., Zhu, H.: PORE: Positive-Only Relation Extraction from Wikipedia Text. In: Aberer, K., Choi, K.-S., Noy, N., Allemang, D., Lee, K.-I., Nixon, L.J.B., Golbeck, J., Mika, P., Maynard, D., Mizoguchi, R., Schreiber, G., Cudré-Mauroux, P. (eds.) ASWC 2007 and ISWC 2007. LNCS, vol. 4825, pp. 580–594. Springer, Heidelberg (2007)
122. Watanabe, H., Muggleton, S.: Can ILP Be Applied to Large Datasets? In: De Raedt, L. (ed.) ILP 2009. LNCS, vol. 5989, pp. 249–256. Springer, Heidelberg (2010)
123. Winkler, W.: The state of record linkage and current research problems. Technical report, Statistical Research Division, U.S. Bureau of the Census (1999)
124. Winkler, W.: Overview of record linkage and current research directions. Technical report, Bureau of the Census - Research Report Series (2006)
125. Wu, H., Zubair, M., Maly, K.: Harvesting social knowledge from folksonomies. In: Proceedings of the seventeenth conference on Hypertext and hypermedia, HYPERTEXT 2006, pp. 111–114. ACM, New York (2006)
126. Yan, Y., Okazaki, N., Matsuo, Y., Yang, Z., Ishizuka, M.: Unsupervised relation extraction by mining wikipedia texts using information from the web. In: ACL 2009, pp. 1021–1029 (2009)
127. Zhou, G., Su, J.: Named entity recognition using an hmm-based chunk tagger. In: Proceedings of the 40th Annual Meeting on Association for Computational Linguistics, ACL 2002. Association for Computational Linguistics, Morristown (2002)

# Foundations of Description Logics

Sebastian Rudolph

Institute AIFB, Karlsruhe Institute of Technology, DE
rudolph@kit.edu

**Abstract.** This chapter accompanies the foundational lecture on Description Logics (DLs) at the 7th Reasoning Web Summer School in Galway, Ireland, 2011. It introduces basic notions and facts about this family of logics which has significantly gained in importance over the recent years as these logics constitute the formal basis for today's most expressive ontology languages, the OWL (Web Ontology Language) family.

We start out from some general remarks and examples demonstrating the modeling capabilities of description logics as well as their relation to first-order predicate logic. Then we begin our formal treatment by introducing the syntax of DL knowledge bases which comes in three parts: RBox, TBox and ABox. Thereafter, we provide the corresponding standard model-theoretic semantics and give a glimpse of the alternative way of defining the semantics via an embedding into first-order logic with equality.

We continue with an overview of the naming conventions for DLs before we delve into considerations about different notions of semantic alikeness (concept and knowledge base equivalence as well as emulation). These are crucial for investigating the expressivity of DLs and performing normalization. We move on by reviewing knowledge representation capabilities brought about by different DL features and their combinations as well as some model-theoretic properties associated thereto.

Subsequently, we consider typical reasoning tasks occurring in the context of DL knowledge bases. We show how some of these tasks can be reduced to each other, and have a look at different algorithmic approaches to realize automated reasoning in DLs.

Finally, we establish connections between DLs and OWL. We show how DL knowledge bases can be expressed in OWL and, conversely, how OWL modeling features can be translated into DLs.

In our considerations, we focus on the description logic $\mathcal{SROIQ}$ which underlies the most recent and most expressive yet decidable version of OWL called OWL 2 DL. We concentrate on the logical aspects and omit data types as well as extralogical features from our treatise. Examples and exercises are provided throughout the chapter.

A. Polleres et al. (Eds.): Reasoning Web 2011, LNCS 6848, pp. 76–136, 2011.

# 1    Introduction

*Come join the DL vaudeville show!*
*It's variable-free, although*
*With quantifiers, not, and, or*
*Quite deeply rooted in FOLklore.*
*Still, curing the first-order ailment*
*We sport decidable entailment!*

While formal, logic-based approaches to representing and working with knowledge occur throughout human history, the advent and widespread adoption of programmable computing devices in the 20th century has led to intensified studies of both theoretical and practical aspects of knowledge representation and automated reasoning. Rooted in early AI approaches, Description Logics (DLs) have developed into one of the main knowledge representation formalisms. The maturity of the field is also reflected by the adoption of description logics as prior specification paradigm

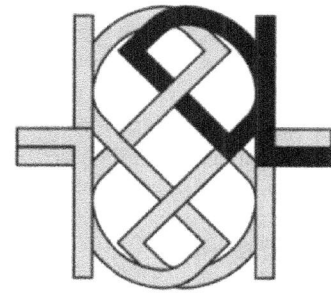

**Fig. 1.** The DL logo

for ontological descriptions – culminating in the standardization of the OWL web ontology language by the World Wide Web Consortium (W3C) – as well as the availability of highly optimized and readily deployable (yet open source) tools for automated inferencing. Thanks to this "dissemination path," DLs constitute the theoretical backbone for information systems in many disciplines, among which life sciences can be seen as the "early adopters" [Sidhu *et al.*, 2005; Wolstencroft *et al.*, 2005; Golbreich *et al.*, 2006].

## 1.1    Outlook

***What is in this Lecture.*** This document is supposed to give a gentle introduction into state-of-the-art description logics. Before going into technicalities the remainder of this section will briefly discuss how DLs are positioned in the landscape of knowledge representation formalisms, provide some examples for modeling features of DLs, and sketch the most prominent application context: the Semantic Web.

Section 2 starts the formal treatment by introducing the syntax of knowledge bases of the description logic $\mathcal{SROIQ}$. Section 3 provides the corresponding model-theoretic semantics and substantiates the claimed connection between DLs and first-order predicate logic (FOL) by giving a translation from $\mathcal{SROIQ}$ into FOL with equality.

Section 4 reviews the naming scheme for DLs between the basic DL $\mathcal{ALC}$ and the high-end DL $\mathcal{SROIQ}$. Section 5 provides several notions that capture that different syntactic specifications may have the same (or "alike") semantical impact. The motivation of Section 6 is to give a feeling for the modeling power provided by different constructs and the according model-theoretic consequences.

Subsequently, Section 7 considers typical reasoning tasks normally occurring in the context of DL-based knowledge representation and discusses the mutual reducibility of these tasks. In Section 8, we give a shallow overview over different algorithmic paradigms for automated inferencing with DLs. Finally, in Section 9, we provide a way to translate $\mathcal{SROIQ}$ knowledge bases into OWL ontologies and, conversely, show how OWL axioms can be translated into DLs.

**What is _not_ in this Lecture.** Due to space limitations, we have to restrict this lecture in many respects. We will focus on the core logical aspects of description logics and hence omit datatypes, keys, etc. despite their obvious practical importance for knowledge representation. Likewise, this is not supposed to be an introduction into OWL nor any other Semantic Web specification language. Thus, we will only briefly state how DL knowledge bases can be translated into OWL such that OWL reasoning tools can be harnessed to perform DL reasoning tasks. Moreover, we will refrain from looking into sub-Boolean fragments of DLs, even though they are practically important for serving as theoretical basis for the tractable profiles of the latest version of OWL. On the theoretical side, we will omit considerations about computational complexity of reasoning tasks.

**Required Previous Knowledge.** This lecture is meant to be introductory and foundational. Consequently, we tried to make it as self-contained as feasibly possible and hope that it is comprehensible even without any background in formal logics, although it can do no harm either. We presume, however, a certain familiarity with basic concepts and notations of naïve set theory. We do not expect prior knowledge about Semantic Web formalisms like the Resource Description Framework (RDF) or OWL, still it would come handy to fully comprehend the comments about the connections between DLs and OWL.

## 1.2   DLs in the Context of Other Formalisms

Historically, DLs have emerged from semantic networks [Quillian, 1968] and frame-based systems [Minsky, 1974]. These early knowledge representation approaches had the advantage of being rather intuitively readable and comprehensible. On the downside, it turned out that the understanding of the precise meaning of these diagrammatic representations differed widely amongst humans. This also became apparent by the heterogeneous behavior of tools implemented to reason with these structures. Under a plethora of names (among them *terminological systems* and *concept languages*), description logics developed out of the attempt to endow these intuitive representations with a formal semantics to establish a common ground for human and tool interoperability.

With the formal semantics introduced it was rather immediately clear that – abstracting from the syntax used – DLs can be seen as a fragment of first-order predicate logic (short: FOL), many of them even as a fragment of FOL's two-variable fragment [Borgida, 1996] in cases extended with counting quantifiers [Pratt-Hartmann, 2005]. As opposed to general FOL where logical inferencing is undecidable, DL research has been focusing on decidable fragments to such an extent that today, decidability is almost conceived as a necessary condition to call a formalism a DL.

**Remark 1.** Recap that in theoretical computer science, a class of problems is called decidable, if there is a generic algorithm that can take any problem instance from this class as an input and provide a yes-or-no answer to it after finite time. In the context of logics, the generic problem normally investigated is whether a given set of statements logically entails another statement. In case there is no danger of confusion about the type of problem considered, sometimes the logic itself is called decidable or undecidable.

In contrast to the well-known correspondence to FOL, it took some time to discover the close relation of DLs to modal logics [Schild, 1991]; in fact, the basic description logic $\mathcal{ALC}$ is just a syntactic variant of the multi-modal logic $\mathbf{K}_m$. As a consequence of this, there is also a close relationship of DLs to the Guarded Fragment [Andréka et al., 1998], a very expressive fragment of FOL which is still decidable.

For application purposes, DLs can be tailored to the specific requirements of a concrete usage scenario. To this end, a set of modeling features is selected such that the resulting logic has sufficient expressivity for the intended purpose while still being manageable in terms of the inferencing needed. This strategy has led to thorough investigations and finally a deeper understanding of the impact of the diverse standard modeling features on decidability and complexity of reasoning.

**Remark 2.** Thereby, the boundaries of the above mentioned fragments are sometimes crossed. For instance, functionality statements and cardinality constraints in general are not supported by the Guarded Fragment, the same holds for transitivity statements, which also lie outside the two-variable fragment. DLs featuring regular expressions on roles [Calvanese et al., 2009] even go beyond FOL with equality, but we will not discuss them here.

Beyond decidability, a crucial design principle in DLs is to establish favorable trade-offs between expressivity and scalability. On the theoretical side, establishing complexity results for inferencing problems (a tradition started by Brachman and Levesque [1984] and meanwhile widely accepted as central part of the DL research methodology) helps to roughly estimate how scalable and how "implementable" reasoning methods are likely to be. Of course, for the deployment in practice, many engineering and optimization considerations are necessary even if they do not influence the worst-case complexities. Today, there exist several highly optimized and efficient systems for reasoning in DL-based formalisms [Motik et al., 2009c; Sirin et al., 2007; Tsarkov and Horrocks, 2006].

## 1.3   DL Modeling in a Nutshell

This section provides an informal introduction of the most common modeling features in DLs. For the interested reader with some background in logics, we will relate them to FOL with equality by giving the corresponding terms and logical translations in square brackets.

All DLs are based on a vocabulary [signature] containing individual names [constants], concept names [unary predicates] and role names [binary predicates].

Two specific class names, $\top$ and $\bot$, denote the concept containing all individuals and the empty concept, respectively. Usually, a DL knowledge base [theory] is partitioned into an assertional part, called ABox and a terminological part, which is further subdivided into TBox and RBox. The ABox contains assertional knowledge [ground facts], the notation of which coincides with FOL: there are concept assertions such as

$$\text{Actor(angelina)}$$

(indicating that the individual named `angelina` belongs to the set of all actors) and role assertions like

$$\text{married(angelina,brad)}$$

(stating that the individuals named `angelina` and `brad` are in the relation of being married). The TBox contains universal statements. The notation used in DLs does not need variables and is inspired by set theory. We can specify subsumptions, e.g. by expressing that every actor is an artist via

$$\text{Actor} \sqsubseteq \text{Artist}$$

$[\forall x(\text{Actor}(x) \rightarrow \text{Artist}(x))]$. A specific feature of DLs is that concept names can be combined into complex concepts by Boolean operators, as in

$$\text{Actor} \sqcap \text{USGovernor} \sqsubseteq \text{Bodybuilder} \sqcup \neg\text{Austrian}$$

$[\forall x(\text{Actor}(x) \wedge \text{USGovernor}(x) \rightarrow \text{Bodybuilder}(x) \vee \neg\text{Austrian}(x))]$, expressing that every actor who is a US governor is also a bodybuilder or not Austrian. Another way to define complex concepts is by quantifying over roles, as for instance in

$$\exists\text{knows}.\text{Actor} \sqsubseteq \forall\text{hasfriend}.\text{Envious}$$

$[\forall x(\exists y(\text{knows}(x,y) \wedge \text{Actor}(y)) \rightarrow \forall z(\text{hasfriend}(x,z) \rightarrow \text{Envious}(z)))]$, which states that everybody knowing some actor has only envious friends.

The modeling features introduced above constitute $\mathcal{ALC}$ (*attributive language with complements*, [Schmidt-Schauß and Smolka, 1991]), the smallest DL that is Boolean-closed (i.e. it allows Boolean operators to be applied to concepts without restriction).

As stated before, in order to satisfy requirements emerging from practical modeling scenarios, these basic modeling features have been enriched by more and more expressive features for specifying and querying knowledge. In DLs, this development has led from the basic $\mathcal{ALC}$ to more expressive formalisms. *Role inverses* can be used to "traverse" roles backward e.g. in

$$\exists\text{HasChild}.\top \sqsubseteq \forall\text{hasChild}^{-}.\text{Grandparent}$$

$[\forall x(\exists y(\text{hasChild}(x,y)) \rightarrow \forall z(\text{hasChild}(z,x) \rightarrow \text{Grandparent}(x)))]$, expressing that everybody having a child is the child of only grandparents. *Cardinality constraints* allow for specifying the number of related instances:

$$\text{Polygamist} \sqsubseteq \geqslant 2.\text{Married}.\top$$

$[\forall x(\mathtt{Polygamist}(x) \rightarrow \exists y \exists z (\mathtt{Married}(x,y) \land \mathtt{Married}(x,z) \land y \neq z))]$ states that a polygamist is married to at least two distinct individuals. By means of *nominals*, classes can be defined by enumerating their instances: the axiom

$$\exists \mathtt{Married}.\{\mathtt{brad}\} \sqsubseteq \{\mathtt{angelina}\}$$

$[\exists x(\mathtt{Married}(x, \mathtt{brad}) \rightarrow x = \mathtt{angelina})]$ claims that being married to Brad is a property only applying to Angelina.

The RBox of a DL knowledge base allows for further, role-centric modeling features. These include role inclusion statements as for instance:

$$\mathtt{married} \sqsubseteq \mathtt{loves}$$

$[\forall x \forall y(\mathtt{married}(x,y) \rightarrow \mathtt{loves}(x,y))]$, which states that being married to somebody implies loving them. A more general and expressive variant of role inclusions are role-chain axioms as in

$$\mathtt{hasChild}^- \circ \mathtt{hasChild} \sqsubseteq \mathtt{hasSibling}$$

$[\forall x \forall y \forall z(\mathtt{hasChild}(y,x) \land \mathtt{hasChild}(y,z) \rightarrow \mathtt{hasSibling}(x,z))]$, saying that the child of somebody I am a child of is my sibling.

### 1.4   The Semantic Web

The rise of the World Wide Web as a large body of digitally accessible knowledge has inspired a plethora of research related to the question how to organize and formalize knowledge on the Web in order to allow for automated, intelligent retrieval and combination of the stored information. The term *Semantic Web* stands for a variety of research and standardization efforts towards this goal, and DLs constitute a crucial part of this endeavor. The underlying idea of the Semantic Web is to provide information on the Web in a sufficiently formal and structured way to enable "intelligent" processing by machines. To this end, several key requirements can be identified: First, it is necessary to agree on common and open standards for representing information, in order to enable the exchange of information between diverse applications and platforms and subsequently the combination of pieces of information from different origins. Such standards have to be defined in a clear formal way but at the same time, they need to be flexible and extendable.

In fact, the World Wide Web Consortium (W3C) has fostered and approved the definition of the basic Semantic Web standards. The ontology languages RDF and its extension RDF Schema (RDFS) as well as OWL have been deliberately developed for a deployment in the Semantic Web.[1]

---

[1] Originating from philosophy, the term *ontology* is not precisely defined in the computer science context either and a lot of deviating definitions can be found throughout the literature. In this treatise, we will use the term to simply refer to a document created in RDF(S) or OWL, modeling knowledge of an application domain. Thereby, we will consider it to be equivalent with the arguably more appropriate term *knowledge base*.

As the second key ingredient for the Semantic Web, methods are needed which automatically infer new knowledge from given knowledge. In order to maximally benefit from specified knowledge, it must be possible to obtain information that is not explicitly given but constitutes a logical consequence of what is known. This directly leads to the multifarious field of formal logic, and in particular to the area of automated reasoning. A significant portion of DL research has been spawned by problems and usage scenarios from the Semantic Web area.

## 2   Syntax

*Deluxe DL delivery*
*Will come in boxes (number: three),*
*Precisely marked with A, T, R.*
*The first exhibits solid grounding,*
*The next allows for simple counting,*
*The third one's strictly regular.*

In this section, we provide the definition of the expressive description logic $\mathcal{SROIQ}$ [Horrocks *et al.*, 2006] which serves as the logical basis for OWL 2 DL, the most expressive member of the OWL family where inferencing is still decidable. Most of today's mainstream DLs are, in fact, sublanguages of $\mathcal{SROIQ}$.

DLs are based on three disjoint sets of primal elements:

- The set $N_I$ of *individual names* contains all names used to denote singular entities (be they persons, objects or anything else) in our domain of interest. Examples would be brad, excalibur, rhine, or sun.
- The set $N_C$ of *concept names* contains names that refer to types, categories, or classes of entities, usually characterized by common properties. Typical concept names are Mammal, Country, Organization, but also Yellow or English.
- The set $N_R$ of *role names* contains names that denote binary relationships which may hold between individuals of a domain, for instance: marriedWith, fatherOf, likes, or locatedIn.

**Remark 3.** There are no mandatory rules for writing and typography of vocabulary elements. According to a convention most widely adopted, we capitalize concept names whereas individual and role names are written in lower case. Moreover, camel case is used for names corresponding to multi word units in natural language.

Having these name sets at hand (they are usually jointly referred to as *vocabulary* or *signature*), we can now turn to the three building blocks of $\mathcal{SROIQ}$ knowledge bases: RBox, TBox and ABox.

### 2.1   RBox

A $\mathcal{SROIQ}$ RBox captures interdependencies between the roles of the considered knowledge base. Given the set $N_R$ of role names, a *role* is either the *universal role u* or it has the form $r$ or $r^-$ for any role name $r$. The set of roles will be denoted by **R**. For convenience, we introduce the function *Inv* that "inverts" roles, i.e. we set $Inv(r) := r^-$ and $Inv(r^-) := r$ in order to simplify notation. In the sequel, we will use the symbols $r, s$, possibly with subscripts, to denote roles.

A *role inclusion axiom* (RIA, sometimes also referred to as *role chain axiom*) is a statement of the form $r_1 \circ \ldots \circ r_n \sqsubseteq r$ where $r_1, \ldots, r_n, r$ are roles. As a special case thereof (for $n = 1$), we obtain *simple role inclusions* $r \sqsubseteq s$. Typical examples of role inclusion axioms are `owns` $\circ$ `partOf` $\sqsubseteq$ `owns` or `fatherOf` $\sqsubseteq$ `childOf`$^-$. A finite set of such RIAs is called a *role hierarchy*.

Given a role hierarchy, it is useful to distinguish the roles that can be "created" by role chains of length greater than one from those which cannot. Consequently, we define *non-simple roles* as follows:

S1. Every role $r$ occurring in a RIA $r_1 \circ \ldots \circ r_n \sqsubseteq r$ where $n > 1$ is non-simple.
S2. Every role $r$ occurring in a simple role inclusion $s \sqsubseteq r$ with a non-simple $s$ is itself non-simple.
S3. If $r$ is non-simple then so is $Inv(r)$.
S4. No other role is non-simple.

We let $\mathbf{R}^n$ denote the set of all non-simple roles of a role hierarchy and call all the other roles *simple* denoted by $\mathbf{R}^s = \mathbf{R} \setminus \mathbf{R}^n$.

**Example 4.** Consider the following role hierarchy:

$$\texttt{motherOf} \sqsubseteq \texttt{parentOf} \tag{1}$$
$$\texttt{parentOf} \sqsubseteq \texttt{ancestorOf} \tag{2}$$
$$\texttt{ancesterOf} \circ \texttt{ancestorOf} \sqsubseteq \texttt{ancestorOf} \tag{3}$$
$$\texttt{ancestorOf} \sqsubseteq \texttt{descendantOf}^- \tag{4}$$

Then we can use S1. to find that `ancestorOf` is non-simple due to (3). This allows us to conclude that `descendantOf`$^-$ is non-simple via (4) and S2.. From the above follows via S3. that also `ancestorOf`$^-$ and `descendantOf` must be non-simple. Finally, S4. ensures that `motherOf`, `motherOf`$^-$, `parentOf`, and `parentOf`$^-$ are simple.

In order to ensure decidability of the ensuing logic, we cannot allow arbitrary role hierarchies but have to restrict to those which have the property of being *regular*. Formally, a role hierarchy is regular if there is a strict partial order $\prec$ on the non-simple roles $\mathbf{R}^n$ such that

- $S \prec R$ iff $Inv(S) \prec R$, and
- every RIA is of one of the forms
    R1 $r \circ r \sqsubseteq r$,
    R2 $Inv(r) \sqsubseteq r$,
    R3 $s_1 \circ \ldots \circ s_n \sqsubseteq r$,
    R4 $r \circ s_1 \circ \ldots \circ s_n \sqsubseteq r$,
    R5 $s_1 \circ \ldots \circ s_n \circ r \sqsubseteq r$,
    such that $r \in \mathsf{N}_R$ is a (non-inverse) role name $r$, and $s_i \prec r$ for $i = 1, \ldots, n$ whenever $s_i$ is non-simple.

**Example 5.** Consider the following role hierarchy containing the RIAs: $s \circ s \sqsubseteq s$, $r \circ s \sqsubseteq r$, and $r \circ s \circ r \sqsubseteq t$. First observe that all involved atomic roles are non-simple. If we define $\prec$ such that $s^- \prec s \prec r^- \prec r \prec t^- \prec t$, then all the above criteria are satisfied: the first RIA is an instance of R1, the second is an instance of R4, and the third is an instance of R3. Hence this role hierarchy is regular.

**Example 6.** Assume a role hierarchy containing $r \circ t \circ s \sqsubseteq t$ as the only axiom. Only $t$ is non-simple here, still this role hierarchy is not regular, as the RIA does not fit any of the allowed forms R1–R5 (to see this, note that $\prec$ is required to be strict, therefore $t \not\prec t$ must always be the case, irrespective of the concrete choice of $\prec$).

**Example 7.** Let a role hierarchy contain the two RIAs $r \circ s \sqsubseteq s$, and $s \circ r \sqsubseteq r$. While each of these RIAs alone would be acceptable as a role hierarchy, they do not go well together: the first requires $r \prec s$ (due to R5) whereas the second enforces $s \prec r$ (due to R4) which as a whole violates the condition of $\prec$ being a strict order. Thus the considered role hierarchy is not regular.

A *role characteristic* is a statement of the form $\mathsf{Ref}(r)$ (*reflexivity*), $\mathsf{Asy}(s)$ (*asymmetry*), or $\mathsf{Dis}(s, s')$ (*role disjointness*), where $s$ and $s'$ are simple roles while $r$ may be simple or non-simple. A $\mathcal{SROIQ}$ RBox (usually denoted by $\mathcal{R}$) is the union of a finite set of role characteristics together with a role hierarchy. A $\mathcal{SROIQ}$ RBox is regular if its role hierarchy is regular.

## 2.2   TBox

Given a $\mathcal{SROIQ}$ RBox $\mathcal{R}$ as defined in the previous section, we now inductively define *concept expressions* (also simply called *concepts*) as follows:

- every concept name $\mathsf{C} \in \mathsf{N}_C$ is a concept expression,
- $\top$ and $\bot$ are concept expressions, called *top concept* and *bottom concept*, respectively,
- $\{a_1, \ldots, a_n\}$ is a concept expression for every finite set $\{a_1, \ldots, a_n\} \subseteq \mathsf{N}_I$ of individual names; concepts of this type are called *nominal concepts*,
- if $C$ and $D$ are concept expressions then so are $\neg C$ (*negation*), $C \sqcap D$ (*intersection*), $C \sqcup D$ (*union*),
- if $r$ is a role and $C$ is a concept expression, then $\exists r.C$ (*existential quantification*) and $\forall r.C$ (*universal quantification*) are also concept expressions,
- if $r$ is a simple role, $n$ is a natural number and $C$ is a concept expression, then $\exists r.\mathsf{Self}$ (*self restriction*), $\geqslant nr.C$ (*at-least restriction*), and $\leqslant nr.C$ (*at-most restriction*) are also concept expressions. The latter two are also jointly referred to as *qualified number restrictions* or *cardinality constraints*.

We will denote the set of all concept expressions thus defined by **C**. Throughout this chapter, the symbols $C$, $D$ will be used to denote concept expressions.

**Remark 8.** Note that the definition of concept expressions depends on the underlying RBox due to the restriction of some concept expressions to contain only simple roles.

A *general concept inclusion axiom* (short: GCI) has the form $C \sqsubseteq D$ where $C$ and $D$ are concepts. This kind of statement is also sometimes called *subsumption* axiom, as $C \sqsubseteq D$ is often read "$C$ is subsumed by $D$." Sometimes, this axiom type is also referred to as *is-a relationship*, inspired by the often chosen wording for this type of statement (e.g. "a cat is a mammal" would be a typical verbalization of Cat $\sqsubseteq$ Mammal).

**Remark 9.** Sometimes, $C \sqsubseteq D$ is also called a *subconcept* statement with $C \sqsubseteq D$ being read "$C$ is a subconcept of $D$." While this is well justified by standard formal theories of (human) conceptual thinking where concepts are hierarchically ordered by subconcept-superconcept relationships [Ganter and Wille, 1997], this naming is unfortunate in the DL setting since it can also be understood syntactically to mean subformula of a concept term. Thus we do not use this term and whenever referring to the latter meaning, we speak of *subexpressions* of a concept.

Finally, a $\mathcal{SROIQ}$ TBox (usually denoted by $\mathcal{T}$) is a finite set of GCIs.

## 2.3   ABox

The *ABox* of a knowledge base contains information that applies to single individuals as opposed to the GCIs in the TBox, which represent statements which are generally true for all individuals alike.

An *individual assertion* can have any of the following forms:

- $C(\mathsf{a})$, called *concept assertion*,
- $r(\mathsf{a}, \mathsf{b})$, called *role assertion*,
- $\neg r(\mathsf{a}, \mathsf{b})$, called *negated role assertion*,
- $\mathsf{a} \approx \mathsf{b}$, called *equality statement*, or
- $\mathsf{a} \not\approx \mathsf{b}$, called *inequality statement*,

with $\mathsf{a}, \mathsf{b} \in \mathsf{N}_I$ individual names, $C \in \mathbf{C}$ a concept expression, and $r \in \mathbf{R}$ a role.

**Remark 10.** Of course, also the form $\neg C(\mathsf{a})$ is captured by the above definition since $\neg C$ is again a concept expression, as opposed to roles, which do not allow for negation (note that the inverse of a role is something quite different from its negation).

A $\mathcal{SROIQ}$ ABox (usually denoted by $\mathcal{A}$) is a finite set of individual assertions. We call an ABox *extensionally reduced* if the only concepts and roles occurring therein are concept names and roles names, respectively.

**Remark 11.** It should be noted that the separation between ABox and TBox – originally conceived for less expressive DLs – becomes less sharp once nominal concepts are allowed, since nominal concepts allow for referring to single individuals in the TBox as well. In fact, every of the different types of individual assertions can be expressed by a GCI featuring nominals: $C(a)$ becomes $\{a\} \sqsubseteq C$, $(\neg)r(a, b)$ is equivalent to $\{a\} \sqsubseteq (\neg)\exists r.\{b\}$, $a \approx b$ can be rewritten into $\{a\} \sqsubseteq \{b\}$, and $a \not\approx b$ into $\{a\} \sqsubseteq \neg\{b\}$. Still the distinction is not entirely meaningless even for DLs featuring nominals as soon as *data complexity* of reasoning is investigated.

A $\mathcal{SROIQ}$ *knowledge base* $\mathcal{KB}$ is the union of a regular RBox $\mathcal{R}$ and a TBox $\mathcal{T}$ as well as an ABox $\mathcal{A}$ for $\mathcal{R}$. The elements of $\mathcal{KB}$ are referred to as *axioms*. Given a knowledge base $\mathcal{KB}$ we write $\mathsf{N}_I(\mathcal{KB})$, $\mathsf{N}_C(\mathcal{KB})$, and $\mathsf{N}_R(\mathcal{KB})$ to denote those individual names, concept names, and role names which occur in $\mathcal{KB}$, respectively.

**Example 12.** As an example consider the following knowledge base $\mathcal{KB}$:

| |
|---|
| RBox $\mathcal{R}$ |
| $\qquad\qquad$ owns $\sqsubseteq$ caresFor |
| $\qquad\qquad\qquad$ "If somebody owns something, they care for it." |
| TBox $\mathcal{T}$ |
| $\qquad\quad$ Healthy $\sqsubseteq$ ¬Dead |
| $\qquad\qquad\qquad$ "Healthy beings are not dead." |
| $\qquad\qquad$ Cat $\sqsubseteq$ Dead $\sqcup$ Alive |
| $\qquad\qquad\qquad$ "Every cat is dead or alive." |
| HappyCatOwner $\sqsubseteq$ $\exists$owns.Cat $\sqcap$ $\forall$caresFor.Healthy |
| $\qquad\qquad\qquad$ "A happy cat owner owns a cat and all beings he cares for are healthy." |
| ABox $\mathcal{A}$ |
| HappyCatOwner (schrödinger) |
| $\qquad\qquad\qquad$ "Schrödinger is a happy cat owner." |

## 3  Semantics

*Semantics has wide applications*
*To relationship-based altercations,*
*For semantics unveils*
*What a statement entails*
*Depending on interpretations.*

Like for any other logic, the definition of a formal semantics for DLs boils down to providing a consequence relation that determines whether an axiom logically follows from (also: is entailed by) a given set of axioms. The semantics of description logics is defined in a model-theoretic way. Thereby, one central notion is that of an interpretation. Interpretations might be conceived as potential "realities" or "worlds." In particular, interpretations need in no way comply with the actual reality.

### 3.1  Interpretations

In the case of DLs, an interpretation, normally denoted with $\mathcal{I}$, provides

- a nonempty set $\Delta^{\mathcal{I}}$, called the *domain* or also *universe of discourse* which can be understood as the entirety of individuals or things existing in the "world" that $\mathcal{I}$ represents, and

- a function $\cdot^{\mathcal{I}}$, called *interpretation function* which connects the vocabulary elements (i.e., the individual, concept, and role names) to $\Delta^{\mathcal{I}}$, by providing
  - for each individual name $\mathsf{a} \in \mathsf{N}_I$ a corresponding individual $\mathsf{a}^{\mathcal{I}} \in \Delta^{\mathcal{I}}$ from the domain,
  - for each concept name $\mathsf{A} \in \mathsf{N}_C$ a corresponding set $\mathsf{A}^{\mathcal{I}} \subseteq \Delta^{\mathcal{I}}$ of domain elements (as opposed to the domain itself, $\mathsf{A}^{\mathcal{I}}$ is allowed to be empty), and
  - for each role name $\mathsf{r} \in \mathsf{N}_R$ a corresponding (also possibly empty) set $\mathsf{r}^{\mathcal{I}} \subseteq \Delta^{\mathcal{I}} \times \Delta^{\mathcal{I}}$ of ordered pairs of domain elements.

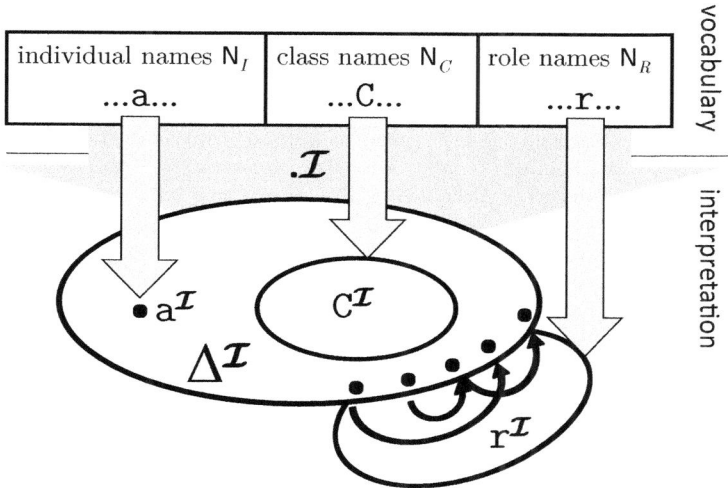

**Fig. 2.** Structure of DL interpretations

Figure 2 depicts this definition graphically. For domain elements $\delta, \delta' \in \Delta$, the intuitive meaning of $\delta \in \mathsf{A}^{\mathcal{I}}$ is that the individual $\delta$ belongs to the class described by the concept name $\mathsf{A}$, while $\langle \delta, \delta' \rangle \in \mathsf{r}$ means that $\delta$ is connected to $\delta'$ by the relation denoted by the role name $\mathsf{r}$.

**Remark 13.** To avoid confusion, it is important to strictly separate syntactic notions (referring to the vocabulary and axioms) from the semantic notions (referring to the domain and domain elements). Individual names, concept names and role names are syntactic entities and so are roles and concepts. Individuals are elements of $\Delta^{\mathcal{I}}$ and hence semantic entities. In order to refer to the semantic counterparts of concepts and roles, one would use the terms *concept extension* or *role extension*, respectively. Single elements of the extension of a concept or role are also called *concept instances* or *role instances*.

**Example 14.** Consider the following signature:

- $\mathsf{N}_I = \{\texttt{sun}, \texttt{morning_star}, \texttt{evening_star}, \texttt{moon}, \texttt{home}\}$.
- $\mathsf{N}_C = \{\texttt{Planet}, \texttt{Star}\}$.
- $\mathsf{N}_R = \{\texttt{orbitsAround}, \texttt{shinesOn}\}$.

We now define an interpretation $\mathcal{I} = (\Delta^\mathcal{I}, \cdot^\mathcal{I})$ as follows: Let our domain $\Delta^\mathcal{I}$ contain the following elements: ☉, ☿, ♀, ♁, ☾, ♂, ♃, ♄, ♅, ♆, ♇. We define the interpretation function by

$$
\begin{aligned}
\texttt{sun}^\mathcal{I} &= \text{☉} & \texttt{Planet}^\mathcal{I} &= \{\text{☿}, \text{♀}, \text{♁}, \text{♂}, \text{♃}, \text{♄}, \text{♅}, \text{♆}\} \\
\texttt{morning_star}^\mathcal{I} &= \text{♀} & \texttt{Star}^\mathcal{I} &= \{\text{☉}\} \\
\texttt{evening_star}^\mathcal{I} &= \text{♀} & \texttt{orbitsAround}^\mathcal{I} &= \{\langle \text{☿}, \text{☉}\rangle, \langle \text{♀}, \text{☉}\rangle, \langle \text{♁}, \text{☉}\rangle, \langle \text{♂}, \text{☉}\rangle, \langle \text{♃}, \text{☉}\rangle, \\
\texttt{moon}^\mathcal{I} &= \text{☾} & &\quad \langle \text{♄}, \text{☉}\rangle, \langle \text{♅}, \text{☉}\rangle, \langle \text{♆}, \text{☉}\rangle, \langle \text{♇}, \text{☉}\rangle, \langle \text{☾}, \text{♁}\rangle\} \\
\texttt{home}^\mathcal{I} &= \text{♁} & \texttt{shinesOn}^\mathcal{I} &= \{\langle \text{☉}, \text{☿}\rangle, \langle \text{☉}, \text{♀}\rangle, \langle \text{☉}, \text{♁}\rangle, \langle \text{☉}, \text{☾}\rangle, \langle \text{☉}, \text{♂}\rangle, \\
& & &\quad \langle \text{☉}, \text{♃}\rangle, \langle \text{☉}, \text{♄}\rangle, \langle \text{☉}, \text{♅}\rangle, \langle \text{☉}, \text{♆}\rangle, \langle \text{☉}, \text{♇}\rangle\}
\end{aligned}
$$

For a better understanding, it is often helpful to display an interpretation as a directed graph with labeled nodes and arcs. Thereby, the nodes correspond to the domain individuals $\Delta^\mathcal{I}$ where a node $\delta \in \Delta^\mathcal{I}$ gets labeled by the individual names assigned to it (i.e. those $\texttt{a} \in \mathsf{N}_I$ for which $\texttt{a}^\mathcal{I} = \delta$) as well as the concept names $\texttt{A}$ in the extensions of which $\delta$ lies (i.e. $\delta \in \texttt{A}^\mathcal{I}$). Moreover, whenever a pair of two domain individuals $\delta, \delta' \in \Delta^\mathcal{I}$ is in the extension of a role name $\texttt{r}$ (that is, if $\langle \delta, \delta' \rangle \in \texttt{r}^\mathcal{I}$), a directed arc is drawn from $\delta$ to $\delta'$ and labeled with $\texttt{r}$. The graphical representation of the interpretation $\mathcal{I}$ defined above would then look like this (where we abbreviate `orbitsAround` by `o` and `shinesOn` by `s`):

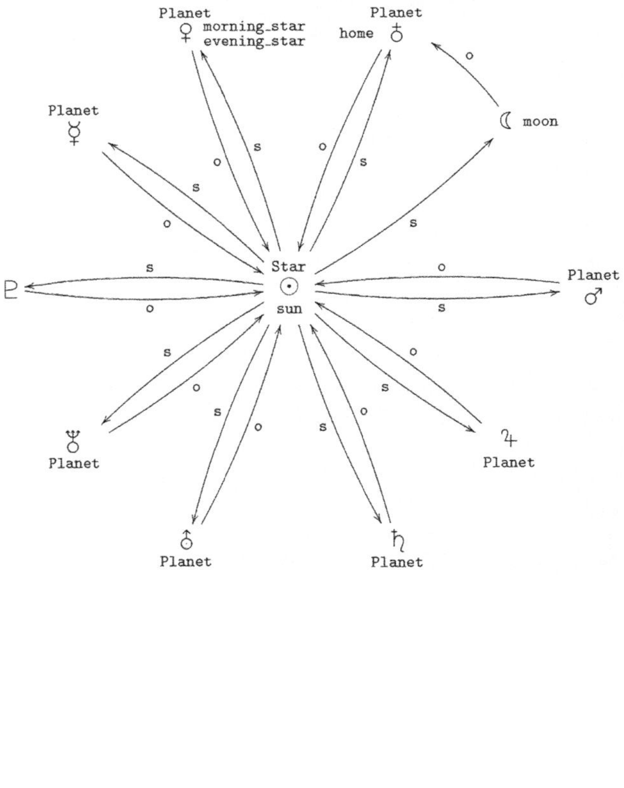

**Remark 15.** One should keep in mind that the domain $\Delta^{\mathcal{I}}$ is not required to be finite, but can also be an infinite set. It is also possible to consider only interpretations with finite domains, but then one explicitly talks about *finite models* or *finite satisfiability*. There are logics where infinite interpretations are "dispensable" as there are always finite ones that do the same job, these logics are said to have the *finite model property*. $\mathcal{SROIQ}$ does not have this property. However, since DLs are normally fragments of first-order logic, we can safely restrict our attention to interpretations with countable domains (that is, domains having at most as many individuals as there are natural numbers). This is a consequence of the downward part of the Theorem of Löwenheim-Skolem, according to which every FOL theory that has an arbitrary infinite model also has a countable one.

**Example 16.** As an example of an interpretation, this time with an infinite domain, consider the following vocabulary:

- $N_I = \{\texttt{zero}\}$.
- $N_C = \{\texttt{Prime}, \texttt{Positive}\}$.
- $N_R = \{\texttt{hasSuccessor}, \texttt{lessThan}, \texttt{multipleOf}\}$.

Now, we define $\mathcal{I}$ as follows: let $\Delta^{\mathcal{I}} = \mathbb{N} = \{0, 1, 2, \ldots\}$, i.e., the set of all natural numbers including zero. Furthermore, we let $\texttt{zero}^{\mathcal{I}} = 0$, as well as $\texttt{Prime}^{\mathcal{I}} = \{n \mid n \text{ is a prime number}\}$ and $\texttt{Positive}^{\mathcal{I}} = \{n \mid n > 0\}$. For the roles, we define

- $\texttt{hasSuccessor}^{\mathcal{I}} = \{\langle n, n+1 \rangle \mid n \in \mathbb{N}\}$
- $\texttt{lessThan}^{\mathcal{I}} = \{\langle n, n' \rangle \mid n < n',\ n, n' \in \mathbb{N}\}$
- $\texttt{multipleOf}^{\mathcal{I}} = \{\langle n, n' \rangle \mid \exists k.n = k \cdot n',\ n, n', k \in \mathbb{N}\}$

Note that this interpretation is well defined, although it has an infinite domain. For space reasons, we refrain from providing the corresponding graph representation.

**Remark 17.** Note that the definition of an interpretation does not require that different individual names denote different individuals, that is, it may happen that for two individual names $\mathbf{a}$ and $\mathbf{b}$, we have $\mathbf{a}^{\mathcal{I}} = \mathbf{b}^{\mathcal{I}}$. A stronger definition of DL interpretations that excludes such cases is usually referred to as *unique name assumption* (short: UNA). Note also, that not every domain element $\delta \in \Delta$ needs to be named, i.e., there may be $\delta$ for which no individual name $\mathbf{a}$ with $\mathbf{a}^{\mathcal{I}} = \delta$ exists. For obvious reasons, such individuals are usually referred to as *anonymous individuals*.

## 3.2    Satisfaction of Axioms

By now, we have seen that an interpretation determines the semantic counterparts of vocabulary elements. However, in order to finally determine the truth of complex axioms, it is necessary to also find the counterparts of complex concepts and roles. We provide a definition according to which the semantics of a complex language construct can be obtained from the semantics of its constituents (thereby following the principle of *compositional semantics*). Formally, this is done by "lifting" the interpretation function $\cdot^{\mathcal{I}}$ to these complex expressions.

First we extend the interpretation function from role names to roles by letting $u^{\mathcal{I}} = \Delta^{\mathcal{I}} \times \Delta^{\mathcal{I}}$ (that is: the universal role interconnects any two individuals of the domain and also every individual with itself), and assigning to inverted role names $\mathbf{r}^-$ the set of all pairs $\langle \delta, \delta' \rangle$ of domain elements for which $\langle \delta', \delta \rangle$ is contained in $\mathbf{r}^{\mathcal{I}}$.

Next we define the interpretation function for concepts:

- $\top$ is the concept which is true for every individual of the domain, hence $\top^{\mathcal{I}} = \Delta^{\mathcal{I}}$.
- $\bot$ is the concept which has no instances, hence $\bot^{\mathcal{I}} = \emptyset$.
- $\{\mathsf{a}_1, \ldots, \mathsf{a}_n\}$ is the concept containing exactly the individuals denoted by $\mathsf{a}_1, \ldots, \mathsf{a}_n$, therefore $\{\mathsf{a}_1, \ldots, \mathsf{a}_n\}^{\mathcal{I}} = \{\mathsf{a}_1^{\mathcal{I}}, \ldots, \mathsf{a}_n^{\mathcal{I}}\}$
- $\neg C$ is supposed to denote the set of all those domain individuals that are not contained in the extension of $C$, i.e., $(\neg C)^{\mathcal{I}} = \Delta^{\mathcal{I}} \setminus C^{\mathcal{I}}$.
- $C \sqcap D$ is the concept comprising all individuals that are simultaneously in $C$ and $D$, thus we define $(C \sqcap D)^{\mathcal{I}} = C^{\mathcal{I}} \cap D^{\mathcal{I}}$.
- $C \sqcup D$ contains individuals being present in $C$ or $D$ (or both), therefore we let $(C \sqcup D)^{\mathcal{I}} = C^{\mathcal{I}} \cup D^{\mathcal{I}}$.
- $\exists r.C$ is the concept that holds for an individual $\delta \in \Delta^{\mathcal{I}}$ exactly if there is some domain individual $\delta' \in \Delta^{\mathcal{I}}$ such that $\delta$ is connected to $\delta'$ via the relation denoted by $r$ and $\delta'$ belongs to the extension of the concept $C$, formally: $(\exists r.C)^{\mathcal{I}} = \{\delta \in \Delta^{\mathcal{I}} \mid \exists \delta' \in \Delta^{\mathcal{I}}.(\langle \delta, \delta' \rangle \in r^{\mathcal{I}} \wedge \delta' \in C^{\mathcal{I}})\}$.
- $\forall r.C$ denotes the set of individuals $\delta \in \Delta^{\mathcal{I}}$ with the following property: whenever $\delta$ is connected to some domain individual $\delta' \in \Delta^{\mathcal{I}}$ via the relation denoted by $r$, then $\delta'$ belongs to the extension of the concept $C$, formally: $(\forall r.C)^{\mathcal{I}} = \{\delta \in \Delta^{\mathcal{I}} \mid \forall \delta' \in \Delta^{\mathcal{I}}.(\langle \delta, \delta' \rangle \in r^{\mathcal{I}} \rightarrow \delta' \in C^{\mathcal{I}})\}$.
- $\exists r.\mathsf{Self}$ comprises those domain individuals which are $r$-related to themselves, thus we let $(\exists r.\mathsf{Self})^{\mathcal{I}} = \{x \in \Delta^{\mathcal{I}} \mid \langle x, x \rangle \in r^{\mathcal{I}}\}$.
- $\leqslant n\, r.C$ refers to the domain elements $\delta \in \Delta^{\mathcal{I}}$ for which no more than $n$ individuals exist to which $\delta$ is $r$-related and that are in the extension of $C$, formally: $(\leqslant n\, r.C)^{\mathcal{I}} = \{\delta \in \Delta^{\mathcal{I}} \mid \#\{\delta' \in \Delta^{\mathcal{I}} \mid \langle \delta, \delta' \rangle \in r^{\mathcal{I}} \wedge \delta' \in C^{\mathcal{I}}\} \leq n\}$ (thereby $\#S$ is used to denote the cardinality of a set $S$, i.e., the number of its elements).
- $\geqslant n\, r.C$, dual to the case before, denotes those domain elements having at least $n$ such $r$-related elements: $(\geqslant n\, r.C)^{\mathcal{I}} = \{\delta \in \Delta^{\mathcal{I}} \mid \#\{\delta' \in \Delta^{\mathcal{I}} \mid \langle \delta, \delta' \rangle \in r^{\mathcal{I}} \wedge \delta' \in C^{\mathcal{I}}\} \geq n\}$.

**Remark 18.** The reader should be aware that by the above definition, the extension of the concept $\forall r.C$ contains every domain individual $\delta \in \Delta^{\mathcal{I}}$ that is not $r$-connected to any $\delta'$. For instance, the concept $\forall \mathtt{hasChild}.\mathtt{Happy}$ comprises all individuals all of whose children are happy (alternatively, and arguably less confusing: all individuals that do not have children which are not happy). This includes those individuals not having children at all. In fact, when modeling with DLs, the concept $\forall r.\bot$ is often used to refer to all individuals not being $r$-connected to any other individual (nor to themselves).

**Example 19.** Consider the interpretation $\mathcal{I}$ from Example 14. With the lifting of the interpretation function just defined, we are able to determine the extension of concepts and roles as follows:

$$
\begin{aligned}
&\texttt{orbitsaround}^{-\mathcal{I}}\\
&\quad = \{\langle\odot,\text{☿}\rangle,\langle\odot,\text{♀}\rangle,\langle\odot,\text{♁}\rangle,\langle\odot,\text{♂}\rangle,\langle\odot,\text{♃}\rangle,\langle\odot,\text{♄}\rangle,\langle\odot,\text{♅}\rangle,\langle\odot,\text{♆}\rangle,\langle\odot,\text{♇}\rangle\}
\end{aligned}
$$

$$
\begin{aligned}
&(\forall\texttt{orbitsAround}.(\neg\texttt{Star}))^{\mathcal{I}}\\
&\quad = \{\delta \mid \forall\delta'.(\langle\delta,\delta'\rangle \in \texttt{orbitsAround}^{\mathcal{I}} \rightarrow \delta' \in (\neg\texttt{Star})^{\mathcal{I}})\}\\
&\quad = \{\delta \mid \forall\delta'.(\langle\delta,\delta'\rangle \in \texttt{orbitsAround}^{\mathcal{I}} \rightarrow \delta' \in \Delta^{\mathcal{I}} \setminus \texttt{Star}^{\mathcal{I}})\}\\
&\quad = \{\delta \mid \forall\delta'.(\langle\delta,\delta'\rangle \in \texttt{orbitsAround}^{\mathcal{I}} \rightarrow \delta' \in \{\text{☿},\text{♀},\text{♁},\text{☾},\text{♂},\text{♃},\text{♄},\text{♅},\text{♆},\text{♇}\})\}\\
&\quad = \{\odot,\text{☾}\}
\end{aligned}
$$

$$
\begin{aligned}
&((\neg\texttt{Planet}) \sqcap \exists\texttt{orbitsAround}.\texttt{Star})^{\mathcal{I}}\\
&\quad = (\neg\texttt{Planet})^{\mathcal{I}} \cap (\exists\texttt{orbitsAround}.\texttt{Star})^{\mathcal{I}}\\
&\quad = \Delta^{\mathcal{I}} \setminus \texttt{Planet}^{\mathcal{I}} \cap \{\delta \mid \exists\delta'.(\langle\delta,\delta'\rangle \in \texttt{orbitsAround}^{\mathcal{I}} \wedge \delta' \in \texttt{Star}^{\mathcal{I}})\}\\
&\quad = \{\odot,\text{☾},\text{♇}\} \cap \{\text{☿},\text{♀},\text{♁},\text{♂},\text{♃},\text{♄},\text{♅},\text{♆},\text{♇}\}\\
&\quad = \{\text{♇}\}
\end{aligned}
$$

$$
\begin{aligned}
&({\geqslant}2\texttt{shinesOn}.\{\texttt{morning_star},\texttt{evening_star}\})^{\mathcal{I}}\\
&\quad = \{\delta \mid \#\{\delta' \mid \langle\delta,\delta'\rangle \in \texttt{shinesOn}^{\mathcal{I}} \wedge \delta' \in \{\texttt{morning_star},\texttt{evening_star}\}^{\mathcal{I}}\} \geq 2\}\\
&\quad = \{\delta \mid \#\{\delta' \mid \langle\delta,\delta'\rangle \in \texttt{shinesOn}^{\mathcal{I}} \wedge \delta' \in \{\texttt{morning_star}^{\mathcal{I}},\texttt{evening_star}^{\mathcal{I}}\} \geq 2\}\\
&\quad = \{\delta \mid \#\{\delta' \mid \langle\delta,\delta'\rangle \in \texttt{shinesOn}^{\mathcal{I}} \wedge \delta' \in \{\text{♀}\}\} \geq 2\}\\
&\quad = \emptyset
\end{aligned}
$$

**Exercise 1.** *Describe – both verbally and formally – the extension of the following concepts with respect to the interpretation $\mathcal{I}$ defined in Example 16:*

(a) $\forall\texttt{hasSuccessor}^{-}.\texttt{Positive}$
(b) $\exists\texttt{multipleOf}.\mathsf{Self}$
(c) $\exists\texttt{multipleOf}.\exists\texttt{hasSuccessor}^{-}.\exists\texttt{hasSuccessor}^{-}.\{\texttt{zero}\}$
(d) ${\geqslant}10\,\texttt{lessThan}^{-}.\texttt{Prime}$
(e) $\neg\texttt{Prime} \sqcap {\leqslant}2\,\texttt{multipleOf}.\top$
(f) $\exists\texttt{lessThan}.\texttt{Prime}$
(g) $\forall\texttt{multipleOf}.(\exists\texttt{hasSuccessor}^{-}.\{\texttt{zero}\} \sqcup \exists\texttt{multipleOf}.\exists\texttt{hasSuccessor}^{-}.\exists\texttt{hasSuccessor}^{-}.\{\texttt{zero}\})$

The final purpose of the (lifted) interpretation function is to determine satisfaction of axioms. In the following, we define when an axiom $\alpha$ is true (also: holds), given a specific interpretation $\mathcal{I}$. If this is the case, we also say that $\mathcal{I}$ is a *model* of $\alpha$ or that $\mathcal{I}$ *satisfies* $\alpha$ and write $\mathcal{I} \models \alpha$.

- A role inclusion axiom $r_1 \circ \ldots \circ r_n \sqsubseteq r$ holds in $\mathcal{I}$ if for every sequence $\delta_0,\ldots,\delta_n \in \Delta^{\mathcal{I}}$ for which holds $\langle\delta_0,\delta_1\rangle \in r_1^{\mathcal{I}}$, ..., $\langle\delta_{n-1},\delta_n\rangle \in r_n^{\mathcal{I}}$, also $\langle\delta_0,\delta_n\rangle \in r^{\mathcal{I}}$ is satisfied. Figuratively, this means that every path in $\Delta^{\mathcal{I}}$ that traverses the roles $r_1,\ldots,r_n$ (in the given order) must have a direct

$r$-"shortcut." When using $\circ$ as symbol for the relation product, we can write down this condition as $r_1^{\mathcal{I}} \circ \ldots \circ r_n^{\mathcal{I}} \subseteq r^{\mathcal{I}}$.

- A role disjointness statement $\mathsf{Dis}(r, s)$ is true in $\mathcal{I}$ if every two domain individuals $\delta, \delta' \in \Delta^{\mathcal{I}}$ that are connected via an $r$-relation are not connected via an $s$-relation. In other words, we can say that the two roles are mutually exclusive which can be formally expressed by the condition $r^{\mathcal{I}} \cap s^{\mathcal{I}} = \emptyset$.
- A general concept inclusion $C \sqsubseteq D$ is satisfied by $\mathcal{I}$, if every instance of $C$ is also an instance of $D$. An alternative wording of this would be that the extension of $C$ is contained in the extension of $D$, formally $C^{\mathcal{I}} \subseteq D^{\mathcal{I}}$.
- A concept assertion $C(\mathsf{a})$ holds in $\mathcal{I}$ if the individual with the name $\mathsf{a}$ is an instance of the concept $C$, that is $\mathsf{a}^{\mathcal{I}} \in C^{\mathcal{I}}$.
- A role assertion $r(\mathsf{a}, \mathsf{b})$ is true in $\mathcal{I}$ if the individual denoted by $\mathsf{a}$ is $r$ connected to the individual denoted by $\mathsf{b}$, i.e. the extension of $r$ contains the corresponding pair of domain elements: $\langle \mathsf{a}^{\mathcal{I}}, \mathsf{b}^{\mathcal{I}} \rangle \in r^{\mathcal{I}}$.
- $\mathcal{I}$ is a model of $\neg r(\mathsf{a}, \mathsf{b})$ exactly if it is not a model of $r(\mathsf{a}, \mathsf{b})$.
- The equality statement $\mathsf{a} \approx \mathsf{b}$ holds in $\mathcal{I}$ if the individual names $\mathsf{a}$ and $\mathsf{b}$ refer to the same domain individual, i.e. $\mathsf{a}^{\mathcal{I}} = \mathsf{b}^{\mathcal{I}}$.
- $\mathcal{I}$ is a model of $\mathsf{a} \not\approx \mathsf{b}$ exactly if it is not a model of $\mathsf{a} \approx \mathsf{b}$.

**Example 20.** We now check for some example axioms whether interpretation $\mathcal{I}$ from Example 14 satisfies them.

- $\mathtt{morning\_star} \approx \mathtt{evening\_star}$ is true, since $\mathtt{morning\_star}^{\mathcal{I}} = \female =$ $\mathtt{evening\_star}^{\mathcal{I}}$, i.e. the two names denote the same domain individual.
- $\mathtt{orbitsAround} \circ \mathtt{orbitsAround} \sqsubseteq \mathtt{shinesOn}^{-}$ is also true: The only chain of domain individuals $\delta_1, \delta_2, \delta_3$ with $\langle \delta_1, \delta_2 \rangle \in \mathtt{orbitsAround}^{\mathcal{I}}$ and $\langle \delta_2, \delta_3 \rangle \in$ $\mathtt{orbitsAround}^{\mathcal{I}}$ is $\delta_1 = \mathbb{C}$, $\delta_2 = \oplus$, $\delta_3 = \odot$. Therefore, we obtain $\mathtt{orbitsAround}^{\mathcal{I}} \circ$ $\mathtt{orbitsAround}^{\mathcal{I}} = \{\langle \mathbb{C}, \odot \rangle\}$. On the other hand, due to $\langle \odot, \mathbb{C} \rangle \in \mathtt{shinesOn}^{\mathcal{I}}$ we obtain $\langle \mathbb{C}, \odot \rangle \in \mathtt{shinesOn}^{-\mathcal{I}}$.
- $\mathtt{Star}(\mathtt{evening\_star})$ is false since the domain element $\mathtt{evening\_star}^{\mathcal{I}} = \female$ is not contained in $\mathtt{Star}^{\mathcal{I}} = \{\odot\}$.
- $\mathtt{Planet} \sqsubseteq \neg\{\mathtt{sun}, \mathtt{moon}\}$ is valid in $\mathcal{I}$ as we get $(\neg\{\mathtt{sun}, \mathtt{moon}\})^{\mathcal{I}} = \Delta^{\mathcal{I}} \setminus$ $(\{\mathtt{sun}, \mathtt{moon}\})^{\mathcal{I}} = \Delta^{\mathcal{I}} \setminus \{\odot, \mathbb{C}\} = \{\female, \female, \oplus, \male, \mathtt{4}, \hbar, \delta, \mathtt{8}, \mathtt{P}\}$ which is a superset of $\mathtt{Planet}^{\mathcal{I}} = \{\female, \female, \oplus, \male, \mathtt{4}, \hbar, \delta, \mathtt{8}\}$.
- $\mathtt{shinesOn}(\mathtt{moon}, \mathtt{earth})$ does not hold in $\mathcal{I}$ since the pair of the respective individuals is not contained in the extension of the $\mathtt{shinesOn}$ role: $\langle \mathtt{moon}^{\mathcal{I}}, \mathtt{earth}^{\mathcal{I}} \rangle = \langle \mathbb{C}, \oplus \rangle \notin \mathtt{shinesOn}^{\mathcal{I}}$.
- $\top \sqsubseteq \forall\mathtt{shinesOn}^{-}.\{\mathtt{sun}\}$ is true. To see this we first need to find $(\forall\mathtt{shinesOn}^{-}.\{\mathtt{sun}\})^{\mathcal{I}}$. In words, this concept comprises those objects that are shone upon by nothing but the sun (if they are shone upon by anything at all). Formally, to check whether a domain individual $\delta$ is in the extension of that concept, we have to verify that every individual $\delta'$ with $\langle \delta, \delta' \rangle \in \mathtt{shinesOn}^{-\mathcal{I}}$ (which is equivalent to $\langle \delta', \delta \rangle \in \mathtt{shinesOn}^{\mathcal{I}}$) also satisfies $\delta' \in \{\mathtt{sun}\}^{\mathcal{I}}$ which just means $\delta' = \odot$. Scrutinizing all elements of $\Delta^{\mathcal{I}}$, we find this condition satisfied for each, therefore we have $\top^{\mathcal{I}} = \Delta^{\mathcal{I}} \subseteq \Delta^{\mathcal{I}} = (\forall\mathtt{shinesOn}^{-}.\{\mathtt{sun}\})^{\mathcal{I}}$.
- $\mathsf{Dis}(\mathtt{orbitsAround}, \mathtt{shinesOn})$ is satisfied by $\mathcal{I}$ since no pair $\langle \delta, \delta' \rangle$ is contained in the extensions of both $\mathtt{orbitsAround}$ and $\mathtt{shinesOn}$ and therefore $\mathtt{orbitsAround}^{\mathcal{I}} \cap \mathtt{shinesOn}^{\mathcal{I}} = \emptyset$.

**Exercise 2.** *Decide whether the following axioms are satisfied by the interpretation $\mathcal{I}$ from Example 16.*

(a) $\texttt{hasSuccessor} \sqsubseteq \texttt{lessThan}$

(b) $\exists\texttt{hasSuccessor}^-.\exists\texttt{hasSuccessor}^-.\{\texttt{zero}\} \sqsubseteq \texttt{Prime}$

(c) $\top \sqsubseteq \forall\texttt{multipleOf}^-.\{\texttt{zero}\}$

(d) $\texttt{Dis}(\texttt{divisileBy}, \texttt{lessThan}^-)$

(e) $\texttt{multipleOf} \circ \texttt{multipleOf} \sqsubseteq \texttt{multipleOf}$

(f) $\top \sqsubseteq {\leqslant}1\texttt{hasSuccessor}.\texttt{Positive}$

(g) $\texttt{zero} \not\approx \texttt{zero}$

(h) ${\leqslant}1\texttt{multipleOf}^-.\top(\texttt{zero})$

(i) $\top \sqsubseteq \forall\texttt{lessThan}.\exists\texttt{lessThan}.(\texttt{Prime} \sqcap \exists\texttt{hasSuccessor}.\exists\texttt{hasSuccessor}.\texttt{Prime})$

Now that we have defined when an interpretation $\mathcal{I}$ is a model of an axiom, we can easily extend this notion to whole knowledge bases: we say that $\mathcal{I}$ is a *model* of a given knowledge base $\mathcal{KB}$ (also: $\mathcal{I}$ *satisfies* $\mathcal{KB}$, written $\mathcal{I} \models \mathcal{KB}$), if it satisfies all the axioms of $\mathcal{KB}$, i.e., if $\mathcal{I} \models \alpha$ for every $\alpha \in \mathcal{KB}$. Moreover, a knowledge base $\mathcal{KB}$ is called *satisfiable* or *consistent* if it has a model, and it is called *unsatisfiable* or *inconsistent* or *contradictory* otherwise.

**Example 21.** The following knowledge base is inconsistent.

$$\texttt{Reindeer} \sqcap \exists\texttt{hasNose}.\texttt{Red}(\texttt{rudolph}) \qquad \texttt{Reindeer} \sqsubseteq \texttt{Mammal}$$
$$\forall\texttt{worksFor}^-.(\neg\texttt{Reindeer} \sqcup \texttt{Flies})(\texttt{santa}) \qquad \texttt{Mammal} \sqcap \texttt{Flies} \sqsubseteq \texttt{Bat}$$
$$\texttt{worksFor}(\texttt{rudolph}, \texttt{santa}) \qquad \texttt{Bat} \sqsubseteq \forall\texttt{worksFor}.\{\texttt{batman}\}$$
$$\texttt{santa} \not\approx \texttt{batman}$$

**Remark 22.** Note that, for determining whether a knowledge base satisfies an interpretation $\mathcal{I}$, only the value of $\cdot^{\mathcal{I}}$ for those individual, concept, and role names are relevant, that occur in $\mathcal{KB}$. All vocabulary elements not contained in $\mathsf{N}_I(\mathcal{KB}) \cup \mathsf{N}_C(\mathcal{KB}) \cup \mathsf{N}_R(\mathcal{KB})$ can be mapped arbitrarily and do not influence the semantics.

## 3.3 Logical Consequence

So far, we have defined a "modelhood" relation, which for a given interpretation and a given set of axioms determines whether the axiom is true with respect to the interpretation. Remember that the actual purpose of a formal semantics is to provide a consequence relation, which tells us whether an axiom is a logical consequence of a knowledge base. This consequence relation is commonly also denoted by $\models$ and defined as follows: an axiom $\alpha$ is a *consequence* of (also *entailed by*) a knowledge base $\mathcal{KB}$ (written: $\mathcal{KB} \models \alpha$) if every model of $\mathcal{KB}$ is also a model of $\alpha$, i.e. for every $\mathcal{I}$ with $\mathcal{I} \models \mathcal{KB}$ also holds $\mathcal{I} \models \alpha$.

**Remark 23.** As a straightforward consequence of this model-theoretic definition of consequences we obtain the fact that an inconsistent knowledge base entails any axiom, since the considered set of models which have to satisfy the axiom is empty and hence the condition is vacuously true. This effect, well-known in many logics, is called the *principle of explosion* according to which "anything follows from a contradiction."

**Exercise 3.** *Decide whether the following propositions about the knowledge base* $\mathcal{KB}$ *from Example 12 are true and give evidence:*

(a) $\mathcal{KB}$ *is satisfiable,*
(b) $\mathcal{KB} \models$ Alive(schrödinger),
(c) $\mathcal{KB} \models$ Dead $\sqcap$ Alive $\sqsubseteq \bot$,
(d) $\mathcal{KB} \models$ Alive $\sqsubseteq$ Healthy.

**Exercise 4.** *Decide whether the following statements are true or false and justify your decision. For arbitrary* $\mathcal{SROIQ}$ *knowledge bases* $\mathcal{KB}$ *and* $\mathcal{KB}'$ *holds:*

(a) *If an axiom* $\alpha$ *is a logical consequence of the empty knowledge base, i.e.* $\emptyset \models \alpha$, *then it is the consequence of any other knowledge base* $\mathcal{KB}$.
(b) *The larger a knowledge base, the more models it has. That is, if* $\mathcal{KB} \subseteq \mathcal{KB}'$ *then every model of* $\mathcal{KB}$ *is also a model of* $\mathcal{KB}'$.
(c) *The larger a knowledge base, the more consequences it has. That is, if* $\mathcal{KB} \subseteq \mathcal{KB}'$ *then every logical consequence from* $\mathcal{KB}$ *is a logical consequence from* $\mathcal{KB}'$.
(d) *If* $\neg C(a) \in \mathcal{KB}$, *then* $\mathcal{KB} \models C(a)$ *can never hold (for arbitrary concepts* $C$*).*
(e) *If two knowledge bases are different (*$\mathcal{KB} \neq \mathcal{KB}'$*), then they also differ in terms of logical consequences, i.e., there is an axiom* $\alpha$ *such that* $\mathcal{KB} \models \alpha$ *and* $\mathcal{KB}' \not\models \alpha$ *or vice versa.*

## 3.4  Excursus: Semantics via Embedding into FOL

As mentioned before, it is often said that most description logics, including $\mathcal{SROIQ}$, are fragments of first-order predicate logic (FOL). Technically, this statement is somewhat misleading since, from a syntax point of view, most DL axioms are not FOL formulae. What is rather meant by this statement is the following: It is obvious that DL interpretations have the same structure as FOL interpretations if one conceives individual names as constants, concept names as unary predicates and role names as binary predicates. Under this assumption, one can define an easy syntactic translation $\tau$ which, applied to a DL axiom $\alpha$, yields a FOL sentence $\tau(\alpha)$ such that the model sets of $\alpha$ and $\tau(\alpha)$ coincide, that is an interpretation $\mathcal{I}$ is a model of $\alpha$ exactly if it is a model of $\tau(\alpha)$. Consequently, every reasoning problem in a DL is easily transferrable to an equivalent reasoning problem in FOL, whence the semantics of description logics could – as an alternative to the previously introduced way – be defined by reducing it to the semantics of FOL via the mentioned translation.

**Remark 24.** Obviously, the converse cannot be the case, for any decidable DL: supposing it were, we could decide any FOL reasoning problem by translating it to the DL and then deciding the DL version. This clearly contradicts the well-known undecidability of FOL.

We provide here a definition of $\tau$ but omit a proof of its correctness. More precisely, the translation outputs first-order predicate logic with equality, a mild generalization of pure first-order predicate logic featuring an equality predicate

=. Every $\mathcal{SROIQ}$ knowledge base $\mathcal{KB}$ thus translates via $\tau$ to a theory $\tau(\mathcal{KB})$ in first-order predicate logic with equality. We define

$$\tau(\mathcal{KB}) = \bigcup_{\alpha \in \mathcal{KB}} \tau(\alpha),$$

i.e., we translate every axiom of the knowledge base separately into a FOL sentence. How exactly $\tau(\alpha)$ is defined depends on the type of the axiom $\alpha$.

However, first we have to define auxiliary translation functions $\tau_\mathbf{R} : \mathbf{R} \times$ Var $\times$ Var $\to$ FOL for roles and $\tau_\mathbf{C} : \mathbf{C} \times$ Var $\to$ FOL for concepts (where Var $= \{x_0, x_1, \ldots\}$ is a set of variables):

$$\tau_\mathbf{R}(u, x_i, x_j) = \textbf{true}$$
$$\tau_\mathbf{R}(\mathbf{r}, x_i, x_j) = \mathbf{r}(x_i, x_j)$$
$$\tau_\mathbf{R}(\mathbf{r}^-, x_i, x_j) = \mathbf{r}(x_j, x_i)$$

$$\tau_\mathbf{C}(\mathsf{A}, x_i) = \mathsf{A}(x_i)$$
$$\tau_\mathbf{C}(\top, x_i) = \textbf{true}$$
$$\tau_\mathbf{C}(\bot, x_i) = \textbf{false}$$
$$\tau_\mathbf{C}(\{\mathsf{a_1}, \ldots, \mathsf{a_n}\}, x_i) = \bigvee_{1 \le j \le n} x_i = \mathsf{a}_j$$
$$\tau_\mathbf{C}(\neg C, x_i) = \neg \tau_\mathbf{C}(C, x_i)$$
$$\tau_\mathbf{C}(C \sqcap D, x_i) = \tau_\mathbf{C}(C, x_i) \wedge \tau_\mathbf{C}(D, x_i)$$
$$\tau_\mathbf{C}(C \sqcup D, x_i) = \tau_\mathbf{C}(C, x_i) \vee \tau_\mathbf{C}(D, x_i)$$
$$\tau_\mathbf{C}(\exists r.C, x_i) = \exists x_{i+1}.(\tau_\mathbf{R}(r, x_i, x_{i+1}) \wedge \tau_\mathbf{C}(C, x_{i+1}))$$
$$\tau_\mathbf{C}(\forall r.C, x_i) = \forall x_{i+1}.(\tau_\mathbf{R}(r, x_i, x_{i+1}) \to \tau_\mathbf{C}(C, x_{i+1}))$$
$$\tau_\mathbf{C}(\exists r.\mathsf{Self}, x_i) = \tau_\mathbf{R}(r, x_i, x_i)$$
$$\tau_\mathbf{C}(\geqslant nr.C, x_i) = \exists x_{i+1} \ldots x_{i+n}.\Big(\bigwedge_{i+1 \le j < k \le i+n}(x_j \neq x_k)$$
$$\wedge \bigwedge_{i+1 \le j \le i+n}(\tau_\mathbf{R}(r, x_i, x_j) \wedge \tau_\mathbf{C}(C, x_j))\Big)$$
$$\tau_\mathbf{C}(\leqslant nr.C, x_i) = \neg \tau_\mathbf{C}(\geqslant (n+1)r.C, x_i)$$

Obviously, the translation assigns to a role a FOL formula with (at most) two free variables and to a concept a FOL formula with (at most) one free variable. Now we are ready to translate axioms:

$$\tau(r_1 \circ \ldots \circ r_n \sqsubseteq r) = \forall x_0 \ldots x_n (\bigwedge_{1 \le i \le n} \tau_\mathbf{R}(r_i, x_{i-1}, x_i)) \to \tau_\mathbf{R}(r, x_0, x_n)$$
$$\tau(\mathsf{Dis}(r, r')) = \forall x_0 x_1 (\tau_\mathbf{R}(r, x_0, x_1) \to \neg \tau_\mathbf{R}(r', x_0, x_1))$$
$$\tau(C \sqsubseteq D) = \forall x_0 (\tau_\mathbf{C}(C, x_0) \to \tau_\mathbf{C}(D, x_0))$$
$$\tau(C(\mathsf{a})) = \tau_\mathbf{C}(C, x_0)[x_0/\mathsf{a}]$$
$$\tau(r(\mathsf{a}, \mathsf{b})) = \tau_\mathbf{R}(C, x_0, x_1)[x_0/\mathsf{a}][x_1/\mathsf{b}]$$
$$\tau(\neg r(\mathsf{a}, \mathsf{b})) = \neg \tau(r(\mathsf{a}, \mathsf{b}))$$
$$\tau(\mathsf{a} \approx \mathsf{b}) = \mathsf{a} = \mathsf{b}$$
$$\tau(\mathsf{a} \not\approx \mathsf{b}) = \neg(\mathsf{a} = \mathsf{b})$$

**Exercise 5.** *Translate the axioms from Example 20 and Exercise 2 into first-order logic with equality.*

**Remark 25.** The considerations in this section do not apply to all DLs, since also extensions of DLs with non-first-order features have been defined and investigated such as non-monotonic features, regular expressions as role constructors or fixpoint operators. However, the mainstream DLs for which mature reasoners exist and which have been used as a basis for OWL are all first-order-embeddable.

# 4    Description Logics Nomenclature

> *What's in a name? That which we call, say, $\mathcal{SHIQ}$,*
> *By any other name would do the trick.*
> *While DL names might leave the novice $\mathcal{SHOQ}$ed,*
> *Some principles of $\mathcal{ALCH}$emy unlocked*
> *Enable understanding in a minute:*
> *Though it be madness, yet there's method in it.*

There is a well-established naming convention for DLs. The naming scheme for mainstream DLs can be summarized as follows:

$$((\,\mathcal{ALC}\,|\,\mathcal{S}\,)[\,\mathcal{H}\,]|\,\mathcal{SR}\,)[\,\mathcal{O}\,][\,\mathcal{I}\,][\,\mathcal{F}\,|\,\mathcal{N}\,|\,\mathcal{Q}\,]$$

The meaning of the name constituents is as follows:

- $\mathcal{ALC}$ is an abbreviation for *attributive language with complements* [Schmidt-Schauß and Smolka, 1991]. This DL disallows RBox axioms as well as the universal role, role inverses, cardinality constraints, nominal concepts, and self concepts.
- By $\mathcal{S}$ we denote $\mathcal{ALC}$ where we additionally allow transitivity statements, i.e., specific role chain axioms of the shape $r \circ r \sqsubseteq r$ for $r \in \mathsf{N}_R$. The name goes back to the name of a modal logic called S.
- $\mathcal{ALC}$ and $\mathcal{S}$ can be extended by role hierarchies (obtaining $\mathcal{ALCH}$ or $\mathcal{SH}$) which allow for simple role inclusions, i.e., role chain axioms of the shape $r \sqsubseteq s$.
- $\mathcal{SR}$ denotes $\mathcal{ALC}$ extended with all kinds of RBox axioms as well as self concepts.
- The letter $\mathcal{O}$ in the name of a DL indicates that nominal concepts are supported.
- When a DL contains $\mathcal{I}$ then it features role inverses.
- The letter $\mathcal{F}$ at the end of a DL name enables support for role functionality statements which can be expressed as $\top \sqsubseteq \leqslant 1.\top$.
- $\mathcal{N}$ at the end of a DL name allows for unqualified number restrictions, i.e., concepts of the shape $\geqslant nr.\top$ and $\leqslant nr.\top$.
- $\mathcal{Q}$ indicates support for arbitrary qualified number restrictions.

As becomes clear from the previous descriptions, $\mathcal{S}$ contains $\mathcal{ALC}$. Moreover $\mathcal{SR}$ subsumes all of $\mathcal{ALC}$, $\mathcal{ALCH}$, $\mathcal{S}$, and $\mathcal{SH}$. Finally $\mathcal{F}$ becomes obsolete once $\mathcal{N}$ is present and both are superseded by $\mathcal{Q}$.

**Exercise 6.** *Come up with a partial order diagram displaying syntactic containment of all DLs that match the above naming scheme and do not contain $\mathcal{F}$ or $\mathcal{N}$.*

**Exercise 7.** *Name, for each of the following knowledge bases, the "smallest" DL that contains it:*

(a) *the knowledge base from Example 12,*
(b) *the knowledge base from Example 21,*
(c) *the knowledge base consisting of the axioms (a), (b) and (e) from Exercise 2,*
(d) *the knowledge base containing the axioms*

$$\top \sqsubseteq \exists\mathsf{sameAs}.\mathsf{Self} \qquad \top \sqsubseteq\, \leqslant 1\mathsf{sameAs}.\top \qquad \mathsf{batman} \sqsubseteq \neg\exists\mathsf{sameAs}^-.\{\mathsf{santa}\}.$$

# 5   Equivalences, Emulation, Normalization

*Don't give told consequences lip,*
*Nor 'bout equivalences quip,*
*'Cause often it's the formal norm*
*That statements be in normal form.*

The language of the DL $\mathcal{SROIQ}$ is rather redundant, that is, a matter can be formulated in in many ways that are syntactically different but semantically the same. In the following, we will survey different kinds of "semantical alikeness." Moreover we also discuss how this "syntactic redundancy" can be reduced by reverting to so-called normal forms, which come handy for preprocessing knowledge bases before performing actual automated reasoning, but are also useful to alleviate proof work when certain meta-logical properties have to be shown.

## 5.1   Concept Equivalences

A very basic form of "semantical alikeness" is *concept equivalence.* Two concepts $C, D \in \mathbf{C}$ are called equivalent – which is usually denoted by $C \equiv D$ – if they have the same extension in any interpretation $\mathcal{I}$, i.e. $C^{\mathcal{I}} = D^{\mathcal{I}}$. Note that this notion does not presume a fixed knowledge base, thus it really refers to all possible interpretations $\mathcal{I}$.

**Remark 26.** It is easy to see that the definition of concept equivalence can be reformulated in terms of axiom entailment: $C \equiv D$ holds exactly if the empty knowledge base entails both $C \sqsubseteq D$ and $D \sqsubseteq C$, i.e. $\emptyset \models C \sqsubseteq D$ and $\emptyset \models D \sqsubseteq C$. In fact, sometimes in the literature, statements of the form $C \equiv D$ are allowed to occur in knowledge bases as TBox axioms.

**Exercise 8.** *Contemplate whether the condition from Remark 26 can be captured by just one axiom, i.e. whether there is an axiom $\alpha$ such that $\emptyset \models \alpha$ if and only if $C \equiv D$. If this question cannot be answered right now, you may revisit it after having read this section.*

Quite a few basic concept equivalences (which are normally simply taken for granted without further consideration) can be directly traced back to the semantics definition for concepts. To recognize and memorize the equivalences it is quite helpful that the syntactical notation of concept constructors ($\sqcup$, $\sqcap$) is inspired by the associated set-theoretical interpretation ($\cup$, $\cap$) and is also very related to the corresponding notation in propositional logic ($\vee$,$\wedge$). First, we find that both concept intersection and union are commutative, associative and idempotent.

$$
\begin{array}{lll}
C \sqcap D \equiv D \sqcap C & C \sqcup D \equiv D \sqcup C & \text{commutativity} \\
(C \sqcap D) \sqcap E \equiv C \sqcap (D \sqcap E) & (C \sqcup D) \sqcup E) \equiv C \sqcup (D \sqcup E) & \text{associativity} \\
C \sqcap C \equiv C & C \sqcup C \equiv C & \text{idempotency}
\end{array}
$$

The law of associativity alone already releases us from the duty to put parentheses if the union or intersection of more than two concepts is written down, this allows us to write $C \sqcup D \sqcup D$ or $C \sqcap D \sqcap D$ without causing semantical ambiguity due to the missing precedence information. By virtue of the laws of commutativity, associativity, and idempotency together, we can even conceive unions and intersections of many concepts as sets and write for concept sets $\{C_1, \ldots, C_n\} = \mathfrak{C} \subseteq \mathbf{C}$

$$
\bigsqcup_{C \in \mathfrak{C}} C \qquad \text{or} \qquad \bigsqcap_{C \in \mathfrak{C}} C
$$

instead of $C_1 \sqcup \ldots \sqcup C_n$ or $C_1 \sqcap \ldots \sqcap C_n$, respectively.

While the aforementioned laws deal with semantical properties of $\sqcap$ and $\sqcup$ separately, the following cope with their mutual interactions. On the right hand side, we see that the two connectives are distributive over each other, while the equivalences on the right are usually referred to as absorption laws.

$$
\begin{array}{ll}
(C \sqcup D) \sqcap E \equiv (C \sqcap E) \sqcup (D \sqcap E) & (C \sqcup D) \sqcap C \equiv C \\
(C \sqcap D) \sqcup E \equiv (C \sqcup E) \sqcap (D \sqcup E) & (C \sqcap D) \sqcup C \equiv C
\end{array}
$$

Next, we investigate equivalence correspondences involving negation and are certainly not too surprised to find that double negation can be removed and also that the laws of de Morgan are valid in the DL setting:

$$
\begin{array}{c}
\neg\neg C \equiv C \\
\neg(C \sqcap D) \equiv \neg D \sqcup \neg C \\
\neg(C \sqcup D) \equiv \neg D \sqcap \neg C
\end{array}
$$

Beyond but similar to the de Morgan laws, negation can be shifted past quantifiers or be "absorbed" by number restrictions and we obtain:

$$
\begin{array}{c}
\neg \exists r.C \equiv \forall r.\neg C \\
\neg \forall r.C \equiv \exists r.\neg C \\
\neg {\leqslant} nr.C \equiv {\geqslant}(n+1)r.C \\
\neg {\geqslant}(n+1)r.C \equiv {\leqslant} nr.C
\end{array}
$$

The above laws provide a lot of leeway to move negation around. In particular, they ensure that for every concept there exists a concept in *negation normal form*. A concept is said to be in negation normal form (short: NNF), if the only negation symbols in it occur in front of concept names, nominal concepts or self concepts. Given a concept $C$, we determine the concept $nnf(C)$ which is in negation normal form and satisfies $C \equiv nnf(C)$ by applying the recursive function $nnf$:

$$nnf(C) := C \text{ if } C \in \{A, \neg A, \{a_1, ..., a_n\}, \neg\{a_1, ..., a_n\}, \exists r.\mathsf{Self}, \neg \exists r.\mathsf{Self}, \top, \bot\}$$

$$nnf(\neg\neg C) := nnf(C)$$

$$
\begin{aligned}
nnf(\neg\top) &:= \bot & nnf(\neg\bot) &:= \top \\
nnf(C \sqcap D) &:= nnf(C) \sqcap nnf(D) & nnf(\neg(C \sqcap D)) &:= nnf(\neg C) \sqcup nnf(\neg D) \\
nnf(C \sqcup D) &:= nnf(C) \sqcup nnf(D) & nnf(\neg(C \sqcup D)) &:= nnf(\neg C) \sqcap nnf(\neg D) \\
nnf(\forall r.C) &:= \forall r.nnf(C) & nnf(\neg\forall r.C) &:= \exists r.nnf(\neg C) \\
nnf(\exists r.C) &:= \exists r.nnf(C) & nnf(\neg\exists r.C) &:= \forall r.nnf(\neg C) \\
nnf(\leqslant n\, r.C) &:= \leqslant n\, r.nnf(C) & nnf(\neg \leqslant n\, r.C) &:= \geqslant (n+1)\, r.nnf(C) \\
nnf(\geqslant n\, r.C) &:= \geqslant n\, r.nnf(C) & nnf(\neg \geqslant n\, r.C) &:= \leqslant (n-1)\, r.nnf(C)
\end{aligned}
$$

The following equivalences show that $\geqslant 0$ cardinality constraints are vacuously true and that existential and universal quantification can be seen as a special case of number restrictions.

$$\geqslant 0 r.C \equiv \top$$
$$\geqslant 1 r.C \equiv \exists r.C$$
$$\leqslant 0 r.C \equiv \forall r.\neg C$$

**Exercise 9.** *Argue that for every $\mathcal{ALCQ}$ concept $C$, there exists a concept $C'$ with $C \equiv C'$ containing (next to concept and role names) only the connectives $\neg, \sqcup$, and $\geqslant n$. Provide a function that computes $C'$.*

We finish our enumeration of concept equivalences with some correspondences showing, next to some interactions of quantifiers with $\top$ and $\bot$, that quantifiers may distribute over corresponding connectives, that nominal concepts can be "split" into unions of singleton nominal concepts, and that in self concepts, inverses don't make a difference.

$$\exists r.\bot \equiv \bot$$
$$\forall r.\top \equiv \top$$
$$\exists r.(C \sqcup D) \equiv \exists r.C \sqcup \exists r.D$$
$$\forall r.(C \sqcap D) \equiv \forall r.C \sqcap \forall r.D$$
$$\{a_1, \ldots, a_n\} \equiv \{a_1\} \sqcup \ldots \sqcup \{a_n\}$$
$$\exists r^-.\mathsf{Self} \equiv \exists r.\mathsf{Self}$$

**Exercise 10.** *Give formal proofs for all concept equivalences established in this section.*

**Exercise 11.** *Show that the following equivalences are not valid:*

(a)  $\exists r.(C \sqcap D) \equiv \exists r.C \sqcap \exists r.D,$

(b)  $C \sqcap (D \sqcup E) \equiv (C \sqcap D) \sqcup E,$

(c)  $\exists r.\{a\} \sqcap \exists r.\{b\} \equiv \geqslant 2.\{a,b\},$

(d)  $\exists r.\top \sqcap \exists s.\top \equiv \exists r.\exists r^{-}.\exists s.\top.$

## 5.2   Knowledge Base Equivalences

Another notion of semantical alikeness is *axiom* or *knowledge base equivalence*. Two knowledge bases $\mathcal{KB}_1$ and $\mathcal{KB}_2$ are called equivalent (which we will write $\mathcal{KB}_1 \Longleftrightarrow \mathcal{KB}_2$), if their model sets coincide, i.e. if an interpretation $\mathcal{I}$ is a model of $\mathcal{KB}_1$ exactly if it is a model of $\mathcal{KB}_2$. As a special case, we obtain axiom equivalence: $\alpha_1$ and $\alpha_2$ are equivalent (written $\alpha_1 \Longleftrightarrow \alpha_2$) if the two singleton knowledge bases $\{\alpha_1\}$ and $\{\alpha_2\}$ are equivalent.

In the following, we will review some of the most important knowledge base equivalences which are e.g. used to define knowledge base normal forms. The first two equivalences show that unions on the left hand side as well as intersections on the right hand side of a GCI can be "taken apart" into several axioms. These correspondences are also well known in the logic programming field where they are usually referred to as Lloyd-Topor transformations [Lloyd and Topor, 1984].

$$\{A \sqcup B \sqsubseteq C\} \Longleftrightarrow \{A \sqsubseteq C,\ B \sqsubseteq C\}$$
$$\{A \sqsubseteq B \sqcap C\} \Longleftrightarrow \{A \sqsubseteq B,\ A \sqsubseteq C\}$$

An axiom equivalence also often used for normalization purposes is the following:

$$C \sqsubseteq D \Longleftrightarrow \top \sqsubseteq \neg C \sqcup D$$

This allows to transform arbitrary GCIs into the statement that a certain concept (in our case $\neg C \sqcup D$) is "universal", i.e., that its extension is the whole domain. Moreover, this transformation together with a reverse Lloyd-Topor modification allows to transform an entire TBox into one single universal concept statement.

**Example 27.** Considering the TBox of the knowledge base from Example 12, we can first perform the following transformations:

- Healthy $\sqsubseteq \neg$Dead        becomes    $\top \sqsubseteq \neg$Healthy $\sqcup \neg$Dead
- Cat $\sqsubseteq$ Dead $\sqcup$ Alive    becomes    $\top \sqsubseteq \neg$Cat $\sqcup$ Dead $\sqcup$ Alive
- HappyCatOwner $\sqsubseteq \exists$owns.Cat $\sqcap \forall$caresFor.Healthy    becomes
  $\top \sqsubseteq \neg$HappyCatOwner $\sqcup (\exists$owns.Cat $\sqcap \forall$caresFor.Healthy$)$

Finally, due to the coinciding left hand side of the created GCIs, we can put them together to obtain

$$\top \sqsubseteq (\neg\text{Healthy} \sqcup \neg\text{Dead}) \sqcap (\neg\text{Cat} \sqcup \text{Dead} \sqcup \text{Alive})$$
$$\sqcap(\neg\text{HappyCatOwner} \sqcup (\exists\text{owns.Cat} \sqcap \forall\text{caresFor.Healthy}))$$

As already mentioned before, ABox statements can be translated into equivalent TBox statements in any DL that allows for nominals, according to the following equivalences:

$$C(\mathsf{a}) \Longleftrightarrow \{\mathsf{a}\} \sqsubseteq C$$
$$r(\mathsf{a}, \mathsf{b}) \Longleftrightarrow \{\mathsf{a}\} \sqsubseteq \exists r.\{\mathsf{b}\}$$
$$\neg r(\mathsf{a}, \mathsf{b}) \Longleftrightarrow \{\mathsf{a}\} \sqsubseteq \neg \exists r.\{\mathsf{b}\}$$
$$\mathsf{a} \approx \mathsf{b} \Longleftrightarrow \{\mathsf{a}\} \sqsubseteq \{\mathsf{b}\}$$
$$\mathsf{a} \not\approx \mathsf{b} \Longleftrightarrow \{\mathsf{a}\} \sqsubseteq \neg\{\mathsf{b}\}$$

**Exercise 12.** *It might come as a surprise that the GCI $\{\mathsf{a}\} \sqsubseteq \{\mathsf{b}\}$ is sufficient to express $\mathsf{a} \approx \mathsf{b}$. Argue why the converse inclusion $\{\mathsf{b}\} \sqsubseteq \{\mathsf{a}\}$ is redundant given $\{\mathsf{a}\} \sqsubseteq \{\mathsf{b}\}$.*

In turn this allows to transfer any knowledge base consisting only of an ABox and a TBox into a singular universal concept statement.

**Exercise 13.** *Consider whether there is a way to also translate RBox axioms into GCIs by a similar technique.*

**Example 28.** The said equivalences can also be applied reversely and thus used to remove axioms containing nominal concepts from TBoxes. This may be worthwhile doing as nominals in TBoxes normally lead to worse runtimes of reasoning algorithms. Give examples of GCIs containing nominals where this removal is not possible.

The following two equivalences may take a moment to verify intuitively. The essential idea here is to transfer the "standpoint" from the source to the target of a role. These correspondences can be used to remove some inverses from a knowledge base.

$$\exists \mathsf{r}^-.C \sqsubseteq D \Longleftrightarrow C \sqsubseteq \forall \mathsf{r}.D$$
$$C \sqsubseteq \forall \mathsf{r}^-.D \Longleftrightarrow \exists \mathsf{r}.C \sqsubseteq D$$

**Example 29.** Give a formal proof for the two preceding axiom equivalences.

**Exercise 14.** *Consider whether the inverse can be removed in axioms of the shape $C \sqsubseteq \exists \mathsf{r}^-.D$.*

Inverses also give rise to an equivalence between role chain axioms. Intuitively, all roles on both sides of the statement have to be inverted and (which is not really a big surprise) additionally the order of the roles in the chain has to be reverted.

$$r_1 \circ \ldots \circ r_n \sqsubseteq r \Longleftrightarrow Inv(r_n) \circ \ldots \circ Inv(r_1) \sqsubseteq Inv(r)$$

**Exercise 15.** *In the light of this section, revisit Exercise 7 and discuss how the knowledge bases there could be equivalently rewritten to fit an even "smaller" DL.*

## 5.3    Emulation

In the previous sections, we considered very strong notions of semantic alikeness based on the equality of extensions or model sets, respectively. These notions are symmetric (i.e. they hold both ways) and presume that the signatures used are the same. However, there are certain modeling tasks and certain normalization requirements that can be accomplished only by virtue of additional vocabulary (i.e. auxiliary individual, concept and role names; often those signature elements are called *fresh* in order to indicate that they must not have been used in the knowledge base before).

**Example 30.** As an easy example, consider the $\mathcal{SROIQ}$ axiom $\top \sqsubseteq \exists u.C$, which specifies that the concept $C$ is non-empty, i.e. in every model $\mathcal{I}$, there must be some individual $\delta \in \Delta^{\mathcal{I}}$ for which $\delta \in C^{\mathcal{I}}$ holds. While we cannot express this equivalently in any DL not featuring the universal role, it is rather easy to do so in an emulating way: we introduce a new individual name c which is meant to denote $\delta$ and specify that it denotes an instance of $C$ by the ABox statement $C(\mathsf{c})$. Note that this example also represents a simple form of Skolemisation (which is not the case for all examples of emulation).

This kind of semantic similarity that allows for introducing additional vocabulary is referred to as *(semantic) emulation*. Formally, a knowledge base $\mathcal{KB}'$ semantically emulates a knowledge base $\mathcal{KB}$ if the two following conditions hold:

- Every model of $\mathcal{KB}'$ is a model of $\mathcal{KB}$, formally: given an interpretation $\mathcal{I}$, we have that $\mathcal{I} \models \mathcal{KB}'$ implies $\mathcal{I} \models \mathcal{KB}$.
- For every model $\mathcal{I}$ of $\mathcal{KB}$ there is a model $\mathcal{I}'$ of $\mathcal{KB}'$ that has the same domain as $\mathcal{I}$, and coincides with $\mathcal{I}$ on the vocabulary used in $\mathcal{KB}$. In other words $\Xi^{\mathcal{I}'} = \Xi^{\mathcal{I}}$ for every $\Xi \in \mathsf{N}_I(\mathcal{KB}) \cup \mathsf{N}_C(\mathcal{KB}) \cup \mathsf{N}_R(\mathcal{KB})$.

**Remark 31.** Note that knowledge base equivalence is a special case of emulation. In particular, every knowledge base emulates itself. Moreover, emulation is transitive: if $\mathcal{KB}''$ emulates $\mathcal{KB}'$ and $\mathcal{KB}'$ emulates $\mathcal{KB}$, then $\mathcal{KB}''$ emulates $\mathcal{KB}$.

Another common wording for expressing that $\mathcal{KB}'$ emulates $\mathcal{KB}$ is saying that $\mathcal{KB}'$ is *conservative over* $\mathcal{KB}$. The semantic correspondence between two knowledge bases $\mathcal{KB}'$ and $\mathcal{KB}$ where the former emulates the latter is still quite tight: $\mathcal{KB}'$ is satisfiable exactly if $\mathcal{KB}$ is, the two knowledge bases coincide in terms of entailment for every axiom $\alpha$ which does not use any name from the auxiliary vocabulary used in $\mathcal{KB}'$, i.e. in this case, we have $\mathcal{KB} \models \alpha$ exactly if $\mathcal{KB}' \models \alpha$. In fact, we even obtain that $\mathcal{KB} \cup \mathcal{KB}_1 \models \mathcal{KB}_2$ exactly if $\mathcal{KB}' \cup \mathcal{KB}_1 \models \mathcal{KB}_2$ for any knowledge bases $\mathcal{KB}_1, \mathcal{KB}_2$ that do not contain any of $\mathcal{KB}'$s auxiliary vocabulary. Thus, $\mathcal{KB}'$ can do the same job as $\mathcal{KB}$ in many respects while the possible usage of auxiliary signature elements provides quite some freedom in terms of normalization possibilities.

**Example 32.** Remember that we call an ABox of a knowledge base *extensionally reduced* if the only concepts and roles occurring therein are concept names and roles names, respectively. While it is easy to convert an ABox into one not containing statements of the form $\mathbf{r}^-(\mathbf{a}, \mathbf{b})$ (as they can be equivalently expressed by $\mathbf{r}(\mathbf{b}, \mathbf{a})$), concept assertions of the form $C(\mathbf{a})$ where $C$ is not a concept name cannot be removed by equivalent transformations in general. However, by making use of an additional, newly introduced concept name $\mathbf{A}_C$, we can rewrite $C(\mathbf{a})$ into the two axioms $\mathbf{A}_C(\mathbf{a})$ and $\mathbf{A}_C \sqsubseteq C$ which together do the same job as the original axiom. Thereby, the complex concept is shifted from the ABox into the TBox, whence an exhaustive application of this step to all concept assertions results in a knowledge base $\mathcal{KB}'$ which is extensionally reduced and emulates $\mathcal{KB}$.

**Exercise 16.** *Prove that* $\{\mathbf{A}_C(\mathbf{a}), \mathbf{A}_C \sqsubseteq C\}$ *indeed emulates* $\{C(\mathbf{a})\}$.

One normalization being of particular importance for many reasoning algorithms is known under the name *structural reduction*. Essentially, structural reduction aims at reducing the complex structure of axioms by means of introducing concept names for substructures and substituting them. This allows us to omit nestings of role restrictions and boolean operators. Technically, the idea works as follows: let $C[D]$ be a complex concept containing $D$ as a subexpression. Then, we can introduce a fresh concept name $\mathbf{A}_D$ and force it to extensionally coincide with $D$ by adding the two axioms $\mathbf{A}_D \sqsubseteq D$ and $D \sqsubseteq \mathbf{A}_D$ to the knowledge base. This enables us to exchange all occurrences of $D$ in $C[D]$ by $\mathbf{A}_D$, obtaining $C[\mathbf{A}_D]$.

**Example 33.** Consider the axiom

$$\exists\texttt{livesAt}.\{\texttt{northPole}\} \sqsubseteq \exists\texttt{worksFor}^-.(\texttt{Reindeer} \sqcap \exists\texttt{hasNose}.(\texttt{Red} \sqcap \texttt{Shiny})).$$

Performing structural reduction (and using $\equiv$ as a shortcut for mutual $\sqsubseteq$) we obtain

$$\mathbf{A}_{\exists\texttt{livesAt}.\{\texttt{northPole}\}} \sqsubseteq \mathbf{A}_{\exists\texttt{worksFor}^-.(\texttt{Reindeer}\sqcap\exists\texttt{hasNose}.(\texttt{Red}\sqcap\texttt{Shiny}))}$$
$$\mathbf{A}_{\exists\texttt{livesAt}.\{\texttt{northPole}\}} \equiv \exists\texttt{livesAt}.\mathbf{A}_{\{\texttt{northPole}\}}$$
$$\mathbf{A}_{\{\texttt{northPole}\}} \equiv \{\texttt{northPole}\}$$
$$\mathbf{A}_{\exists\texttt{worksFor}^-.(\texttt{Reindeer}\sqcap\exists\texttt{hasNose}.(\texttt{Red}\sqcap\texttt{Shiny}))} \equiv \exists\texttt{worksFor}^-.\mathbf{A}_{\texttt{Reindeer}\sqcap\exists\texttt{hasNose}.(\texttt{Red}\sqcap\texttt{Shiny})}$$
$$\mathbf{A}_{\texttt{Reindeer}\sqcap\exists\texttt{hasNose}.(\texttt{Red}\sqcap\texttt{Shiny})} \equiv \texttt{Reindeer} \sqcap \mathbf{A}_{\exists\texttt{hasNose}.(\texttt{Red}\sqcap\texttt{Shiny})}$$
$$\mathbf{A}_{\exists\texttt{hasNose}.(\texttt{Red}\sqcap\texttt{Shiny})} \equiv \exists\texttt{hasNose}.\mathbf{A}_{\texttt{Red}\sqcap\texttt{Shiny}}$$
$$\mathbf{A}_{\texttt{Red}\sqcap\texttt{Shiny}} \equiv \texttt{Red} \sqcap \texttt{Shiny}$$

**Remark 34.** There are other, more elaborate and space-saving ways to perform structural reduction. In fact normally only one of the two axioms $\mathbf{A}_D \sqsubseteq D$ or $D \sqsubseteq \mathbf{A}_D$ is necessary to achieve emulation. Which one depends on the position of $D$ inside an axiom related to scopes of negation and other junctors. This position information is captured by the notion of *polarity* of a subexpression.

**Exercise 17.** *Using the technique of structural reduction and other semantic alikeness correspondences introduced above, argue that any knowledge base $\mathcal{KB}$ can be emulated by a knowledge base $\mathcal{KB}'$ the TBox of which contains only GCIs of the form*

$$\bigsqcap_{C \in \mathfrak{C}} C \sqsubseteq \bigsqcup_{D \in \mathfrak{D}} D$$

*where $\mathfrak{C} \cup \mathfrak{D}$ contains only concepts of the forms $\{a\}$, $A$, $\exists r.\mathsf{Self}$, $\leq nr.A$, or $\geq nr.A$ with $a \in \mathsf{N}_I$, $A \in \mathsf{N}_C$ and $r \in \mathbf{R}$ (note that no negation is allowed, whatsoever).*

**Example 35.** Given a concept expression of the form $A \sqsubseteq \geq nr.B$, the cardinality constraint can be removed as follows: We introduce fresh role names $\mathbf{r}_1, \ldots \mathbf{r}_n$ which we specify as subroles of $r$ (by the axioms $\mathbf{r}_i \sqsubseteq r$ for all $1 \leq i \leq n$) and as pairwise disjoint (i.e. we add $\mathsf{Dis}(\mathbf{r}_i, \mathbf{r}_j)$ for all $1 \leq i < j \leq n$). With that background axiomatization, the above statement can be rewritten into $A \sqsubseteq \bigsqcap_{1 \leq i < j \leq n} \exists \mathbf{r}_i.B$.

Emulation techniques can also be used to show that a number of statements which can be directly expressed in other logics (such as FOL) but not in DL, are nevertheless expressible by using some "makros" involving auxiliary vocabulary. In the following, we give some examples for this.

***The universal role.*** The universal role $u$ connects all individuals of the described domain. In a DL where this feature is not built in, we may want to introduce a new role $u'$ and write down statements which force $u'$ to behave like the universal role (by making sure that $u'$ must be interpreted as $\Delta^{\mathcal{I}} \times \Delta^{\mathcal{I}}$ in every model $\mathcal{I}$). Note that this can be easily done in FOL by the statement $\forall x, y(u'(x, y))$. However, if a DL supports transitivity and nominal concepts, we can obtain the same by introducing a new nominal $\mathbf{a}_{aux}$ and specify the axioms $\top \sqsubseteq \exists u'.\{\mathbf{a}_{aux}\}$ and $\top \sqsubseteq \exists u'^{-}.\{\mathbf{a}_{aux}\}$ and $u' \circ u' \sqsubseteq u'$. The only downside to this is that $u'$ is then necessarily non-simple whence it cannot be used in all places where $u$ could.

***Concept products.*** Sometimes, there are situations where one wants to express that any instance of a concept $C$ is connected with any instance of a concept $D$ via a role $r$. In fact, *concept product* statements of the form $C \times D \sqsubseteq r$ which express exactly that have been introduced into description logics rather early but never found their way into the mainstream.

**Example 36.** As an example, the fact that alkaline solutions neutralize acid solutions could expressed by the concept product axiom $\mathsf{AlkalineSolution} \times \mathsf{AcidSolution} \sqsubseteq \mathsf{neutralises}$.

Again, it is rather easy to find that the FOL statement $\forall x, y(C(x) \wedge D(y) \rightarrow r(x, y))$ realizes this (where we for the sake of simplicity assume that $C, D$ are concept names and $r$ is a role name). However, $\mathcal{SROIQ}$ provides enough modeling capabilities to emulate this situation as well via the GCIs $C \sqsubseteq \exists \mathbf{r}_{aux}.\mathsf{Self}$ and $D \sqsubseteq \exists \mathbf{r}'_{aux}.\mathsf{Self}$ as well as the complex role inclusion $\mathbf{r}_{aux} \circ u \circ \mathbf{r}'_{aux} \sqsubseteq r$. Concept products and their impact on reasoning complexity have e.g. been considered by Rudolph *et al.* [2008].

**Qualified role inclusion.** Likewise, the specialization of roles due to concept memberships of the two involved individuals seems to surpass the modeling capabilities of the DLs treated here. The FOL statement $\forall x, y(C(x) \wedge r(x, y) \wedge D(y) \rightarrow s(x, y))$ (expressing that any $C$-instance and $D$-instance that are interconnected by $r$ are also interconnected by $s$) can be emulated by a DL axiomatization in a similar way as discussed above: Introduce the GCIs $C \sqsubseteq \exists r_{aux}.\mathsf{Self}$ and $D \sqsubseteq \exists r'_{aux}.\mathsf{Self}$ as well as the complex role inclusion $r_{aux} \circ r \circ r'_{aux} \sqsubseteq s$.

**Exercise 18.** *Use this technique to express the proposition "any person of age having signed a contract which is legal is bound to that contract." Use the concept names* OfAge, Contract, Legal *and the role names* hasSigned *and* boundTo.

Qualified role inclusions and concept products constitute special cases of the more general framework of *description logic rules* as described by Krötzsch *et al.* [2008].

**Boolean Combination of Axioms.** From the point of view of FOL, it seems quite straightforward that any statement can be negated or any two statements can be connected by disjunction and conjunction, obtaining a new statement inside the logic. In other words, FOL is Boolean-closed on the sentence level. In DLs, the situation is quite different: there is no direct way to, for instance, say that one of the two GCIs $A \sqsubseteq B$ and $C \sqsubseteq D$ must hold. This is, roughly speaking, due to the fact that DL axioms can be understood as "element-wise" propositions (the verbalization of which starts "for each element of the domain holds..."), whereas the above statement gives an alternative choice concerning all individuals at once. Fortunately, $\mathcal{SROIQ}$ provides a way to handle this by virtue of the universal role. We first recap that the above axioms can be rewritten into $\top \sqsubseteq \neg A \sqcup B$ and $\top \sqsubseteq \neg C \sqcup D$ respectively. Then we axiomatize the following statement: "every domain element is an instance of $A \sqcup B$ or every domain element is an instance of $C \sqcup D$." To this end we exploit the fact that every individual is connected to every individual via the universal role, whence we can formally express the above wording by the axiom $\top \sqsubseteq \forall u.(\neg A \sqcup B) \sqcup \forall u.(\neg C \sqcup D)$.

**Exercise 19.** *In fact, the encoding introduced above doesn't need any auxiliary vocabulary. However, arbitrary Boolean combinations of axioms can also be emulated in $\mathcal{SHOIQ}$. In that case, the vocabulary must be extended. Explain how this can be done. Hint: try using a "hub nominal."*

**Exercise 20.** *Find a way to emulate $C(\mathsf{a}) \vee D(\mathsf{b})$ in $\mathcal{SHIQ}$.*

**Exercise 21.** *Consider whether it is possible to emulate ABox statements of the shape $\neg r(\mathsf{a}, \mathsf{b})$, $\mathsf{a} \approx \mathsf{b}$, and $\mathsf{a} \not\approx \mathsf{b}$ with an $\mathcal{ALCHIQ}$ knowledge base by using only ABox statements of the form $C(\mathsf{a})$ and $r(\mathsf{a}, \mathsf{b})$.*

# 6  Modeling with DLs

*While frowning on plurality,*
*The pope likes cardinality:*
*It can enforce infinity,*
*And hence endorse divinity.*
*But, theologically speaking,*
*The papal theory needs tweaking*
*For it demands divine assistance*
*to prove "the three are one"-consistence.*

In this section, we will discuss the added value brought about by certain DL modeling features. We will also discuss specific types of statements for which some formalisms provide dedicated modeling primitives, although they are just "syntactic sugar," that is they can be expressed by virtue of the modeling features already introduced. Moreover, we will provide some insight about model-theoretic consequences that arise from using or not using certain constructs.

**Remark 37.** Thereby, one can see that the expressive power of a logic can be characterized by its capability to "distinguish" interpretations. That is, a "stronger" logic might be able to distinguish two interpretations $\mathcal{I}_1$ and $\mathcal{I}_2$ meaning that there is a knowledge base $\mathcal{KB}$ such that $\mathcal{I}_1 \models \mathcal{KB}$ but $\mathcal{I}_2 \not\models \mathcal{KB}$ (or vice versa), whereas a "weaker" logic may not have this capability. In many cases, this indistinguishability can be cast into statements of the following type: given any knowledge base $\mathcal{KB}$ in a certain DL and a (set of) model(s) of $\mathcal{KB}$ then performing a certain operation or manipulation on that model(s) will inevitably result in an interpretation which is again a model of $\mathcal{KB}$. We then say the set of models of $\mathcal{KB}$ is *closed under* the considered operation.

## 6.1  A Lot Can Be Done in $\mathcal{ALC}$

Already $\mathcal{ALC}$ features many modeling capabilities usually found in knowledge representation languages. Beyond the ones explicitly introduced, quite some more correspondences can be expressed indirectly. We will tackle the most important ones.

*Concept Disjointness.* Two concepts $C$ and $D$ are disjoint with respect to an interpretation $\mathcal{I}$, if their extensions do not overlap, i.e. $C^{\mathcal{I}} \cap D^{\mathcal{I}} = \emptyset$. It is straightforward that this semantic condition can be cast into the GCI $C \sqcap D \sqsubseteq \bot$. Equivalently, this can be expressed by $C \sqsubseteq \neg D$ or $D \sqsubseteq \neg C$. Disjointness information is often neglected when doing logical modeling. It can, however, be very useful to derive negative information, e.g., the guarantee that some individual is *not* an instance of a concept.

*Domain and Range of Roles.* Given a role $r$, we may want to make statements about the source and target individuals for the respective relation. We say that the role $r$ has *domain* $C$ in an interpretation $\mathcal{I}$ if any source individual of the relation associated with $r$ is an instance of $C$, in other words, for every $\langle \delta, \delta' \rangle \in r^{\mathcal{I}}$, we have $\delta \in C^{\mathcal{I}}$. Likewise, we say that $r$ has *target* $D$ if for every

$\langle \delta, \delta' \rangle \in r^{\mathcal{I}}$, also $\delta' \in D^{\mathcal{I}}$ is satisfied. The standard DLs covered here do not provide modeling primitives for specifying domain or range of a role, but they can be easily expressed with the means already present in $\mathcal{ALC}$. The above domain statement is equivalent to the GCI $\exists r. \top \sqsubseteq C$ whereas the range statement can be written as $\top \sqsubseteq \forall r. D$.

**The Empty Role.** It might seem a bit peculiar that, while $\mathcal{SROIQ}$ supports both the universal and the empty concept ($\top$ and $\bot$, respectively), it features only the universal role $u$ whereas the empty role is not part of the definition. This is, however, not a severe omission as the empty role can be easily axiomatized: for a new role name emptyRole we can use the GCI $\top \sqsubseteq \forall$emptyRole.$\bot$ to force the extension of emptyRole to be empty. An alternative axiom (beyond $\mathcal{ALC}$) with the same effect is $\mathsf{Dis}(u, \mathtt{emptyRole})$.

> **Exercise 22.** *Come up with an $\mathcal{ALC}$ GCI that expresses the following statement: "If an academic supervises a project, then he is a project leader and the project is a research project." Use the role name* **supervises** *as well as the concept names* **Academic**, **Project**, **ProjectLeader**, *and* **ResearchProject**.

## 6.2   Looking Back: Inverse Roles

Inverses allow for traversing roles in reverse direction. While DLs without inverses only allow for describing domain individuals by means of their "outgoing" roles, by means of inverses, "incoming" roles can be taken into account as well.

> **Example 38.** Consider the interpretation $\mathcal{I}$ from Example 16. It is rather easy to see that the domain individuals 3 and 5 (as well as any other prime number) are not distinguishable by $\mathcal{ALC}$ concepts (in fact, not even by $\mathcal{SROQ}$ concepts), that is, there is no concept $C$ having 3 as an instance but not 5, or vice versa. On the other hand, the $\mathcal{ALCI}$ concept $\exists \mathsf{succ}^-.\exists \mathsf{succ}^-.\exists \mathsf{succ}^-.\neg\exists \mathsf{succ}^-.\top$ does the job.

Moreover, some rather natural properties of relations can be expressed by means of inverses. A role $\mathbf{r}$ is called *symmetric* if for any $\langle \delta, \delta' \rangle \in \mathbf{r}^{\mathcal{I}}$ also $\langle \delta', \delta \rangle \in \mathbf{r}^{\mathcal{I}}$ holds, that is, relatedness via $\mathbf{r}$ always holds both ways. On the other hand it is called *asymmetric* if for all $\langle \delta, \delta' \rangle \in \mathbf{r}^{\mathcal{I}}$ satisfy $\langle \delta', \delta \rangle \notin \mathbf{r}^{\mathcal{I}}$ holds, this means that $\mathbf{r}$-relatedness never holds both ways. Sometimes, symmetry or asymmetry of a role $\mathbf{r}$ is included in a DL as a separate axiom type, denoted by $\mathsf{Sym}(\mathbf{r})$ or $\mathsf{Asy}(\mathbf{r})$, respectively. The former can be easily expressed by stating that $\mathbf{r}$ has its own inverse as a subrole: $\mathbf{r}^- \sqsubseteq \mathbf{r}$. The latter can be characterized by stating that $\mathbf{r}$ and its inverse are disjoint: $\mathsf{Dis}(\mathbf{r}, \mathbf{r}^-)$.

## 6.3   Model Manipulation Part I: Filtration

Now we will turn our attention to our first model transformation. Given a set $\mathfrak{C}$ of concepts and an interpretation $\mathcal{I}$, we can obtain the *filtration* of $\mathcal{I}$ with respect to $\mathfrak{C}$ as follows: First, we define an equivalence relation $\simeq$ on the domain elements of $\mathcal{I}$ by letting $\delta \simeq \delta'$ for anonymous $\delta, \delta' \in \Delta^{\mathcal{I}}$ whenever $\delta$ and $\delta'$ coincide in terms of concept memberships for concepts from $\mathfrak{C}$, that is, for every $C \in \mathfrak{C}$ we have $\delta \in C^{\mathcal{I}}$ exactly if $\delta \in C^{\mathcal{I}}$. Then, for some $\delta \in \Delta^{\mathcal{I}}$ we let $[\delta]_{\simeq} = \{\delta' \mid \delta \simeq \delta'\}$ and $\Delta^{\mathcal{I}}_{/\simeq} = \{[\delta]_{\simeq} \mid \delta \in \Delta\}$. Verbally, the set $\Delta^{\mathcal{I}}_{/\simeq}$ consists of "bags" of domain elements from $\mathcal{I}$ where all elements in one bag coincide on the concepts from $\mathfrak{C}$ they satisfy. The filtration of $\mathcal{I}$ is the interpretation $\mathcal{J}$ with

- $\Delta^{\mathcal{J}} = \Delta^{\mathcal{I}}_{/\simeq}$
- for each $\mathbf{a} \in \mathsf{N}_I$, set $\mathbf{a}^{\mathcal{J}} = [\mathbf{a}^{\mathcal{I}}]_{\simeq}$;
- for each concept name $\mathsf{A} \in \mathsf{N}_C$, set $\mathsf{A}^{\mathcal{J}} = \{[\delta]_{\simeq} \mid \delta \in \mathsf{A}^{\mathcal{I}}\}$;
- for each role name $\mathbf{r} \in \mathsf{N}_R$, set $\mathbf{r}^{\mathcal{J}} = \{\langle [\delta]_{\simeq}, [\delta]_{\simeq} \rangle \mid \langle \delta, \delta' \rangle \in \mathbf{r}^{\mathcal{I}}\}$;

Intuitively, this means, that the filtration is obtained by collapsing domain elements which are not distinguishable by virtue of concepts from $\mathfrak{C}$ (nor by individual names) into one.

**Example 39.** Let $\mathcal{I}$ be the interpretation from Example 14 and let $\mathfrak{C}$ contain all $\mathcal{ALC}$ concepts. Then the according filtration can be sketched as follows.

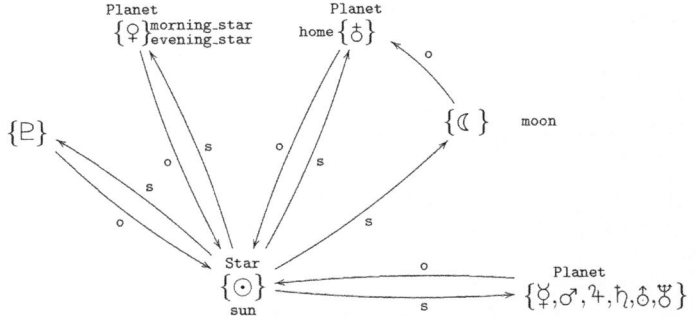

If, for a given $\mathcal{SROI}$ knowledge base $\mathcal{KB}$, we let $\mathfrak{C}$ be all concepts occurring in $\mathcal{KB}$ (including the subexpressions of concepts) then the filtration of a model of $\mathcal{KB}$ will again be a model of $\mathcal{KB}$. On the other hand, since in this case, $\mathfrak{C}$ is finite, there can be only finitely many "bags" in $\Delta^{\mathcal{I}}_{/\simeq}$ which means that the filtration will even be a finite model of $\mathcal{KB}$. This allows to conclude that every satisfiable $\mathcal{SROI}$ knowledge base has a finite model.

**Remark 40.** In general, logics for which the existence of an arbitrary model implies the existence of a model where $\Delta^{\mathcal{I}}$ is a finite set (usually briefly called *finite model*) are said to have the *finite model property*. This is a rather convenient property, since one may disregard infinite representations when looking for models of a knowledge base. Moreover, for any logic that has the finite model property and that can be embedded into FOL, the problem of knowledge base satisfiability is decidable.

Concluding, we can state that filtrations are quite stable in terms of model-hood preservation, however they fail as soon as cardinality constraints come into play.

**Exercise 23.** *Consider Example 39 and find an $\mathcal{ALCQ}$ axiom which is not satisfied in the interpretation given there although it is satisfied in the original interpretation from Example 14.*

## 6.4  Up to Infinity: Cardinality Constraints

By means of cardinality constraints, precise statements about the number of individuals related to a certain individual via a role can be made. This kind of modeling features is of obvious practical value and wide-spread in other knowledge specification formalisms such as entity-relationship modeling or UML. Cardinality constraints also naturally capture certain role characteristics.

For instance, *role functionality* can be seen and treated as a special case of cardinality constraints. In words, a role is functional, if every domain individual is connected to at most one domain individual via the relation associated to that role. Formally, a role $r$ is functional, if for every domain individual $\delta \in \Delta^{\mathcal{I}}$ there is at most one individual $\delta' \in \Delta^{\mathcal{I}}$ satisfying $\langle \delta, \delta' \rangle \in r^{\mathcal{I}}$. This condition can be enforced by the axiom $\top \sqsubseteq {\leqslant} 1.\top$. Sometimes, in DLs which do not support number restrictions in general, the according axiom is noted as $\mathsf{Fun}(r)$. Typical examples for functional roles are `hasFather`, `marriedWith`, or `locatedInCountry`.

**Remark 41.** Note that by definition, a role can be functional and still not start from every domain individual, as in the case of `marriedWith`. Thus the term "functional" may be misleading as it may cause the erroneous impression that the role extension is a (total) function. Rather, functional roles semantically correspond to partial functions.

In fact, in the presence of cardinality constraints allows to enforce that a knowledge base has only models the domain of which is infinite. Consider the following knowledge base:

$$(\forall \mathsf{succ}^-.\top)(\mathsf{zero}) \qquad \top \sqsubseteq \exists \mathsf{succ}.\top \qquad \top \sqsubseteq {\leqslant} 1.\mathsf{succ}^-.\top$$

It is not to difficult to find a model for this knowledge base which has an infinite domain: in fact the interpretation described in Example 16 is such a model. On the other hand the knowledge base cannot have a model with finite domain.

**Exercise 24.** *Prove this. Hint: assume a finite number of domain elements and count sources and targets for `succ`.*

Note that we have just shown that any extension of $\mathcal{ALCIF}$ does not have the finite model property.

## 6.5   Model Manipulation Part II: Unraveling

However, another nice property still holds in the presence of number restrictions. Roughly speaking, this property states that we can take an arbitrary model and "unfold" or "unroll" it such that all the parts of the model not containing named individuals are tree-like (i.e., cycle-free). More formally, the *unraveling* of an interpretation $\mathcal{I}$ is an interpretation that is obtained from $\mathcal{I}$ as follows: First, we define the set $S \subseteq (\Delta^{\mathcal{I}})^*$ of *paths* to be the smallest set of sequences of domain elements such that

- for every $\mathsf{a} \in \mathsf{N}_I$, $\mathsf{a}^{\mathcal{I}}$ is a path;
- $\delta_1 \cdots \delta_n \cdot \delta_{n+1}$ is a path, if
    - $\delta_2 \neq \mathsf{a}^{\mathcal{I}}$ for all $\mathsf{a} \in \mathsf{N}_I$,
    - $\delta_1 \cdots \delta_n$ is a path,
    - $\delta_{i+1} \neq \delta_{i-1}$ for all $i = 2, \ldots, n$,
    - $\langle \delta_n, \delta_{n+1} \rangle \in r^{\mathcal{I}}$ for some $r \in \mathbf{R}$.

For each $w = \delta_1 \cdots \delta_n \in S$, set $\mathsf{last}(w) = \delta_n$. Now, we define the unraveling of $\mathcal{I}$ as the interpretation $\mathcal{J} = \langle \Delta^{\mathcal{J}}, \cdot^{\mathcal{J}} \rangle$ with $\Delta^{\mathcal{J}} = S$ and, for each sequence $w \in \Delta^{\mathcal{J}}$, we define the interpretation of concept and role names as follows:

(a) for each $\mathsf{a} \in \mathsf{N}_I$, set $\mathsf{a}^{\mathcal{J}} = \mathsf{a}^{\mathcal{I}}$;

(b) for each concept name $\mathsf{A} \in \mathsf{N}_C$, set $w \in \mathsf{A}^{\mathcal{J}}$ iff $\mathsf{last}(w) \in \mathsf{A}^{\mathcal{I}}$;

(c) for each role name $r \in \mathsf{N}_R$, set $\langle w, w' \rangle \in r^{\mathcal{J}}$ iff

  - $w' = w\delta$ for some $\delta \in \Delta^{\mathcal{I}}$ and $\langle \mathsf{last}(w), \delta \rangle \in r^{\mathcal{I}}$ or
  - $w = w'\delta$ for some $\delta \in \Delta^{\mathcal{I}}$ and $\langle \delta, \mathsf{last}(w') \rangle \in r^{\mathcal{I}}$ or
  - $w = \mathsf{a}^{\mathcal{I}}$, $w' = \mathsf{b}^{\mathcal{I}}$ for some $\mathsf{a}, \mathsf{b} \in \mathsf{N}_I$ and $\langle \mathsf{a}^{\mathcal{I}}, \mathsf{b}^{\mathcal{I}} \rangle \in r^{\mathcal{I}}$.

With this notion of unraveling we find that for any $\mathcal{ALCHIQ}$ knowledge base $\mathcal{KB}$, an interpretation $\mathcal{I}$ is a model exactly if its unraveling is. This correspondence has some practical consequences: First it guarantees that $\mathcal{ALCHIQ}$ has the *forest model property*. That means that every satisfiable $\mathcal{ALCHIQ}$ knowledge base $\mathcal{KB}$ has a model with a particular shape: there is a "root tangle" of named elements from which trees of anonymous elements grow. This property is for instance of particular interest to prove the completeness of tableau algorithms.

**Example 42.** To demonstrate what happens during the unraveling of an interpretation, consider this small example interpretation (where mbt is intended to mean "more beats than"):

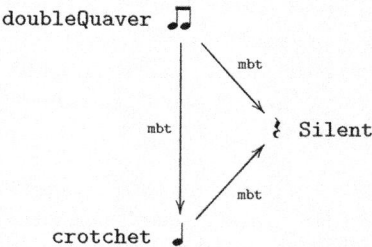

In order to unravel this interpretation, intuitively, we first pick all named individuals (i.e., ♫ and ♩) and keep them as well as their mutual relationships. Then, in the original interpretation, we walk along the (incoming and outgoing) role links to anonymous elements to find the named individuals' role neighbors, these neighbors are (as well as the corresponding role links) reproduced in the unraveling. Even if the neighbors are the same, we introduce separate copies in the unraveling, using the "origin element" as a prefix to distinguish them. In our example, we introduce ♫𝄽 as the mbt-neighbor of ♫ (caused by 𝄽 in the original interpretation) and ♩𝄽 as the mbt-neighbor of ♩ (caused by the same 𝄽). We then proceed to neighbors of neighbors and so forth. Thereby, we exclude the elements that we "just came from" in the previous step. We may, however, traverse elements of the original interpretation several times, we will however disregard their names and create anonymous copies of them in the unraveling. In our case, the result of this procedure is an infinite interpretation which is partially depicted below.

**Exercise 25.** *Sketch or formally describe the unravelings of the interpretations from Example 14 and Example 16. For the latter and for Example 42, give one axiom from the DL $\mathcal{S}$ and one from the DL $\mathcal{ALCOI}$, either of which hold in the interpretation but not the according unraveling.*

**Remark 43.** In fact, variants of the forest model property also hold for some DLs containing role chain axioms and/or nominal concepts, requiring also to modify the employed unraveling technique. In the presence of role chain axioms, one usually defines a "skeleton" of the model via unraveling into a forest structure and thereafter adds further "role links" the presence of which is enforced by the RIAs. In the presence of nominals, one has to allow so-called "backlinks" i.e. tree individuals are allowed to have role links back into the root tangle (but not into other trees).

## 6.6   Far Far Away: Transitivity

Transitivity of a role $r$ is expressible by the complex role inclusion $r \circ r \sqsubseteq r$. In DLs that do not feature any complex role inclusions but transitivity this axiom is often alternatively written as $\mathsf{Tra}(r)$. Role transitivity statements come about quite naturally for a variety of relations that are to be modeled. Typical examples for transitive roles are $\mathsf{ancestorOf}$, $\mathsf{superiorOf}$, $\mathsf{partOf}$, $\mathsf{greaterThan}$, etc. Role transitivity declarations allow for a more succinct modeling and better querying capabilities via entailment checks.

> **Example 44.** Envisioning a company and a knowledge base containing employee data, it would of course be possible to explicitly add all superior relations as ABox role assertions $\mathsf{superiorOf}(\mathsf{a},\mathsf{b})$. On the other hand, the same can be achieved (in terms of inferrable superior information) by only adding role assertions for the cases of where $\mathsf{a}$ denotes a direct superior of $\mathsf{b}$, if we additionally state that $\mathsf{superiorOf}$ is transitive. Moreover this second version is advantageous in terms of maintenance: whenever a new employee joins the company, only their direct superior(s) and inferior(s) need to be explicitly specified.

However, what can be expressed in terms of transitivity in standard DLs is limited. Thereby the limitations are inherited from FOL. What cannot be done in the DLs treated here is to precisely talk about the transitive closure of a given role. In other words, there is no way to axiomatize the condition that one role $r$ is the transitive closure of another role $s$ (formally, this condition can be expressed by $r^{\mathcal{I}} = (s^{\mathcal{I}})^*$). What can be done is to say that the extension of $r$ contains the transitive closure of $s$ (i.e. $(s^{\mathcal{I}})^* \subseteq r^{\mathcal{I}}$) by specifying $s \sqsubseteq r$ and $r \circ r \sqsubseteq r$. Presuming this axiomatization of an upper bound for the transitive closure, we can e.g. check whether there is an "$s$-path" of arbitrary length from an individual $\mathsf{a}$ to an individual $\mathsf{b}$ in every model of the knowledge base by checking whether the knowledge base entails the role assertion $s(\mathsf{a},\mathsf{b})$. Still, there is no way to check for the necessary absence of such a path in all models of the knowledge base.

## 6.7   Model Manipulation Part III: Disjoint Union

We now consider a transformation which, roughly speaking, takes two interpretations and puts them side by side. More formally, given two interpretations $\mathcal{I} = (\Delta^{\mathcal{I}}, \cdot^{\mathcal{I}})$ and $\mathcal{J} = (\Delta^{\mathcal{J}}, \cdot^{\mathcal{J}})$, assuming that $\Delta^{\mathcal{I}} \cap \Delta^{\mathcal{J}} = \emptyset$, we define the *disjoint union* of $\mathcal{I}$ with $\mathcal{J}$ denoted by $\mathcal{I}{+}\mathcal{J} = (\Delta^{\mathcal{I}{+}\mathcal{J}}, \cdot^{\mathcal{I}{+}\mathcal{J}})$ as follows: $\Delta^{\mathcal{I}{+}\mathcal{J}} = \Delta^{\mathcal{I}} \cup \Delta^{\mathcal{J}}$, $\mathsf{a}^{\mathcal{I}{+}\mathcal{J}} = \mathsf{a}^{\mathcal{I}}$, $\mathsf{A}^{\mathcal{I}{+}\mathcal{J}} = \mathsf{A}^{\mathcal{I}} \cup \mathsf{A}^{\mathcal{J}}$ and $\mathsf{r}^{\mathcal{I}{+}\mathcal{J}} = \mathsf{r}^{\mathcal{I}} \cup \mathsf{r}^{\mathcal{J}}$. Note that, unlike most definitions of disjoint unions, this definition is asymmetric since, for the mapping of the individuals, preference is given to $\mathcal{I}$. One can show that whenever $\mathcal{I}$ and $\mathcal{J}$ are models of a $\mathcal{SHIQ}$ knowledge base $\mathcal{KB}$ then so is their disjoint union $\mathcal{I}{+}\mathcal{J}$.

> **Exercise 26.** *Prove the claim above. Hint: An intermediate lemma showing $C^{\mathcal{I}{+}\mathcal{J}} = C^{\mathcal{I}} \cup C^{\mathcal{J}}$ will come handy for that. This will require a structural induction over the concepts.*

**Example 45.** Given the interpretation $\mathcal{I}$ from Example 14, let $\mathcal{I}'$ denote $\mathcal{I}$ where every domain element $\delta$ has been renamed into $\delta'$. Then the interpretation $\mathcal{I} + \mathcal{I}'$ can be displayed as follows:

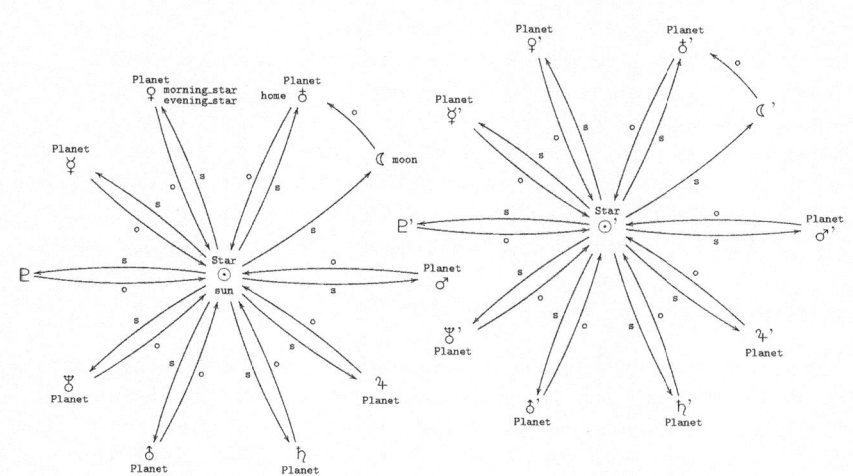

In fact, the above result can be generalized to disjoint unions of infinitely many models. This gives rise to a property which could be called the *infinite model property*: whenever there is an arbitrary model for a $\mathcal{SHIQ}$ knowledge base $\mathcal{KB}$, then there is also an infinite one.

**Remark 46.** More generally, these properties even hold for all $\mathcal{SRIQ}$ knowledge bases not containing the universal role.

**Exercise 27.** *Consider Example 45 and find an $\mathcal{ALCO}$ axiom which is not satisfied in the interpretation given there although it is satisfied in the original interpretation from Example 14.*

Wrapping up, what we have learned about model manipulations, their range of applicability, and the model properties they give rise to can be summarized in the following table.

| manipulation | preserves models for | associated property |
|---|---|---|
| filtration | $\mathcal{SROI}$ | finite model property |
| unraveling | $\mathcal{ALCHIQ}$ | forest model property |
| disjoint union | $\mathcal{SRIQ}\backslash u$ | infinite model property |

## 6.8    Know Your Bounds: Nominal Concept and Universal Role

The modeling power brought about by nominal concepts and universal roles is quite similar. For instance, having the universal role at disposal, we can remove all nominal concepts from a $\mathcal{SROIQ}$ knowledge base as follows: first, rewrite every nominal concept $\{a_1, \ldots, a_n\}$ into $\{a_1\} \sqcup \ldots \sqcup \{a_n\}$ according to the equivalence given in Section 5. Next, introduce fresh concept names $A_{\{a\}}$ for all singleton nominal concepts thus obtained and substitute every occurrence of any $\{a\}$ by the according $A_{\{a\}}$. Finally, add the concept assertion $A(a)$ as well as the GCI $\top \sqsubseteq \,\leqslant 1u.A_{\{a\}}$ for any introduced $A_{\{a\}}$.

On the other hand, the universal role can be emulated once nominal concepts are allowed: we introduce a fresh individual name center and a new role name toCenter and force every individual to have a toCenter relation to the individual denoted by center by means of the axiom $\top \sqsubseteq \exists$toCenter.$\{$center$\}$. Now we can get from every domain individual to every other by a two-hop travel along toCenter and toCenter$^-$. Thus we can replace every $\varXi u.C$ with $\varXi \in \{\forall, \exists, \leqslant n, \geqslant n\}$ by the concept expression $\exists$toCenter.$\varXi$toCenter$^-.C$.

**Exercise 28.** *Find a way to remove the RBox occurrences of $u$ as well.*

A crucial feature showing the added expressivity obtained from nominal concepts or the universal role is the capability to bound or fix the number of individuals in the extension of a class or even in the whole domain. Both the GCIs AtMostTwo $\sqsubseteq \{$one, two$\}$ and $\top \sqsubseteq \,\leqslant 2u.$AtMostTwo specify that the concept AtMostTwo has at most two instances in every model. In order to cause the extension size to be exactly two, we would have to add one $\not\approx$ two or $\top \sqsubseteq \,\geqslant 2u.$AtMostTwo, respectively. Likewise, we can enforce the whole domain to contain at most (or exactly) two individuals by imposing these axiom with AtMostTwo substituted by $\top$.

**Remark 47.** These considerations show that as soon as nominal concepts or the universal role is involved, models of knowledge bases need not be closed under disjoint union as it was the case for e.g. $\mathcal{SHIQ}$.

**Exercise 29.** *As we have seen, $\mathcal{SROIQ}$ allows to enforce that the domain size (i.e. the number of its elements) is at most $n$ for any given $n \in \mathbb{N}$. Contemplate whether there is a knowledge base $\mathcal{KB}_{\text{fin}}$ that emulates finite models, i.e., for every knowledge base $\mathcal{KB}$ not using vocabulary from $\mathcal{KB}_{\text{fin}}$ the models of $\mathcal{KB} \cup \mathcal{KB}_{\text{fin}}$ are exactly those models of $\mathcal{KB}$ with finite domain, if one abstracts from the vocabulary of $\mathcal{KB}_{\text{fin}}$.*

**Exercise 30.** *Is it possible to create a $\mathcal{SHIQ}$ knowledge base $\mathcal{KB}$ such that every model contains one individual which is connected via a role $r$ to infinitely many other individuals? Can the same be achieved in $\mathcal{ALCHOIQ}$? What about $\mathcal{ALCHIQ}$? For each of the cases either provide such a knowledge base or argue why this is not possible.*

## 6.9  Selfishness

The self concept enables to speak about "role loops", i.e. situations where an individual is simultaneously source and target of the same relation, or in other words the individual is connected to it*self*. This allows to define concepts based on such situations, for instance we could define `PersonCommittingSuicide` ≡ ∃`kills`.Self or `Narcissist` ≡ ∃`loves`.Self. Beyond that, this feature comes handy when global properties of roles are to be enforced. A role $r$ is said to be *reflexive* if its associated relation is, i.e. if $\langle \delta, \delta \rangle \in r^{\mathcal{I}}$ for all $\delta \in \Delta^{\mathcal{I}}$. Conversely, it is called *irreflexive* if $\delta \neq \delta'$ for all $\langle \delta, \delta' \rangle \in r^{\mathcal{I}}$. In some places, the definition of $\mathcal{SROIQ}$ includes additional RBox axioms of the form $\mathsf{Ref}(r)$ or $\mathsf{Irr}(r)$ to specify reflexivity or irreflexivity of $r$, respectively. However, these role characteristics can be equivalently expressed by the GCIs $\top \sqsubseteq \exists r.\mathsf{Self}$ or $\exists r.\mathsf{Self} \sqsubseteq \bot$, respectively.

**Exercise 31.** *If one has a closer look into the literature, these additional axiom types require $r$ to be simple in the case of irreflexivity but not in the case of reflexivity statements. In our current translation, role simplicity would be required in both cases. How can this restriction be circumvented by an alternative translation of the reflexivity statement?*

### 6.10    Open World vs. Closed World

A useful distinction often made in the context of logic-based information systems is that between *closed-world* and *open-world* reasoning. Essentially, this distinction is concerned with the question how missing information is treated. Under the *closed-world assumption* (CWA) facts which cannot be deduced from a knowledge base are supposed to be *false* whereas under the *open-world assumption* the truth of these facts is simply *unknown*. Expert or database systems often implement the CWA. Opposed to this, as a consequence of the semantics introduced in Section 3, DLs follow the OWA. This is also implied by the fact, that (most) DLs are fragments of first-order logic, which also adheres to the OWA.

**Example 48.** Consider the following knowledge base $\mathcal{KB}$ containing merely ABox statements:

$$\mathsf{Planet(home)} \qquad \mathsf{orbitsAround(home, sun)}$$
$$\mathsf{Planet(morning\_star)} \qquad \mathsf{orbitsAround(moon, home)}$$
$$\mathsf{Star(sun)} \qquad \mathsf{orbitsAround(morning\_star, sun)}$$
$$\mathsf{evening\_star} \approx \mathsf{morning\_star}$$

While `Planet(evening_star)` and `orbitsAround(evening_star, sun)` are consequences of $\mathcal{KB}$, negated statements like ¬`Star(home)` or ¬`orbitsAround(sun, moon)` or moon $\not\approx$ home are not due to the OWA. This can be explained by the fact, that there are models for the $\mathcal{KB}$ where these statements do not hold (but rather their unnegated variants). In order to enforce these negated statements they would heave to be explicitly added to the knowledge base.

While the OWA is commonly argued to be the right perspective in the context of the Semantic Web where completeness seems to be hard to achieve, there are cases, where e.g. the extension of a concept or a role is entirely known and one wants to express this information in order to guarantee that the according additional consequences can be drawn. To a certain extent, this can be implemented by virtue of nominal concepts.

> **Example 49.** Revisiting Example 48, to obtain the consequence ¬Star(home) we could alternatively state that sun is the name of the only individual belonging to the concept Star by adding the TBox axiom Star ⊑ {sun}. This has the advantage that also the concept membership of anonymous individuals is thereby excluded which cannot be achieved by ABox statements. Yet, in order to get the above consequence we still have to additionally assert sun ≉ home, thereby excluding the case that home and sun refer to the same individual. In the same way, we can treat roles. For example, the axiom {home} ⊑ ∀orbitsAround.{sun} expresses that home is orbitsAround-connected to nothing but (possibly) sun.

While nominals come handy for making "nothing but" statements, they cannot fully simulate closed-world behavior. Therefore (local) closed-world extensions to DLs have been investigated. Notable approaches in that direction are *(auto)epistemic DLs*, and *circumscriptive DLs*.

## 7    Reasoning Tasks and Their Reducibility

*A knowledge base with statements in it*
*Seeks a model sound and nice*
*No matter, finite or infinite,*
*It asks a hermit for advice.*
*Yet, shattering is the reaction:*
*"Inconsistency detection,*
*You can't get no satisfaction."*

It is one of the major selling points of logic-based knowledge representation in general and of DLs in particular that, once a body of knowledge has been accumulated and transferred into a logical representation, this knowledge can be queried and worked with in an intelligent way which goes well beyond what can be done with traditional information systems such as databases. In this section we will review typical tasks that can be performed with DL knowledge bases and that require elaborate inferencing. We can see that some of those tasks can be reduced to others which alleviates the task of creating tools performing those tasks.

### 7.1    Knowledge Base Satisfiability

Remember that a knowledge base $\mathcal{KB}$ is called satisfiable (also: consistent) if it has a model, i.e., there is an interpretation $\mathcal{I}$ with $\mathcal{I} \models \mathcal{KB}$, otherwise it is called unsatisfiable, inconsistent, or contradictory. Deciding whether a knowledge base is consistent is important in its own right, as knowledge base inconsistency often hints at severe modeling errors: since knowledge bases are supposed to describe real state of affairs, they should not be contradictory. Moreover, due to the principle of explosion, an inconsistent knowledge base entails every statement which renders any derived information useless. Additionally, as we will see in a bit, axiom entailment checks can be reduced to detecting inconsistency of knowledge bases.

## 7.2  Axiom Entailment

We remember that a knowledge base $\mathcal{KB}$ entails a DL axiom $\alpha$ if every model of $\mathcal{KB}$ is also a model of $\alpha$. Axiom entailment can be seen as the prototypical reasoning task for querying knowledge: given a body of knowledge formally specified in a knowledge base, this knowledge is to be "logically queried" by checking whether some statement is necessarily true, presuming the statements of the knowledge base.

The problem of checking axiom entailment can be reduced to deciding knowledge base satisfiability. The idea behind this reduction is proof by contradiction: we show that something holds by assuming the opposite and deriving a contradiction from that assumption. Suppose $\alpha$ and $\beta$ are axioms claiming the opposite of each other. Then every interpretation (hence in particular every model of the knowledge base $\mathcal{KB}$) satisfies either $\alpha$ or $\beta$, but not both. Now, if $\alpha$ is a consequence of $\mathcal{KB}$, we know that every model of $\mathcal{KB}$ is a model of $\alpha$. This means that no model of $\mathcal{KB}$ can be a model of $\beta$. In other words, the extended knowledge base $\mathcal{KB}' = \mathcal{KB} \cup \{\beta\}$ can have no model which just means that $\mathcal{KB}'$ is unsatisfiable. Thus the axiom entailment problem can be easily recast into a knowledge base unsatisfiability problem, provided we find such an "opposite" axiom for the given $\alpha$. In $\mathcal{SROIQ}$ this is obvious for some cases. In some other cases, we have to revert to finding an axiom or a set of axioms emulating the opposite of $\alpha$, which works just as well. We give the correspondences for all types of $\mathcal{SROIQ}$ axioms in Table 1.

**Table 1.** Definition of axiom sets $\mathcal{A}_\alpha$ such that $\mathcal{KB} \models \alpha$ exactly if $\mathcal{KB} \cup \mathcal{A}_\alpha$ is unsatisfiable. Individual names c with possible subscripts are supposed to be fresh. For GCIs (third line), the first variant is normally employed, however, we also give a variant which is equivalent instead of just emulating.

| $\alpha$ | $\mathcal{A}_\alpha$ |
|---|---|
| $r_1 \circ \ldots \circ r_n \sqsubseteq r$ | $\{\neg r(c_0, c_n),\ r_1(c_0, c_1), \ldots, r_n(c_{n-1}, c_n)\}$ |
| $\mathrm{Dis}(r, r')$ | $\{r(c_1, c_2),\ r'(c_1, c_2)\}$ |
| $C \sqsubseteq D$ | $\{(C \sqcap \neg D)(c)\}$ or: $\{\top \sqsubseteq \exists u(C \sqcap \neg D)\}$ |
| $C(a)$ | $\{\neg C(a)\}$ |
| $r(a, b)$ | $\{\neg r(a, b)\}$ |
| $\neg r(a, b)$ | $\{r(a, b)\}$ |
| $a \approx b$ | $\{a \not\approx b\}$ |
| $a \not\approx b$ | $\{a \approx b\}$ |

## 7.3   Concept Satisfiability

Given a knowledge base $\mathcal{KB}$, a concept $C \in \mathbf{C}$ is called *satisfiable* with respect to $\mathcal{KB}$, if it may contain individuals, i.e. there is a model $\mathcal{I}$ of $\mathcal{KB}$ that maps $C$ to a nonempty set, formally: $C^{\mathcal{I}} \neq \emptyset$. Obviously, there are concepts which are unsatisfiable irrespective of the underlying knowledge base, like $\mathsf{A} \sqcap \neg \mathsf{A}$ or simply $\bot$. If, however some atomic concept $\mathsf{A} \in N_I$ is unsatisfiable, this may as well indicate modeling errors. A knowledge base where all atomic concepts are satisfiable is usually called *coherent*. Note that a knowledge base can be incoherent but satisfiable. Like knowledge base satisfiability and axiom entailment, concept satisfiability is a decision problem, i.e. we get *yes* or *no* as an answer.

The problem of deciding concept satisfiability can be reduced to axiom entailment. An unsatisfiable concept $C$ is necessarily empty in any model $\mathcal{I}$, i.e., $C^{\mathcal{I}} = \emptyset$. This can be rewritten into $C^{\mathcal{I}} \subseteq \emptyset$ (since the other direction is trivial), and further (using the fact that $\bot^{\mathcal{I}} = \emptyset$) into $C^{\mathcal{I}} \subseteq \bot^{\mathcal{I}}$. However this means $\mathcal{I} \models C \sqsubseteq \bot$ for every model $\mathcal{I}$ of $\mathcal{KB}$, therefore $\mathcal{KB} \models C \sqsubseteq \bot$. Hence, unsatisfiability of of a concept $C$ with respect to some knowledge base $\mathcal{KB}$ can be decided by checking whether $\mathcal{KB}$ entails the GCI $C \sqsubseteq \bot$.

## 7.4   Instance Retrieval

Given a knowledge base $\mathcal{KB}$ and a concept $C$, it is a rather natural desire to ask for $C$'s instances. However, there are two issues with that: First, a knowledge base usually has many models and a specific individual may be instance of $C$ in one model but not in another. So, one typically asks for individuals which are instances of $C$ in *every* model. The other problem is that from model to model, the domain $\Delta^{\mathcal{I}}$ may vary and does not need to contain the same individual. The only way to refer to individuals in a sensible, cross-domain way is via their names. This is why one restricts to named individuals for the instance retrieval task. Consequently, the task could be formulated as follows: given a knowledge base $\mathcal{KB}$ and a concept $C$, give me all individual names $\mathsf{a} \in \mathsf{N}_I$ for which $\mathsf{a}^{\mathcal{I}} \in C^{\mathcal{I}}$ for every model $\mathcal{I}$ of $\mathcal{KB}$.

> **Remark 50.** This definition of instance retrieval may even lead to the peculiar case that one can infer from a knowledge base that a concept $C$ is nonempty in every model (which can e.g. be tested by asking whether $\mathcal{KB} \models \top \sqsubseteq \exists u.C$) while the instance retrieval for $C$ yields nothing. A simple example for this would be the knowledge base containing only the axiom $(\exists \mathsf{r}.C)(\mathsf{a})$.

Given the definition of instance retrieval above, it is obvious that an individual name $\mathsf{a}$ will be delivered as part of the answer of an instance retrieval with respect to a concept $C$ precisely if $\mathcal{KB} \models C(\mathsf{a})$. Therefore, instance retrieval can be performed by successively checking whether the considered knowledge base entails $C(\mathsf{a})$ for every individual name $\mathsf{a}$. This takes $|\mathsf{N}_I(\mathcal{KB})|$ entailment checks. Depending on what concrete reasoning algorithm is employed, fewer calls to the reasoning procedure may be required since it might be possible to retrieve many instances at once. This particularly applies to reasoning methods based on logic programming and/or database systems.

Sometimes, the term instance retrieval is also used for roles. In that case we are looking for all pairs $\langle \mathsf{a}, \mathsf{b} \rangle$ of individual names $\mathsf{a}, \mathsf{b} \in \mathsf{N}_I$ for which $\langle \mathsf{a}^{\mathcal{I}}, \mathsf{b}^{\mathcal{I}} \rangle \in r^{\mathcal{I}}$ for every model $\mathcal{I}$ of $\mathcal{KB}$. This can be easily checked by asking for the entailment $\mathcal{KB} \models r(\mathsf{a}, \mathsf{b})$ for every combination of individual names.

## 7.5    Classification

Given a knowledge base $\mathcal{KB}$, the concept names occurring therein can be put into a hierarchy according to their subsumption relationships. More precisely, if we define a relation $\sqsubseteq_{\mathcal{KB}}$ on the set $\mathsf{N}_C$ of concept names by $\mathsf{A} \sqsubseteq_{\mathcal{KB}} \mathsf{B}$ iff $\mathcal{KB} \models \mathsf{A} \sqsubseteq \mathsf{B}$, we find that this relation is a *preorder*, that is, we have $\mathsf{A} \sqsubseteq_{\mathcal{KB}} \mathsf{A}$ for all $\mathsf{A} \in \mathsf{N}_C$ and from $\mathsf{A} \sqsubseteq_{\mathcal{KB}} \mathsf{B}$ and $\mathsf{B} \sqsubseteq_{\mathcal{KB}} \mathsf{C}$ follows $\mathsf{A} \sqsubseteq_{\mathcal{KB}} \mathsf{C}$.

**Exercise 32.** *Prove that $\sqsubseteq_{\mathcal{KB}}$ is indeed a preorder.*

Classification of a knowledge base is the task of entirely determining $\sqsubseteq_{\mathcal{KB}}$. This task is practically important due to several reasons: During the knowledge base modeling process, the modeler has an overview over the hierarchical structure of the used concept names which can be diagrammatically visualized in a nice, intuitive way. On the other hand, classification can serve as a preprocessing step that speeds up subsequently performed reasoning tasks with respect to the underlying knowledge base.

Obviously, classification of a knowledge base can be performed by checking the entailment $\mathcal{KB} \models \mathsf{A} \sqsubseteq \mathsf{B}$ for any pair $\mathsf{A}, \mathsf{B}$ of class names, which amounts to $|\mathsf{N}_C| \cdot (|\mathsf{N}_C| - 1)$ separate entailment checks. However, exploiting the properties of preorders and concept subsumption statements explicitly given by GCIs, the number of such checks can be drastically reduced [Shearer and Horrocks, 2009].

## 7.6    Conjunctive Query Answering

*Conjunctive queries* (CQs) and *unions of conjunctive queries* (UCQs) are well known in the database community [Chandra and Merlin, 1977] and constitute an expressive query language with capabilities that go well beyond standard reasoning tasks in DLs. In terms of first-order logic, these CQs and UCQs are formulae from the positive existential fragment. Free variables in a query (not bound by an existential quantifier) are also called *answer variables* or *distinguished variables*, whereas existentially quantified variables are called *non-distinguished*. As an example, $\exists y \exists z (\mathtt{childOf}(x, y) \wedge \mathtt{childOf}(x, z) \wedge \mathtt{married}(y, z))$ with distinguished variable $x$ and non-distinguished variables $y$ and $z$ represents a conjunctive query asking for all children whose parents are married with each other. If all variables in the query are non-distinguished, the query answer is just *true* or *false* and the query is called a *Boolean query*. Given a knowledge base $\mathcal{KB}$ and a Boolean UCQ $q$, the query entailment problem is deciding whether $q$ is *true* or *false* w.r.t. $\mathcal{KB}$, i.e., we have to decide whether each model of $\mathcal{KB}$ provides for a suitable assignment for the variables in $q$.[2] For a query with distinguished variables, the answers to the query are those tuples of individual names (constants) for which the knowledge base entails the query that is obtained by replacing the free variables with the individual names in the answer tuple. The problem of finding all answer tuples is known as query answering.

In general, conjunctive query answering or checking Boolean conjunctive query entailment are not easily (more precisely: polynomially) reducible to any of the other standard reasoning tasks treated above, which can be concluded from the fact that the worst-case complexities for these problems are usually way harder than the complexities of the other tasks [Lutz, 2008]. Conversely, it is trivial to reduce the task of checking knowledge base consistency to checking conjunctive query entailment: for instance, $\mathcal{KB}$ is inconsistent exactly if for fresh concept names $A_{aux}$ and $B_{aux}$ the knowledge base $\mathcal{KB} \cup \{A_{aux} \sqcap B_{aux} \sqsubseteq \bot\}$ satisfies the conjunctive query $\exists x (A_{aux}(x) \wedge B_{aux}(x))$.

**Exercise 33.** *A conjunctive query is called tree-shaped if for any two query variables $x, y$ there is exactly one sequence of pairwise different query variables $z_0, \ldots, z_n$ and exactly one sequence $r_1, \ldots r_n$ of role names such that $z_0 = x$, $z_n = y$, and for every $1 \leq i \leq n$ either $r_i(z_{i-1}, z_i) \in q$ or $r_i(z_i, z_{i-1}) \in q$. Argue that query answering for a tree-shaped conjunctive query with one distinguished variable can be reduced to (concept) instance retrieval.*

## 7.7 Other Reasoning Tasks

The reasoning tasks described above, excluding conjunctive query answering, are often referred to as *standard reasoning tasks*. Still, conjunctive query answering is conceptually in line with those, since it can be formulated as entailment checking. Beyond those *deductive* tasks which are all concerned with determining logical consequences, there are several *non-standard reasoning tasks* where the goal is somewhat different. In the following, we will briefly go through a selection of these.

***Induction.*** As opposed to the aforementioned deductive methods, inductive approaches[3] usually take an amount of factual (assertional) data and try to generalize therefrom by generating hypotheses expressed as terminological axioms or complex concepts. This sort of reasoning tasks are related to data mining problems and respective approaches draw their inspiration from machine learning and in particular inductive logic programming (ILP, [Lehmann, 2009]). Since inductive reasoning is not truth-preserving (i.e. hypotheses which are generated may be falsified), also interactive methods with human expert involvement have been proposed [Rudolph, 2004].

---

[2] Note that in general, solving this task is way harder than querying a classical database, as the considered models may be infinite in both size and number.

[3] Not to be confused with the mathematical proof technique of induction.

***Abduction.*** Like induction and unlike deduction, abduction is an inferencing method which is not truth-preserving. Roughly speaking, abduction could be described as "premise guessing." More precisely, given a knowledge base $\mathcal{KB}$ in some DL and an axiom $\alpha$ such that $\alpha$ is not entailed by $\mathcal{KB}$, abductive reasoning is concerned with finding a knowledge base $\mathcal{KB}'$ with specific properties such that $\alpha$ is a logical consequence of $\mathcal{KB} \cup \mathcal{KB}'$. In ontology engineering, abductive reasoning services come handy when a wanted consequence is not entailed and one wants to determine what information is missing [Noia *et al.*, 2009].

***Explanation.*** If results of automated reasoning are to be shared with human users, it is often not sufficient to just display the result. Often it is also desirable to give an account on the cause why some axiom is entailed by the knowledge base, in other words to give an *explanation* for it. In most cases, only few axioms actually contribute to an entailment. Thus it is already quite helpful to find a minimal subset of a knowledge base for which the entailment still holds. More precisely, given a knowledge base $\mathcal{KB}$ and an axiom $\alpha$ with $\mathcal{KB} \models \alpha$, a *justification* for the entailment is a knowledge base $\mathcal{KB}' \subseteq \mathcal{KB}$ such that $\mathcal{KB}' \models \alpha$ but for every $\mathcal{KB}'' \subset \mathcal{KB}'$ holds $\mathcal{KB}'' \not\models \mathcal{KB}$. In general, a justification does not need to be unique, there might be more than one justification for an entailment [Horridge *et al.*, 2008].

***Module extraction.*** When confronted with large knowledge bases, it might be worthwhile to identify natural partitions of them which logically interact which each other only in a restricted way, such that they can be handled independently when it comes to query answering or reasoning in general. In other cases, one may be interested only in a part of the knowledge specified in a knowledge base which is expressible in a certain fraction of the vocabulary. In general, the task of identifying or computing such knowledge base parts is referred to as *module extraction* [Stuckenschmidt *et al.*, 2009].

# 8    Algorithmic Approaches to DL Reasoning

*Is it consequence-driven*
*Automatically given*
*What we base our system upon?*
*Or do, fueled by Rousseau,*
*we say "Guerre aux tableaux!*
*Et vive la resolution!"?*

Various reasoning paradigms have been investigated with respect to their applicability to DLs. Most of them originate from well-known approaches for theorem proving in a first-order logic setting. However, in contrast to the unavoidable downside that reasoning methods for first-order logic cannot be sound, complete, and terminating, approaches to reasoning in DLs aim at a sound and complete decision procedures, whence the adopted reasoning techniques have to be adapted in order to guarantee termination.

The majority of state-of-the art OWL reasoners, such as Pellet [Sirin *et al.*, 2007], FaCT++ [Tsarkov and Horrocks, 2006], or RacerPro

[Haarslev and Möller, 2001] use tableau methods with good performance results, but even those successful systems are not applicable in all practical scenarios. This motivates the search for alternative reasoning approaches that employ different methods in order to address cases where tableau algorithms exhibit certain weaknesses. Successful examples in this respect are the works based on resolution [Motik and Sattler, 2006] and hyper-tableaux [Motik et al., 2009c] as well as consequence-based approaches [Kazakov, 2009].

As we have seen in the previous section, many important reasoning tasks can be reduced to checking knowledge base satisfiability, hence we will focus on this specific task. In general, reasoning methods can be subdivided into *model-theoretic methods* on one hand and *proof-theoretic methods* on the other.

Model-theoretic methods essentially try to construct models of a given knowledge base in an organized way. If this succeeds, the knowledge base has obviously been shown to be satisfiable, if it can be shown that the construction must necessarily fail, unsatisfiability has been established. Typical reasoning paradigms of that sort are tableau procedures and automata-based approaches.

**Remark 51.** If models are represented explicitly (i.e., for an interpretation $\mathcal{I} = (\Delta^{\mathcal{I}}, \cdot^{\mathcal{I}})$ both $\Delta_{\mathcal{I}}$ and $\cdot^{\mathcal{I}}$ are stored in some data structure), a naïve model construction strategy can only arrive at finite models, obviously. While this may be enough for logics that satisfy the finite model property, it is insufficient in the general case. However, this problem can be circumvented if one reverts to *finite model representations*, which only store a finite part of the model explicitly and provide additional (finite) information how this partial model could be deterministically extended into a real model. Intuitively, this can be compared to the decimal representation of rational numbers: while the correct value of $\frac{13}{11} = 1.18181818\ldots$ needs infinitely many digits to be precisely noted down, it is not hard to come up with a finite representation, namely $1.\overline{18}$ which, by virtue of the additional extra information provided by the overline, shows how the infinite "pure" representation could be constructed (if one had infinite time and memory). Of course, when working with finite representations, it is crucial that these allow for effective detection of axiom satisfaction.

As opposed to model-theoretic reasoning methods, proof-theoretic approaches operate more on the syntactic side: starting out from a normalized version of the knowledge base, deduction rules are applied to derive further logical statements about a potential model. If, in the course of these derivations an overt contradiction is derived, the considered knowledge base has shown to be unsatisfiable. In order to guarantee a termination of the procedure also in the case of satisfiability it is crucial that in the course of derivation, some sort of saturation will be reached in finite time. This can e.g. be achieved by restricting the relevant propositions (which may or may not be derived) to a finite set.

In the following, we will survey some well-known reasoning paradigms for DLs without going into technical details.

## 8.1 Tableau

Tableau procedures aim at constructing a model that satisfies all axioms of the given knowledge base. The strategy here is to maintain a set $D$ of elements representing domain individuals (including anonymous ones) and acquire information about their concept memberships and role interrelatedness. $D$ is initialized by all the individual names and the according ABox facts. Normally, the partial model thus constructed does not satisfy all the TBox and RBox axioms. Thus, the intermediate model is "repaired" as required by the axioms. This may mean to establish new concept membership or role interrelatedness information about the maintained elements, yet sometimes it may also be necessary to extend the set of considered domain individuals. Now and again, it might be required to make case distinctions and backtrack later. If we arrive at a state, where the intermediate model satisfies all the axioms and hence does not need to be repaired further, the knowledge base is satisfiable. If the intermediate model contains overt contradictions (such as an element marked as instance of a concept $C$ and its negation $\neg C$ or an element marked as an instance of $\bot$), we can be sure that repairing it further by adding more information will never lead to a proper model, hence we are in a "dead end" need to backtrack. If every alternative branch thus followed leads into such a "dead end", we can be sure that no model can exist.

**Example 52.** Omitting a lot of technical details, we shortly explain how the satisfiability of the knowledge base from Example 12 would be established by a tableau algorithm. For better reference, we first recap the knowledge base.

$$\text{owns} \sqsubseteq \text{caresFor} \tag{5}$$

$$\text{Healthy} \sqsubseteq \neg\text{Dead} \tag{6}$$

$$\text{Cat} \sqsubseteq \text{Dead} \sqcup \text{Alive} \tag{7}$$

$$\text{HappyCatOwner} \sqsubseteq \exists\text{owns}.\text{Cat} \sqcap \forall\text{caresFor}.\text{Healthy} \tag{8}$$

$$\text{HappyCatOwner}(\text{schrödinger}) \tag{9}$$

As explained we first initialize the set of domain elements by letting $D = \{\text{schrödinger}\}$, moreover, due to the only ABox axiom (9) we mark schrödinger with HappyCatOwner. Inspecting the axioms, we find that (8) is not satisfied by the current representation. Thus, we repair it as required by (8), "inventing" a new element, say 🐱, and adding it to $D$. Accordingly, we stipulate that schrödinger is connected to 🐱 by an owns relation and marking 🐱 with Cat. We find that, as a consequence of these changes, (8) is satisfied (for the moment). However, the changes have invalidated axioms (5) and (7). We account for the former by introducing a caresFor connection from schrödinger to 🐱. The latter essentially leaves us with two options: we need to mark 🐱 either by Dead or by Alive. This means, we have to make a case distinction and investigate each option separately.

- Let us try and pick Dead. Again, examining the axioms, we find (8) violated due to the second part of its consequence. Repairing this requires to mark 😈 with Healthy which in turn invalidates (6). Hence we have to mark 😈 by ¬Dead. Unfortunately, we now observe that 😈 is marked both by Dead and ¬Dead, thus we have reached a "dead end" and need to backtrack.
- So, we mark 😈 by Alive. Also here, we find (8) violated and repair it by marking 😈 with Healthy, obtaining invalidation of (6) and coping with it by marking 😈 by ¬Dead. We have thus arrived at a state where our intermediate model satisfies all axioms. Hence, we have obtained a proper model of $\mathcal{KB}$ and conclude that the knowledge base is satisfiable.

However, note that the continued "repairing" performed in a tableau procedure does not necessarily terminate, since performing one repair might cause the need for another repair and so forth ad infinitum.

**Example 53.** Consider the knowledge base containing the single axiom $\top \sqsubseteq \exists\mathtt{succ}.\top$, which forces every domain element to have a successor. Applying the naïve repair approach from above we will need to introduce a successor for every individual, then successors of successors etc.

Therefore, in order to be applicable as a decision procedure, these infinite computations must be prevented to ensure termination. This can be achieved by a strategy called *blocking*, where certain domain elements are "blocked" (which essentially means that they are exempt from the necessity of being repaired) by other domain individuals which "look the same" in terms of concept memberships. For more advanced DLs, more complicated blocking strategies are needed.

A tableau algorithm for $\mathcal{SHOIQ}$ is described by Horrocks and Sattler [2007]. A refinement of the tableau technique, called *hypertableau* is at the core of the OWL 2 DL reasoner HermiT [Motik *et al.*, 2009c].

## 8.2    Automata

As mentioned earlier, most DLs satisfy some sort of tree-model property. On the other hand, families of trees (in other words: tree languages) can be represented by appropriate tree-automata. Thus, given an automaton that characterizes the tree models of a knowledge base, the problem of knowledge base satisfiability can be rephrased into the question whether the tree language represented by this corresponding automaton is non-empty. This line of research has been followed by several investigations targeted at standard reasoning as well as conjunctive query answering. Approaches along those lines are e.g. described by Glimm *et al.* [2008a] and Calvanese *et al.* [2009].

**Exercise 34.** *To get a feeling for the relatedness between automata and DL reasoning, try to design an $\mathcal{ALC}$ knowledge base $\mathcal{KB}$ with the property that for any $r_1, r_2, \ldots, r_n \in \mathsf{N}_R$ we have that $\mathcal{KB} \models A \sqsubseteq \exists r_1 \exists r_2 \ldots \exists r_n.B$ exactly if the word $r_1 r_2 \ldots r_n$ matches the regular expression $\mathtt{s}^*(\mathtt{rs}|\mathtt{srr})^*$.*

## 8.3   Consequence-Based Reasoning

As suggested by their name, consequence-based (also: consequence-driven) reasoning approaches start from the given knowledge base and derive logical consequences of it by means of applying *deduction rules*. A deduction rule has the shape

$$\textbf{name}\ \frac{\alpha_1\ \cdots\ \alpha_n}{\alpha}$$

with $\alpha, \alpha_1, \ldots, \alpha_n$ being axioms of the underlying logic. To apply a deduction rule means to add $\alpha$ to the set of statements known to be true if truth is already established for $\alpha_1, \ldots, \alpha_n$ (be it due to their presence in the knowledge base or because they have been derived by an earlier application of a deduction rule). If, given a set $\mathbb{D}$ of deduction rules, an axiom $\beta$ can be generated like this from an axiom set $\{\beta_1, \ldots, \beta_k\}$ by (possibly manifold) applications of deduction rules, we say that $\beta$ is *derivable* from $\{\beta_1, \ldots, \beta_k\}$ and write $\{\beta_1, \ldots, \beta_k\} \vdash \beta$.

In order to be of proper use for the reasoning, the used set $\mathbb{D}$ of deduction rules (also jointly called a *deduction calculus*) has to mimic the logical entailment as defined by the formal semantics. That means that on one hand, $\beta$ must be a logical consequence of $\{\beta_1, \ldots, \beta_k\}$ whenever $\beta$ is derivable therefrom (in short: $\{\beta_1, \ldots, \beta_k\} \vdash \beta$ implies $\{\beta_1, \ldots, \beta_k\} \models \beta$) – a property called *soundness* of the deduction calculus. On the other hand, we require its *completeness*, i.e. that every logical consequence of $\{\beta_1, \ldots, \beta_k\}$ can also be derived from it (in short: $\{\beta_1, \ldots, \beta_k\} \models \beta$ implies $\{\beta_1, \ldots, \beta_k\} \vdash \beta$). Sometimes, completeness is constrained to specific axiom types $\beta$, e.g. a deduction calculus is called refutationally complete, if inconsistency of a knowledge base implies derivability of $\top \sqsubseteq \bot$.

**Example 54.** The following deduction calculus is sound and refutationally complete for $\mathcal{ALC}$ TBoxes in an appropriate normal form (for details see Simancik *et al.* [2011]). Thereby A and B denote concept names, $H$ and $K$ are conjunctions of negated and unnegated concept names, whereas $M$, $N$, and $N_i$ are disjunctions of concept names.

$$\mathbf{R_A^+}\ \frac{}{\textsf{A} \sqcap H \sqsubseteq \textsf{A}}$$

$$\mathbf{R_A^-}\ \frac{\neg\textsf{A} \sqcap H \sqsubseteq N \sqcup \textsf{A}}{\neg\textsf{A} \sqcap H \sqsubseteq N}$$

$$\mathbf{R_\sqcap^n}\ \frac{H \sqsubseteq N_1 \sqcup \textsf{A}_1\ \cdots\ H \sqsubseteq N_n \sqcup \textsf{A}_n\quad \bigsqcap_{i=1}^{n} \textsf{A}_i \sqsubseteq M}{H \sqsubseteq M \sqcup \bigsqcup_{i=1}^{n} N_i}$$

$$\mathbf{R_\exists^+}\ \frac{H \sqsubseteq N \sqcup \textsf{A}\quad \textsf{A} \sqsubseteq \exists r.\textsf{B}}{H \sqsubseteq N \sqcup \exists r.\textsf{B}}$$

$$\mathbf{R_\exists^-}\ \frac{H \sqsubseteq N \sqcup \exists r.K\quad K \sqsubseteq N \sqcup \textsf{A}\quad \exists r.\textsf{A} \sqsubseteq \textsf{B}}{H \sqsubseteq M \sqcup \textsf{B} \sqcup \exists r.(K \sqcap \neg\textsf{A})}$$

$$\mathbf{R_\exists^\bot}\ \frac{H \sqsubseteq N \sqcup \exists r.K\quad K \sqsubseteq \bot}{H \sqsubseteq M}$$

$$\mathbf{R_\forall}\ \frac{H \sqsubseteq N \sqcup \exists r.K\quad H \sqsubseteq N \sqcup \textsf{A}\quad \textsf{A} \sqsubseteq \forall r.\textsf{B}}{H \sqsubseteq M \sqcup N \sqcup \exists r.(K \sqcap \textsf{B})}$$

**Exercise 35.** *Using the above deduction calculus, show that the axiom* D $\sqsubseteq$ G *can be derived from the knowledge base containing the axioms*

A $\sqsubseteq$ B $\sqcup$ C        D $\sqsubseteq$ $\forall$r.A        $\exists$r.B $\sqsubseteq$ E        D $\sqsubseteq$ F $\sqcup$ $\exists$r.$\neg$C        E $\sqcap$ F $\sqsubseteq$ G.

However, just a sound and complete deduction calculus is not sufficient for a decision procedure (note that FOL itself has such a calculus while being undecidable). In addition to that, one has to ensure that the "enrichment process" of adding more and more derived consequences to the set of true statements will terminate at some point. One way to guarantee this is to make sure that only finitely many (syntactically) different axioms can be derived. Consequence-driven approaches are described e.g. by Kazakov [2009] and Simancik *et al.* [2011].

## 8.4   Resolution

Resolution is a technique prominently used in first-order logic theorem proving. At the core of reasoning via resolution is the resolution rule which looks as follows:

$$\textbf{Res } \frac{A_1 \vee \ldots \vee A_i \vee \ldots A_n \qquad B_1 \vee \ldots \vee B_j \vee \ldots B_m}{A_1 \vee \ldots \vee A_{i-1} \vee A_{i+1} \vee \ldots A_n \vee B_1 \vee \ldots \vee B_{j-1} \vee B_{j+1} \vee \ldots B_m}$$

Thereby, $A_k$ and $B_k$ denote literals, i.e. negated or unnegated FOL atoms and the two literals $A_i$ and $B_j$ are assumed to be complements of each other (i.e. $A_i = \neg B_j$ or $B_j = \neg A_i$). As the resolution rule is a deduction rule, resolution can be seen as a variant of consequence-based reasoning. One of the differences is that resolution is performed not on DL knowledge bases directly but on a FOL translation thereof. Resolution-based methods have been described for DLs up to $\mathcal{SHOIQ}$ [Motik and Sattler, 2006; Kazakov and Motik, 2008].

## 9   Description Logics and OWL

*In fact, in terms of syntax, OWL*
*Just tends to be a bulky fowl,*
*However, if it mates with Turtle*
*This union turns out rather fertile;*
*I deem the offspring of this love*
*As graceful as a turtledove.*

As mentioned before, the web ontology language OWL is based on Description Logics but also features additional types of extra-logical information, concerning, e.g., ontology versioning information and annotations. Moreover, OWL supports modeling and reasoning with datatypes which we omitted from our consideration. Likewise, *keys* are supported in OWL but not discussed here.

In this section, we will see how any OWL DL compliant reasoning tool can be used to decide $\mathcal{SROIQ}$ knowledge base satisfiability as well as any other reasoning task which can be reduced to it.

"OWL speak" differs partially from the terms normally used in description logics. The following table gives a synopsis of the corresponding terms used in the OWL vs. the DL community as well as in the domain of classical first-order logic.

| OWL | DL | FOL |
|-----|-----|-----|
| class name | concept name | unary predicate |
| class | concept | formula with one free variable |
| object property name | role name | binary predicate |
| object property | role | formula with two free variables |
| ontology | knowledge base | theory |
| axiom | axiom | sentence |
| vocabulary | vocabulary / signature | signature |

In the next two sections, we briefly explain how a $\mathcal{SROIQ}$ knowledge base can be translated into an OWL 2 DL ontology such that satisfiability and entailment checks can be performed by OWL reasoning engines.

## 9.1    Translating DL KBs into OWL

For translating a $\mathcal{SROIQ}$ knowledge base into an OWL ontology, some technical issues need to be taken care of. First of all, both the used vocabulary as well as the constructors have to be URIs (i.e. uniform resource identifiers, that is, terms following the prescribed naming scheme prevalent in the Semantic Web). The URIs for the used individual, concept, and role names can be chosen rather arbitrarily, while the URIs for constructors etc. are prescribed and associated to specific namespaces usually associated to the prefixes `owl:`, `rdfs:`, `rdf:`, and `xsd:`. For the sake of simplicity, we will assume that all used individual, concept and role names from the DL knowledge base are syntactically well-formed URIs.

Second, the mainly used encoding of OWL is as an RDF document [Manola and Miller, 2004]. One one hand, this is advantageous from a downward compatibility and tool interoperability point of view; in fact the encoding of concept and role assertions in OWL and RDF coincide and some other RDFS statements are available in OWL as well with a similar semantics. On the other hand, the encoding as RDF also imposes some restrictions on the way logical axioms can be encoded. As RDF is a graph-based formalism consisting of node-edge-node triples, DL axioms and complex concepts have to be transformed into a graph-like representation. This is done by virtue of the typical means used to encode complex structures in RDF: structural bnodes and graph-based encoding of lists.

For our treatise, we will use the Turtle [Beckett and Berners-Lee, 14 January 2008] notation, which seems most appropriate as it illustrates the underlying RDF triple structure while at the same time hiding the very low-level details (such as the tripli-fication of the list structures employed for the RDF encoding of OWL).

The translation of a $\mathcal{SROIQ}$ knowledge base $\mathcal{KB}$ contains three parts: a preamble containing the definition of namespaces, declarations of the used concept (resp. class) and role (resp. object property) names, and finally a part containing the OWL counterparts of the axioms from $\mathcal{KB}$. Hence, we let

$$[\![\mathcal{KB}]\!] = \mathrm{Pre} + \mathrm{Dec}(\mathcal{KB}) + \sum_{\alpha \in \mathcal{KB}} [\![\alpha]\!]$$

where $+$ denotes concatenation of strings. Thereby the preamble is defined by

$$\mathrm{Pre} = \left\{ \begin{array}{l} \texttt{@prefix owl: <http://www.w3.org/2002/07/owl\#> .} \\ \texttt{@prefix rdfs: <http://www.w3.org/2000/01/rdf-schema\#> .} \\ \texttt{@prefix rdf: <http://www.w3.org/1999/02/22-rdf-syntax-ns\#> .} \\ \texttt{@prefix xsd: <http://www.w3.org/2001/XMLSchema\#> .} \end{array} \right.$$

whereas the declarations are expressed by according typing statements:

$$\mathrm{Dec}(\mathcal{KB}) = \sum_{A \in \mathsf{N}_C(\mathcal{KB})} A \;\; \texttt{rdf:type owl:Class .} \\ + \sum_{r \in \mathsf{N}_R(\mathcal{KB})} r \;\; \texttt{rdf:type owl:ObjectProperty .}$$

As displayed above, the actual knowledge base is translated axiom-wise via the function $[\![\cdot]\!]$ defined on the next page. The latter makes calls to the functions $[\![\cdot]\!]_\mathbf{C}$ and $[\![\cdot]\!]_\mathbf{R}$ given further below, which are used to decompose and recursively translate complex concepts and roles, respectively.

$$[\![r_1 \circ \ldots \circ r_n \sqsubseteq r]\!] = [\![r]\!]_\mathbf{R} \;\; \texttt{owl:propertyChainAxiom} \; ([\![r_1]\!]_\mathbf{R} \cdots [\![r_n]\!]_\mathbf{R}) \; .$$

$$[\mathsf{Dis}(r, r')] = [\![r]\!]_\mathbf{R} \;\; \texttt{owl:propertyDisjointWith} \; [\![r']\!]_\mathbf{R} \; .$$

$$[\![C \sqsubseteq D]\!] = [\![C]\!]_\mathbf{C} \;\; \texttt{rdfs:subClassOf} \; [\![D]\!]_\mathbf{C} \; .$$

$$[\![C(a)]\!] = a \;\; \texttt{rdf:type} \; [\![C]\!]_\mathbf{C} \; .$$

$$[\![r(a, b)]\!] = a \;\; r \;\; b \; .$$

$$[\![r^-(a, b)]\!] = b \;\; r \;\; a \; .$$

$$[\![\neg r(a, b)]\!] = \texttt{[] rdf:type owl:NegativePropertyAssertion ;} \\ \texttt{owl:assertionProperty} \; [\![r]\!]_\mathbf{R} \; \texttt{;} \\ \texttt{owl:sourceIndividual} \; a \; \texttt{; owl:targetValue} \; b \; .$$

$$[\![a \approx b]\!] = a \;\; \texttt{owl:sameAs} \; b \; .$$

$$[\![a \not\approx b]\!] = a \;\; \texttt{owl:differentFrom} \; b \; .$$

$$\llbracket u \rrbracket_{\mathbf{R}} = \texttt{owl:topObjectProperty}$$
$$\llbracket r \rrbracket_{\mathbf{R}} = r$$
$$\llbracket r - \rrbracket_{\mathbf{R}} = [\ \texttt{owl:inverseOf} \ :r \ ]$$

$$\llbracket A \rrbracket_{\mathbf{C}} = A$$
$$\llbracket \top \rrbracket_{\mathbf{C}} = \texttt{owl:Thing}$$
$$\llbracket \bot \rrbracket_{\mathbf{C}} = \texttt{owl:Nothing}$$
$$\llbracket \{a_1, \ldots, a_n\} \rrbracket_{\mathbf{C}} = [\ \texttt{rdf:type owl:Class ; owl:oneOf (} \ :a_1 \ \ldots \ :a_n \ )]$$
$$\llbracket \neg C \rrbracket_{\mathbf{C}} = [\ \texttt{rdf:type owl:Class ; owl:complementOf} \ \llbracket C \rrbracket_{\mathbf{C}} \ ]$$
$$\llbracket C_1 \sqcap \ldots \sqcap C_n \rrbracket_{\mathbf{C}} = [\ \texttt{rdf:type owl:Class ; owl:intersectionOf (} \llbracket C_1 \rrbracket_{\mathbf{C}} \ \ldots \ \llbracket C_n \rrbracket_{\mathbf{C}})]$$
$$\llbracket C_1 \sqcup \ldots \sqcup C_n \rrbracket_{\mathbf{C}} = [\ \texttt{rdf:type owl:Class ; owl:unionOf (} \llbracket C_1 \rrbracket_{\mathbf{C}} \ \ldots \ \llbracket C_n \rrbracket_{\mathbf{C}})]$$
$$\llbracket \exists r.C \rrbracket_{\mathbf{C}} = [\ \texttt{rdf:type owl:Restriction ;}$$
$$\texttt{owl:onProperty} \ \llbracket r \rrbracket_{\mathbf{R}} \ \texttt{; owl:someValuesFrom} \ \llbracket C \rrbracket_{\mathbf{C}} \ ]$$
$$\llbracket \forall r.C \rrbracket_{\mathbf{C}} = [\ \texttt{rdf:type owl:Restriction ;}$$
$$\texttt{owl:onProperty} \ \llbracket r \rrbracket_{\mathbf{R}} \ \texttt{; owl:allValuesFrom} \ \llbracket C \rrbracket_{\mathbf{C}} \ ]$$
$$\llbracket \exists r.\mathsf{Self} \rrbracket_{\mathbf{C}} = [\ \texttt{rdf:type owl:Restriction ;}$$
$$\texttt{owl:onProperty} \ \llbracket r \rrbracket_{\mathbf{R}} \ \texttt{; owl:hasSelf "true"\^{}\^{}xsd:boolean} \ ]$$
$$\llbracket \geqslant n\,r.C \rrbracket_{\mathbf{C}} = [\ \texttt{rdf:type owl:Restriction ;}$$
$$\texttt{owl:minQualifiedCardinality} \ n\texttt{\^{}\^{}xsd:nonNegativeInteger ;}$$
$$\texttt{owl:onProperty} \ \llbracket r \rrbracket_{\mathbf{R}} \ \texttt{; owl:onClass} \ \llbracket C \rrbracket_{\mathbf{C}} \ ]$$
$$\llbracket \leqslant n\,r.C \rrbracket_{\mathbf{C}} = [\ \texttt{rdf:type owl:Restriction ;}$$
$$\texttt{owl:maxQualifiedCardinality} \ n\texttt{\^{}\^{}xsd:nonNegativeInteger ;}$$
$$\texttt{owl:onProperty} \ \llbracket r \rrbracket_{\mathbf{R}} \ \texttt{;} \quad \texttt{owl:onClass} \ \llbracket C \rrbracket_{\mathbf{C}} \ ]$$

**Example 55.** For the knowledge base $\mathcal{KB}$ from Example 12, the translation $\llbracket \mathcal{KB} \rrbracket$ looks as follows (for better readability, we use the namespace http://www.example.org/# for individual, concept, and role names and abbreviate it by the empty prefix as shown in the first line of the translation):

```
@prefix :     <http://www.example.org/#> .
@prefix owl:  <http://www.w3.org/2002/07/owl#> .
@prefix rdfs: <http://www.w3.org/2000/01/rdf-schema#> .
@prefix rdf:  <http://www.w3.org/1999/02/22-rdf-syntax-ns#> .
@prefix xsd:  <http://www.w3.org/2001/XMLSchema#> .


:cwns          rdf:type owl:ObjectProperty .
:caresFor      rdf:type owl:ObjectProperty .
:Cat           rdf:type owl:Class .
:Dead          rdf:type owl:Class .
:Alive         rdf:type owl:Class .
:Healthy       rdf:type owl:Class .
:HappyCatOwner rdf:type owl:Class .
```

:owns          rdfs:subPropertyOf :caresFor .

:Healthy       rdfs:subClassOf [ owl:complementOf :Dead ] .
:Cat           rdfs:subClassOf [ owl:unionOf (:Dead :Alive) ] .
:HappyCatOwner rdfs:subClassOf
      [ owl:intersectionOf
          ( [ rdf:type owl:Restriction ;
              owl:onProperty :owns ; owl:someValuesFrom :Cat ]
            [ rdf:type owl:Restriction ;
              owl:onProperty :caresFor ; owl:allValuesFrom :Healthy] )
      ] .

:schrödinger   rdf:type :HappyCatOwner .

To give an idea, how the RDF graph representation of an OWL ontology looks
like, the last TBox axiom is displayed graphically in the following picture.

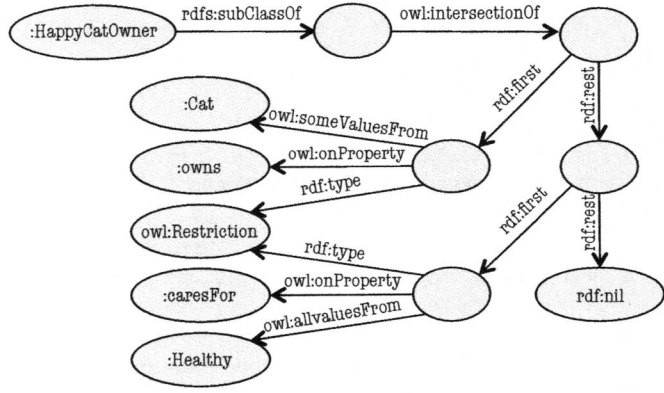

**Exercise 36.** *Translate the knowledge base from Example 21 and the initial axiom
from Example 33 into OWL ontologies in Turtle syntax.*

## 9.2   Expressing OWL Axioms in $\mathcal{SROIQ}$

In fact, the OWL specification features much more axiom types than the ones
used above to translate $\mathcal{SROIQ}$ knowledge bases. As far as the purely logical
axioms are concerned (i.e. excluding everything referring to datatypes, keys,
annotations, or the like), all these axioms can be considered as *syntactic sugar*,
i.e., they can be conceived as shortcuts for other axioms expressed in the "core"
OWL language used in the definitions above. In the sequel, we give the DL
paraphrases of these axioms

| Axiom type | Turtle notation | DL paraphrase |
|---|---|---|
| Class Equivalence | $[\![C]\!]_\mathbf{C}$ owl:equivalentClass $[\![D]\!]_\mathbf{C}$ . | $C \sqsubseteq D,\ D \sqsubseteq C$ |
| Class Disjointness | $[\![C]\!]_\mathbf{C}$ owl:disjointWith $[\![D]\!]_\mathbf{C}$ . | $C \sqcap D \sqsubseteq \bot$ |
| Disjoint Classes | [] rdf:type owl:AllDisjointClasses ; owl:members ($[\![C_1]\!]_\mathbf{C}$ ... $[\![C_n]\!]_\mathbf{C}$) . | $C_i \sqcap C_j \sqsubseteq \bot$ for all $1 \leq i < j \leq n$ |
| Disjoint Union | $[\![C]\!]_\mathbf{C}$ owl:disjointUnionOf ($[\![C_1]\!]_\mathbf{C}$ ... $[\![C_n]\!]_\mathbf{C}$) . | $\bigsqcup_{i<j} C_i \sqsubseteq C$ $C_i \sqcap C_j \sqsubseteq \bot$ for all $1 \leq i < j \leq n$ |
| Property Equivalence | $[\![r]\!]_\mathbf{R}$ owl:equivalentProperty $[\![s]\!]_\mathbf{R}$ . | $r \sqsubseteq s,\ s \sqsubseteq r$ |
| Disjoint Properties | [] rdf:type owl:AllDisjointProperties ; owl:members ($[\![r_1]\!]_\mathbf{R}$ ... $[\![r_n]\!]_\mathbf{R}$) . | $\mathrm{Dis}(r_i, r_j)$ for all $1 \leq i < j \leq n$ |
| Inverse Properties | $[\![r]\!]_\mathbf{R}$ owl:inverseOf $[\![s]\!]_\mathbf{R}$ . | $Inv(r) \sqsubseteq s$ |
| Property Domain | $[\![r]\!]_\mathbf{R}$ rdfs:domain $[\![C]\!]_\mathbf{C}$ . | $\exists r.\top \sqsubseteq C$ |
| Property Range | $[\![r]\!]_\mathbf{R}$ rdfs:range $[\![C]\!]_\mathbf{C}$ . | $\top \sqsubseteq \forall r.C$ |
| Functional Property | $[\![r]\!]_\mathbf{R}$ rdf:type owl:FunctionalProperty . | $\top \sqsubseteq\ \leqslant 1 r.\top$ |
| Inverse Functional Property | $[\![r]\!]_\mathbf{R}$ rdf:type owl:InverseFunctionalProperty . | $\top \sqsubseteq\ \leqslant 1 Inv(r).\top$ |
| Reflexive Property | $[\![r]\!]_\mathbf{R}$ rdf:type owl:ReflexiveProperty . | $\top \sqsubseteq \exists r.\mathsf{Self}$ |
| Irreflexive Property | $[\![r]\!]_\mathbf{R}$ rdf:type owl:IrreflexiveProperty . | $\exists r.\mathsf{Self} \sqsubseteq \bot$ |
| Symmetric Property | $[\![r]\!]_\mathbf{R}$ rdf:type owl:SymmetricProperty . | $Inv(r) \sqsubseteq r$ |
| Asymmetric Property | $[\![r]\!]_\mathbf{R}$ rdf:type owl:AsymmetricProperty . | $\mathrm{Dis}(Inv(r), r)$ |
| Transitive Property | $[\![r]\!]_\mathbf{R}$ rdf:type owl:TransitiveProperty . | $r \circ r \sqsubseteq r$ |
| Different Individuals | [] rdf:type owl:AllDifferent ; owl:members ( $a_1$ ... $a_n$ ) . | $a_i \not\approx a_j$ for all $1 \leq i < j \leq n$ |

## 10   Further Reading

At the end of this chapter, we give a few pointers to further reading with respect to different aspects of the contents presented here. Note that this list is certainly incomplete and subject to personal inclinations.

**Description Logics.**  As central reference to the area of Description Logics, the primary resource is certainly the Description Logic Handbook [Baader *et al.*, 2007], providing an overview of the subject, introductory parts as well as parts dedicated to advanced issues. The description logic $\mathcal{SROIQ}$ which, together with its sublogics, was the main subject of our treatise is introduced by Horrocks *et al.* [2006], the according reasoning complexity results are established by Kazakov [2008].

**Conjunctive Queries in Description Logics.**  While the principled problem of reasoning in DLs up to $\mathcal{SROIQ}$ can considered to be solved, conjunctive query answering is still a subject of active research and only preliminary decidability and complexity results are available. Most notably, decidability of $\mathcal{SROIQ}$ and even of $\mathcal{SHOIQ}$ is unsolved. On the other hand, the problem is settled for $\mathcal{SHIQ}$ [Glimm *et al.*, 2008c] and $\mathcal{SHOQ}$ [Glimm *et al.*, 2008b]. Moreover, Calvanese *et al.* [2009] captured additionally $\mathcal{SHOI}$ and extended the results to regular path queries. The most expressive Boolean-closed DL simultaneously featuring nominal concepts, inverses and number restrictions (i.e., $\mathcal{O}$, $\mathcal{I}$, and $\mathcal{Q}$) for which decidability is known is $\mathcal{ALCHOIQb}$ [Rudolph and Glimm, 2010].

***Relations to Logics in General.*** For foundations of logics, the textbook by Schöning [2008] is certainly a good starting point in particular for computer scientists, whereas Ebbinghaus *et al.* [1996] capture mathematical aspects. Model theory is treated in depth by Chang and Keisler [1990]. For an introduction into the area of theorem proving in a first-order logic setting, we recommend the textbook by Fitting [1996]. We suggest to consult Papadimitriou [1994] for the study of algorithmic complexity theory.

The correspondence of DLs and first-order logic (in particular the 2-variable fragment) has e.g. been described by Borgida [1996], the complexity treatment on the 2-variable fragment of FOL with counting quantifiers by Pratt-Hartmann [2005] has served as basis for a row of DL complexity results. The relatedness of DLs with modal logics (see the textbook by Blackburn *et al.* [2006] for a thorough introduction) is treated by Schild [1991]. As another closely related logic, the guarded fragment of FOL is described by Andréka *et al.* [1998].

***AI and Knowledge Representation.*** A central reference for a comprehensive overview of the area of AI as a whole is the seminal textbook by Russell and Norvig [2003]. Knowledge Representation in particular is treated by Sowa [1984] and van Harmelen *et al.* [2008].

***Semantic Web and OWL.*** The Semantic Web vision is described in the seminal paper by Berners-Lee *et al.* [2001]. In order to get an overview over all aspects of (Web) ontologies, the Ontology Handbook by Staab and Studer [2009] is a central reference.

As far as technical questions about syntax and semantics of OWL is concerned, the primary resource are the W3C Recommendation Documents. Next to an overview [OWL Working Group, 2009], syntax and semantics are treated by Motik *et al.* [2009b] and Motik *et al.* [2009a], respectively, whereas Patel-Schneider and Motik [2009] tackle the RDF serialization of OWL. The OWL 2 Primer by Hitzler *et al.* [2009a] gives an informal introduction into the use of OWL. A thorough treatment of all the standardized Semantic Web formalisms is provided by the textbook Foundations of Semantic Web Technologies [Hitzler *et al.*, 2009b].

**Acknowledgements.** I thank all people who helped me in one or the other way to accumulate the knowledge about DLs which I gave a partial overview of in this lecture. I am grateful to the organizers of the Reasoning Web Summer School 2011 for giving me the opportunity to teach. I thank the anonymous reviewers for their comments on an earlier version of this document. I am indebted to Anees ul Mehdi, Nadeschda Nikitina and Jens Wissmann for their thorough proofreading.

Special thanks go to Ian Horrocks for his valuable feedback in terms of poetic quality assurance. The DL logo displayed at the beginning of the chapter goes back to Enrico Franconi, the deduction calculus for $\mathcal{ALC}$ comes from Yevgeny Kazakov. Further inspiration was drawn from (in alphabetical order) Benedict XVI, Nicolas Chamfort, Edward Lear, the Rolling Stones, William Shakespeare and W.A. Spooner.

# References

[Andréka *et al.*, 1998] Andréka, H., van Benthem, J.F.A.K., Németi, I.: Modal languages and bounded fragments of predicate logic. Journal of Philosophical Logic 27(3), 217–274 (1998)

[Baader *et al.*, 2007] Baader, F., Calvanese, D., McGuinness, D., Nardi, D., Patel-Schneider, P. (eds.): The Description Logic Handbook: Theory, Implementation, and Applications, 2nd edn. Cambridge University Press, Cambridge (2007)

[Beckett and Berners-Lee, 14 January 2008] Beckett, D., Berners-Lee, T.: Turtle – Terse RDF Triple Language. W3C Team Submission (January 14, 2008), http://www.w3.org/TeamSubmission/turtle/

[Berners-Lee *et al.*, 2001] Berners-Lee, T., Hendler, J., Lassila, O.: The Semantic Web. In: Scientific American, pp. 96–101 (May 2001)

[Blackburn *et al.*, 2006] Blackburn, P., van Benthem, J.F.A.K., Wolter, F. (eds.): Handbook of Modal Logic. Studies in Logic and Practical Reasoning, vol. 3. Elsevier Science, Amsterdam (2006)

[Borgida, 1996] Borgida, A.: On the relative expressiveness of description logics and predicate logics. Artificial Intelligence 82(1–2), 353–367 (1996)

[Brachman and Levesque, 1984] Brachman, R.J., Levesque, H.J.: The tractability of subsumption in frame-based description languages. In: Brachman, R.J. (ed.) Proceedings of the 4th National Conference on Artificial Intelligence (AAAI 1984), pp. 34–37. AAAI Press, Menlo Park (1984)

[Calvanese *et al.*, 2009] Calvanese, D., Eiter, T., Ortiz, M.: Regular path queries in expressive description logics with nominals. In: Boutilier, C. (ed.) Proceedings of the 21st International Conference on Artificial Intelligence (IJCAI 2009), pp. 714–720 (2009)

[Chandra and Merlin, 1977] Chandra, A.K., Merlin, P.M.: Optimal implementation of conjunctive queries in relational data bases. In: Hopcroft, J.E., Friedman, E.P., Harrison, M.A. (eds.) Proceedings of the 9th Annual ACM Symposium on Theory of Computing (STOC 1977), pp. 77–90. ACM Press, New York (1977)

[Chang and Keisler, 1990] Chang, C.C., Jerome Keisler, H.: Model Theory, 3rd edn. Studies in Logic and the Foundations of Mathematics, vol. 73. North Holland, Amsterdam (1990)

[Ebbinghaus *et al.*, 1996] Ebbinghaus, H.-D., Flum, J., Thomas, W.: Mathematical Logic. Springer, Heidelberg (1996)

[Fitting, 1996] Fitting, M.: First-Order Logic and Automated Theorem Proving, 2nd edn. Springer, Heidelberg (1996)

[Ganter and Wille, 1997] Ganter, B., Wille, R.: Formal Concept Analysis: Mathematical Foundations. Springer, Heidelberg (1997)

[Glimm *et al.*, 2008a] Glimm, B., Horrocks, I., Sattler, U.: Deciding $\mathcal{SHOQ}^{\cap}$ knowledge base consistency using alternating automata. In: Baader, F., Lutz, C., Motik, B. (eds.) Description Logics. CEUR Workshop Proceedings, vol. 353 (2008), CEUR-WS.org

[Glimm *et al.*, 2008b] Glimm, B., Horrocks, I., Sattler, U.: Unions of conjunctive queries in $\mathcal{SHOQ}$. In: Brewka, G., Lang, J. (eds.) Proceedings of the 11th International Conference on Principles of Knowledge Representation and Reasoning (KR 2008), pp. 252–262. AAAI Press, Menlo Park (2008)

[Glimm *et al.*, 2008c] Glimm, B., Lutz, C., Horrocks, I., Sattler, U.: Answering conjunctive queries in the SHIQ description logic. Journal of Artificial Intelligence Research 31, 150–197 (2008)

[Golbreich *et al.*, 2006] Golbreich, C., Zhang, S., Bodenreider, O.: The foundational model of anatomy in OWL: Experience and perspectives. Journal of Web Semantics 4(3) (2006)

[Haarslev and Möller, 2001] Haarslev, V., Möller, R.: RACER System Description. In: Goré, R.P., Leitsch, A., Nipkow, T. (eds.) IJCAR 2001. LNCS (LNAI), vol. 2083, pp. 701–705. Springer, Heidelberg (2001)

[Hitzler *et al.*, 2009a] Hitzler, P., Krötzsch, M., Parsia, B., Patel-Schneider, P.F., Rudolph, S. (eds.): OWL 2 Web Ontology Language: Primer. W3C Recommendation (2009), http://www.w3.org/TR/owl2-primer/

[Hitzler *et al.*, 2009b] Hitzler, P., Krötzsch, M., Rudolph, S.: Foundations of Semantic Web Technologies. Chapman & Hall/CRC (2009)

[Horridge *et al.*, 2008] Horridge, M., Parsia, B., Sattler, U.: Laconic and Precise Justifications in OWL. In: Sheth, A.P., Staab, S., Dean, M., Paolucci, M., Maynard, D., Finin, T., Thirunarayan, K. (eds.) ISWC 2008. LNCS, vol. 5318, pp. 323–338. Springer, Heidelberg (2008)

[Horrocks and Sattler, 2007] Horrocks, I., Sattler, U.: A tableau decision procedure for $\mathcal{SHOIQ}$. Journal of Automated Reasoning 39(3), 249–276 (2007)

[Horrocks *et al.*, 2006] Horrocks, I., Kutz, O., Sattler, U.: The even more irresistible $\mathcal{SROIQ}$. In: Doherty, P., Mylopoulos, J., Welty, C.A. (eds.) Proceedings of the 10th International Conference on Principles of Knowledge Representation and Reasoning (KR 2006), pp. 57–67. AAAI Press, Menlo Park (2006)

[Kazakov and Motik, 2008] Kazakov, Y., Motik, B.: A resolution-based decision procedure for $\mathcal{SHOIQ}$. Journal of Automated Reasoning 40(2-3), 89–116 (2008)

[Kazakov, 2008] Kazakov, Y.: $\mathcal{RIQ}$ and $\mathcal{SROIQ}$ are harder than $\mathcal{SHOIQ}$. In: Brewka, G., Lang, J. (eds.) Proceedings of the 11th International Conference on Principles of Knowledge Representation and Reasoning (KR 2008), pp. 274–284. AAAI Press, Menlo Park (2008)

[Kazakov, 2009] Kazakov, Y.: Consequence-driven reasoning for horn $\mathcal{SHIQ}$ ontologies. In: Boutilier, C. (ed.) Proceedings of the 21st International Conference on Artificial Intelligence (IJCAI 2009), pp. 2040–2045 (2009)

[Krötzsch *et al.*, 2008] Krötzsch, M., Rudolph, S., Hitzler, P.: Description logic rules. In: Ghallab, M., Spyropoulos, C.D., Fakotakis, N., Avouris, N. (eds.) Proceedings of the 18th European Conference on Artificial Intelligence (ECAI 2008), pp. 80–84. IOS Press, Amsterdam (2008)

[Lehmann, 2009] Lehmann, J.: Dl-learner: Learning concepts in description logics. Journal of Machine Learning Research 10, 2639–2642 (2009)

[Lloyd and Topor, 1984] Lloyd, J.W., Topor, R.W.: Making prolog more expressive. Journal of Logic Programming 1(3), 225–240 (1984)

[Lutz, 2008] Lutz, C.: The complexity of conjunctive query answering in expressive description logics. In: Armando, A., Baumgartner, P., Dowek, G. (eds.) IJCAR 2008. LNCS (LNAI), vol. 5195, pp. 179–193. Springer, Heidelberg (2008)

[Manola and Miller, 2004] Manola, F., Miller, E. (eds.): Resource Description Framework (RDF): Primer. W3C Recommendation (2004), http://www.w3.org/TR/rdf-primer/

[Minsky, 1974] Minsky, M.: A framework for representing knowledge. Artificial intelligence memo, A.I. Laboratory. Massachusetts Institute of Technology, Cambridge (1974)

[Motik and Sattler, 2006] Motik, B., Sattler, U.: A Comparison of Reasoning Techniques for Querying Large Description Logic ABoxes. In: Hermann, M., Voronkov, A. (eds.) LPAR 2006. LNCS (LNAI), vol. 4246, pp. 227–241. Springer, Heidelberg (2006)

[Motik et al., 2009a] Motik, B., Patel-Schneider, P.F., Grau, B.C. (eds.): OWL 2 Web Ontology Language: Direct Semantics. W3C Recommendation (2009), http://www.w3.org/TR/owl2-direct-semantics/

[Motik et al., 2009b] Motik, B., Patel-Schneider, P.F., Parsia, B. (eds.): OWL 2 Web Ontology Language: Structural Specification and Functional-Style Syntax. W3C Recommendation (2009), http://www.w3.org/TR/owl2-syntax/

[Motik et al., 2009c] Motik, B., Shearer, R., Horrocks, I.: Hypertableau reasoning for description logics. Journal of Artificial Intelligence Research (JAIR) 36, 165–228 (2009)

[Noia et al., 2009] Di Noia, T., Di Sciascio, E., Donini, F.M.: A tableaux-based calculus for abduction in expressive description logics: Preliminary results. In: Grau, B.C., Horrocks, I., Motik, B., Sattler, U. (eds.) Description Logics. CEUR Workshop Proceedings, vol. 477 (2009), CEUR-WS.org

[OWL Working Group, 2009] W3C OWL Working Group. OWL 2 Web Ontology Language: Document Overview. W3C Recommendation (2009) http://www.w3.org/TR/owl2-overview/

[Papadimitriou, 1994] Papadimitriou, C.H.: Computational Complexity. Addison-Wesley, Reading (1994)

[Patel-Schneider and Motik, 2009] Patel-Schneider, P.F., Motik, B. (eds.): OWL 2 Web Ontology Language: Mapping to RDF Graphs. W3C Recommendation (2009), http://www.w3.org/TR/owl2-mapping-to-rdf/

[Pratt-Hartmann, 2005] Pratt-Hartmann, I.: Complexity of the two-variable fragment with counting quantifiers. Journal of Logic, Language and Information 14, 369–395 (2005)

[Quillian, 1968] Ross Quillian, M.: Semantic memory. In: Minsky, M. (ed.) Semantic Information Processing, ch. 4, pp. 227–270. MIT Press, Cambridge (1968)

[Rudolph and Glimm, 2010] Rudolph, S., Glimm, B.: Nominals, inverses, counting, and conjunctive queries or: Why infinity is your friend! Journal of Artificial Intelligence Research (JAIR) 39, 429–481 (2010)

[Rudolph et al., 2008] Rudolph, S., Krötzsch, M., Hitzler, P.: All elephants are bigger than all mice. In: Baader, F., Lutz, C., Motik, B. (eds.) Proceedings of the 21st International Workshop on Description Logics (DL 2008). CEUR Workshop Proceedings, vol. 353 (2008), CEUR-WS.org

[Rudolph et al., 2008b] Rudolph, S., Krötzsch, M., Hitzler, P.: Terminological reasoning in SHIQ with ordered binary decision diagrams. In: Pro- ceedings of the 23rd AAAI Conference on Artificial Intelligence (AAAI 2008), pp. 529–534. AAAI Press, Menlo Park (2008)

[Rudolph, 2004] Rudolph, S.: Exploring relational structures via FLE. In: Wolff, K.E., Pfeiffer, H.D., Delugach, H.S. (eds.) ICCS 2004. LNCS (LNAI), vol. 3127, pp. 196–212. Springer, Heidelberg (2004)

[Russell and Norvig, 2003] Russell, S., Norvig, P.: Artificial Intelligence: A Modern Approach, 2nd edn. Prentice Hall, Englewood Cliffs (2003)

[Schild, 1991] Schild, K.: A correspondence theory for terminological logics: Preliminary report. In: Mylopoulos, J., Reiter, R. (eds.) Proceedings of the 12th International Joint Conference on Artificial Intelligence (IJCAI 1991), pp. 466–471. Morgan Kaufmann, San Francisco (1991)

[Schmidt-Schauß and Smolka, 1991] Schmidt-Schauß, M., Smolka, G.: Attributive concept descriptions with complements. Journal of Artificial Intelligence 48, 1–26 (1991)

[Schöning, 2008] Schöning, U.: Logic for Computer Scientists. Birkhäuser, Basel (2008)

[Shearer and Horrocks, 2009] Shearer, R., Horrocks, I.: Exploiting Partial Information in Taxonomy Construction. In: Bernstein, A., Karger, D.R., Heath, T., Feigenbaum, L., Maynard, D., Motta, E., Thirunarayan, K. (eds.) ISWC 2009. LNCS, vol. 5823, pp. 569–584. Springer, Heidelberg (2009)

[Sidhu et al., 2005] Sidhu, A., Dillon, T., Chang, E., Sidhu, B.S.: Protein ontology development using OWL. In: Proceedings of the 1st OWL Experiences and Directions Workshop (OWLED 2005). CEUR Workshop Proceedings, vol. 188 (2005), http://ceur-ws.org/

[Simancik et al., 2011] Simancik, F., Kazakov, Y., Horrocks, I.: Consequence-based reasoning beyond horn ontologies. In: Walsh, T. (ed.) Proceedings of the 22nd International Conference on Artificial Intelligence, IJCAI 2011 (2011)

[Sirin et al., 2007] Sirin, E., Parsia, B., Grau, B.C., Kalyanpur, A., Katz, Y.: Pellet: A practical OWL-DL reasoner. Journal of Web Semantics 5(2), 51–53 (2007)

[Sowa, 1984] Sowa, J.F.: Conceptual Structures: Information Processing in Mind and Machine. Addison-Wesley, Reading (1984)

[Staab and Studer, 2009] Staab, S., Studer, R. (eds.): Handbook on Ontologies, 2nd edn. International Handbooks on Information Systems. Springer, Heidelberg (2009)

[Stuckenschmidt et al., 2009] Stuckenschmidt, H., Parent, C., Spaccapietra, S. (eds.):Modular Ontologies: Concepts, Theories and Techniques for Knowledge Modularization. LNCS, vol. 5445. Springer, Heidelberg (2009)

[Tsarkov and Horrocks, 2006] Tsarkov, D., Horrocks, I.: FaCT++ Description Logic Reasoner: System Description. In: Furbach, U., Shankar, N. (eds.) IJCAR 2006. LNCS (LNAI), vol. 4130, pp. 292–297. Springer, Heidelberg (2006)

[van Harmelen et al., 2008] van Harmelen, F., Lifschitz, V., Porter, B.: Handbook of Knowledge Representation. Foundations of Artificial Intelligence. Elsevier, Amsterdam (2008)

[Wolstencroft et al., 2005] Wolstencroft, K., Brass, A., Horrocks, I., Lord, P., Sattler, U., Turi, D., Stevens, R.: A little semantic web goes a long way in biology. In: Gil, Y., Motta, E., Benjamins, V.R., Musen, M.A. (eds.) ISWC 2005. LNCS, vol. 3729, pp. 786–800. Springer, Heidelberg (2005)

# Using SPARQL with RDFS and OWL Entailment

Birte Glimm

The University of Oxford, Department of Computer Science, UK

**Abstract.** This chapter accompanies the lecture on SPARQL with entailment regimes at the 7th Reasoning Web Summer School in Galway, Ireland, 2011. SPARQL is a query language and protocol for data specified in the Resource Description Format (RDF). The basic evaluation mechanism for SPARQL queries is based on subgraph matching. The query criteria are given in the form of RDF triples possibly with variables in place of the subject, object, or predicate of a triple, called basic graph patterns. Each instantiation of the variables that yields a subgraph of the queried RDF graph constitutes a solution. The query language further contains capabilities for querying for optional basic graph patterns, alternative graph patterns etc. We first introduce the main features of SPARQL as a query language. In order to define the semantics of a query, we show how a query can be translated to an abstract query, which can then be evaluated according to SPARQL's query evaluation mechanism. Apart from the features of SPARQL 1.0, we also briefly introduce the new features of SPARQL 1.1, which is currently being developed by the Data Access Working Group of the World Wide Web Consortium.

In the second part of these notes, we introduce SPARQL's extension point for basic graph pattern matching. We illustrate how this extension point can be used to define a semantics for basic graph pattern evaluation based on more elaborate semantics such as RDF Schema (RDFS) entailment or OWL entailment. This allows for solutions to a query that implicitly follow from an RDF graph, but which are not necessarily explicitly present. We illustrate what constitutes an extension point and how problems that arise from using a semantic entailment relation can be addressed. We first introduce SPARQL in combination with the RDFS entailment relation and then move on to the more expressive Web Ontology Language OWL. We cover OWL's Direct Semantics, which is based on Description Logics, and the RDF-Based Semantics, which is an extension of the RDFS semantics. For the RDF-Based Semantics we mainly focus on the OWL 2 RL profile, which allows for an efficient implementation using rule engines.

We assume that readers have a basic knowledge of RDF and Turtle, which we use in examples. For the OWL parts, we assume some background in OWL or Description Logics (see lecture notes *Foundations of Description Logics*). The examples for the OWL part are given in Turtle, OWL's functional-style syntax and Description Logics syntax. Although the inferences that are relevant for the example queries are explained, a basic idea about OWL's modeling constructs and their semantics are certainly helpful.

A. Polleres et al. (Eds.): Reasoning Web 2011, LNCS 6848, pp. 137–201, 2011.

# 1  Introduction

Query answering is important in the context of the Semantic Web, since it provides a mechanism via which users and applications can interact with ontologies and data. Several query languages have been designed for this purpose, including RDQL, SeRQL and, most recently, SPARQL. We consider the SPARQL [26] query language (pronounce *sparkle*) here, which was standardized in 2008 by the World Wide Web Consortium (W3C) and which is now supported by most RDF triple stores. Currently, the next SPARQL standard is being developed by W3C, named SPARQL 1.1 [13]. Apart from being a query language, the W3C standard also defines a protocol for communicating queries between client and server [11] and a results format for representing query results in XML [5].

The main mechanism for computing query results in SPARQL is subgraph matching: RDF triples in both the queried RDF data and the query pattern are interpreted as nodes and edges of directed graphs, and the resulting query graph is matched to the data graph using variables as wild cards.

In this section, we give some simple examples of SPARQL queries and the query evaluation process. We further introduce the basic ideas behind entailment regimes. In Section 2, we introduce the general features of SPARQL in more detail and explain more formally how a SPARQL query is evaluated. We then introduce SPARQL entailment regimes and explain the design rationals behind the RDFS entailment regime in Section 3. Next, we clarify the relationship between OWL's structural specification and RDF graphs in and introduce the OWL Direct Semantics Entailment Regime in Section 4. Finally, we give some exercises and pointers to additional literature for further reading.

## 1.1  SPARQL Query Examples

We start with a simple example that illustrates SPARQL's standard query evaluation mechanism, which is based on sub-graph matching. We use Turtle [4] to write down RDF data throughout this chapter and we use the RDF triples shown in Table 1 throughout this and the next section.

**Example 1.** We consider the following SPARQL query over that data from Table 1:

```
PREFIX foaf: <http://xmlns.com/foaf/0.1/>
SELECT ?name ?mbox
WHERE { ?x foaf:name ?name . ?x foaf:mbox ?mbox }
```

For the query, we start by declaring a prefix that allows for abbreviating otherwise long IRIs in the query body. In the remainder we omit this prefix declaration, but assume that the foaf prefix is declared as above. The main part of the query starts with a *select clause* that specifies which variables should be returned with their bindings as part of the result. The *where clause* specifies the conditions that answers have to satisfy. Note that the where clause is still written as a set of triples as in the data above, but the subject and the object are now variables. In order to evaluate such a *basic graph pattern* (BGP), we substitute the variables with terms from the data and if the substitution yields a subgraph of the queried graph, the substituted values are called a solution. The BGP from the above query yields the solutions:

**Table 1.** Example data used in Section 1 and 2

| @prefix | foaf: | <http://xmlns.com/foaf/0.1/>. |
|---|---|---|
| _:a | foaf:name | "Birte Glimm". |
| _:a | foaf:mbox | "b.glimm@googlemail.com". |
| _:a | foaf:icqChatID | "b.glimm". |
| _:a | foaf:aimChatID | "b.glimm". |
| _:b | foaf:name | "Sebastian Rudolph". |
| _:b | foaf:mbox | <mailto:rudolph@kit.edu>. |
| _:c | foaf:name | "Pascal Hitzler". |
| _:c | foaf:aimChatID | "phi". |
| foaf:icqChatID | rdfs:subPropertyOf | foaf:nick. |
| foaf:name | rdfs:domain | foaf:Person. |

| ?x | ?name | ?mbox |
|---|---|---|
| _:a | "Birte Glimm" | "b.glimm@googlemail.com" |
| _:b | "Sebastian Rudolph" | <mailto:rudolph@kit.edu> |

Since the select clause only specifies ?name and ?mbox as output variables, a further projection step is required to evaluate the complete query, which only leaves the values for ?name and ?mbox in the query solutions.

## 1.2 RDF Datasets

Some might be surprised by the absence of a from clause in the query from Example 1, which specifies which data is to be queried. This is because SPARQL queries are executed against an *RDF dataset*, which represents a collection of graphs, and each SPARQL query engine has a default dataset that is normally used. An RDF dataset comprises one graph, called the *default graph*, which does not have a name, and zero or more *named graphs*, where each named graph is identified by an IRI. Unless we change the so called *active graph* for the BGP evaluation with the GRAPH keyword to one of the named graphs, the query is executed against the default graph. In Example 1 the active graph is the default graph, which we have implicitly assumed to contain the given set of triples. Alternatively, a SPARQL query may specify a custom dataset that is to be used for matching by using the FROM and FROM NAMED keyword. In this case, the dataset used for the query consists of a default graph, which is obtained by merging all graphs referred to in a from clause, and a set of (IRI, graph) pairs, one from each from named clause.

**Example 2.** For an example with custom datasets, let us assume that the data from Table 1 is available under the IRI <http://example.org/foaf/myFoaf>. The following query creates a custom dataset with an empty default graph (no FROM clause) and one named graph.

```
SELECT        ?name ?mbox
FROM NAMED <http://example.org/foaf/myFoaf>
WHERE         { GRAPH <http://example.org/foaf/myFoaf>
                    { ?x foaf:name ?name. ?x foaf:mbox ?mbox }
                }
```

Since we used the GRAPH keyword, the active graph for the BGP evaluation is the given named graph. Alternatively, we could use a variable instead of the IRI for the GRAPH keyword, which would evaluate the BGP once over each named graph, binding the graph variable to the corresponding IRI of the named graph. The query answer is the same as for Example 1.

### 1.3   Blank Nodes in Queries and Query Results

Finally, we want to point out that in our intermediate results for the query from Example 1 we use exactly the same blank node names as in the data from Table 1. This does not have to be the case. Blank nodes just denote the existence of something and we cannot rely on a label being used consistently, it can change when a graph is reloaded or during a merge operation. Furthermore, the query is in fact evaluated against the *scoping graph*, which is equivalent to the active graph, but allows for renaming of blank nodes. Thus, evaluating the BGP could equally result in the solutions where _:a is renamed into _:x and _:b is consistently renamed into _:y. Note, however, that we cannot rename _:a and _:b into the same blank node, nor can we rename the first occurrence of _:b different from the second occurrence.

Since blank node merely denote the existence of something, we can also not understand a blank node in the query as referring to an element with exactly that blank node label in the queried graph. Blank nodes in a BGP can rather be understood as variables, which are immediately projected out after BGP matching, i.e., a blank node cannot occur in the SELECT clause. Since in Example 1 we are not interested in the concrete value that ?x is mapped to, we could equally replace the query pattern with

{ _:x foaf:name ?name . _:x foaf:mbox ?mbox }

which exactly the same results.

### 1.4   SPARQL Entailment Regimes

Various W3C standards, including RDF and OWL, provide semantic interpretations for RDF graphs that allow additional RDF statements to be inferred from explicitly given assertions. The entailment regimes in SPARQL 1.1 [12] define how basic graph pattern matching can be defined using semantic entailment relations instead subgraph matching.

**Example 3.** We again use the data from Table 1 to illustrate the use of inference with the query:

```
SELECT ?name ?nick
WHERE { ?x foaf:name ?name . ?x foaf:nick ?nick }
```

Using subgraph matching, we do not get an answer. The triple _:a foaf:nick "b.glimm" is, however, entailed by the given triples under RDFS semantics since any subject related with the property foaf:icqChatID is necessarily related to that object also with the property foaf:nick in any RDFS-interpretation that satisfies the data. Under the RDFS entailment regime we expect, therefore, to get "Birte Glimm" as binding for ?name and "b.glimm" as binding for ?nick in the solution.

The entailment regimes developed by the W3C specify exactly what answers we get for several common entailment relations such as RDFS entailment or OWL Direct Semantics entailment. Aspects that have to be addressed include:

- How are the infinitely many axiomatic triples under the RDF(S) semantics handled? These are entailed even by an empty graph and infinite answers, at least due to such axiomatic triples, are rarely desirable.
- How are entailed triples handled that just differ in blank node labels?
- How are inconsistent graphs handled?
- What happens in case of errors?

For OWL's Direct Semantics we further have to address the issue that the semantics is not defined in terms of triples, but in terms of structural objects, which correspond to Description Logic constructs.

### 1.5   SPARQL as a Protocol

The SPARQL Protocol for RDF defines how a SPARQL query can be conveyed from a query client to a query processor. The protocol is firstly described in an abstract interface independent of any concrete realization, implementation, or binding to another protocol; but a HTTP and SOAP binding of the interface is also provided. We do not further explain the protocol and instead focus on the semantics of SPARQL queries either with subgraph matching (aka simple entailment) or with more elaborate entailment regimes.

## 2   SPARQL Basics

In the previous examples, we have already seen some basic SPARQL queries. We now make it more precise what parts belong to a query and which choices we have in selecting the data that is returned as the query answer.

### 2.1   Graph Patterns

The basic selection criteria are specified in the WHERE clause, but before we describe this in more detail, we first recall some basic notions from RDF.

**Definition 1.** *We write* I *for the set of all* International Resource Identifiers *(IRIs),* L *for the set of all* RDF literals, *and* B *for the set of all* blank nodes. *The set of* RDF terms, *denoted* T, *is* I ∪ L ∪ B.

*An* RDF graph *is a set of* RDF triples *of the form* (subject, predicate, object) ∈ (I ∪ B) × I × T. *We normally omit "RDF" in our terminology if no confusion is likely, and we use Turtle syntax [4] for all examples. The* vocabulary Voc(G) *of a graph* G *is the set of all terms that occur in* G.

*Queries are built using a countably infinite set* V *of* query variables *disjoint from* T. *A variable v* ∈ V *is prefixed by the variable identifier* ? *or* $ . *The outer-most graph pattern in a query is called the* query pattern.

The variable identifier is not part of the variable name, e.g., $x and ?x denotes the same variable even if both prefixes are used within one query. The variable name itself can contain numbers, letters, and various other admissible symbols [26].

We generally abbreviate IRIs using prefixes rdf, rdfs, owl, xsd, and foaf to refer to the RDF, RDFS, OWL, XML Schema Datatypes, and FOAF namespaces, respectively. We further use the prefix ex for an imaginary example namespace. Prefix declarations for these namespaces are generally omitted in example data and queries.

The simplest form of a WHERE clause consists of a basic graph pattern (BGP), but we can also construct more complex graph patterns by combining smaller patterns in various ways that are described in detail within this section.

**Basic Graph Patterns.** As we have seen in the introductory section, basic graph patterns are the basic building blocks for building a SPARQL query and we can define these formally as follows:

**Definition 2.** *A triple pattern is member of the set* (T ∪ V) × (I ∪ V) × (T ∪ V), *and a basic graph pattern (BGP) is a set of triple patterns.*

According to the above definition, variables can, thus, occur in place of a subject, predicate, or an object. It is also worth pointing out, that the subject of a triple pattern can be a literal although this is not allowed in RDF. This is meant to support (possible future) extensions of RDF. At the moment queries with literals in the subject position can simply not have an answer.

So far we have not seen the effect of blank nodes in BGPs. Since blank nodes do not really refer to a particular resource in the graph, but only denote the existence of something, we cannot expect that the blank node _:a in a BGP is mapped to exactly the blank node _:a in the data as foaf:name in the query is mapped to foaf:name in the queried graph. Instead, blank nodes act similar to variables with the difference that they cannot be selected in the SELECT clause.

**Example 4.** Since we are not interested in the mappings for ?x in Example 1, we could achieve the same result as with that query pattern:

{ _:a foaf:name ?name . _:a foaf:mbox ?mbox }

Note that we have deliberately used _:a, which is a blank node label that occurs in the data. The blank node _:a in the BGP acts, however, as a variable and can be substituted any of the blank nodes from the data. Thus, we get the same result with the above query pattern as in Example 1.

As we have seen in Section 1, a **WHERE** clause that consists of a BGP requires that the set of triple patterns that make up the BGP must all match. In the following, we introduce more complex patters:

- *Group Graph Patterns*, where a set of graph patterns must all match,
- *Optional Graph Patterns*, where additional patterns may extend the solution,
- *Alternative Graph Patterns*, where two or more possible patterns are tried, and
- *Patterns on Named Graphs*, where patterns are matched against named graphs.

**Group Graph Patterns.** Group graph patterns combine patterns conjunctively, similarly to BGPs that combine triple patterns conjunctively. In order to create a group, SPARQL uses curly braces. Grouping by itself is not very useful unless if we only work with basic graph patterns, but it becomes useful when we consider further constructors. For example, we can combine two groups with the **UNION** keyword, which means that solutions are obtained by matching one or the other group. Before we come to that, we first introduce grouping in more detail.

**Example 5.** Using groups, patterns of the query from Example 1 can equivalently be written as:

```
{ { ?x foaf:name ?name}
  { }
  { ?x foaf:mbox ?mbox } }
```

We now separated the BGP from Example 1 into three groups. The first and the third group now consists of a single triple pattern. The second group is the *empty* group pattern, which matches to any data. The inclusion or omission of the empty pattern has, therefore, no effect here. A query with only the empty pattern returns always one solution in which any variable that is selected is unbound. Omitting the braces around the two triple patterns as illustrated below:

```
{ ?x foaf:name ?name
  { }
  ?x foaf:mbox ?mbox }
```

leaves us with a group of three elements: a BGP of one triple pattern, an empty group, and again a BGP of one triple pattern.

**Alternative Patterns.** Now that we can group patterns, we can combine groups with other constructors, e.g., the **UNION** constructor for specifying alternative restrictions. The UNION constructor is a binary operator, i.e., it is used as **pattern UNION pattern**.

**Example 6.** We assume that the default graph contains the data from Table 1 and that the SPARQL query is:

```
SELECT ?mbox
WHERE { { ?x foaf:name "BirteGlimm". ?x foaf:mbox ?mbox }
        UNION
        { ?x foaf:name "SebastianRudolph" . ?x foaf:mbox ?mbox }
      }
```

The result for this query consists of the two email addresses from the queried data. The first email address matches the first BGP and the one for Sebastian Rudolph matches the second BGP, and the results from both BGPs contribute to the final answer due to the UNION keyword.

It is worth noting that the UNION keyword is not denoting an exclusive or and that SPARQL does not have a set semantics as, for example, SQL, so we can have duplicate results in the answer.

**Example 7.** We illustrate the fact that UNION does not represent an exclusive or and that SPARQL queries can have duplicate results by means of the following query again over the data from Table 1:

```
SELECT ?name ?chatID
WHERE  { ?x foaf:name ?name .
         { ?x foaf:icqChatID ?chatID } UNION
         { ?x foaf:aimChatID ?chatID } }
```

The results for the query are as follows:

| ?name | ?chatID |
|---|---|
| "Birte Glimm" | "bgl" |
| "Birte Glimm" | "bgl" |
| "Pascal Hitzler" | "phi" |

where the first solution results from matching the first alternative and the latter two result from matching the second alternative. The solutions for this query are in fact computed by building the union (without duplicate elimination) from the results of evaluating two graph patters:

```
{ ?x foaf:name ?name.              and     { ?x foaf:name ?name.
  { ?x foaf:icqChatID ?chatID } }            { ?x foaf:aimChatID ?chatID } }
```

In case we use multiple unions, e.g., of the form pattern UNION pattern UNION pattern, this is equivalent to writing: { pattern UNION pattern } UNION pattern, i.e., the UNION operator is left-associative.

**Optional Patterns.** Apart from using the UNION keyword, we can declare some parts as optional using the OPTIONAL keyword, i.e., we allow for only retrieving bindings for these optional parts when these are available.

**Example 8.** We assume that the default graph contains the data from Table 1 and that the SPARQL query is:

```
SELECT ?name ?mbox
WHERE  { ?x foaf:name ?name
         OPTIONAL { ?x foaf:mbox ?mbox } }
```

The result for this query now consists of one additional solution compared to the result for Example 1 in which ?name is bound to "Pascal Hitzler" and ?mbox is unbound. We indicate unbound values by simply leaving the entry in the results table empty:

| ?name | ?mbox |
|-------|-------|
| "Birte Glimm" | "b.glimm@googlemail.com" |
| "Sebastian Rudolph" | \<mailto:rudolph@kit.edu\> |
| "Pascal Hitzler" | |

The OPTIONAL operator is again binary, i.e., it is used pattern OPTIONAL pattern. Although { OPTIONAL pattern } is syntactically valid, it just abbreviates { {} OPTIONAL pattern }. Like UNION, OPTIONAL is left-associative, i.e., pattern OPTIONAL pattern OPTIONAL pattern is equivalent to: { pattern OPTIONAL pattern } OPTIONAL pattern.

**Mixing Optional and Alternative Patterns.** We can, of course, also mix the use of OPTIONAL and UNION. In this case, left-associativity still applies.

**Example 9.** The following query over the data from Table 1 gives three results given the left-associativity of UNION and OPTIONAL:

> SELECT ?name ?chatID ?mbox
> WHERE { ?x foaf:name ?name .
>             { ?x foaf:icqChatID ?chatID } UNION
>             { ?x foaf:aimChatID ?chatID } OPTIONAL
>             { ?x foaf:mbox ?mbox } }

| ?name | ?chatID | ?mbox |
|-------|---------|-------|
| "Birte Glimm" | "b.glimm" | "b.glimm@googlemail.com" |
| "Birte Glimm" | "b.glimm" | "b.glimm@googlemail.com" |
| "Pascal Hitzler" | "phi" | |

After matching the first triple pattern, the union is evaluated. Finally, the optional part is applied to enrich the so far computed solutions. If we were to make the operator preference explicit, we get the following equivalent pattern:

> { { ?x foaf:name ?name .
>     { { ?x foaf:icqChatID ?chatID } UNION { ?x foaf:aimChatID ?chatID } }
>   }
>   OPTIONAL { ?x foaf:mbox ?mbox }
> }

If we do not want the standard left-associate behavior of SPARQL, we have to use braces to enforce a different grouping.

**Filters.** SPARQL *filters* restrict solutions to those for which the filter evaluates to true. The FILTER keyword is followed by a Boolean filter function that evaluates to true or false. Only if the filter function evaluates to true is the solution to be included in the solution sequence.

**Example 10.** In order to illustrate the use of a filters, we employ the isIRI filter function to filter out results in which the foaf:mbox is not given as an IRI. We use again the data from Table 1 and the query:

```
SELECT ?name ?mbox
WHERE { ?x foaf:name ?name . ?x foaf:mbox ?mbox
          FILTER isIRI(?mbox)
        }
```

Since the isIRI function evaluates to false when ?mbox is bound to the plain literal "b.glimm@googlemail.com", the match cannot be included in the solutions and we get just one result:

| ?name | ?mbox |
|---|---|
| "Sebastian Rudolph" | \<mailto:rudolph@kit.edu\> |

There are quite a range of filter functions, e.g., functions for comparing numerical values or date, filtering strings according to a regular expression, test whether a binding is a blank node, or whether a variable is bound at all. For more details, we refer to the SPARQL Query specification [26].

**Literals.** The general syntax for literals in a SPARQL query is a string enclosed in either double quotes ("...") or single quotes ('...') with either an optional language tag (introduced by @) or an optional datatype IRI or prefixed name (introduced by ^^).

For convenience, integers can be written without quotation marks and an explicit datatype IRI. Such literals are interpreted as typed literals of datatype xsd:integer, xsd:decimal if there is no '.' in the number; otherwise the number is interpreted as xsd:decimal if no exponent is given and as xsd:double otherwise. Literals of type xsd:boolean can also be written as true or false.

To facilitate writing literal values which themselves contain quotation marks or which are long and contain newline characters, SPARQL provides an additional quoting construct in which literals are enclosed in three single- or double-quotation marks.

Examples of literal syntax in SPARQL include:

- "chat" or 'chat'
- "chat"@fr with language tag "fr"
- "xyz"^^\<http://example.org/ns/userDatatype\>
- "abc"^^appNS:appDataType
- "'The librarian said, "Perhaps you would enjoy 'War and Peace'." "'
- 1, which is the same as "1"^^xsd:integer
- 1.3, which is the same as "1.3"^^xsd:decimal
- 1.300, which is the same as "1.300"^^xsd:decimal
- 1.0e6, which is the same as "1.0e6"^^xsd:double
- true, which is the same as "true"^^xsd:boolean
- false, which is the same as "false"^^xsd:boolean

## 2.2   Result Formats

SPARQL has four query forms. These query forms use the solutions from pattern matching to form result sets or RDF graphs. The query forms are:

- SELECT returns all, or a subset of, the variables bound in a query pattern match.
- CONSTRUCT returns an RDF graph constructed by substituting variables in a set of triple templates.
- ASK returns a boolean indicating whether a query pattern matches or not.
- DESCRIBE returns an RDF graph that describes the resources found.

The XML results format of SPARQL can be used to serialize the result set from a select query or the boolean result of an ask query.

**SELECT**   queries return variables and their bindings directly. The syntax SELECT * is an abbreviation that selects all of the variables in a query that are in scope. The definition of scope becomes relevant when making use of the SPARQL 1.1 sub-select feature in which case only variables that are projected in the sub-query are visible in the enclosing query.

**CONSTRUCT**   queries return a single RDF graph. The result is an RDF graph formed by taking the specified graph template and by instantiating it with each query solution in the solution sequence. The triples obtained from each solution are combining into a single RDF graph by set union.

   If any such instantiation produces a triple containing an unbound variable or an illegal RDF construct, e.g., containing a literal in subject or predicate position, then that triple is not included in the output RDF graph. The graph template can contain triples with no variables (known as ground or explicit triples), and these also appear in the output RDF graph returned by the CONSTRUCT query form.

   A template can create an RDF graph containing blank nodes. The blank node labels are scoped to the template for each solution. If the same label occurs twice in a template, then there will be one blank node created for each query solution, but there will be different blank nodes for triples generated by different query solutions.

**Example 11.** In order to see an example for construct and the effect of blank nodes in the template, we se the following query over the data from Table 1:

```
CONSTRUCT { ?x rdf:type foaf:Person . ?x foaf:givenName _:x }
WHERE        { ?x foaf:name ?name }
```

Evaluating the query pattern yields the following bindings:

| ?x | ?name |
|----|-------|
| _:a | "Birte Glimm" |
| _:b | "Sebastian Rudolph" |
| _:c | "Pascal Hitzler" |

Instantiating and building the set union of the template then results in:

| _:u | rdf:type | foaf:Person. |
|---|---|---|
| _:u | foaf:givenName | _:$x_1$ |
| _:v | rdf:type | foaf:Person. |
| _:v | foaf:givenName | _:$x_2$ |
| _:w | rdf:type | foaf:Person. |
| _:w | foaf:givenName | _:$x_3$ |

Note that the variable binds to blank nodes in the data and there is not even a guarantee that in the intermediate results the same blank node labels are used. In the constructed data a different blank node label is created. Similarly, the blank node in the template just causes a different blank node label to be created each time the template is instantiated.

**ASK** queries test whether or not a query pattern has a solution. No information is returned about the possible query solutions, just whether or not a solution exists.

**Example 12.** The following queries illustrates the use of the ASK query form for the data from Table 1:

ASK { ?x foaf:name "Birte Glimm" }

The result to this query is true or yes since there is a possible binding for ?x in the data.

**DESCRIBE** queries return a single RDF graph containing RDF data about resources.

**Example 13.** Possible examples for DESCRIBE are the following:

DESCRIBE foaf:Person   or
DESCRIBE ?x WHERE { ?x foaf:name "Birte Glimm" }

The exact output is not prescribed by the SPARQL Query specification, i.e., results depend on the SPARQL query processor and can vary between systems. The resulting RDF graph can be complex and can, for example, mention other resources that are somehow related to the given resource. For example, whether a property denotes an inverse or an inverse functional property, or the name and mbox if the resource that is to be described is from a FOAF file.

### 2.3   Solution Modifiers

Query patterns generate a multiset of solutions, each solution being a partial function from variables to RDF terms. These solutions are then treated as a sequence (a solution sequence), initially in no specific order; any sequence modifiers are then applied to create another sequence. Finally, the sequence is used to generate one of the results of a SPARQL query form. A *solution sequence modifier* is one of:

- *Order modifier* to put the solutions in order
- *Projection modifier* to choose certain variables and eliminate others from the solutions
- *Distinct modifier* to eliminate duplicate solutions
- *Reduced modifier* to permit elimination of some non-unique solutions
- *Offset modifier* to control where the solutions start from in the overall sequence of solutions
- *Limit modifier* to restrict the number of solutions

**ORDER BY** is a keyword that establishes an order within a solution sequence. It is followed by a sequence of order comparators, composed of an expression and an optional order modifier (either ASC($\cdot$) or DESC($\cdot$)). Each ordering comparator is either ascending (indicated by the ASC($\cdot$) modifier or by no modifier) or descending (indicated by the DESC($\cdot$) modifier).

**Example 14.** We illustrate the use ORDER BY with the following query over the data from Table 1:

```
SELECT    ?name
WHERE     { ?x foaf:name ?name }
ORDER BY ?name
```

Since ascending is the default ordering, the query is equivalent to:

```
SELECT    ?name
WHERE     { ?x foaf:name ?name }
ORDER BY ASC(?name)
```

The results are now ordered according to the bindings for the variable ?name:

| ?name |
| --- |
| "Birte Glimm" |
| "Pascal Hitzler" |
| "Sebastian Rudolph" |

With the DESC($\cdot$) keyword we would get the exact opposite order.

The ascending order of two solutions with respect to an ordering comparator is established by substituting the solution bindings into the expressions and comparing them with the < operator, which is defined by the SPARQL Query specification for numerics, simple literals, xsd:string, xsd:boolean, and xsd:dateTime. Descending order is the reverse of the ascending order. Ordering never changes the cardinality of solutions.

Pairs of IRIs are ordered by comparing them as simple literals. SPARQL also fixes an order between some kinds of RDF terms that would not otherwise be ordered (given here from lowest in the order):

1. no value assigned to the variable or expression in this solution,
2. blank nodes,
3. IRIs,
4. RDF literals.

A plain literal is lower than an RDF literal with type xsd:string of the same lexical form. Note that SPARQL does not define a total ordering over all possible RDF terms; a few examples of pairs of terms for which the relative order is undefined are:

- a simple literal and a literal with a language tag, e.g., "a" and "a"en_gb,
- two literals with language tags, e.g., "a"en_gb and "b"en_gb,
- a simple literal and an xsd:string, e.g., "a" and "a"^^xsd:string,
- a simple literal and a literal with a supported data type, e.g., "a" and "1"^^xsd:integer,
- two unsupported data types, e.g., "1"^^my:integer and "2"^^my:integer
- a supported data type and an unsupported data type, e.g., "1"^^xsd:integer and "2"^^my:integer.

The ORDER BY clause can also contain a LIMIT n and an OFFSET m condition to limit the number of returned results to $n$ and to start only with the $m^{th}$ result. Using LIMIT and OFFSET to select different subsets of the query solutions will not be useful unless the order is made predictable by using ORDER BY.

Using ORDER BY on a solution sequence for a CONSTRUCT or DESCRIBE query has no direct effect because only SELECT returns a sequence of results. Used in combination with LIMIT and OFFSET, ORDER BY can be used to return results generated from a different slice of the solution sequence. An ASK query does not include ORDER BY, LIMIT or OFFSET.

**Projection** can be used to transform the solution sequence into one involving only a subset of the variables. For each solution in the sequence, a new solution is formed using a specified selection of the variables as specified in the SELECT clause.

We have used projection already any many previous examples, so do not give another example here.

**DISTINCT** is a solution modifier that causes the elimination of duplicate solutions. Specifically, each solution that binds the same variables to the same RDF terms as another solution is eliminated from the solution set.

**REDUCED** is not as strong as DISTINCT because it permits duplicate elimination, but does not enforce it. Thus, some duplicates might be eliminated, whereas other remain in the solution sequence.

## 2.4   SPARQL Algebra Processing

So far we have mainly used examples to illustrate the effect of different SPARQL operators. In this section, we precisely define how a the result for a SPARQL query is computed, which requires transforming a query string into a SPARQL algebra object that is then evaluated in order to compute the query result.

The first step towards obtaining an algebra object for a query string is parsing. In the parsing process, we expanding abbreviations for IRIs and triple patterns, e.g., for triple patterns that use Turtle's comma or semicolon abbreviations.

**Table 2.** SPARQL 1.0 grammar elements that make up a query pattern

| | |
|---|---|
| `GroupGraphPattern` | `::= '{' TriplesBlock?` |
| | `( ( GraphPatternNotTriples | Filter ) '.'?` |
| | `TriplesBlock? )*` |
| | `'}'` |
| `GraphPatternNotTriples` | `::= OptionalGraphPattern | GroupOrUnionGraphPattern` |
| | `| GraphGraphPattern` |
| `OptionalGraphPattern` | `::= 'OPTIONAL' GroupGraphPattern` |
| `GraphGraphPattern` | `::= 'GRAPH' VarOrIRIref GroupGraphPattern` |
| `GroupOrUnionGraphPattern` | `::= GroupGraphPattern ( 'UNION' GroupGraphPattern )*` |
| `Filter` | `::= 'FILTER' Constraint` |

**Translating SPARQL Query Patterns to Algebra Expressions.** Parsing of a SPARQL query string involves expanding abbreviations for IRIs and triple patterns and the construction of an abstract syntax tree that can then be transformed into a SPARQL algebra expressions. The semantics of a query is then given based on the algebra expression. After parsing, we have an abstract syntax tree that represents the expanded query string and which is then converted into an algebra object.

**Example 15.** The algebra expression corresponding to the simple query

SELECT ?s WHERE { ?s :p ?o }

is Project(Bgp(?s <http://example.org#p> ?o), {?s}) if we assume that the empty prefix is defined as <http://example.org#>. The algebra expression is evaluated inside out, i.e., we first evaluate the pattern ?s <http://example.org#p> ?o given as parameter to the Bgp algebra expression; then a projection is performed so that only the values for the variable ?s remain.

We start by looking into the translation of a query pattern before we come to solution modifiers and other parts that are not related to the query pattern. We restrict our explanation to SPARQL 1.0 elements; SPARQL 1.1 works quite similar, but due to the much increased features the translation gets far more involved.

The translation is defined in terms of objects from the SPARQL grammar, i.e., one has to understand which part of a query pattern has been produced by a certain grammar object. The grammar defines a query pattern as a `GroupGraphPattern` and we give the relevant grammar part in Table 2, where `TriplesBlock` denotes a basic graph pattern, `VarOrIRIref` denotes a variable or an IRI, and `Constraint` represents a filter expression.

We first define a function algbr that takes a query pattern and inductively translates it into a SPARQL algebra expression. The algebra objects that we encounter in the translation process are

- Bgp, for expressing that a BGP has to be evaluated, e.g., by performing subgraph matching,
- Join, for joining results, e.g., from different groups,
- LeftJoin, for combining results with optional values,

– Filter, for filtering results according to a filter expression,
– Union, for combining results from alternatives,
– Graph, for evaluating a query part on a named graph.

We further encounter the empty pattern, denoted $Z$, which is a basic graph pattern that evaluates to an empty solution mapping, i.e., to a solution mapping that does not map any variable to a value. Thus, the empty pattern can be joined with any other pattern without any effect, i.e., $Z$ is the identity for join.

**Definition 3.** *We define the function* algbr(P) *as follows: If* P *is* `TriplesBlock`, *then*

$$\text{algbr}(P) := \begin{cases} Z & \textit{if } P \textit{ is empty} \\ \text{Bgp}(P) & \textit{otherwise.} \end{cases}$$

*If* P *is* `GroupOrUnionGraphPattern` *with elements* $e_1, \ldots, e_n$, *then*

$$\text{algbr}(P) := \begin{cases} \text{algbr}(e_1) & \textit{for } n = 1 \textit{ and} \\ \text{Union}(\text{algbr}(e_1), \text{algbr}(e_2 \text{ UNION} \ldots \text{UNION } e_n)) & \textit{otherwise.} \end{cases}$$

*If* P *is* `GraphGraphPattern` *of the form* term GRAPH P′, *then*

$$\text{algbr}(P) := \text{Graph}(\text{term}, \text{algbr}(P')).$$

*If* P *is* `GroupGraphPattern` *containing filter elements* $f_1, \ldots, f_n$ *and other elements* $e_1, \ldots, e_m$ *then*

$$\text{algbr}(P) := \begin{cases} \text{Filter}(f_1 \&\& \ldots \&\& f_n, \text{translateGroup}(e_1, \ldots, e_m)) & \textit{if } n > 0 \textit{ and} \\ \text{translateGroup}(e_1, \ldots, e_m) & \textit{otherwise,} \end{cases}$$

*where* translateGroup *is described in Algorithm 1 and* && *is SPARQL's conjunction operator for filter expressions.*

The resulting algebra objects can be *simplified* by exploiting the join identity property of the empty pattern $Z$: we can replace $\text{Join}(Z, A)$ by $A$ and $\text{Join}(A, Z)$ by $A$.

**Example 16.** For example, the simple query pattern
{ ?s ?p ?o }      is translated to      Join(Z, Bgp(?s ?p ?o))
since { ?s ?p ?o } is an instance of `GroupGraphPattern` (as every query pattern), which is translated according to Algorithm 1 (line 10), where the triple pattern itself is translated as `TriplesBlock`. According to the join identity simplification, the expression can be simplified to Bgp(?s ?p ?o).

In order to see some examples of the translation, we go through several of the example query patterns that we have encountered so far.

---

**Algorithm 1.** Translation of non-filter elements in group graph patterns

---

**Algorithm:** translateGroup($e_1, \ldots, e_n$)
**Input:** $e_1, \ldots, e_n$: the list of non-filter elements in a group pattern
**Output:** a SPARQL algebra expression A
 1: A := Z {the empty pattern}
 2: **for** $i = 1, \ldots, n$ **do**
 3:     **if** $e_i$ is of the form OPTIONAL pattern **then**
 4:         **if** algbr(pattern) is of the form Filter(F, A′) **then**
 5:             A := LeftJoin(A, A′, F)
 6:         **else**
 7:             A := LeftJoin(A, algbr(pattern), true)
 8:         **end if**
 9:     **else**
10:         A := Join(A, algbr($e_i$))
11:     **end if**
12: **end for**
13: **return**  A

---

**Example 1:**
> *Query pattern:* Join(Z, Bgp(?x foaf:name ?name. ?x foaf:mbox ?mbox)),
> *Simplification:* Bgp(?x foaf:name ?name. ?x foaf:mbox ?mbox).

**Example 2:**
> *Query pattern:* Join(Z, Graph(*iri*, Join(Z, Bgp(BGP$_1$)))),
> *Simplification:* Graph(*iri*, Bgp(BGP$_1$)).
> *We use* iri *and* BGP$_1$ *instead of the given graph IRI and BGP.*

**Example 3:**
> *Query pattern:* Join(Z, Bgp(?x foaf:name ?name. ?x foaf:nick ?nick)),
> *Simplification:* Bgp(?x foaf:name ?name. ?x foaf:nick ?nick).
> *Note that the example was used to illustrate the effects of entailment regimes, but this does not influence the conversion to algebra objects. The only effect is that the* Bgp($\cdot$) *algebra objects are evaluated differently.*

**Example 5:**
> *Query pattern:* Join(Join(Join(Z, Bgp(TP$_1$)), Z), Bgp(TP$_2$)),
> *Simplification:* Join(Bgp(?x foaf:name ?name), Bgp(?x foaf:mbox ?mbox)).
> *We abbreviate the two triples patterns with* TP$_1$ *and* TP$_2$, *respectively. Although the query from Example 5 yields the same results as the query from Example 1 on any data, its algebra version is different due to the use of groups. We first used Algorithm 1 on the three elements of the query pattern, which is a group graph pattern. Each element itself is then translated as a* GroupOrUnionGraphPattern *using the case for single elements. A query optimizer might further modify this algebra object so that it becomes the same as the simplified version of the algebra expression for Example 1.*

**Example 6:**
> *Query pattern:* Join(Z, Union(Bgp(*BGP$_1$*), Bgp(*BGP$_2$*))),
> *Simplification:* Union(Bgp(*BGP$_1$*), Bgp(*BGP$_2$*)).
> *We abbreviate the two BGPs from the union pattern with* BGP$_1$ *and* BGP$_2$, *respectively.*

**Example 7:**
> *Query pattern:* Join(Join(Z, Bgp(tp$_1$)), Union(Bgp(tp$_2$), Bgp(tp$_3$))),

*Simplification:* Join(Bgp(tp$_1$), Union(Bgp(tp$_2$), Bgp(tp$_3$))).
*We abbreviate the three triple patterns from the example with* tp$_1$ *to* tp$_3$, *respectively.*

**Example 8:**

*Query pattern:* LeftJoin(Join(Z, Bgp(tp$_1$)), Bgp(tp$_2$), true),
*Simplification:* LeftJoin(Bgp(tp$_1$), Bgp(tp$_2$), true).
*The triple patterns from the example are abbreviated with* tp$_1$ *and* tp$_2$, *respectively.*

**Example 9:**

*Query pattern:*
LeftJoin(Join(Join(Z, Bgp(tp$_1$)), Union(Bgp(tp$_2$), Bgp(tp$_3$))), Bgp(tp$_4$), true),
*Simplification:*
LeftJoin(Join(Bgp(tp$_1$), Union(Bgp(tp$_2$), Bgp(tp$_3$))), Bgp(tp$_4$), true).
*We again abbreviate the triple patterns from the example with* tp$_1$ *to* tp$_4$, *respectively.*

**Example 10:**

*Query pattern:* Filter(*isIRI*(?mbox), Join(Z, Bgp(bgp$_1$))),
*Simplification:* Filter(*isIRI*(?mbox), Bgp(bgp$_1$)).
*For the translation of the query pattern we use the first case of translating group graph patterns. We abbreviate the basic graph pattern from the example with* bgp$_1$.

We omit the examples for NOT EXISTS, EXISTS, and MINUS since these are SPARQL 1.1 features, for which we do not go into details in the algebra translation.

**Evaluating Algebra Expressions for Query Patterns.** In order to define the evaluation of an algebra object for a query pattern, we first define the most basic operation, i.e., the evaluation of a basic graph pattern.

**Definition 4.** *Evaluating a SPARQL graph pattern results in a* solution sequence *that lists possible bindings of query variables to RDF terms in the active graph. Such bindings are represented by partial functions $\mu$ from* V *to* T, *called* solution mappings. *For a solution mapping $\mu$ – and more generally for any (partial) function – the set of elements on which $\mu$ is defined is the* domain dom($\mu$) *of $\mu$, and the set* ran($\mu$) := $\{\mu(x) \mid x \in$ dom($\mu$)$\}$ *is the* range *of $\mu$. For a graph pattern* GP, *we use $\mu$(GP) to denote the pattern obtained by applying $\mu$ to all elements of* GP *in* dom($\mu$).

This convention is extended in the obvious way to filter expressions, and to all functions that are defined on variables or terms.

The order of solution sequences is relevant for later processing steps in SPARQL, but not for obtaining the solutions for a graph pattern. Thus, we obtain a *solution multiset* when evaluating a basic graph pattern, or, more generally, any SPARQL graph patterns.

**Definition 5.** *A multiset* over an *underlying set $S$ is a total function $M : S \to \mathbf{N}^+ \cup \{\omega\}$ where $\mathbf{N}^+$ are the positive natural numbers, and $\omega > n$ for all $n \in \mathbf{N}^+$. The value $M(s)$ is the* multiplicity *of $s \in S$, and $\omega$ denotes a countably infinite number of occurrences.*

Infinitely many occurrences of individual solution mappings are indeed possible when considering SPARQL entailment regimes, although a major concern when defining extensions to basic graph pattern matching is how to avoid sources of infinite solutions.

We often represent a multiset $M$ with underlying set $S$ by the set $\{(s, M(s)) \mid s \in S\}$. Accordingly, we may write $(s, n) \in M$ if $M(s) = n$. Also, we assume that $M(s)$ denotes

**Table 3.** Evaluation of algebraic operators for query patterns in SPARQL over a dataset D, where the multiplicity functions $M, M_1$, and $M_2$ are assumed to be those for the multisets $[\![GP]\!]_{D,G}$, $[\![GP_1]\!]_{D,G}$, and $[\![GP_2]\!]_{D,G}$, v is a variable, and iri is an IRI

$$[\![\mathsf{Union}(GP_1, GP_2)]\!]_{D,G} := \big\{(\mu, n) \mid n = M_1(\mu) + M_2(\mu) > 0\big\}$$

$$[\![\mathsf{Join}(GP_1, GP_2)]\!]_{D,G} := \big\{(\mu, n) \mid n = \textstyle\sum_{(\mu_1, \mu_2) \in J(\mu)} (M_1(\mu_1) * M_2(\mu_2)) > 0\big\} \text{ where}$$
$$J(\mu) := \{(\mu_1, \mu_2) \mid \mu_1, \mu_2 \text{ compatible and } \mu = \mu_1 \cup \mu_2\}$$

$$[\![\mathsf{Filter}(F, GP)]\!]_{D,G} := \big\{(\mu, n) \mid M(\mu) = n > 0 \text{ and } [\![\mu(F)]\!] = \mathsf{true}\big\}$$

$$[\![\mathsf{LeftJoin}(GP_1, GP_2, F)]\!]_{D,G} := [\![\mathsf{Filter}(F, \mathsf{Join}(GP_1, GP_2))]\!]_{D,G} \cup$$
$$\big\{(\mu_1, M_1(\mu_1)) \mid \text{for all } \mu_2 \text{ with } M_2(\mu_2) > 0 : \mu_1 \text{ and } \mu_2 \text{ are}$$
$$\text{incompatible or } [\![(\mu_1 \cup \mu_2)(F)]\!] = \mathsf{false}\big\}$$

$$[\![\mathsf{Graph}(\mathsf{iri}, GP)]\!]_{D,G} := \begin{cases} [\![GP]\!]_{D,G_{\mathsf{iri}}} & \text{if iri is an IRI with } (\mathsf{iri}, G_{\mathsf{iri}}) \in D \\ \{\} & \text{otherwise} \end{cases}$$

$$[\![\mathsf{Graph}(v, GP)]\!]_{D,G} := [\![\mathsf{Union}(\mathsf{Join}(\mathsf{Graph}(\mathsf{iri}_n, GP), \{\mu : v \mapsto \mathsf{iri}_n\}),$$
$$\mathsf{Union}(\mathsf{Join}(\mathsf{Graph}(\mathsf{iri}_{n-1}, GP), \{\mu : v \mapsto \mathsf{iri}_{n-1}\}),$$
$$\mathsf{Union}(\dots,$$
$$\mathsf{Join}(\mathsf{Graph}(\mathsf{iri}_1, GP), \{\mu : v \mapsto \mathsf{iri}_1\})))]\!]_{D,G}$$
$$\text{for } \mathsf{iri}_1, \dots, \mathsf{iri}_n \text{ the IRIs of the named graphs in D}$$

$0$ whenever $s \notin S$. In some cases, it is also convenient to use a set-like notation where repeated elements are allowed, e.g. writing $\{a, b, b\}$ for the multiset $M$ with underlying set $\{a, b\}$, $M(a) = 1$, and $M(b) = 2$.

To define the solution multiset for a BGP under the simple semantics, we still need to consider the effect of blank nodes. Intuitively, these act like variables that are projected out of a query result, and thus they may lead to duplicate solution mappings. This is accounted for using RDF instance mappings as follows:

**Definition 6.** *An* RDF instance mapping *is a partial function* $\sigma \colon B \to T$ *from blank nodes to RDF terms. We extend* $\sigma$ *to pattern graphs and filters as done for solution mappings above. The* solution multiset $[\![BGP]\!]_{D,G}$ *for a basic graph pattern* BGP *over the dataset* D *with active graph* G *is the following multiset of solution mappings:*

$\{(\mu, n) \mid \mathrm{dom}(\mu) = V(BGP), \text{ and } n \text{ is the maximal number such that}$
$\quad \sigma_1, \dots, \sigma_n \text{ are distinct RDF instance mappings such that, for all } 1 \le i \le n,$
$\quad \mathrm{dom}(\sigma_i) = B(BGP) \text{ and } \mu(\sigma_i(BGP)) \text{ is a subgraph of } G\}.$

Note that the number $n$ in the definition of $[\![BGP]\!]_{D,G}$ is always finite.

The algebraic operators that are required for evaluating non-basic graph patterns correspond to operations on multisets of solution mappings. This remains unchanged even if we use an entailment regime different from SPARQL's standard simple entailment. To take infinite multiplicities into account, which can occur in some entailment regimes, we assume $\omega + n = n + \omega = \omega$ for all $n \ge 0$, $\omega * n = n * \omega = \omega$ for all $n > 0$ and $\omega * 0 = 0 * \omega = 0$. We denote the truth value from evaluating a filter F by $[\![F]\!]$.

**Definition 7.** *Two solution mappings $\mu_1$ and $\mu_2$ are* compatible *if $\mu_1(v) = \mu_2(v)$ for all $v \in \mathsf{dom}(\mu_1) \cap \mathsf{dom}(\mu_2)$. If this is the case, a solution mapping $\mu_1 \cup \mu_2$ is defined by setting $(\mu_1 \cup \mu_2)(v) := \mu_1(v)$ if $v \in \mathsf{dom}(\mu_1)$, and $(\mu_1 \cup \mu_2)(v) := \mu_2(v)$ otherwise.*

*The* evaluation *of a graph pattern over* G, *denoted* $[\![ \cdot ]\!]_{D,G}$, *is defined as in Table 3, where the multiplicity functions $M$ / $M_1$ / $M_2$ are assumed to be those for the multisets* $[\![GP]\!]_{D,G} / [\![GP_1]\!]_{D,G} / [\![GP_2]\!]_{D,G}$.

Note that, for brevity, we join an algebra object with a multiset in the evaluation of Graph(v, GP) with v a variable, which is strictly speaking not possible since both of the joined elements should be algebra objects that are then evaluated in the join evaluation.

**Translating and Evaluating SPARQL Queries.** Apart from the query pattern itself, the solution modifiers of a query are also translated into corresponding algebra objects. The resulting algebra object together with a dataset for the query and a query form defines a SPARQL abstract query.

**Definition 8.** *Given a SPARQL query* Q *with query pattern* P, *we step by step construct an algebra expression* E *as follows:*

1. E := ToList(algbr(P)), *where* ToList *turns a multiset into a sequence with the same elements and cardinality. There is no implied ordering to the sequence; duplicates need not be adjacent.*
2. E := OrderBy(E, $(c_1, \ldots, c_n)$) *if the query string has an* ORDER BY *clause, where $c_1, \ldots, c_n$ the order comparators in* Q.
3. E := Project(E, vars) *if the query form is* SELECT, *where* vars *is the set of variables mentioned in the* SELECT *clause or all named variables in the query if* SELECT * *is used.*[1]
4. E := Distinct(E) *if the query form is* SELECT *and the query contains the* DISTINCT *keyword.*
5. E := Reduced(E) *if the query form is* SELECT *and the query contains the* REDUCED *keyword.*
6. E := Slice(E, start, length) *if the query contains* OFFSET start *or* LIMIT length, *where* start *defaults to 0 and* length *defaults to* $(\mathrm{size}(E) - \mathrm{start})$ *with* $\mathrm{size}(E)$ *denoting the cardinality of $E$.*
7. E := Construct(E, templ) *if the query form is* CONSTRUCT *and* templ *is the template of the query.*
8. E := Describe(E, VarsRes) *if the query form is* DESCRIBE, *where* VarsRes *is the set of variables and resources mentioned in the* DESCRIBE *clause or all named variables in the query if* DESCRIBE * *is used.*

*The algebra expression for* Q, *denoted* algbr(Q), *is* E.

*We define the* RDF dataset for Q *as follows: if* Q *contains a* FROM *or* FROM NAMED *clause, then the* RDF dataset D for Q *is* $\{G, (iri_1, G_1), \ldots, (iri_n, G_n)\}$ *where the default*

---

[1] Note that for SPARQL 1.1, * only refers to variables that are in scope, e.g., if the query contains a sub-query, then only variables that are selected in the sub-query are in scope for the enclosing query.

**Table 4.** Evaluation of algebraic operators for queries over a dataset D with default graph G

$$[\![\mathsf{ToList}(\mathsf{E})]\!]_{\mathsf{D},\mathsf{G}} := (\mu_1, \ldots, \mu_n) \text{ for } \{\mu_1, \ldots, \mu_n\} = [\![\mathsf{E}]\!]_{\mathsf{D},\mathsf{G}}$$

$$[\![\mathsf{OrderBy}(\mathsf{E}, (c_1, \ldots, c_m))]\!]_{\mathsf{D},\mathsf{G}} := (\mu_1, \ldots, \mu_n) \text{ such that } (\mu_1, \ldots, \mu_n) \text{ satisfies } (c_1, \ldots, c_m)$$
$$\text{and } \{\mu_1, \ldots, \mu_n\} = \{\mu \mid \mu \in [\![\mathsf{E}]\!]_{\mathsf{D},\mathsf{G}}\}$$

$$[\![\mathsf{Project}(\mathsf{E}, \mathsf{vars})]\!]_{\mathsf{D},\mathsf{G}} := (\mu_1', \ldots, \mu_n') \text{ with } (\mu_1, \ldots, \mu_n) = [\![\mathsf{E}]\!]_{\mathsf{D},\mathsf{G}}, \mathsf{dom}(\mu_i') = \mathsf{vars} \subseteq \mathsf{dom}(\mu_i),$$
$$\text{and } \mu_i' \text{ is compatible with } \mu_i \text{ for } 1 \le i \le n$$

$$[\![\mathsf{Distinct}(\mathsf{E})]\!]_{\mathsf{D},\mathsf{G}} := (\mu_1, \ldots, \mu_m) \text{ with } \{\mu_1, \ldots, \mu_n\} = \{\mu \mid \mu \in \{\mu \mid \mu \in [\![\mathsf{E}]\!]_{\mathsf{D},\mathsf{G}}\}$$
$$\text{and } (\mu_1, \ldots, \mu_m) \text{ preserves the order of } [\![\mathsf{E}]\!]_{\mathsf{D},\mathsf{G}}$$

$$[\![\mathsf{Reduced}(\mathsf{E})]\!]_{\mathsf{D},\mathsf{G}} := (\mu_1, \ldots, \mu_m) \text{ with } \{\mu_1, \ldots, \mu_n\} \subseteq \{\mu \mid \mu \in [\![\mathsf{E}]\!]_{\mathsf{D},\mathsf{G}}\},$$
$$\{\mu_1, \ldots, \mu_m\} = \{\mu \mid \mu \in [\![\mathsf{E}]\!]_{\mathsf{D},\mathsf{G}}\}$$
$$\text{and } (\mu_1, \ldots, \mu_m) \text{ preserves the order of } [\![\mathsf{E}]\!]_{\mathsf{D},\mathsf{G}}$$

$$[\![\mathsf{Slice}(\mathsf{E}, \mathsf{start}, \mathsf{length})]\!]_{\mathsf{D},\mathsf{G}} := (\mu_{start}, \ldots, \mu_{start+length}) \text{ for } (\mu_1, \ldots, \mu_m) = [\![\mathsf{E}]\!]_{\mathsf{D},\mathsf{G}}$$

$$[\![\mathsf{Construct}(\mathsf{E}, \mathsf{templ})]\!]_{\mathsf{D},\mathsf{G}} := \{\mu_i(\mathsf{templ}_i) \mid \{\mu_1, \ldots, \mu_n\} = [\![\mathsf{E}]\!]_{\mathsf{D},\mathsf{G}}, 1 \le i \le n, \mathsf{templ}_i \text{ is graph}$$
$$\text{equivalent to } \mathsf{templ}, \mu_i(\mathsf{templ}_i) \text{ is valid RDF, and}$$
$$B(\mathsf{templ}_i) \cap \bigcup_{1 \le j \le n, j \ne i}(B(\mathsf{templ}_j) \cup \mathsf{ran}(mu_i)) = \emptyset\}$$

$$[\![\mathsf{Describe}(\mathsf{E}, \mathsf{VarsRes})]\!]_{\mathsf{D},\mathsf{G}} := \{\mathsf{desc}(\mu_i(\mathsf{VarsRes})) \mid \{\mu_1, \ldots, \mu_n\} = [\![\mathsf{E}]\!]_{\mathsf{D},\mathsf{G}}, 1 \le i \le n$$
$$\text{where } \mathsf{descr} \text{ generates a system-dependent description}$$
$$\text{for the given resources}$$

*graph* G *is the RDF merge of the graphs referred to in the* **FROM** *clauses and each pair* $(iri_i, G_i)$ *results from a* **FROM NAMED** *clause with IRI* $iri_i$ *where* $iri_i$ *identifies a resource that serializes the graph* $G_i$; *otherwise the dataset for* Q *is the dataset used by the queried service.*

*The SPARQL abstract query for* Q *is a tuple* $(\mathsf{algbr}(Q), \mathsf{D}, \mathsf{F})$ *where* D *is the RDF dataset for* Q, *and* F *is the query form for* Q.

*We extend the evaluation of algebra expressions as defined in Table 4. To evaluate* $(\mathsf{E}, \mathsf{D}, \mathsf{F})$, *one first computes* $[\![ E ]\!]_{\mathsf{D},\mathsf{G}}$. *If the query form is* **SELECT**, **CONSTRUCT**, *or* **DESCRIBE**, *the query answer is* $[\![ E ]\!]_{\mathsf{D},\mathsf{G}}$, *which is a solution sequence for* **SELECT** *queries and a set of RDF triples otherwise. If the query form is* **ASK** *the query answer is* $\sharp([\![ E ]\!]_{\mathsf{D},\mathsf{G}}) > 0$.

Note that in case of **DESCRIBE** queries, the concrete result is not normatively defined and depends on the implementation.

**Example 17.** In order to see some examples of abstract queries, we go through several of the examples again, in particular those with solution modifiers or query forms other than **SELECT**.

**Example 1**

Abstract Query: (Project(ToList(algbr(P)), {?name, ?mbox}), D, SELECT)

with P the query pattern from Example 1 and D the dataset of the SPARQL query processor.

**Example 2**

Abstract Query:

(Project(ToList(algbr(P)), {?name, ?mbox}), {∅, (iri, $G_{iri}$)}, SELECT)
with P the query pattern from Example 2, iri the IRI in the FROM NAMED clause, and $G_{iri}$ the graph serialized by iri.

**Example 11**

Abstract Query: (Construct(ToList(algbr(P)), templ), D, CONSTRUCT)
with P the query pattern from Example :ex : construct, templ the template from the query, and D the dataset of the SPARQL query processor.

**Example 12**

Abstract Query: (ToList(algbr(P)), D, ASK)
with P the query pattern from Example :ex : ask and D the dataset of the SPARQL query processor.

**Example 13**

Abstract Query (a): (Describe(ToList(Z), {foaf:Person}), D, DESCRIBE)
Abstract Query (b): (Describe(ToList(algbr(P)), {?x}), D, DESCRIBE)
with P the query pattern from the second query in Example :ex : describe and D the dataset of the SPARQL query processor.

**Example 14**

Abstract Query:
(Project(OrderBy(ToList(algbr(P)), (ASC(?name))), {?name}), D, SELECT)
with P the query pattern from Example :ex : order and D the dataset of the SPARQL query processor.

## 2.5   SPARQL 1.1 Features

The W3C Data Access Working Group is currently in the process of specifying the next version of SPARQL, which is named SPARQL 1.1. The new version adds several features to the query language [13], which we briefly introduce in this section. We do not provide an algebra translation for these features and only give examples of how these features can be used. We further give a brief overview of the new parts in SPARQL 1.1 apart from the query language and the entailment regimes.

**Aggregates.** Aggregates allow for counting the number of answers, computing average, minimal, or maximal values from solutions by applying expressions ver groups of solutions.

**Example 18.** One intuitive way of aggregates is counting, which we illustrate with the data from Table 1 and the query:

```
SELECT (COUNT(?x) AS ?num)
WHERE { ?x foaf:name ?name }
```

We obtain one solution with binding 3 for ?num.

**Example 19.** In order to select variables to which no aggregate is applied, one has to group the solutions accordingly. We illustrate this with the data from Table 5 and the query:

**Table 5.** Triples used in the examples for the new features of SPARQL 1.1

```
:auth1 :writes :book1 .
:auth1 :writes :book2 .
:auth2 :writes :book3 .
:auth3 :writes :book4 .
:book1 :costs  9 .
:book2 :costs  5 .
:book3 :costs  11 .
:book4 :costs  2 .
```

```
SELECT      ?auth (AVG(?price)AS?avgPrice)
WHERE       { ?auth :writes ?book . ?book:costs ?price }
GROUP BY ?auth
```

Without grouping by author, we were not able to have ?auth in the SELECT clause. We obtain:

| ?auth | ?avgPrice |
|-------|-----------|
| :auth1 | 7 |
| :auth2 | 11 |
| :auth2 | 2 |

We can further extend the query with a HAVING clause to filter some of the aggregated values. For example, adding

$$\text{HAVING (AVG(?price)} > 5)$$

results in the last solution being filtered out.

**Subqueries.** Subqueries provide a way to embed SPARQL queries within other queries, normally to achieve results which cannot otherwise be achieved

**Example 20.** We use again the triples from Table 5. We use a subquery to answer the query "Which author has a book that costs more than the most expensive book of :auth1?". For such query, two queries have to be executed: the first query finds the most expensive book of :auth1 and the second finds those authors who have a book more expensive than that. Using subqueries we can directly embed the first query into the second:

```
SELECT ?auth
WHERE { ?auth :writes ?book . ?book :costs ?price
            {
                SELECT (MAX(?price) AS ?max)
                WHERE { :auth1 :writes ?book . ?book :costs ?price }
            }
            FILTER (?price > ?max)
        }
```

Evaluating the inner query yields 9 as binding for ?max. Note that only ?max is projected and, therefore, available to the outer query. The variable ?price from the inner query is not visible for the outer query (it is *out of scope*) and the variable ?price from the outer query unrelated to it. The filter applies, as usual, to all elements in the group, which makes sure that the two triple patterns and the subquery are evaluated and the results are joined before the filter is applied.

**Negation** comes in two styles, one is based on filtering out results that do not match a given graph pattern using filers with the NOT EXISTS keyword, and the other way is to directly remove solutions related to another pattern with MINUS.

Filtering of query solutions is done within a FILTER expression using NOT EXIST and EXISTS.

**Example 21.** We illustrate the use of filtering combined with NOT EXISTS (EXISTS) using the data from Table 1 and the query:

```
SELECT ?name
WHERE { ?x foaf:name ?name
              FILTER NOT EXISTS { ?x foaf:mbox ?mbox }
       }
```

Since only for ?x bound to "Pascal Hitzler" the NOT EXISTS filter evaluates to true, we get just one solution:

| ?name |
| --- |
| "Pascal Hitzler" |

We can similarly test for the existence of a match for the pattern by using FILTER EXISTS instead of FILTER NOT EXISTS, which would yield one solution with ?name once bound to "Birte Glimm" and once to "Sebastian Rudolph".

A different form of negation is supported via the MINUS keyword, which takes the form pattern MINUS pattern.

**Example 22.** We illustrate the use MINUS with the following query over the data from Table 1:

```
SELECT ?name ?mbox
WHERE { ?x foaf:name ?name . ?x foaf:mbox ?mbox
              MINUS { ?x foaf:name "Birte Glimm" }
       }
```

In this case, the left-hand side consists of the two triple patterns, which yield the two solutions:

| ?x | ?name | ?mbox |
| --- | --- | --- |
| _:a | "Birte Glimm" | "b.glimm@googlemail.com" |
| _:b | "Sebastian Rudolph" | <mailto:rudolph@kit.edu> |

The right-hand side of the MINUS operator yields one solution in which ?name and ?mbox are unbound (since they do not occur in the pattern and are, therefore, not matched):

| ?x | ?name | ?mbox |
| --- | --- | --- |
| _:a | | |

In order to compute the query result, we keep only solutions for the left-hand side pattern if they are not "compatible" with the solutions for the right-hand side. Two mappings are compatible if whenever they both map a variable, then they map it to the same value. We will refer to the two solutions for the left-hand side pattern as $l_1$ and $l_2$, respectively, and to the solution for the right-hand side pattern as $r_1$. In our case, we have that $l_1$ is compatible with $r_1$ since the mappings for ?x is the same and since ?name and ?mbox are unbound in $r_1$, which does not contradict the mapping for ?name and ?mbox in $l_1$. Since $l_1$ and $r_1$ are compatible, $l_1$ is removed from the solutions. For $l_2$ and $r_1$, it is, however, clear that the mappings are not compatible since $l_2$ maps ?x to _:b whereas $r_1$ maps ?x to _:a. This means that $l_2$ remains in the solutions for the whole pattern, which gives the following overall result:

| ?name | ?mbox |
|---|---|
| "Sebastian Rudolph" | <mailto:rudolph@kit.edu> |

**SELECT Expressions** can be used in the SELECT clause to combine variable bindings already in the query solution, or defined earlier in the SELECT clause to produce a binding in the query solution, e.g., SELECT ?net ((?net * 1.2) AS ?gross) will return bindings for ?net and 120% of that value as binding for ?gross.

**Property Paths** allow for specifying a possible route through a graph between two graph nodes through property path expressions that apply to the predicate of a triple. A trivial case is a property path of length exactly 1, which is a triple pattern. Property paths allow for more concise expression of some SPARQL basic graph patterns and also add the ability to match arbitrary length paths. For example, the BGP

?x foaf:name "Birte Glimm" . ?x foaf:knows+/foaf:name ?name

The query starts from an element which is associated with the name Birte Glimm and then follows a path of length one or more (+) along the property foaf:knows followed by (/) a path of length one along the foaf:name property. Thus, the query finds the names of all people that Birte knows directly or indirectly. The / operator can always be eliminated, e.g., by rewriting the second triple into ?x foaf:knows + ?y . ?y foaf:name ?name. The arbitrary length paths, however, cannot be eliminated in this way and a SPARQL query processor has to implement them natively in order to fully support property paths.

**Assignments** can be used in addition to SELECT expressions in order to add bindings. Whereas SELECT expressions are limited to the SELECT clause, on can use the BIND keyword followed by an expression to add bindings already in the WHERE clause. The BINDINGS keyword can further be used to provide a solution sequence that is to be joined with the query results. It can used by an application to provide specific requirements on query results and also by SPARQL query engine implementations that provide federated query through the SERVICE keyword to send a more constrained query to a remote query service.

**CONSTRUCT Short Forms** allow for abbreviating CONSTRUCT queries provided the template and the pattern are the same and the pattern is just a basic graph pattern (i.e., no FILTERs and no complex graph patterns are allowed in the short form). The keyword WHERE is required in the short form.

Furthermore, SPARQL 1.1 provides an expanded set of functions and operators.

**SPARQL 1.1 Update** [28] provides a way of modifying a graph store by inserting or deleting triples from graphs. It provides the following facilities:

– Insert new triples into an RDF graph.
– Delete triples from an RDF graph.
– Perform a group of update operations as a single action.
– Create a new RDF graph in a graph store.
– Delete an RDF graph from a graph store.

The behavior of update queries in a system that uses entailment regimes is left open in SPARQL 1.1. A straightforward way of implementing updates under entailment regimes would be to interpret such queries under simple entailment semantics.

The SPARQL protocol has also been extended to allow for an exchange of update requests between a client and a SPARQL endpoint [22].

**Service Descriptions** [30] have been added SPARQL 1.1 as a method for discovering and vocabulary for describing SPARQL services made available via the SPARQL Protocol for RDF [11]. Such a description is intended to provide a mechanism by which a client or end user can discover information about the SPARQL implementation/service such as supported extension functions and details about the available dataset or the used entailment regime.

**Federation Extensions** are currently under development as part of SPARQL 1.1 to express queries across distributed data sources. At the time of writing a first public working draft is available at http://www.w3.org/TR/sparql11-federated-query/.

**JSON Result Format** is so far a working group note available at http://www.w3.org/TR/rdf-sparql-json-res/. The working group intends to bring this to recommendation status, but at the time of writing no public working draft is available.

## 3  SPARQL Entailment Regimes

In the previous section, we have defined the syntax of SPARQL queries and how such queries are evaluated with subgraph matching as means of evaluating basic graph patterns. This form of basic graph pattern evaluation is also called simple entailment since it can equally be defined in terms of the simple entailment relation between RDF graphs. In order to use more elaborate entailment relations, which also allow for retrieving solutions that implicitly follow from the queried graph, we now look at so-called *entailment regimes*. An entailment regime specifies how an entailment relation such as RDF

Schema entailment can be used to redefine the evaluation of basic graph patterns from a SPARQL query making use of SPARQL's extension point for basic graph pattern matching. In order to satisfy the conditions that SPARQL places on extensions to basic graph pattern matching, an entailment regimes specifies conditions that limit the number of entailments that contribute solutions for a basic graph pattern. For example, only a finite number of the infinitely many axiomatic triples can contribute solutions under RDF Schema (RDFS) entailment. In this section, we introduce the RDFS entailment regime and explain the design rationale behind the regime. In Section 4, we then show how the OWL 2 Direct Semantics entailment relation can be used.

Each entailment regime is characterized by a set of properties:

**Name:**      A name for the entailment regime, usually the same as the entailment relation used to define the evaluation of a basic graph pattern.

**IRI:**      The IRI for the regime. This IRI can be used in the service description for a SPARQL endpoint, which is an RDF graph that describes the functionality and the features that it provides.

**Legal Graphs:**      Describes which graphs are legal for the regime.

**Legal Queries:**      Describes which queries are legal for the regime.

**Illegal Handling:** Describes what happens in case of an illegal graph or query.

**Entailment:**      Specifies which entailment relation is used in the evaluation of basic graph patterns.

**Inconsistency:**      Defines what happens if the queried graph is inconsistent under the used semantics.

**Query Answers:** Defines how a basic graph pattern is evaluated, i.e., what the solutions are for a given graph and basic graph pattern of a query.

Before we start describing a concrete entailment regime, we first analyze what conditions an entailment regime has to satisfy. These conditions also motivate the choice of the above properties that are defined for each entailment regime.

### 3.1 Conditions on Extensions of Basic Graph Pattern Matching

In order to extend SPARQL for an entailment relation $E$ such as RDFS or OWL Direct Semantics entailment, it suffices to modify the evaluation of BGPs accordingly, while the remaining algebra operations can still be evaluated as in Definition 7. When considering $E$-entailment, we thus define solution multisets $[\![BGP]\!]_{D,G}^E$.

The SPARQL Query 1.0 specification [26] already envisages the extension of the BGP matching mechanism, and provides a set of conditions for such extensions that we recall in Table 6. These conditions can be hard to interpret since their terminology is not aligned well with the remaining specification. In the following, we discuss our reading of these conditions, leading to a revised clarified version presented in Table 7.[2]

Condition (1) forces an entailment regime to specify a scoping graph based on which query answers are computed instead of using the active graph directly. Since an entailment regime's definition of BGP matching is free to refer to such derived graph

---

[2] The SPARQL 1.1 Query working draft has been updated to contain the revised conditions.

**Table 6.** Conditions for extending BGP matching to E-entailment (quoted from [26])

1. The scoping graph SG, corresponding to any consistent active graph AG, is uniquely specified and is E-equivalent to AG.
2. For any basic graph pattern BGP and pattern solution mapping P, P(BGP) is well-formed for E.
3. For any scoping graph SG and answer set $\{P_1, \ldots, P_n\}$ for a basic graph pattern BGP, and where $BGP_1, \ldots, BGP_n$ is a set of basic graph patterns all equivalent to BGP, none of which share any blank nodes with any other or with SG

$$SG \models_E (SG \cup P_1(BGP_1) \cup \ldots \cup P_n(BGP_n)).$$

4. Each SPARQL extension must provide conditions on answer sets which guarantee that every BGP and AG has a finite set of answers which is unique up to RDF graph equivalence.

**Table 7.** Clarified conditions for extending BGP matching to E-entailment

An entailment regime E must provide conditions on basic graph pattern evaluation such that for any basic graph pattern BGP, any RDF graph G, and any evaluation $[\![\cdot]\!]_G^E$ that satisfies the conditions, the multiset of graphs $\{(\mu(BGP), n) \mid (\mu, n) \in [\![BGP]\!]_G^E\}$ is uniquely determined up to RDF graph equivalence. An entailment regime must further satisfy the following conditions:

1. For any consistent active graph AG, the entailment regime E uniquely specifies a *scoping graph* SG that is E-equivalent to AG.
2. A set of *well-formed* graphs for E is specified such that, for any basic graph pattern BGP, scoping graph SG, and solution mapping $\mu$ in the underlying set of $[\![BGP]\!]_{SG}^E$, the graph $\mu(BGP)$ is well-formed for E.
3. For any basic graph pattern BGP, and scoping graph SG, if $\{\mu_1, \ldots, \mu_n\} = [\![BGP]\!]_{SG}^E$ and $BGP_1, \ldots, BGP_n$ are basic graph patterns all equivalent to BGP but not sharing any blank nodes with each other or with SG, then

$$SG \models_E SG \cup \bigcup_{1 \leq i \leq n} \mu_n(BGP_n).$$

4. Entailment regimes *should* provide conditions to prevent trivial infinite solution multisets.

structures anyway, the additional use of a scoping graph does not increase the freedom of potential extensions. We assume, therefore, that the scoping graph is the active graph in the remainder. If the active graph is E-inconsistent, entailment regimes specify the intended behavior directly, e.g., by requiring that an error is reported.

Condition (2) refers to a "pattern solution mapping" though what is probably meant is a *pattern instance mapping* P, defined in [26] as the combination of an RDF instance mapping $\sigma$ and a solution mapping $\mu$ where $P(x) = \mu(\sigma(x))$. We assume, however, that (2) is actually meant to refer to all solution mappings in $[\![BGP]\!]_{D,G}^E$. Indeed, even for simple entailment where well-formedness only requires P(BGP) to be an RDF graph, condition (2) would be violated when using *all* pattern instance mappings. To see this, consider a basic graph pattern

$$\{ \_:a \ :b \ :c \ \}.$$

Clearly, there is a pattern instance mapping P with $P(\_:a) = $ "1"$^{\wedge\wedge}$xsd:int, but P(BGP) = {"1"$^{\wedge\wedge}$xsd:int :b :c} is not an RDF graph. Similar problems occur when using all solution mappings. Hence we assume (2) to refer to elements of the computed solution multiset $[\![BGP]\!]^E_{D,G}$. The notion of *well-formedness* in turn needs to be specified explicitly for entailment regimes.

Condition (3) uses the term "answer set" to refer to the results computed for a BGP. To match the rest of [26], this has to be interpreted as the solution multiset $[\![BGP]\!]^E_{D,G}$. This also means mappings $P_i$ are solution mappings (not pattern instance mappings as their name suggests). The purpose of (3), as noted in [26], is to ensure that if blank node names are returned as bindings for a variable, then the same blank node name occurs in different solutions only if it corresponds to the same blank node in the graph.

**Example 23.** To illustrate the problem, consider the following graphs:

> $G$ : :a :b _:c.        $G_1$ : :a :b _:b$_1$.        $G_2$ : :a :b _:b$_2$.        $G_3$ : :a :b _:b$_1$.
> _:d :e :f.        _:b$_2$ :e :f.        _:b$_1$ :e :f.        _:b$_1$ :e :f.

Clearly, G simply entails $G_1$ and $G_2$, but not $G_3$ where the two blank nodes are identified. Now consider a basic graph pattern BGP

$$\{ \text{:a :b ?x. ?y :e :f} \}.$$

A solution multiset for BGP could comprise two mappings

$$\mu_1: \text{?x} \mapsto \_\text{:b}_1, \text{?y} \mapsto \_\text{:b}_2 \text{ and}$$
$$\mu_2: \text{?x} \mapsto \_\text{:b}_2, \text{?y} \mapsto \_\text{:b}_1.$$

We then have $\mu_1(BGP) = G_1$ and $\mu_2(BGP) = G_2$, and both solutions are entailed. Condition (3) requires, however, that $G \cup \mu_1(BGP) \cup \mu_2(BGP)$ is also entailed by G, and this is not the case in our example since this union contains $G_3$.

The reason is that our solutions have unintended co-references of blank nodes that (3) does not allow. SPARQL's basic subgraph matching semantics respects this condition by requiring solution mappings to refer to blank nodes that actually occur in the active graph, so blank nodes are treated like (Skolem) constants.[3] The revised condition in Table 7 has further been modified to not implicitly require finite solution multisets which may not be appropriate for all entailment regimes. In addition, we use RDF instance mappings for renaming blank nodes instead of requiring renamed variants of the BGP.

Finally, condition (4) requires that solution multisets are finite and uniquely determined up to RDF graph equivalence, again using the "answer set" terminology. Our revised condition clarifies what it means for a solution multiset to be "unique up to RDF graph equivalence." We move the uniqueness requirement above all other conditions, since (2) and (3) do not make sense if the solution multiset was not defined in this sense. The rest of the condition was relaxed since entailment regimes may inherently require infinite solution multisets, e.g., in the case of the Rule Interchange Format RIF [17]. It is desirable that this only happens if there are infinite solutions that are "interesting," so the condition has been weakened to merely recommend the elimination of infinitely many "trivial" solution mappings in solution multisets. The requirement thus

---

[3] Yet, SPARQL allows blank nodes to be renamed when loading documents, so there is no guarantee that blank node IDs used in input documents are preserved.

is expressed in an informal way, leaving the details to the entailment regime. Within this paper, we will make sure that the solution multisets are in fact finite (both regarding the size of the underlying set, and regarding the multiplicity of individual elements).

## 3.2   Addressing the Extension Point Conditions

Before coming to OWL, we introduce the RDFS entailment regime since RDFS is well-known and simpler than OWL while the regime still illustrates the main points in which an entailment regime differs from SPARQL's standard query evaluation. The major problem for RDFS entailment is to avoid trivially infinite solution multisets as suggested by Table 7 (4), where three principal sources of infinite query results have to be addressed:

1. An RDF graph can be inconsistent under the RDFS semantics in which case it RDFS-entails all (infinitely many) conceivable triples.
2. The RDFS semantics requires all models to satisfy an infinite number of *axiomatic triples* even when considering an empty graph.
3. Every non-empty graph entails infinitely many triples obtained by using arbitrary blank nodes in triples.

We now discuss each of these problems, and derive a concrete definition for BGP matching in the proposed entailment regime at the end of this section.

**Treatment of Inconsistencies.** SPARQL does not require entailment regimes to yield a particular query result in cases where the active graph is inconsistent. As stated in [26], "[the] effect of a query on an inconsistent graph [...] must be specified by the particular SPARQL extension." One could simply require that implementations of the RDFS entailment report an error when given an inconsistent active graph. However, a closer look reveals that inconsistencies are extremely rare in RDFS, so that the requirement of checking consistency before answering queries would impose an unnecessary burden on implementations.

Indeed, graphs can only be RDFS-inconsistent due to improper use of the datatype rdf:XMLLiteral.

**Example 24.** A typical example for this is the following graph:

:a :b "<"^^rdf:XMLLiteral.       :b rdfs:range rdfs:Literal.

The literal in the first triple is *ill-typed* as it does not denote a value of rdf:XMLLiteral. This does not cause an inconsistency yet but forces "<"^^rdf:XMLLiteral to be interpreted as a resource that is not in the extension of rdfs:Literal, which in turn cannot be the case in any model that satisfies the second triple.

Ill-typed literals are the only possible cause of inconsistency in RDFS and as such not a frequent problem.[4] Moreover, inconsistencies of this type are inherently "local" as

---

[4] Implementations may support additional datatypes that can lead to similar problems. Such extensions go beyond the RDFS semantics we consider here, yet inconsistencies remain rare even in these cases.

they are based on individual ill-typed literals that could easily be ignored if not related to a given query.

It has thus been decided in the SPARQL working group that systems only have to report an error if they actually detect an inconsistency. Until this happens, queries can be answered as if all literals were well-typed. Our exact formalization corresponds to a behavior where tools simply assume that all strings are well-typed for rdf:XMLLiteral, and hence does not put additional burden on implementers.

**Treatment of Axiomatic Triples.** Every RDFS model is required to satisfy an infinite number of *axiomatic triples*. The reason is that the RDF vocabulary for encoding lists includes property names rdf:_i for all $i \geq 1$, with several (RDFS) axiomatic triples for each rdf:_i. For instance, we find a triple rdf:_i rdf:type rdf:Property for all $i \in \mathbf{N}$. Thus, the query ?x rdf:type rdf:Property could have infinitely many results. We consider such results trivial in the sense of Table 7 (4), and thus we want avoid them in the RDFS entailment regime.

We therefore propose that axiomatic triples with a subject of the form rdf:_i are only taken into account if the subject's IRI explicitly occurs in the active graph. This ensures that only finitely many axiomatic triples are considered, since there is only a finite number of axiomatic triples whose subjects do not have the form rdf:_i. To conveniently formalize this, Definition 10 below still refers to the standard RDFS entailment, but restricts the range of solution mappings to a *finite* vocabulary, which consists of terms from the queried graph and from terms of the RDFS vocabulary apart from those of the form rdf:_i.

**Treatment of Blank Nodes.** Even if condition (3) in Table 7 holds, solution multisets could include infinitely many results that only differ in the identifiers for blank nodes. Simple entailment avoids this problem by restricting results to blank nodes that occur in the active graph. For entailment regimes, however, one must take entailed triples into account. This already leads to triples with different blank nodes, as illustrated in the graphs $G_1$ and $G_2$ in Example 23.

Restricting the range of solution mappings to blank nodes in the active graph would ensure finiteness but is not a satisfactory solution.

**Example 25.** To see why restricting the range of solution mappings to blank nodes in the active graph is not a satisfactory, consider the graph

$$G : \text{:a :b :c.} \quad \text{:d :e \_:f.}$$

The query pattern BGP = { :a :b ?x } yields only one solution mapping $\mu$ : ?x $\mapsto$ :c under simple entailment. Yet, the mapping $\mu'$ : ?x $\mapsto$ \_:f uses only blank nodes from G, and satisfies $G \models \mu'(\text{BGP})$ even under simple semantics.

This shows that the latter two conditions are not sufficiently specific for handling blank nodes in entailment regimes. A more adequate approach is the use of *Skolemization*:

**Definition 9.** *Let the prefix* skol *refer to a namespace IRI that does not occur as the prefix of any IRI in the active graph or query. The* Skolemization sk(\_:b) *of a blank* *node* _:b *is defined as* sk(_:b) := skol:b. *We extend* sk(·) *to graphs and filters just like other (partial) functions on RDF terms.*

Intuitively, Skolemization changes blank nodes into resource identifiers that are not affected by entailment. Clearly, we do not want Skolemized blank nodes to occur in query results, but it is useful to restrict to solution mappings $\mu$ for which sk(G) $\models$ sk($\mu$(BGP)). In Example 25 above, this condition is indeed satisfied by $\mu$ but not by $\mu'$.

In order to illustrate the effect, we use an RDF graph that does not make use of any special RDFS terms, i.e., simple entailment would result in the same solutions. Let G, sk(G), and BGP be as follows:

$$G : \text{:a :b :c.} \qquad sk(G) : \text{:a :b :c.} \qquad BGP : \text{?x :b \_:d}$$
$$\text{\_:a :b \_:c.} \qquad \text{skol:a :b skol:c.}$$

Here the Skolem function sk maps _:a to skol:a and _:c to skol:c for skol defined as some imaginary prefix not used anywhere in G or BGP. We can now return only those solutions $\mu$ for which applying the Skolem function to blank nodes in the range of $\mu$ and some RDF instance mapping $\sigma$ yields ground triples that are entailed by sk(G). For example, all the mappings below yield entailed triples, but only the first two satisfy the stated requirement because applying sk to _:a and _:c yields a ground triple that is entailed by sk(G):

$$\mu_1 : \text{?x} \mapsto \text{:a} \qquad \sigma_1 : \text{\_:d} \mapsto \text{:c}$$
$$\mu_2 : \text{?x} \mapsto \text{\_:a} \qquad \sigma_2 : \text{\_:d} \mapsto \text{\_:c}$$
$$\mu_3 : \text{?x} \mapsto \text{\_:b}_1 \qquad \sigma_3 : \text{\_:d} \mapsto \text{\_:b}_2$$

### 3.3   The RDFS Entailment Regime

The set of *well-formed* graphs for the RDFS entailment regime is simply the set of all RDF graphs. BGP matching for RDFS is defined as follows.

**Definition 10.** *Let* G *be an RDF graph,* BGP *a basic graph pattern,* V(BGP) *the set of variables in* BGP, B(BGP) *the set of blank nodes in* BGP, sk *a Skolemization function as in Definition 9 such that* ran(sk) $\cap$ (Voc(G) $\cup$ Voc(BGP)) = $\emptyset$. *Let* Voc(RDFS) *be the RDFS vocabulary and* Voc⁻(RDFS) = Voc(RDFS) \ {rdf:_i | i $\in$ **N**}.

*We write* $\models_{RDFS}$ *for the RDFS entailment relation and define the* evaluation of BGP over G under RDFS entailment, $[\![BGP]\!]_{D,G}^{RDFS}$, *as the solution multiset*

$\{(\mu, n) \mid \text{dom}(\mu) = V(BGP), \text{ and } n \text{ is the maximal number of distinct RDF instance}$
$\qquad mappings \ such \ that, for \ each \ 1 \leq i \leq n,$
$\qquad (i) \ \text{dom}(\text{sigma}_i) = B(BGP),$
$\qquad (ii) \ \mu(\sigma_i(BGP)) \ are \ well-formed \ RDF \ triples,$
$\qquad (iii) \ \text{sk}(\mu(\sigma_i(BGP))) \ are \ ground \ RDF \ triples,$
$\qquad (iv) \ \text{sk}(G) \models_{RDFS} \text{sk}(\mu(\sigma_i(BGP))), \ and$
$\qquad (v) \ \text{ran}(\mu) \subseteq \text{Voc}(G) \cup \text{Voc}^-(RDFS)\}.$

Other types of graph patterns are evaluated as in Definition 7. If the active graph is RDFS-inconsistent, implementations may compute solution multisets based on the assumption that all literals of type rdf:XMLLiteral are well-typed, so that no inconsistency occurs. When the inconsistency is detected, implementations should report an error. We summarize the RDFS entailment regime in Table 8.

Condition (*i*) ensures that only RDF instance mappings that map all and only the blank nodes of BGP can increase the multiplicity of a solution mapping. Condition (*ii*) ensures that the instantiated triples are well-formed, e.g., variables tat occur in the subject position cannot be mapped to a literal by a solution mapping. Similarly, variables in the predicate position cannot be mapped to blank nodes. Condition (*iii*) then ensures that all blank nodes are indeed Skolemized by sk, resulting in ground RDF triples. Condition (*iv*) and (*v*) ensure that blank nodes and the axiomatic triples are handled as described in the previous section, therefore, avoiding infinitely many answers.

The definition might look quite complicated, but has the advantage that we can simply swap in another entailment relation and vocabulary to get another entailment regime. For example, when we use the simple entailment relation in place of the RDFS entailment relation and the empty set instead of Voc(RDFS) (as there are no special terms for simple interpretations), then we get exactly the behavior of subgraph matching (aka simple entailment) described in Definition 6. Furthermore, we can also swap the RDFS entailment relation for RDF or the OWL RDF-Based Semantics entailment relation and get a valid entailment regime. The OWL Direct Semantics needs some minor tweaks as the Direct Semantics is not defined in terms of triples, but based on Description Logics.

**Example 26.** In order to see why the range of a solution mapping can also use terms from Voc(RDFS), we consider the data from Table 1 and the query:

> SELECT ?name
> WHERE { ?x foaf:name ?name . ?x rdf:type foaf:Person }

Under RDFS entailment, the queried graph entails

> _:a   foaf:name   "Birte Glimm"
> _:a   rdf:type    foaf:Person

Thus, $\mu_1$: ?name $\mapsto$ "Birte Glimm" is a solution. Note, however, that rdf:type is not part of the vocabulary of the graph, and the solution is only part of the result since we include the RDFS vocabulary. Overall, we get the following three solutions:

| ?name |
| --- |
| "Birte Glimm" |
| "Sebastian Rudolph" |
| "Pascal Hitzler" |

Furthermore, in order to implement the regime, we can simply materialize all RDFS inferences and use subgraph matching on the extended graph. We illustrate this with the next example.

**Example 27.** In order to get an idea of how we can implement the RDFS entailment regime via materialization, we consider again the data from Table 1 and the query from the previous example.

**Table 8.** The RDFS entailment regime

| Name | RDFS |
|---|---|
| **IRI** | http://www.w3.org/ns/entailment/RDFS |
| **Legal Graphs** | Any legal RDF graph |
| **Legal Queries** | Any legal SPARQL query |
| **Illegal Handling** | In case the query is illegal (syntax errors), the system must raise a MalformedQuery fault. In case the queried graph is illegal (syntax errors), the system must raise a QueryRequestRefused fault. |
| **Entailment** | RDFS Entailment |
| **Inconsistency** | The scoping graph is graph-equivalent to the active graph even if the active graph is RDFS-inconsistent. If the active graph is RDFS-inconsistent, an implementation may raise a QueryRequestRefused fault or issue a warning and it should generate such a fault or warning if, in the course of processing, it determines that the data or query is not compatible with the request. In the presence of an inconsistency the conditions on solutions still guarantee that answers are finite. |
| **Query Answers** | Basic Graph Patterns are evaluated as in Definition 10 |

In order to materialize all RDFS inferences, we add triples that are RDFS entailed and obtain a graph G′, which contains (among other triples):

    _:a   rdf:type   foaf:Person
    _:b   rdf:type   foaf:Person
    _:c   rdf:type   foaf:Person

due to the triple foaf:name rdfs:domain foaf:Person combined with the three triples with the predicate foaf:name. Furthermore, we would add

    _:a   foaf:nick   foaf:b.glimm
    _:c   foaf:nick   foaf:phi

due to the fact that foaf:icqChatID is a subproperty of foaf:nick. Furthermore, the full materialization would also contain triples such as t rdf:type rdfs:Resource, for each term t in subject or object position plus other triples (cf. [14], [15]).

For evaluation the query, we do not have to make the Skolemization explicit, instead, we can just consider the blank nodes in G′ as constants. However, if a blank node occurs in the query that occurs also in the graph, we have to keep in mind that the blank node from the query cannot only map to that very blank node in the graph, but it still acts like a variable. Thus, if _:x in our query were _:a, it could still match to _:b in G′. Hence, we get the same three solution by performing subgraph matching on G′ as in the previous example.

Since computing the required partial RDFS closure (partial, since we do not require all axiomatic triples) can be done in polynomial time [15] and BGP evaluation then amounts to subgraph matching over the partial closure, it follows that the complexity of the evaluation problem under the RDFS regime is the same as for standard SPARQL. For set semantics instead of multiset semantics this is known to be PSPACE-complete [24].

# 4    The OWL Entailment Regimes

In contrast to the RDFS semantics, a graph does no longer admit a unique canonical model that can be used to compute answers under the RDF-Based Semantics (RBS) and Direct Semantics (DS) of OWL, i.e., we can no longer imagine queries to act on a unique "completed" version of the active graph. This affects reasoning algorithms, but has only little effect on our definitions. The main new challenges for OWL are its expressive datatype constructs that may lead to infinite answers, and the fact that the OWL DS is defined in terms of OWL objects to which a given RDF graph and query must first be translated. The problems discussed for RDF(S) also require slightly different solutions for OWL:

1. Inconsistent input ontologies are required to be rejected with an error.
2. The axiomatic triples of RDFS are used only by the RBS and can again be handled by suitably restricting solutions to terms from a finite vocabulary.
3. The problem of blank nodes occurs for both semantics and can again be addressed by Skolemization, but for DS the blank nodes that are used to encode OWL objects must not be Skolemized.

The main difference to RDFS is the stricter first item which no longer permits deferred inconsistency detection. Inconsistencies in RDFS were easy to ignore since they always related to single literals. Neither OWL semantics suggests such simple reasoning under inconsistencies. Although proposals exists for addressing this, they disagree on the inferred entailments and tend to require complex computations. On the other hand, typical OWL reasoning algorithms are model building procedures which detect inconsistencies as part of their normal operation. Hence, reporting errors in this case can usually be done without additional effort.

## 4.1    Mapping from RDF Graphs to OWL Structural Objects

For the OWL 2 Direct Semantics entailment regime, semantic conditions are defined with respect to ontology structures (i.e., instances of the Ontology class as defined in the OWL 2 structural specification [21]). Given an RDF graph G, the ontology structure for G, denoted $O_G$, is obtained by mapping the queried RDF graph into an OWL 2 ontology [23]. This mapping is only defined for OWL 2 DL ontologies, i.e., ontologies that satisfy certain syntactic conditions.

In this section, we use both Turtle and OWL's functional-style syntax (FSS) that is used in the OWL 2 structural specification [21]. We further provide a Description Logic (DL) syntax version for those with a background in DLs.

For many triples that use as predicate a special term from the RDFS vocabulary, the mapping to OWL structural objects is straightforward.

**Example 28.** For example a subclass statement in RDFS has a straightforward representation in OWL's FSS:

|  |  |
|---|---|
| Turtle: | foaf:Person rdfs:subClassOf foaf:Agent . |
| FSS: | SubClassOf(foaf:Person foaf:Agent) |
| DL: | Person    ⊑    Agent |

Note that DLs have no notion of IRIs, namespaces, or prefix declaration and we just write the short name without any prefix in the DL syntax. It is also characteristic that several terms of the specialized RDFS and OWL vocabulary in the Turtle syntax are translated to constructors in the FSS, e.g., rdfs:subClassOf is mapped into a SubClassOf constructor.

Similarly, the translation of domains and ranges is relatively straightforward.

**Example 29.** For example, the following domain and range statements translate straightforwardly to the FSS, but the DL syntax is slightly more involved:

| Turtle: | foaf:knows rdfs:range   foaf:Person . |
| | foaf:knows rdfs:domain foaf:Person . |
| FSS: | ObjectPropertyRange(foaf:knows foaf:Person) |
| | ObjectPropertyDomain(foaf:knows foaf:Person) |
| DL: | $\top \sqsubseteq \forall$ knows Person |
| | $\exists$ knows.$\top \sqsubseteq$ Person |

First, it can be noted that in the FSS the term rdfs:range becomes ObjectPropertyRange. The counterpart to ObjectPropertyRange is DataPropertyRange range, which is used for properties that relate individuals (such as instances of the class foaf:Person) to concrete data values. For example, the property foaf:name relates an individual to a string, i.e., an element from xsd:String. Since OWL supports very expressive reasoning with datatypes, which requires different algorithms from reasoning with abstract (non-datatype) elements, every property in OWL DL must be typed. Thus, we would have that foaf:knows is of type owl:ObjectProperty whereas foaf:name is of type owl:DataProperty.

In the DL syntax, there is no direct constructor for domains and ranges. The above statements are, however, logically equivalent. The first axiom uses on the left-hand side the special symbol $\top$, which corresponds to owl:Thing and is always true. Thus, the axiom can be read as "It is always implied that all ($\forall$) knows-successors of an element are instances of the class Person," which is exactly what a range axiom specifies. The second axiom can be read as "If an element has some ($\exists$) knows-successor, then it is an instance of the class Person."

**Elements of an OWL 2 DL Ontology.** Now that we have seen some examples of the mapping from RDF triples to OWL axioms, we introduce the basic elements in an OWL 2 DL ontology. An OWL 2 DL ontology consists of an *ontology header* and a set of *axioms*. The ontology header specifies the IRI of the ontology and which other ontologies are imported by it.

**Example 30.** The following set of RDF triples constitute a valid OWL 2 DL ontology.

| Turtle: | @prefix foaf: <http://xmlns.com/foaf/0.1/> . |
| | <http://example.org/ont1> rdf:type       owl:Ontology . |
| | <http://example.org/ont1> owl:imports <http://example.org/ont2> . |
| FSS: | Prefix(foaf:= <http://xmlns.com/foaf/0.1/>) |
| | Ontology(<http://example.org/ont1> |
| | Import(<http://example.org/ont2>) |
| | ) |

The ontology header has no representation in Description Logic syntax and it has no direct influence on the logical consequences of the ontology other than through imports, which instruct an OWL parser to additionally include the triples that are obtained from parsing the imported ontology.

The axioms in an ontology are used to describe a domain of interest, e.g., in the previous section we described people, their names, email addresses and chat IDs making use of terms from the FOAF (Friend of a Friend) ontology. Within the axioms, we distinguish between *logical* and *non-logical* axioms. As the ontology header, non-logical axioms carry no semantics, i.e., they do not influence the consequences of an ontology, and include:

- Annotations,
- Entity Declarations

With ontology annotations, one can describe properties of the ontology, e.g., who created it, which version of the ontology this is and other things. Similarly, one can annotate other axioms, e.g., with a comment or with provenance information, and one can even annotate annotations themselves. Entity declarations specify the types of terms. For example, we have learned above that foaf:knows is an object property whereas foaf:name is a data property. In addition to object and data properties, OWL also provides recognizes annotation properties, e.g., rdfs:label or rdfs:comment are built-in annotation properties, but one can define additional custom ones too. Similarly one can declare classes and custom datatypes (ones that are not defined in the OWL 2 datatype map) and named individuals. Such declarations are required to allow for an unambiguous parsing process.

**Example 31.** We can extend the ontology from Example 30 with the following annotations and declaration. Since the axioms are non-logical, the extended ontology still only entails tautological statements under the Direct Semantics.

Turtle:  `<http://example.org/ont1> owl:priorVersion <http://example.org/ont0> .`
`foaf:knows           rdf:type          owl:ObjectProperty .`
`<http://example.org/ont1> rdfs:label     "An example" .`

FSS:  `Annotation(owl:priorVersion <http://example.org/ont0>)`
`Annotation(rdfs:label "An example")`
`Declaration(ObjectProperty(foaf:knows))`

The first annotation gives the IRI of a previous version for the current ontology and the second annotation just provides a label for the ontology. The declaration axiom specifies foaf:knows as an object property.

In the remainder we frequently omit type declarations. Unless otherwise specified, examples assume that properties are object properties and that terms refer to classes rather than data ranges.

**Complex Classes and Axioms.**  So far we always had a straightforward correspondence between one triple and one OWL axiom. A FSS axiom can, however, correspond to several RDF triples, and the RDF triples might contain auxiliary blank nodes that are not part of the corresponding OWL objects and are not visible in the corresponding FSS axiom. This is usually the case if we want to represent complex OWL classes in RDF triples. In most cases, we can "hide" the blank nodes and obtain a slightly more readable Turtle format by making use of Turtles's abbreviations: [...] implicitly introduces a blank node, ";" can be used if the following triple has the same subject, which is them omitted, "," acts as ";" but for the case where triples share subject and object, the (...) constructor abbreviates lists of terms, and a abbreviates rdf:type.

**Example 32.** The first class assertion uses just a class name, which requires a single RDF triple, but the second assertion uses a complex class, which requires several RDF triples with auxiliary blank nodes.

| | | |
|---|---|---|
| Turtle: | :Peter rdf:type | :Person . |
| | :Peter rdf:type | _:x . |
| | _:x    rdf:type | owl:Restriction . |
| | _:x    owl:onProperty | :hasFather . |
| | _:x    owl:someValuesFrom | :Person . |

Turtle (abbr.):  :Peter a :Person .
:Peter a [ a  owl:Restriction ;
           owl:onProperty  :hasFather ;
           owl:someValuesFrom   :Person ] .

FSS:    ClassAssertion(:Person :Peter)
        ClassAssertion(ObjectSomeValuesFrom(:hasFather :Person) :Peter)

DL:     Person(Peter)
        (∃ hasFather.Person)(Peter)

The first axiom just states that the individual :Peter is an instance of the class :Person. The second axiom states that :Peter belongs to the class of things that have a :hasFather-successor which is an instance of the class :Person.

**Example 33.** Disjunctions and conjunctions in the FSS similarly require several triples in RDF:

| | | |
|---|---|---|
| Turtle: | :Birte rdf:type | _:x . |
| | _:x    rdf:type | owl:Class . |
| | _:x    owl:unionOf | _:l$_1$ . |
| | _:l$_1$    rdf:first | :Vegetarian . |
| | _:l$_1$    rdf:next | _:l$_2$ . |
| | _:l$_2$    rdf:first | :Vegan . |
| | _:l$_2$    rdf:rest | rdf:nil . |

Turtle (abbr.):   :Birte a [ a owl:Class ; owl:unionOf ( :Vegetarian :Vegan ) ] .

FSS:    ClassAssertion(ObjectUnionOf(:Vegetarian :Vegan) :Birte)
DL:     Birte ⊑ Vegetarian ⊔ Vegan

The typing as owl:Class is required since owl:unionOf can equally be used to build the union of two datatypes or data ranges (i.e., complex datatypes that are already obtained by combining datatypes). Axiom states that the individual :Birte is a vegan or a vegetarian, i.e., an instance of the class ObjectUnionOf(:Vegan :Vegetarian).

**Blank Nodes and Anonymous Individuals.** Although in the above examples it was always the case that the blank nodes disappeared in the FSS, this is not always the case. The FSS may still contain blank nodes, but these correspond to OWL individuals that have no explicit names and are called *anonymous individuals*.

**Example 34.** The following axiom uses anonymous individuals:

> Turtle:    :Peter :hasBrother _:y .
> FSS:    ObjectPropertyAssertion(:hasBrother :Peter _:y)

The meaning of the axiom is exactly the same as the meaning of the second axiom from Example 32, i.e., we say that :Peter is related to *some* element with the property :hasBrother. Note that in DL notation there is no counterpart to anonymous individuals and one always has to use existential quantifiers ($\exists$) as in the first version of this axiom. For RDF graphs that can be mapped into OWL 2 DL ontologies, it is, however, guaranteed that an according DL version always exists.

While parsing an input document (containing RDF triples) into an OWL ontology, it can be necessary to rename blank nodes/anonymous individuals and there is no guarantee that the blank node identifier _:y from the above triple is used as an identifier for Peter's brother in the ontology structure. Thus, the latter axiom from Example 34 could also be parsed as the OWL axiom

> ObjectPropertyAssertion(:hasBrother :Peter _:somethingelse)

## 4.2    Introduction to the OWL Direct Semantics for SPARQL

Having introduced the basic ideas of how we get from an RDF graph to an ontology that can be interpreted under OWL's Direct Semantics, we now turn our attention to the issue of deciding what is a consequence of an OWL ontology and how we can query for such consequences with SPARQL.

**OWL Entailment.** OWL reasoners are tools that decide OWL entailment. In order to decide whether an RDF graph G entails an RF graph G' under OWL 2 Direct Semantic entailment, we can proceed as follows:

1. We compute the imports closure clos(G) of G by enriching G with directly and indirectly imported triples and then we transform clos(G) into $O_G$ using the mapping process as defined in the OWL 2 Mapping to RDF Graphs specification. If the mapping fails, then G is not well-formed and, thus, cannot be used under the OWL 2 Direct Semantics.
2. We proceed similarly for G', obtaining $O_{G'}$.
3. We check whether $O_G \models O_{G'}$, where $\models$ denotes the OWL Direct Semantics entailment relation. Most commonly OWL reasoners do this by searching for a counter-model, i.e., a model $\mathcal{I}$ that satisfies $O_G$ and the negation of $O_{G'}$. A problem is that not all axioms can be negated in OWL. Thus, it is usually required to reformulate the reasoning problem and deal with each axiom in $O_{G'}$ separately.

**Table 9.** RDF data for Example 35

(1)     <http://example.org/myOntology> a owl:Ontology

(2)     :eats a owl:ObjectProperty
(3)     :contains a owl:ObjectProperty
(4)     :Vegetarian a owl:Class
(5)     :Vegan a owl:Class
(6)     :MilkProduct a owl:Class

(7)     :Birte a [ a owl:Class ; owl:unionOf ( :Vegetarian :Vegan ) ] .
(8)     :Birte :eats :Yoghurt .
(9)     :Yoghurt :contains :Milk .
(10)    :Milk a :MilkProduct .
(11)    [ a owl:Restriction ; owl:onProperty :contains ; owl:someValuesFrom :MilkProduct ]
            rdfs:subClassOf :MilkProduct .
(12)    :Vegan rdfs:subClassOf
            [ a owl:Restriction ; owl:onProperty :eats ; owl:allValuesFrom
                [a owl:Class ; owl:complementOf :MilkProduct ]
            ]

We illustrate some of the problems that have to be addressed in an OWL DS entailment regime in Example 35 below.

**Example 35.** We consider the query:

SELECT ?ind
WHERE { ?ind rdf:type :Vegetarian }

We assume that the default (and, hence, the active graph for the query) contains the triples from Table 9. Since the Direct Semantics is defined in terms of OWL structural objects, we first have to map the triples from Table 9 into OWL objects. The result of the mapping is shown in Table 10. Triple (1) results in the ontology header (1'). This triple does not contribute anything towards the logical consequences of the ontology, but is required to satisfy the constraints of OWL 2 DL. Similarly, Triples (2) to (6) result in the non-logical axioms (2') to (6'), which declare terms as classes or object properties. Such declarations are required to allow for an unambiguous parsing process. The remaining triples lead to logical axioms: Triple (7) is the same as in Example 33 and states that the individual :Birte is a vegan or a vegetarian.

Note that in the FSS version of (7') we have ObjectUnionOf whereas in the RDF triples, we just have unionOf. This is because the FSS makes it explicit whether the element is a class or a data range. In case of a data range DataUnionOf would be used. In order to be able to decide what applies, the declarations are used, e.g., from (4) and (5) (in FSS (4') and (5'), respectively), we know that :Vegetarian and :Vegan are classes. Triple (8) translates into an assertion saying that the individual :Birte :eats the individual :Yoghurt. In order to see whether this is a data or an object property assertion in the FSS, we can again use the declarations. Axiom (9') is obtained similarly. From (11), we obtain a more complicated axiom that states: if an element has a :contains relationship with something that is an instance of :MilkProduct, then this element is itself an instance of :MilkProduct. Finally, (12) translates into a statement that says that instances of the class :Vegan can only be related with the property :eats to something that is not an instance of :MilkProduct. For those more familiar with Description Logic syntax,

**Table 10.** FSS version of the triples for Example 35

(1')     Ontology(<http://example.org/myOntology>  
(2')     Declaration(ObjectProperty(:eats))  
(3')     Declaration(ObjectProperty(:contains))  
(4')     Declaration(Class(:Vegetarian))  
(5')     Declaration(Class(:Vegan))  
(6')     Declaration(Class(:MilkProduct))  

(7')     ClassAssertion(ObjectUnionOf(:Vegetarian :Vegan) :Birte)  
(8')     ObjectPropertyAssertion(:eats :Birte :Yoghurt)  
(9')     ObjectPropertyAssertion(:contains :Yoghurt :Milk)  
(10')    ClassAssertion(:MilkProduct :Milk)  
(11')    SubClassOf(ObjectSomeValuesFrom(:contains :MilkProduct) :MilkProduct)  
(12')    SubClassOf(:Vegan ObjectAllValuesFrom(:eats ObjectComplementOf(:MilkProduct)))  
       )

Table 11 shows the logical axioms into Description Logic syntax with (7*), (11*), and (12*) terminological (TBox) axioms and (8*), (9*), and (10*) assertional (ABox) axioms.

In order to find the answers for the query under OWL DS entailment, we also need a version of the BGP that can be interpreted according to the OWL structural specification. One way of doing this would be to replace the variables with terms from the ontology, then map the resulting triples to OWL axioms, and check entailment. This would, however, require frequent parsing/mapping attempts that frequently will fail because we substituted a variable with a value that violates the OWL 2 DL constraints, e.g., when we replace the variable ?ind with a class name, e.g., :Vegan, we obtain a triple that cannot be mapped since :Vegan rdf:type :Vegetarian is not allowed in OWL 2 DL, i.e., rdf:type cannot be used to relate two classes. Since we know that :Vegetarian is a class from (4), we know that ?ind has to be instantiated with individual names. In order to avoid a parsing attempt for each possible assignment of variables, the choice has been made to extend OWL's structural specification to allow for variables in place of atomic objects such as individuals, classes, properties, or literals. We can then simply map a BGP into axioms from the extended specification. This yields:

$$\text{ClassAssertion(:Vegetarian ?ind)}$$

For this axiom it is clear that ?ind occurs in an individual position and, therefore, has to be replaced with individual names from the queried ontology. For this example, we only have to substitute ?ind with :Birte. We could also use dedicated reasoner methods to retrieve instances of the class :Vegetarian without iterating over all individual names to obtain the query result:

$$\frac{\textbf{?ind}}{\text{:Birte}}$$

Note that the class used in the query pattern could equally be a class expression such as

$$\text{ObjectUnionOf(:Vegetarian :Vegan ObjectAllValuesFrom(:eats:MilkProduct)),}$$

although that last disjunct is somehow far-fetched as a class of things that only eat milk products. Assume further that we extend the ontology with:

(13)    ClassAssertion(ObjectUnionOf(:Vegetarian :Vegan) :Ian)  
(14)    SubclassOf(:Vegetarian :HasSpecialMealRequest)  
(15)    SubclassOf(:Vegan :HasSpecialMealRequest)

**Table 11.** Description Logic version of the logical axioms for Example 35

$$(7^*) \quad (\text{Vegetarian} \sqcup \text{Vegan})(\text{Birte})$$
$$(8^*) \quad \text{eats}(\text{Birte, Yoghurt})$$
$$(9^*) \quad \text{contains}(\text{Yoghurt, Milk})$$
$$(10^*) \quad \text{MilkProduct}(\text{Milk})$$
$$(11^*) \quad \exists\text{contains.MilkProduct} \sqsubseteq \text{MilkProduct}$$
$$(12^*) \quad \text{Vegan} \sqsubseteq \forall\text{eats.}(\neg\text{MilkProduct})$$

Clearly :Ian belongs to the above stated disjunction, so should be returned as query answer although membership in that class is not explicitly stated nor can we foresee all such classes and extend the queried ontology accordingly. Furthermore, we might have to do case-based reasoning. In this case, we can neither extend the ontology with a statement that :Ian belongs to the class :Vegetarian nor with one that establishes that :Ian belongs to :Vegan. Nevertheless, we know that :Ian belongs to the extension of the class :HasSpecialMealRequest.

### 4.3   Mapping BGPs to Extended OWL Objects

Note that in the above example, it was clear from the queried ontology that ?ind rdf:type :Vegetarian corresponds to a class assertion with ?ind mapping to individual names since :Vegetarian was declared as a class in $O_G$. In some cases, however, the variables in a BGP do no longer allow for an unambiguous mapping, which is addressed by variable typing triples.

**Variable Typing.** In order to have an unambiguous correspondence between BGPs and extended OWL objects, the Direct Semantics entailment regime requires for some cases extra triples in a basic graph pattern that give typing information for the variables.

**Example 36.** In order to see why this is required, consider the following query:

SELECT ?s ?p ?o WHERE { ?s ?p ?o }

Without any restrictions this query could be a query for

- declarations, i.e., the BGP maps to a declaration such as Declaration(Class(?s)) where ?p binds to rdf:type, ?o to owl:Class, and bindings for ?s have to be computed or Declaration(ObjectProperty(?s)) where ?p binds to rdf:type and ?o to owl:ObjectProperty, or any other type of declaration,
- inverse object properties, i.e., the BGP maps to ObjectInverseOf(?o) where ?s maps to a blank node and ?p to owl:inverseOf,
- subclasses, i.e., the BGP maps to SubClassOf(?s ?o) with rdfs:subClassOf as binding for ?p,
- equivalent classes, i.e., the BGP maps to EquivalentClasses(?s ?o) where ?p binds to owl:equivalentClass,
- disjoint classes, i.e., the BGP maps to DisjointClasses(?s ?o) where ?p binds to owl:disjointWith,
- . . .

In order to answer the query without any typing constraints, all possible ways of mapping the BGP into ontology structures have to be considered. Even if variables can only occur in the position of function parameters of the functional-style syntax, the BGP from the above query can still be mapped to ObjectPropertyAssertion(?p ?s ?o), DataPropertyAssertion(?p ?s ?o), or AnnotationAssertion(?p ?s ?o) without variable typing information.

The inclusion of type declarations from the queried ontology means that at least the non-variable terms in the query can be disambiguated without additional typing information in the query. Typically, variables have to be declared if they represent classes, properties, or datatypes, whereas individual variables do not need declarations for an unambiguous mapping process. This is similar to typing in ontologies, where typing of individuals is optional, but typing for properties, classes, and non-OWL 2 datatypes is mandatory.

**Example 37.** The BGP of the query

SELECT ?x WHERE { ?x :p ?y }

is parsed into (a) or (b) depending on whether :p is declared as an object or a data property in the queried ontology

(a) ObjectPropertyAssertion(:p ?x ?y)        (b) DataPropertyAssertion(:p ?x ?y)

If :p is changed into the variable ?p, we need an extra typing triple, e.g.,

Declaration(ObjectProperty(?p))

to allow for an unambiguous mapping process.

**Definition 11.** *Let* BGP *be a basic graph pattern with* ?x *a variable occurring in* BGP. *If* BGP *contains a triple*

?x rdf:type TYPE,

*where* TYPE *is one of*

- owl:Class,
- owl:ObjectProperty,
- owl:DataProperty,
- owl:Datatype, *or*
- owl:NamedIndividual,

*then* ?x *is declared* *to be of type* TYPE.

**From BGPs to Extended OWL Objects.** We now formally define how BGPs are mapped into OWL axioms extended to contain variables, i.e., the result of the mapping yields rather axiom *templates* than axioms.

The BGP of the query is mapped into an OWL 2 DL ontology, extended to allow variables in place of class names, object property names, datatype property names, individual names, or literals. Table 12 shows how productions of the OWL 2 functional-style

**Table 12.** Grammar extension for extended OWL objects

Class := IRI | Var      ObjectProperty := IRI | Var      DataProperty := IRI | Var
Individual := NamedIndividual | AnonymousIndividual | Var
Literal := typedLiteral | stringLiteralNoLanguage | stringLiteralWithLanguage | Var

syntax grammar [21] are extended to allow variables as defined by the Var production from the SPARQL grammar [26]. If BGP contains no ontology header, i.e., a triple of the form x rdf:type owl:Ontology with x ∈ I ∪ B, we assume that BGP is extended with _:o rdf:type owl:Ontology for _:o a blank node name not occurring in BGP or the active graph before parsing BGP into extended OWL objects. Solution mappings in a query result are applied to such extended ontologies to obtain a set of OWL DL axioms that is compatible with the queried ontology and also entailed by it under DS.

**Definition 12.** *An extended ontology* $O_{BGP}^G$ *is constructed for a basic graph pattern* BGP *and graph* G *using the parsing process for RDF graphs as defined in [23] with three modifications:*

1. *variable identifiers are allowed in place of IRIs and literals in all parsing steps,*
2. *an ontology header may be added to* BGP *if not given, and*
3. *the type declarations given in* BGP *are augmented with the declarations in* G *and those obtained from graphs imported by* G *(denoted* AllDecl(G) *in [23]).*

*The complete parsing process is detailed in the latest entailment regimes working draft.[5] A basic graph pattern* BGP *satisfies the* typing constraints *of the entailment regime if*

- *no variable is declared as being of more than one type,*
- *variables without a type declaration occur either only in individual positions or only in literal positions, and*
- *it is possible to disambiguate all types of IRIs and variables when parsing* BGP *into extended OWL objects taking the typing information from* $O_G$ *and from* BGP *into account;*

*A basic graph pattern* BGP *is* well-formed for the OWL DS entailment regime and a graph G *if* $O_{BGP}^G$ *can be obtained in this way and is an extended OWL DL ontology. An RDF graph* G *is* well-formed for the OWL DS entailment regime *if is mapping to structural OWL objects [23], resulting in an ontology* $O_G$, *is defined.*

**SPARQL Syntax Extensions for BGPs.** Considering the fact that each BGP has to be mapped to structural OWL objects anyway in order to use the OWL DS, it seems natural to directly allow for specifying BGPs in other OWL syntaxes, e.g., the FSS. Such an extension has not been specified by the W3C as part of the entailment regimes document, but it seems likely that implementations of the OWL DS regime might also accept other syntaxes for the BGP.

---

[5] http://www.w3.org/TR/2010/WD-sparql11-entailment-20100601/

**Table 13.** A query with infinitely many entailed solutions

G : :Peter a [ a owl:Restriction;          BGP : :Peter a [ a owl:Restriction;
    owl:onProperty :dp;                              owl:onProperty :dp;
    owl:allValuesFrom [ a rdfs:Datatype;            owl:allValuesFrom [ a rdfs:Datatype;
      owl:oneOf ("5"^^xsd:integer)]]                    owl:datatypeComplementOf [
                                   a rdfs:Datatype; owl:oneOf (?x)]]]

$O_G$    : ClassAssertion(DataAllValuesFrom(:dp DataOneOf("5"^^xsd:integer)) :Peter)
$O_{BGP}^G$ : ClassAssertion(DataAllValuesFrom(:dp DataComplementOf(DataOneOf(?lit))) :Peter)

## 4.4 Infinite Entailments in Datatype Reasoning

**Example 38.** In order to see how datatype reasoning in OWL can cause infinite entailments, consider the graph and query in Table 13. The graph G states that all data values to which Peter is related via :dp are in the singleton set of the integer 5. The BGP asks for all data values to which :Peter cannot be related with :dp. Without suitable restrictions, all (infinitely many) integers other than 5 could be used in solution mappings for ?x. Moreover, it is currently unknown how to compute all mappings for literal variables even for cases where there number is finite – testing all literals is clearly not an option.[6]

We will again use the vocabulary of the queried graph to include only literals that are explicitly mentioned in the input graph for the OWL entailment regimes. Like for the IRIs rdf:_i, this may lead to unexpected behavior, since mentioning a literal in the input may lead to new query results even for queries not directly related to this literal. Yet, this problem seems so rare in practice that a more detailed analysis of the problematic datatype expressions is not worthwhile, even if it could further limit unintuitive behavior.

## 4.5 The OWL 2 Direct Semantics Entailment Regime

We now define the evaluation of graph patterns. For the Direct Semantics, Skolemization is applied to $O_G$, which ensures that only blank nodes that represent anonymous OWL individuals are Skolemized, not blank nodes used for encoding complex OWL syntax in RDF.

**Definition 13.** *Let* G *be an RDF graph that is well-formed for the OWL 2 DS entailment regime,* BGP *a basic graph pattern that is well-formed for DS and* G*,* $V(O_{BGP}^G)$ *the set of variables in* $O_{BGP}^G$*,* $B(O_{BGP}^G)$ *the set of blank nodes in* $O_{BGP}^G$*,* sk *a Skolemization function for the blank nodes in* $O_{BGP}^G$ *as in Definition 9 such that* $ran(sk) \cap (Voc(O_G) \cup Voc(O_{BGP}^G)) = \emptyset$*.*

    *We write* $\models_{DS}$ *for the OWL 2 Direct Semantics entailment relation and define the* evaluation of BGP over G under OWL 2 Direct Semantics entailment, $[\![BGP]\!]_{D,G}^{DS}$, as *the solution multiset*

---

[6] Hence one cannot call such solutions "trivial" in the sense of Table 7. Indeed, our restrictions are motivated by pragmatic considerations, not by formal requirements of SPARQL.

$\{(\mu, n) \mid \mathsf{dom}(\mu) = \mathsf{V}(\mathsf{BGP}),$ *and $n$ is the maximal number of distinct RDF instance*
*mappings such that, for each* $1 \leq i \leq n,$
   *(i)* $\mathsf{dom}(\mathsf{sigma_i}) = \mathsf{B}(\mathsf{BGP}),$
   *(ii)* $\mu(\sigma_i(\mathsf{O}^\mathsf{G}_\mathsf{BGP})) \cup \mathsf{O}^\mathsf{G}$ *is an OWL 2 DL ontology,*
   *(iii)* $\mathsf{sk}(\mu(\sigma_i(\mathsf{O}^\mathsf{G}_\mathsf{BGP})))$ *are ground RDF triples,*
   *(iv)* $\mathsf{sk}(\mathsf{O}_\mathsf{G}) \models_\mathsf{DS} \mathsf{sk}(\mu(\sigma_i(\mathsf{O}^\mathsf{G}_\mathsf{BGP}))),$ *and*
   *(v)* $\mathsf{ran}(\mu) \subseteq \mathsf{Voc}(\mathsf{O}_\mathsf{G})\}.$

If $\mathsf{O}_\mathsf{G}$ is inconsistent, queries must be rejected with an error.

**Restrictions on Solutions.** Since solutions can only bind to terms from a finite vocabulary, clearly the solution multiset and each multiplicity is finite too. Although this avoids infinite results as discussed in Section 4.4, reasoners may have to consider a large number of literals as potential variable bindings and we expect that not all systems will provide a complete implementation for queries with literal variables.

Note that for the OWL DS regime no vocabulary other than that of the graph itself is required since there are no axiomatic triples and variables can only bind to built-in terms that are also built-in entities. Built-in entities such as owl:Thing are, however, assumed to be present in any ontology [21, Table 5], i.e., $\mathsf{O}_\mathsf{G}$ automatically includes declarations for these built-in entities. Thus, we have omitted any OWL 2 specific vocabulary from condition $(v)$.

Compared to the RDFS regime, condition (ii) requires $\mu(\sigma_i(\mathsf{O}^\mathsf{G}_\mathsf{BGP})) \cup \mathsf{O}^\mathsf{G}$ to be an OWL 2 DL ontology. Thus, the axioms from the instantiated BGP together with the axioms from the queried ontology must satisfy the restrictions for OWL 2 DL ontologies. These restrictions are in place to guarantee that the key reasoning tasks in OWL 2 with Direct Semantics are decidable. For example, for owl:topDataProperty, the following requirement has to be met in OWL 2 DL:

> The owl:topDataProperty property occurs in a SubDataPropertyOf axiom only in the position of the super-property.

The condition guarantees that these restrictions are equally applied to the query. Furthermore, the condition prevents that the BGP uses a property in a number restriction that is declared as transitive in the queried ontology since transitive properties cannot occur in number restrictions in OWL 2 DL.

The complexity of standard reasoning problems in OWL are well-understood and BGP evaluation can be implemented using the standard reasoning techniques. The complexity of OWL reasoning usually outweighs that of the SPARQL algebra operations, i.e., checking whether a solution mapping is a solution is complete for nondeterministic double exponential time in OWL 2 DL.

**Higher Order Queries.** The Direct Semantics entailment regime allows for certain (but not all) forms of higher order queries.

**Example 39.** The BGP

$$?x \text{ rdfs:subClassOf } ?y$$

can be used to query for pairs of sub- and super-classes. This means that variables can bind to classes (representing sets of individuals) and not just to individuals or data values.

Queries in which variables are used in positions of a First-Order Logic quantifier, will, however, be illegal since such queries cannot be mapped to OWL objects as required.

**Example 40.** The following (illegal) query asks whether some or all brothers of Peter are persons:

```
SELECT ?x
WHERE { :Peter a [
                a owl:Restriction ;
                owl:onProperty :hasBrother ;
                ?x :Person
           ]
       }
```

In FSS the BGP of the query corresponds to the axiom:

ClassAssertion(?x(:hasBrother :Person) :Peter)

Here the variable occurs in the position of a quantifier (ObjectSomeValuesFrom or ObjectAllValuesFrom, i.e., ∃ and ∀ in Description Logics) and not just in the position of OWL entities such as class names or individual names.

## 4.6  The OWL 2 RDF-Based Semantics Entailment Regime

The OWL 2 RDF-Based Semantics is a direct extension of the RDFS semantics, which means that it interprets RDF triples directly without the need of mapping an RDF graph into structural objects. Compared to the Direct Semantics, the RDF-Based Semantics treats classes as individuals that refer to elements of the domain. Each such element is then associated with a subset of the domain, called the class extension. This means that semantic conditions on class extensions are only applicable to those classes that are actually represented by an element of the domain which can lead to less consequences than expected. An example is given by the following graph and BGP:

G : :a rdf:type :C    BGP : ?x rdf:type [ rdf:type owl:Class ;
                                          owl:unionOf ( :C :D ) ]

G states that :a has type :C, while BGP asks for instances of the complex class denoting the union of :C and :D. One might expect $\mu : ?x \mapsto :a$ to be a solution, but this is not the case under the OWL 2 RDF-Based Semantics (see also [29, Sec. 7.1]). It is guaranteed that the union of the class extensions for :C and :D exists as a subset of the domain; no statement in G implies, however, that this union is the class extension of any domain element. Thus, $\mu(\text{BGP})$ is not entailed by G.

The entailment holds, however, when the statement :E owl:unionOf ( :C :D ) is added to G. In the OWL Direct Semantics, in contrast, classes denote sets and not domain elements, so G entails $\mu(\text{BGP})$ under DS where, formally, G must first be extended with an ontology header to become well-formed for DS. Note that a similar situation occurs for Example 38, but the problem of infinitely many answers occurs if the necessary expressions are introduced.

Summing up, the RBS handles blank nodes just like RDFS, even in cases where they are needed for encoding OWL class expressions. This allows us to use Skolemization just like in the case of RDFS in the next definition. The expressive datatype reasoning is again addressed as for the DS using the answer domain.

**Definition 14.** *Let* G *be an RDF graph,* BGP *a basic graph pattern,* V(BGP) *the set of variables in* BGP, B(BGP) *the set of blank nodes in* BGP, sk *a Skolemization function as in Definition 9 such that* ran(sk) $\cap$ (Voc(G) $\cup$ Voc(BGP)) = $\emptyset$. *Let* Voc(OWL2RB) *be the OWL 2 RDF-Based vocabulary and* Voc$^-$(OWL2RB) = Voc(OWL2RB) \ {rdf:_i | $i \in \mathbf{N}$}.*

*We write* $\models_{\text{RBS}}$ *for the OWL 2 RDF-Based Semantics entailment relation and define the evaluation of* BGP *over* G *under OWL 2 RDF-Based Semantics entailment,* $[\![\text{BGP}]\!]^{\text{RBS}}_{\text{D,G}}$, *as the solution multiset*

$\{(\mu, n) \mid \text{dom}(\mu) = \text{V(BGP)}$, *and* n *is the maximal number of distinct RDF instance*
  *mappings such that, for each* $1 \leq i \leq n$,
  *(i)* dom(sigma$_i$) = B(BGP),
  *(ii)* $\mu(\sigma_i(\text{BGP}))$ *are well-formed RDF triples,*
  *(iii)* sk($\mu(\sigma_i(\text{BGP}))$) *are ground RDF triples,*
  *(iv)* sk(G) $\models_{\text{RBS}}$ sk($\mu(\sigma_i(\text{BGP}))$), *and*
  *(v)* ran($\mu$) $\subseteq$ Voc(G) $\cup$ Voc$^-$(OWL2RB)}.

### 4.7    OWL 2 Profiles

OWL 2 DL is decidable, but computationally hard and not scalable enough for many applications. OWL Full is not even decidable and, consequently, not many implementations that support all of OWL Full are available. Thus, OWL 2 identifies subsets of OWL 2, called *profiles*, which are sufficiently expressive, but of lower complexity (tractable) and tailored to specific reasoning services (see also Figure 1):

- Terminological/schema reasoning: OWL 2 EL
- Query Answering via database engines: OWL 2 QL
- Assertional/data reasoning with rule engines: OWL 2 RL

The OWL 2 QL and EL profiles further restrict the allowed inputs compared to OWL 2 DL, but equally use the Direct Semantics. The OWL 2 RL profile, in principle, can be used with both semantics, but for the Direct Semantics the input RDF graph has to satisfy some constrains. The RDF-Based semantics can be use with any RDF graph but under the OWL 2 RL profile one derives only certain consequences.

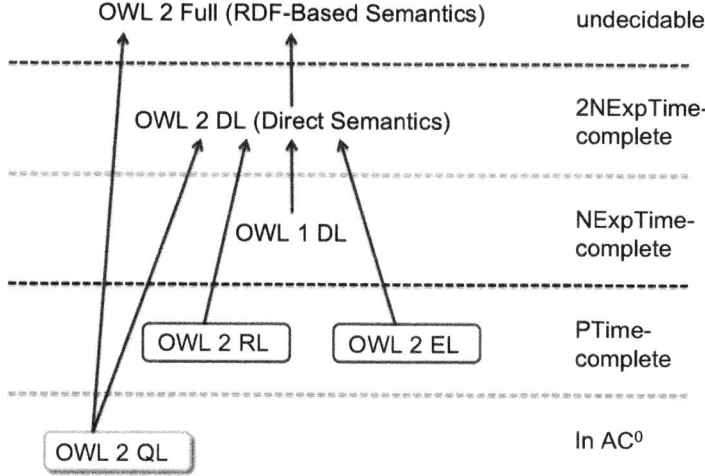

**Fig. 1.** An overview for the complexity of reasoning in OWL and its profiles

**OWL 2 DL** is the largest subset of RDF graphs for which the OWL 2 Direct Semantics is defined. Systems that support OWL 2 DL can also handle ontologies that satisfy the restrictions of the OWL 2 EL and QL profiles because these profiles are even more restrictive.

**The OWL 2 EL Profile** is particularly useful in applications employing ontologies that contain very large numbers of properties and/or classes. The profile captures the expressive power used by many ontologies and is a subset of OWL 2 DL for which the basic reasoning problems can be performed in time that is polynomial with respect to the size of the ontology. Worth mentioning is that the class hierarchy (all subclass relations between classes) can be computed in "one pass", whereas OWL 2 DL reasoner typically have to check each pair of classes separately. The one-pass classification exploits saturation-based techniques developed for $\mathcal{EL}$ Description Logics [2, 1, 7, 3, 7] and can be extended to the Horn (non-disjunctive) fragment of OWL DL [16].

**The OWL 2 QL Profile** is aimed at applications that use very large volumes of instance data, and where query answering is the most important reasoning task. In OWL 2 QL, conjunctive query answering can be implemented using conventional relational database systems [9, 10, 25]. Using query rewriting techniques, sound and complete conjunctive query answering can be performed in LogSpace with respect to the size of the data (assertions) using standard database management systems. Recently developed techniques prevent an exponential blowup from query rewriting [18, 27]. As in OWL 2 EL, polynomial time algorithms can be used to implement the ontology consistency and class expression subsumption reasoning problems.

Note that OWL 2 QL implementations most commonly will only support conjunctive queries, i.e., queries where the BGP consists only of axioms of the following type:

- ClassAssertion,
- ObjectPropertyAssertion, and
- DataPropertyAssertion.

With the additional restriction that variables can only occur in the position of individuals and literals (if datatype reasoning is supported). Future versions of SPARQL could define further entailment regimes, e.g., one that defines a dedicated conjunctive query regime. Since an implementations is, however, free to reject any unsupported query anyway, the currently defined OWL regime can still be used.

**The OWL 2 RL Profile** defines a syntactic subset of OWL 2, which is amenable to implementation using rule-based technologies. For RDF graphs that fall into this syntactic subset, reasoning is sound and complete and both semantics of OWL can be used yielding the same results. Outside of this syntactic fragment, the RDF-Based Semantics can still be used, but reasoning can be incomplete. The main reasoning in the RL profile are PTime-complete (ontology consistency, class expression satisfiability, class expression subsumption, instance checking, and conjunctive query answering). Reasoning can be implemented in a rule engine (with equality support) by materializing schema inferences for facts.

### 4.8   Implementing the OWL 2 RL Profile via Rules

The OWL 2 RL specification provides a complete rule set that can be used to materialize all OWL 2 RL inferences. Each RDF triple is encoded via a ternary predicate $T(\_, \_, \_)$. A given set of rules is then applied to the ternary predicates.

**Example 41.** Subproperty reasoning is, for example, handled via the rule `prp-spo1`:

$$\text{prp-spo1: } T(?p_1, \text{rdfs:subPropertyOf}, ?p_2) \wedge T(?x, ?p_1, ?y) \rightarrow T(?x, ?p_2, ?y)$$

Given the first two triples below (as ternary predicates), we can derive the third one by applying the above rule:

$$T(\text{:hasSister, rdfs:subPropertyOf, :hasSibling})$$
$$T(\text{:Peter, :hasSister, :Mary})$$

$$\Rightarrow T(\text{:Peter, :hasSibling, :Mary})$$

Functionality for properties is taken into account via the `prp-fp` rule:

$$\text{prp-fp: } T(?p, \text{rdf:type, owl:FunctionalProperty}) \wedge T(?x, ?p, ?y_1) \wedge T(?x, ?p, ?y_2)$$
$$\rightarrow T(?y_1, \text{owl:sameAs}, ?y_2)$$

Given the first three triples, we can then apply the rule to derive the forth triple:

$$T(\text{:hasMother, rdf:type, owl:FunctionalProperty})$$
$$T(\text{:John, :hasMother, :Anna})$$
$$T(\text{:John, :hasMother, :Ann})$$

$$\Rightarrow T(\text{:Anna, owl:sameAs, :Ann})$$

**Table 14.** Data used to illustrate subclass reasoning with complex class expressions in OWL RL

Turtle:    (1a) :Person rdfs:subClassOf _:c
             (1b) _:c rdf:type owl:Restriction
             (1c) _:c owl:allValuesFrom :Person
             (1d) _:c owl:onProperty :hasChild
             (2) :Anna :hasChild :Mary
             (3) :Anna rdf:type :Person

FSS:     (1) SubClassOf(:Person ObjectAllValuesFrom(:hasChild :Person))
            (2) ObjectPropertyAssertion(:hasChild :Anna :Mary)
            (3) ClassAssertion(:Person :Anna)

DL:      (1) $Person \sqsubseteq \forall hasChild.Person$
            (2) hasChild(Anna, Mary)
            (3) Person(Mary)

We illustrate how subclass reasoning with complex class expressions can be performed using the data from Table 14.

cax-sco: $T(?c_1, \text{rdfs:subClassOf}, ?c_2) \wedge T(?x, \text{rdf:type}, ?c_1) \rightarrow T(?x, \text{rdf:type}, ?c2)$
cls-avf: $T(?x, \text{owl:allValuesFrom}, ?y) \wedge T(?x, \text{owl:onProperty}, ?p) \wedge$
           $T(?u, \text{rdf:type}, ?x) \wedge T(?u, ?p, ?v) \rightarrow T(?v, \text{rdf:type}, ?y)$

The rule cax-sco can be applied to the ternary form of triple (1a) and (3) to derive the first of the two triples below. Then, the ternary form of triples (1c), (1d), (4), and (2) can be used to satisfy the body of the rule cls-avf binding ?x to _:c, ?y to :Person, ?p to :hasChild, ?u to :Anna, and ?v to :Mary, to derive triple (5).

$\Rightarrow$ (4) :Anna rdf:type _:c
$\Rightarrow$ (5) :Mary rdf:type :Person

Note that triple (4) has no representation in FSS or DL notation and would not be derived by a non-rule-based OWL reasoner that uses the Direct Semantics. The triple is rather an intermediate consequence with the purpose of deriving the class assertion (5).

After exhaustively applying the OWL RL rules [20] to a set of RDF triples, the resulting extended graph contains triples that state the (atomic) types for all individuals as well as the relationships between individuals. Schema reasoning is, however, not performed by applying the OWL 2 RL rules, i.e., we do not have all triples :$c_1$ rdfs:subClassOf :$c_2$ for :$c_1$ a subclass of :$c_2$ under the Direct or RDF-Based semantics.

In order to evaluate BGP over an active graph G using the OWL 2 RL profile one can proceed as follows:
1. Saturate G using the OWL 2 RL rule to obtain G′.
2. Evaluate BGP over G′ using sub-graph matching (i.e., via any standard SPARQL implementation).

More optimized implementation than via the fixed OWL 2 RL rule set are possible [19]. It is further possible to implement the RL profile in any rule engine that supports the RIF Core dialect [6, 8] either as fixed or ontology-specific rule set.

# 5   Exercises

We provide a couple of exercises in this section that can be used to test the understanding of several aspects that have been presented in the previous sections. Solutions to the exercises are provided in the following section.

## 5.1   Mapping to the SPARQL Algebra

**Exercise 1.** *Translate the following SPARQL query into an abstract query:*

$$SELECT \; ?mbox$$
$$WHERE \; \{ \; ?x \; foaf{:}mbox \; ?mbox \; \}$$

**Exercise 2.** *Translate the following SPARQL query into an abstract query:*

$$SELECT \; DISTINCT \; ?name$$
$$WHERE \; \{ \; ?x \; foaf{:}name \; ?name \; FILTER \; regex(?name, \text{"ian"}) \; \}$$

**Exercise 3.** *Translate the following SPARQL query into an abstract query:*

```
SELECT ?mbox
WHERE { { ?x foaf:name "Birte Glimm". ?x foaf:mbox ?mbox }
         UNION
         { ?x foaf:name ?name . ?x foaf:mbox ?mbox
           FILTER regex(?name, "ian") }
       }
```

**Exercise 4.** *Translate the following SPARQL query into an abstract query:*

```
SELECT ?name ?id
WHERE { { ?x foaf:name ?name OPTIONAL { ?x foaf:icqChatID ?id } }
         UNION { ?x foaf:name ?name . ?x foaf:mbox <mailto:rudolph@kit.edu> }
       } ORDER BY ?name
```

## 5.2   Query Evaluation

For the query evaluation in this section we assume simple entailment, i.e., subgraph matching.

**Exercise 5.** *Illustrate the evaluation of the query from Exercise 3 including its intermediate results assuming the default graph contains the triples from Table 1.*

**Table 15.** RDF triples for Exercise 7

```
@prefix  : <http://example.org/> .
@prefix  w3c: <http://www.w3.org/> .
@prefix  iswc2010: <http://data.semanticweb.org/conference/iswc/2010/> .
```

| | | | |
|---|---|---|---|
| (1) | iswc2010:paper/280 | rdf:type | :ConferencePaper. |
| (2) | iswc2010:paper/280 | :authors | _:l1. |
| (3) | _:l1 | rdf:type | rdf:Seq. |
| (4) | _:l1 | rdf:_1 | "Birte Glimm". |
| (5) | _:l1 | rdf:_2 | "Markus Krötzsch". |
| (6) | w3c:TR/rdf-sparql-query | rdf:type | :W3CStandard. |
| (7) | w3c:TR/rdf-sparql-query | :writtenBy | _:l2. |
| (8) | _:l2 | rdf:type | rdf:Seq. |
| (9) | _:l2 | rdf:_1 | "Eric Prud'hommeaux". |
| (10) | _:l2 | rdf:_2 | "Andy Seaborne". |
| (11) | :ConferencePaper | rdfs:subClassOf | :Publication. |
| (12) | :W3CStandard | rdfs:subClassOf | :Publication. |
| (13) | :writtenBy | rdfs:subPropertyOf | :authors. |

**Exercise 6.** *Illustrate the evaluation of the query from Exercise 4 including its intermediate results assuming the default graph contains the triples from Table 1.*

## 5.3   RDFS Semantics Queries

In this section we assume RDFS entailment, i.e., we use the RDFS entailment regime.

**Exercise 7.** *We assume a graph with the triples from Table 15 and the query:*

> SELECT ?auth ?pub
> WHERE { ?pub rdf:type :Publication . ?pub :authors ?seq . ?seq ?ind ?auth }

*List the query results under the RDFS entailment regime and argue, for each solution, why the solution follows.*

**Exercise 8.** *You might have noticed that the query from Exercise 7 has two answers in which the binding for ?auth is not an author name. How can we modify the query to query for solutions in which ?auth binds to an author name?*

**Exercise 9.** *We again assume a graph with the triples from Table 15 and the query:*

> SELECT ?type
> WHERE { iswc2010:paper/280 rdf:type ?type }

*Which answers does the query have under RDFS entailment and why?*

**Exercise 10.** *We again assume a graph with the triples from Table 15. Is the triple* iswc2010:paper/280 :authors _:x *entailed under RDFS entailment? What is then the answer to the following query?*

$$ASK \{ \text{ iswc2010:paper/280 :authors \_:x } \}$$

### 5.4   OWL Direct Semantics Queries

**Exercise 11.** *We assume that the queried ontology contains the axioms from Table 9. Map the following BGP into an extended OWL axiom, list the results of evaluating the BGP under OWL Direct Semantics, and explain, for each solution, why the solution is entailed:*

$$\text{?mp rdf:type :MilkProduct}$$

**Exercise 12.** *Map the query pattern of the following query into extended OWL objects and illustrate the evaluation of the query over the ontology from Table 9:*

> *SELECT* ?sup
> *WHERE* { :MilkProduct rdfs:subClassOf ?sup. ?sup rdf:type owl:Class }

**Exercise 13.** *We consider the ontology from Table 9. Why is the query*

> *SELECT* ?rel
> *WHERE* { :Vegetarian ?rel :Vegan }

*not a well-formed query under the OWL 2 Direct Semantics?*

**Exercise 14.** *What query can one use to retrieve a list of all classes tat occur in the ontology?*

**Exercise 15.** *A typical reasoning tasks in OWL is the classification of classes, i.e., the computation of all pairs $\langle C, D \rangle$ such that $C$ is a direct sub-class of $D$ or $C$ is equivalent to $D$. Can a SPARQL query be used to retrieve the subsumption hierarchy?*

**Exercise 16.** *Can the OWL Direct Semantics entailment regime be implemented via materialization, as sketched for the RDFS regime? If so, sketch what one would have to do. If not, why is it no possible and would it possible for subsets of the language?*

## 6   Solutions to the Exercises

In this section, we provide the solution for the exercises from the previous section.

## 6.1   Mapping to the SPARQL Algebra

**Solution 1.** We start with the query pattern, which is, as every query pattern, a group graph pattern here consisting of one element, which is a `TriplesBlock`. Since Definition 3 defines the translation for `GroupGraphPattern` according to Algorithm 1 (we have no filter, but one other element, which is the BGP), we get Join(Z, algbr(bgp)) with bgp the BGP of the query. Going back to Definition 3 for the translation of the BGP, we can now use the first case for `TriplesBlock` and we obtain Join(Z, Bgp(?x foaf:mbox ?mbox)). The object can be simplified to just Bgp(?x foaf:mbox ?mbox).

Now that we have the algebra translation for the query pattern, which we denote with E, we can obtain the algebra translation for the whole query and then the abstract query as described in Definition 8. We first obtain ToList(E), then go on to Project(ToList(E), {?mbox}). Finally, we obtain the abstract query (assuming D is the dataset):

$$(\text{Project}(\text{ToList}(E), \{?mbox\}), D, \text{SELECT})$$

**Solution 2.** We start again with the query pattern, which is again a group graph pattern this time consisting of an element (a `TriplesBlock`) with a filter. We translate according to Algorithm 1 and then according to the case for `GroupGraphPattern` with one filter and one element. We have to apply TranslateGroup and obtain, as in the previous exercise, Join(Z, Bgp(?x foaf:name ?name)), which we simplify to Bgp(?x foaf:name ?name). Together with the filter translation, this results in

$$\text{Filter}(\text{regex}(?name, \text{"ian"}), \text{Bgp}(?x \text{ foaf:name } ?name)).$$

Now that we have the algebra translation for the query pattern, which we denote with E, we can obtain the algebra translation for the whole query according to Definition 8. After applying ToList(E) and Project(ToList(E), {?name}) as above, we further translate the DISTINCT keyword and obtain:

$$(\text{Distinct}(\text{Project}(\text{ToList}(\text{Filter}(\text{regex}(?name, \text{"ian"}), \text{Bgp}(?x \text{ foaf:name } ?name )))$$
$$\{?mbox\})), D, \text{SELECT})$$

**Solution 3.** We start with the query pattern, which is, as every query pattern, a group graph pattern consisting of one element, which is a `GroupOrUnionGraphPattern` of the form

GroupGraphPattern UNION GroupGraphPattern

as can be seen from the grammar in Table 2. Thus, we start with a translation according to Algorithm 1 and then according to the case for `GroupOrUnionGraphPattern` from Definition 3 obtaining: Join(Z, Union(algbr($G_1$), algbr($G_2$))) with $G_1$ and $G_2$ denoting the first and the second group of the union, respectively. For $G_1$ we again use Algorithm 1 followed by the case for `TriplesBlock` from Definition 3, leading to

$$\text{Join}(Z, \text{Bgp}(?x \text{ foaf:name "Birte Glimm" . } ?x \text{ foaf:mbox } ?mbox)).$$

Since $G_2$ has a filter, we obtain

$$\text{Filter}( \text{regex}(?name, \text{"ian"}),$$
$$\text{Join}(Z, \text{Bgp}(?x \text{ foaf:name } ?name . ?x \text{ foaf:mbox } ?mbox))).$$

Putting all together, we get:

Join(Z, Union( Join(Z, Bgp(?x foaf:name "Birte Glimm" . ?x foaf:mbox ?mbox)),
  Filter( regex(?name, "ian"),
    Join(Z, Bgp(?x foaf:name ?name . ?x foaf:mbox ?mbox)))))

which can be simplified to

Union( Bgp(?x foaf:name "Birte Glimm" . ?x foaf:mbox ?mbox),
  Filter(regex(?name, "ian"), Bgp(?x foaf:name ?name . ?x foaf:mbox ?mbox)))

Now that we have the algebra translation for the query pattern, which we denote with E, we can obtain the algebra translation for the whole query and then the abstract query as described in Definition 8:

$$(\mathsf{Project}(\mathsf{ToList}(E), \{?\mathsf{mbox}\}), D, \mathsf{SELECT})$$

**Solution 4.** We again translate the query pattern first obtaining:

Union( LeftJoin(Join(Z, Bgp(?x foaf:name ?name)), Bgp(?x foaf:icqChatID ?id), true),
  Bgp(?x foaf:name ?name . ?x foaf:mbox <mailto:rudolph@kit.edu>))

The expression can be simplified to:

Union(LeftJoin(Bgp(?x foaf:name ?name), Bgp(?x foaf:icqChatID ?id), true),
  Bgp(?x foaf:name ?name . ?x foaf:mbox <mailto:rudolph@kit.edu>))

We refer to the simplified expression as E and obtain the abstract query:

$$(\mathsf{Project}(\mathsf{OrderBy}(\mathsf{ToList}(E), (\mathsf{ASC}(?\mathsf{name})), \{?\mathsf{name}, ?\mathsf{id}\}))$$

## 6.2  Query Evaluation

**Solution 5.** We evaluate the algebra expression inside out, starting with the BGPs. The evaluation of Bgp(?x foaf:name "Birte Glimm" . ?x foaf:mbox ?mbox) yields $\Omega_1 = \{\mu_1\}$ with

$$\mu_1 : ?\mathsf{x} \mapsto \_{:}\mathsf{a}, ?\mathsf{mbox} \mapsto \text{"b.glimm@googlemail.com"}.$$

The evaluation of Bgp(?x foaf:name ?name . ?x foaf:mbox ?mbox) yields $\Omega_2 = \{\mu_2, \mu_3\}$ with

$$\mu_2 : ?\mathsf{x} \mapsto \_{:}\mathsf{a}, ?\mathsf{name} \mapsto \text{"Birte Glimm"}, ?\mathsf{mbox} \mapsto \text{"b.glimm@googlemail.com"},$$
$$\mu_3 : ?\mathsf{x} \mapsto \_{:}\mathsf{b}, ?\mathsf{name} \mapsto \text{"Sebastian Rudolph"}, ?\mathsf{mbox} \mapsto <\text{mailto:rudolph@kit.edu}>.$$

We next evaluate Filter(regex(?name, "ian"), $\Omega_2$) obtaining $\Omega_2' = \{\mu_3\}$. We can now evaluate the union operator, which yields $\Omega = \{\mu_1, \mu_3\}$, which is then turned into a list by the ToList operator. Applying the projection operator yields the final solution sequence: $(\mu_1', \mu_3')$ with

$$\mu_1' : ?\mathsf{mbox} \mapsto \text{"b.glimm@googlemail.com"},$$
$$\mu_3' : ?\mathsf{mbox} \mapsto <\text{mailto:rudolph@kit.edu}>.$$

**Solution 6.** We again evaluate the algebra expression inside out, starting with the BGPs. The evaluation of Bgp(?x foaf:name ?name) yields $\Omega_1 = \{\mu_1^1, \mu_1^2, \mu_1^3\}$ with

$$\mu_1^1 \; : \; ?x \mapsto \_{:}a, ?name \mapsto \text{"Birte Glimm"},$$
$$\mu_1^2 \; : \; ?x \mapsto \_{:}b, ?name \mapsto \text{"Sebastian Rudolph"},$$
$$\mu_1^3 \; : \; ?x \mapsto \_{:}c, ?name \mapsto \text{"Pascal Hitzler"}.$$

The evaluation of Bgp(?x foaf:icqChatID ?id) yields $\Omega_2 = \{\mu_2^1\}$ with

$$\mu_2^1 : \; ?x \mapsto \_{:}a, ?id \mapsto \text{"b.glimm"}.$$

For Bgp(?x foaf:name ?name . ?x foaf:mbox <mailto:rudolph@kit.edu>) we obtain $\Omega_3 = \{\mu_3^1\}$ with

$$\mu_3^1 : \; ?x \mapsto \_{:}b, ?name \mapsto \text{"Sebastian Rudolph"}.$$

In order to evaluate LeftJoin($\Omega_1, \Omega_2,$ true), we first compute Filter(true, Join($\Omega_1, \Omega_2$)) which yields $\Omega_4 = \{\mu_4^1\}$ with

$$\mu_4^1 : \; ?x \mapsto \_{:}a, ?name \mapsto \text{"Birte Glimm"}, ?id \mapsto \text{"b.glimm"}.$$

The mappings $\mu_1^2$ and $\mu_1^3$ cannot be joined with $\mu_2^1$ since they are not compatible. Due to the incompatibility, both these mapping participate, however, in the union and are part of the solution for LeftJoin($\Omega_1, \Omega_2,$ true) due to the second part of the LeftJoin definition. Evaluating LeftJoin($\Omega_1, \Omega_2,$ true) yields $\Omega_5 = \{\mu_4^1\} \cup \{\mu_1^2, \mu_1^3\} = \{\mu_5^1, \mu_5^2, \mu_5^3\}$ with

$$\mu_5^1 = \mu_4^1 \; : \; ?x \mapsto \_{:}a, ?name \mapsto \text{"Birte Glimm"}, ?id \mapsto \text{"b.glimm"},$$
$$\mu_5^2 = \mu_1^2 \; : \; ?x \mapsto \_{:}b, ?name \mapsto \text{"Sebastian Rudolph"},$$
$$\mu_5^3 = \mu_1^3 \; : \; ?x \mapsto \_{:}c, ?name \mapsto \text{"Pascal Hitzler"}.$$

We can now evaluate Union($\Omega_5, \Omega_3$), which yields $\Omega_6 = \{\mu_6^1, \mu_6^2, \mu_6^3, \mu_6^4\}$ with

$$\mu_6^1 \; : \; ?x \mapsto \_{:}a, ?name \mapsto \text{"Birte Glimm"}, ?id \mapsto \text{"b.glimm"},$$
$$\mu_6^2 \; : \; ?x \mapsto \_{:}b, ?name \mapsto \text{"Sebastian Rudolph"},$$
$$\mu_6^3 \; : \; ?x \mapsto \_{:}c, ?name \mapsto \text{"Pascal Hitzler"},$$
$$\mu_6^4 \; : \; ?x \mapsto \_{:}b, ?name \mapsto \text{"Sebastian Rudolph"}$$

The multiset $\Omega_6$ is then turned into a list by the ToList operator. Applying the OrderBy operator yields the list $(\mu_6^1, \mu_6^3, \mu_6^2, \mu_6^4)$. Finally, applying the projection operator yields: $(\mu_7^1, \mu_7^2, \mu_7^3, \mu_7^4)$ with

$$\mu_7^1 \; : \; ?name \mapsto \text{"Birte Glimm"}, ?id \mapsto \text{"b.glimm"},$$
$$\mu_7^2 \; : \; ?name \mapsto \text{"Pascal Hitzler"},$$
$$\mu_7^3 \; : \; ?name \mapsto \text{"Sebastian Rudolph"},$$
$$\mu_7^4 \; : \; ?name \mapsto \text{"Sebastian Rudolph"}.$$

## 6.3   RDFS Semantics Queries

**Solution 7.** We first list triples that are entailed under RDF semantics that are contributing solutions. The entailment follows from the RDFS entailment rules [14]. The relevant rule and the triples to which the rule is applied are indicated in the left-hand side column.

rdfs9 + (1) + (11) → (14)  iswc2010:paper/280     rdf:type  :Publication.
rdfs9 + (6) + (11) → (15)  w3c:TR/rdf-sparql-query rdf:type  :Publication.
rdfs7 + (7) + (13) → (15)  w3c:TR/rdf-sparql-query :authors _:l2.

If we were to materialize all RDFS-entailed triples, there would be several additional triples, but we focus here on the relevant ones. Although the above RDFS-entailed triples do not contain freshly generated blank nodes, we want to point out that sometimes blank nodes have to be introduced in the rule application process, but such freshly introduced blank nodes cannot be returned in a solution since they are not part of the answer domain. We obtain the following solutions from evaluating the BGP:

|  | ?pub | | ?seq | ?ind | ?auth |
|---|---|---|---|---|---|
| $\mu_1$ : | iswc2010:paper/280 | _:l1 | rdf:type | rdf:Seq | |
| $\mu_2$ : | iswc2010:paper/280 | _:l1 | rdf:_1 | "Birte Glimm" | |
| $\mu_3$ : | iswc2010:paper/280 | _:l1 | rdf:_2 | "Markus Krötzsch" | |
| $\mu_4$ : | w3c:rdf-sparql-query | _:l2 | rdf:type | rdf:Seq | |
| $\mu_5$ : | w3c:rdf-sparql-query | _:l2 | rdf:_1 | "Andy Seaborne" | |
| $\mu_6$ : | w3c:rdf-sparql-query | _:l2 | rdf:_2 | "Eric Prud'hommeaux" | |

Computing the projection is then straightforward.

**Solution 8.** One possibility would be to apply a filter to ?auth that only permits literals as binding:

```
SELECT ?auth ?pub
WHERE { ?pub rdf:type :Publication . ?pub :authors ?seq . ?seq ?ind ?auth
        FILTER isLiteral(?auth) }
```

Other solutions with different filters are equally possible.

**Solution 9.** We first list triples that are entailed under RDF semantics that are contributing solutions. The entailment follows from the RDFS entailment rules [14]. The relevant rule and the triples to which the rule is applied are indicated in the left-hand side column.

rdfs4a + (1) → (14)   iswc2010:paper/280 rdf:type rdfs:Resource.
rdfs9 + (1) + (11) → (15)   iswc2010:paper/280 rdf:type :Publication.

thus, the query has two answers. The first inference might be surprising, but under RDFS entailment, we derive several such triples. If such triples are not desired, a filter can again be used to filter them out.

**Solution 10.** The triple iswc2010:paper/280 :authors _:x is indeed entailed under RDFS semantics since entailment treats black nodes as existential variables. According to triple (2), iswc2010:paper/280 is related via the property :authors to *some* element, witnessed by the blank node _:l1 in the data. Since the actual names of variables do not matter, i.e., the only question to decide is whether there is some element such that iswc2010:paper/280 is related to this element with the property :authors, which is the case.

Regarding the Boolean query (here we only have a blank node, no variable), we have two possible outcomes: there is a solution sequence containing a mapping ( $\mu$ ) where $\mu$ has an empty domain (it does not map any variable to anything) or there is only an empty solution sequence ( ). In the first case, the query answer is yes (true), whereas in the second case the query answer is no (false).

For the RDFS entailment regime, we work with a Skolem function that maps blank nodes from the active graph to constants, i.e., to fresh terms that occur neither in the query nor in the active graph. Let us assume that _:l1 is mapped to sk(l1). Since the query contains a blank node, we have to find an RDF instance mapping such that when we apply the mapping and then use the same Skolem function, the triples are entailed and ground. Thus, let $\mu$ be the mapping with an empty domain and $\sigma$: _:x $\mapsto$ _:l1, then

$$\mu(\sigma(\text{iswc2010:paper/280 :authors \_:x)}) = \text{iswc2010:paper/280 :authors sk(l1)},$$

which is a ground triple that is entailed by sk(G) (even contained in sk(G)). Thus, the query answer is true.

## 6.4 OWL Direct Semantics Queries

**Solution 11.** The BGP is mapped into

> FSS: ClassAssertion(:MilkProduct ?mp)        DL: MilkProduct(?mp)

using the declaration axiom (6). Evaluating the BGP yields two solutions:

$$\mu_1 : \text{?mp} \mapsto \text{:Yoghurt}$$
$$\mu_2 : \text{?mp} \mapsto \text{:Milk}$$

where $\mu_2$ is a direct consequence of Axiom (10) and $\mu_1$ follows from Axiom (9) and (11).

**Solution 12.** The BGP of the query pattern is mapped into SubClassOf(:MilkProduct ?sup) (DL: MilkProduct $\sqsubseteq$ ?sup). Evaluating the mapped BGP yields $\Omega = \{\mu_1, \mu_2\}$ with

$$\mu_1 : \text{?sup} \mapsto \text{:MilkProduct}$$
$$\mu_2 : \text{?mp} \mapsto \text{owl:Thing}.$$

The solution $\mu_1$ follows since each class is a subclass of itself under the DS (the subclass relation is reflexive) and $\mu_2$ follows since owl:Thing is a superclass of every class. Applying ToList and Project yields the solution sequence $(\mu_1, \mu_2)$.

**Solution 13.** The first problem is that ?rel is not typed. This makes it difficult to map the query pattern into an extended OWL object. Even worse, no matter what type we could add, the query cannot be fixed. Two classes, such as :Vegetarian and :Vegan, can only be related with terms from the special vocabulary, e.g., by saying that :Vegan is a subclass of :Vegetarian (in Turtle: :Vegan rdfs:subClassOf :Vegetarian) or by saying that the two classes are disjoint (in Turtle: :Vegetarian owl:disjointWith :Vegan). However, since terms of the special vocabulary do not have any of the types that variables can take, the query pattern cannot be fixed.

**Solution 14.** The query

> SELECT ?class
> WHERE { ?class rdfs:subClassOf owl:Thing . ?class rdf:type owl:Class }

would retrieve all classes of the ontology since any class is a subclass of owl:Thing (in DL: $\top$) under OWL's semantics. The typing triple is not necessary in this case since the parsing is unambiguous given that owl:Thing assumed to be declared as a class in any ontology even if such a declaration is not explicitly present.

**Solution 15.** A SPARQL query cannot distinguish between direct and indirect subclasses. Thus, a single query can, in general, not be used to retrieve all and only the required pairs. One would also get the indirect subclasses and it would be difficult to filter them out, at least in a single query.

**Solution 16.** If completeness is required, i.e., we want to return all solutions that are solutions, then materialization cannot be used as a general implementation technique. One of the problems are disjunctions, i.e., there is not just one canonical model of an OWL ontology that represents all relevant possible states of the world. One could argue that we could just include facts that hold in every model, e.g., if we have

```
:a rdf:type :C .
:b rdf:type [ rdf:type owl:Class ; owl:unionOf ( :D :E :F ) ].
:F rdfs:subClassOf owl:Nothing .
```

which is

$$C(a)$$
$$(D \sqcup E \sqcup F)(b)$$
$$F \sqsubseteq \bot$$

in DL notation, then we could argue that we add

```
:b rdf:type [ rdf:type owl:Class ; owl:unionOf ( :D :E ) ]
```

which is

$$(D \sqcup E)(b)$$

in DL notation to obtain a "canonical" model (since :F is a subclass of owl:Nothing it cannot have any instances). However, a BGP such as

```
?ind rdf:type [ rdf:type owl:Class ; owl:unionOf ( :C :D :F ) ]
```

would still have :a and :b as solutions (:a since it belongs to :C and :b as it belongs to the union of :D and :E). It would be impossible to foresee all such queries and materialize the required axioms in a finite ontology.

This is different for the OWL RL profile. The semantics of OWL 2 RL is defined such that certain consequences have to be derived, e.g., one materializes only (named) classes to which an individual belongs. The OWL 2 RL specification includes a set of rules that materialize all such consequences. Under certain restrictions for the ontology, the OWL RL rules derive all consequences that one would derive under the Direct Semantics. If the ontology violates the restrictions, then one might miss some answers that a tool that implements OWL 2 with its Direct Semantics could derive.

## 7    Links and Further Reading

The following list of references is not meant to be complete and is a subjective selection by the author. References that are not listed can equally be relevant and students are encouraged to look for references that most closely fit with their interests.

A text book covering the topics relevant for this summer school is: *Foundations of Semantic Web Technologies* Hitzler, P., Krötzsch, M., Rudolph, S. CRC Press 2009

## 7.1   Public SPARQL Endpoints

*data.gov.uk* . The UK Government makes over 5,400 datasets publicly available, from all central government departments and a number of other public sector bodies and local authorities. The site also includes links to SPARQL tutorials and examples:
`http://data.gov.uk/sparql`

*DBPedia* contains structured information from Wikipedia (> 100 million triples):
`http://dbpedia.org/sparql`, see `http://www.dbpedia.com` for further information and documentation

*DBTune* provides access to music-related structured data with more than 14 billion RDF triples. The interface also allows for selecting an entailment regime that is to be used (RDF, RDFS, plus the non-standardized RDFSLite and p2r) :
`http://dbtune.org/jamendo/store/user/query`

*CKAN* is a platform to share, use, and find data that is publicly available
`http://semantic.ckan.net/sparql/`

*Linked Movie Database* . A semantic web database for movies, including a large number of interlinks to several datasets on the open data cloud and references to related webpages
`http://data.linkedmdb.org/` and `http://data.linkedmdb.org/sparql`

*SPARQL Editor* . Talis hosts a SPARQl Editor with Examples for Space Data `http://api.talis.com/stores/space/items/tutorial/spared.html`

*Semantic Web Dog Food* contains data about authors and publications for several conferences:
`http://data.semanticweb.org/snorql`

*SPARQL Endpoint Status* collects uptime information for SPARQL endpoints from CKAN
`http://labs.mondeca.com/sparqlEndpointsStatus/index.html`

## 7.2   RDFS

*Completeness, decidability and complexity of entailment for RDF Schema and a semantic extension involving the OWL vocabulary* ter Horst, H.J.: Journal of Web Semantics 3(2–3), 79–115 (2005).

## 7.3  OWL and OWL Reasoning

*OWL 2 Web Ontology Language: Primer*  Hitzler, P., Krötzsch, M., Parsia, B., Patel-Schneider, P.F., Rudolph, S. (eds.). W3C Recommendation (2009), available at http://www.w3.org/TR/owl2-primer/

*OWL 2: The next step for OWL*  Cuenca Grau, B. Horrocks, I., Motik, B., Parsia, B., Patel-Schneider, P., and Sattler, U.: Journal of Web Semantics: Science, Services and Agents on the World Wide Web, 6(4):309–322, 2008.

*From $\mathcal{SHIQ}$ and RDF to OWL: The Making of a Web Ontology Language*  Horrocks, I., Patel-Schneider, P.F., van Harmelen, F.: Journal of Web Semantics 1(1), 7–26 (2003)

*The Description Logic Handbook*  Baader, F., Calvanese, D., McGuinness, D.L., Nardi, D., Patel-Schneider, P.F. Cambridge University Press (2003)

*Hypertableau Reasoning for Description Logics*  Motik, B., Shearer, R., Horrocks, I.: Journal of Artificial Intelligence Research 173(14), 1275–1309 (2009)

*The even more irresistible $\mathcal{SROIQ}$*  Horrocks,I., Kutz,O., Sattler,U.: Proceedings of the 10th International Conference on the Principles of Knowledge Representation and Reasoning (KR 2006). pp. 57–67 (2006)

*$\mathcal{RIQ}$ and $\mathcal{SROIQ}$ are harder than $\mathcal{SHOIQ}$*  Kazakov, Y.: Proceedings of the 11th International Conference on the Principles of Knowledge Representation and Reasoning (KR 2008). AAAI Press/The MIT Press (2008)

*A Tableau Decision Procedure for $\mathcal{SHOIQ}$*  Horrocks, I., Sattler, U.: Journal of Automated Reasoning 39(3), 249–276 (2007)

*Reasoning in Description Logics using Resolution and Deductive Databases*  Motik, B.: Ph.D. thesis, Univesität Karlsruhe (TH), Karlsruhe, Germany (2006)

*A practical OWL-DL Reasoner*  Sirin, E., Parsia, B., Cuenca Grau, B., Kalyanpur, A., Katz, Y.: Pelle. Journal of Web Semantics 5(2) (2007)

*Reducing OWL Entailment to Description Logic Satisfiability*  Horrocks, I., Patel-Schneider, P.: Journal of Web Semantics 1(4), 345–357 (2004)

*Rules and Ontologies for the Semantic Web*  Eiter, T., Ianni, G., Krennwallner, T., Polleres, A. Reasoning Web, Fourth International Summer School 2008 Springer, 2008.

*Scalable Authoritative OWL Reasoning for the Web*  Hogan, A., Harth, A., Polleres, A.: IJSWIS "Semantic Services, Interoperability and Web Applications: Emerging Concepts". Journal Summation Volume. To appear, 2011.

*Dynamic Querying of Mass-Storage RDF Data with Rule-Based Entailment Regimes* Ianni, G., Krennwallner, K., Martello, A., Polleres, A.: Proceedings of the 8th International Semantic Web Conference (ISWC 2009), LNCS, Springer-Verlag 2009.

*From SPARQL to Rules (and back)* Polleres, A.: Proceedings of the 16th International World Wide Web Conference, 2007.

*Scalable Authoritative OWL Reasoning on a Billion Triples* Hogan, A., Harth, A., Polleres, A.: Proceedings of Billion Triple Semantic Web Challenge Workshop at 7th International Semantic Web Conference, 2008.

## 7.4  SPARQL

*Semantics and complexity of SPARQL* Pérez, J., Arenas, M., Gutierrez, C. ACM Transactions on Database Systems 34(3), 1–45 (2009)

*Search RDF data with SPARQL* McCarthy, P.:
`http://www.ibm.com/developerworks/xml/library/j-sparql/`

*SPARQL Tutorial* – Jena/ARQ `http://jena.sourceforge.net/ARQ/Tutorial/`

*SPARQL by Example* – Cambridge Semantics `http://www.cambridgesemantics.com/2008/09/sparql-by-example/`

*Data Extraction & Exploration with SPARQL & the Talis platform* `http://www.slideshare.net/ldodds/sparql-tutorial`

*Introducing SPARQL: Querying the Semantic Web* Dodds, L.: `http://www.xml.com/pub/a/2005/11/16/introducing-sparql-querying-semantic-web-tutorial.html`

## 7.5  SPARQL over OWL Ontologies

*SPARQL Beyond Subgraph Matching* Glimm, B., Krötzsch, M.: In: Proceedings of the 9th International Semantic Web Conference (ISWC 2010). vol. 6496, pp. 241–256. Springer-Verlag (2010)

*SPARQL-DL: SPARQL Query for OWL-DL* Sirin,E., Parsia,B.: Proceedings of the 3rd OWL Experiences and Directions Workshop (OWLED 2007) (2007)

*Optimizations for Answering Conjunctive ABox Queries* Sirin, E., Parsia, B.: Proceedings of the 2006 Description Logic Workshop (DL 2006) (2006)

# References

1. Baader, F.: Terminological cycles in a description logic with existential restrictions. In: Proceedings of the 18th International Joint Conference on Artificial Intelligence (IJCAI 2003), pp. 325–330 (2003)
2. Baader, F., Brandt, S., Lutz, C.: Pushing the $\mathcal{EL}$ envelope. In: Proceedings of the 19th International Joint Conference on Artificial Intelligence (IJCAI 2005), vol. 19, pp. 364–369 (2005)
3. Baader, F., Lutz, C., Suntisrivaraporn, B.: Efficient reasoning in $\mathcal{EL}^+$. In: Proceedings of the 2006 Description Logic Workshop (DL 2006). CEUR Workshop Proceedings (2006)
4. Beckett, D., Berners-Lee, T.: Turtle – Terse RDF Triple Language. W3C Team Submission (January 14, 2008), http://www.w3.org/TeamSubmission/turtle/
5. Beckett, D., Broekstra, J. (eds.): SPARQL Query Results XML Format (January 15, 2008), http://www.w3.org/TR/rdf-sparql-XMLres/
6. Boley, H., Hallmark, G., Kifer, M., Paschke, A., Polleres, A., Reynolds, D. (eds.): RIF Core Dialect. W3C Recommendation (2010), http://www.w3.org/TR/rif-core/
7. Brandt, S.: Polynomial time reasoning in a description logic with existential restrictions, GCI axioms, and–what else? In: de Mantáras, R.L., Saitta, L. (eds.) Proceedings of the 16th European Conference on Artificial Intelligence (ECAI 2004), pp. 298–302. IOS Press, Amsterdam (2004)
8. de Bruijn, J. (ed.): RIF RDF and OWL Compatibility. W3C Recommendation (2010), http://www.w3.org/TR/rif-rdf-owl/
9. Calvanese, D., De Giacomo, G., Lembo, D., Lenzerini, M., Rosati, R.: DL-Lite: Tractable description logics for ontologies, pp. 602–607 (2005)
10. Calvanese, D., De Giacomo, G., Lembo, D., Lenzerini, M., Rosati, R.: Tractable reasoning and efficient query answering in description logics: The DL-Lite family 39(3), pp. 385–429 (2007)
11. Charboneau, D., Feigenbaum, L. (eds.): SPARQL 1.1 Protocol for RDF. W3C Working Draft (January 26, 2010), http://www.w3.org/TR/sparql11-protocol/
12. Glimm, B., Ogbuji, C. (eds.): SPARQL 1.1 Entailment Regimes. W3C Working Draft (October 14, 2010), http://www.w3.org/TR/sparql11-entailment/
13. Harris, S., Seaborne, A. (eds.): SPARQL 1.1 Query Language. W3C Working Draft (October 14, 2010), http://www.w3.org/TR/sparql11-query/
14. Hayes, P.: RDF semantics (2004), http://www.w3.org/TR/rdf-mt/
15. ter Horst, H.J.: Completeness, decidability and complexity of entailment for RDF Schema and a semantic extension involving the OWL vocabulary. Journal of Web Semantics 3(2-3), 79–115 (2005)
16. Kazakov, Y.: Consequence-driven reasoning for horn SHIQ ontologies. In: Proceedings of the 21st International Joint Conference on Artificial Intelligence (IJCAI 2009), pp. 2040–2045 (2009)
17. Kifer, M., Boley, H. (eds.): RIF Overview. W3C Working Group Note (2010), http://www.w3.org/TR/rif-overview/
18. Kontchakov, R., Lutz, C., Toman, D., Wolter, F., Zakharyaschev, M.: The combined approach to query answering in DL-Lite. In: Proceedings of the 12th International Conference on the Principles of Knowledge Representation and Reasoning (KR 2010). AAAI Press,The MIT Press (2010)
19. Krötzsch, M.: Efficient inferencing for OWL EL. In: Janhunen, T., Niemelä, I. (eds.) JELIA 2010. LNCS, vol. 6341, pp. 234–246. Springer, Heidelberg (2010)
20. Motik, B., Cuenca Grau, B., Horrocks, I., Wu, Z., Fokoue, A., Lutz, C. (eds.): OWL 2 Web Ontology Language: Profiles. W3C Recommendation (2009), http://www.w3.org/TR/owl2-profiles/

21. Motik, B., Patel-Schneider, P.F., Parsia, B. (eds.): OWL 2 Web Ontology Language: Structural Specification and Functional-Style Syntax. W3C Recommendation (2009), http://www.w3.org/TR/owl2-syntax/
22. Ogbuji, C. (ed.): SPARQL 1.1 Uniform HTTP Protocol for Managing RDF Graphs. W3C Working Draft (October 14, 2010),
http://www.w3.org/TR/sparql11-http-rdf-update/
23. Patel-Schneider, P.F., Motik, B. (eds.): OWL 2 Web Ontology Language: Mapping to RDF Graphs. W3C Recommendation (2009), http://www.w3.org/TR/owl2-mapping-to-rdf/
24. Pérez, J., Arenas, M., Gutierrez, C.: Semantics and complexity of SPARQL. ACM Trans. Database Syst. 34(3), 1–45 (2009)
25. Pérez-Urbina, H., Horrocks, I., Motik, B.: Efficient query answering for OWL 2. In: Bernstein, A., Karger, D.R., Heath, T., Feigenbaum, L., Maynard, D., Motta, E., Thirunarayan, K. (eds.) ISWC 2009. LNCS, vol. 5823, pp. 489–504. Springer, Heidelberg (2009)
26. Prud'hommeaux, E., Seaborne, A. (eds.): SPARQL Query Language for RDF. W3C Recommendation (January 15, 2008), http://www.w3.org/TR/rdf-sparql-query/
27. Rosati, R., Almatelli, A.: Improving query answering over DL-Lite ontologies. In: Rosati, R., Almatelli, A. (eds.) Proceedings of the 12th International Conference on the Principles of Knowledge Representation and Reasoning (KR 2010). AAAI Press, The MIT Press (2010)
28. Schenk, S., Gearon, P., Passant, A. (eds.): SPARQL 1.1 Update. W3C Working Draft (October 14, 2010), http://www.w3.org/TR/sparql11-update/
29. Schneider, M. (ed.): OWL 2 Web Ontology Language: RDF-Based Semantics. W3C Recommendation (2009), http://www.w3.org/TR/owl2-rdf-based-semantics/
30. Williams, G.T. (ed.): SPARQL 1.1 Service Description. W3C Working Draft (October 14, 2010), http://www.w3.org/TR/sparql11-service-description/

# Database Foundations for Scalable RDF Processing

Katja Hose[1], Ralf Schenkel[1,2], Martin Theobald[1], and Gerhard Weikum[1]

[1] Max-Planck-Institut für Informatik, Saarbrücken, Germany
[2] Saarland University, Saarbrücken, Germany

**Abstract.** As more and more data is provided in RDF format, storing huge amounts of RDF data and efficiently processing queries on such data is becoming increasingly important. The first part of the lecture will introduce state-of-the-art techniques for scalably storing and querying RDF with relational systems, including alternatives for storing RDF, efficient index structures, and query optimization techniques. As centralized RDF repositories have limitations in scalability and failure tolerance, decentralized architectures have been proposed. The second part of the lecture will highlight system architectures and strategies for distributed RDF processing. We cover search engines as well as federated query processing, highlight differences to classic federated database systems, and discuss efficient techniques for distributed query processing in general and for RDF data in particular. Moreover, for the last part of this chapter, we argue that extracting knowledge from the Web is an excellent showcase – and potentially one of the biggest challenges – for the scalable management of uncertain data we have seen so far. The third part of the lecture is thus intended to provide a close-up on current approaches and platforms to make reasoning (e.g., in the form of probabilistic inference) with uncertain RDF data scalable to billions of triples.

## 1   RDF in Centralized Relational Databases

The increasing availability and use of RDF-based information in the last decade has led to an increasing need for systems that can store RDF and, more importantly, efficiently evaluate complex queries over large bodies of RDF data. The database community has developed a large number of systems to satisfy this need, partly reusing and adapting well-established techniques from relational databases [122]. The majority of these systems can be grouped into one of the following three classes:

1. *Triple stores* that store RDF triples in a single relational table, usually with additional indexes and statistics,
2. *vertically partitioned tables* that maintain one table for each property, and
3. Schema-specific solutions that store RDF in a number of *property tables* where several properties are jointly represented.

A. Polleres et al. (Eds.): Reasoning Web 2011, LNCS 6848, pp. 202–249, 2011.

```
<Katja,teaches,Databases>
<Katja,works_for,MPI Informatics>
<Katja,PhD_from,TU Ilmenau>
<Martin,teaches,Databases>
<Martin,works_for,MPI Informatics>
<Martin,PhD_from,Saarland University>
<Ralf,teaches,Information Retrieval>
<Ralf,PhD_from,Saarland University>
<Ralf,works_for,Saarland University>
<Saarland University,located_in,Germany>
<MPI Informatics,located_in,Germany>
```

**Fig. 1.** Running example for RDF data

In the following sections, we will describe each of these classes in detail, focusing on two important aspects of these systems: *storage and indexing*, i.e., how are RDF triples mapped to relational tables and which additional support structures are created; and *query processing*, i.e., how SPARQL queries are mapped to SQL, which additional operators are introduced, and how efficient execution plans for queries are determined. In addition to these purely relational solutions, a number of specialized RDF systems has been proposed that built on non-relational technologies, we will briefly discuss some of these systems. Note that we will focus on SPARQL[1] processing, which is not aware of underlying RDF/S or OWL schema and cannot exploit any information about subclasses; this is usually done in an additional layer on top.

We will explain especially the different storage variants with the running example from Figure 1, some simple RDF *facts* from a university scenario. Here, each line corresponds to a fact (*triple*, *statement*), with a *subject* (usually a resource), a *property* (or *predicate*), and an *object* (which can be a resource or a constant). Even though resources are represented by URIs in RDF, we use string constants here for simplicity. A collection of RDF facts can also be represented as a graph. Here, resources (and constants) are nodes, and for each fact <s,p,o>, an edge from s to o is added with label p. Figure 2 shows the graph representation for the RDF example from Figure 1.

### 1.1 Triple Stores

Triple stores keep RDF triples in a simple relational table with three or four attributes. This very generic solution with low implementation overhead has been very popular, and a large number of systems based on this principle are available. Prominent examples include 3store [56] and Virtuoso [41] from the Semantic Web community, and RDF-3X [101] and HexaStore [155] that were developed by database groups.

**Storage.** RDF facts are mapped to a generic three-attribute table of the form (subject,property,object), also know as *triple table*; for simplicity, we will abbreviate the attributes by S, P, and O. To save space (and to make access structures

---

[1] http://www.w3.org/TR/rdf-sparql-query/

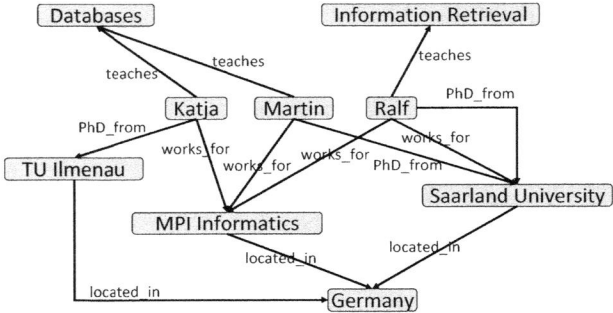

**Fig. 2.** Graph representation for the RDF example from Figure 1

| subject | property | object |
|---------|----------|--------|
| Katja | teaches | Databases |
| Katja | works_for | MPI Informatics |
| Katja | PhD_from | TU Ilmenau |
| Martin | teaches | Databases |
| Martin | works_for | MPI Informatics |
| Martin | PhD_from | Saarland University |
| Ralf | teaches | Information Retrieval |
| Ralf | PhD_from | Saarland University |
| Ralf | works_for | Saarland University |
| Ralf | works_for | MPI Informatics |
| Saarland University | locatedIn | Germany |
| MPI Informatics | located_in | Germany |

**Fig. 3.** Triple store representation for the running example from Figure 1

more efficient), most systems convert resource identifiers, properties and constants to numeric ids before storing them in the relation, for example by hashing. The resulting map is usually stored in an additional table, sometimes separately for resource ids and constants. If a system stores data from more than one source (or more than one RDF graph), the relation is often extended by a fourth numeric attribute, the **graph id** (abbreviated as **G**), that uniquely identifies the source of a triple. In this case, the relation is also called a *quadruple table*.

Figure 3 shows the resulting three-attribute relation for the example from Figure 1.

For efficient query processing, indexes on (a subset of) all combinations of S, P, and OS are maintained. This allows to efficiently retrieve all matches for a triple pattern of a SPARQL query. We will often refer to indexes with the sequence of the abbreviations of the indexed attributes (such as SPO). Since each index has approximately the size of the relation, the number of combinations for which indexes are kept is usually limited, or indexes are stored in a compressed way.

*Virtuoso* [40] comes with a space-optimized way of mapping resources, predicates and constants to numeric ids (IRI ID). These strings are mapped to numeric ids only if they are long (which means at least 9 bytes long), otherwise, they are stored as text in the quadruple relation (this saves for short objects over a solution that maps everything to ids). It uses a standard quadruple table (G,S,P,O) with a primary key index on all four attributes together. In addition, it uses a bitmap index on (O,G,P,S): This index maintains, for each combination of (O,G,P) in the data, a bit vector. Each subject is assigned a bit position, and for a quadruple (g,s,p,o) in the data, the bit position for s is set to 1 in the bit vector for (o,g,p). Virtuoso stores distinct values within a page only once and eliminates common prefixes of strings. An additional compression of each page with gzip yields a compression from 8K to 3K for most pages.

*RDF-3X* uses a standard triple table and does not explicitly support multiple graphs. However, RDF-3X never actually materializes this table, but instead keeps clustered B+ tree indexes on all six combinations of (S,P,O). Additionally, RDF-3X includes aggregated indexes for each possible pair of (S,P,O) and each order, resulting in six additional indexes. The index on (S,O), for example, stores for each pair of subject and object that occurs in the data the number of triples with this subject and this object, we'll refer to this index as SO*. Such an index allows to efficiently answer queries like select ?s ?o where {?s ?p ?o}. We could process this by scanning the SOP index, but we don't need the exact bindings for ?p to generate the result, so we read many index entries that don't add new results. All we need is, for each binding of ?s and ?o, the number of triples with this subject and this object, so that we can generate the right number of results for this binding (including duplicates). The SO* index can help a lot here. Finally, RDF-3X maintains aggregated indexes for each single attribute, again plus triple counts.

To reduce space requirements of these indexes, RDF-3X stores the leaves of the indexes in pages and compresses them. Since these leaves contain triples that often share some attribute values and all attributes are numeric, it uses delta encoding for compression. This, together with encoding strings into comparably short numbers, helps to keep the overall size of the database comparable or even slightly smaller than the size of the uncompressed RDF triples in textual representation. The original RDF-3X paper [101] includes a discussion of space-time tradeoffs for compression and shows that, for example, compression with LZ77 generates more compact indexes, but requires significantly more time to decompress.

**Query Processing and Optimization.** Query execution on a quadruple store is done in two steps, *converting* the SPARQL query into an equivalent SQL query, and creating and executing a *query plan* for this SQL query.

*Step 1.* The conversion of a SPARQL query to an equivalent SQL query on the triple/quadruple table is a rather straight-forward process; we'll explain it now for triple tables. For each triple pattern in the SPARQL query, a copy of the triple relation is added to the query. Whenever a common variable is used in two patterns, a join between the corresponding relation instances is created on the attributes where the variable occurs. Any constants are directly mapped to conditions on the corresponding relation's attribute.

As an example, consider the SPARQL query

```
SELECT ?a ?b WHERE
  {?a works_for ?u.
   ?b works_for ?u.
   ?a phd_from ?u. }
```

which selects people who work at the same place where they got their phd, together with their coworkers. This is mapped to the SQL query

```
SELECT t1.s, t2.s FROM triple t1, triple t2, triple t3
WHERE t1.p='works_for'
  AND t2.p='works_for'
  AND t3.p='phd_from'
  AND t1.o=t2.o
  AND t1.o=t3.o
  AND t1.s=t3.s
```

Note that in a real system, the string constants would usually be mapped to the numeric id space first. Further note that we can optimize away one join here (t2.o=t3.o) since it is redundant.

*Step 2.* Now that we have a standard SQL query, it is tempting to simply rely on the existing relational backends for optimizing and processing this query. This is actually done in many systems, and even those systems which implement their own backend system use the same operators used in relational databases. Converting the SQL query into an equivalent abstract operator tree, for example an expression in relational algebra, is again straight-foward.

Once this is done, the next step is creating an efficient physical execution plan, i.e., decide how the abstract operators (joins, projection, selection) are mapped to physical implementations in a way that the resulting execution is as cheap (in terms of I/O and CPU usage) and as fast (in terms of processing times) as possible. The choice of implementations is rather huge, for example a join operator can be implemented with merge joins, hash joins, or nested loop joins. Additionally, a number of specialized joins exist (such as outer joins and semi joins) that can further improve efficiency. An important physical operator in many systems are index lookups and scans which exploit the numerous indexes that systems keep. Often, each triple pattern in the original SPARQL query corresponds to an index scan in the corresponding index if that index is available, for example, the triple pattern ?a works_for ?b could be mapped on a scan of the PSO index, if that exists. If the optimal index does not exist, scans of less specific indexes can be used, but some information from that index must be skipped. For example, if the system provides only an O index, a pattern ?a works_for MPI Informatics can be mapped to a scan of the O index, starting at MPI Informatics and skipping all entries that do not match the predicate constraint.

Finding the most efficient plan now includes considering possible variants of physical plans (such as different join implementations, different join orders, etc.) and selecting the most efficient plan. This, in turn, requires that the execution cost for each plan is estimated. It turns out that off-the-shelf techniques implemented in current relational databases, for example attribute-level histograms to represent the distribution. These techniques were not built for dealing with a single, large table. The main problem is that they ignore correlation of attributes, since statistics are available only separately for each attribute. Estimates (for example how many results a join will have, or how many results a selection will have) are therefore often way off, which can lead to arbitrarily bad execution plans. Multi-dimensional histograms, on the other hand, could capture this correlation, but can easily get too large for large-scale RDF data.

RDF-3X [101, 100] comes with specializes data structures for maintaining statistics. It uses histograms that can handle any triple pattern and any join, but assume independence of different patterns, and it comes with optimizations for frequent join paths. To further speed up processing, it applies sideway information passing between operators [100]. It also includes techniques to deal with unselective queries which return a large fraction of the database.

Virtuoso [40] exploits the bit vectors in its indexes for simple joins, which can be expressed as a conjunction of thse sparse bit vectors. As an example, consider the SPARQL query

```
select ?a
where {?a works_for Saarland University.
       ?a works\for MPI Informatics.}
```

To execute this, it is sufficient to load the bit vectors for both triple patterns and intersect them. For cost estimation, Virtuoso does not rely on per-attribute histogram, but uses query-time sampling: If a triple pattern has constants for p and o and the graph is fixed, it loads the first page of the bit vector for that pattern from the index, and extrapolates selectivity from the selectivity of this small sample.

Further solutions on the problem of selectivity estimation for graph queries were proposed by [91, 90] outside the context of an RDF system; Stocker et al. [135] consider the problem of query optimization with graph patterns.

## 1.2   Vertically Partitioned Tables

The vast majority of triple patterns in queries from real applications has fixed properties. To exploit this fact for storing RDF, one table with two attributes, one for storing subjects and one for storing objects, is created for each property in the data; if quadruples are stored, a third attribute for the graphid is added. An RDF triple is now stored in the table for its property. Like in the Triple table solution, string literals are usually encoded as numeric ids. Figure 4 shows how our example data from Figure 1 is represented with vertically partitioned tables.

Sinces tables have only two columns, this idea can be further pushed by not storing them in a traditional relational system (a *row store*), but in a *column*

| teaches | |
|---------|--------|
| subject | object |
| Katja | Databases |
| Martin | Databases |
| Ralf | Information Retrieval |

| works_for | |
|-----------|--------|
| subject | object |
| Katja | MPI Informatics |
| Martin | MPI Informatics |
| Ralf | MPI Informatics |
| Ralf | Saarland University |

| PhD_from | |
|----------|--------|
| subject | object |
| Katja | TU Ilmenau |
| Martin | Saarland University |
| Ralf | Saarland University |

| located_in | |
|------------|--------|
| subject | object |
| Saarland University | Germany |
| MPI Informatics | Germany |

**Fig. 4.** Representation of the running example from Figure 1 with vertically partitioned tables

*store.* A column store does not store tables as collections of rows, but as collection of columns, where each entry of a column comes with a unique ID that allows to reconstruct the rows at query time. This has the great advantage that all entries within a column have the same type and can therefore be compressed very efficiently. The idea of using column stores for RDF was initially proposed by Abadi et al. [2], Sidirourgos et al. [130] pointed out advantages and disadvantages of this technique.

Regarding query processing, it is evident that triple patterns with a fixed property can be evaluated very efficiently, by simply scanning the table for this property (or, in a column store, accessing the columns of this table). Query optimization can also be easier as per-table statistics can be maintained. On the other hand, triple patterns with a property wildcard are very expensive since they need to access all two-column tables and form the union of the results.

## 1.3 Property Tables

In many RDF data collections, a large number of subjects have the same or at least are largely overlapping set of properties, and many of these properties will be accessed together in queries (like in our example above that asked for people that did their PhD at the same place where they are working now). Combining all properties of a subject in the same table makes processing such queries much faster since there is no need for a join to combine the different properties. Property tables do exactly this: Groups of subjects with similar properties are represented by a single table where each attribute corresponds to a property. A set of facts for one of these subjects is then stored as one row in that table, where one column represents the subjects, and the other columns store objects for the properties that correspond to that column, or NULL if no such property exists. The most prominent example for this storage structure is Jena [158,23]. Chong et al. [27] proposed property tables as external view (which can be materialized) to simplify access to triple stores. Levandoski et al [80] demonstrate that property tables can outperform triple stores and vertically partitioned tables for RDF data collections with regular structure, such as DBLP or DBPedia.

| People | | |
|--------|--------|--------|
| subject | teaches | PhD_from |
| Katja | Databases | TU Ilmenau |
| Martin | Databases | Saarland University |
| Ralf | Information Retrieval | Saarland University |

| Institutions | |
|--------|--------|
| subject | located_in |
| Martin | Saarland University |
| Ralf | Saarland University |

| Remainder | | |
|--------|--------|--------|
| subject | predicate | object |
| Katja | works_for | MPI Informatics |
| Martin | works_for | MPI Informatics |
| Ralf | works_for | Saarland University |
| Ralf | works_for | MPI Informatics |

**Fig. 5.** Representation of the running example from Figure 1 with property tables

This table layout comes with two problems to solve: First, there should not be too many NULL values since they increase storage space, so storing the whole set of facts in a single table is not a viable solution. Instead, the set of subjects to store in one table can be determined by clustering subjects by the set of their properties, or subjects of the same type can be stored in the same table if schema information is available. Second, multi-valued properties, i.e., properties that can have more than one object for the same subject, cannot be stored in this way without breaking the relational paradigm. In our example, people can work for more than one institution at the same time. To solve this, one can either create multiple attributes for the same property, but this works only if the maximal number of different objects for the same property and the same subject is rather small. Alternatively, one can store facts with these properties in a standard triple table.

Figure 5 shows the representation of the example from Figure 1 with property tables. We grouped information about people in the **People** table and information about Institutions in the **Institutions** table. As the **works_for** property can have multiple objects per subject, we store facts with this property in the **Remainder** triple table.

## 1.4   Specialized Systems

Beyond the three classes of systems we presented so far, there are a number of systems that don't fit into these categories. We will shortly sketch these systems here without giving much detail, and refer the reader to the referenced original papers for more information.

Atre et al. [9] recently proposed to store RDF data in matrix form. Compressed bit vectors help to make query processing efficient. Zhou and Wu [160] propose to split RDF data into XML trees, rewrite SPARQL as XPath and XQuery expressions, and implement an RDF system on top of an XML database. Fletcher and Beck [42] propose to index not triples, but atoms, and introduce the Three-Way Triple Tree, a disk-based index.

A number of proposals aims at representing and indexing RDF as graphs. Baolin and Bo [85] combine in their HPRD system a triple index with a path index and a content index. Liu and Hu [84] propose to use a dedicated path index to improve efficiency of RDF query processing. Grin [149] explicitly uses a graph index for RDF processing. Matono et al. [92] propose to use a path index based on suffix arrays. Bröcheler et al propose to use DOGMA, a disk-based graph index, for RDF processing.

## 2    RDF in Distributed Setups

Managing and querying RDF data efficiently in a centralized setup is an important issue. However, with the ever-growing amount of RDF data published on the Web, we also have to pay attention to relationships between web-accessible knowledge bases. Such relationships arise when knowledge bases store semantically similar data that overlaps. For example, a source storing extracted information from Wikipedia (DBpedia [10]) and a source providing geographical information about places (GeoNames[2]) might both provide information about the same city, e.g., Berlin.

Such relationships are often expressed explicitly in the form of RDF links, i.e., subject and object URIs refer to different namespaces and therefore establish a semantic connection between entities contained in different knowledge bases. Consider again the above mentioned sources (DBpedia and GeoNames) [15], both provide information about the same entity (e.g. Berlin) but use different identifiers. Thus, the following RDF triple links the respective URIs and expresses that both sources refer to the same entity: (`<http://dbpedia.org/resource/Berlin>`, `<owl:sameAs>`, `http://sws.geonames.org/2950159/`).

The exploitation of these relationships and links offers users the possibility to obtain a wide variety of query answers that could not be computed considering a single knowledge base but require the combination of knowledge provided by multiple sources. Computing such query answers requires sophisticated reasoning and query optimization techniques that also take the distribution into account. An important issue in this context that needs to be considered is the interface that is available to access a knowledge base [63]; some sources provide SPARQL endpoints that can answer SPARQL queries, whereas other sources offer RDF dumps – an RDF dump corresponds to a large RDF document containing the RDF graph representing a source's complete dataset.

The *Linking Open Data initiative*[3] is trying to enforce the process of establishing links between web-accessible RDF data sources – resulting in *Linked Open Data* [15], a detailed introduction to Linked Data is provided in [11]. The Linked Open Data cloud resembles the structure of the World Wide Web (web pages connected via hyperlinks) and relies on the so-called Linked Data principles [14]:

– Use URIs as names for things.
– Use HTTP URIs so that people can look up those names.

---

[2] `http://www.geonames.org/ontology/`
[3] `http://esw.w3.org/SweoIG/TaskForces/CommunityProjects/LinkingOpenData`

- When someone looks up a URI, provide useful information, using the standards (RDF, SPARQL).
- Include links to other URIs, so that they can discover more things.

For a user there are several ways to access knowledge bases and exploit links between them. A basic solution is *browsing* [148,38,57]: the user begins with one data source and progressively traverses the Web by following RDF links to other sources. Browsers allow users to navigate through the sources and are therefore well-suited for non-experts. However, for complex information needs, formulating queries in a structured query language and executing them on the RDF sources is much more efficient. Thus, in the remainder of this section we will discuss the main approaches for distributed query processing on RDF data sources – a topic, which has also been considered in recent surveys [51,63]. We begin with an overview of search engines in Section 2.1 and proceed with approaches adhering to the principles of data warehouses (Section 2.2) and federated systems (Section 2.3). We proceed with approaches that discover new sources during query processing (Section 2.4) and end with approaches applying the P2P principle (Section 2.5).

## 2.1   Search Engines

Before discussing how traditional approaches known from distributed database systems can be applied in the RDF and Linked Data context, let us discuss an alternative to find linked data on the Web: search engines.

The main idea is to crawl RDF data from the Web and create a centralized index based on the crawled data. The original data is not stored permanently but dropped once the index is created. Since the data provided by the original sources changes, indexes need to be recreated from time to time. In order to answer a user query, all that needs to be done is to perform a lookup operation on the index and to determine the set of relevant sources. In most cases user queries consist of keywords that the search engine tries to find matching data for. The so found results and relevant sources are output to the user with some additional information about the results.

The literature proposes several central index search engines, some of them are discussed in the following. We can distinguish local-copy approaches [58,34,26] that collect local copies of the crawled data and index-only approaches [37,108] that only hold local indexes for the data found on the Web. Another distinguishing characteristic is what search engines find: RDF documents [37,108] or RDF objects/entities [58,34,26].

**Swoogle.** One of the most popular search engines for the Semantic Web was Swoogle [37]. It was designed as a system to automatically discover Semantic Web documents, index their metadata, and answer queries about them. Using this engine, users can find ontologies, documents, terms, and other data published on the Web.

Swoogle uses an SQL database to store metadata about the documents, i.e., information about encoding, language, statistics, ontology annotations, relationships among documents (e.g., one ontology imports another one), etc.

A crawler-based approach is used to discover RDF documents, metadata is extracted, and relationships between documents are computed. Swoogle uses two kinds of crawlers: a Google crawler using the Google web service to find relevant Semantic Web documents and a crawler based on Jena2 [157], which identifies Semantic Web content in a document, analyzes the content, and discovers new documents through semantic relations.

Discovered documents are indexed by an information retrieval system, which can use either character N-Gram or URIrefs as keywords to find relevant documents and to compute the similarity among a set of documents. Swoogle computes ranks for documents in a similar way as the PageRank [109] algorithm used by Google.

Thus, a user can formulate a query using keywords and Swoogle will report back a list of documents matching those keywords in a ranked order. Optionally, a user might also define content-based constraints to a general SQL query on the underlying database, e.g., the type of the document, the number of defined classes, language, encoding, etc.

**Semantic Web Search Engine (SWSE).** In contrast to other search engines (document-centric), which given a keyword look for relevant sources and documents, SWSE [58] is entity-centric, i.e., given a keyword it looks for relevant entities.

SWSE uses a hybrid approach towards data integration: first, it uses YARS2 [60] as an internal RDF store to manage the crawled data (data warehousing) and second, it applies virtual integration using wrappers for external sources. In order to increase linkage between data from different sources, SWSE applies entity consolidation. The goal is to identify URIs that actually refer to the same real-world entity by finding matches analyzing values of inverse functional properties. In addition, existing RDF entities are linked to Web documents (e.g., HTML) using an inverted index over the text of documents and specific properties such as `foaf:name` to identify entities.

YARS2 uses three different index types to index the data: (i) a keyword index using Apache Lucene[4] as an inverted index – this index maps terms occurring in an RDF object of a triple to the subject, (ii) quad[5] indexes in six different orderings of triple components – distributed over a set of machines according to a hash function, and (iii) join indexes to speed up queries containing combinations of values or paths in the graph.

In the first step of searching, the user defines a keyword query. The result of the search is a list of all entities matching the keyword together with a summary description for the entities. The results are ranked by an algorithm similar to PageRank combining ranks from the RDF graph with ranks from the data source graph [65]. To refine the search and filter results, the user can specify a specific type/class, e.g., person, document, etc. By choosing a specific result entity, the user can obtain additional information. These additional pieces of information

---

[4] `http://lucene.apache.org/java/docs/fileformats.html`

[5] A quad is a triple extended by a fourth value indicating the origin.

might originate from different sources. Users can then continue their exploration by following semantic links that might also lead to documents related to the queried entity.

**WATSON.** WATSON [34] is a tool and an infrastructure that automatically collects, analyzes, and indexes ontologies and semantic data available online in order to provide efficient access to this knowledge for Semantic Web users and applications.

The first step, of course, is crawling. WATSON tries to discover locations of semantic documents and collects them when found. The crawling process exploits well-known search engines and libraries, such as Swoogle and Google. The retrieved ontologies are inspected for links to other ontologies, e.g., exploiting `owl:import`, `rdfs:seeAlso`, dereferenceable URIs, etc.

Afterwards, the semantic content is validated, indexed, and metadata is generated. WATSON collects metadata such as the language, information about contained entities (classes, properties, literals), etc. By exploiting semantic relations (e.g. `owl:import`), implicit links between ontologies can be computed in order to detect and remove duplicate information and so that storing redundant information and presenting duplicated results to the user can be avoided.

WATSON supports queries based on keywords similar to Swoogle to retrieve and access semantic content including a particular search phrase (multiple keywords are supported). Keywords are matched against the local names, labels, comments and/or literals occurring in ontologies. WATSON returns URIs of matching entities, which can serve as an entry point for iterative search and exploration.

**Sindice.** Sindice [108] crawls RDF documents (files and SPARQL endpoints) from the Semantic Web and uses three indexes for resource URIs, Inverse Functional Properties (IFPs)[6], and keywords. Consequently, the user interface allows users to search for documents based on keywords, URIs, or IFPs. To process the query, the query only needs to be passed to the relevant index, results need to be gathered, and the output (HTML page) needs to be created.

The indexes correspond to inverted indexes of occurrences in documents. Thus, the URI index has one entry for each URI. The entry contains a list of document URLs that mention the corresponding URI. The structure of the keyword index is the same, the only difference is that it does not consider URIs but tokens (extracted from literals in the documents), the IFP index uses the uniquely identifying pair (property, value).

When looking for the URI of a specific entity (e.g. Berlin), Sindice provides the user with several documents that mention the searched URI. For each result some further information is given (human description, date of last update) to enable users to choose the best suitable source. The results are ranked in order of relevance, which is determined based on the TF/IDF relevance metric [43] in

---

[6] `http://www.w3.org/TR/owl-ref/#InverseFunctionalProperty-def`: If a property is declared to be inverse-functional, then the object of a property statement uniquely determines the subject (some individual).

information retrieval, i.e., sources that share rare terms (URIs, IFPs, keywords) are preferred. In addition, the ranking prefers sources whose URLs a similar to queried URIs, i.e., containing the same hostnames.

**Falcons.** In contrast to other search engines, Falcons [26] is a keyword-based search engine that focuses on finding linked objects instead of RDF documents or ontologies. In this sense, it is similar to SWSE.

Just like other systems, Falcons crawls RDF documents, parses them using Jena[7], and follows URIs discovered in the documents for further crawling. To store the RDF triples, Falcons creates quadruples and uses a quadruple store (MySQL).

To provide detailed information about objects, the system constructs a virtual document containing literals associated with the object, e.g., human-readable names and descriptions (`rdfs:label`, `rdfs:comment`).

Falcons creates an inverted index based on the terms in the virtual documents and uses this index later on for keyword-based search. A second inverted index is built based on the objects' classes – it is used to perform filtering based on the objects' classes/types. For a query containing both, keywords and class restrictions, Falcons computes the intersection between the result sets returned by both indexes.

Thus, given a keyword query, Falcons uses the index to find virtual documents (and therefore objects) containing the keywords. Falcons supports class-based (typing) query refinement by employing class-inclusion reasoning, which is the main difference to SWSE, which does not allow such refinements and reasoning. The so-found result objects are ranked by considering both their relevance to the query (similarity between the virtual documents and the keyword query) and their popularity (a measure based on the number of documents referring to the object).

Each presented result object is accompanied with a snippet that shows associated literals and linked objects matched with the query. – also shows detailed RDF descriptions loaded from the quadruple store.

**More systems.** There are many more search engines such as OntoSearch2 [111], which stores a copy of an ontology in a tractable description logic and supports SPARQL as a query language to find, for instance, all the instances of a given class or the relations occurring between two instances. Several other systems aim at providing efficient access to ontologies and semantic data available online. For example, OntoKhoj [112] is an ontology portal that crawls, classifies, ranks, and searches ontologies. For ranking they use an algorithm similar to PageRank. Oyster [110] is different in the sense that it focuses on ontology sharing: users register ontologies and their metadata, which can afterwards accessed over a peer-to-peer network of local registries. Finally let us mention OntoSelect [19] is an ontology library that focuses on providing natural language based access to ontologies.

---

[7] http://jena.sourceforge.net

## 2.2   Data Warehousing

So far, we have discussed several techniques to search for documents and entities using search engines. However, the problem of answering queries based on data provided by multiple sources has been known in the database community since the 80s. Thus, from a database point of view the scenario we face with distributed RDF processing is similar to *data integration.*

Figure 6 shows a categorization of data integration systems from a classical database point of view. The same categorization can be applied to the problem of processing Linked Data because, despite some differences, similar problems have to be solved and similar techniques have already been applied by different systems for Linked Data processing.

The first and most important characteristic that distinguishes data integration systems is whether they copy the data into a central database, or storage system respectively, or leave the data at the sources and only work with statistical information about the sources, e.g., indexes. Approaches falling into the former category are generally referred to as *materialized data integration systems* with *data warehouses* as a typical implementation. Approaches of the second category are referred to as *virtual data integration systems* as they only virtually integrate the data without making copies.

The process of integrating data into a data warehouse is referred to as the ETL process (Extract-Transform-Load). First, the data is extracted from the sources, then it is transformed, and finally loaded into the data warehouse. This workflow can also be applied in the Semantic Web context.

*Extract.* As many knowledge bases are available as dumps for download, it is possible to download a collection of interesting linked datasets and import them into a data warehouse. This warehouse resembles a centralized RDF storage system that can be queried and optimized using the techniques discussed in Section 1. If a data source is not available for download, the data can be crawled by looking up URIs or accessing a SPARQL endpoint. As the data originates from different sources, the system should keep track of provenance, e.g., by using named graphs [22] or quads.

*Transform.* Data warehouses in the database world usually provide the data in an aggregated format, e.g., in a data warehouse storing information about sales we do not keep detailed information about all transactions (as provided by the sources) but, for instance, the volume of sales per day (computing aggregated values based on the data provided by the sources). For linked data, this roughly corresponds to running additional analyses on the crawled data for duplicate detection/removal or entity consolidation as applied, for instance, by WATSON and SWSE.

*Load.* Finally, the transformed data is loaded into the data warehouse. In dependence on aspects such as the update frequency of the sources and the size of the imported datasets, it might be difficult to keep the data warehouse up-to-date. In any case the data needs to be reloaded or recrawled so that the data warehouse can be updated accordingly. It is up to the user, or the user application respectively, to decide if and to what extent out-of-date data is acceptable or not.

Reconsidering the search engines we have already discussed above, we see that some of these approaches actually fall into this category: the search engines using a central repository to store a local copy of the RDF data they crawled from the web, e.g., SWSE [58], WATSON [34], and Falcons [26].

In summary, the data warehouse approach has some advantages and disadvantes. The biggest advantage, of course, is that all information is available locally, which allows for efficient query processing, optimization, and therefore low query response times. On the other hand, there is no guarantee that the data loaded into the data warehouse is up-to-date and we have to update it from time to time. From the perspective of a single user or application, the data warehouse contains a lot of data that is not queried and unnecessarily increases storage space consumption. So, the data warehouse solution is only suitable if we have a sufficiently high number of queries and diverse applications.

### 2.3   Federated Systems

As mentioned above, database literature proposes two main classes of data integration approaches (Figure 6): materialized and virtual data integration. In contrast to materialized data integration approaches, *virtual data integration systems* do not work with copies of the original data but only virtually integrate the data. If we are only interested in answering unstructured queries based on keywords, the search engine approach, as a variant of virtual integration systems, is an appropriate solution. We have already discussed several examples of Semantic Web search engines in Section 2.1. Some of them are pure search engines that only work with indexes and output pointers to relevant data at the sources, whereas other Semantic Web search engines rather correspond to data warehouses.

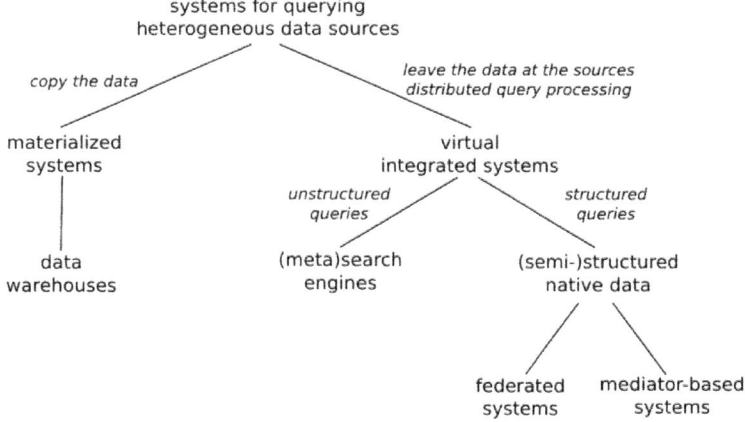

**Fig. 6.** Classification of data integration approaches

When we are interested in answering structured queries on the most recent version of the data, we need to consider approaches that integrate sources in a way that preserves the sources' autonomy but at the same time allow for evaluating queries based on the data of multiple sources: we distinguish between *mediator-based systems* and *federated systems*.

*Mediator-based systems.* Classic mediator-based systems provide a service that integrates data from a selection of independent sources. In such a scenario, sources are often unaware that they are participating in an integration system. A mediator provides a common interface to the user that is used to formulate queries. Based on the global catalog (statistics about the data available at the sources), the mediator takes care of rewriting and optimizing the query, i.e., the mediator determines which sources are relevant with respect to the given query and creates subqueries for the relevant sources. In case the sources manage their local data using data models or query languages different from what the mediator uses, so-called wrappers rewrite the subqueries in a way that makes them "understandable" for the sources, e.g., SQL to XQuery. Wrappers also take care of transforming the data that the sources produce as answers to the query into the mediator's model. In practice, the task of the mediator/wrapper architecture is to overcome heterogeneity on several levels: data model, query language, etc.

*Federated systems.* The second variant of virtual integration systems from a database point of view is *federated systems*. A federation is a consolidation of multiple sources providing a common interface and therefore very similar to the mediator-based approach. The main difference to mediator-based systems is that sources are aware that they are part of the federation because they actively have to support the data model and the query language that the federation agreed upon. In addition, sources might of course support other query languages.

However, from a user's point-of-view there is no difference between both architectures as both provide transparent access to the data. Likewise, in general the Semantic Web community does not distinguish between these two architectures either so that both approaches are referred to as federated systems, virtual integration, and/or federated query processing [51,54,63]. Thus, in the following we adopt the notion of a federation from the Semantic Web community and do not distinguish between federated and mediated architectures nor between federators and mediators.

There are some peculiarities unique to the web of linked data that we need to deal with in a distributed setup. Some sources, for instance, do not provide a query interface that a federated architecture could be built upon, i.e., some sources only provide information to dereferenceable HTTP URIs, some sources are only available as dumps, and some sources provide access via SPARQL endpoints.

For the first two cases, we can crawl the data and load it into a central repository (data warehouse) or multiple repositories that can be combined in a federation – comparable to using wrappers for different data sources. If the data is already available via SPARQL endpoints, we only need to register them in the federation and their data can be accessed by the federation to answer future

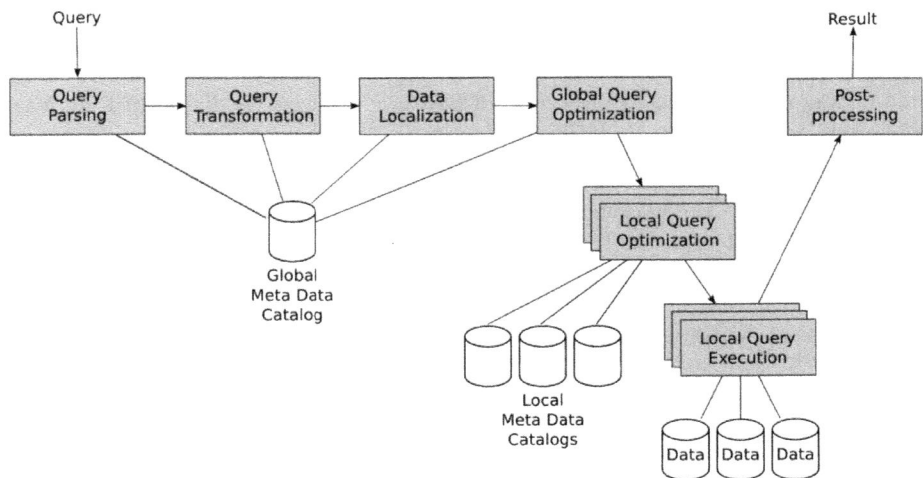

**Fig. 7.** Steps of distributed query processing

queries. In addition, if we adopt SPARQL as the common protocol and global query language, then no wrappers are necessary for SPARQL endpoints because they already support the global query language and data format.

Another interesting difference between classic federated database systems and federated RDF systems is that for systems using the relational data model, answering a query involves only a few joins between different datasets defined on attributes of the relations. For RDF data, this is more complicated because the triple format requires much more (self) joins. Moreover, some aspects of RDF, e.g., explicit links (`owl:sameAs`), are also aspects that need to be considered during query optimization and execution.

Query processing in distributed database systems roughly adheres to the steps highlighted in Figure 7: query parsing, query transformation, data localization, global query optimization, local query optimization, local query execution, and post-processing. In the following we discuss each of these steps in detail; we first describe each step with respect to classic distributed databases and then compare it to distributed RDF systems.

*Query parsing.* After the query has been formulated by the user, it has to be parsed, i.e., the query formulated in a declarative query language is represented using an internal format that facilitates optimization. For example, a query formulated in SQL is transformed into an algebra operator tree.

Likewise, SPARQL queries can be parsed into directed query graphs: SQGM (SPARQL Query Graph Model) [62]. In any case, after parsing we have a graph consisting of operators that need to be evaluated in order to answer the query – a connection between operators indicates the data flow, i.e., the exchange of intermediate results, between them.

*Query transformation.* In this step, simple and straightforward optimizations are applied to the initial query that do not require any sophisticated considerations but mostly rely on heuristics and straightforward checks. One aspect is to perform a semantic analysis, e.g., to check whether relations and attributes referred to in the query do actually exist in the schema. The schema information necessary for this step are stored in the global catalog. In addition, query predicates are transformed into a canonical format (normalization) to facilitate further optimization, e.g., to identify contradicting query predicates that would result in empty result sets. Moreover, simple algebraic rewriting is applied using heuristics to eliminate bad query plans, e.g., replacing crossproducts followed by selections with join operators, redundant predicates are removed, expressions are simplified, unnesting of subqueries and views, etc.

Similar transformations can be done for SPARQL queries. For example, we can check for contradictions of conditions in the query that would lead to an empty result set. We can also apply some basic optimizations such as removing redundant predicates and simplifying expressions. In addition, we can also check if the predicates referenced in the query do actually exist, i.e., if there is any source providing corresponding triples. However, this is only useful for distributed RDF systems that do not traverse links to discover new sources during query processing. As discussed below, the information necessary to perform these checks are part of the global catalog.

*Data localization.* In the classic data localization step, the optimizer replaces references to global relations with references to the sources' fragment relations that contribute to the global relation. The optimizer simply needs to access the global catalog and make use of reconstruction expressions: algebraic expressions that, when executed, reconstruct the global relations by combining the sources' fragments in an appropriate manner. In consideration of predicates and joins contained in the query, a subset of fragments and therefore sources can be eliminated from the query because the data they provide cannot contribute to the query result.

In case of Linked Data and systems for distributed RDF processing, this step entails the identification of sources relevant to a given query. Whereas we have reconstruction expressions for relational databases, it is more complicated for RDF data because there are no strict rules and or restrictions on which source uses which vocabulary. Thus, what an appropriate optimizer for distributed RDF systems needs to do in this step is to go through the list of basic graph patterns contained in the query and identify relevant sources. The information necessary to perform this task are contained in the global catalg, which we will discuss in detail below.

*Global query optimization.* In principle, the goal of this step is to find an efficient query execution plan. In classic distributed systems, the optimizer can optimize for total query execution time or for response time. The former is a measure for the total research consumption in the network, i.e., it sums up the duration of all operations necessary to answer the query over all nodes altogether. Response time takes parallel execution into account and measures the time until the results are presented to the user. Optimizing response time can exploit all different flavors of parallelization only when the mediator has some "control" over the sources, i.e., considering direct interaction, communication, and data exchange between any two nodes for optimization.

In federated RDF systems, however, sources are still more autonomous so that some of the options available to classic optimizers cannot be considered, e.g., we cannot decide on a query plan that requires one source to send an intermediate result directly to another source that uses the received data to compute a local join. Thus, techniques commonly referred to as data, query, or hybrid shipping [76] are not applicable in the RDF context. Likewise, the option of pipelining[8] results between operators executed at different sources is hardly applicable.

In general, however, it is efficient in terms of load at the mediator and communication costs when joins can directly be executed at the sources. The problem is that with the restrictions mentioned above joins in distributed RDF systems can only be executed in a completely decentralized fashion at a source $s_1$ if there is no additional data at any other source $s_2$ that could produce additional results when joined with part of the data provided by $s_1$. Thus, even though $s_2$ can process the join locally it has to return, in addition to the join result, all data that could be joined with other sources' data. Hence, in most cases joins have to be computed in a centralized fashion by the mediator [139, 51]. In that case, the optimizer applies techniques for local query optimization as discussed in Section 1.

Another option to process joins efficiently is to use a semijoin [76] algorithm to compute a join between an intermediate result at the mediator and a data source [51]. The mediator needs to compute a projection on the join variables and extract all variable bindings present in the intermediate result. For all these variable bindings, the mediator needs to query the remote data source with respect to the join definition and retrieve the result, which consists of all triples matching the join condition. So, the mediator retrieves all relevant data to perform the join locally. Unfortunately, SPARQL does not support the inclusion of variable bindings in a query so that the only alternative is to include variable bindings as filter expressions in a query, which for a large number of bindings might blow up the size of the query message. The alternative of sending separate messages for each binding [115] results in a high number of messages and therefore increases network load and query execution costs.

A heuristic to minimize query execution costs is to minimize the size of intermediate results. Hence, optimizers for federated RDF systems also try to find a plan that minimizes the size of intermediate results. Theresfore, the optimizer considers additional statistics about selectivities and cardinalities provided by the global catalog (see below).

---

[8] Result items (tuples) are propagated through a hierarchy of operators (e.g., joins) so that when a tuple is output by an operator on a lower level, it *immediately* serves as input to the upper operator – before all input tuples have been processed by the lower operator.

Another standard technique for every optimizers is to apply heuristics to re-order query operators in a way that intermediate results are small, i.e., pushing highly selective operators downwards in the query graph so that they are executed first. Considering RDF data, this means to push down value constraints – ideally into subqueries executed at the sources to reduce the amount of exchanged data.

Another important and related issue for standard optimizers is join order optimization. The goal is to find the order of joins that minimizes execution costs. This process heavily relies on statistics (global catalog) to estimate the size of intermediate results. Obviously, it is beneficial to execute joins first that produce small outputs so that all other joins can be computed more efficiently. For optimization in distributed RDF systems, join order optimization corresponds to determining the order of evaluating basic graph patterns. To estimate the size of intermediate results, the optimizer needs detailed statistics that allow for cardinality estimation. Histograms are very common for relational distributed systems and can also be useful in the context of federated RDF systems. A simple heuristic that can be applied without detailed statistics is variable counting [136], which estimates the selectivity of basic graph patterns in dependence on the type and number of unbound components.

In consideration of all these different possibilities for query optimization, the optimizer has to explore and evaluate a potentially high number of alternative query plans that all compute the same result. For exploration, a common technique is the application of dynamic programming for plan enumeration, which enables an exhaustive search of all query plans. In order to decide on the benefit of each plan, the optimizer has to determine its costs using an appropriate cost model [105, 51, 139]. For this purpose, the optimizer again needs statistics about the data, selectivity, cardinality estimations about intermediate results. In addition, it is worthwhile to consider network latencies as well.

The current standard for distributed RDF optimizers is to optimize a query and to determine a query plan for each query individually at runtime. This can be improved by adopting more techniques from classic distributed database systems that pre-optimize (partial) execution plans for frequently issued queries or subqueries contained in many queries, e.g., two-step-plans [76].

As a high number of sources provide potentially relevant data for a given query, approximation by reducing the number of queried sources is an appropriate technique to reduce the overall execution costs. This can be achieved by ranking sources based on triple and join pattern cardinality estimations and prune all but the top-ranked sources from consideration [59].

*Global catalog and indexes.* As we have seen above, the global catalog plays an important role for query optimization. It might contain information about specific sources such as vocabulary, supported predicates, network latencies, etc. In addition, for the distributed setup and especially for cardinality estimation, the catalog has to contain statistics about the data, e.g., in the form of indexes. In constrast to indexes that are applied in the context of centralized systems (complete indexes), these indexes cannot provide the same level of details because of the amount of data from all sources altogether. Therefore indexes and statistics about the data suitable for distributed optimization have to abstract the data in a way that allows for index size compression. The goal is to find a suitable trade-off between the level of detail and memory space consumption – the general rule is the more detailed an index is, the more accurate are cardinality estimations but the higher is the memory consumption.

A widely used approach for indexing is *schema-level indexing*, i.e., predicate URIs and the types of instances. Whereas types and predicates that occur only rarely represent good discriminators to detect relevant sources for a given query, frequently used types and predicates that almost every source provides are only of little use for optimization, e.g., `rdfs:label`. The disadvantage is that this index cannot be used for basic graph patterns with variables in the predicate position. An alternative to indexing predicates and types detached from the structure of the overall RDF graph is indexing paths of predicates [139]. Moreover, when considering the RDF data of a source as a graph, it might also be useful to index frequent subgraphs [51,89,146].

Another kind of indexes are *inverted URI indexes*. They index the data on instance level by indexing all URIs occuring in a data source. This kind of index allows the query processor to identify all sources which contain a given URI and thus potentially contribute bindings for a triple pattern containing that URI.

There are also indexes indexing data on both instance-level and schema-level [59]. They make use of data structures known from classic relational database: histograms. As these data structures have originally been developed to summarize informtion about numerical data, hash functions are applied to all elements of a triple so that triples are transformed into numerical space. One-dimensional histograms index each component (subject, predicate, object) in separate. But when considering each of these three dimensions in separate, the optimizer looses a great potential for optimization because the combination of instances is much more selective and therefore more useful for query optimization. Thus, multidimensional histograms are applied, i.e., three-dimensional histograms. An alternative to histograms, other structures that efficiently summarize multidimensional data can be applied, e.g., QTrees [59].

*Local query optimization and execution.* When the global optimizer has decided for a specific query execution plan, subqueries are extracted and sent to the sources for local execution. They apply the same optimization techniques as for local queries, so that the received query is treated like a query issued directly at the source. Consequently, it is optimized and executed using the techniques discussed in Section 1.

*Post-processing* In this last step the partial results received from the sources are combined into the final result. For simple distributed queries, the post-processing might simply consist of a union operation. For more complex queries, however, post-processing is much more complex and costly because all operations that could not be executed at the sources have to be processed by the mediator after retrieving the data from the sources. As discussed above, this is particularly true for joins for which multiple sources provide relevant data. In addition, it might be necessary to remove duplicates in the result set, which represents the very last operation for post-processing.

To conclude the section about federated systems and federated RDF processing, we discuss some systems that have been proposed in this context.

**SemWIQ.** The Semantic Web Integrator and Query Engine [79], SemWIQ in short, uses an architecture based on the mediator/wrapper approach, i.e., wrappers are used to enable the participation of sources using other data models, e.g. relational databases. All registered data sources must either be connected by a SPARQL-capable wrapper or support the SPARQL protocol directly.

Queries are formulated in SPARQL, the parser computes a canonical query plan, which is optimized by the federator/mediator. The optimization process only considers very basic statistics, such as a list of classes and the number of instances a data source provides for each class as well as a list of predicates and their occurrences. For query optimization, the federator analyzes the query and scans the catalog for relevant registered data sources. The resulting plan is executed by sending subqueries to the sources (via wrappers or SPARQL endpoints).

The system requires that every data item has an asserted type, i.e., for a query it requires type information for each subject variable in order to be able to retrieve instances. As a consequence, there are some restrictions with respect to query formulation, e.g., all subjects must be variables and for each subject variable its type must be explicitly or implicitly defined. The optimizer then uses this information to look for registered data sources providing instances of the queried types. More sophisticated optimization techniques, such as the push-down of joins, have been proposed as future work.

**DARQ.** DARQ [115] (Distributed ARQ) is a federated system of SPARQL endpoints that allows distributed query processing on the set of registered endpoints. It also adopts a mediator-based approach and assumes that sources that do not support SPARQL themselves are connected to the federation using wrappers.

Data sources are described using service descriptions (represented in RDF). These descriptions contain capabilities, i.e., constraints expressed with regular SPARQL filter expressions. These constraints can express, for instance, that a data source only stores data about specific types of resources. For sources with limitations of access patterns, e.g., allowing lookups on personal data only when the user can specify the name of the person of interest. DARQ supports this by defining patterns in the service descriptions that must be included in the query. To provide the query optimizer with statistics, service descriptions contain the total number provided by a data source and optionally information for each capability, e.g., the number of triples with a specific predicate and the selectivities (bound subject/object) of a triple pattern with a specific predicate.

Queries are formulated in SPARQL, parsed, and handed to the query planning and optimization components. DARQ uses a cost-based query optimization technique relying on the statistical information provided as service descriptions and capabilities.

A SPARQL query contains one or more filtered basic graph patterns (triple patterns). DARQ performs query planning for each basic graph pattern in separate. By comparing the triple patterns of the query against the capabilities of the service descriptions, the system can detect a set of relevant sources for each pattern. As this matching procedure is based on predicates, DARQ only supports queries with bound predicates.

After having determined the relevant sources, subqueries are created, one for each basic graph pattern and data source matching. Thus, the system might create multiple subqueries that are to be sent to the same data source. In that case, the subqueries can be combined into one message.

Based on these subqueries the query optimizer considers limitations on access patterns and tries to find a feasible and efficient query execution plan. For logical query optimization, heuristics are used that try to simplify the query and reduce intermediate result sizes, e.g., push value constraints into subqueries, which corresponds to pushing down selections in classic query optimization. In order to decide for a specific implementation to process joins (nested loop join or bind join), the optimizer estimates the result size of joins based on the statistics given in the service descriptions. The optimizer chooses the implementation with the least estimated transfer costs (computed based on the result estimates).

In the end, subqueries are executed at the sources and remaining operations are executed by the federator.

**Hermes.** Hermes [147] is a system based on a federated architecture that has a slightly different focus than other systems discussed so far. Queries are not formulated using SPARQL. Instead, the user enters a keyword query and Hermes tries to translate the keyword query into a structured query (SPARQL). The query is decomposed into subqueries, which are executed at the sources.

In order to achieve this, a number of indexes are created: a keyword index (indexing terms extracted from the labels of data graph elements), a structure index (information about schema graphs, i.e., relations between classes derived from the connections given in the data graph), and a mapping index (information about mappings on data- and schema-level, pairwise mappings between elements).

After having received the input keywords, relevant sources are determined using the keyword index. The retrieved keyword elements are combined with schema graphs received from the structure index to find substructures that connect all keyword elements. A ranking function is used to rank the computed query graphs so that the user can choose some of them for execution.

The selected query is decomposed into subqueries, each of which is answered by a different data source. Optimization uses the same techniques as DARQ [115]. Before sending the subqueries to the sources, they are mapped to the query format supported by the receiving data source. After execution, the results received from the sources are combined.

**Other systems.** Virtuoso [41], a native quad store, provides the option to consider remote sources for query execution. The system dereferences URIs and holds the retrieved data in a cache for future queries.

Dartgrid [25] is a system for SPARQL queries over multiple relational databases. One of the main components is the semantic registration service which maintains mappings from the schemas of registered data sources to the internal ontology. An query interface based on forms and ontologies helps users to construct semantic queries formulated in SPARQL. Queries are translated using the mapping information into SQL queries that can be executed on the sources.

The networked graphs approach [126] allows users to reuse and integrate RDF content from other sources. It allows users to define RDF graphs by extensionally listing content and by using views on other graphs. These views can be used to include parts of other graphs. Networked graphs are designed for distributed settings and are exchangeable between sources. However, the paper focuses more on semantics and reasoning than on aspects of query execution.

[139] presents an approach for querying distributed RDF data sources and introduces index structures, a cost model, and an algorithm for query answering. The approach supports distributed SeRQL path queries over multiple Sesame RDF repositories using a special index structure to determine the relevant sources for a query.

## 2.4   Discovering New Sources During Query Processing

In contrast to the classic understanding of federated databases, processing Linked Data arises additional challenges that we have not discussed in detail above, e.g., by dereferencing URIs and considering the returned document as a new virtual "data source" not all sources are known in advance and available for indexing. Furthermore, approaches that index a static set of sources, i.e., all approaches we have discussed so far, have to recreate their indexes from time to time in order to reflect recent updates to the sources, which means that some of the information provided by the index might be out-of-date.

**Pure.** Some strategies for query processing over Linked Data rely on the principle of following links between sources [61]. An advantage in comparison to federated architectures is that sources that are not accessible via SPARQL endpoints can be considered. Another advantage is that users can retain complete control over the data they provide.

The query is executed without a previous query planning or optimization step. First, the system retrieves data from the sources mentioned in the query. The data is partially evaluated on the retrieved data so that relevant source URIs and links can be identified. The system uses these URIs to retrieve more data. It iteratively evaluates and discovers further data until all sources found to be relevant have been processed.

The peculiarity of this approach is the intertwining of the two phases: query evaluation and link traversal. Previous work [94,17] kept these two phases in separate by first retrieving the data and then evaluating the query on the retrieved data.

**Hybrid.** It is also possible to combine federated query processing with active discovery of unknown but potentially relevant sources [78]. The main assumption is that knowledge about some sources is available for query planning and optimization. During query execution, sources are retrieved, new sources are iteratively discovered, the query plan is reoptimized and partially executed until all relevant sources have been processed.

## 2.5    Systems Based on the P2P Paradigm

As we have seen, distributed processing of RDF data can be realized by adopting the federated database system architecture and developing efficient solutions for problems that come along with the characteristics of Linked Data. But there is another important class of distributed systems that Linked Data processing can also benefit from: peer-to-peer (P2P) systems.

P2P systems are networks of autonomous peers that are connected through logical links, which express that a pair of peers "know" each other, i.e., they can exchange messages and data and are referred to as *neighbors*. A pair of peers without a link do not know each other and can therefore only contact each other if they find a path of links via intermediate peers that connects them.

In general, sources/peers in a P2P network have a higher degree of autonomy in comparison to sources in distributed database systems. In pure P2P networks there is no central component, i.e., no federator or mediator, that could be used for query planning and optimization. Instead, the behavior of the whole system is the consequence of local interactions between peers.

Whereas sources in the context of distributed databases are considered to be rather stable and available, P2P systems assume a higher degree of dynamic behavior, i.e., they assume that peers might join and leave the network at any time. Even though a peer leaves the network, the system should still be able to answer queries – even though the data provided by the peer that left the network is (momentarily) unavailable. Moreover, each peer in the network might issue queries and participate in answering queries. There are different classes of P2P systems: *centralized P2P, pure P2P, hybrid P2P, structured P2P.*

*Centralized P2P systems.* For centralized P2P systems, like Napster[9], there is a central component providing a centralized index that is used to locate relevant data to a given query. So, a user query issued at a peer is sent to the central component, which uses the index to find peers providing the queried data. This information is sent back to the querying peer, which then directly communicates with other peers to access the queried data.

*Pure/unstructured P2P systems.* As the centralized server represents a bottleneck and a single point of failure, pure P2P system strictly avoid peers with special roles. Instead, all peers are considered equal and query processing is realized by *flooding*, i.e., a peer with a local user query forwards the query to all its neighbors, they proceed the same so that the query finally reaches all the peers in the network. The answers to the query are routed back to the query initiator, which can then directly communicate with peers providing relevant

---

[9] http://www.napster.com

data. In analogy to structured P2P systems, which we will discuss below, pure P2P systems are often referred to as *unstructured P2P systems*.

*Hybrid P2P systems.* The problem with flooding is that query processing consumes much bandwidth and weak/slow peers represent the bottleneck. So, hybrid P2P systems use super-peers to counteract this problem. Super-peers are strong peers that form an unstructured P2P network. Weaker peers are connected to a super-peer as leaf nodes and form a centralized P2P system.

*Structured P2P systems.* All the P2P system discussed so far assume that the data shared in the network remains at the peers that "own" them. However, in structured P2P systems, a global rule is used to redistribute the data among peers in the system. In most cases a hash function is used for this purpose so that each peer is responsible for a specific hash range and therefore for all the data with hash values in that range. Peers are arranged in a logical overlay structure, e.g., a logical ring [137], that is used to organize the peer in way that alleviates efficient lookup. For query processing peers use the globally known rule according to which the data has been distributed in the first place, e.g., peers compute hash values for the queried data and use the overlay network to locate peers responsible for overlapping hash ranges.

After having introduced the basic types of P2P systems, let us now discuss some approaches that apply these concepts to the Semantic Web and RDF data.

**Edutella.** An early approach that combined the two paradigms RDF and P2P is Edutella [98]. Edutella assumes that all resources maintained in the network can be described with metadata in RDF format. All functionality in the Edutella network is mediated through RDF statements and queries on them.

Peers participating in an Edutella network might use different schemas so that Gnutella applies the mediator-wrapper approach based on a common data model and a common query exchange format to overcome heterogeneity. A peer that wants to participate in the network, registers at a so-called mediator peer. Peers register the metadata schemas they support and in this way indicate which queries they can answer. Queries are sent through the Edutella network to the subset of peers that have registered with the corresponding schema. The resulting RDF statements are sent back to the requesting peer. To broaden the search space, mediators provide a service that translates queries over one schema into queries over another schema.

Because of the mediators that mediate between clusters of peers supporting different schemas, the overall architecture can be considered a *hybrid P2P system*.

**GridVine.** GridVine [5] uses P-Grid [4] to organize peers in a *structured P2P* overlay network, which is used for communication and interaction between peers. Data is indexed and stored in the standard way of structured P2P systems, i.e., each peer maintains a local database at the semantic layer to store the triples whose keys are contained in the key range the peer is responsible for. GridVine also supports sharing schemas by associating schemas with unique keys and storing them in the overlay network.

GridVine supports queries based on basic graph patterns and exploits pairwise mappings (OWL statements relating similar classes and properties) between different schemas to overcome schema heterogeneity and evaluate queries against schemas they were originally not formulated against. Mappings are also stored in the network.

In order to locate peers providing relevant data for a given query, or a basic graph pattern respectively, each triple is indexed and stored three times – once for each component of the triple.

**RDFPeers.** RDFPeers [20] is similar to GridVine but uses a different semantic overlay network for storing, indexing, and querying RDF data. To efficiently answer multi-attribute and range queries, RDFPeers relies on a multi-attribute addressable network (MAAN) [21], which extends Chord [137] – *structured P2P*.

Each triple is stored three times applying hash functions to subject, predicate, and object. The system's query processing capabilities are very similar to the ones of GridVine. It supports triple pattern queries, disjunctive and range queries, and conjunctive multi-predicate queries using RDQL.

**QC and SBV.** Another approach for evaluating conjunctive queries of triple patterns over RDF data using *structured* overlay networks is proposed in [83]. It uses Chord [137] and also stores each triple three times (subject, predicate, object).

This approach proposes two algorithms for query processing: *query chain* (QC) and *spread by value* (SBV). The main idea of the query chain algorithm is that intermediate results flow through the nodes of the chain. First, responsible nodes are determined for each triple pattern contained in the query – exploiting constants contained in the patterns and using the overlay structure to identify relevant peers. The found responsible peers form a chain and exchange messages to answer the query.

The spread by value algorithm constructs multiple chains for each query that can be processed in parallel. Query processing starts at a node responsible for the first query pattern, which uses values of matching triples to forward the query to nodes providing data for these values.

**Other approaches.** YARS2 [60] uses the P2P paradigm (*structured P2P*) in a different way than the approaches discussed so far. As mentioned above in Section 2.1, it comes in combination with the search engine SWSE. YARS2 does not distribute the data in the structured overlay network but an index structure. More specifically, YARS2 uses indexes in six different orderings of triple components and distributes this index according to a hash function.

KAONp2p [55] also suggests a P2P-like architecture for query answering over distributed ontologies. Queries are evaluated against resources, which are integrated using a virtual ontology that logically imports all relevant ontologies distributed over the network.

# 3  Scalable Reasoning with Uncertain RDF Data

As we have seen in the previous chapter, state-of-the-art SPARQL engines for RDF data (see, e.g., [3, 99]) primarily focus on conjunctive queries on top of a relational encoding of RDF [1] data. They often employ a so-called "triple-store" technique by indexing or slicing the data according to various permutations of the basic subject-predicate-object (SPO) pattern. These engines generally follow a deterministic data and query processing model and do not have a notion of uncertain reasoning or probabilistic inference.

In this chapter, we aim to devise possible directions for scalable reasoning with uncertain RDF knowledge bases, following ideas from probabilistic databases, logic programming, and recent imperative programming platforms for inference in probabilistic graphical models. We focus on RDF as our basic data model and SPARQL as the default query language for RDF. Consequently, the forms of uncertain reasoning we consider here focus on SQL-style queries in probabilistic databases, Datalog-style reasoning using Horn clauses as rules, as well as some probabilistic extensions to logic programming.

By default, an RDF graph itself is schemaless. RDFS [1] thus introduces basic facilities for imposing schema information in terms of a class hierarchy. Intuitively, entities that belong to the same instance of a class, i.e., entities of the same RDF *type*, share common properties. In RDFS, instances of classes, class memberships and class properties are transitively inherited to resources in subclasses via the *type*, *subClassOf* and *subPropertyOf* relations, respectively. More generally, rule-based reasoning using Datalog-style Horn clauses subsumes *type*, *subclassOf* and *subPropertyOf* inferences in RDF/S and also captures some of the constraints expressible in OWL [7] (e.g., the *transitive* or *functional* properties of predicates). Transitivity, for example, can very easily be expressed via first-order predicate logic (specifically with rules in the form of Horn clauses), which require a from of logical reasoning in order to check consistency or to compute entailment. Considering relations as logical predicates and facts as literals, we thus also briefly investigate the relationship to logic programming and its probabilistic extensions in this chapter.

As a (more popular) counterpart to probabilistic logic programming, probabilistic databases [32] investigate query processing techniques for structured (SQL) queries over relational data with fixed schemata. While the general idea of including probabilistic models into databases is not new and leads back to more than 30 years of research (see, e.g., [24]), in particular recent works in this context have provided us with a rich body of literature and have led to the development of a plethora of systems, which all aim at the scalable management of uncertain relational data. Ultimately, these approaches however need to face the very same scalability and complexity issues as any approach dealing with inference in probabilistic graphical models.

As this part of the lecture focuses on database-style infrastructures for the management of uncertain RDF data, we do not go into details on probabilistic extensions to OWL and its variants based on description logics (OWL-DL). For probabilistic description logics [87] (PDL), we refer the reader to [138], which provides a very good overview of PDL and related approaches, which has found wide-spread acceptance in the semantic web community. PDL generalizes the description logic $\mathcal{SHOIN}^{(D)}$ and can thus express, for example, functional rules. Reasoners such as PRONTO[10] can be used to decide consistency or to compute entailment with probabilistic bounds.

### 3.1 Probabilistic Databases

Approaches for managing uncertainty in the context of probabilistic databases [8, 13,30,31,132] focus on relational data with fixed schemata, and they often employ strong independence assumptions among data objects (the "base tuples"). SQL is used as the default query language for running queries, defining views, or triggering updates to the data.

Most probabilistic databases adopt the *possible worlds* [6] model as basis for their data model and for defining the semantics of queries. Intuitively, every tuple in a probabilistic database corresponds to a binary random variable which may exist in the database with some probability. A probabilistic database thus encodes a large number of possible instances of deterministic databases, where each deterministic instance contains a different combination (i.e., possible world) of tuples. Each such possible world has a probability between 0 and 1. The probabilities of all worlds form a distribution and sum up to 1. The *marginal probability* of a tuple can be obtained by summing up the probabilities of all worlds which contain that tuple, which leaves confidence computations in probabilistic databases #P-complete [120, 30] in the general case. The semantics of queries is then formally defined by running the query against each of these instances individually, and by encoding the results obtained from each of the individual instances back into the probabilistic database. While often the actual data computation can be carried out directly along with the SQL operators, different inference techniques for confidence computations may have to be implemented as separate function calls (e.g., as stored procedures [96]).

In addition to this basic uncertainty model, the ULDB [13] data model (for "Uncertainty and Lineage Database") provides a lineage-based representation formalism for probabilistic data, which has been shown to be *closed and complete* for any combination of SQL-style relational operators and across arbitrary levels of materialized views. Here, the lineage (aka. "history" in [132] or "repair-key" operator in [8]) of a derived tuple (or an entire view) is captured as a Boolean formula which recursively unrolls the logical dependencies from the derived tuple back to the base tuples. The lineage of a derived tuple may never be cyclic, but it may impose a DAG structure over the derivation of a tuple. Thus, being a form of a directed (but acyclic) graphical model, probabilistic inference in ULDBs remains #P-complete for general SQL queries.

On the other hand, considering restricted classes of SQL queries and corresponding query plans, where exact confidence computations remain tractable, has led to a notion of "safe plans" in [30]. Intuitively, an entire query plan is safe, if all query operators take only independent subgoals as their input, which

---

[10] http://pellet.owldl.com/pronto/

guarantees a hierarchical derivation structure (i.e., tree-shaped lineage) of all tuples involved in a query result. Following this idea on a more fine-grained (i.e., on a per-tuple rather than a per-plan) level, [129] considers a class of so-called "read-once" functions, where the Boolean lineage formula can be factorized (in polynomial time) into a form, where every variable appears at most once. Moreover, efficient top-$k$ query processing [116, 134] and unified ranking approaches [81] have investigated different semantics of ranking queries over uncertain data. Recently, the modeling of correlated tuples [127] with the explicit usage of probabilistic graphical models such as Bayesian Nets [18, 151] and Markov Random Fields [128] has found increasing attention also in the database community. In [70], the authors define a class of Markov networks, where query evaluation with soft constraints can be done in polynomial time, while the case with hard constraints is considered separately in [31]. Also in the context of these graphical models, lineage remains the key for a closed representation formalism [71].

While most probabilistic database systems provide extensions to the DDL (data definition language) part of the SQL standard to define dependencies at the schema level, only fairly few works explicitly tackle the DML (data manipulation language) part of SQL, including data modifications such as updates or deletes [66, 125]. Most probabilistic database approaches focus on SELECT-PROJECT-JOIN (SPJ) query patterns, where query processing bears a number of interesting analogies to inference in probabilistic graphical models.

**MystiQ.** The MystiQ system [16, 117, 30] developed at the University of Washington provides support for a wide range of SQL queries over uncertain and inconsistent data sources. On the DML part, MystiQ introduces the notion of an *approximate match* operator between a query attribute and a data attribute, which can be evaluated by data-type-specific built-in functions. On the DDL part, MystiQ supports so-called *predicate functions*, which define how probabilities should be generated for an approximate match on such an attribute. Moreover, in the spirit of functional dependencies in deterministic databases, MystiQ allows for the specification of global constraints among attributes at the schema level. Unlike classic functional dependencies, these constraints can be specified to be either strict (e.g., a person may have only exactly one date-of-birth) or soft (e.g., most people have different names). A violation of a constraint mutually affects the confidences of all tuples involved in the conflict. In further works, the authors present an efficient top-$k$ algorithm (coined "multi-simulations") for a class of SELECT DISTINCT queries which exhibit a DNF structure as lineage. Multi-simulation works by running multiple Monte Carlo simulations [73] in parallel until the lower confidence bounds of the top-$k$ answers to the query have sufficiently converged, in order to distinguish them from the upper confidence bounds of the non-top-$k$ answers.

**Trio.** Based on the ULDB data model, the Trio [96] system developed at Stanford University provides an integrated approach for the management of *data*, *uncertainty*, and *lineage*. Trio is implemented on top of a conventional database system (PostgreSQL) and employs an SQL-based rewriting layer for the Trio query language (TriQL) into a series of relational queries and calls to stored procedures. As core of its data model, Trio adopts the notion of X-tuples [123] for mutually exclusive tuple alternatives, Boolean lineage formulas, *maybe* annotations (which indicate the possible absence of the tuple in the uncertain database), and confidence values that may be attached to each tuple alternative. In Trio (and ULDBs), lineage enables for the complete decoupling of data and confidence computations, which may yield significant efficiency benefits for query processing. Later extensions to Trio have investigated in more detail how to exploit lineage for probabilistic confidence computations [124] and data updates [125].

**MayBMS.** The MayBMS [8,66] system initially developed at Saarland University and then at the Cornell database group is designed as a completely native extension to PostgreSQL. Based on the encoding scheme of conditional tables (C-tables) [67,123], the current MayBMS system employs a compact form of schema decompositions (coined "U-relations"), which results in a succinct encoding of an uncertain relation with independent attributes. Another focal point of the MayBMS project includes the investigation of foundations for query languages in probabilistic databases in analogy to relational algebra and SQL [75]. Ongoing research issues in MayBMS include query optimization, an update language, concurrency control and recovery, and the design of generic APIs for uncertain data. MayBMS is the basis for the SPROUT system discussed in Section 4.

**Orion.** Inspired by large-scale sensor nets, the Orion [132,131] system developed at Purdue University also investigates continuous probability density functions (PDFs) in combination with SQL-style relational query processing techniques. Originally inspired by the application of sensor networks, the current Orion 2.0 prototype has built-in support for a number of continuous (e.g., Gaussian, Uniform) and discrete (e.g., Binomial, Poisson) distributions, which are treated symbolically at query processing time. If the resulting distribution of an SQL-style operation cannot be represented by a standard distribution anymore, Orion switches to an approximate distribution using histograms and sampling. Further features of Orion include the handling of correlations among attributes, which are captured as explicit joint distributions among the correlated attributes, and the handling of missing (i.e., incomplete) data, which is implemented by allowing for partial PDFs whose confidence distributions may sum up to less than 1. Using a form of query history, the Orion data model is closed under common relational operations and consistent with the possible-worlds semantics.

**PrDB.** One of the most active current probabilistic database projects is the PrDB [128] system developed at the University of Maryland. Unlike other probabilistic database approaches, PrDB employs undirected probabilistic graphical models, specifically Markov Random Fields, as basis for handling correlated base tuples. The graphical model is stored directly in the underlying database system in the form of *factor tables*, which capture correlations among tuples and serve as input for the probabilistic inference algorithms. Due to the generality of this data model, PrDB incorporates a large variety and optimizations for both exact and approximate inference, including variable elimination [36], the reuse of shared factors, and Gibbs sampling [47] in the general case. In addition, bisimulation [72] is shown to significantly speed up inference for DAG-shaped queries in this context. Also here, lineage, in the form of Boolean formulas that capture the logical dependencies of derived data objects (tuples) back to the base tuples, is the key for a closed and complete representation model. Moreover, the efficient processing of lineage for probabilistic inference under this data model has been studied in [71].

## 3.2   Logic Programming and Rule-Based Reasoning

The semantic web has led to the development of a plethora of rule-based and description-logic-based reasoning engines, including reasoners like Sesame, Jena, IRIS, Bossam, Prova, and many more[11]. Besides classical logic programming frameworks based on Prolog and Datalog, these engines specifically focus on different ontological reasoning concepts based on the RDF/S standards and the DL (based on description logic), RL (supporting first-order Horn rules) and EL (supporting rules with restricted existential quantifiers) fragments of OWL.

Besides the different grounding techniques and varying expressiveness of the ontological concepts these reasoners support, an important semantic distinction can be made in the way these engines handle negation. *Negation-as-failure* [28] is the most common semantics for handling negation in rule antecedents (bodies), which is also the default semantics used in most Prolog and Datalog engines. Intuitively, a negated literal in the body of the rule is grounded if no proof for the literal can be derived from the knowledge base. As a form of *closed-world assumption* this semantics can lead to non-monotonic inferences when new information (i.e., facts or rules) are entered into the knowledge base, such that the negation of the literal no longer holds. To tackle this issue, the *well-founded negation* and *stable-model semantics* have been proposed to handle negation in rule antecedents. We refer the reader to [143, 45, 102] for details. However, care is advisable when reasoning with recursive rules; already plain Datalog without negation has been shown to be EXPTIME-complete for non-linear, recursive programs (i.e., for predicates with more than one argument and rules with more than one recursive predicate in their antecedents) and still PSPACE-complete for linear, recursive programs, respectively [142].

To evaluate the performance of these engines, a number of benchmarks have been defined, out of which the most prominent ones are probably [145] for RDF/S, as well as the Lehigh University Benchmark (LUBM) [52] and the more recent OpenRuleBench [82] initiative. In the following subsections, we provide a brief overview of the top-performing engines evaluated in OpenRuleBench: OntoBroker, XSB, Yap, and DLV.

**OntoBroker**[12] is designed as a Java-based, object-oriented database system. It has been originally developed at the AIFB Karlsruhe and meanwhile turned into

---

[11] See http://www.w3.org/2007/OWL/wiki/Implementations for a current overview.

[12] http://www.ontoprise.de/

a commercial product. It follows a bottom-up, deductive grounding technique and supports Magic-Sets-based rule rewriting [12], as well as a cost-based query optimizer (similar to a relational database system)—a feature that many other rule engines lack. OntoBroker supports the well-founded negation.

**XSB**[13] is an open-source Prolog engine implemented in native C. In addition to a top-down processing of Prolog or function-free Datalog programs, it can also be used in a deductive (i.e., bottom-up) database fashion, using advanced grounding techniques based on tabling (aka. "memoing") [153]. Through tabling, XSB is able to terminate even for cases when many Prolog engines (based on top-down SLD resolution [77]) run into cycles. Unlike most Prolog engines, XSB supports the well-founded negation and (via a plug-in) also the stable-model semantics.

**Yap**[14] is a highly optimized Prolog engine developed at the Center for Research in Advanced Computing Systems and the University of Porto. Like XSB it supports advanced grounding techniques based on tabling, but (unlike XSB) it can also create indices for faster data access on-the-fly, when it determines that a particular index may speed up the access to a large amount of data. It however supports only negation-as-failure, but not the well-founded negation nor the stable-model semantics.

**DLV**[15] is a bottom-up rule system implemented in C++. It is unique in that it allows for a form of disjunctive Datalog programming with disjunctive rule consequents (heads). Moreover, it is the only system along theses lines which supports the stable-model semantics. As additional feature, DLV has built-in support for propositional reasoning over the grounded model using Max-SAT solving and similar techniques. As most of these engines, it is a pure query engine, i.e., it does not support incremental updates to the knowledge base.

### 3.3    Combining First-Order Logic and Probabilistic Inference

In the following, we assume a knowledge base to consist of a finite set of first-order logical formulas and a finite set of (potentially typed) entities. We focus only on reasoning techniques which work by grounding (i.e., by instantiating) the first-order rules, and which again result in a finite set of propositional formulas. This class of rules conforms to a subset of first-order logic that is generally referred to as the Bernays-Schönfinkel-Ramsey class, which is decidable and can be evaluated by grounding the first-order formulas. In some settings, we might want to distinguish between *soft rules*, which may be violated and thus typically have a confidence weight associated with them, and *hard constraints*, which may not be violated. A wide variety of grounding techniques exist, each leading to a different reasoning semantics and potentially different answers to queries. What all grounding techniques have in common is that they bind the variables in the

---

[13] http://xsb.sourceforge.net/
[14] http://www.dcc.fc.up.pt/~vsc/Yap/
[15] http://www.dbai.tuwien.ac.at/proj/dlv/

rules with the entities contained in the knowledge base in order to obtain a (grounded) set of propositional formulas. In the following, we will consider the actual grounding procedure mostly as a black-box function.

We formally call a knowledge base *inconsistent* if the conjunction of all propositional statements that can be derived from it (e.g., via grounding the rules) evaluates to *false*. Moreover, we call a knowledge base *unsatisfiable* if there exists no truth assignment to variables such that the hard constraints are satisfied.

**Propositional Reasoning.** Classic (deterministic) approaches to handling inconsistencies in a set of propositional formulas are based on the Boolean satisfiability problem, generally known as SAT. Although the general SAT problem is NP-complete, many real-world problems have actually been shown to be "easy" to solve even for thousands of Boolean variables and many tens of thousands of constraints. Moreover, in recent years the field of SAT solving has made great progress in developing strategies, which allow for tackling also non-trivial instances of the SAT problem very efficiently. Introducing soft rules, on the other hand, leads to a weighted form of the satisfiability problem, generally known as the *maximum satisfiability problem* (Max-SAT). Here the goal is to find a truth assignment to variables which maximizes the aggregated weights (typically the sum) of the formulas which are satisfied by this assignment. Finding the optimum solution in Max-SAT solving however is also NP-complete. More specifically, the maximum satisfiability problem over Horn clauses (coined Max-Horn-SAT) has been studied in detail in [69]. In [48], the authors provide a 3/4 approximation algorithm for the weighted Max-SAT problem over Boolean formulas in conjunctive normal form (CNF). None of these classic (deterministic) Max-SAT solvers however considers a distinction between soft and hard constraints. Thus, recently a family of stochastic Max-SAT solvers has been introduced with Walk-SAT [154] and MaxWalkSAT [74], which apply different strategies in order to explore multiple possible worlds to more accurately approximate the optimum solution. A counterpart to Max-SAT solving from a probabilistic perspective is Maximum-a-Posteriori estimation (or "MAP-inference") [133], which selects the most likely mode, i.e., the most likely assignments to variables, according to their posteriori distribution. Recently, for example in the context of information extraction, grounding a set of first-order formulas and post-processing the propositional formulas by a Max-SAT solver has been applied successfully in the SOFIE [141] and PROSPERA [97] projects, in order to automatically populate the YAGO [140] knowledge base[16].

**Markov Logic Networks.** Statistical relational learning (SRL) [46] has been gaining an increasing momentum in the machine learning, database, and semantic web communities, with Markov Logic Networks [118] probably being the most generic approach for combining first-order logic and probabilistic graphical models into a unified representation and reasoning framework. Intuitively, Markov Logic works by grounding a set of first-order logical rules against a

---

[16] http://www.mpi-inf.mpg.de/yago-naga/

knowledge base, and by sampling states ("worlds") over a Markov network that represents the grounded (i.e., propositional) formulas. Inference in probabilistic graphical models in general is #P-complete. Therefore, Markov Chain Monte Carlo (MCMC) [47, 114, 133] denotes a family of efficient sampling algorithms for probabilistic inference in these graphical models, with Gibbs sampling [47] being one of the most widely used sampling technique, which is also employed in Markov Logic.

Markov Logic however does not easily scale to very large knowledge bases. Grounding a first-order Markov network works by binding all entities (constants) to variables in first-order predicates that match the type signature of the predicate. For binary predicates, this results in grounded networks which are often nearly quadratic in the number of entities in the knowledge base. Scaling Markov Logic to large knowledge bases with millions of entities (and hundreds of millions of facts) thus remains all but straightforward. Recently, the Tuffy [103] engine (see Section 4) has been addressing the issue of scaling up Markov Logic by coupling Alchemy[17] with a relational back-end, and by replacing the grounding procedure of Alchemy with a more efficient bottom-up grounding technique.

**MAP-Inference.** More recently, stochastic ways of addressing inference over a combination of deterministic (hard) and probabilistic (soft) dependencies has been addressed also in the context of Markov Logic. Maximum-a-Posteriori (MAP) inference [133] (based on the stochastic weighted Max-SAT solver MaxWalkSAT [74]) and MC-SAT [114] (based on slice sampling [33]) are two approximation algorithms for propositional and probabilistic inference in Markov Logic, respectively. Using a log-linear model for generating the factors of grounded formulas, MAP-inference can be shown to directly correspond to an execution of MaxWalkSAT over a Markov Logic network [119].

**MC-SAT.** Hard constraints may introduce isolated regions of states which cannot easily be overcome by a Gibbs sampler (i.e, by just flipping one variable at a time). MC-SAT [113] thus introduces auxiliary variables which provide the sampler with the ability a "jump" into another (otherwise disconnected) region with some probability. Experimentally, MC-SAT has been shown to outperform Gibbs sampling and simulated tempering by a significant margin, particularly when deterministic dependencies are present. However, allowing arbitrary constraints as hard rules may lead to the formulation of unsatisfiable constraints, which either renders the knowledge base inconsistent (if there is no solution at all) or empty (if the only solution is to set all facts to false). Satisfiability checks, which includes checking whether a derived fact is false in all the possible worlds and thus has a probability of exactly 0, cannot be approximated and thus remain an NP-hard problem.

**Constrained Conditional Models.** Another framework which combines (first-order) logical constraints and probabilistic inference is given by Constrained Conditional Models [88] (CCMs). Intuitively, constraints between input

---

[17] http://alchemy.cs.washington.edu/

variables (observations) and output variables (labels) are encoded into linear weight vectors, which can be solved by Integer Linear Programming. Working with CCMs involves both learning weights for the model and efficient inference. CCMs allow for encoding Markov Random Fields, Hidden Markov Models and Conditional Random Fields [121]. They found strong applications in natural language processing, including tasks like co-reference resolution, semantic role labeling, and information extraction.

**Probabilistic Datalog.** Early probabilistic extensions to Datalog have been studied already in [44] and have later been refined to a number OWL concepts [104]. Although this approach already introduced a notion of lineage (coined "intensional query semantics"), the probabilistic computations are restricted to a class of rules which is guaranteed to provide independent subgoals (similar to the notion of safe plans or read-once functions in [30, 129]), where confidence computations can be propagated "upwards" the lineage tree using the inclusion-exclusion principle (aka. "sieve formula").

### 3.4   Programming Platforms for Probabilistic Inference

The "declarative-imperative" [64], a term coined in the context of the Berkeley Orders of Magnitude (BOOM) project[18], brings two seemingly contracting paradigms in data management to the point: how can we combine the power of an imperative programming language with the convenience of a declarative query language? In the following, we briefly highlight two imperative programming platforms for probabilistic inference: FACTORIE and Infer.NET.

**FACTORIE**[19] is a toolkit for deployable probabilistic modeling developed by the machine learning group at the University of Massachusetts Amherst [93]. It is based on the idea of using an imperative programming language (Scala) to define templates which generate factors between random variables, an approach coined *imperatively defined factor graphs*. Intuitively, when instantiated these templates form a factor graph, where all factors that have been instantiated from the same template also share the same parameters that were used to define the template. For inference, FACTORIE provides a variety of techniques based on MCMC, including Gibbs sampling. FACTORIE has been successfully applied to various inference tasks in natural language processing and information integration. Recently, FACTORIE has also been coupled with a relational back-end and thus potentially scales to probabilistic database settings with billions of variables [156].

**Infer.NET**[20] provided by Microsoft Research in Cambridge provides a rich programming language for modeling Bayesian inference tasks in graphical models and comes with an out-of-the-box selection of inference algorithms. It provides a

---

[18] http://boom.cs.berkeley.edu/

[19] http://code.google.com/p/factorie/

[20] http://research.microsoft.com/en-us/um/cambridge/projects/infernet/

built-in API for defining random variables (binary/multivariate-discrete or continuous), factors, message-passing operators, and other algebraic operators. It has been used in many machine-learning settings, with tasks involving classification or clustering, and in a wide variety of domains, including information retrieval, bio-informatics, epidemiology, vision, and many others.

### 3.5 Distributed Probabilistic Inference

Distribution bears the highest potential to scale-up rule-based reasoning and probabilistic inference, but still is fairly unexplored in the context of uncertain reasoning and probabilistic data management. Although distribution of course cannot tackle the asymptotical runtime issues inherently involved in these algorithms, it bears two key advantages:

- Storing a large data or knowledge base with billions of uncertain data objects in a distributed environment immediately allows for an increased *main-memory locality* of the data, which is a key for both efficient rule-based reasoning and probabilistic inference, with a majority of fine-grained, random-access-style data accesses.

- Running queries over a cluster of machines bears great potential for high-performance *parallel computations*, but clearly also poses major algorithmic challenges in terms of synchronizing these computations and the preservation of approximation guarantees (e.g., convergence guarantees for the MCMC-based sampling techniques).

**MCDB.** The MCDB [68, 159] project at IBM Almaden focuses on supporting Monte Carlo techniques for complex data analysis tasks directly within a database system. MCDB is one of the few database approaches to probabilistic data management that has specifically been adapted to Hadoop[21], a massively parallel, Map-Reduce-like [35] computing environment. MCDB focuses on analytical tasks over a broad range of user-defined stochastic models, e.g., risk analysis with complex analytical queries including grouping and aggregations [53]. Another, recent, application domain of MCDB is declarative information extraction [95].

**Message Passing & Distributed Inference.** Its iterative nature makes the Map-Reduce paradigm not well suitable for inference tasks, which inherently involve many fine-grained updates between states of objects that may be distributed across a compute cluster. For probabilistic inference, two main paradigms for distribution co-exist: the *shared memory* and the *distributed memory* model. In the shared-memory model, every processor has access to all the memory in the cluster; while for the distributed memory model, every processor only has a limited amount of local memory, and each processor can pass "messages" to other processors in the cluster. An alternative to the classic Message

---

[21] http://hadoop.apache.org/

Passing Interface[22] (MPI) is the Internet Communications Engine[23] (ICE). Both are shipped as C++ libraries.

With the ResidualSplash [49] algorithm, the authors present a parallel belief propagation algorithm under the shard memory model, which is shown to achieve optimal runtime compared to a theoretical lower bound for parallel inference on chain graphical models. In their later DBRSplash [50] algorithm, the authors drop the shared memory model and consider parallel inference techniques in generic probabilistic graphical models, which are captured as *distributed factor graphs*. In this setting, a factor graph is distributed into a number of (disjoint or slightly overlapping) partitions, such that the number of partitions matches the number of processors available in the compute cluster. The objective of the partitioning function is to minimize the communication cost among nodes while ensuring load balance. Since computing an optimal partitioning under these constraints is NP-hard, an efficient (linear-time) approximation algorithm is devised as basis for the data partitioning. As for inference, a belief propagation algorithm is employed, with a local priority queue for incoming update messages at each processor. DBRSplash even reports a super-linear performance scale-up compared to a centralized setting. Moreover, GraphLab[24] [86] is a framework for deploying parallel (provably correct) machine learning algorithms. Unlike MapReduce, it focuses on more asynchronous communication protocols with different levels of sequential-consistency guarantees. In the *full consistency* model, during the execution of a function $f(v)$ on a vertex $v$, no other function is allowed to read or write data to any node in the scope (neighborhood) $S(v)$. In the *edge consistency* model, no other function is allowed to read or write data to an edge associated with $v$, while a function $f(v)$ is executed. Finally, in the weakest form of consistency model, the *vertex consistency* model, no other function is allowed to read or write data to the vertex $v$ itself, while a function $f(v)$ is executed.

## 4   New Trends: BayesStore, SPROUT, Tuffy, URDF

SQL-style query processing over uncertain relational data or, respectively, rule-based reasoning with uncertain RDF data is an emerging topic in the database, knowledge management and semantic web communities. Recent trends along these lines include moving away from strict relational data, lifting inference to higher-order constraints, and scaling-up inference for graphical models which combine first-order logic and probabilistic graphical models (in particular Markov Logic). Extracting structured data from the Web is an excellent showcase—and likely one of the biggest challenges—for the scalable management of uncertain data we have seen so far. In the following, we thus briefly highlight a few projects, each of which devises exciting directions that could help making the management of uncertain RDF data scalable to billions of triples.

---

[22] http://www.mcs.anl.gov/research/projects/mpich2/
[23] http://www.zeroc.com/ice.html
[24] http://www.graphlab.ml.cmu.edu/

**BayesStore.** Based on a native extension to the PostgreSQL database system, the BayesStore [151] project at Berkeley aims to bridge the gap between statistical models induced by the data and the uncertainty model supported by the probabilistic database. BayesStore supports statistical models, evidence data, and inference algorithms directly as first-class citizens inside a database management system (DBMS). It combines a probabilistic database system (PDBMS) for relational data with a declarative first-order extension to Bayesian Nets, which allows for capturing complex correlation patterns in the PDBMS in a compact way. Using a combination of propositional and first-order factors, BayesStore supports probabilistic inference for all common database operations, including selection, projection, and joins. Recent extensions to BayesStore include the investigation of probabilistic-declarative information extraction techniques by using Conditional Random Fields (CRFs) for extraction and by implementing the Viterbi algorithm for efficient inference in the CRF directly via SQL [150].

**SPROUT.** Using the MayBMS probabilistic database system (see Section 3.1) as basis, the SPROUT (for "Scalable PROcessing on Uncertain Tables") [106] project at Oxford University aims at the scalable processing of queries over uncertain data. A particular focus of the project lies on tractable classes of queries for which exact probabilistic inference can be done in polynomial time. Similarly to the notion of safe plans [30], a probabilistic query is tractable if it has a hierarchical structure and (in the case of SPROUT) can be decomposed into a binary decision diagram in polynomial time. [107], on the other hand, investigates the decision diagrams for approximate inference for queries where exact probabilistic inference is intractable.

**Tuffy.** Based on the observation that Markov Logic does not easily scale to real-world datasets with millions of data objects, the Tuffy [103] at the University of Wisconsin-Madison investigates pushing Markov Logic Networks (MLNs) directly into a relational database system (RDBMS). In contrast to the open-world grounding strategy followed by the MLN implementation Alchemy, the authors pursue a more efficient bottom-up, closed-world grounding strategy of first-order rules through iterative SQL statements, which is fully supported by the DBMS optimizer. Focusing on MAP-inference, Tuffy provides a partitioning strategy for the search (inference) phase by splitting the grounded network into a number of independent components, which allows for parallel inference and an exponentially faster search compared to running the search on the global problem. For a given classification benchmark, Tuffy (consuming only 15MB of RAM) was reported to produce much better result quality within minutes than Alchemy (using 2.8GB of RAM) even after days of running.

**URDF.** In probabilistic databases, SQL is employed as means for formulating queries and for defining views. While SQL queries and schema-level constraints generally yield "hard" Boolean constraints among tuples, they lack the notion of "soft" dependencies among data items. The URDF [144] project developed at the Max Planck Institute for Informatics specifically focuses on weighted Horn clauses as soft rules and mutual-exclusion constraints as hard rules. Unlike

Markov Logic, URDF follows a Datalog-style, deductive grounding strategy for soft rules in the form of Horn clauses, which typically results in a much smaller grounded network size than for Markov Logic. While URDF originally employed a Max-SAT solver, which had been tailored to a specific class of mutual-exclusion hard rules, URDF currently also employs probabilistic models based on deductive grounding and lineage. Moreover, URDF explores several directions for managing large amounts of web-extracted RDF data in a declarative way, including temporal reasoning extensions [152, 39], as well as inductively learning soft inference rules automatically from a given knowledge base. As exact inference for this class of queries is intractable, URDF investigates various approximation techniques based on MCMC (Gibbs sampling) for approximate inference.

# References

1. RDF Primer & RDF Schema (W3C Rec.2004-02-10), http://www.w3.org/TR/rdf-primer/, http://www.w3.org/TR/rdf-primer/
2. Abadi, D.J., Marcus, A., Madden, S., Hollenbach, K.: SW-Store: a vertically partitioned DBMS for Semantic Web data management. VLDB J. 18(2), 385–406 (2009)
3. Abadi, D.J., Marcus, A., Madden, S., Hollenbach, K.J.: Scalable semantic web data management using vertical partitioning. In: Koch, C., Gehrke, J., Garofalakis, M.N., Srivastava, D., Aberer, K., Deshpande, A., Florescu, D., Chan, C.Y., Ganti, V., Kanne, C.-C., Klas, W., Neuhold, E.J. (eds.) VLDB, pp. 411–422. ACM, New York (2007)
4. Aberer, K., Cudré-Mauroux, P., Datta, A., Despotovic, Z., Hauswirth, M., Punceva, M., Schmidt, R.: P-Grid: a self-organizing structured P2P system. SIGMOD Rec 32, 29–33 (2003)
5. Aberer, K., Cudré-Mauroux, P., Hauswirth, M., Van Pelt, T.: GridVine: Building Internet-Scale Semantic Overlay Networks. In: McIlraith, S.A., Plexousakis, D., van Harmelen, F. (eds.) ISWC 2004. LNCS, vol. 3298, pp. 107–121. Springer, Heidelberg (2004)
6. Abiteboul, S., Kanellakis, P., Grahne, G.: On the representation and querying of sets of possible worlds. Theor. Comput. Sci. 78(1), 159–187 (1991)
7. Antoniou, G., van Harmelen, F.: A Semantic Web Primer (Cooperative Information Systems). MIT Press, Cambridge (2004)
8. Antova, L., Koch, C., Olteanu, D.: MayBMS: Managing incomplete information with probabilistic world-set decompositions. In: ICDE, pp. 1479–1480 (2007)
9. Atre, M., Chaoji, V., Zaki, M.J., Hendler, J.A.: Matrix bit loaded: a scalable lightweight join query processor for RDF data. In: Rappa, M., Jones, P., Freire, J., Chakrabarti, S. (eds.) WWW, pp. 41–50. ACM, New York (2010)
10. Auer, S., Bizer, C., Kobilarov, G., Lehmann, J., Cyganiak, R., Ives, Z.G.: DBpedia: A nucleus for a web of open data. In: Aberer, K., Choi, K.-S., Noy, N., Allemang, D., Lee, K.-I., Nixon, L.J.B., Golbeck, J., Mika, P., Maynard, D., Mizoguchi, R., Schreiber, G., Cudré-Mauroux, P. (eds.) ASWC 2007 and ISWC 2007. LNCS, vol. 4825, pp. 722–735. Springer, Heidelberg (2007)
11. Auer, S., Ngomo, A.-D.N., Lehmann, J.: Introduction to linked data. In: Polleres, A., et al. (eds.) Reasoning Web 2011. LNCS, vol. 6848, pp. 203–250. Springer, Heidelberg (2011)

12. Beeri, C., Ramakrishnan, R.: On the power of magic. J. Log. Program. 10(1/2/3/4), 255–299 (1991)
13. Benjelloun, O., Sarma, A.D., Halevy, A.Y., Widom, J.: ULDBs: Databases with uncertainty and lineage. In: VLDB, pp. 953–964 (2006)
14. Berners-Lee, T.: Linked Data - Design Issues (2006), http://www.w3.org/DesignIssues/LinkedData.html
15. Bizer, C., Heath, T., Berners-Lee, T.: Linked Data – The Story So Far. Int. J. Semantic Web. Inf. Syst. 5(3), 1–22 (2009)
16. Boulos, J., Dalvi, N., Mandhani, B., Mathur, S., Ré, C., Suciu, D.: MystiQ: a system for finding more answers by using probabilities. SIGMOD, 891–893 (2005)
17. Bouquet, P., Ghidini, C., Serafini, L.: Querying the Web of Data: A Formal Approach. In: Gómez-Pérez, A., Yu, Y., Ding, Y. (eds.) ASWC 2009. LNCS, vol. 5926, pp. 291–305. Springer, Heidelberg (2009)
18. Bravo, H.C., Ramakrishnan, R.: Optimizing MPF queries: decision support and probabilistic inference. SIGMOD, 701–712 (2007)
19. Buitelaar, P., Eigner, T., Declerck, T.: OntoSelect: A Dynamic Ontology Library with Support for Ontology Selection. In: Proceedings of the Demo Session at the International Semantic Web Conference (2004)
20. Cai, M., Frank, M.: RDFPeers: a scalable distributed RDF repository based on a structured peer-to-peer network. In: Proceedings of the 13th International Conference on World Wide Web, WWW 2004, pp. 650–657 (2004)
21. Cai, M., Frank, M., Chen, J., Szekely, P.: MAAN: A Multi-Attribute Addressable Network for Grid Information Services. In: Proceedings of the 4th International Workshop on Grid Computing, GRID 2003, p. 184 (2003)
22. Carroll, J.J., Bizer, C., Hayes, P., Stickler, P.: Named graphs. Journal of Web Semantics 3, 247–267 (2005)
23. Carroll, J.J., Dickinson, I., Dollin, C., Reynolds, D., Seaborne, A., Wilkinson, K.: Jena: implementing the Semantic Web recommendations. In: Feldman, S.I., Uretsky, M., Najork, M., Wills, C.E. (eds.) WWW (Alternate Track Papers & Posters), pp. 74–83. ACM, New York (2004)
24. Cavallo, R., Pittarelli, M.: The theory of probabilistic databases. In: VLDB, pp. 71–81. Morgan Kaufmann, San Francisco (1987)
25. Chen, H., Wang, Y., Wang, H., Mao, Y., Tang, J., Zhou, C., Yin, A., Wu, Z.: Towards a Semantic Web of relational databases: A practical semantic toolkit and an in-use case from traditional Chinese medicine. In: Cruz, I., Decker, S., Allemang, D., Preist, C., Schwabe, D., Mika, P., Uschold, M., Aroyo, L.M. (eds.) ISWC 2006. LNCS, vol. 4273, pp. 750–763. Springer, Heidelberg (2006)
26. Cheng, G., Qu, Y.: Searching Linked Objects with Falcons: Approach, Implementation and Evaluation. Int. J. Semantic Web Inf. Syst. 5(3), 49–70 (2009)
27. Chong, E.I., Das, S., Eadon, G., Srinivasan, J.: An efficient SQL-based RDF querying scheme. In: Böhm, K., Jensen, C.S., Haas, L.M., Kersten, M.L., Larson, P.-Å., Ooi, B.C. (eds.) VLDB, pp. 1216–1227. ACM, New York (2005)
28. Clark, K.L.: Negation as failure. In: Logic and Data Bases, pp. 293–322. Plenum Press, New York (1978)
29. Cruz, I.F., Kashyap, V., Decker, S., Eckstein, R. (eds.): Proceedings of SWDB 2003, The first International Workshop on Semantic Web and Databases, Co-located with VLDB 2003, September 7-8. Humboldt-Universität, Berlin (2003)
30. Dalvi, N., Suciu, D.: Efficient query evaluation on probabilistic databases. In: VLDB, pp. 864–875 (2004)
31. Dalvi, N., Suciu, D.: The dichotomy of conjunctive queries on probabilistic structures. In: PODS Conference, pp. 293–302 (2007)

32. Dalvi, N.N., Ré, C., Suciu, D.: Probabilistic databases: diamonds in the dirt. Commun. ACM 52(7), 86–94 (2009)
33. Damlen, P., Wakefield, J., Walker, S.: Gibbs sampling for Bayesian non-conjugate and hierarchical models by using auxiliary variables. Journal of the Royal Statistical Society: Series B (Statistical Methodology) 61(2), 331–344 (1999)
34. d'Aquin, M., Baldassarre, C., Gridinoc, L., Angeletou, S., Sabou, M., Motta, E.: Characterizing Knowledge on the Semantic Web with Watson. In: EON, pp. 1–10 (2007)
35. Dean, J., Ghemawat, S.: MapReduce: simplified data processing on large clusters. Commun. ACM 51, 107–113 (2008)
36. Dechter, R.: Bucket elimination: A unifying framework for reasoning. Artif. Intell. 113(1-2), 41–85 (1999)
37. Ding, L., Finin, T., Joshi, A., Pan, R., Cost, R.S., Peng, Y., Reddivari, P., Doshi, V., Sachs, J.: Swoogle: a search and metadata engine for the semantic web. In: CIKM 2004: Proceedings of the thirteenth ACM International Conference on Information and Knowledge Management, pp. 652–659 (2004)
38. Ding, Y., Sun, Y., Chen, B., Borner, K., Ding, L., Wild, D., Wu, M., DiFranzo, D., Fuenzalida, A.G., Li, D., Milojevic, S., Chen, S., Sankaranarayanan, M., Toma, I.: Semantic web portal: a platform for better browsing and visualizing semantic data. In: Proceedings of the 6th International Conference on Active Media Technology, AMT 2010, pp. 448–460 (2010)
39. Dylla, M., Sozio, M., Theobald, M.: Resolving temporal conflicts in inconsistent rdf knowledge bases. In: BTW, pp. 474–493 (2011)
40. Erling, O., Mikhailov, I.: Towards web-scale rdf, http://virtuoso.openlinksw.com/whitepapers/Web-Scale%20RDF.pdf
41. Erling, O., Mikhailov, I.: RDF Support in the Virtuoso DBMS. In: Pellegrini, T., Auer, S., Tochtermann, K., Schaffert, S. (eds.) Networked Knowledge - Networked Media. SCI, vol. 221, pp. 7–24. Springer, Berlin (2009)
42. Fletcher, G.H.L., Beck, P.W.: Scalable indexing of RDF graphs for efficient join processing. In: Cheung, D.W.-L., Song, I.-Y., Chu, W.W., Hu, X., Lin, J.J. (eds.) CIKM, pp. 1513–1516. ACM, New York (2009)
43. Frakes, W.B., Baeza-Yates, R.A. (eds.): Information Retrieval: Data Structures & Algorithms. Prentice-Hall, Englewood Cliffs (1992)
44. Fuhr, N.: Probabilistic Datalog - a logic for powerful retrieval methods. In: SIGIR, pp. 282–290 (1995)
45. Gelfond, M., Lifschitz, V.: The stable model semantics for logic programming. In: Logic Programming, pp. 1070–1080. MIT Press, Cambridge (1988)
46. Getoor, L., Taskar, B.: An Introduction to Statistical Relational Learning. MIT Press, Cambridge (2007)
47. Gilks, W., Richardson, S., Spiegelhalter, D.J.S.: Markov Chain Monte Carlo in Practice. Chapman and Hall, Boca Raton (1996)
48. Goemans, M.X., Williamson, D.P.: New 3/4-approximation algorithms for the maximum satisfiability problem. SIAM J. Discrete Math. 7(4), 656–666 (1994)
49. Gonzalez, J.E., Low, Y., Guestrin, C.: Residual splash for optimally parallelizing belief propagation. In: Artificial Intelligence and Statistics (AISTATS), pp. 177–184 (2009)
50. Gonzalez, J.E., Low, Y., Guestrin, C., O'Hallaron, D.: Distributed parallel inference on large factor graphs. In: Uncertainty in Artificial Intelligence (UAI), pp. 203–212 (2009)
51. Görlitz, O., Staab, S.: Federated Data Management and Query Optimization for Linked Open Data, ch. 5, pp. 109–137. Springer, Heidelberg (2011)

52. Guo, Y., Pan, Z., Heflin, J.: LUBM: A benchmark for OWL knowledge base systems. J. Web Sem. 3(2-3), 158–182 (2005)
53. Haas, P.J., Jermaine, C.M., Arumugam, S., Xu, F., Perez, L.L., Jampani, R.: MCDB-R: Risk analysis in the database. PVLDB 3(1), 782–793 (2010)
54. Haase, P., Mathäß, T., Ziller, M.: An evaluation of approaches to federated query processing over linked data. In: Proceedings of the 6th International Conference on Semantic Systems, I-SEMANTICS 2010, pp. 5:1–5:9 (2010)
55. Haase, P., Wang, Y.: A decentralized infrastructure for query answering over distributed ontologies. In: Proceedings of the 2007 ACM symposium on Applied computing, SAC 2007, pp. 1351–1356 (2007)
56. Harris, S., Gibbins, N.: 3store: Efficient bulk RDF storage. In: Volz, R., Decker, S., Cruz, I.F. (eds.) PSSS. CEUR Workshop Proceedings, vol. 89 (2003)
57. Harth, A.: VisiNav: Visual web data search and navigation. In: Bhowmick, S.S., Küng, J., Wagner, R. (eds.) DEXA 2009. LNCS, vol. 5690, pp. 214–228. Springer, Heidelberg (2009)
58. Harth, A., Hogan, A., Delbru, R., Umbrich, J., O'Riain, S., Decker, S.: SWSE: Answers Before Links! In: Semantic Web Challenge (2007)
59. Harth, A., Hose, K., Karnstedt, M., Polleres, A., Sattler, K., Umbrich, J.: Data Summaries for On-Demand Queries over Linked Data. In: WWW 2010, pp. 411–420 (2010)
60. Harth, A., Umbrich, J., Hogan, A., Decker, S.: YARS2: A federated repository for querying graph structured data from the web. In: Aberer, K., Choi, K.-S., Noy, N., Allemang, D., Lee, K.-I., Nixon, L.J.B., Golbeck, J., Mika, P., Maynard, D., Mizoguchi, R., Schreiber, G., Cudré-Mauroux, P. (eds.) ASWC 2007 and ISWC 2007. LNCS, vol. 4825, pp. 211–224. Springer, Heidelberg (2007)
61. Hartig, O., Bizer, C., Freytag, J.-C.: Executing SPARQL queries over the web of linked data. In: Bernstein, A., Karger, D.R., Heath, T., Feigenbaum, L., Maynard, D., Motta, E., Thirunarayan, K. (eds.) ISWC 2009. LNCS, vol. 5823, pp. 293–309. Springer, Heidelberg (2009)
62. Hartig, O., Heese, R.: The SPARQL query graph model for query optimization. In: Franconi, E., Kifer, M., May, W. (eds.) ESWC 2007. LNCS, vol. 4519, pp. 564–578. Springer, Heidelberg (2007)
63. Hartig, O., Langegger, A.: A Database Perspective on Consuming Linked Data on the Web. Datenbank-Spektrum 10(2), 57–66 (2010)
64. Hellerstein, J.M.: The declarative imperative: experiences and conjectures in distributed logic. SIGMOD Record 39(1), 5–19 (2010)
65. Hogan, A., Harth, A., Decker, S.: ReConRank: A Scalable Ranking Method for Semantic Web Data with Context. In: 2nd Workshop on Scalable Semantic Web Knowledge Base Systems (2006)
66. Huang, J., Antova, L., Koch, C., Olteanu, D.: MayBMS: a probabilistic database management system. SIGMOD, 1071–1074 (2009)
67. Imielinski, T., Lipski Jr., W.: Incomplete information in relational databases. J. ACM 31(4), 761–791 (1984)
68. Jampani, R., Xu, F., Wu, M., Perez, L.L., Jermaine, C.M., Haas, P.J.: MCDB: a Monte Carlo approach to managing uncertain data. In: Wang, J.T.-L. (ed.) SIGMOD, pp. 687–700. ACM, New York (2008)
69. Jaumard, B., Simeone, B.: On the complexity of the maximum satisfiability problem for Horn formulas. Information Processing Letters 26(1), 1–4 (1987)
70. Jha, A., Rastogi, V., Suciu, D.: Query evaluation with soft-key constraints. In: PODS, pp. 119–128 (2008)

71. Kanagal, B., Deshpande, A.: Lineage processing over correlated probabilistic databases. In: SIGMOD, pp. 675–686 (2010)
72. Kanellakis, P.C., Smolka, S.A.: CCS expressions finite state processes, and three problems of equivalence. Inf. Comput. 86, 43–68 (1990)
73. Karp, R.M., Luby, M.: Monte-Carlo algorithms for enumeration and reliability problems. In: FOCS, pp. 56–64 (1983)
74. Kautz, H., Selman, B., Jiang, Y.: A general stochastic approach to solving problems with hard and soft constraints. In: The Satisfiability Problem: Theory and Applications, pp. 573–586. American Mathematical Society, Providence (1996)
75. Koch, C.: A compositional query algebra for second-order logic and uncertain databases. In: ICDT, pp. 127–140 (2009)
76. Kossmann, D.: The state of the art in distributed query processing. ACM Comput. Surv. 32, 422–469 (2000)
77. Kowalski, R.A., Kuehner, D.: Linear resolution with selection function. Artif. Intell. 2(3/4), 227–260 (1971)
78. Ladwig, G., Tran, T.: Linked Data Query Processing Strategies. In: Patel-Schneider, P.F., Pan, Y., Hitzler, P., Mika, P., Zhang, L., Pan, J.Z., Horrocks, I., Glimm, B. (eds.) ISWC 2010, Part I. LNCS, vol. 6496, pp. 453–469. Springer, Heidelberg (2010)
79. Langegger, A., Wöß, W., Blöchl, M.: A semantic web middleware for virtual data integration on the web. In: Bechhofer, S., Hauswirth, M., Hoffmann, J., Koubarakis, M. (eds.) ESWC 2008. LNCS, vol. 5021, pp. 493–507. Springer, Heidelberg (2008)
80. Levandoski, J.J., Mokbel, M.F.: RDF data-centric storage. In: ICWS, pp. 911–918. IEEE Computer Society Press, Los Alamitos (2009)
81. Li, J., Saha, B., Deshpande, A.: A unified approach to ranking in probabilistic databases. In: PVLDB, vol. 2(1), pp. 502–513 (2009)
82. Liang, S., Fodor, P., Wan, H., Kifer, M.: OpenRuleBench: an analysis of the performance of rule engines. In: WWW, pp. 601–610. ACM, New York (2009)
83. Liarou, E., Idreos, S., Koubarakis, M.: Evaluating Conjunctive Triple Pattern Queries over Large Structured Overlay Networks. In: Cruz, I., Decker, S., Allemang, D., Preist, C., Schwabe, D., Mika, P., Uschold, M., Aroyo, L.M. (eds.) ISWC 2006. LNCS, vol. 4273, pp. 399–413. Springer, Heidelberg (2006)
84. Liu, B., Hu, B.: Path queries based RDF index. In: SKG, p. 91. IEEE Computer Society Press, Los Alamitos (2005)
85. Baolin, L., Bo, H.: HPRD: A high performance RDF database. In: Li, K., Jesshope, C., Jin, H., Gaudiot, J.-L. (eds.) NPC 2007. LNCS, vol. 4672, pp. 364–374. Springer, Heidelberg (2007)
86. Low, Y., Gonzalez, J., Kyrola, A., Bickson, D., Guestrin, C., Hellerstein, J.M.: GraphLab: A new parallel framework for machine learning. In: Conference on Uncertainty in Artificial Intelligence (UAI), Catalina Island, California (2010)
87. Lukasiewicz, T.: Probabilistic description logic programs. Int. J. Approx. Reasoning 45(2), 288–307 (2007)
88. Chang, N.R.M., Ratinov, L., Roth, D.: Learning and inference with constraints. In: AAAI (2008)
89. Maduko, A., Anyanwu, K., Sheth, A.P., Schliekelman, P.: Graph summaries for subgraph frequency estimation. In: Bechhofer, S., Hauswirth, M., Hoffmann, J., Koubarakis, M. (eds.) ESWC 2008. LNCS, vol. 5021, pp. 508–523. Springer, Heidelberg (2008)

90. Maduko, A., Anyanwu, K., Sheth, A.P., Schliekelman, P.: Estimating the cardinality of RDF graph patterns. In: Williamson, C.L., Zurko, M.E., Patel-Schneider, P.F., Shenoy, P.J. (eds.) WWW, pp. 1233–1234. ACM, New York (2007)
91. Maduko, A., Anyanwu, K., Sheth, A.P., Schliekelman, P.: Graph summaries for subgraph frequency estimation. In: Bechhofer, S., Hauswirth, M., Hoffmann, J., Koubarakis, M. (eds.) ESWC 2008. LNCS, vol. 5021, pp. 508–523. Springer, Heidelberg (2008)
92. Matono, A., Amagasa, T., Yoshikawa, M., Uemura, S.: An indexing scheme for RDF and RDF schema based on suffix arrays. In: Cruz, et al [29], pp. 151–168
93. McCallum, A., Schultz, K., Singh, S.: FACTORIE: Probabilistic programming via imperatively defined factor graphs. In: NIPS (2009)
94. Mendelzon, A.O., Milo, T.: Formal models of Web queries. In: Proceedings of the sixteenth ACM SIGACT-SIGMOD-SIGART symposium on Principles of database systems, PODS 1997, pp. 134–143 (1997)
95. Michelakis, E., Krishnamurthy, R., Haas, P.J., Vaithyanathan, S.: Uncertainty management in rule-based information extraction systems. SIGMOD, 101–114 (2009)
96. Mutsuzaki, M., Theobald, M., de Keijzer, A., Widom, J., Agrawal, P., Benjelloun, O., Sarma, A.D., Murthy, R., Sugihara, T.: Trio-One: Layering uncertainty and lineage on a conventional DBMS (demo). In: CIDR, pp. 269–274 (2007)
97. Nakashole, N., Theobald, M., Weikum, G.: Scalable knowledge harvesting with high precision and high recall. In: WSDM, pp. 227–236 (2011)
98. Nejdl, W., Wolf, B., Qu, C., Decker, S., Sintek, M., Naeve, A., Nilsson, M., Palmér, M., Risch, T.: EDUTELLA: a P2P networking infrastructure based on RDF. In: WWW 2002: Proceedings of the 11th International Conference on World Wide Web, pp. 604–615. ACM Press, New York (2002)
99. Neumann, T., Weikum, G.: Rdf-3x: a risc-style engine for rdf. In: PVLDB, vol. 1(1), pp. 647–659 (2008)
100. Neumann, T., Weikum, G.: Scalable join processing on very large RDF graphs. In: Çetintemel, U., Zdonik, S.B., Kossmann, D., Tatbul, N. (eds.) SIGMOD Conference, pp. 627–640. ACM, New York (2009)
101. Neumann, T., Weikum, G.: The RDF-3X engine for scalable management of rdf data. VLDB J 19(1), 91–113 (2010)
102. Niemelä, I., Simons, P.: Smodels - an implementation of the stable model and well-founded semantics for normal logic programs. In: Logic Programming and Nonmonotonic Reasoning, Springer, Heidelberg (1997)
103. Niu, F., Ré, C., Doan, A., Shavlik, J.: Tuffy: scaling up statistical inference in Markov logic networks using an RDBMS. Technical report, University of Wisconsin-Madison (2010)
104. Nottelmann, H., Fuhr, N.: Adding probabilities and rules to OWL lite subsets based on probabilistic Datalog. Int. Journal of Uncertainty, Fuzziness and Knowledge-Based Systems 14(1), 17–41 (2006)
105. Obermeier, P., Nixon, L.: A Cost Model for Querying Distributed RDF-Repositories with SPARQL. In: Workshop on Advancing Reasoning on the Web: Scalability and Commonsense (2008)
106. Olteanu, D., Huang, J., Koch, C.: SPROUT: Lazy vs. eager query plans for tuple-independent probabilistic databases. In: ICDE, pp. 640–651. IEEE, Los Alamitos (2009)
107. Olteanu, D., Huang, J., Koch, C.: Approximate confidence computation in probabilistic databases. In: ICDE, pp. 145–156 (2010)

108. Oren, E., Delbru, R., Catasta, M., Cyganiak, R., Stenzhorn, H., Tummarello, G.: Sindice.com: a document-oriented lookup index for open linked data. Int. J. Metadata Semant. Ontologies 3, 37–52 (2008)

109. Page, L., Brin, S., Motwani, R., Winograd, T.: The PageRank Citation Ranking: Bringing Order to the Web. Technical Report 1999-66, Stanford InfoLab (November 1999)

110. Palma, R., Haase, P.: Oyster - Sharing and Re-using Ontologies in a Peer-to-Peer Community. In: Gil, Y., Motta, E., Benjamins, V.R., Musen, M.A. (eds.) ISWC 2005. LNCS, vol. 3729, pp. 1059–1062. Springer, Heidelberg (2005)

111. Pan, J.Z., Thomas, E., Sleeman, D.: Ontosearch2: Searching and querying web ontologies. In: Proc. of the IADIS International Conference, pp. 211–218 (2006)

112. Patel, C., Supekar, K., Lee, Y., Park, E.K.: OntoKhoj: a semantic web portal for ontology searching, ranking and classification. In: Proceedings of the 5th ACM International Workshop on Web Information and Data Management, WIDM 2003, pp. 58–61 (2003)

113. Poon, H., Domingos, P.: Sound and efficient inference with probabilistic and deterministic dependencies. In: AAAI. AAAI Press, Menlo Park (2006)

114. Poon, H., Domingos, P., Sumner, M.: A general method for reducing the complexity of relational inference and its application to MCMC. In: AAAI, pp. 1075–1080 (2008)

115. Quilitz, B., Leser, U.: Querying distributed RDF data sources with SPARQL. In: Bechhofer, S., Hauswirth, M., Hoffmann, J., Koubarakis, M. (eds.) ESWC 2008. LNCS, vol. 5021, pp. 524–538. Springer, Heidelberg (2008)

116. Re, C., Dalvi, N., Suciu, D.: Efficient top-k query evaluation on probabilistic data. In: ICDE, pp. 886–895 (2007)

117. Re, C., Suciu, D.: Managing probabilistic data with mystiQ: The can-do, the could-do, and the can't-do. In: Greco, S., Lukasiewicz, T. (eds.) SUM 2008. LNCS (LNAI), vol. 5291, pp. 5–18. Springer, Heidelberg (2008)

118. Richardson, M., Domingos, P.: Markov logic networks. Machine Learning 62(1-2) (2006)

119. Riedel, S.: Cutting plane MAP inference for Markov Logic. In: International Workshop on Statistical Relational Learning, SRL (2009)

120. Roth, D.: On the hardness of approximate reasoning. Artif. Intell. 82, 273–302 (1996)

121. Roth, D., Yih, W.: Integer linear programming inference for conditional random fields. In: Proc. of the International Conference on Machine Learning (ICML), pp. 737–744 (2005)

122. Sakr, S., Al-Naymat, G.: Relational processing of rdf queries: a survey. SIGMOD Record 38(4), 23–28 (2009)

123. Sarma, A.D., Benjelloun, O., Halevy, A.Y., Widom, J.: Working models for uncertain data. In: ICDE, p. 7 (2006)

124. Sarma, A.D., Theobald, M., Widom, J.: Exploiting lineage for confidence computation in uncertain and probabilistic databases. In: ICDE, pp. 1023–1032 (2008)

125. Das Sarma, A., Theobald, M., Widom, J.: LIVE: A lineage-supported versioned DBMS. In: Gertz, M., Ludäscher, B. (eds.) SSDBM 2010. LNCS, vol. 6187, pp. 416–433. Springer, Heidelberg (2010)

126. Schenk, S., Staab, S.: Networked graphs: a declarative mechanism for SPARQL rules, SPARQL views and RDF data integration on the Web. In: Proceeding of the 17th International Conference on World Wide Web, WWW 2008, pp. 585–594 (2008)

127. Sen, P., Deshpande, A.: Representing and querying correlated tuples in probabilistic databases. In: ICDE, pp. 596–605 (2007)
128. Sen, P., Deshpande, A., Getoor, L.: PrDB: managing and exploiting rich correlations in probabilistic databases. VLDB J. 18(5), 1065–1090 (2009)
129. Sen, P., Deshpande, A., Getoor, L.: Read-once functions and query evaluation in probabilistic databases. PVLDB 3(1), 1068–1079 (2010)
130. Sidirourgos, L., Goncalves, R., Kersten, M.L., Nes, N., Manegold, S.: Columnstore support for RDF data management: not all swans are white. PVLDB 1(2), 1553–1563 (2008)
131. Singh, S., Mayfield, C., Mittal, S., Prabhakar, S., Hambrusch, S.E., Shah, R.: Orion 2.0: native support for uncertain data. SIGMOD, 1239–1242 (2008)
132. Singh, S., Mayfield, C., Shah, R., Prabhakar, S., Hambrusch, S.E., Neville, J., Cheng, R.: Database support for probabilistic attributes and tuples. In: ICDE, pp. 1053–1061 (2008)
133. Singla, P., Domingos, P.: Memory-efficient inference in relational domains. In: AAAI (2006)
134. Soliman, M.A., Ilyas, I.F., Chang, K.C.: URank: formulation and efficient evaluation of top-k queries in uncertain databases. SIGMOD, 1082–1084 (2007)
135. Stocker, M., Seaborne, A., Bernstein, A., Kiefer, C., Reynolds, D.: SPARQL basic graph pattern optimization using selectivity estimation. In: Huai, J., Chen, R., Hon, H.-W., Liu, Y., Ma, W.-Y., Tomkins, A., Zhang, X. (eds.) WWW, pp. 595–604. ACM, New York (2008)
136. Stocker, M., Seaborne, A., Bernstein, A., Kiefer, C., Reynolds, D.: SPARQL basic graph pattern optimization using selectivity estimation. In: Proceeding of the 17th International Conference on World Wide Web, WWW 2008, pp. 595–604 (2008)
137. Stoica, I., Morris, R., Liben-Nowell, D., Karger, D.R., Kaashoek, M.F., Dabek, F., Balakrishnan, H.: Chord: a scalable peer-to-peer lookup protocol for internet applications. IEEE/ACM Trans. Netw. 11, 17–32 (2003)
138. Straccia, U.: Managing Uncertainty and Vagueness in Description Logics, Logic Programs and Description Logic Programs. In: Baroglio, C., Bonatti, P.A., Małuszyński, J., Marchiori, M., Polleres, A., Schaffert, S. (eds.) Reasoning Web. LNCS, vol. 5224, pp. 54–103. Springer, Heidelberg (2008)
139. Stuckenschmidt, H., Vdovjak, R., Houben, G.-J., Broekstra, J.: Index structures and algorithms for querying distributed RDF repositories. In: Proceedings of the 13th International Conference on World Wide Web, WWW 2004, pp. 631–639 (2004)
140. Suchanek, F.M., Kasneci, G., Weikum, G.: YAGO: a core of semantic knowledge. In: WWW, pp. 697–706 (2007)
141. Suchanek, F.M., Sozio, M., Weikum, G.: SOFIE: a self-organizing framework for information extraction. In: WWW, pp. 631–640 (2009)
142. Systeme, A.W., Gottlob, G., Voronkov, A., Dantsin, E., Dantsin, E., Eiter, T., Eiter, T.: Complexity and expressive power of logic programming (1999)
143. Terracina, G., Leone, N., Lio, V., Panetta, C.: Experimenting with recursive queries in database and logic programming systems. Theory Pract. Log. Program. 8, 129–165 (2008)
144. Theobald, M., Sozio, M., Suchanek, F., Nakashole, N.: URDF: Efficient reasoning in uncertain RDF knowledge bases with soft and hard rules. Technical Report MPII20105-002, Max Planck Institute Informatics, MPI-INF (2010)

145. Theoharis, Y., Christophides, V., Karvounarakis, G.: Benchmarking database representations of RDF/S stores. In: Gil, Y., Motta, E., Benjamins, V.R., Musen, M.A. (eds.) ISWC 2005. LNCS, vol. 3729, pp. 685–701. Springer, Heidelberg (2005)

146. Tran, T., Haase, P., Studer, R.: Semantic search – using graph-structured semantic models for supporting the search process. In: Rudolph, S., Dau, F., Kuznetsov, S.O. (eds.) ICCS 2009. LNCS, vol. 5662, pp. 48–65. Springer, Heidelberg (2009)

147. Tran, T., Wang, H., Haase, P.: Hermes: Data Web search on a pay-as-you-go integration infrastructure. Web Semant. 7, 189–203 (2009)

148. Tummarello, G., Cyganiak, R., Catasta, M., Danielczyk, S., Delbru, R., Decker, S.: Sig.ma: live views on the web of data. In: Proceedings of the 19th International Conference on World Wide Web, WWW 2010, pp. 1301–1304 (2010)

149. Udrea, O., Pugliese, A., Subrahmanian, V.S.: GRIN: A graph based RDF index. In: AAAI, pp. 1465–1470. AAAI Press, Menlo Park (2007)

150. Wang, D.Z., Michelakis, E., Franklin, M.J., Garofalakis, M.N., Hellerstein, J.M.: Probabilistic declarative information extraction. In: ICDE, pp. 173–176 (2010)

151. Wang, D.Z., Michelakis, E., Garofalakis, M.N., Hellerstein, J.M.: BayesStore: managing large, uncertain data repositories with probabilistic graphical models. PVLDB 1(1), 340–351 (2008)

152. Wang, Y., Yahya, M., Theobald, M.: Time-aware reasoning in uncertain knowledge bases. In: Workshop on Management of Uncertain Data, MUD (2010)

153. Warren, D.S.: Memoing for logic programs. Commun. ACM 35, 93–111 (1992)

154. Wei, W., Erenrich, J., Selman, B.: Towards efficient sampling: Exploiting random walk strategies. In: AAAI, pp. 670–676 (2004)

155. Weiss, C., Karras, P., Bernstein, A.: Hexastore: sextuple indexing for Semantic Web data management. PVLDB 1(1), 1008–1019 (2008)

156. Wick, M.L., McCallum, A., Miklau, G.: Scalable probabilistic databases with factor graphs and mcmc. PVLDB 3(1), 794–804 (2010)

157. Wilkinson, K., Sayers, C., Kuno, H., Reynolds, D.: Efficient RDF Storage and Retrieval in Jena2. In: First International Workshop on Semantic Web and Databases (SWDB 2003), pp. 131–150 (2003)

158. Wilkinson, K., Sayers, C., Kuno, H.A., Reynolds, D.: Efficient RDF storage and retrieval in Jena2. In: Cruz, et al [29], pp. 131–150

159. Xu, F., Beyer, K.S., Ercegovac, V., Haas, P.J., Shekita, E.J.: $E = MC^3$: managing uncertain enterprise data in a cluster-computing environment. SIGMOD, 441–454 (2009)

160. Zhou, M., Wu, Y.: XML-based RDF data management for efficient query processing. In: Dong, X.L., Naumann, F. (eds.) WebDB (2010)

# Scalable OWL 2 Reasoning for Linked Data⋆

Aidan Hogan[1], Jeff Z. Pan[2], Axel Polleres[1,3], and Yuan Ren[2]

[1] Digital Enterprise Research Institute, National University of Ireland, Galway
{aidan.hogan,axel.polleres}@deri.org
[2] Department of Computing Science, University of Aberdeen
{jeff.z.pan,y.ren}@abdn.ac.uk
[3] Siemens AG Österreich, Siemensstrasse 90, 1210 Vienna, Austria

**Abstract.** The goal of the Scalable OWL 2 Reasoning for Linked Data lecture is twofold: first, to introduce scalable reasoning and querying techniques to Semantic Web researchers as powerful tools to make use of Linked Data and large-scale ontologies, and second, to present interesting research problems for the Semantic Web that arise in dealing with TBox and ABox reasoning in OWL 2. The lecture consists of three parts. The first part will begin with an introduction and motivation for reasoning over Linked Data, including a survey of the use of RDFS and OWL on the Web. The second part will present a scalable, distributed reasoning service for instance data, applying a custom subset of OWL 2 RL/RDF rules (based on a tractable fragment of OWL 2). The third part will present recent work on faithful approximate reasoning for OWL 2 DL. The lecture will include our implementation of the mentioned techniques as well as their evaluations. These notes provide complimentary reference material for the lecture, and follow the three-part structure and content of the lecture.

## 1 Introduction

Over the past few years, various Web publishers have turned to RDF and Linked Data principles as a means of disseminating information in a machine-interpretable way, resulting in a burgeoning Web of Data which now includes interlinked content provided by corporate bodies (e.g., BBC [53], BestBuy [35], New York Times[1], Freebase[2]), community-driven efforts (e.g., WIKIPEDIA/DBpedia[3] [15]), social networking sites (e.g., hi5[4], LiveJournal[5]), biomedical datasets (e.g., DrugBank[6], Linked

---

⋆ The work presented in this paper has been funded in part by Science Foundation Ireland under Grant No. SFI/08/CE/I1380 (Lion-2), by an IRCSET Scholarship, by the EU MOST project, and by the EPSRC LITRO project.

[1] http://data.nytimes.com/; retr. 15/10/2010
[2] http://www.freebase.com/; retr. 15/10/2010
[3] http://dbpedia.org/; retr. 15/10/2010
[4] http://api.hi5.com/; retr. 15/10/2010
[5] http://livejournal.com; retr. 15/10/2010
[6] http://www.drugbank.ca/; retr. 15/10/2010

A. Polleres et al. (Eds.): Reasoning Web 2011, LNCS 6848, pp. 250–325, 2011.
© Springer-Verlag Berlin Heidelberg 2011

Clinical Trials[7]), governmental entities (e.g., `data.gov.uk`, `data.gov`), academia (e.g., DBLP[8], UniProt[9]), as well as some esoteric corpora (e.g., Poképédia[10], Linked Open Numbers[11] [91]). See `http://lod-cloud.net` (retr. 15/10/2010) for Cyganiak and Jentzsch's Linked Open Data cloud diagram which illustrates the datasets comprising the current (and past) Web of Data. As such, there now exists a rich vein of heterogeneous, structured and interlinked data on the Web. (*Please see [2] in these proceedings for extensive introduction and general discussion relating to Linked Data and the Web of Data; herein, we focus on reasoning issues.*)

Linked Data's successes are perhaps due to its bottom-up approach to bootstrapping Semantic *Web* publishing, best epitomised by the Linked Data 5-star scheme which can be summarised as follows:

⋆ PUBLISH DATA ON THE WEB UNDER AN OPEN LICENSE
⋆⋆ PUBLISH STRUCTURED DATA
⋆⋆⋆ USE NON-PRIORIETARY FORMATS
⋆⋆⋆⋆ USE URIs TO IDENTIFY THINGS
⋆⋆⋆⋆⋆ LINK YOUR DATA TO OTHER DATA

**—paraphrased from [10]**

Here, each additional star is promoted as representing a tangible, incremental step towards increasing the potential reusability and interoperability of the publishers' data.[12] As part of this bottom-up approach, Linked Data downplays higher levels of the traditional Semantic Web stack, viz., *ontologies*, *logic*, *proof*, *trust* and *cryptography* [34]. One may note that many of these layers relate to reasoning in one form or another, where—as we will discuss in the next section—the reasoning research literature (and arguably, the RDFS and OWL standards) have often focussed on scenarios and goals orthogonal to the Linked Data use-case.

Still, the original challenges envisaged for the traditional Semantic Web are now being realised on the Web of Data: consumers wishing to process data offered by many different publishers will encounter problems with respect to integrating and making meaningful use of the resulting corpus. In fact, as we will see, Linked Data publishers already use lightweight subsets of the RDFS and OWL standards to provide "mappings" which help consumers overcome this problem of heterogeneity; thereafter, we argue that many Linked Data consumers (will) often need some lightweight forms of reasoning when dealing with heterogeneous Linked Data sourced from different domains. Looking further into the future, the demand for more and more expressive reasoning services—which allow further machine automation of typical Web tasks—is bound to grow.

---

[7] `http://linkedct.org/`; retr. 15/10/2010

[8] `http://www4.wiwiss.fu-berlin.de/dblp/`; retr. 15/10/2010

[9] `http://www.uniprot.org/`; retr. 15/10/2010

[10] `http://www.pokepedia.net/`; retr. 15/10/2010

[11] `http://km.aifb.kit.edu/projects/numbers/`; retr. 15/10/2010

[12] Please see `http://lab.linkeddata.deri.ie/2010/star-scheme-by-example/` (retr. 2011/01/22) for the rationale behind these stars. Note that although the final star does not explicitly mention Linked Data or RDF, use of these technologies is implied.

Along these lines, herein we give an overview of the state-of-the-art with respect to
Linked Data reasoning. In particular, we present the following discussion:

- we begin in § 2 by motivating reasoning in the context of Linked Data, presenting
  some concrete use-cases, and discussing the use of the RDFS and OWL standards
  on the Web of Data.

We then detail two complementary reasoning approaches which may be applicable in
such scenarios:

- in § 3, we look at SAOR: a lightweight, pragmatic rule-based materialisation engine which applies a scalable subset of OWL 2 RL/RDF, is designed to operate
  over a cluster of commodity hardware in a distributed setting with little or no coordination between machines, and conservatively discards terminological knowledge given by unverifiable sources;
- in § 4, we describe a system for performing approximative, but relatively expressive TBox reasoning with respect to OWL 2 DL, where $\mathcal{SROIQ}$ ontologies are
  simplified into $\mathcal{EL}^{++}$, and where the approximate representation is then classified
  using scalable techniques which typically demonstrate high recall when compared
  with the non-approximative classification.

The rule-based reasoning described in § 3 offers linear scalable with respect to assertional (instance) data, but is rather inexpressive, especially when dealing with terminological knowledge; this approach is well-suited to lightweight reasoning over the large
amounts of assertional knowledge which constitute the bulk of RDF data found on the
Web. Conversely, the approximative reasoning of § 4 offers expressive PTime-complete
reasoning over terminological knowledge, but is still a memory-based approach; this
approach is well-suited to expressive reasoning over the smaller amounts of terminological knowledge found on the Web.

In summary, we aim to provide the reader with insights into why Linked Data reasoning is useful, what the use-cases are, what the main challenges are, how RDFS and
OWL are being used on the Web, and what techniques can be used to realise reasoning
in such scenarios.

## 2   Linked Data: RDFS, OWL and Reasoning

We begin this section by highlighting three potential problems faced by consumers of
Linked Data; for each problem, we present a motivating example which helps illustrate
why we believe that Linked Data needs reasoning (§ 2.1). Thereafter, we discuss what
kind of reasoning is possible over Linked Data where we look at the use of RDFS and
OWL on the Web of Data (§ 2.2). Finally, we give a brief overview of how Linked
Data reasoning can be realised (§ 2.3) before moving onto the concrete proposals for
reasoning systems in the later sections.

Note that throughout this section, we present examples and analyses extracted from a
real-world Linked Data corpus of 1.118 billion quadruples (965 million unique triples)
extracted from 3.985 million RDF/XML documents through an open-domain crawl

conducted in May 2010. The corpus consists of data from 783 different pay-level domains (direct subdomains of top-level domains, such as `dbpedia.org`), and thus
represents a domain-agnostic *sample* of the Web of Data. We refer the reader to [39,
§ 4] for more details on this dataset.

## 2.1  Why Does Linked Data Need Reasoning?

We begin by introducing three potential problems faced by consumers of Linked Data
which can be addressed through reasoning techniques:

1. heterogeneity in terminology;
2. heterogeneity in naming resources;
3. contradictions given by inconsistency.

**Heterogeneity in terminology.** To enable interoperability and subsequent data integration, Linked Data literature encourages reuse of URIs—particularly those referential to
classes and properties (schema-level *terminology*)—across data sources: in the ideal
case, a Linked Data consumer can perform a simple (RDF-)merge of datasets, where
consistent use of terminology ensures that resources are described uniformly and thus
can be accessed and queried uniformly. Although this ideal is achievable in part by community agreement and self-organising phenomena such as preferential attachment [6]—
whereby, for example, the most popular classes and properties would become the de-
facto consensus and thus more widely used—given the ad-hoc decentralised nature of
the Web, complete and appropriate agreement upon the broad spectrum of terminology
needed to fully realise the Web of Data is probably infeasible.

Instead, Linked Data publishers may use different but analogous terminology to describe their data: competing vocabularies may offer different levels of granularity or expressivity more suitable to a given publisher's needs, may be popular at different times
or within different communities, etc. For example, one publisher may chose the property `foaf:maker` whereas another publisher may chose the property `dc:creator`,
where both properties serve the same purpose. Publishers may not only choose different
vocabularies, but may also choose alternate terms within a given vocabulary to model
analogous information. For example, vocabularies may offer pairs of *inverse properties*—e.g., `foaf:made`/`foaf:maker`—which poses the publisher with two options
for stating the same information. Further still, Linked Data best-practices encourage
publisher to "cherry-pick" different vocabularies, choosing a heterogeneous "bag of
terms" to describe their data [14].

This becomes a significant obstacle for applications consuming a sufficiently heterogeneous corpus: for example, queries posed against the data must emulate the various
terminological permutations possible to achieve (more) complete answers. Here, we
take the motivating example of a simple query described in prose as:

*What are the webpages related to* **`ex:resource`***?*

Knowing that the property `foaf:page` is commonly used in Linked Data to define the
relationship from resources to the documents somehow concerning them, we can formulate a simple structured query in SPARQL [73]—the W3C standardised RDF query

language—as given in Listing 1. (*Please see [24] in these proceedings for an introduction to SPARQL, and further discussion on combining RDFS and OWL entailment with SPARQL.*)

**Listing 1.** Simple query for all pages relating to ex:resource

```
SELECT ?page
WHERE {
  ex:resource foaf:page ?page .
}
```

However, within Linked Data, there exist various other, more fine-grained properties for relating a resource to specific types of pages—these properties are not only given by FOAF, but also by remote vocabularies. Thus, to ensure more complete answers, the SPARQL query must use disjunction (UNION clauses) to reflect the possible triples which may answer the query; we give such an example in Listing 2 involving properties we found in our one-billion-triple Linked Data corpus, where we additionally annotate each pattern with the total number of triples found in our corpus for the respective predicate; this gives a *rough* indicator of the relative likelihood of finding additional answers with each additional pattern (prefix values are enumerated in Appendix B).

Not only is the resulting query much more cumbersome to formulate, but it also requires a much more in-depth knowledge of the various vocabularies in the corpus. In reality, people querying the data will often end up omitting many (if not all) of the UNION clauses and simply accept partial results, perhaps only querying for those properties they know about. Similar (albeit less convincing) examples can be shown for classes. Considering that the highlighted example might only be one part of a much larger query, the inherent difficulties in "adequately" querying a heterogeneous corpus of Linked Data become clear.[13]

One possible solution to ease the burden on consumers (and one *sometimes* promoted in the Linked Data community) is for publishers to provide some redundancy in their data, and explicitly repeat the same information in whatever different ways they foresee would be useful to consumers. Taking such a publishing pattern to its extremes, to express that the person Tim Berners-Lee (whose known URI is timblfoaf:i) has the homepage timblfoaf:hp, a publisher might explicitly state the following set of triples:

```
timblfoaf:i  foaf:homepage timblfoaf:hp ;
             foaf:isPrimaryTopicOf timblfoaf:hp ;
             foaf:page timblfoaf:hp .
timblfoaf:hp foaf:primaryTopic timblfoaf:i ;
             foaf:topic timblfoaf:i .
```

---

[13] Of course, this problem of *heterogeneity* is not purely specific to SPARQL querying, but likewise extends to many other consumer applications or techniques operating over a corpus of Linked Data.

thus covering all possible ways in which the data could be queried or otherwise consumed. However, clearly such a pattern would not be sustainable as the number of publishers grows, as the amount of data grows, and as the variety of relevant vocabularies and terms grows.

Thus, we propose that Linked Data needs some means of translating assertional data between different terminology, allowing for consumers to enjoy the same answers for the simple query in Listing 2 as if they had formulated and asked the extended query given in Listing 1 (or, as if publishers had gone to the bother of publishing and maintaining redundant data in all pertinent terminological combinations).

Of course, the Linked Data community are not oblivious to these problems. Vocabulary maintainers commonly publish machine-interpretable RDFS and OWL definitions of their local terms, as well as mappings to related terms in remote vocabularies. In fact, since foaf:page is relatively well-known within the Linked Data community, all of the properties appearing in the example extended query are (possibly indirectly) related to foaf:page using RDFS and OWL connectives in their respective vocabularies: all properties referenced in Listing 2 are chosen on the basis that they are directly or indirectly related to foaf:page by rdfs:subPropertyOf or owl:inverse-Of, where relations using these properties can be used to infer foaf:page answers —note that we italicise patterns for properties which have an *inverse* (sub-)relation to foaf:page in Listing 2. (We will look in more detail at the use of RDFS and OWL within Linked Data later in § 2.2.)

Thus, given these ad-hoc mappings provided by the Linked Data publishers themselves, we have the necessary formal knowledge to be able to answer the former simple query with all of the answers given by the latter elaborate query. In order to exploit this knowledge and realise this goal in the general case, we require some form of *reasoning*.

**Heterogeneity in naming resources.** A similar problem is posed by the lack of agreement on identifiers assigned to assertional resources. Complete agreement upon a single URI for each possible resource of interest is unrealistic, and would require either a centralised naming registry to corroborate name proposals, or agreement upon some universal bijective naming scheme compatible with any arbitrary resource. Aside from feasibility, having one agreed-upon URI to identify each resource may not even be desirable, and would in fact contradict one of the core Linked Data principles: that URIs should be made *dereferenceable* such that when a HTTP lookup is performed, useful information should be returned (in RDF) [34]. A URI can only dereference to a single document, and thus if there were a one-to-one relationship bewteen resources and URIs, the description of individual resources would be in the control of individual publishers who mint the pertinent URI. Instead, Linked Data best-practices often encourage minting novel, locally dereferenceable URIs for resources, even if legacy compatible URIs are (somehow) known. Finally, many publishers (esp. older FOAF exporters) may forego using URIs at all, instead using default blank-nodes to "identify" resources.

Thus, we can expect resources to be identified using different naming schemes, or— worse still—to be represented as blank-nodes. Thus, the total knowledge contribution on that resource is fractured by the disparity in naming across sources. Consumers of

**Listing 2.** Extended query for all pages relating to ex:resource

```
SELECT ?page
WHERE {
  { ex:resource foaf:page ?page . }                          #4,923,026
  UNION { ex:resource foaf:weblog ?page . }                  #10,061,003
  UNION { ex:resource foaf:homepage ?page . }                #9,522,912
  UNION { ?page foaf:topic ex:resource . }                   #6,163,769
  UNION { ?page foaf:primaryTopic ex:resource . }            #3,689,888
  UNION { ex:resource mo:musicbrainz ?page . }               #399,543
  UNION { ex:resource foaf:openid ?page . }                  #100,654
  UNION { ex:resource foaf:isPrimaryTopicOf ?page . }        #92,927
  UNION { ex:resource mo:wikipedia ?page . }                 #55,228
  UNION { ex:resource mo:myspace ?page . }                   #28,197
  UNION { ex:resource po:microsite ?page . }                 #15,577
  UNION { ex:resource mo:amazon_asin ?page . }               #14,790
  UNION { ex:resource mo:imdb ?page . }                      #9,886
  UNION { ex:resource mo:fanpage ?page . }                   #5,135
  UNION { ex:resource mo:biography ?page . }                 #4,609
  UNION { ex:resource mo:discogs ?page . }                   #1,897
  UNION { ex:resource rail:arrivals ?page . }                #347
  UNION { ex:resource rail:departures ?page . }              #347
  UNION { ex:resource mo:musicmoz ?page . }                  #227
  UNION { ex:resource mo:discography ?page . }               #195
  UNION { ex:resource mo:review ?page . }                    #46
  UNION { ex:resource mo:freedownload ?page . }              #37
  UNION { ex:resource mo:mailorder ?page . }                 #35
  UNION { ex:resource mo:licence ?page . }                   #28
  UNION { ex:resource mo:paiddownload ?page . }              #13
  UNION { ex:resource foaf:tipjar ?page . }                  #8
  UNION { ex:resource doap:homepage ?page . }                #1
  UNION { ex:resource doap:old-homepage ?page . }            #1
  UNION { ex:resource mo:download ?page . }                  #0
  UNION { ex:resource mo:event_homepage ?page . }            #0
  UNION { ex:resource mo:free_download ?page . }             #0
  UNION { ex:resource mo:homepage ?page . }                  #0
  UNION { ex:resource mo:paid_download ?page . }             #0
  UNION { ex:resource mo:preview_download ?page . }          #0
  UNION { ex:resource mo:olga ?page . }                      #0
  UNION { ex:resource mo:onlinecommunity ?page . }           #0
  UNION { ex:resource plink:addFriend ?page . }              #0
  UNION { ex:resource plink:atom ?page . }                   #0
  UNION { ex:resource plink:content ?page . }                #0
  UNION { ex:resource plink:foaf ?page . }                   #0
  UNION { ex:resource plink:profile ?page . }                #0
  UNION { ex:resource plink:rss ?page . }                    #0
  UNION { ex:resource xfn:mePage ?page . }                   #0
}
```

such a corpus may struggle to achieve complete answer for their queries; again, consider a simple example query:

**What are the webpages related to Tim Berners-Lee?**

Knowing that Tim uses the URI `timblfoaf:i` to refer to himself in his personal FOAF profile document, and again knowing that the property `foaf:page` defines the relationship from resources to the documents somehow concerning them, we can formulate the SPARQL query given in Listing 3.

**Listing 3.** Simple query for pages relating to Tim Berners-Lee

```
SELECT ?page
WHERE {
  timbl\-foaf:\-i foaf:page ?page .
}
```

However, other publishers use different URIs to identify Tim, where to get more complete answers across these naming schemes, the SPARQL query must (as per the previous example) use disjunctive `UNION` clauses for each known URI; we give an example in Listing 4 using identifiers from a recent Linked Data corpus (see Appendix B for prefix mappings), where we see disparate URIs not only across data publishers, but also within the same namespace. Thus (again), the expanded query quickly becomes extremely cumbersome. Combined with the terminological permuations for `foaf:page` exemplified earlier, this query again illustrates the difficulties posed to consumers of Linked Data.

Again, one possible solution would be to encourage publishers to provide redundant data for all commonly known URIs of the given resource, but as before, this again becomes infeasible as Linked Data diversifies and expands.

Consequently, we propose that Linked Data needs some means of (i) *resolving coreferent identifiers* which signify the same thing; (ii) *handling coreferent identifiers* such that consumers can access and process a heterogeneous corpus as if (more) complete agreement on identifiers was present. Without this, the information about all resources in the Linked Data corpus will be fractured across naming schemes, and a fundamental goal of the Web of Data—to attenuate the traditional barriers between data publishers—will be compromised.

Again, Linked Data publishers acknowledge and take steps to counter-act such problems and help consumers. Firstly, in [14, § 6], Bizer et al. state that `owl:sameAs` should be used to interlink coreferent resources in remote datasets:

> *"It is common practice to use the owl:sameAs property for stating that another data source also provides information about a specific non-information resource."*

—**[14, § 6]**

Thus, the `owl:sameAs` property can be used to relate locally defined (and ideally dereferenceable) identifiers to external legacy identifiers which signify the same thing.

**Listing 4.** Extended query for pages relating to Tim Berners-Lee (sic.)

```
SELECT ?page
WHERE {
  UNION { timbl\-foaf:\-i foaf:page ?page . }
  UNION { identicauser:45563 foaf:page ?page . }
  UNION { dbpedia:Berners-Lee foaf:page ?page . }
  UNION { dbpedia:Dr._Tim_Berners-Lee foaf:page ?page . }
  UNION { dbpedia:Dr._Tim_Berners_Lee foaf:page ?page . }
  UNION { dbpedia:Sir_Timothy_John_Berners-Lee foaf:page ?page . }
  UNION { dbpedia:Tim-Berners_Lee foaf:page ?page . }
  UNION { dbpedia:TimBL foaf:page ?page . }
  UNION { dbpedia:Tim_Berners-Lee foaf:page ?page . }
  UNION { dbpedia:Tim_Bernes-Lee foaf:page ?page . }
  UNION { dbpedia:Tim_Bernes_Lee foaf:page ?page . }
  UNION { dbpedia:Tim_Burners_Lee foaf:page ?page . }
  UNION { dbpedia:Tim_berners-lee foaf:page ?page . }
  UNION { dbpedia:Timbl foaf:page ?page . }
  UNION { dbpedia:Timothy_Berners-Lee foaf:page ?page . }
  UNION { dbpedia:Timothy_John_Berners-Lee foaf:page ?page . }
  UNION { yagor:Tim_Berners-Lee foaf:page ?page . }
  UNION { fb:en.tim_berners-lee foaf:page ?page . }
  UNION { fb:guid.9202a8c04000641f800000000003b0a foaf:page ?page . }
  UNION { swid:Tim_Berners-Lee foaf:page ?page . }
  UNION { dblpperson:100007 foaf:page ?page . }
  UNION { avtimbl:me foaf:page ?page . }
  UNION { bmpersons:Tim+Berners-Lee foaf:page ?page . }
  ...
}
```

This approach offers two particular advantages: (i) publishers can define an ad-hoc local naming scheme for their resources—thus reducing the initial inertia for Linked Data publishing—and thereafter, incrementally provide mappings to external coreferent identifiers as desirable; (ii) multiple dereferenceable identifiers can implicitly provide alternative sources of information for a given resource, useful for discovery.

Furthermore, OWL provides the class `owl:InverseFunctionalProperty`: properties contained within this class have values unique to a given resource—loosely, these can be thought of as *key* values where two resources sharing identical values for some such property are, by OWL semantics, coreferent. Along these lines, inverse-functional properties can be used in conjunction with existing identification schemes—such as ISBNs for books, EAN·UCC-13 or MPN for products, MAC addresses for network-enabled devices, etc.—to bootstrap identity on the Web of Data within certain domains; such identification values can be encoded as simple datatype strings, thus bypassing the requirement for bespoke agreement or mappings between URIs. Similar other constructs are available in OWL for resolving coreference, such as `owl:-FunctionalProperty`, `owl:maxCardinality`, and `owl:hasKey` (the latter was introduced in the updated OWL 2 standard). Note that these OWL constructs require agreement on terminology—for example, agreement on a given property term to denote the ISBN attribute—without which, coreference cannot be established. (We will look in more detail at such use of OWL later in § 2.2.)

As per the previous examples, all of the identifiers given by the extended query in Listing 3 are linked (either directly or indirectly) by explicit `owl:sameAs` on the Web of Data. Thus, we again have the necessary formal knowledge to be able to answer the former simple query with all of the answers given by the latter elaborate query. In order to exploit this knowledge and realise this goal in the general case, Linked Data consumers again require some form of *reasoning*.

**Contradictions given by inconsistency.** A more speculative use-case for Linked Data relates to the identification and treatment of noisy data in the corpus as symptomised by inconsistency. Given that Linked Data and OWL operate under the Open World Assumption and without a Unique Name Assumption, OWL ontologies are not naturally suited to specifying constraints, particular those which check for missing information—e.g., that all `Persons` have at least one parent—or which verify that a property has not been "overloaded" with too many values—e.g., that no `Person` has more than two biological parents. In the former case, the Open World Assumption (OWA) states that data should be assumed to be incomplete; in the latter case, the lack of a Unique Name Assumption (UNA) means that if a `Person` has three parents, one or more such values might (by default) be referring to the same real-world parent.

However, even with the OWA and without the UNA, OWL still provides some limited machinery for checking the validity of data, in the form of consistency checking. Consistency checking ensures that there are no formal conflicts in the presented data. With respect to Linked Data, the most prevalent forms of inconsistency stem from (i) lexically-invalid (aka. ill-typed) datatype literals: e.g., `"true"^^<xsd:boolean>`; and (ii) disjoint classes, where the OWL property `owl:disjointWith` can be used to relate two classes whose intersection is empty, and where it is inconsistent for a resource to be a member of both. In fact, inconsistency often occurs as the result of reasoning, where we provide a real-world example in Listing 5 caused as follows:

1. the FOAF vocabulary assigns the property `foaf:knows` the domain `foaf:Person`, defines the property `foaf:homepage` to be inverse-functional, and states that the classes `foaf:Person` and `foaf:Organization` are disjoint;
2. the W3C has a profile on the "identi.ca" micro-blogging platform, which uses `foaf:knows` to assert follower/followee relationships: in these data, the W3C is identified by the `identicauser:48404` URI, is said to "know" various other users (such that it is entailed to be a `foaf:Person`), and is given the `foaf:-homepage` `<http://www.w3.org/>`;
3. various sites state that the W3C is an `foaf:Organization` with the `foaf:-homepage` `<http://www.w3.org/>`.

Thus, we see an example of inconsistency caused by merging and performing inference over data collected from three different publishers. The entailment that the W3C is a person can, of course, be considered noise, where (in this case) the problem causes an inconsistency. Once inconsistencies are detected by way of reasoning, algorithms can be devised to repair such problems; for example, the claim that the W3C is a person can be removed since more sources claim it to be an organisation [39].

Repairing the inconsistencies present in a Linked Data corpus can thus improve the quality of the data by resolving formal conflicts (perhaps) based on analyses of the trustworthiness of the sources involved. However, we note that the granularity of consistency checking offered by OWL means that only a small subset of the noise in a dataset may be symptomised as inconsistency [39]. Note that OWL 2 has added numerous more features which allow for performing consistency checks—e.g., `owl:propertyDisjointWith`, `owl:AsymmetricProperty`, `owl:IrreflexiveProperty`, etc.—but (as we will see in the next section) these features have yet to gain significant adoption on the Web.

**Listing 5.** The W3C is inconsistent

```
# http://xmlns.com/foaf/spec (FOAF Vocabulary)
foaf:Person owl:disjointWith foaf:Organization .                        #1
foaf:homepage a owl:InverseFunctionalProperty .                         #2
foaf:knows rdfs:domain foaf:Person .                                    #3

# http://identi.ca/w3c/foaf (W3C's identi.ca profile)
identicauser:48404 foaf:knows ident\-ica\-user:\-45563 .                #4
identicauser:48404 foaf:name "W3C" .                                    #5
identicauser:48404 foaf:homepage <http://www.w3.org/> .                 #6

# Entailed by #2/#3 & OWL 2 RL/RDF rule prp-dom
identicauser:48404 a foaf:Person .                                      #7

# http://data.semanticweb.org/organization/w3c/rdf (Sem. Web Dogfood Server)
sworg:w3c a foaf:Organization .                                         #8
sworg:w3c foaf:name "W3C" .                                             #9
sworg:w3c foaf:homepage <http://www.w3.org/> .                         #10

# Entailed by #6/#10 & OWL 2 RL/RDF rule prp-ifp
identicauser:48404 owl:sameAs sworg:w3c .                              #11

# Entailed by #7/#11 & OWL 2 RL/RDF rule eq-rep-s
sworg:w3c a foaf:Person .                                              #12

#        Inconsistency entailed by #1/#8/#12 & OWL 2 RL/RDF rule cax-dw
```

## 2.2  What Reasoning Does Linked Data Need?

Having looked at some scenarios and motivating examples which help demonstrate why Linked Data might need reasoning, we now briefly survey the use of RDFS and OWL in our Linked Data corpus (see § 2) and thus present insights into the use of these standards on the Web of Data.

We acknowledge that presenting raw counts of the use of RDFS and OWL primitives in our corpus could be heavily influenced by a small number of documents which publish large amounts of schema data, or a small number of domains which publish many documents using these primitives [39]: in other words, a single (obscure) publisher could greatly influence such results. We thus also characterise such primitives in terms of the perceived *importance* of all documents using them. Borrowing from work presented in [32], we rank RDF documents in the corpus using the prominent PageRank links-analysis technique [71]: PageRank calculates a variant of the Eigenvector centrality of nodes (e.g., documents) in a graph: given the intuition of directed links as "positive votes", the resulting scores help characterise the relative prominence

**Table 1.** Top ten ranked documents and other notable ranks

| # | Document | Rank |
|---|----------|------|
| 1 | http://www.w3.org/1999/02/22-rdf-syntax-ns | 0.112 |
| 2 | http://www.w3.org/2000/01/rdf-schema | 0.104 |
| 3 | http://dublincore.org/2008/01/14/dcelements.rdf | 0.089 |
| 4 | http://www.w3.org/2002/07/owl | 0.067 |
| 5 | http://www.w3.org/2000/01/rdf-schema-more | 0.045 |
| 6 | http://dublincore.org/2008/01/14/dcterms.rdf | 0.032 |
| 7 | http://www.w3.org/2009/08/skos-reference/skos.rdf | 0.028 |
| 8 | http://www.w3.org/2003/g/data-view | 0.014 |
| 9 | http://xmlns.com/foaf/spec/ | 0.014 |
| 10 | http://www.w3.org/2000/01/combined-ns-translation.rdf.fr | 0.010 |
| 73 | http://www.w3.org/People/Berners-Lee/card | 2.55E-4 |
| 116 | http://www.w3.org/2006/03/wn/wn20/schemas/wnfull.rdfs | 1.65E-4 |
| 120 | http://www.w3.org/2006/time | 1.60E-4 |
| 150 | http://motools.sourceforge.net/timeline/timeline.rdf | 1.25E-4 |
| 337 | http://rdf.geospecies.org/ont/geospecies | 5.24E-5 |

of particular nodes on the Web. For more information on this ranking scheme, we refer the reader to [39]; here we focus on the results. Table 1 presents the top ten ranked documents, which are dominated by core meta-vocabularies, documents linked therefrom, and other popular vocabularies; we also present the ranks of other notable documents mentioned in this section.

Given these source-level ranks, we then rank RDFS and OWL primitives in our corpus based on the summation of the ranks of documents featuring them. Note that we focus on features of RDFS and OWL which are supported by OWL 2 RL/RDF.

Table 2 lists the results, where for each primitive, we list the sum of the ranks of all documents in which it is used ($\sum$ RANK), its position based on that summation of ranks (#), the value of the highest ranked document in which it is used (max RANK), the relative rank position of that document out of the 3.985 million surveyed (max POS), as well as the total number of quads (QUAD), the total number of *unique* triples (TRIPLE), the total number of documents (DOC), and the total number of pay-level domains (DOM) in which that primitive appeared. We additionally *italicise* primitives new to OWL 2. Note again that Table 1 provides a legend for notable documents (POS>337).

From Table 2, we make the following high-level observations:

1. The top four primitives equate to the core of RDFS (aka. $\rho$DF [68]).
2. Ten of the bottom twelve primitives are new to OWL 2; the exceptions are owl:-differentFrom and owl:complementOf which are eleventh and ninth from bottom resp. (the former is also expressible without blank-nodes). Our crawl was conducted seven months after OWL 2 became a W3C Recommendation (Oct. 2009) —by means of a quick scan of the max POS column of Table 2, we note that new OWL 2 features had little penetration in the popular Web vocabularies in that interim.
3. The most widespread primitive is owl:sameAs, used in 122 of the 783 domains contributing to our corpus (15.6%) and 1.346 million of the 3.985 million

**Table 2.** Ranks of RDFS/OWL primitives supported by OWL 2 RL/RDF rules

| # | Primitive | $\sum$ Rank | max Rank | max Pos | Quad | Triple | Doc | Dom |
|---|---|---|---|---|---|---|---|---|
| 1 | rdfs:subClassOf | 4.15E-1 | 1.12E-1 | 1 | 334,589 | 307,170 | 60,368 | 63 |
| 2 | rdfs:range | 4.12E-1 | 1.12E-1 | 1 | 91,068 | 13,947 | 1,317 | 61 |
| 3 | rdfs:domain | 4.11E-1 | 1.12E-1 | 1 | 164,891 | 16,213 | 1,270 | 63 |
| 4 | rdfs:subPropertyOf | 2.16E-1 | 1.04E-1 | 2 | 12,898 | 9,160 | 504 | 52 |
| 5 | owl:Class | 1.38E-1 | 6.73E-2 | 4 | 217,334 | 173,542 | 42,986 | 54 |
| 6 | owl:DatatypeProperty | 1.34E-1 | 6.73E-2 | 4 | 82,866 | 4,484 | 9,464 | 41 |
| 7 | owl:ObjectProperty | 1.33E-1 | 6.73E-2 | 4 | 64,199 | 8,760 | 17,417 | 48 |
| 8 | owl:sameAs | 1.24E-1 | 2.55E-4 | 73 | 11,928,308 | 3,769,898 | 1,346,218 | 122 |
| 9 | owl:FunctionalProperty | 7.41E-2 | 2.82E-2 | 7 | 4,345 | 496 | 239 | 21 |
| 10 | owl:disjointWith | 5.92E-2 | 2.82E-2 | 7 | 1,860 | 917 | 145 | 16 |
| 11 | owl:inverseOf | 5.82E-2 | 2.82E-2 | 7 | 2,146 | 992 | 234 | 29 |
| 12 | owl:unionOf | 3.46E-2 | 2.82E-2 | 7 | 4,168 | 4,168 | 109 | 18 |
| 13 | owl:SymmetricProperty | 3.36E-2 | 2.82E-2 | 7 | 657 | 128 | 64 | 15 |
| 14 | owl:equivalentClass | 3.21E-2 | 1.41E-2 | 9 | 23,398 | 22,929 | 22,781 | 18 |
| 15 | owl:TransitiveProperty | 3.03E-2 | 2.82E-2 | 7 | 259 | 140 | 76 | 16 |
| 16 | owl:InverseFunctionalProperty | 2.96E-2 | 1.41E-2 | 9 | 966 | 62 | 118 | 17 |
| 17 | owl:equivalentProperty | 2.94E-2 | 1.41E-2 | 9 | 618 | 173 | 176 | 16 |
| 18 | owl:someValuesFrom | 1.79E-2 | 1.42E-2 | 8 | 465 | 465 | 48 | 6 |
| 19 | owl:oneOf | 3.36E-4 | 1.60E-4 | 120 | 121 | 121 | 19 | 9 |
| 20 | owl:hasValue | 2.86E-4 | 1.60E-4 | 120 | 14 | 13 | 6 | 3 |
| 21 | owl:maxCardinality | 2.81E-4 | 1.60E-4 | 120 | 297 | 297 | 18 | 9 |
| 22 | owl:allValuesFrom | 2.62E-4 | 1.65E-4 | 116 | 258 | 257 | 35 | 7 |
| 23 | owl:intersectionOf | 1.73E-4 | 1.25E-4 | 150 | 136 | 136 | 12 | 6 |
| 24 | owl:AllDifferent | 9.94E-5 | 5.24E-5 | 337 | 68 | 68 | 8 | 3 |
| 25 | owl:differentFrom | 4.11E-6 | 1.43E-7 | 439,559 | 337 | 335 | 98 | 4 |
| 26 | owl:AllDisjointClasses | 4.20E-7 | 6.17E-8 | 1,888,937 | 9 | 9 | 9 | 1 |
| 27 | owl:complementOf | 1.62E-7 | 9.10E-8 | 916,865 | 7 | 7 | 2 | 2 |
| 28 | owl:IrreflexiveProperty | 1.10E-7 | 5.94E-8 | 2,014,801 | 20 | 10 | 2 | 1 |
| "" | owl:AsymmetricProperty | 1.10E-7 | 5.94E-8 | 2,014,801 | 18 | 9 | 2 | 1 |
| 30 | owl:hasKey | 4.99E-8 | 4.99E-8 | 2,818,089 | 1 | 1 | 1 | 1 |
| 31 | owl:propertyChainAxiom | 4.48E-8 | 4.48E-8 | 3,206,125 | 6 | 6 | 1 | 1 |
| 32 | owl:maxQualifiedCardinality | — | — | — | — | — | — | — |
| "" | owl:propertyDisjointWith | — | — | — | — | — | — | — |
| "" | owl:sourceIndividual (NPAs) | — | — | — | — | — | — | — |
| "" | owl:AllDisjointProperties | — | — | — | — | — | — | — |

documents (33.7%)—owl:sameAs relations are typically asserted between individuals, making them much more numerous than the terminological-focused features. Relatedly, owl:FunctionalProperty and owl:InverseFunctionalProperty—which can be used to infer owl:sameAs relations—were used in various prominent vocabularies (e.g., SKOS and FOAF).

4. With regards to consistency checking, owl:disjointWith was the only such feature with prominent use.

Another interesting side observation is that OWL features which require blank-nodes to express in RDF (e.g., those requiring lists or blank-node subjects in the mapping to RDF [26]) appear to have less adoption. The subset of OWL (1) *without* blank-nodes—which has higher adoption—corresponds closely with various proposals to extend RDFS, including RDFS Plus [1] and RDFS 3.0 [7]. The highest ranked primitive to buck this trend was owl:unionOf in twelfth position which requires use of RDF lists.

Regarding the OWL primitives with no support in OWL 2 RL which we omitted from the table, owl:cardinality would be the highest ranked (position 19 of the current table); resp., owl:maxCardinality, owl:ReflexiveProperty, owl:qualifiedCardinality and owl:minQualifiedCardinality also saw sparse use, and would fill various positions below 25. The `owl:hasSelf` and `owl:disjointUnionOf` primitives did not appear.

To wrap up, we acknowledge that such a survey of RDFS and OWL cannot give a universal or definitive indication of the most important RDFS/OWL features for the Web—for example, based on our subjective experience, we would consider `owl:sameAs` as more important than many of the primitives ranked above it. However, the results offer useful insights with respect to *trends* of adoption on the Web, and what kind of reasoning the RDFS and OWL provided by Linked Data enables.

### 2.3    How Can We Reason over Linked Data?

Reasoning over Linked Data—and over RDF Web data in general—is a relatively young research topic, but one which is attracting more and more attention.

The traditional reasoning research literature has mostly focussed on issues relating to expressivity, tractability, soundness and completeness—particularly in closed domains—with evaluation demonstrated over various curated, vetted ontologies [37].

However, the Linked Data use-case has an orthogonal set of requirements: in particular, the two biggest challenges faced in such a scenario are (i) *scalability*, where reasoning may be required over billions of facts; and (ii) *robustness*, where reasoning may be required over noisy, impudent and/or inconsistent Web data. Conversely, (i) expressivity is often not such a problem for Linked Data where (as we will see) only lightweight subsets of the RDFS and OWL standards are used to describe vocabularies; (ii) formal tractability (i.e., polynomial complexity) often does not translate into scalability; (iii) soundness is still an important goal of reasoning, but guarantees of soundness are not critical since the base knowledge (sourced from the Web) cannot itself be guaranteed to be sound; and (iv) guarantees of completeness are also often not critical given that the base knowledge itself cannot be characterised as complete [37].

In terms of scalability, a lot of work has been done recently with respect to large-scale rule-based reasoning, typically where inferences are materialised and written to storage [20,40,93,89,88,62,42,54,13]. Various authors have suggested optimisations and distribution schemes based on a separation of terminological knowledge from assertional data during the inferencing process [93,89,88,42], enabling rule-based reasoning over datasets comprising of hundreds of millions [93], billions [89,42], or hundreds of billions [88] of triples. We discuss systems which provide scalable, distributed, rule-based materialisation in § 3; in particular, we detail one such proposal called SAOR, discussing related approaches later in the section.

However, despite the recent upsurge in research pertaining to scalable rule-based reasoning, few authors have demonstrated reasoning over realistic Linked Data [20,40,89,42,13], and even fewer have tackled issues of robustness or provenance of data [20,40]. Two main approaches have been proposed based on analysis of the source of data: (i) *authoritative reasoning*, which views the corpus as one big graph for reasoning, but only considers terminological data from sources which it deems can be trusted based on the dereferenceability of terms involved [40,39]; and (ii) *quarantined reasoning*, which performs a large-number of smaller "per-document" closures, where for each document under analysis, documents providing related terminology are imported with reasoning performed over these closed "quarantined" subsets of documents. We discuss authoritative reasoning in detail in § 3; we also briefly discuss quarantined reasoning later in that section, where we refer the interested reader to [20] for further detail.

However, we note that all of these proposals for reasoning are rule-based, and thus are limited in terms of their expressivity: although we have seen that Linked Data primarily uses lightweight features of OWL, rule-based reasoning is also limited with regards coverage for terminological reasoning [62]. Thus, an interesting direction for a more "holistic" approach to reasoning is to use tableau-based systems for expressive reasoning over the (typically smaller) terminological data, and rule-based systems for lightweight reasoning over the (potentially massive) assertional data. Such an approach has been recently proposed for the DLEJena system [62], where the Pellet DL reasoner is leveraged for more complete T-Box reasoning, and the Jena rule-engine is leveraged for more scalable A-Box reasoning. This hybrid approach is practically appealing in that it exploits the complementary strengths of rule-based and tableau-based approaches.

However, scalability and efficiency are still a concern for tableau-based reasoners in such scenarios, where although the terminological segment of a typical Linked Data corpus will be relatively small, it may still be too large for many existing tableau-based reasoners to cope with. One promising avenue of research to improve the scalability of tableau-based reasoning is to apply approximative reasoning, where soundness and/or completeness are traded for significant gains in terms of scalable and efficient execution; this is perhaps a more mid-to-long-term research direction with regards reasoning over Linked Data, but one which could enable much more expressive reasoning in such scenarios than the current rule-based proposals. Along these lines, in § 4 we present a proposal for performing tractable OWL 2 DL reasoning using sound but (possibly) incomplete approximations.

Finally, we note that the proposals presented hereafter mainly focus on the first use-case for reasoning over Linked Data: handling heterogeneity in the terminology used to describe assertional data. For discussion of proposals which target the latter two use-cases—handling coreference and inconsistency—we refer the interested reader to [39].

## 3   Scalable, Incomplete, OWL 2 RL/RDF Rule-Based Reasoning

Herein, we detail our system for performing rule-based reasoning over large-scale Linked Data corpora, which we call the *Scalable Authoritative OWL Reasoner* (SAOR) [40,42,39]. Our use-case scenario is the Semantic Web Search Engine (SWSE) [41],[14] which offers search and browsing over datasets consisting of approximately one billion Linked Data triples crawled from the Web: we want to use reasoning to materialise inferences and make them available in the SWSE results presented to users. In particular, we will be designing and evaluating our approach for performing materialisation over the heterogeneous ∼billion triple Linked Data corpus introduced at the outset of § 2.

Thus, in this section we look at applying incomplete OWL 2 RL/RDF materialisation in a manner sympathetic with our use-case, over static corpora of unverified Linked Data collected from millions of sources, consisting of approximately one billion input facts. In particular:

---

[14] Prototype available at `http://swse.deri.org/`

- we begin by discussing the requirements of our system for performing Linked Data/Web reasoning and high-level design decisions (§ 3.2);
- we continue by discussing a separation of terminological data during reasoning, providing soundness and conditional completeness results (§ 3.3);
- we subsequently detail the generic optimisations that a separation of terminological data allows (§ 3.4);
- we then look at identifying a subset of OWL 2 RL/RDF rules suitable for scalable materialisation, and discuss our method for performing distributed *authoritative* reasoning over our Linked Data corpus (§ 3.5);
- finally, we provide an overview of related work (§ 3.7) and give general discussion (§ 3.8).

## 3.1 Preliminaries

Before we continue, we briefly give some necessary preliminaries relating to (i) RDF, (ii) Linked Data principles, and (iii) rules.

**RDF.** We briefly give some necessary notation relating to RDF constants and RDF triples; see [33].

*RDF Constant.* Given the set of URI references U, the set of blank nodes B, and the set of literals L, the set of *RDF constants* is denoted by $C := U \cup B \cup L$. We interpret blank-nodes as skolem constants signifying particular individuals, as opposed to existential variables as prescribed by the RDF Semantics [33]; also, we rewrite blank-node labels when merging documents to ensure uniqueness of labels across those documents [33]. Finally, note that we may use 'a' as a shortcut for $rdf:type$, following convention in Turtle [8].

*RDF Triple.* A triple $t := (s, p, o) \in (U \cup B) \times U \times C$ is called an *RDF triple*, where $s$ is called subject, $p$ predicate, and $o$ object. A triple $t := (s, p, o) \in G, G := C \times C \times C$ is called a *generalised triple* [27], which allows any RDF constant in any triple position: henceforth, we assume generalised triples unless explicitly stated otherwise. We call a finite set of triples $G \subset G$ a *graph*.

*RDF Variable/RDF Triple Pattern.* We denote the set of all *RDF variables* as V; we call a generic member of the set $V \cup C$ an *RDF term*. Again, we denote RDF variables as alphanumeric strings with a '?' prefix. We call a triple of RDF terms—where variables are allowed in any position—an *RDF triple pattern*.

*Variable Substitution* We call a mapping from the set of variables to the set of constants $\theta : V \to C$ a *variable substitution*; we denote the set of all such substitutions by $\Theta$.

**Linked Data Principles and Provenance.** In order to cope with the unique challenges of handling diverse and unverified Web data, many of our components and algorithms require inclusion of a notion of provenance: consideration of the source of RDF data found on the Web. Thus, herein we provide some formal preliminaries for the Linked Data principles, and HTTP mechanisms for retrieving RDF data. (*Please see [2] in these proceedings for extensive introduction and discussion relating to Linked Data and the Web of Data.*)

*Linked Data Principles.* Herein, we will refer to the four best practices of Linked Data as follows [10]:

- (**LDP1**) use URIs as names for things;
- (**LDP2**) use HTTP URIs so those names can be dereferenced;
- (**LDP3**) return useful information upon dereferencing of those URIs; and
- (**LDP4**) include links using externally dereferenceable URIs.

*Data Source.* We define the *http-download* function get $:$ $\mathsf{U} \to 2^{\mathsf{G}}$ as the mapping from a URI to an RDF graph it provides by means of a given HTTP lookup [22] which directly returns status code 200 OK and data in a suitable RDF format, or to the empty set in the case of failure; this function also performs a rewriting of blank-node labels (based on the input URI) to ensure uniqueness when merging RDF graphs [33]. We define the set of *data sources* $\mathsf{S} \subset \mathsf{U}$ as the set of URIs $\mathsf{S} := \{s \in \mathsf{U} \mid \text{get}(s) \neq \emptyset\}$.

*RDF Triple in Context/RDF Quadruple.* An ordered pair $(t, c)$ with a triple $t :=$ $(s, p, o)$, and with a context $c \in \mathsf{S}$ and $t \in \text{get}(c)$ is called a *triple in context c.* We may also refer to $(s, p, o, c)$ as an *RDF quadruple* or quad $q$ with context $c$.

*HTTP Redirects/Dereferencing.* A URI may provide a HTTP redirect to another URI using a 30x response code [22]; we denote this function as redir $:$ $\mathsf{U} \to \mathsf{U}$ which may map a URI to itself in the case of failure (e.g., where no redirect exists). We denote the fixpoint of redir as redirs, denoting traversal of a number of redirects (a limit may be set on this traversal to avoid cycles and artificially long redirect paths). We define *dereferencing* as the function deref $:=$ get $\circ$ redirs which maps a URI to an RDF graph retrieved with status code 200 OK *after* following redirects, or which maps a URI to the empty set in the case of failure.

**Atoms and Rules.** In this section, we briefly introduce some notation as familiar particularly from the field of Logic Programming [57], which eventually gives us our notion of a *rule*. As such, much of the notation in this section serves as a generalisation of the RDF notation already presented; we will discuss this relation as pertinent.

*Atom* An *atomic formula* or *atom* is a formula of the form $p(e_1, \ldots, e_n)$, where all such $e_1, \ldots, e_n$ are terms (like Datalog, function symbols are disallowed) and where $p$ is a *predicate* of arity $n$—we denote the set of all such atoms by Atoms. As such, this notation can be thought of as generalising that of RDF triples, where we use a standard RDF *ternary predicate* $T$ to represent RDF triples in the form $T(s, p, o)$—for example, $T(\text{Fred, age, } 56)$—where we will typically leave $T$ implicit.

Note that a term $e_i$ can also be a variable, and thus RDF triple patterns can also be represented directly as atoms. Atoms not containing variables are called *ground atoms* or simply *facts*, denoted as the set Facts (a generalisation of G); a finite set of facts $I$ is called a (Herbrand) *interpretation* (a generalisation of a graph). Letting $A$ and $B$ be two atoms, we say that $A$ *subsumes* $B$—denoted $A \rhd B$—if there exists a substitution $\theta \in \Theta$ of variables such that $A\theta = B$ (applying $\theta$ to the variables of $A$ yields $B$);

we may also say that $B$ is an *instance* of $A$; if $B$ is ground, we say that it is a *ground instance*. Similarly, if we have a substitution $\theta \in \Theta$ such that $A\theta = B\theta$, we say that $\theta$ is a *unifier* of $A$ and $B$; we denote by $\mathrm{mgu}(A, B)$ the *most general unifier* of $A$ and $B$ which provides the "minimal" variable substitution (up to variable renaming) required to unify $A$ and $B$.

*Rule.* A *rule* $R$ is given as follows:

$$H \leftarrow B_1, \ldots, B_n (n \geq 0), \tag{1}$$

where $H, B_1, \ldots, B_n$ are atoms, $H$ is called the *head* (conclusion/consequent) and $B_1, \ldots, B_n$ the *body* (premise/antecedent). We use $\mathrm{Head}(R)$ to denote the head $H$ of $R$ and $\mathrm{Body}(R)$ to denote the body $B_1, \ldots, B_n$ of $R$.[15] The variables of our rules are *range restricted*, also known as *safe* [87]: like Datalog, the variables appearing in the head of each rule must also appear in the body, which means that a substitution which grounds the body must also ground the head. We denote the set of all such rules by Rules. A rule with an empty body is considered a fact; a rule with a non-empty body is called a *proper-rule*. We call a finite set of such rules a *program $P$*.

Like before, a ground rule is one without variables. We denote with $\mathrm{Ground}(R)$ the set of ground instantiations of a rule $R$ and with $\mathrm{Ground}(P)$ the ground instantiations of all rules occurring in a program $P$. Again, an *RDF rule* is a specialisation of the above rule, where atoms strictly have the ternary predicate $T$ and contain RDF terms; an *RDF program* is one containing RDF rules, etc.

Note that we may find it convenient to represent rules as having multiple atoms in the head, such as:

$$H_1, \ldots, H_m (m \geq 1) \leftarrow B_1, \ldots, B_n (n \geq 0),$$

where we imply a *conjunction* between the head atoms, such that this can be equivalently represented as the set of rules:

$$\{H_i \leftarrow B_1, \ldots, B_n \mid (1 \leq i \leq m)\}.$$

*Immediate Consequence Operator.* We give the immediate consequence operator $\mathfrak{T}_P$ of a program $P$ under interpretation $I$ as:[16]

$$\mathfrak{T}_P : 2^{\mathsf{Facts}} \rightarrow 2^{\mathsf{Facts}}$$

$$I \mapsto \left\{ \mathrm{Head}(R)\theta \mid R \in P \wedge \exists I' \subseteq I \text{ s.t. } \theta = \mathrm{mgu}\big(\mathrm{Body}(R), I'\big) \right\}$$

Intuitively, the immediate consequence operator maps from a set of facts $I$ to the set of facts it directly entails with respect to the program $P$—note that $\mathfrak{T}_P(I)$ will retain the facts in $P$ since facts are rules with empty bodies and thus unify with any interpretation, and note that $\mathfrak{T}_P$ is *monotonic*—the addition of facts and rules to a program can only lead to the same or additional consequences. We may refer to the application of a single rule $\mathfrak{T}_{\{R\}}$ as a *rule application*.

---

[15] Such a rule can be represented as a definite Horn clause.

[16] Note that in our Herbrand semantics, an interpretation $I$ can be thought of as simply a set of facts.

Since our rules are a syntactic subset of Datalog, $\mathfrak{T}_P$ has a *least fixpoint*—denoted $\mathsf{lfp}(\mathfrak{T}_P)$—which can be calculated in a bottom-up fashion, starting from the empty interpretation $\Delta$ and applying iteratively $\mathfrak{T}_P$ [94] (here, convention assumes that $P$ contains the set of input facts as well as proper rules). Define the iterations of $\mathfrak{T}_P$ as follows: $\mathfrak{T}_P \uparrow 0 = \Delta$; for all ordinals $\alpha$, $\mathfrak{T}_P \uparrow (\alpha + 1) = \mathfrak{T}_P(\mathfrak{T}_P \uparrow \alpha)$; since our rules are Datalog, there exists an $\alpha$ such that $\mathsf{lfp}(\mathfrak{T}_P) = \mathfrak{T}_P \uparrow \alpha$ for $\alpha < \omega$, where $\omega$ denotes the least infinite ordinal—i.e., the immediate consequence operator will reach a fixpoint in countable steps [87]. Thus, $\mathfrak{T}_P$ is also *continuous*. We call $\mathsf{lfp}(\mathfrak{T}_P)$ *the least model*, or the *closure* of $P$, which is given the more succinct notation $\mathsf{lm}(P)$.

### 3.2 Linked Data Reasoning: Overview

Performing reasoning over large amounts of arbitrary RDF data sourced from the Web implies unique challenges which have not been significantly addressed by the literature. Given that we will be dealing with a corpus in the order of a billion triples collected from millions of unvetted sources, we must acknowledge two primary challenges:

- **scalability**: the reasoning approach must scale to billion(s) of statements;
- **robustness**: the reasoning approach should be tolerant to noisy, impudent and inconsistent data.

These requirements heavily influence the design choices of our reasoning approach, where in particular we (must) opt for performing reasoning which is incomplete with respect to OWL semantics.

**Incomplete Reasoning: Rationale.** As alluded to in § 2.3, current standard RDF-S/OWL reasoning approaches are not naturally suited to meet the aforementioned challenges.

Firstly, standard RDFS entails infinite triples, although implementations commonly support a decidable (finite) subset [83,67,68,93]. In any case, RDFS does not support reasoning over OWL axioms commonly provided by Linked Data vocabularies.

With respect to OWL, reasoning with respect to OWL (2) Full is known to be undecidable. Reasoning with standard dialects such as OWL (2) DL or OWL Lite have more than exponential worst-case complexity, and are typically implemented using tableau-based algorithms which have yet to demonstrate scalability for reasoning over assertional data which would be propitious to our scenario: certain reasoning tasks may require satisfiability checking which touch upon a large proportion of the individuals in the knowledgebase, and may have to operate over a large, branching search space [4]. Similarly, although certain optimisation techniques may make the performance of such tableau-reasoning sufficient for certain *reasonable* inputs and use-cases, guarantees of such *reasonability* do not extend to a Web corpus like ours. Reasoning with respect to the new OWL 2 profiles—viz., OWL 2 EL/QL/RL—have polynomial runtime, which although an improvement, may still be prohibitively expensive for reasoning over the assertional data of a corpus such as ours.

Aside from complexity considerations, most OWL documents on the Web are in any case OWL Full: "syntactic" assumptions made in DL-based profiles are violated by even very commonly used ontologies. For example, the FOAF vocabulary knowingly falls into OWL Full since, e.g., `foaf:name` is defined as a sub-property of the core RDFS property `rdfs:label`, and `foaf:mbox_sha1sum` is defined as a member of both `owl:InverseFunctionalProperty` and `owl:DatatypeProperty`: such axioms are disallowed by OWL (2) DL (and by extension, disallowed by the sub-dialects and profiles).

Finally, OWL semantics prescribe that anything can be entailed from an inconsistency, following the *principle of explosion* in classical logics. This is not only true of OWL (2) Full semantics, but also of those sub-languages rooted in Description Logics, where reasoners check entailment by reduction to satisfiability—if the original graph is inconsistent, it is already in itself unsatisfiable, and the entailment check will return true for any arbitrary graph. Given that consistency cannot be expected on the Web, we wish to avoid the arbitrary entailment of all possible triples from our knowledge-base. Along these lines, a number of paraconsistent reasoning approaches have been defined in the literature (see, e.g., [46,60,96,61,56]) typically relying upon four-valued logic [9]—however, again, these approaches have yet to demonstrate the sort of performance required for our scenario.

Thus, due to the prohibitive computational complexity involved, complete reasoning with respect to the standardised RDFS/OWL (sub-)languages is infeasible for our scenario, esp. given the volume of assertional data involved. We instead argue that completeness (with respect to the language) is not a requirement for our use-case, particularly given that the corpus itself represents incomplete knowledge (cf. [21,37]).

Moving forward, we opt for sound but incomplete support of OWL Full semantics such that entailment is axiomatised by a set of rules which are applicable to arbitrary RDF graphs (no syntactic restrictions) and which do not rely on satisfiability checking (are so not bound by the principle of explosion).

**Rule-based Reasoning.** Predating OWL 2—and in particular the provision of the OWL 2 RL/RDF ruleset—numerous rule-based entailment regimes were proposed in the literature to provide a partial axiomatisation of OWL's semantics. These regimes included Description Logic Programs (DLP) [29,63], pD* [82,83], RDFS-Plus [1], etc. Recognising the evident demand for rule-based support of OWL, in 2009, the W3C OWL Working Group standardised the OWL 2 RL profile and accompanying OWL 2 RL/RDF ruleset [27].The OWL 2 RL profile is a syntactic subset of OWL 2 which is implementable through translation to the Direct Semantics (DL-based semantics) or the RDF-Based Semantics (OWL 2 Full semantics). As such, the OWL 2 RL/RDF ruleset comprises a partial-axiomatisation of the OWL 2 RDF-Based Semantics which is applicable for arbitrary RDF graphs, and thus is is compatible with RDF Semantics [33]. We thus select OWL 2 RL/RDF as the most comprehensive, standard means of supporting RDFS *and* OWL entailment using rules, which largely subsumes the entailment possible through RDFS, DLP, pD*, RDFS-Plus, etc. For reference, we provide the OWL 2 RL/RDF ruleset in Appendix A, highlighting various characteristics which we will discuss as appropriate. (*Please also see [36] in these proceedings for discussion on the combination of rules and ontologies.*)

**Forward Chaining.** We opt to perform *forward-chaining materialisation* of inferred data with respect to (a subset of) OWL 2 RL/RDF rules—i.e., we aim to make explicit the implicit data inferable through these rules (as opposed to, e.g., rewriting/extending queries and posing them against the original data in situ).

A materialisation approach offers two particular benefits:

- *pre-runtime execution*: materialisation can be conducted off-line (or, more accurately while loading data) avoiding the run-time expense of query-specific backward-chaining techniques which may adversely affect query response times;
- *consumer independent*: the inferred data provided by materialisation can subsequently be consumed in the same manner as explicit data, without the need for integrating a reasoning component into the runtime engine.

Note that in the spirit of *one size does not fit all*, forward-chaining materialisation is not a "magic-bullet": backward-chaining may be more propitious to support inferences where the amount of data involved is prohibitively expensive to materialise and index, and where these inferred data are infrequently required by the consumer application. Herein, we focus on materialisation, but grant that the inherent trade-off between off-line forward-chaining and runtime backward-chaining warrants further investigation in another scope.

Alongside **scalability** and **robustness**, we identify two further requirements for our materialisation approach:

- **efficiency**: the reasoning algorithm must not only be able to process large-amounts of data, but should do so in as little computation time as possible;
- **terseness**: to reduce the burden on the consumer system—e.g., with respect to indexing or query-processing—we wish to keep a succinct volume of materialised data and aim instead for "reasonable" completeness.

Both of these additional requirements are intuitive, but also non-trivial, and so will provide important input for our design decisions.

**OWL 2 RL/RDF Scalability** Full materialisation with respect to the entire set of OWL 2 RL/RDF rules is infeasible for our use-case. First, a subset of OWL 2 RL/RDF rules are expressed informally—i.e., they are not formalised by means of Horn clauses—and may introduce new terms as a consequence of the rule, which in turn affects decidability (i.e., the achievability of a finite fixpoint). For example, OWL 2 RL/RDF rule dt-eq is specified as:

$$\forall lt_1, lt_2 \in L \; with \; the \; same \; data \; value, \; infer \; (lt_1, owl:sameAs, lt_2).$$

Note that this rule does not constrain $lt_1$ or $lt_2$ to be part of any graph or interpretation under analysis. Similarly, rule dt-diff entails pairwise owl:differentFrom relations between all literals with different data values, and rule dt-type2 entails an explicit membership triple for each literal to its datatype. These rules applied to, e.g., the value set of decimal-expressible real numbers (denotable by the datatype xsd:decimal) entail infinite triples.[17]

---

[17] Typically, materialisation engines support non-standard versions of these rules using heuristics such as canonicalisation of datatype literals, or only applying the rules over literals that appear in the ruleset or in the data under analysis.

Aside from these datatype rules, the worst-case complexity of applying OWL 2 RL/RDF rules is cubic with respect to the known set of constants (a.k.a. the Herbrand universe); for example, consider the following two triples:

$$(\texttt{owl:sameAs}, \texttt{owl:sameAs}, \texttt{rdf:type})$$
$$(\texttt{owl:sameAs}, \texttt{rdfs:domain}, \texttt{bad:Hub})$$

Adding these two triples to any arbitrary RDF graph will lead to the inference of all possible (generalised) triples by the OWL 2 RL/RDF rules: i.e., the inference of $C \times C \times C$ (a.k.a. the Herbrand base), where $C \subset \mathsf{C}$ is the set of RDF constants (§ 3.1) mentioned in the OWL 2 RL/RDF ruleset and the graph (a.k.a. the Herbrand universe). The process involves the OWL 2 RL/RDF "equality rules" (eq-*) and the rule for supporting rdfs:domain (prp-dom), which lead to the inference of $|C|^3$ triples, as such *emulating* the explosive nature of inconsistency without actually requiring any inconsistency—we leave the details of the inferencing as an exercise for the reader (available in [39]).

This simple example raises concerns with respect to all of our defined requirements: materialising the required entailments for a large graph will be neither **scalable** nor **efficient**; even assuming that materialisation were possible, the result would not be **terse** (or be of any use at all to the consumer system); given that a single remote publisher can arbitrarily make such assertions in any location they like, such reasoning is clearly not **robust**.

Even for reasonable inputs, the result size and expense of OWL 2 RL/RDF materialisation can be prohibitive for our scenario. For example, chains of transitive relations of length $n$ mandate quadratic ($\frac{n^2-n}{2}$) materialisation. Large equivalence classes (sets of individuals who are pairwise related by owl:sameAs) similarly mandate the materialisation of $n^2$ pairwise symmetric, reflexive and transitive owl:sameAs relations. Given our input sizes and the distinct possibility of such phenomena in our corpus, such quadratic materialisation quickly infringes upon our requirements for **scalability**, **efficiency** and arguably **terseness**.

Moreover, certain rules can materialise inferences which hold for every term in the graph—we call these inferences *tautological*. For example, the OWL 2 RL/RDF rule eq-ref materialises a reflexive owl:sameAs statement for every known term, reflecting the fact that everything is the same as itself. Thus, we omit such tautological rules (eq-ref for OWL 2 RL/RDF), viewing them as contrary to our requirement for **terseness**.

As such, the OWL 2 RL/RDF ruleset—and application thereof—requires significant tailoring to meet our requirements; we begin with our first non-standard optimisation in the following section.

## 3.3   Distinguishing Terminological Data

As previously described, RDFS/OWL allow for disseminating terminological data— loosely schema-level data—which provide definitions of classes and properties. Given a sufficiently large corpora collected from the Web, the percentage of terminological data is relatively small when compared to the volume of assertional data: typically— and as we will see in § 3.6—terminological data represent less than one percent of such a corpus [40,42]. Assuming that the proportion of terminological data is quite small—and given that these data are among the most commonly accessed during reasoning—we formulate an approach around the assumption that such data can be efficiently handled and processed independently of the main bulk of assertional data. First, we provide some preliminaries relating to our notion of terminological data.

*Meta-class.* We consider a *meta-class* as a class specifically of classes or properties; i.e., the members of a meta-class are themselves either classes or properties. *Herein, we restrict our notion of meta-classes to the set defined in RDF(S) and OWL specifications*, where examples include `rdf:Property`, `rdfs:Class`, `owl:Class`, `owl:FunctionalProperty`, `owl:Restriction`, `owl:DatatypeProperty`, etc.; note that `rdfs:Resource`, `rdfs:Literal`, e.g., are not considered meta-classes.

*Meta-property.* A *meta-property* is one which has a meta-class as its domain; *again, we restrict our notion of meta-properties to the set defined in RDF(S) and OWL specifications*, where examples include `rdfs:domain`, `rdfs:subClassOf`, `owl:has-Key`, `owl:inverseOf`, `owl:oneOf`, `owl:onProperty`, `owl:unionOf`, etc.; note that `rdf:type`, `owl:sameAs`, `rdfs:label`, e.g., do *not* have a meta-class as domain and so are *not* considered meta-properties.

*Terminological triple.* We define the set of *terminological triples* as the union of the following sets of triples:

1. triples with `rdf:type` as predicate and a meta-class as object;
2. triples with a meta-property as predicate;
3. triples forming a *valid* RDF list whose head is the object of a meta-property (e.g., a list used for `owl:unionOf`, `owl:intersectionOf`, etc.);

Our approach for separating terminological data is related to the area of partial evaluation and program specialisation of Logic Programs [55,58,49]: we take a generic (meta) program—such as RDFS, pD*, OWL 2 RL/RDF, etc.—and partially evaluate this program with respect to terminological knowledge. The result of this partial evaluation is a set of terminological inferences and a residual program which can be applied over the assertional data; this specialised *assertional program* is then primed using further optimisation before application over the bulk of assertional data.

Towards this goal, we begin by formalising the notion of a T-split rule which distinguishes between terminological and assertional atoms (T-atoms/A-atoms).

**Definition 1 (T-split rule).** *A T-split rule $R$ is given as follows:*

$$H \leftarrow A_1, \ldots, A_n, T_1, \ldots, T_m \quad (n, m \geq 0), \tag{2}$$

*where the $T_i, 0 \leq i \leq m$ atoms in the body (T-atoms) are all those that can only have terminological ground instances, whereas the $A_i, 1 \leq i \leq n$ atoms (A-atoms), can have arbitrary ground instances. We use $\mathsf{TBody}(R)$ and $\mathsf{ABody}(R)$ to respectively denote the set of T-atoms and the set of A-atoms in the body of $R$.*

Henceforth, we assume rules are T-split such that T-atoms and A-atoms can be referenced using the functions TBody and ABody when necessary.

*Example 1.* Let $R_{EX}$ denote the following rule:

$$(?\text{x}, \text{a}, ?\text{c2}) \leftarrow \underline{(?\text{c1}, \text{rdfs:subClassOf}, ?\text{c2})}, (?\text{x}, \text{a}, ?\text{c1})$$

When writing T-split rules, we denote TBody($R_{EX}$) by underlining: the underlined T-atom can only be bound by a triple with the meta-property rdfs:subClassOf as RDF predicate, and thus can only be bound by a terminological triple. The second atom in the body can be bound by assertional or terminological triples, and so is considered an A-atom. ◊

The notion of a *T-split program*—containing T-split rules and ground T-atoms and A-atoms—follows naturally. Distinguishing terminological atoms in rules enables us to define a form of stratified program execution, whereby a terminological fixpoint is reached first, and then the assertional data is reasoned over; we call this the *T-split least fixpoint*. Before we formalise this alternative fixpoint procedure, we must first describe our notion of a T-ground rule, where the variables appearing in T-atoms of a rule are grounded separately by terminological data:

**Definition 2 (T-ground rule).** *A T-ground rule is a set of rule instances for the T-split rule R given by grounding* TBody($R$) *and the variables it contains across the rest of the rule. We denote the set of such rules for a program P and a set of facts I as* Ground$^T(P, I)$, *defined as:*

$$\text{Ground}^T(P, I) := \{\text{Head}(R)\theta \leftarrow \text{ABody}(R)\theta \mid R \in P, \exists I' \subseteq I \text{ s.t. } \theta = \text{mgu}(\text{TBody}(R), I')\}$$

The result is a set of rules whose T-atoms are grounded by the terminological data in $I$.

*Example 2.* Consider the T-split rule $R_{EX}$ as before:

$$(?\text{x}, \text{a}, ?\text{c2}) \leftarrow \underline{(?\text{c1}, \text{rdfs:subClassOf}, ?\text{c2})}, (?\text{x}, \text{a}, ?\text{c1})$$

Now let

$$I_{EX} := \{\, (\text{foaf:Person}, \text{rdfs:subClassOf}, \text{foaf:Agent}),$$
$$(\text{foaf:Agent}, \text{rdfs:subClassOf}, \text{dc:Agent}) \,\}$$

Here,

$$\text{Ground}^T(\{R_{EX}\}, I_{EX}) = \{\, (?\text{x}, \text{a}, \text{foaf:Agent}) \leftarrow (?\text{x}, \text{a}, ?\text{foaf:Person});$$
$$(?\text{x}, \text{a}, \text{dc:Agent}) \leftarrow (?\text{x}, \text{a}, ?\text{foaf:Agent}) \,\}.$$

◊

We can now formalise our notion of the T-split least fixpoint, where a terminological least model is determined, T-atoms of rules are grounded against this least model, and the remaining (proper) assertional rules are applied against the bulk of assertional data in the corpus. (In the following, we recall from § 3.1 the notions of the immediate consequence operator $\mathfrak{T}_P$, the least fixpoint lfp($\mathfrak{T}_P$), and the least model lm($P$) for a program $P$.)

**Definition 3 (T-split least fixpoint).** *The* T-split least fixpoint *for a program P is broken up into two parts: (i) the* terminological least fixpoint, *and (ii) the* assertional least fixpoint. *Let* $P^F := \{R \in P \mid \mathsf{Body}(R) = \emptyset\}$ *be the set of facts in* $P$,[18] *let* $P^{T\emptyset} := \{R \in P \mid \mathsf{TBody}(R) \neq \emptyset, \mathsf{ABody}(R) = \emptyset\}$, *let* $P^{\emptyset A} := \{R \in P \mid \mathsf{TBody}(R) = \emptyset, \mathsf{ABody}(R) \neq \emptyset\}$, *and let* $P^{TA} := \{R \in P \mid \mathsf{TBody}(R) \neq \emptyset, \mathsf{ABody}(R) \neq \emptyset\}$. *Note that* $P = P^F \cup P^{T\emptyset} \cup P^{\emptyset A} \cup P^{TA}$. *Now, let*

$$TP := P^F \cup P^{T\emptyset}$$

*denote the initial (terminological) program containing ground facts and T-atom only rules, and let* $\mathsf{lm}(TP)$ *denote the least model for the terminological program. Let*

$$P^{A+} := \mathsf{Ground}^T(P^{TA}, \mathsf{lm}(TP))$$

*denote the set of (proper) rules achieved by grounding rules in* $P^{TA}$ *with the terminological atoms in* $\mathsf{lm}(TP)$: *Now, let*

$$AP := \mathsf{lm}(TP) \cup P^{\emptyset A} \cup P^{A+}$$

*denote the second (assertional) program containing all available facts and proper assertional rules. Finally, we can give the least model of the T-split program P as* $\mathsf{lm}(AP)$ *for AP derived from P as above—we more generally denote this by* $\mathsf{lm}^T(P)$.

An important question thereafter is how the standard fixpoint of the program $\mathsf{lm}(P)$ relates to the T-split fixpoint $\mathsf{lm}^T(P)$. Firstly, we show that the latter is *sound* with respect to the former:

**Theorem 1 (T-split soundness).** *For any program P, it holds that* $\mathsf{lm}^T(P) \subseteq \mathsf{lm}(P)$.

*Proof available in [39].*

Thus, for any given program containing rules and facts (as we define them), the T-split least fixpoint is necessarily a subset of the standard least fixpoint. Next, we look at characterising the *completeness* of the former with respect to the latter; beforehand, we need to define our notion of a T-Box:

**Definition 4 (T-Box).** *We define the* T-Box *of an interpretation I with respect to a program P as the subset of facts in I that are an instance of a T-atom of a rule in P:*

$$\mathsf{TBox}(P, I) := \{F \in I \mid \exists R \in P, \exists T \in \mathsf{TBody}(R) \text{ s.t. } T \triangleright F\} .$$

(Here we recall the $\triangleright$ notation of an instance [§ 3.1] whereby $A \triangleright B$ iff $\exists \theta$ s.t. $A\theta = B$.) Thus, our T-Box is precisely the set of terminological triples in a given interpretation (i.e., graph) that can be bound by a terminological atom of a rule in the program.

We now give a conditional proposition of completeness which states that if no new T-Box facts are produced during the execution of the assertional program, the T-split least model is equal to the standard least model.

---

[18] Of course, $P^F$ can refer to axiomatic facts and/or the initial facts given by an input knowledge-base.

**Theorem 2 (T-split conditional completeness).** *For any program $P$, its terminological program $TP$ and its assertional program $AP$, if it holds that* $\mathsf{TBox}(P, \mathsf{Im}(TP)) = \mathsf{TBox}(P, \mathsf{Im}(AP))$, *then it holds that* $\mathsf{Im}(P) = \mathsf{Im}^T(P)$.

**Corollary 1 (Rephrased condition for T-split completeness).** *For any program $P$, if a rule with non-empty* $\mathsf{ABody}$ *does not infer a terminological fact, then* $\mathsf{Im}(P) = \mathsf{Im}^T(P)$.

*Proofs available in [39].*

So one may wonder when this condition of completeness is broken—i.e., when do rules with assertional atoms infer terminological facts? Analysis of how this can happen must be applied per rule-set, but for OWL 2 RL/RDF, we conjecture that such a scenario can only occur through (i) so called *non-standard use* of the set of RDFS/OWL meta-classes and meta-properties required by the rules, *or*, (ii) by the semantics of replacement for `owl:sameAs` (supported by OWL 2 RL/RDF rules **eq-rep-\***).[19]

We first discuss the effects of non-standard use for T-split reasoning over OWL 2 RL/RDF, starting with a definition.

**Definition 5 (Non-standard triples).** *With respect to a set of meta-properties $MP$ and meta-classes $MC$, a* non-standard triple *is a terminological triple (T-fact wrt. $MP/MC$) where additionally:*

- *a meta-class in $MC$ appears in a position other than as the value of* `rdf:type`; or
- *a property in $MP \cup \{$`rdf:type`,`rdf:first`,`rdf:rest`$\}$ appears outside of the RDF predicate position.*

We call the set $MP \cup MC \cup \{$`rdf:type`,`rdf:first`,`rdf:rest`$\}$ the *restricted vocabulary*. (Note that restricting the use of `rdf:first` and `rdf:rest` would be superfluous for RDFS and pD\* which do not support terminological axioms containing RDF lists.)

Now, before we formalise a proposition about the incompleteness caused by such usage, we provide an intuitive example thereof:

*Example 3.* As an example of incompleteness caused by non-standard use of the meta-property `owl:InverseFunctionalProperty`, consider:

$1_a.$    `(ex:KeyProperty, rdfs:subClassOf, owl:InverseFunctionalProperty)`
$2_a.$    `(ex:isbn13, a, ex:KeyProperty)`
$3_a.$    `(ex:The_Road, ex:isbn13, "978-0307265432")`
$4_a.$    `(ex:Road%2C_The, ex:isbn13, "978-0307265432")`

where triple ($1_a$) is considered non-standard use. The static T-Box in the terminological program will include the first triple, and, through the assertional rule **cax-sco** and triples ($1_a$) and ($2_a$) will infer:

---

[19] We note that the phrase "non-standard use" has appeared elsewhere in the literature with the same intuition, but with slightly different formulation and intention; e.g., see [18].

$5_a.|$                    $(\texttt{ex:isbn13}, \texttt{a}, \texttt{owl:InverseFunctionalProperty})$

but this T-fact will not be considered by the pre-ground T-atoms of the rules in the assertional program. Thus, the inferences:

$6_a.|$                    $(\texttt{ex:The\_Road}, \texttt{owl:sameAs}, \texttt{ex:Road\%2C\_The})$
$7_a.|$                    $(\texttt{ex:Road\%2C\_The}, \texttt{owl:sameAs}, \texttt{ex:The\_Road})$

which should hold through rule **prp-ifp** and triples $(3_a)$, $(4_a)$ and $(5_a)$ will not be made.
   A similar example follows for non-standard use of meta-classes; e.g.:

$1_b.\|$                 $(\texttt{ex:inSubFamily}, \texttt{rdfs:subClassOf}, \texttt{rdfs:subClassOf})$
$2_b.\|$                    $(\texttt{ex:Bos}, \texttt{ex:inSubFamily}, \texttt{ex:Bovinae})$
$3_b.\|$                      $(\texttt{ex:Daisy}, \texttt{a}, \texttt{ex:Bos})$

which through the assertional rule **prp-spo1** and triples $(1_b)$ and $(2_b)$ will infer:

$4_b.\|$                 $(\texttt{ex:Bos}, \texttt{rdfs:subClassOf}, \texttt{ex:Bovinae})$,

but not:

$5_b.\|$                    $(\texttt{ex:Daisy}, \texttt{a}, \texttt{ex:Bovinae})$

since triple $(4_b)$ is not included in the terminological program.                           $\Diamond$

**Theorem 3 (Conditional completeness for standard use).** *Let $\mathcal{O}2\mathcal{R}'$ denote the set of (T-split) OWL 2 RL/RDF rules excluding **eq-rep-s**, **eq-rep-p** and **eq-rep-o**; let I be any interpretation not containing any non-standard use of the restricted vocabulary which contains (i) meta-classes or meta-properties appearing in the T-atoms of $\mathcal{O}2\mathcal{R}'$, and (ii) $\texttt{rdf:type}, \texttt{rdf:first}, \texttt{rdf:list}$; and let $P := \mathcal{O}2\mathcal{R}' \cup I$; then, it holds that $\mathsf{lm}(P) = \mathsf{lm}^T(P)$.*

*Sketch of proof involving inspection of OWL 2 RL/RDF rules available in [39].*

Briefly, we note that [93] have given a similar result for RDFS by inspection of the rules, and that pD* inference relies on non-standard axiomatic triples whereby the above results do not translate naturally.

   With respect to rules **eq-rep-\*** (which we have thus far omitted), new terminological triples can be inferred from rules with non-empty ABody through the semantics of $\texttt{owl:sameAs}$, breaking the condition for completeness from Theorem 2. However, with respect to the T-split inferencing procedure, we conjecture that incompleteness can only be caused if $\texttt{owl:sameAs}$ affects some constant in the TBody of an OWL 2 RL/RDF rule; we refer the interested reader to [39] for some examples and further discussion. In any case, note that (i) in our intended use-case, we do not apply rules $\texttt{eq-rep-*}$ in our inferencing procedure due to scalability concerns—this will be discussed further in § 3.5; and (ii) we believe that in practice, T-split incompleteness through such $\texttt{owl:sameAs}$ relations would only occur for rare corner cases. (For more detailed work looking at scalable "equality reasoning" for Linked Data, please see [39, § 7].)

Conceding the possibility of incompleteness—in particular in the presence of non-standard triples or `owl:sameAs` relations affecting certain terminological constants—we proceed by describing our implementation of the T-split program execution, how it enables unique optimisations, and how it can be used to derive a subset of OWL 2 RL/RDF rules which are linear with respect to assertional knowledge.

**Implementing T-split Inferencing.** Given that the T-Box remains static during the application of the assertional program, our T-split algorithm enables a partial-indexing approach to reasoning, whereby only a subset of assertional triples—in particular those required by rules with multiple A-atoms in the body—need be indexed. Thus, the T-split closure can be achieved by means of two triple-by-triple scans of the corpus:

1. the first scan **identifies and separates out the T-Box and applies the terminological program**:
   (a) during the scan, any triples that are instances of a T-atom of a rule are indexed in memory;
   (b) after the scan, rules with only T-atoms in the body are applied over the in-memory T-Box until the terminological least model is reached, and rules with T-atoms and A-atoms in the body have their T-atoms grounded by these data;
   (c) novel inferences in the terminological least model are written to an on-disk file (these will later be considered as part of the inferred output, and as input to the assertional program);
2. the second scan **applies the assertional program** over the main corpus and the terminological inferences;
   (a) each triple is individually checked to see if it unifies with an atom in an assertional rule body;
      i. if it unifies with a single-atom rule body, the inference is immediately applied;
      ii. if it unifies with a multi-atom rule body, the triple is indexed and the index is checked to determine whether the other atoms of the rule can be instantiated by previous triples—if so, the inference is applied;
   (b) inferred triples are immediately put back into step (2a), with an in-memory cache avoiding cycles and (partially) filtering duplicates.

The terminological program is applied using standard *semi-naïve evaluation* techniques, whereby only instances of rule bodies involving novel data will fire, ensuring that derivations are not needlessly and endlessly repeated (see, e.g., [87]).

We give a more formal break-down of the application of the assertional program in Algorithm 3.1. For our purposes, the A-Box input is the set of axiomatic statements in the rule fragment, the set of novel terminological inferences, and the entire corpus; i.e., we consider terminological data as also being assertional in a unidirectional form of punning [25].

---

**Algorithm 3.1.** Reason over the A-Box

---

**Require:** ABOX: A                                                                    /* $\{t_0 \dots t_m\}$ */
**Require:** ASSERTIONAL PROGRAM: $AP$              /* $\{R_0 \dots R_n\}$, $\mathsf{TBody}(R_i) = \emptyset$ */
 1: Index := $\{\}$                                                           /* triple index */
 2: LRU := $\{\}$                                      /* fixed-size, least recently used cache */
 3: **for all** $t \in$ A **do**
 4:    $G_0 := \{\}, G_1 := \{t\}, i := 1$
 5:    **while** $G_i \neq G_{i-1}$ **do**
 6:      **for all** $t_\delta \in G_i \setminus G_{i-1}$ **do**
 7:        **if** $t_\delta \notin$ LRU **then**                       /* if $t_\delta \in$ LRU, make $t_\delta$ most recent entry */
 8:          add $t_\delta$ to LRU                           /* remove eldest entry if necessary */
 9:          $\mathrm{output}(t_\delta)$
10:          **for all** $R \in AP$ **do**
11:            **if** $|\mathsf{Body}(R)| = 1$ **then**
12:              **if** $\exists \theta$ s.t. $\{t_\delta\} = \mathsf{Body}(R)\theta$ **then**
13:                $G_{i+1} := G_{i+1} \cup \mathsf{Head}(R)\theta$
14:              **end if**
15:            **else**
16:              **if** $\exists \theta$ s.t. $t_\delta \in \mathsf{Body}(R)\theta$ **then**
17:                $card = |\text{Index}|$
18:                Index := Index $\cup \{t_\delta\}$
19:                **if** $card \neq |\text{Index}|$ **then**
20:                  **for all** $\theta$ s.t. $\mathsf{Body}(R\theta) \subseteq$ Index, $t_\delta \in \mathsf{Body}(R\theta)$ **do**
21:                    $G_{i+1} := G_{i+1} \cup \mathsf{Head}(R\theta)$
22:                  **end for**
23:                **end if**
24:              **end if**
25:            **end if**
26:          **end for**
27:        **end if**
28:      **end for**
29:      $i\text{++}$
30:      $G_{i+1} := \mathrm{copy}(G_i)$                    /* copy inferences to new set to avoid cycles */
31:    **end while**
32: **end for**
33: **return** output                                                              /* on-disk inferences */

---

First note that duplicate inference steps may be applied for rules with only one atom in the body (Lines 11–14): one of the main optimisations of our approach is that it minimises the amount of data that we need to index, where we only wish to store triples which may be necessary for later inference, and where triples *only* grounding single atom rule bodies need not be indexed. To provide *partial* duplicate removal, we instead use a Least-Recently-Used (LRU) cache over a sliding window of recently encountered triples (Lines 7 & 8)—outside of this window, we may not know whether a triple has been encountered before or not, and may repeat inferencing steps.

Thus, in this *partial-indexing* approach, we need only index those triples which are matched by a rule with a multi-atom body (Lines 15–25). For indexed triples, aside from the LRU cache, we can additionally check to see if that triple has been indexed before (Line 19) and we can apply a semi-naïve check to ensure that we only materialise inferences which involve the current triple (Line 20). We note that as the assertional index is required to store more data, the two-scan approach becomes more inefficient than the "full-indexing" approach; in particular, a rule with a body atom containing all variable terms will require indexing of all data, negating the benefits of the approach; e.g., if the rule OWL 2 RL/RDF rule **eq-rep-s**:

$$(?s', ?p, ?o) \leftarrow (?s, \mathtt{owl:sameAs}, ?s'), (?s, ?p, ?o)$$

is included in the assertional program, the entire corpus of assertional data must be indexed (in this case according to subject) because of the latter "open" atom. We emphasise that our partial-indexing performs well if the assertional index remains small and performs best if every proper rule in the assertional program has only one A-atom in the body—in the latter case, no assertional indexing is required. We will use this observation to identify a subset of T-split OWL 2 RL/RDF rules which are linear with respect to the assertional knowledge in § 3.5, but first we look at some generic optimisations for the assertional program.

### 3.4  Optimising the Assertional Program

Note that in Algorithm 3.1 Line 10, all rules are checked for all triples to see if an inference should take place. Given that (i) the assertional program will be applied over a corpus containing in the order of a billion triples; (ii) the process of grounding the T-atoms of T-split rules may lead to a large volume of assertional rules given a sufficiently complex terminology; we deem it worthwhile to investigate some means of optimising the execution of the assertional program. Herein, we discuss such optimisations and provide initial evaluation thereof—note that since our assertional program contains only assertional atoms, we herein omit the T-split notation where $\mathrm{Body}(R)$ always refers to a purely assertional body.

**Merging Equivalent T-ground Rules.** Applying the T-grounding of rules to derive purely assertional rules may generate "equivalent rules": rules which can be unified by an *bijective variable rewriting*. Similarly, there may exist T-ground rules with "equivalent bodies" which can be *merged* into one rule. To formalise these notions, we first define the bijective variable rewriting function used to determine equivalence of atoms.

**Definition 6 (Variable rewriting).** *A bijective variable rewriting function is an automorphism on the set of variables, given simply as:*

$$\nu : \mathsf{V} \mapsto \mathsf{V}$$

As such, this function is a specific form of variable substitution, where two atoms which are unifiable by such a rewriting are considered *equivalent*:

**Definition 7 (Equivalent atoms).** *Two atoms are* equivalent *(denoted $A_1 \lhdrhd A_2$ reflecting the fact that both atoms are instances of each other) iff they are unifiable by a bijective variable rewriting:*[20]

$$A_1 \lhdrhd A_2 \Leftrightarrow \exists \nu \text{ s.t. } A_1 \nu = A_2$$

*Equivalence of a set of atoms follows naturally. Two rules are* body-equivalent *($R_1 \lhdrhd_b R_2$) iff their bodies are equivalent:*

$$R_1 \lhdrhd_b R_2 \Leftrightarrow \mathsf{Body}(R_1) \lhdrhd \mathsf{Body}(R_2) \Leftrightarrow \exists \nu \text{ s.t. } \mathsf{Body}(R_1)\nu = \mathsf{Body}(R_2)$$

*Two rules are considered* fully-equivalent *if their bodies and heads are unifiable by the same variable rewriting:*

$$R_1 \lhdrhd_r R_2 \Leftrightarrow \exists \nu \text{ s.t. } \Big( \mathsf{Body}(R_1)\nu = \mathsf{Body}(R_2) \wedge \mathsf{Head}(R_1)\nu = \mathsf{Head}(R_2) \Big)$$

Note that fully-equivalent rules are considered redundant, and all but one can be removed without affecting the computation of the least model. Using these equivalence relations, we can now define our *rule-merge* function (again recall from § 3.1 our interpretation of multi-atom heads as being conjunctive, and a convenient representation of the equivalent set of rules):

**Definition 8 (Rule merging).** *Given an equivalence class of rules $[R]_{\lhdrhd_b}$—a set of rules between which $\lhdrhd_b$ holds—select a* canonical rule $R \in [R]_{\lhdrhd_b}$; *we can now describe the* rule-merge *of the equivalence class as*

$$\mathsf{merge}([R]_{\lhdrhd_b}) := \mathsf{Head}_{[R]_{\lhdrhd_b}} \leftarrow \mathsf{Body}(R)$$

*where*

$$\mathsf{Head}_{[R]_{\lhdrhd_b}} := \bigcup_{R_i \in [R]_{\lhdrhd_b}} \mathsf{Head}(R_i)\nu_i \text{ s.t. } \mathsf{Body}(R_i)\nu_i = \mathsf{Body}(R)$$

*Now take a program $P$ and let:*

$$P/\lhdrhd_b := \{ [R]_{\lhdrhd_b} \mid R \in P \}$$

*denote the quotient set of $P$ given by $\lhdrhd_b$: the set of all equivalent classes $[R]_{\lhdrhd_b}$ wrt. the equivalence relation $\lhdrhd_b$ in $P$. We can generalise the* rule merge *function for a set of rules as*

$$\mathsf{merge} : 2^{\mathsf{Rules}} \to 2^{\mathsf{Rules}}$$

$$P \mapsto \bigcup_{[R]_{\lhdrhd_b} \in P/\lhdrhd_b} \mathsf{merge}([R]_{\lhdrhd_b})$$

---

[20] Note that in the unification, only the variables in the left atom are rewritten and not both; otherwise two atoms such as (?a, foaf:knows, ?b) and (?b, foaf:knows, ?c) would not be equivalent: they could not be aligned by any (necessarily *injective*) rewriting function $\nu$.

*Example 4.* Take three T-ground rules:

$$(\text{?x}, \text{a}, \text{foaf:Person}) \leftarrow (\text{?x}, \text{foaf:img}, \text{?y})$$
$$(\text{?s}, \text{foaf:depicts}, \text{?o}) \leftarrow (\text{?s}, \text{foaf:img}, \text{?o})$$
$$(\text{?a}, \text{foaf:depicts}, \text{?b}) \leftarrow (\text{?a}, \text{foaf:img}, \text{?b})$$

The second rule can be merged with the first using $\nu_1 = \{\text{?s/?x}, \text{?o/?y}\}$, which gives:

$$(\text{?x}, \text{a}, \text{foaf:Person}),(\text{?x}, \text{foaf:depicts}, \text{?y}) \leftarrow (\text{?x}, \text{foaf:img}, \text{?y})$$

The third rule can be merged with the above rule using $\nu_1 = \{\text{?a/?x}, \text{?b/?y}\}$ to give:

$$(\text{?x}, \text{a}, \text{foaf:Person}),(\text{?x}, \text{foaf:depicts}, \text{?y}) \leftarrow (\text{?x}, \text{foaf:img}, \text{?y})$$

...the same rule. This demonstrates that the merge function removes redundant fully-equivalent rules. ◇

Merging the rules thus removes redundant rules, and reduces the total number of rule applications required for each triple without affecting the final least model:

**Proposition 1.** *For any program* $P$, $\text{lm}(P) = \text{lm}(\text{merge}(P))$.

*Sketch of proof available in [39].*

**Rule Index.** We have reduced the amount of rules in the assertional program through merging; however, given a sufficiently complex T-Box, we may still have a prohibitive number of rules for efficient recursive application. We now look at the use of a rule index which maps a fact to rules containing a body atom for which that fact is an instance, thus enabling the efficient identification and application of only relevant rules for a given triple.

**Definition 9 (Rule lookup).** *Given a fact $F$ and program $P$, the* rule lookup function *returns all rules in the program containing a body atom for which $F$ is an instance:*

$$\text{lookup} : \text{Facts} \times 2^{\text{Rules}} \rightarrow 2^{\text{Rules}}$$
$$(F, P) \mapsto \left\{ R \in P \mid \exists B_i \in \text{Body}(R) \text{ s.t. } B_i \rhd F \right\}$$

Now, instead of attempting to apply all rules, for each triple we can perform the above lookup function and return only triples from the assertional program which could potentially lead to a successful rule application.

*Example 5.* Given a triple:

$$t := (\text{ex:me}, \text{a}, \text{foaf:Person})$$

and a simple example ruleset:

$$P := \{(\text{?x}, \text{a}, \text{foaf:Person}) \leftarrow (\text{?x}, \text{foaf:img}, \text{?y}),$$
$$(\text{?x}, \text{a}, \text{foaf:Agent}) \leftarrow (\text{?x}, \text{a}, \text{foaf:Person}),$$
$$(\text{?y}, \text{a}, \text{rdfs:Class}) \leftarrow (\text{?x}, \text{a}, \text{?y})\}$$

lookup($t, P$) returns a set containing the latter two rules.                    ◊

With respect to implementing this lookup function, we require a rule index. A triple pattern has $2^3 = 8$ possible forms: $(?, ?, ?)$, $(s, ?, ?)$, $(?, p, ?)$, $(?, ?, o)$, $(s, p, ?)$, $(?, p, o)$, $(s, ?, o)$, $(s, p, o)$. Thus, we require eight indices for indexing body patterns, and eight lookups to perform lookup($t, P$) and find all relevant rules for a triple. We use seven in-memory hashtables storing the constants of the rule antecedent patterns as key, and a set of rules containing such a pattern as value; e.g., $\{(?\texttt{x}, \texttt{a}, \texttt{foaf:Person})\}$ is put into the $(?, p, o)$ index with $(\texttt{a}, \texttt{foaf:Person})$ as key. Rules containing $(?, ?, ?)$ patterns without constants are stored in a set, as they are relevant to all triples—they are returned for all lookups.

We further optimise the rule index by linking dependencies between rules, such that once one rule fires, we can determine which rules should fire next without requiring an additional lookup. This is related to the notion of a rule graph in Logic Programming (see, e.g., [74]):

**Definition 10 (Rule graph).** *A* rule graph *is defined as a directed graph:*

$$\Gamma := (P, \hookrightarrow)$$

*such that:*[21]

$$R_i \hookrightarrow R_j \Leftrightarrow \exists B \in \mathsf{Body}(R_j), \exists H \in \mathsf{Head}(R_i) \text{ s.t. } B \rhd H$$

*where $R_i \hookrightarrow R_j$ is read as "$R_j$ follows $R_i$".*

By building and encoding such a rule graph into our index, we can "wire" the recursive application of rules for the assertional program. However, from the merge function (or otherwise) there may exist rules with large sets of head atoms. We therefore extend the notion of the rule graph to a directed labelled graph with the inclusion of a labelling function

**Definition 11 (Rule-graph labelling).** *Let $\Lambda$ denote a labelling function as follows:*

$$\Lambda : \mathsf{Rules} \times \mathsf{Rules} \to 2^{\mathsf{Atoms}}$$

$$(R_i, R_j) \mapsto \left\{ H \in \mathsf{Head}(R_i) \mid \exists B \in \mathsf{Body}(R_j) \text{ s.t. } B \rhd H \right\}$$

*A* labelled rule graph *is thereafter defined as a directed labelled graph:*

$$\Gamma^\Lambda := (P, \hookrightarrow, \Lambda)$$

Each edge in the rule graph is labelled with $\Lambda(R_i, R_j)$, denoting the set of atoms in the head of $R_i$ that, when grounded, would be matched by atoms in the body of $R_j$.

*Example 6.* Take the two rules:

---

[21] Here, we recall from § 3.1 the '$\rhd$' notation for an instance.

$R_i.\Big|$     (y, a, foaf:Image),(?x, a, foaf:Person) ← (?x, foaf:img, ?y)

$R_j.\Big|$          (s, a, foaf:Agent) ← (?s, a, foaf:Person)

We say that $R_i \overset{\lambda}{\hookrightarrow} R_j$, where $\lambda = \Lambda(R_i, R_j) = \{(\text{?x, a, foaf:Person})\}$.          ◇

In practice, our rule index stores sets of elements of a linked list, where each element contains a rule and links to rules which are relevant for the atoms in that rule's head. Thus, for each input triple, we can retrieve all relevant rules for all eight possible patterns, apply those rules, and if successful, follow the respective labelled links to recursively find relevant rules without re-accessing the index until the next input triple.

**Rule Saturation.** We briefly describe the final optimisation technique we investigated, but which later evaluation demonstrated to be mostly disadvantageous: *rule saturation*. We say that a subset of dependencies in the rule graph are *strong dependencies*, where the successful application of one rule *will* always lead to the successful application of another. Now, we can saturate rules with single-atom bodies by pre-computing the recursive rule application of its dependencies; we give the gist with an example:

*Example 7.* Take rules

$R_i.\Big|$     (?x, a, foaf:Person),(?y, a, foaf:Image) ← (?x, foaf:img, ?y)

$R_j.\Big|$          (?s, a, foaf:Agent) ← (?s, a, foaf:Person)

$R_k.\Big|$          (?y, a, rdfs:Class) ← (?x, a, ?y)

We can see that $R_i \hookrightarrow R_j$, $R_i \hookrightarrow R_k$, $R_j \hookrightarrow R_k$ as before. Now, we can remove the links from $R_i$ to $R_j$ and $R_k$ by saturating $R_i$ to:

$R_i'.\Big|$     (?x, a, foaf:Person),(?y, a, foaf:Image),(?x, a, foaf:Agent),
      (foaf:Person, a, rdfs:Class),(foaf:Image, a, rdfs:Class),
           (foaf:Agent, a, rdfs:Class) ← (?x, foaf:img, ?y)

and, analogously, we can remove the links from $R_j$ to $R_k$ by saturating $R_j$ to:

$R_j'.\Big|$   (?s, a, foaf:Agent),(foaf:Agent, a, rdfs:Class) ← (?s, a, foaf:Person)

Thus, the index now stores $R_i', R_j', R_k$, but without the links between them.          ◇

However, as we will see in § 3.4, our empirical analysis found rule saturation to be mostly disadvantageous: although it decreases the number of necessary rule applications, as a side-effect, saturated rules can immediately produce a large batch of duplicates which would otherwise have halted a traversal of the rule graph early on. Using the above example, consider encountering the following sequence of input triples:

1.$\Big|$          (ex:Fred, a, foaf:Person)

2.$\Big|$     (ex:Fred, foaf:img, ex:FredsPic)

The first triple will fire rule $R'_j$ and $R_k$; the second triple will subsequently fire rule $R'_i$, and in so doing, will produce a superset of inferences already given by its predecessor. Without saturation, the second triple would fire $R_i$, identify (ex:Fred, a, foaf:-Person) as a duplicate, and instead only fire $R_k$ for (ex:FredsPic, a, foaf:Image).

**Preliminary Performance Evaluation.** We now perform some (relatively) small-scale experiments to empirically (in)validate our optimisations for the assertional program execution. Experiments are run on a 2.2GHz Opteron x86-64, 4GB main memory, 160GB SATA hard-disks, running Java 1.6.0_12 on Debian 5.0.4.

We applied reasoning for RDFS (minus the infinite rdf:_n axiomatic triples [33]), pD* and OWL 2 RL/RDF over LUBM(10) [30], consisting of about 1.27 million assertional triples and 295 terminological triples.[22] For each rule profile, we applied the following configurations:

1. N: **no partial evaluation**: T-Box atoms are bound at runtime from an in-memory triple-store;
2. NI: **no partial evaluation with linked (meta-)rule index**;
3. P: **partial-evaluation**: generating and applying an assertional program;
4. PI: **partial evaluation with linked rule index**;
5. PIM: **partial evaluation with linked rule index and rule merging**;
6. PIMS: **partial evaluation with linked rule index, rule merging and rule saturation**.

Table 3 enumerates the results for each profile, with a breakdown of (i) the number of inferences made, (ii) the total number of assertional rules generated, (iii) the total number of merged rules; and for each of the six configurations; (iv) the time taken, (v) the total number of attempted rule applications—i.e., the total number of times a triple is *checked* to see if it grounds a body atom of a rule to produce inferences—and the percent of rule applications which generated inferences, and (vi) the number of duplicate triples filtered out by the LRU cache (Lines 7 & 8, Algorithm 3.1).

In all approaches, applying the non-optimised partially evaluated (assertional) program takes the longest: although the partially evaluated rules are more efficient to apply, this approach requires an order of magnitude more rule applications than directly applying the meta-program, and so applying the unoptimised residual assertional program takes approximately $2\times$ to $4\times$ longer than the baseline.

With respect to rule indexing, the technique has little effect when applying the meta-program directly—many of the rules contain open patterns in the body. Although the number of rule applications diminishes somewhat, the expense of maintaining and accessing the rule index actually worsens performance by between 10% and 20%. However, with the partially evaluated rules, more variables are bound in the body of the rules, and thus triple patterns offer more selectivity and, on average, the index returns fewer rules. We see that for PI and for each profile respectively, the rule index sees a 78%, 69% and 72% reduction in the equivalent runtime (P) without the rule index; the

---

[22] Note that we exclude lg/gl rules for RDFS/pD* since we allow generalised triples [27]. We also restrict OWL 2 RL/RDF datatype reasoning to apply only to literals in the program.

**Table 3.** Details of reasoning for LUBM(10)—containing 1.27M assertional triples and 295 terminological triples—given different reasoning configurations (the most favourable result for each row is highlighted in bold)

| RDFS | | | | | | |
|---|---|---|---|---|---|---|
| inferred | 0.748 million | | | | | |
| T-ground rules | 149 | | | | | |
| after merge | 87 | | | | | |
| config. | N | NI | P | PI | PIM | PIMS |
| time (s) | 99 | 117 | 404 | 89 | 81 | **69** |
| rule apps (m) | 16.5 | 15.5 | 308 | 11.3 | 9.9 | **7.8** |
| % success | 43.4 | 46.5 | 2.4 | **64.2** | 62.6 | 52.3 |
| cache hits (m) | 10.8 | 10.8 | 8.2 | 8.2 | 8.2 | **8.1** |

| pD* | | | | | | |
|---|---|---|---|---|---|---|
| inferred | 1.328 million | | | | | |
| T-ground rules | 175 | | | | | |
| after merge | 108 | | | | | |
| config. | N | NI | P | PI | PIM | PIMS |
| time (s) | 365 | 391 | 734 | 227 | **221** | 225 |
| rule apps (m) | 62.5 | 50 | 468 | 22.9 | 21.1 | **13.9** |
| % success | 18.8 | 23.4 | 2.6 | 51.5 | 48.7 | **61.3** |
| cache hits (m) | 19.1 | 19.1 | 15.1 | 15.1 | **14.9** | 38.7 |

| OWL 2 RL/RDF | | | | | | |
|---|---|---|---|---|---|---|
| inferred | 1.597 million | | | | | |
| T-ground rules | 378 | | | | | |
| after merge | 119 | | | | | |
| config. | N | NI | P | PI | PIM | PIMS |
| time (s) | 858 | 940 | 1,690 | 474 | **443** | 465 |
| rule apps (m) | 149 | 110 | 1,115 | 81.8 | 78.6 | **75.6** |
| % success | 4.2 | 5.6 | 0.8 | 10.5 | 6.8 | **15** |
| cache hits (m) | 16.5 | 16.5 | 13.1 | 13 | **12.7** | 34.4 |

reduction in rule applications (73%, 80%, 86% reduction resp.) is significant enough to more than offset the expense of maintaining and using the index. With respect to the baseline (N), PI makes a 10%, 38% and 45% saving respectively; notably, for RDFS, the gain in performance over the baseline is less pronounced, where, relative to the more complex rulesets, the number of rule applications is not signficantly reduced by partial evaluation and indexing.

Merging rules provided a modest saving across all rulesets, with PIM giving a 9%, 3% and 6.5% saving in runtime and a 12%, 8% and 4% saving in rule applications over PI respectively for each profile. Note that although OWL 2 RL/RDF initially creates more residual rules than pD* due to expanded T-Box level reasoning, these are merged to a number just above pD*: OWL 2 RL supports intersection-of inferencing used by LUBM and not in pD*. LUBM does not contain OWL 2 constructs, but redundant meta-rules are factored out during the partial evaluation phase.

Finally, we look at the effect of saturation and approach PIMS. For RDFS, we encountered a 15% reduction in runtime over PIM, with a 21% reduction in rule applications required. However, for pD* we encountered a 2% *increase* in runtime over that of PIM despite a 34% reduction in rule applications: as previously alluded to, the cache was burdened with 2.6× more duplicates, negating the benefits of fewer rule applications. Similarly, for OWL 2 RL/RDF, we encountered a 4% increase in runtime over that of PIM despite a 4% reduction in rule applications: again, the cache encountered 2.7× more duplicates.

The purpose of this evaluation is to give a granular analysis and empirical justification for our optimisations for different rule-based profiles: one might consider different scenarios (such as a terminology-heavy corpus) within which our optimisations may not work. However, we will later demonstrate these optimisations—with the exception of rule saturation—to be propitious for our scenario of reasoning over Linked Data.

It is worth noting that—aside from reading input and writing output—we performed the above experiments almost entirely in-memory. Given the presence of (pure) assertional rules which have multi-atom bodies where one such atom is "open" (all terms are variables)—viz., pD* rule rdfp11 and OWL 2 RL/RDF rules eq-rep-*—we currently must naïvely store *all* data in memory, and cannot scale much beyond LUBM(10).[23]

### 3.5   Towards Linked Data Reasoning

With the notions of a T-split program, partial evaluation and assertional program optimisations in hand, we now reunite with our original use-case of Linked Data reasoning, for which we move our focus from clean corpora in the order of a million statements to our corpus in the order of a billion statements collected from almost four million sources—we will thus describe some trade-offs we make in order to shift up (at least) these three orders of magnitude in scale, and to be tolerant to noise and impudent data present in the corpus. More specifically, we:

1. first describe, motivate and characterise the scalable subset of OWL 2 RL/RDF that we implement based partially on the discussion in the previous section;
2. introduce and describe *authoritative reasoning*, whereby we include cautious consideration of the source of terminology into the reasoning process;
3. outline our distribution strategy for reasoning;
4. evaluate our methods by applying reasoning over our Linked Data evaulation corpus of 1.12 billion quadruples crawled from 4 million RDF/XML documents.

**"A-linear" OWL 2 RL/RDF.** Again, for a generic set of RDF rules (which do not create new terms in the head), the worst case complexity is cubic—in § 3.2 we have already demonstrated a simple example which instigates cubic reasoning for OWL 2 RL/RDF rules, and discussed how, for many reasonable inputs, rule application is quadratic.

---

[23] We could consider storing data in an on-disk index with in-memory caching; however, given the morphology and volume of the assertional data, and the frequency of lookups required, we believe that the cache hit rate would be low, and that the naïve performance of the on-disk index would suffer heavily from hard-disk latency, becoming a severe bottleneck for the reasoner.

Given our use-case, we want to define a profile of rules which will provide linear complexity with respect to the assertional data in the corpus: what we call "A-linearity".

In fact, in the field of Logic Programming (and in particular Datalog) the notion of a linear program refers to one which contains rules with no more than one recursive atom in the body—a recursive atom being one which cannot be instantiated from an inference (e.g., see [17]).[24] For Datalog, recursiveness is typically defined on the level of predicates using the notion of *intensional predicates*, which represent facts that can (only) be inferred by the program, and *extensional predicates*, which represent facts in the original data; atoms with intensional predicates are non-recursive [17]. Since we deal with a single ternary predicate, such a predicate-level distinction does not apply, but the general notion of recursiveness does. This has a notable relationship to our distinction of terminological knowledge—which we deem to be recursive only within itself (assuming standard use of the meta-vocabulary and "well-behaved equality" involving `owl:sameAs`)—and assertional knowledge which *is* recursive.

Based on these observations, we identify an A-linear subset of OWL 2 RL/RDF rules which contain only one recursive/assertional atom in the body, and apply only these rules. Taking this subset as our "meta-program", after applying our T-grounding of meta-rules during partial evaluation, the result will be a set of facts and proper rules with only one assertional atom in the body. The resulting linear assertional program can then be applied without any need to index the assertional data (other than for the LRU duplicates soft-cache); also, since we do not need to compute assertional *joins*— i.e., to find the most general unifier of multiple A-atoms in the data—we can employ a straightforward distribution strategy for applying the program.

**Definition 12 (A-linear program).** *Let $P$ be any T-split (a.k.a. meta) program. We denote the* A-linear program *of $P$ by $P^{\propto A}$ defined as follows:*

$$P^{\propto A} := \{R \in P : |\mathsf{ABody}(R)| \leq 1\}$$

*(Note that by the above definition, $P^{\propto A}$ also includes the pure-terminological rules and the facts of P.)*

Thus, the proper rules of the assertional program $AP^{\propto A}$ generated from an A-linear meta-program $P^{\propto A}$ will only contain one atom in the head. For convenience, we denote the A-linear subset of OWL 2 RL/RDF by $\mathcal{O}2\mathcal{R}^{\propto A}$, which consists of rules in Tables 15– 17 (Appendix A).

Thereafter, the assertional program demonstrates two important characteristics with respect to scalability: (i) the assertional program can be independently applied over subsets of the assertional data, where a subsequent union of the resultant least models will represent the least model achievable by application of the program over the data in whole; (ii) the volume of materialised data and the computational expense of applying the assertional program are linear with respect to the assertional data.

---

[24] There is no relation between a linear program in our case, and the field of Linear Programming [90].

**Proposition 2 (Assertional partitionability).** *Let $I$ be any interpretation, and $\{I_1, \ldots, I_n\}$ be any set of interpretations such that:*

$$I = \bigcup_{i=1}^{n} I_i$$

*Now, for any meta-program $P$, its A-linear subset $P^{\times A}$, and the assertional program $AP^{\times A}$ derived therefrom, it holds that:*

$$\mathsf{Im}(AP^{\times A} \cup I) = \bigcup_{i=1}^{n} \mathsf{Im}(AP^{\times A} \cup I_i)$$

*Proof. (Sketch)* Follows naturally from the fact that rules in $AP^{\times A}$ (i) are monotonic and (ii) only contain single-atom bodies. □

Thus, deriving the least model of the assertional program can be performed over any partition of an interpretation; the set union of the resultant least models is equivalent to the least model of the unpartitioned interpretation. Aside from providing a straightforward distribution strategy, this result allows us to derive an upper-bound on the cardinality of the least model of an assertional program.

**Proposition 3 (A-Linear least model size).** *Let $AP^{\times A}$ denote any A-linear assertional program composed of RDF proper rules and RDF facts composed of ternary-arity atoms with the ternary predicate $T$. Further, let $I^{\times A}$ denote the set of facts in the program and $PR^{\times A}$ denote the set of proper rules in the program (here, $AP^{\times A} = I^{\times A} \cup PR^{\times A}$). Also, let the function Const denote the Herbrand universe of a set of atoms (the set of RDF constants therein), and let $\tau$ denote the cardinality of the Herbrand universe of the heads of all rules in $PR^{\times A}$ (the set of RDF constants in the heads of the proper T-ground rules of $AP^{\times A}$) as follows:*

$$\tau = \left| \mathsf{Const}\left( \bigcup_{R \in PR^{\times A}} \mathsf{Head}(R) \right) \right|$$

*Finally, let $\alpha$ denote the cardinality of the set of facts:*

$$\alpha = |I^{\times A}|$$

*Then it holds that:*

$$|\mathsf{Im}(AP^{\times A})| \leq \tau^3 + \alpha(9\tau^2 + 27\tau + 27)$$

*Proof given in [39].*

Note that $\tau$ is given by the terminology (more accurately the T-Box) of the data and the terms in the heads of the original meta-program. Considering $\tau$ as a constant, we arrive at the maximum size of the least model as $c + c\alpha$: i.e., the least model is linear with respect to the assertional data. In terms of rule applications, the number of rules is again a function of the terminology and meta-program, and the maximum number of rule applications is the product of the number of rules (considered a constant) and the maximum size of the least model. Thus, the number of rule applications remains linear with respect to the assertional data. This is a tenuous result with respect to scalability, and constitutes a refactoring of the cubic complexity to separate out a static terminology. Thereafter, assuming the terminology to be small, the constant $c$ will be small and the least model will be **terse**; however, for a sufficiently complex terminology, obviously the $\tau^3$ and $\alpha\tau^2$ factors begin to dominate—for a terminology heavy program, the worst-case complexity again approaches $\tau^3$. Thus, applying an A-linear subset of a program is again not a "magic bullet" for scalability, although it should demonstrate scalable behaviour for small terminologies (i.e., where $\tau$ is small) and/or other reasonable inputs.

Moving forward, we select an A-linear subset of the OWL 2 RL/RDF ruleset for application over our ruleset. This subset is enumerated in Appendix A, with rule tables categorised by terminological and assertional arity of rule bodies. Again, we also make some other amendments to the ruleset:

1. we omit datatype rules which lead to the inference of (near-)infinite triples;
2. we omit inconsistency checking rules;
3. for reasons of **terseness**, we omit rules which infer 'tautologies'—statements that hold for every term in the graph, such as reflexive `owl:sameAs` statements (we also filter these from the output).

**Authoritative Reasoning.** In preliminary evaluation of our Linked Data reasoning [40], we encountered a puzzling deluge of inferences: We found that remote documents sometimes cross-define terms resident in popular vocabularies, changing the inferences *authoritatively* mandated for those terms. For example, we found one document[25] which defines `owl:Thing` to be an element (i.e., a subclass) of 55 union class descriptions—thus, materialisation wrt. OWL 2 RL/RDF rule **cls-uni** [27, Table 6] over any member of `owl:Thing` would infer 55 additional memberships for these obscure union classes. We found another document[26] which defines nine *properties* as the domain of `rdf:type`—again, anything defined to be a member of any class would be inferred to be a member of these nine *properties* by rules **prp-dom**. Even aside from "cross-defining" core RDF(S)/OWL terms, popular vocabularies such as FOAF were also affected (we will see more in the evaluation presented in § 3.6).

In order to curtail the possible side-effects of open Web data publishing (as also exemplified by the two triples which cause cubic reasoning in § 3.2), we include the source of data in inferencing. Our methods are based on the view that a publisher instantiating a vocabulary's term (class/property) thereby accepts the inferencing mandated by that vocabulary (and recursively referenced vocabularies) for that term. Thus, once a publisher instantiates a term from a vocabulary, only that vocabulary and its references should influence what inferences are possible through that instantiation. As such, we ignore unvetted terminology at the potential cost of discounting serendipitous mappings provided by independent parties, since we currently have no means of distinguishing "good" third-party contributions from "bad" third-party contributions. We call this more conservative form of reasoning *authoritative reasoning*, which only considers

---

[25] http://lsdis.cs.uga.edu/ oldham/ontology/wsag/wsag.owl; retr. early 2010, offline 2011/01/13.

[26] http://www.eiao.net/rdf/1.0; retr. 2011/01/13.

authoritatively published terminological data, and which we now describe. (*Please also see [84] in these proceedings for discussion on trust models for the Web.*)

Firstly, we must define the relationship between a class/property term and a vocabulary, and give the notion of *term-level authority*. We view a term as an RDF constant, and a vocabulary as a Web document: from § 3.1, we recall the get mapping from a URI (a Web location) to an RDF graph it may provide by means of a given HTTP lookup, and the redirs mapping for traversing the HTTP redirects given for a URI.

**Definition 13 (Authoritative sources for terms).** *Letting* $B(G)$ *denote the set of blank-nodes appearing in the graph* $G$, *we denote a mapping from a source URI to the set of terms it speaks* authoritatively *for as follows:*[27]

$$\mathsf{auth} : \mathsf{S} \to 2^{\mathsf{C}}$$
$$s \mapsto \{c \in \mathsf{U} \mid \mathsf{redirs}(c) = s\} \cup \mathsf{B}(\mathsf{get}(s))$$

Thus, a Web source is authoritative for URIs which dereference to it and the blank nodes it contains; for example, the FOAF vocabulary is authoritative for terms in its namespace since it follows best-practices and makes its class/property URIs dereference to an RDF/XML document defining the terms. Note that we consider all documents to be non-authoritative for all literals.

To negate the effects of non-authoritative terminological axioms on reasoning over Web data, we add an extra condition to the T-grounding of a rule (see Definition 2): in particular, we only require amendment to rules where both $\mathsf{TBody}(R) \neq \emptyset$ and $\mathsf{ABody}(R) \neq \emptyset$.

**Definition 14 (Authoritative T-ground rule instance).** *Let* $\mathsf{TAVars}(R) \subset \mathsf{V}$ *denote the set of variables appearing in both* $\mathsf{TBody}(R)$ *and* $\mathsf{ABody}(R)$, *let* $G$ *denote a graph, and let* $s$ *denote the source of that graph. Now, we define the set of authoritative T-ground rule instances for a program* $P$ *in the graph* $G$ *as:*
$$\widehat{\mathsf{Ground}}^T(P, G, s) :=$$
$$\{ \mathsf{Ground}_\theta^T(\{R\}, G) \mid R \in P^{T\emptyset} \cup P^{\emptyset A} \vee \left( R \in P^{TA} \wedge \exists v \in \mathsf{TAVars}(R) \text{ s.t. } \theta(v) \in \mathsf{auth}(s) \right) \}$$
*where* $\mathsf{Ground}_\theta^T$ *is the T-grounding (as per Definition 2) using the the most general unifier* $\theta$, *and where we recall the* $P^{TA}$, $P^{T\emptyset}$, $P^{\emptyset A}$ *conventions from Definition 3.*

The additional condition for authoritativeness states that if $\mathsf{ABody}(R) \neq \emptyset$ and $\mathsf{TBody}(R) \neq \emptyset$, then the unifier $\theta$ must substitute at least one variable appearing in both $\mathsf{ABody}(R)$ and $\mathsf{TBody}(R)$ for an authoritative term (wrt. source $s$)—i.e., source $s$ must speak authoritatively for a term that necessarily appears in each instance of $\mathsf{ABody}(R)$, and cannot create rule instances which could apply over arbitrary assertional data not mentioning any of its terms. We now formalise this notion:

**Theorem 4 (Authoritative reasoning guarantee).** *Let* Const *denote a function which returns the Herbrand universe of a set of rules (including facts): i.e., a function which*

---

[27] Even predating Linked Data, dereferencable vocabulary terms were encouraged; cf. `http://www.w3.org/TR/2006/WD-swbp-vocab-pub-20060314/`; retr. 2011/01/13.

*returns the set of RDF constants appearing in a program P or a graph G. Next, let G'*
*be any graph, let s' be the source of graph G' such that* get$(s') = G'$, *and let P be any*
*(T-split) program and G be any graph such that*

$$\mathsf{Const}(P \cup G) \cap \mathsf{auth}(s') = \emptyset\,;$$

*i.e., neither P nor G contain any terms for which s' speaks authoritatively. Finally, let*
*P' be the set of partially evaluated rules derived from G with respect to P, where:*

$$P' := \{R \in \widehat{\mathsf{Ground}}^T(P, G', s') | \mathsf{Body}(R) \neq \emptyset\}$$

*Now, it holds that* $\mathsf{lm}(P \cup G) = \mathsf{lm}(P \cup P' \cup G)$.

**Corollary 2.** *Given the same assumption(s) as Theorem 4, it also holds that* $\mathsf{lm}^T(P \cup G) = \mathsf{lm}^T(P \cup P' \cup G)$.

*Proofs available in [39].*

*Example 8.* Take the T-split rule $R_{EX}$ as before:

$$(?\mathsf{x}, \mathsf{a}, ?\mathsf{c2}) \leftarrow \underline{(?\mathsf{c1}, \mathsf{rdfs:subClassOf}, ?\mathsf{c2})}, (?\mathsf{x}, \mathsf{a}, ?\mathsf{c1})$$

and let $G_{EX}$ be the graph from source $s$:

$$G_{EX} := \{\,(\mathsf{foaf:Person}, \mathsf{rdfs:subClassOf}, \mathsf{foaf:Agent}),$$
$$(\mathsf{foaf:Agent}, \mathsf{rdfs:subClassOf}, \mathsf{dc:Agent})\,\}$$

Here, $\mathsf{TAVars}(R_{EX}) = \{?\mathsf{c1}\}$. Now, for each substitution $\theta$, there must exist $v \in \mathsf{TAVars}(R_{EX})$ such that $s$ speaks authoritatively for $\theta(v)$. In this case, $s$ must speak authoritatively for the $?\mathsf{c1}$ substitution $\mathsf{foaf:Person}$ for the rule:

$$(?\mathsf{x}, \mathsf{a}, \mathsf{foaf:Agent}) \leftarrow (?\mathsf{x}, \mathsf{a}, \mathsf{foaf:Person})$$

to be an authoritatively T-ground rule instance, and speak authoritatively for the $?\mathsf{c1}$ substitution $\mathsf{foaf:Agent}$ for:

$$(?\mathsf{x}, \mathsf{a}, \mathsf{dc:Agent}) \leftarrow (?\mathsf{x}, \mathsf{a}, \mathsf{foaf:Agent})$$

to be authoritative. In other words, for these T-ground rules to be authoritative, $G_{EX}$ must be served by the document referenced by the FOAF terms—i.e., the FOAF vocabulary. Note that this authoritatively ground rule contains the term $\mathsf{foaf:Agent}$ in the body, and thus can only generate inferences over graphs containing this term (for which $s$ is authoritative). ◇

For reference, we highlight variables in $\mathsf{TAVars}(R)$ with boldface in the rule tables of Appendix A (only applies to rules with A-atoms *and* T-atoms in the body).

It is worth noting that for rules where $\mathsf{ABody}(R)$ and $\mathsf{TBody}(R)$ are both non-empty, authoritative instantiation of the rule will only consider unifiers for $\mathsf{TBody}(R)$ which come from one source: however, in practice for OWL 2 RL/RDF this is not so restrictive: although $\mathsf{TBody}(R)$ may contain multiple atoms, in such rules $\mathsf{TBody}(R)$ usually refers to an atomic axiom which requires multiple triples to represent—indeed, the OWL 2 Structural Specification [64] enforces usage of blank-nodes and cardinalities on such constructs to ensure that the constituent triples of the multi-triple axiom appear in one source. To take an example, for the T-atoms:

$$(?x, owl:hasValue, ?y)$$
$$(?x, owl:onProperty, ?p)$$

we would expect ?x to be ground by a blank-node skolem and thus expect the instance to come from one graph. Although it should be noted that such restrictions do not carry over for OWL 2 Full—which is applicable for arbitrary RDF graphs—it still seems reasonable for us to restrict those OWL 2 Full terminological axioms which require multiple triples to express to be given entirely within one Web document (here, perhaps even making our reasoning more robust). In any case, as we say in § 2.2, such features of OWL are not so commonly adopted for Linked Data.

Note finally that terminological inferences—produced by rules with only T-atoms—are never considered authoritative. Thus, by applying authoritative reasoning, we do not T-ground rules from such facts. For OWL 2 RL/RDF, this only has a "minor" effect on the least model computation since OWL 2 RL/RDF (intentionally) contains redundant rules [27], which allow for deriving the same inferences on a purely assertional level. Along these lines, in Appendix A, Table 22, we list all of the T-atom only rules; assuming that the inferences given by each rule are not considered terminological, we show how the omissions are covered by the recursive application of other assertional rules. We note that we may miss some inferences possible through inference of rdfs:subClassOf relations between owl:someValuesFrom restriction classes, and also between owl:allValuesFrom restriction classes, since we do not support the respective assertional rules cls-svf1 and cls-avf.

**Distributed Reasoning.** As previously mentioned, Proposition 2 lends itself to a straightforward distribution strategy for applying our A-linear OWL 2 RL/RDF subset. We briefly discuss our distribution strategy, where we use one master machine to compute and coordinate "global knowledge" and use several slave machines to perform tasks in parallel over the bulk of the corpus. We assume that all machines are in a shared-nothing configuration [79]; we also assume that the corpus is evenly split over the slave machines in preparation for reasoning (in our setting, this is the direct result of our distributed crawler), and that the slave machines have roughly even specifications. For more information about our distribution architecture, we refer the interested reader to [39, § 3.6].

As a first step for the distributed reasoning, we extract the T-Box data from each machine, use the master machine to execute the terminological program and create the residual assertional program, and then distribute this assertional program (the proper rules) to each slave machine and let it apply the program independently (and in parallel) over its local segment of the corpus. This process is summarised as follows:

1. **parallel:** identify and separate out the T-Box from the main corpus in parallel on the slave machines;
2. **local:** the master machine then
   (a) gathers and merges the T-Box segments from the slave machines;
   (b) generates axiomatic triples from the meta-program and applies T-atom only rules over the T-Box;
   (c) authoritatively grounds the T-atoms in rules with one A-atom, thus generating the A-linear assertional program;

(d) optimises the assertional program by merging rules and building a linked rule index;

3. **parallel:** send the assertional linked rule index to all slave machines and reason over the main corpus in parallel on each machine.

The results of the above three-step operation are: (a) axiomatic triples and terminological inferences resident on the master machine; and (b) assertional inferences split over the slave machines. Note further that the output of this process may contain (both local and global) duplicates.

### 3.6 Linked Data Reasoning Evaluation

We now give evaluation of applying our subset of OWL 2 RL/RDF over the 1.12b quads (947m unique triples) of Linked Data crawled in the previous section. Note that we also require information about redirects encountered in the crawl to reconstruct the redirs function required for authoritative reasoning (see Definition 13) and that we output a flat file of G-Zipped triples. All of our evaluation is based on nine machines (1 master/8 slaves) connected by Gigabit ethernet[28], each with uniform specifications; viz.: 2.2GHz Opteron x86-64, 4GB main memory, 160GB SATA hard-disks, running Java 1.6.0_12 on Debian 5.0.4. Please note that much of the evaluation presented in these notes assumes that the slave machines have roughly equal specifications in order to ensure that tasks finish in roughly the same time, assuming even data distribution.

**Survey of Terminology.** In [39], we presented an analysis—similar to that presented in § 2.3—of the use of RDFS and OWL in the terminology given by our corpus, but specifically with respect to the ruleset we apply. We refer [39] for the details, but in summary we found that our A-linear rules support 99.3% of the total T-ground rules generated from the terminology in the Linked Data corpus, and *authoritative* reasoning with respect to these rules supports 81.7% of the total; excluding one document from the ontologydesignpatterns.org domain which publishes 61,887 non-authoritative axioms, the latter percentage increases to 95.1%. Our authoritative A-linear rules fully support (with respect to OWL 2 RL/RDF rules) 90.6% of the documents containing unique terminology, and partially support 99% of these documents. The summation of the ranks of documents fully supported by our A-linear rules was 77% of the total, and the analogous percentage for documents supported by authoritative reasoning over these rules was 70.3% of the total. We found that the top-ranked document serving non-authoritative axioms was FOAF (#7), which recently added an owl:equivalentClass assertion between foaf:Agent and dct:Agent, and an owl:equivalentProperty assertion between foaf:maker and dct:creator (in effect, our authoritative reasoning algorithm would treat axioms these as uni-directions sub-class/-property mappings from FOAF to DC).

**Authoritative Reasoning.** In [39], we also compared the effects of authoritative vs. non-authoritative reasoning for our corpus. We refer [39] for the details, but in summary

---

[28] We observe, e.g., a max FTP transfer rate of 38MB/sec between machines.

we found that for the instance data of the top five most popular classes and properties, non-authoritative inference sizes are on average 55.46× larger than the authoritative equivalent. Much of this is attributable to noise in and around core RDF(S)/OWL terms, in particular `rdf:type`, `owl:Thing` and `rdfs:Resource`;[29] without these core terms, non-authoritative inferencing creates 12.74× more inferences than the authoritative equivalent.

We present a selected example for the most popular class in our data: `foaf:Person`. Excluding the top-level concepts `rdfs:Resource` and `owl:Thing`, and the inferences possible therefrom, each `rdf:type` triple with `foaf:Person` as value leads to (at least) five authoritative inferences and twenty-six *additional* non-authoritative inferences (all class memberships). Of the latter twenty-six, fourteen are anonymous classes. Table 4 enumerates the five authoritatively-inferred class memberships and the remaining twelve non-authoritatively inferred *named* class memberships; also given are the occurrences of the class as a value for `rdf:type` in the raw data. Although we cannot claim that all of the additional classes inferred non-authoritatively are *noise*—although classes such as `b2r2008:Controlled_vocabularies` appear to be—we can see that they are infrequently used and arguably *obscure*. Although some of the inferences we omit may of course be serendipitous—e.g., perhaps `po:-Person`—again we currently cannot distinguish such cases from noise or blatant spam; for reasons of **robustness** and **terseness**, we conservatively omit such inferences.

**Single-machine Reasoning.** We first applied authoritative reasoning on one machine: reasoning over the dataset described inferred 1.58 billion raw triples, which were filtered to 1.14 billion triples removing non-RDF generalised triples and tautological statements (see § 3.2)—post-processing revealed that 962 million (∼61%) were unique and had not been asserted (roughly a 1:1 *inferred:asserted* ratio). The first step—extracting 1.1 million T-Box triples from the dataset—took 8.2 h.

Subsequently, Table 5 gives the results for reasoning on one machine for each approach outlined in § 3.4. T-Box level processing—e.g., applying terminological rules, partially evaluation, rule indexing, etc.—took roughly the same time (∼9 min) for each approach. During the partial evaluation of the meta-program, 301 thousand assertional rules were created with 2.23 million links; these were subsequently merged down to 216 thousand (71.8%) with 1.15 million (51.6%) links. After saturation, each rule has an average of 6 atoms in the head and all links are successfully removed; however, the saturation causes the same problems with extra duplicate triples as before, and so the fastest approach is PIM, which takes ∼15% of the time for the baseline N algorithm. Note that with 301 thousand assertional rules and without indexing, applying all rules to all statements—roughly 750 trillion rule applications—would take approximately 19 years. In Figure 1, we also show the linear performance of the fastest approach: PIM (we would expect all methods to be similarly linear).

---

[29] We note that much of the noise is attributable to 107 terms from the `opencalais.com` domain; cf.
`http://d.opencalais.com/1/type/em/r/PersonAttributes.rdf`     (retr. 2011/01/22) and `http://groups.google.com/group/pedantic-web/browse_thread/thread/5e5bd42a9226a419` (retr. 2011/01/22).

**Table 4.** Breakdown of non-authoritative and authoritative inferences for `foaf:Person`, with number of appearances as a value for `rdf:type` in the raw data

| Class | (Raw) Count |
|---|---|
| ***Authoritative*** | |
| `foaf:Agent` | 8,165,989 |
| `wgs84:SpatialThing` | 64,411 |
| `contact:Person` | 1,704 |
| `dct:Agent` | 35 |
| `contact:SocialEntity` | 1 |
| ***Non-Authoritative*** (additional) | |
| `po:Person` | 852 |
| `wn:Person` | 1 |
| `aifb:Kategorie-3AAIFB` | 0 |
| `b2r2008:Controlled_vocabularies` | 0 |
| `foaf:Friend_of_a_friend` | 0 |
| `frbr:Person` | 0 |
| `frbr:ResponsibleEntity` | 0 |
| `pres:Person` | 0 |
| `po:Category` | 0 |
| `sc:Agent_Generic` | 0 |
| `sc:Person` | 0 |
| `wn:Agent-3` | 0 |

**Distributed Reasoning.** We also apply reasoning over 1, 2, 4 and 8 slave machines using the distribution strategy outlined in § 3.5; Table 6 gives the performance. Note that the most expensive aspects of the reasoning process—extracting the T-Box from the dataset and executing the assertional program—can be executed in parallel by the slave machines without coordination. The only communication required between the machines is during the aggregation of the T-Box and the subsequent partial evaluation and creation of the shared assertional-rule index: this takes ∼10 min, and becomes the lower bound for time taken for distributed evaluation with arbitrary machine count.

In summary, taking our best performance, we apply reasoning over 1.12 billion Linked Data triples in 3.35 h using 9 machines (1 master/8 slaves), deriving 1.58 billion inferred triples, of which 962 million are novel and unique.

## 3.7   Related Work

Herein, we discuss related works specifically in the field of scalable and distributed reasoning as well as works in the area of robust Web reasoning.

**Scalable/Distributed Reasoning.** From the perspective of scalable RDF(S)/OWL reasoning, one of the earliest engines to demonstrate reasoning over datasets in the order of a billion triples was the commercial system BigOWLIM [13], which is based on a scalable and custom-built database management system over which a rule-based materialisation layer is implemented, supporting fragments such as RDFS and pD*, and

**Table 5.** Performance for reasoning over 1.1 billion statements on one machine for all approaches

|        | T-Box (min) | A-Box (hr)         |
|--------|-------------|--------------------|
| N      | 8.9         | 118.4              |
| NI     | 8.9         | 121.3              |
| P      | 8.9         | 171609[a]          |
| PI     | 8.9         | 22.1               |
| PIM    | 8.9         | 17.7               |
| PIMS   | 8.9         | 19.5               |

[a] Estimated as a linear product from one day of reasoning.

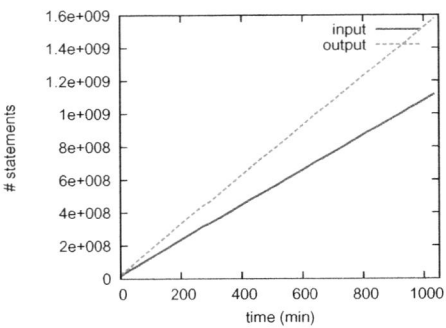

**Fig. 1.** Detailed throughput performance for application of assertional program using the fastest approach: PIM

**Table 6.** Distributed reasoning in *minutes* using PIM for 1, 2, 4 & 8 slave machines

| Machines | Extract T-Box | Build T-Box | Reason A-Box | Total |
|----------|---------------|-------------|--------------|-------|
| 1        | 492           | 8.9         | 1062         | 1565  |
| 2        | 240           | 10.2        | 465          | 719   |
| 4        | 131           | 10.4        | 239          | 383   |
| 8        | 67            | 9.8         | 121          | 201   |

more recently OWL 2 RL/RDF. Most recent results claim to be able to load 12 billion statements of the LUBM synthetic benchmark, and 20.5 billion statements statements inferrable by pD* rules on a machine with 2x Xeon 5430 (2.5GHz, quad-core), and 64GB (FB-DDR2) of RAM.[30] We note that this system has been employed for relatively high-profile applications, including use as the content management system for a live BBC World Cup site.[31] BigOWLIM features distribution, but only as a replication strategy for fault-tolerance and supporting higher query load.

A number of scalable and distributed reasoners adopt a similar approach to SAOR.

Weaver and Hendler [93] discuss a similar approach for distributed materialisation with respect to RDFS—they also describe a separation of terminological (what they call ontological) data from assertional data. Thereafter, they identify that all RDFS rules have only one assertional atom and, like us, use this as the basis for a scalable distribution strategy: they flood the ontological data and split the assertional data over their machines. They demonstrate the completeness of their approach—arriving to a similar conclusion to us—but by inspection of the RDFS fragment. Inferencing is done over an in-memory RDF store. They evaluate their approach over a LUBM-generated synthetic corpus of 345.5 million triples using a maximum of 128 machines (each with

---

[30] http://www.ontotext.com/owlim/benchmarking/lubm.html; retr. 2011/01/22

[31] http://www.readwriteweb.com/archives/bbc_world_cup_website_semantic_technology.php; retr. 2012/01/22

two dual-core 2.6 GHz AMD Opteron processors and 16 GB memory); with this setup, reasoning in memory takes just under 5 minutes, producing 650 million triples.

Similarly following our earlier work on SAOR, Urbani et al. [89] use MapReduce [19] for distributed RDFS materialisation over 850m Linked Data triples. They also consider a separation of terminological (what they call schema) data from assertional data as a core optimisation of their approach, and—likewise with [93]—identify that RDFS rules only contain one assertional atom. As a pre-processing step, they sort their data by subject to reduce duplication of inferences. Based on inspection of the rules, they also identify an ordering (stratification) of RDFS rules which (again assuming standard usage of the RDFS meta-vocabulary) allows for completeness of results without full recursion—unlike us, they do reasoning on a per-rule basis as opposed to our per-triple basis. Unlike us, they also use a 8-byte dictionary encoding of terms. Using 32 machines (each with 4 cores and 4 GB of memory) they infer 30 billion triples from 865 million triples in less than one hour; however, they do not materialise or *decode* the output—a potentially expensive process. Note that they do not include any notion of authority (although they mention that in future, they may include such analysis): they attempted to apply pD* on 35 million Web triples and stopped after creating 3.8 billion inferences in 12 h, lending strength to our arguments for authoritative reasoning.

In more recent work, (approximately) the same authors [88] revisit the topic of materialisation with respect to pD*. They again use a separation of terminological data from assertional data, but since pD* contains rules with multiple assertional atoms, they define bespoke MapReduce procedures to handle each such rule, some of which are similar in principle to those presented in [40] (and later on) such as canonicalisation of terms related by `owl:sameAs`. They demonstrate their methods over three datasets; (i) 1.51 billion triples of UniProt data, generating 2.03 billion inferences in 6.1 h using 32 machines; (ii) 0.9 billion triples of LDSR data, generating 0.94 billion inferences in 3.52 h using 32 machines; (iii) 102.5 billion triples of LUBM, generating 47.6 billion inferences in 45.7 h using 64 machines. The latter experiment is two orders of magnitude above our current experiments, and features rules which require A-Box joins; however, the authors do not look at open Web data, stating that:

> "[...] reasoning over arbitrary triples retrieved from the Web would result in useless and unrealistic derivations."

—[88]

They do, however, mention the possibility of including our authoritative reasoning algorithm in their approach, in order to prevent such adverse affects.

In very recent work, [54] have presented an (Oracle) RDBMS-based OWL 2 RL/RDF materialisation approach. They again use some similar optimisations to the scalable reasoning literature, including parallelisation, canonicalisation of `owl:sameAs` inferences, and also partial evaluation of rules based on highly selective patterns—from discussion in the paper, these selective patterns seem to correlate with the terminological patterns of the rule. They also discuss many low-level engineering optimisations and Oracle tweaks to boost performance. Unlike the approaches mentioned thus far, [54] tackle the issue of updates, proposing variants of semi-naïve evaluation to avoid rederivations. The authors evaluate their work for a number of different datasets and hardware configurations; the largest scale experiment they present consists of applying OWL 2 RL/RDF materialisation over 13 billion triples of LUBM using 8 nodes (Intel Xeon 2.53 GHz CPU, 72GB memory each) in just under 2 hours.

**Web Reasoning.** As previously mentioned, [89] discuss reasoning over 850m Linked Data triples—however, they only do so over RDFS and do not consider any issues relating to provenance.

Kiryakov et al. [52] apply reasoning over 0.9 billion Linked Data triples using the aforementioned BigOWLIM reasoner; however, this dataset is manually selected as a merge of a number of smaller, known datasets as opposed to an arbitrary corpus—they do not consider any general notions of provenance or Web tolerance. (Again, Urbani et al. [88] also apply reasoning over the LDSR dataset.)

Related to the idea of authoritative reasoning is the notion of "conservative extensions" described in the Description Logics literature (see, e.g., [23,59,48]). However, the notion of a "conservative extension" was defined with a slightly different objective in mind: according to the notion of deductively conservative extensions, a dataset $G_a$ is only considered malicious towards $G_b$ if it causes additional inferences with respect to the intersection of the *signature*—loosely, the set of classes and properties defined in the dataset's namespace—of the original $G_b$ with the newly inferred statements. Thus, for example, defining ex:moniker as a *super*-property of foaf:name outside of the FOAF spec would be "disallowed" by our authoritative reasoning: however, this would still be a conservative extension since no new inferences using FOAF terms can be created.

Work presented by [16] use a notion of an *authoritative description* which aligns very much with our notion of authority. They use their notion of authority to do reasoning over class hierarchies, but only include custom support of rdfs:subClassOf and owl:equivalentClass, as opposed to our general framework for authoritative reasoning over arbitrary T-split rules.

A viable alternative approach—which looks more generally at provenance for Web reasoning—is that of "quarantined reasoning", described by Delbru et al.[20] and employed by Sindice [69]. The core intuition is to consider applying reasoning on a per-document basis, taking each Web document and its recursive (implicit and explicit) imports and applying reasoning over the union of these documents. The reasoned corpus is then generated as the merge of these per-document closures. Their evaluation was performed in parallel using three machines (quad-core 2.33GHz CPU with 8GB memory each); they reported loading, on average, 40 documents per second.

### 3.8    Critical Discussion and Future Directions

Herein, we have demonstrated that materialisation with respect to a carefully selected—but still inclusive—subset of OWL 2 RL/RDF rules is currently feasible over large corpora (in the order of a billion triples) of arbitrary RDF data collected from the Web; in order to avoid creating a massive bulk of inferences and to protect popular vocabularies from third-party interference, we include analyses of the source of terminological data into our reasoning, conservatively ignoring third-party contributions and only considering first-party definitions and alignments. Referring back to our motivating foaf:page example in the introduction, we can now get the same answers for the simple query if posed over the union of the input and inferred data as for the extended query posed over only the input data.

We do however identify some shortcomings of our approach. Firstly, the scalability of our approach is predicated on the assumption that the terminological fragment of the corpus remain relatively small and simple—as we have seen in § 3.6, this holds true for our current Linked Data corpus. The further from this assumption we get, the closer we get to quadratic (and possibly cubic) materialisation on a terminological level, and a high $\tau$ "multiplier" for the assertional program. Thus, the future feasibility of our approach for the Web (in its current form) depends on the assumption that assertional data dwarves terminological data. We note that almost all highly-scalable approaches in the literature currently rely on a similar premise to some extent, especially for partial-evaluation and distribution strategies.

Secondly, we adopt a very conservative authoritative approach to reasoning which may miss some interesting inferences given by independently published mappings: although we still allow one vocabulary to map its local terms to those of an external vocabulary, we thus depend on each vocabulary to provide all useful mappings in the dereferenced document. In future work, it would be worthwhile to investigate identifying "trusted" third-party mappings in the wild, perhaps based on links-analysis or observed adoption.

Thirdly, thus far we have not considered rules with more than one A-atom—rules which could, of course, lead to useful inferences for our query-answering use-case. Many such rules—for example supporting property-chains, transitivity or equality—can naïvely lead to quadratic inferencing with respect to many reasonable corpora of assertional data. As previously discussed, a backward-chaining or hybrid approach may often make more sense in cases where materialisation produces too many inferences; in fact, we discuss such an approach for equality reasoning in [39]. Note however that not all multiple A-atom rules can produce quadratic inferencing with respect to asser-tional data: some rules (such as cls-int1, cls-svf1) are what we call *A-guarded*, whereby (loosely) the head of the rule contains only one variable not ground by partial evalua-tion with respect to the terminology, and thus we posit that such rules also abide by our maximum least-model size for A-linear programs (these are highlighted in Table 20). Despite this, such rules would not fit neatly into our distribution framework (would not be conveniently partitionable), where assertional data must then be coordinated be-tween machines to ensure correct computation of joins (such as in [88]); similarly, some variable portion of assertional data must also be indexed to compute these joins.

Finally, despite our authoritative analysis, reasoning may still introduce significant noise and produce unwanted or unintended consequences; in particular, publishers of assertional data are sometimes unaware of the precise semantics of the vocabulary terms they use. An interesting avenue to explore would be non-standard reasoning approaches (e.g., using statistical models or inductive reasoning) as an alternative or complement to the standard approaches presented herein. (*Please see [80] in these proceedings for discussion on combining probabalistic and logical reasoning for Web data; see [11] in these proceedings for an introduction to scalable non-standard reasoning for the Semantic Web; also, see [31] in these proceedings for discussion on building models for the Web of Data*).

Along similar lines, in [39], we looked at a use-case for annotated reasoning whereby we rank triples in the input data (based on a PageRank analysis of the sources of data) and propagate these ranks to inferences through the annotated reasoning framework. Thereafter, we perform a granular repair of inconsistencies, with the core approach being to removing the weakest triple causing the underlying inconsistency. We refer the interested reader to [39] for more detail.

## 4 Scalable Approximative OWL 2 DL Reasoning

Ontologies have been so phenomenally successful, as a machine-understandable compilation of human knowledge, that OWL2 (the second version of OWL) is recently standardised by W3C. As more and more large ontologies become available [70], there is a pressing need for efficient and robust reasoning services. Such reasoning services will help us gain insight of the semantic relations among vocabularis of ontologies and facilitate further processings such as the materialisation that we introduced in § 3.1.

Expressive Description Logics (DLs) [5] have high worst case computational complexity. For example, TBox (terminological box) reasoning in the DL $\mathcal{SROIQ}$ [44], the adjacent logic of OWL2-DL, is N2ExpTime-complete [50]. Mainstream reasoners for expressive DLs provide reasoning services, such as classification (computing subsumption relations among all the named concepts), based on tableau [45] and hypertableau [65] algorithms. Such *model constructing* algorithms classify an ontology, in general, by iterating all necessary pairs of concepts, and trying to construct a model of the ontology that violates the subsumption relation between them [51]. On the other hand, light-weight DLs can have very efficient reasoning algorithms. For example, TBox reasoning in $\mathcal{EL}^{++}$ [3], the logic underpinning of an OWL2 tractable profile OWL2-EL, is PTime-complete. However, they only provide limited expressive power.

This brings a new challenge: can users use OWL2-DL to build their ontologies and still enjoy the efficient reasoning as in tractable profiles? For example, the Foundational Model of Anatomy ontology (FMA) , which is built in $\mathcal{ALCOIF}$, beyond any tractable DLs, can hardly be classified by any mainstream DL reasoners [66]. Given the current efforts of ontology construction, it might not take long before many other FMA-like (or even larger and more complicated) ontologies appear and go beyond the capability of existing DL reasoners.

Approximation [81,28,38,92,72] has been identified as a potential way to reduce the complexity of ontology reasoning. However, many of these approximation approaches still rely on the reasoners of the more expressive DLs. For example, [28] replaces certain parts of a concept expression with $\top$ or $\bot$ to obtain a simpler expression that can be classified more easily with a tableau reasoner. [72] requires the use of reasoners of the more expressive DLs to pre-compute the entailments to achieve efficient online performance. Furthermore, most of the above approaches are on ABox reasoning and query answering. To the best of our knowledge, the only approach on TBox reasoning is [28], which presents an overview of approximation approaches (including language weakening, knowledge compilation and approximate deduction), as well as investigating and reporting negative results of the approximate deduction approach – a problematic side effect of using their approximate deduction approach is that the collapsing of concept expressions leading to many unnecessary approximation steps.

In this section, we propose to combine the idea of language weakening and approximate deduction [28] into soundness preserving approximation for ontology TBox reasoning.

1. After an informative discussion of the technical challenges (§ 4.1), we propose a syntactic language weakening approach (§ 4.3, § 4.4 and § 4.5) to approximating an arbitrary $\mathcal{SROIQ}$ TBox with a corresponding $\mathcal{EL}^{++}$ TBox and additional data structures maintaining the complementary information and cardinality information. It is shown that the proposed approximation is in linear time (Lemma 1, 2 and 3).
2. We present soundness-guaranteed approximate deduction rules to classify the approximated TBox (§ 4.4 and § 4.5). In contrast to the twisted trade-off between tractability and expressiveness, our approach compromises the completeness of reasoning to yield large portion of logical consequences in polynomial time while imposing no restrictions on expressivity of the language used in source ontologies and preserve correctness of results (§ 4.6).
3. We present our implementation and preliminary evaluations (§ 4.7). Evaluation against a set of real world ontologies [70] suggested that, a naive implementation of our approach can (i) outperform existing OWL2-DL reasoners such as Pellet and FaCT++, and (ii) provide rather complete results with high recall (over 95% for $\mathcal{EL}_C^{++}$ and over 99% for $\mathcal{EL}_{CQ}^{++}$, where $\mathcal{EL}_C^{++}$ and $\mathcal{EL}_{CQ}^{++}$ are two more and more fine-grained approximation).

Proofs of all propositions, lemmas and theorems can be found in tech report available at http://www.box.net/shared/nm913g22ie.

## 4.1   Technical Motivations

In order to motivate our investigation on syntactic approximation of $\mathcal{SROIQ}$ ontologies to $\mathcal{EL}^{++}$ ontologies, this section first briefly introduces $\mathcal{SROIQ}$ and $\mathcal{EL}^{++}$ and then illustrates the technical challenges in their TBox reasoning and approximation.

In $\mathcal{SROIQ}$, concept $C$, $D$ can be inductively composed with the following constructs:

$$\top \mid \bot \mid A \mid C \sqcap D \mid \exists R.C \mid \{a\} \mid \neg C \mid \ \geq nR.C \mid \exists R.Self$$

where $\top$ is the top concept, $\bot$ the bottom concept, $A$ atomic concept, $n$ an integer number, $a$ an individual, $\exists R.Self$ the self-restriction and $R$ a role that can be either an atomic role $r$ or the inverse of another role ($R^-$). Conventionally, $C \sqcup D, \forall R.C$ and $\leq nR.C$ are used to abbreviate $\neg(\neg C \sqcap \neg D)$, $\neg \exists R.\neg C$ and $\neg \geq (n+1)R.C$, respectively. $\{a_1, a_2, \ldots, a_n\}$ can be regarded as abbreviation of $\{a_1\} \sqcup \{a_2\} \sqcup \cdots \sqcup \{a_n\}$. Without loss of generality, in what follows, we assume all the concepts to be in their negation normal forms (NNF)[32] and use ~$C$ to denote the NNF of $\neg C$. We also call $\top, \bot, A, \{a\}$ *basic concepts* because they are not composed by other concepts or roles. Given a TBox $T$, we use $CN_T$ ($RN_T$) to denote the set of basic concepts (atomic roles) in $T$. The $\mathcal{EL}$ family is dedicated for large TBox reasoning and has been widely applied in some largest ontologies, e.g. SNOMED [78]. $\mathcal{EL}^{++}$ supports

---

[32] An $\mathcal{SROIQ}$ concept is in NNF *iff* negation is applied only to atomic concepts, nominals or Self-restriction. NNF of a given concept can be computed in linear time[43].

$$\top \mid \bot \mid A \mid C \sqcap D \mid \exists r.C \mid \{a\}.$$

Both $\mathcal{SROIQ}$ and $\mathcal{EL}^{++}$ support concept inclusions (CIs, e.g. $C \sqsubseteq D$) and role inclusions (RIs, e.g. $r \sqsubseteq s$, $r_1 \circ \cdots \circ r_n \sqsubseteq s$). $\mathcal{SROIQ}$ supports also other axioms such asymmetric of roles. If $C \sqsubseteq D$ and $D \sqsubseteq C$, we write $C \equiv D$. If $C$ is non-atomic, $C \sqsubseteq D$ is a general concept inclusion (GCI). For more details about syntax and semantics of DLs, we refer the readers to [76] in these proceedings and [5].

A TBox is a set of concept and role axioms. TBox reasoning services include concept subsumption checking, concept satisfiability checking (to check if a given concept is instantiatable ) and classification (to compute the concept hierarchy). For example, given the following TBox $\mathcal{T}_1$ (in $\mathcal{ALC}$), we can infer $Koala \sqsubseteq Herbivore$.

*Example 9.* An example TBox $\mathcal{T}_1$.

- $\alpha_1 : Koala \sqsubseteq \forall eat.(\exists partof.Eucalypt)$
- $\alpha_2 : Eucalypt \sqsubseteq Plant$
- $\alpha_3 : Plant \sqcup \exists partof.Plant \sqsubseteq VegeFood$
- $\alpha_4 : \forall eat.VegeFood \sqsubseteq Herbivore$

The tableau algorithm [45] constructs a tableau (as a witness of a model of the TBox $\mathcal{T}_1$) as a graph in which each node $x$ represents an individual and is labelled with a set of concepts it must satisfy, each edge $\langle x, y \rangle$ represents a pair of individuals satisfying a role that labels the edge. Subsumption checking $C \sqsubseteq D$ can be reduced to unsatisfiability checking $C \sqcap \neg D \sqsubseteq \bot$. To test this, a tableau is initialised with a single node labelled with $C \sqcap \neg D$, and is then expanded by repeatedly applying the completion rules [45].

One of the major difficulties for tableau algorithms is the high degree of non-determinism introduced by GCIs. For each GCI $C \sqsubseteq D$ in the ontology, the algorithm generates a meta-constraint $\neg C \sqcup D$ for each node of the tableau. The algorithm first extends a node with $\neg C$. If it finds a clash, it backtracks and extends the node with $D$. If there are $n$ GCIs, this expands to $2^n$ combinations for each node of the tableau. This significantly enlarges the search space.

Some techniques have been developed to deal with GCIs. *Absorption* [86] can reduce, e.g. a GCI $A \sqcap C \sqsubseteq D$, where $A$ is a named concept, into non-GCI $A \sqsubseteq \neg C \sqcup D$; however, it is only applicable for GCIs whose LHS is a conjunction with a named concept as conjunct or whose RHS is a negated named concept or a disjunction with a negated named concept as disjunct. *(Extended) Role Absorption* [85, Sec.4.1] can absorb GCIs of form $\exists r.C \sqsubseteq D$ $(C \sqsubseteq \forall r.D)$ into domain (range) constrains. For example $\alpha_3$ can be decomposed into $\exists partof.Plant \sqsubseteq VegeFood$ and thus absorbed as $Domain(partof, VegeFood \sqcup \neg \exists partof.Plant)$. But its applicability is still limited and it still contains a disjunction in the domain. *Binary Absorption* [47] tries to rewrite GCIs into form $A_1 \sqcap A_2 \sqsubseteq C$ where $A_1$ and $A_2$ are named concepts. To sum up, the above absorptions can only be applied to a limited patterns of GCIs; e.g., $\alpha_4$ can not be dealt with by any absorption optimisation.

**Table 7.** $\mathcal{EL}^{++}$ completion rules (no datatypes)

| | |
|---|---|
| **R1** | If $A \in S(X)$, $A \sqsubseteq B \in \mathcal{T}$ and $B \notin S(X)$ <br> then $S(X) := S(X) \cup \{B\}$ |
| **R2** | If $A_1, A_2, \ldots, A_n \in S(X)$, <br> $A_1 \sqcap A_2 \sqcap \cdots \sqcap A_n \sqsubseteq B \in \mathcal{T}$ and $B \notin S(X)$ <br> then $S(X) := S(X) \cup \{B\}$ |
| **R3** | If $A \in S(X)$, $A \sqsubseteq \exists r.B \in \mathcal{T}$ and $(X, B) \notin R(r)$ <br> then $R(r) := R(r) \cup \{(X, B)\}$ |
| **R4** | If $(X, A) \in R(r)$ $A' \in S(A)$, $\exists r.A' \sqsubseteq B \in \mathcal{T}$ <br> and $B \notin S(X)$ <br> then $S(X) := S(X) \cup \{B\}$ |
| **R5** | If $(X, A) \in R(r)$, $\bot \in S(A)$ and $\bot \notin S(X)$ <br> then $S(X) := S(X) \cup \{\bot\}$ |
| **R6** | If $\{a\} \in S(X) \cap S(A)$, $X \rightsquigarrow_R A$ and $S(A) \nsubseteq S(X)$ <br> then $S(X) := S(X) \cup S(A)$ |
| **R7** | If $(X, A) \in R(r)$, $r \sqsubseteq s \in \mathcal{T}$ and $(X, A) \notin R(s)$ <br> then $R(s) := R(s) \cup \{(X, A)\}$ |
| **R8** | If $(X, A) \in R(r_1)$, $(A, B) \in R(r_2)$, $r_1 \circ r_2 \sqsubseteq r_3 \in \mathcal{T}$, <br> and $(X, B) \notin R(r_3)$ <br> then $R(r_3) := R(r_3) \cup \{(X, B)\}$ |

Reasoning with $\mathcal{EL}^{++}$ is more efficient. [3] presents a set of completion rules (Table 7) [33] to compute, given a normalised $\mathcal{EL}^{++}$ TBox $\mathcal{T}$, for each $A \in CN_T$, a subsumer set $S(A) \subseteq CN_T \cup \{\bot\}$ in which for each $B \in S(A)$, $\mathcal{T} \models A \sqsubseteq B$, and for each $r \in RN_T$, a relation set $R(r) \subseteq CN_T \times CN_T$ in which for each $(A, B) \in R(r)$, $\mathcal{T} \models A \sqsubseteq \exists r.B$.

Reasoning with rules **R1-R8** is tractable. However, these rules can not handle $\mathcal{T}_1$ because the ontology is in a language beyond the $\mathcal{EL}^{++}$.

Groot et al. [28] attempt to speed up concept unsatisfiability checking via approximation. Given a concept $C$, it constructs a sequence of $C_i^\top$ such that $C \sqsubseteq \cdots \sqsubseteq C_1^\top \sqsubseteq C_0^\top$, and a sequence of $C_i^\bot$ such that $C_0^\bot \sqsubseteq C_1^\bot \sqsubseteq \ldots C$ by replacing all existential restrictions ($\exists R.D$) after $i$ universal quantifiers ($\forall$) inside $C$ with $\top$ and $\bot$ respectively. Then $C$ is unsatisfiable (satisfiable) if some $C_i^\top$ ($C_i^\bot$) is unsatisfiable (satisfiable). In case $C_i^\top$ ($C_i^\bot$) is usually simpler than $C$, its (un)satisfiability checking should also be easier. For example, a concept $C \equiv \neg Herbivore \sqcap \forall eat.(VegeFood \sqcap \exists partof.Plant)$ can be approximated to $C_1^\top \equiv \neg Herbivore \sqcap \forall eat.(VegeFood \sqcap \top) \equiv \neg Herbivore \sqcap \forall eat.VegeFood$, which is unsatisfiable in $\mathcal{T}_1$. Thus $C$ is unsatisfiable. However, this approach has several limitations when applied to TBox reasoning:

1. It only approximates the tested concept, but not the ontology, thus the unsatisfiability checking still requires reasoners for the original language of the ontology. In other words, it does not reduce the complexity of reasoning.
2. Similar to the Tableau algorithms, to classify an ontology, one has to reduce concept subsumption $C \sqsubseteq D$ to unsatisfiability of $C \sqcap \neg D$ for each necessary pair of $C, D$.

---

[33] In **R6** $X \rightsquigarrow_R A$ iff there exists $C_1, \ldots, C_k \in CN_T$ s.t. $C_1 = X$ or $C_1 = \{b\}$, $(C_j, C_{j+1}) \in R(r_j)$ for some $r_j \in RN_T (1 \leq j \leq k)$ and $C_k = A$.

3. When the test concept subsumption contains no existential restriction, such as $Koala \sqsubseteq Herbivore$, this approach can not help. Hence, it does not help for classification (subsumption checking among named concepts).

Due to the above reasons, this approximation technology is not suitable for TBox Reasoning, especially computing the atomic concept hierarchies.

To sum up, tableau algorithms have difficulties to handle complex structured axioms; tractable DL algorithms can not support more expressive languages; while traditional approximation approach lacks usability in TBox reasoning. In what follows, we present our approach which is motivated and inspired by these works, and show that it overcomes these difficulties with evaluations.

### 4.2   Approach Overview and Preliminary

Different from Groot et al.'s approximation approach, we approximate both the ontology and the tested concept (if needed) by replacing concept sub-expressions (role expressions) that are not in the target DL, e.g. $\mathcal{EL}^{++}$, with atomic concepts (atomic roles) and rewrite axioms accordingly (§ 4.3). Then, additional data structures and completion rules (§ 4.4 and § 4.5) are used to maintain and restore some semantic relations among basic concepts, respectively. We show that all these approachs are tractable and soundness-guaranteed (§ 4.6).

In approximation, we only consider concepts corresponding to the particular TBox in question. We use the notion *term* to refer to these "interesting" concept expressions. More precisely, a term is:

1. a concept expression on the LHS or RHS of any CI, or
2. the singleton of any individual in the ontology, or
3. the syntactic sub-expression of a term, or
4. the complement of a term.

In order to represent all these terms and role expressions that will be used in $\mathcal{EL}^{++}$ reasoning, we first assign names to them.

**Definition 15.** *(Name Assignment)* *Given $S$ a set of concept expressions, $E$ a set of role expressions, a* name assignment *$fn$ is a function as for each $C \in S$ ($R \in E$), $fn(C) = C$ ($fn(R) = R$) if $C$ is a basic concept ($R$ is atomic), otherwise $fn(C)$ ($fn(R)$) is a fresh name.*

As an example, name assignments of some terms in Example 9 are illustrated in Table 8.

From § 4.1 we can see that there is an expressivity gap between $\mathcal{SROIQ}$ and $\mathcal{EL}^{++}$, especially in concept constructs. In the rest of this section, we present 3 stages of approximation to (partially) bridge this gap.

### 4.3   $\mathcal{EL}^{++}$ Approximation

A naive $\mathcal{EL}^{++}$ approximation is to approximate an arbitrary TBox into an $\mathcal{EL}^{++}$ TBox.

**Table 8.** Name Assignment

| Term | Name |
|------|------|
| $\forall eat.\exists partof.Eucalypt$ | $C_1$ |
| $\exists eat.\forall partof.\neg Eucalypt$ | $nC_1$ |
| $\forall partof.\neg Eucalypt$ | $C_2$ |
| $\exists partof.Eucalypt$ | $nC_2$ |
| $Plant \sqcup \exists partof.Plant$ | $C_3$ |
| $\neg Plant \sqcap \forall partof.\neg Plant$ | $nC_3$ |
| $\forall partof.\neg Plant$ | $C_4$ |
| $\exists partof.Plant$ | $nC_4$ |
| $\forall eat.VegeFood$ | $C_5$ |
| $\exists eat.\neg VegeFood$ | $nC_5$ |
| $\neg Plant$ | $nPlant$ |
| $\neg VegeFood$ | $nVegeFood$ |

**Definition 16.** ($\mathcal{EL}^{++}$ **Transformation**) *Given a TBox $\mathcal{T}$ and a name assignment $fn$, its $\mathcal{EL}^{++}$ transformation $A_{fn,\mathcal{EL}^{++}}(\mathcal{T})$ is a set of axiom $T$ constructed as follows:*

1. *$T$ is initialised as $\emptyset$.*
2. *for each $C \sqsubseteq D$ ($C \equiv D$) in $\mathcal{T}$, $T = T \cup \{fn(C) \sqsubseteq fn(D)\}$ ($T = T \cup \{fn(C) \equiv fn(D)\}$).*
3. *for each $\mathcal{EL}^{++}$ role axiom $\beta \in \mathcal{T}$, add $\beta_{[R/fn(R)]}$ into $T$.*
4. *for each term $C$ in $\mathcal{T}$,*
   *(a) if $C$ is the form $C_1 \sqcap \cdots \sqcap C_n$, then $T = T \cup \{fn(C) \equiv fn(C_1) \sqcap \cdots \sqcap fn(C_n)\}$,*
   *(b) if $C$ is the form $\exists R.D$, then $T = T \cup \{fn(C) \equiv \exists fn(R).fn(D)\}$,*
   *(c) otherwise $T = T \cup \{fn(C) \sqsubseteq \top\}$.*

In the above definition, Step 2 rewrites all the concept axioms; Step 3 rewrites all the $\mathcal{EL}^{++}$ role axioms; Step 4 rewrites all the $\mathcal{EL}^{++}$ terms. We call this procedure an $\mathcal{EL}^{++}$ *approximation*.

**Lemma 1.** *For a TBox $\mathcal{T}$ and a name assignment $fn$, let $A_{fn,\mathcal{EL}^{++}}(\mathcal{T}) = T$. Then $T$ is an $\mathcal{EL}^{++}$ TBox and $|T| \leq n_{\mathcal{T}} + |\mathcal{T}|$ where $n_{\mathcal{T}}$ is the number of terms in $\mathcal{T}$ and $|T|$ ($|\mathcal{T}|$) is the number of axioms in $T$ ($\mathcal{T}$) .*

According to Table 8, we can transform the TBox $\mathcal{T}_1$ into $T_{Koala}$ as follows:

*Example 10.* $T_{Koala}$ contains axioms generated by Step 2 and 4, the most important ones include:

- $\alpha_1 \rightarrow Koala \sqsubseteq C_1$, $nC_1 \equiv \exists eat.C_2$, $nC_2 \equiv \exists partof.Eucalypt$;
- $\alpha_2$ is preserved;
- $\alpha_3 \rightarrow C_3 \sqsubseteq VegeFood$, $nC_3 \equiv nPlant \sqcap C_4$, $nC_4 \equiv \exists partof.Plant$;
- $\alpha_4 \rightarrow C_5 \sqsubseteq Herbivore$, $nC_5 \equiv \exists eat.nVegeFood$.

## 4.4   Complement-Enriched $\mathcal{EL}_\mathcal{C}^{++}$ Approximation

In Example 10, reasoning can be performed directly with the completion rules **R1-R8** presented in Table 7. However, $T_{Koala} \not\models Koala \sqsubseteq Herbivore$ because the relations between a term and its complement, e.g. $C_1$ and $nC_1$, can not be directly represented in $\mathcal{EL}^{++}$. To solve this problem, we maintain such relations in a separate *complement table (CT)*, and apply additional completion rules in reasoning.

**Approximate Complement.** We first extend the naive $\mathcal{EL}^{++}$ approximation with a complement table $(CT)$.

**Definition 17.** *($\mathcal{EL}_\mathcal{C}^{++}$ Transformation) Given a TBox $T$ and a name assignment $fn$, its complement-enriched $\mathcal{EL}_\mathcal{C}^{++}$ transformation $A_{fn,\mathcal{EL}_\mathcal{C}^{++}}(T)$ is a pair $(T, CT)$ constructed as follows:*

1. *$T = A_{fn,\mathcal{EL}^{++}}(T)$ (Ref. Def. 16).*
2. *$CT$ is initialised as $\emptyset$.*
3. *for each term $C$ in $T$, $CT = CT \cup \{(fn(C), fn(\sim C))\}$.*

We call this procedure an $\mathcal{EL}_\mathcal{C}^{++}$ *approximation*. The following proposition shows the structure of the approximation results:

**Proposition 4.** *($\mathcal{EL}_\mathcal{C}^{++}$ Approximation) For a TBox $T$, let $A_{fn,\mathcal{EL}_\mathcal{C}^{++}}(T) = (T, CT)$, we have:*

1. *$T$ is an $\mathcal{EL}^{++}$ TBox*
2. *for each $A \in CN_T$, there exists $(A, B) \in CT$*
3. *if $(A, B) \in CT$ then $A, B \in CN_T$ and $(B, A) \in CT$*

This indicates that, by Def.17, a TBox can be syntactically transformed into an $\mathcal{EL}^{++}$ TBox with a table maintaining complementary relations for all names in the $\mathcal{EL}^{++}$ TBox.

*Example 11.* The $\mathcal{EL}_\mathcal{C}^{++}$ approximation of $T_1$ in Example 9 is $(T_{Koala}, CT_{Koala})$, where $T_{Koala}$ is the same as in Example 10, and $CT_{Koala}$ contains pairs such as $(C_1, nC_1), (C_2, nC_2), (C_3, nC_3), (C_4, nC_4), (C_5, nC_5), (Plant, nPlant), (VegeFood, nVegeFood)$, etc.

**Lemma 2.** *For any TBox $T$ and $(T, CT)$ its $\mathcal{EL}_\mathcal{C}^{++}$ approximation, if $T$ contains $n_T$ terms, then $|T| \leq n_T + |T|$ and $|CT| = n_T$, where $|T|(|T|)$ is the number of axioms in $T(T)$ and $|CT|$ is the number of pairs in $CT$.*

**Completion Rules for Complement.** Given an $\mathcal{EL}_\mathcal{C}^{++}$ transformation $(T, CT)$, we normalise axioms of form $C \sqsubseteq D_1 \sqcap \cdots \sqcap D_n$ into $C \sqsubseteq D_1, \ldots, C \sqsubseteq D_n$, and recursively normalise role chain $r_1 \circ \cdots \circ r_n \sqsubseteq s$ with $n > 2$ into $r_1 \circ \cdots \circ r_{n-1} \sqsubseteq u$ and $u \sqsubseteq s$. This procedure can be done in linear time. In the following, we assume $T$ to be always normalised. For convenience, we use a *complement function* $fc : CN_T \mapsto CN_T$ as: for each $A \in CN_T$, $fc(A) = B$ such that $(A, B) \in CT$.

**Table 9.** Complement completion rules

| | |
|---|---|
| **R9** | If $A, B \in S(X), A = fc(B)$ and $\bot \notin S(X)$ then $S(X) := S(X) \cup \{\bot\}$ |
| **R10** | If $A \in S(B)$ and $fc(B) \notin S(fc(A))$ then $S(fc(A)) := S(fc(A)) \cup \{fc(B)\}$ |
| **R11** | If $A_1 \sqcap \cdots \sqcap A_i \sqcap \cdots \sqcap A_n \sqsubseteq \bot, A_1, \ldots, A_{i-1}, A_{i+1}, \ldots, A_n \in S(X)$ and $fc(A_i) \notin S(X)$ then $S(X) := S(X) \cup \{fc(A_i)\}$ |

To utilise the complementary relations in $CT$, we propose additional completion rules (Table 9) to $\mathcal{EL}^{++}$.

**R9** realises axiom $A \sqcap {\sim}A \sqsubseteq \bot$. **R10** asserts the reverse subsumption between concepts to supplement the absence of negation, i.e. $A \sqsubseteq B \to {\sim}A \sqsubseteq {\sim}B$. **R11** builds up the relations between conjuncts of a conjunction, e.g. $A \sqcap B \sqsubseteq \bot$ implies $A \sqsubseteq {\sim}B$.

Now we can infer $Koala \sqsubseteq Herbivore$ (Example 11) as follows:

- $\alpha_2 \to nC_2 \sqsubseteq nC_4 \to_{R10} C_4 \sqsubseteq C_2 \to nC_3 \sqsubseteq C_2$
- $C_3 \sqsubseteq VegeFood \to_{R10} nVegeFood \sqsubseteq nC_3$
- $nVegeFood \sqsubseteq nC_3, nC_3 \sqsubseteq C_2 \to nVegeFood \sqsubseteq C_2 \to nC_5 \sqsubseteq nC_1 \to_{R10} C_1 \sqsubseteq C_5 \to Koala \sqsubseteq Herbivore$

where the inferences with $\to_{R10}$ are enabled by **R10**.

### 4.5  Cardinality-Enriched $\mathcal{EL}_{\mathcal{CQ}}^{++}$ Approximation

In Def.17 we presented an extension of the naive approximation which approximates non-$\mathcal{EL}^+$ concept expressions, particularly concepts constructed by $\neg$, $\sqcup$ and $\forall$, by the definition of their complements. With the completion rules in Tab.9, more entablements can be computed.

It is a natural question to ask, is that possible to approximate even more non-$\mathcal{EL}^{++}$ construct, e.g. cardinality, into $\mathcal{EL}^{++}$? In this subsection, we further extend the $\mathcal{EL}_{\mathcal{C}}^{++}$ transformation to yield more complete reasoning results for ontology containing cardinalities.

**Approximating Cardinality.** In $\mathcal{EL}_{\mathcal{C}}^{++}$ approximation, a concept constructed by $\geq$ or $\leq$ can only be represented as a fresh name. In this way, subsumption $X \sqsubseteq \bot$ can not be entailed in $\mathcal{T}_4$ in the following Example 12.

*Example 12.* $\mathcal{T}_4 = \{X \sqsubseteq \geq 4r.A, X \sqsubseteq \leq 2s.B, A \sqsubseteq B, r \sqsubseteq s\}$.
 $X \sqsubseteq \bot$ should be entailed.

This subsumption requires to maintain the relations between the filler concepts (e.g. $A$ and $B$), the role ($r$) and the cardinality values (e.g 4 and 2). We maintain such relations in a (cardinality table) ($QT$) whose elements are tuples $(A, r, n)$, where $A$ is a basic concept denoting a filler name, $r$ is the atomic role denoting the role name and $n$ is the cardinality value.

**Definition 18. (Cardinality-enriched $\mathcal{EL}_{\mathcal{CQ}}^{++}$ Transformation)** *Given a TBox $\mathcal{T}$, a name assignment $fn$, let $A_{fn,\mathcal{EL}_{\mathcal{C}}^{++}}(\mathcal{T}) = (T', CT')$, its cardinality-enriched $\mathcal{EL}_{\mathcal{CQ}}^{++}$ transformation $A_{fn,\mathcal{EL}_{\mathcal{CQ}}^{++}}(\mathcal{T})$ is a tuple $(T, CT, QT)$ constructed as follows:*

1. *$T$ is initialised as $T'$.*
2. *$CT = CT'$.*
3. *$QT$ is initialised as $\emptyset$.*
4. *for each term $C$ that is the form $\geq nR.D$ in $\mathcal{T}$,*
    (a) *if $n = 0, T = T \cup \{\top \sqsubseteq fn(C)\}$*
    (b) *if $n = 1, T = T \cup \{fn(C) \equiv \exists fn(R).fn(D)\}$*
    (c) *otherwise, $T = T \cup \{fn(C) \equiv fn(D)^{fn(R),n}\}$, and $QT = QT \cup \{(fn(C), fn(R), n)\}$.*
5. *for each pair of names $A$ and $r$, if there exist $(A, r, i_1), (A, r, i_2), \ldots, (A, r, i_n) \in QT$ with $i_1 < i_2 < \cdots < i_n, T = T \cup \{A^{r,i_n} \sqsubseteq A^{r,i_{n-1}}, \ldots, A^{r,i_2} \sqsubseteq A^{r,i_1}, A^{r,i_1} \sqsubseteq \exists r.A\}$*

In step 4, $fn(D)^{fn(R),n}$ is a fresh name. For example, $nVegeFood^{eat,3}$ for $\geq 3eat.\neg VegeFood$. Obviously, this is unique for a given tuple of $D$, $R$ and $n$. Similarly, $\leq nR.D$ will be approximated via the approximation of its complement $\geq (n+1)R.D$. In step 5, for each pair of name assignment $A, r$ in $T$, a subsumption chain is added into $T$ because $\geq i_n r.A \sqsubseteq \cdots \sqsubseteq \geq i_2 r.A \sqsubseteq \geq i_1 r.A \sqsubseteq \exists r.A$.

We call this procedure an $\mathcal{EL}_{\mathcal{CQ}}^{++}$ approximation. The following proposition shows the structure of the results:

**Proposition 5. ($\mathcal{EL}_{\mathcal{CQ}}^{++}$ Approximation)**
*For a TBox $\mathcal{T}$, a name assignment $fn$, let $A_{fn,\mathcal{EL}_{\mathcal{CQ}}^{++}}(\mathcal{T}) = (T, CT, QT)$, we have $T$ an $\mathcal{EL}^{++}$ TBox.*

This indicates that, by Def.18 a TBox can be syntactically transformed into a tuple of an $\mathcal{EL}^{++}$ TBox, a complement table and a cardinality table.

Now, in Example 12, $\mathcal{T}_4$ can be approximated into $T_4 \supseteq \{X, \sqsubseteq Y_1, Y_1 \equiv A^{r,4}, X \sqsubseteq Y_2, nY_2 \equiv B^{s,3}, A \sqsubseteq B, r \sqsubseteq s\}$ with $fn(\geq 4r.A) = Y_1$, $fn(\leq 2s.B) = Y_2$ and $fn(\geq 3s.B) = nY_2$, $CT_4 \supseteq \{(Y_1, nY_1), (Y_2, nY_2)\}$, $QT_4 \supseteq \{(A, r, 4), (B, s, 3)\}$.

**Lemma 3.** *For any TBox $\mathcal{T}$, let $(T, CT, QT)$ its $\mathcal{EL}_{\mathcal{CQ}}^{++}$ transformation, if $\mathcal{T}$ contains $n_{\mathcal{T}}$ terms, then $|CN_T| \leq 2 \times n_{\mathcal{T}}, |T| \leq 3 \times n_{\mathcal{T}} + |\mathcal{T}|, |CT| = n_{\mathcal{T}}$ and $|QT| \leq n_{\mathcal{T}}$, where $CN_T$ is the number of basic concepts in $T$, $|T|(|\mathcal{T}|)$ the number of axioms in $T(\mathcal{T})$, $|CT|$ the number of pairs in $CT$ and $|QT|$ the number of tuples in $QT$.*

**Completion rules.** We further extend Tab.9 with Table 10.

**R12**, in which $r \sqsubseteq_* s$ if $r = s$ or $r \sqsubseteq s \in T$, realises inference $A \sqsubseteq B, R \sqsubseteq S, i \geq j \rightarrow \geq iR.A \sqsubseteq \geq jS.B$. **R13** is the extension of **R4** and **R14-16** are extensions of **R8**. Now we can entail $X \sqsubseteq \bot$ in Example 12 as follows:

1. $A \sqsubseteq B, r \sqsubseteq s \rightarrow_{R12} A^{r,4} \sqsubseteq B^{s,3}$,
2. $A^{r,4} \sqsubseteq B^{s,3}, X \sqsubseteq Y_1, Y_1 \equiv A^{r,4}, nY_2 \equiv B^{s,3} \rightarrow X \sqsubseteq nY_2$
3. $X \sqsubseteq nY_2, X \sqsubseteq Y_2, (Y_2, nY_2) \in CT \rightarrow_{R9} X \sqsubseteq \bot$

**Table 10.** Cardinality completion rule

| |
|---|
| **R12** If $B \in S(A), (A, r, i), (B, s, j) \in QT, r \sqsubseteq_* s,$ $i \geq j$ and $B^j \notin S(A^i)$ then $S(A^{r,i}) := S(A^{r,i}) \cup \{B^{s,j}\}$ |
| **R13** If $A^{r,i} \in S(X), A' \in S(A), \exists r.A' \sqsubseteq B \in T$ and $B \notin S(X)$ then $S(X) := S(X) \cup \{B\}$ |
| **R14** If $A^{r_1,i} \in S(X), (A, B) \in R(r_2), r_1 \circ r_2 \sqsubseteq r_3 \in T,$ and $(X, B) \notin R(r_3)$ then $R(r_3) := R(r_3) \cup \{(X, B)\}$ |
| **R15** If $(X, A) \in R(r_1), B^{r_2,i} \in S(A), r_1 \circ r_2 \sqsubseteq r_3 \in T,$ and $(X, B) \notin R(r_3)$ then $R(r_3) := R(r_3) \cup \{(X, B)\}$ |
| **R16** If $A^{r_1,i} \in S(X), B^{r_2,j} \in S(A), r_1 \circ r_2 \sqsubseteq r_3 \in T,$ and $(X, B) \notin R(r_3)$ then $R(r_3) := R(r_3) \cup \{(X, B)\}$ |

## 4.6  Reasoning Properties

In this subsection, we analyze the reasoning complexity of our approximation and reasoning approach.

**Theorem 5.** *(Complexity) For any $\mathcal{EL}_{CQ}^{++}$ transformation $(T, CT, QT)$ $(T$ normalised), TBox reasoning by completion rules **R1-R16** will terminate in polynomial time w.r.t. $|CN_T| + |RN_T|$.*

Similarly, reasoning on the $\mathcal{EL}^{++}$ and $\mathcal{EL}_C^{++}$ approximations are also tractable. Note that, from lemma 1, 2 and 3, the approximation is always linear. To sum up, the approximation-reasoning approach is tractable.

With the approximation and corresponding rules, we can compute concept subsumption in an $\mathcal{SROIQ}$ TBox. The quality of the approximate reasoning is described by the following theorem:

**Theorem 6.** *(Concept Subsumption Checking) Given a TBox $T$, its vocabulary $V_T$ and $A_{fn,\mathcal{EL}_{CQ}^{++}} = (T, CT, QT)$, for any two concepts $C$ and $D$ constructed from $V_T$, if $A_{fn,\mathcal{EL}_{CQ}^{++}}(\{C \sqsubseteq \top, D \sqsubseteq \top\}) = (T', CT', QT')$, then $T \models C \sqsubseteq D$ if $fn(D) \in S(fn(C))$ can be computed by rules **R1-R16** on $(T \cup T', CT \cup CT', QT \cup QT')$.*

The theorem indicates that our $\mathcal{EL}_{CQ}^{++}$ approximate reasoning approach is soundness-preserving. This conclusion holds similarly on $\mathcal{EL}^{++}$ and $\mathcal{EL}_C^{++}$ approximate reasoning.

Particularly, when $C, D$ are terms in $T$, $T \models C \sqsubseteq D$ if $fn(D) \in S(fn(C))$ can be derived from $(T, CT, QT)$.

As in classical reasoning, unsatisfiability checking of a concept $C$ can be reduced to entailment checking of $C \sqsubseteq \bot$; ontology inconsistency checking can be reduced to entailment checking of $\top \sqsubseteq \bot$ or $\{a\} \sqsubseteq \bot$. By applying ABox internaliation, ABox reasoning can be reduced to TBox reasoning, e.g. $a : A$ if $A \in S(\{a\})$ can be computed.

For more optimised approach on ABox reasoning with syntactic approximation, we refer readers to [75].

More extension patterns can be exploit to improve the completeness of the approximate reasoning while keep it tractable. Our framework is flexible and extendible.

Our extra completion rules process each axiom and term in $T$ individually. This helps keeping the reasoning tractable but some information can be lost:

*Example 13.* $T_5 = \{A \sqcap \neg B \sqsubseteq C, A \sqcap B \sqsubseteq C, D \sqsubseteq \exists r.\top, \exists r.C \sqsubseteq E, \exists r.\neg A \sqsubseteq E\}$

Obviously, we have $T_5 \models A \sqsubseteq C$ and thus $D \sqsubseteq E$.

We approximate $T_5$ into $(\{X_1 \equiv A \sqcap nB, X_1 \sqsubseteq C, X_2 \equiv A \sqcap B, X_2 \sqsubseteq C, X_3 \equiv \exists r.nA, X_3 \sqsubseteq E, \dots\}, \{(B, nB), \dots\})$. Our approach will reach $B \sqcap A \sqsubseteq C$ and $nB \sqcap A \sqsubseteq C$. Because $B$ and $nB$ are not subsumers of $A$ thus we can't infer $C \in S(A)$. Even if we can compute it, in order to further infer $D \sqsubseteq F$. A new axiom $\exists r.(C \sqcup \neg A) \equiv \exists r.C \sqcup \exists r.\neg A$ has to be added into $T$ and approximated for incremental reasoning.

Although we do not guarantee completeness, we will see in next section it has high recall for many test ontologies.

### 4.7    Evaluation

We implemented 3 versions of our approach, namely the $\mathcal{EL}^{++}$, $\mathcal{EL}_C^{++}$, $\mathcal{EL}_{CQ}^{++}$ approximation and reasoning systems, in the TrOWL tractable reasoning infrastructure [34]. To evaluate their performance in practice, we compared with mainstream reasoners Pellet 2.0.0, FaCT++ 1.3.0.1 and HermiT 1.1. All experiments were conducted in an environment of Windows XP SP3 with 2.66 GHz CPU and 1G RAM allocated to JVM 1.6.0.07.

Following [66], we examined the most difficult ontologies in the HermiT benchmark [70]. To focus on TBox reasoning, we removed the ABox axioms with care[35] from these ontologies. Most of the remaining TBoxes can be classified easily by all the reasoners and completely by our $\mathcal{EL}_C^{++}$ system. We evaluate the hard ones, results shown in Table 11 and 12. We mainly conducted the evaluations on $\mathcal{EL}_C^{++}$ system. To show the effects of complement-enriched approximate reasoning, we present also the $\mathcal{EL}^{++}$ recall. For those TBoxes that the $\mathcal{EL}_C^{++}$ provides incomplete classification, we further classified them with the $\mathcal{EL}_{CQ}^{++}$ system.

Each reasoner was given 10 min to classify each ontology. We queried for subsumption relations between named concepts (including *owl:Thing* and *owl:Nothing*) and counted the numbers. Recall of our systems is computed against others to measure the completeness. Thus the time shown in our evaluation includes classification time, subsumption retrieval and counting time. Time unit is second.

Results illustrated in Table 12 show that with extension of the approximation, higher and higher recall can be achieved. $\mathcal{EL}^{++}$ is naive and quite incomplete on some ontologies. $\mathcal{EL}_C^{++}$ approximation can significantly improve the recall on some ontologies (such as Cyc and Tambis Full). With further extension to $\mathcal{EL}_C^{++}$ approximation, all

---

[34] http://trowl.eu

[35] ABox axioms involving individuals appearing in the TBox were internalised, e.g. $a : C$ into $\{a\} \sqsubseteq C, a \neq b$ into $\{a\} \sqcap \{b\} \sqsubseteq \bot$, etc.. The others are removed.

**Table 11.** Results of main stream reasoners

| Ontology $\mathcal{O}$ | $|\mathcal{O}|$ | FaCT++ | HermiT | Pellet |
|---|---|---|---|---|
| Biological Process | 32289 | 3.656 | 5.343 | 10.063 |
| Cellular Component | 47348 | 5.872 | 8.077 | 16.966 |
| GO | 32289 | 18.563 | 6.047 | 16.39 |
| Cyc | 11727 | 25.531 | 16.853 | 142.889 |
| FMA Constitutional | 123564 | e/o | e/o | e/o |
| Tambis Full | 606 | 0.375 | 1.063 | 1.343 |
| Wine | 454 | 0.578 | 0.875 | 1.359 |
| DLP | 1216 | 0.219 | 61.948 | 98.024 |

**Table 12.** Results of our systems

| Ontology | $\mathcal{EL}^{++}$ | $\mathcal{EL}_{\mathcal{C}}^{++}$ | | $\mathcal{EL}_{\mathcal{CQ}}^{++}$ | |
|---|---|---|---|---|---|
| | recall | time | recall | time | recall |
| Biological Process | 93.1% | 1.11 | 100% | - | - |
| Cellular Component | 91.9% | 1.359 | 100% | - | - |
| GO | 93.1% | 4.203 | 100% | - | - |
| Cyc | 1.2% | 1.672 | 100% | - | - |
| FMA Constitutional | N/A | 10.062 | N/A | 50.89 | N/A |
| Tambis Full | 7.2% | 0.11 | 99.3% | 0.203 | 100% |
| Wine | 95.8% | 0.078 | 96.8% | 0.156 | 99.4% |
| DLP | 100% | 0.125 | 100% | - | - |

the recalls are over 99% (except FMA). Comparison with results illustrated in Table 11 shows that the efficiency of our systems is in general better than all other reasoners. Even the slowest $\mathcal{EL}_{\mathcal{CQ}}^{++}$ system is faster than all main stream reasoners. Also, our systems are the only reasoners that can return result on the FMA ontology.

We were also interested in the scalability of our approach. Based on Table 11 we chose 3 easiest ontologies and enlarged them by duplicating all the concept names ( but keep the role names). Duplications were distinguished by a subscript. Consequently, all the concept axioms were duplicated. We classified these ontologies using our $\mathcal{EL}_{\mathcal{C}}^{++}$ system, which has a nice balance between efficiency and completeness (Ref. Table 12). It performed quite stable when the quantity of data increased (Table 13). Due to the interactions between duplications through role axioms, our system even gained some recall on Wine.

Due to the lack of OWL2-DL benchmarks, we turned to ontologies generated from realistic use cases. In [95] an approach of using DL to model relation-based access control (relBAC) has been presented. In this paper, a rather expressive DL $\mathcal{ALCQIBO}$ has been employed to encode various access control schemata. For evaluation purpose, we generated 100 TBoxes containing the following patterns:

1. "User in U are allowed to access (with P) at most n objects in O": $U \sqsubseteq\, \leq nP.O$
2. "Users in U have access to at least m objects in O with P": $U \sqsubseteq\, \geq nP.O$
3. "User u is of user type U": $\{u\} \sqsubseteq U$

**Table 13.** Comparison on duplicated TBox

| Size | FaCT++ | HermiT | Pellet | $\mathcal{EL}_C^{++}$ | Recall |
|------|--------|--------|--------|-----------------------|--------|
| | Tambis Full | | | | |
| 5× | 9.125 | 37.922 | 24.25 | 0.719 | 99.3% |
| 10× | 40.577 | 292.481 | 205.192 | 1.985 | 99.3% |
| 20× | e/o | t/o | t/o | 5.671 | N/A |
| 30× | e/o | t/o | t/o | 11.624 | N/A |
| | Wine | | | | |
| 5× | 13.784 | 56.853 | 86.662 | 0.641 | 97.7% |
| 10× | 33.01 | t/o | t/o | 2.188 | 97.9% |
| 20× | 243.496 | t/o | t/o | 10.077 | 98.0% |
| 30× | t/o | t/o | t/o | 27.529 | N/A |
| | DLP | | | | |
| 5× | t/o | e/o | e/o | 3.39 | N/A |
| 10× | t/o | e/o | e/o | 20.827 | N/A |
| 20× | t/o | e/o | e/o | 142.305 | N/A |
| 30× | t/o | e/o | e/o | 450.6 | N/A |

where $U$ is a type of users, $P$ a permission type, $O$ a object type while $u$ a individual user. Each of these 100 ontologies contains 20 user types, 20 object types, 10 permission types, 750 individuals and 20 access control model axioms (axioms of type 1 and 2). Hierarchies among users types (object types) are randomly generated. The numbers in cardinality restrictions are randomly selected. The combinations of user individual, user type class, permission type class and object type class are also random. Obviously, these TBoxes are in DL $\mathcal{ALHOQ}$.

Different from previous evaluations, the TBoxes generated here can be inconsistent. For example, when a particular user belongs to two types $U_1$ and $U_2$ with $U_1 \sqsubseteq\leq mP.O_1$ and $U_2 \sqsubseteq\geq nP.O_2$ where $m < n$ and $O_2 \sqsubseteq O_1$, inconsistency occurs. Thus, in this evaluation, we are particularly interested in whether the inconsistency can be detected instead of the number of subsumptions.

We classified these TBoxes using FaCT++, $\mathcal{EL}_C^{++}$ and $\mathcal{EL}_{CQ}^{++}$ systems. Each reasoner was given 10 minutes. FaCT++ finished 98 of them, failing the other 2. $\mathcal{EL}_C^{++}$ classified all the TBoxes but failed to find any inconsistency, because it does not support cardinality at all. $\mathcal{EL}_{CQ}^{++}$ classified all the TBoxes efficiently and reported all the inconsistencies correctly. The average and maximal time of FaCT++ and $\mathcal{EL}_{CQ}^{++}$ and the precisions of the 98 ontologies are illustrated in Table 14. Time unit is second. Notice that in FaCT++, reasoning is immediately terminated when any inconsistency is detected, which means the reasoning time of inconsistent TBox is shorter. While our $\mathcal{EL}_{CQ}^{++}$ continues to find all the inconsistency. Therefore we separate the results of consistent and inconsistent TBoxes.

The results show that, $\mathcal{EL}_{CQ}^{++}$ system can classify all the ontologies very efficiently and the presicion is 100%. Also, the average and maximal time is quite stable no matter the ontology is consistent or not. While FaCT++ has difficulty in dealing with consistent TBox containning many cardinality restrictions (Max. time is about 10 seconds

**Table 14.** Comparison on relBAC TBox

| Consistency | FaCT++ | | $\mathcal{EL}_{CQ}^{++}$ | | |
|---|---|---|---|---|---|
| | Ave. | Max. | Ave. | Max. | presicion |
| Consistent | 1.226 | 9,984 | 0.021 | 0.047 | 100% |
| Inconsistent | 0.248 | 2.297 | 0.022 | 0.047 | 100% |

and failed on two other TBoxes). For those inconsistent ones, even though FaCT++ terminates earlier, $\mathcal{EL}_{CQ}^{++}$ system can still outperform it.

## 4.8   Discussion

Approximate reasoning has been an important topic for ontology (KR) and AI research. On the one hand, expressive Description Logics (such as those underpin the standard Semantic Web ontology languages) have high worst case computational complexity. Hence, approximate reasoning is an attractive way to provide scalable and efficient reasoning services [72]. On the other hand, it has been argued that [28] while logic has always aimed at modelling idealised forms of reasoning under idealised circumstances, this is not what is required under the practical circumstances in knowledge-based systems. Instead, we also need to consider (i) reasoning under time-pressure, (ii) reasoning with other limited resources besides time and (iii) reasoning that is not perfect but instead good enough for given tasks under given circumstances.

In this section, we address a long-lasting open problem; i.e, effective and efficient approximate TBox reasoning. With their negative results, Groot et al. concluded that traditional approximation method by Cadoli and Schaerf [77] is not suited for ontology reasoning, and that new approximate strategy are needed. In this paper, we propose to combine the ideas of language weakening and approximate deduction to provide soundness preserving TBox reasoning for expressive Description Logics. We apply our idea to approximate OWL2-DL ontologies to $\mathcal{EL}^{++}$ ones, preliminary evaluation results showed that our approach performs effectively and efficiently on real world ontologies.

In the approximate deduction step, instead of simplifying a model constructing algorithm (such as tableau algorithm), we enrich the existing $\mathcal{EL}^{++}$ reasoning algorithm with some deterministic completion rules (for complement and cardinality). $\mathcal{EL}^{++}$ retain tractability by imposing strict syntactic restriction. However these restrictions are not always necessary. For example, if we rewrite each axiom $C \sqsubseteq D$ of an $\mathcal{EL}^{++}$ ontology into $\neg D \sqsubseteq \neg C$, the language appears to be $\mathcal{ALC}$, but the complexity does not essentially change. Our approximation can naturally cover these situations. As our evaluation shows, it helps increase the recall.

This piece of work is also related to Horn $\mathcal{SHIQ}$, which has an even more complicated set of syntactic restrictions, which can not be satisfied by our Koala example (more precisely, axioms $\alpha_4$). In [51], *structure transformation* is applied in a similar manner as our approximation to facilitate reasoning. However *structure transformation* still preserves the syntactic structure of the axioms, while our approximation actually changed the structures and hence ontology such as the Koala example can be classified with a more efficient algorithm.

## 5  Conclusion

In these lecture nodes, we focused on the investigation of Scalable OWL 2 Reasoning for Linked Data. Along these lines, we briefly introduced Linked Data and its relationship with ontologies and then presented two scalable approaches for OWL 2 RL and OWL 2 DL, respectively. The first approach utilises a rule-based engine to materialise inferences for a scalable subset of OWL 2 RL/RDF, enabling distributed processing over a cluster of commodity hardware that is demonstrated to be feasible for a billion triples of open-domain Linked Data—the first approach is designed to handle large amounts of assertional data. The second approach deals with much more complex ontology languages—as such, the second approach is designed to handle expressive reasoning over terminological data. It approximates reasoning in OWL 2 DL into a tractable profile OWL 2 EL and substantially reduces reasoning complexity from 2NEXPTIME-complete to PTIME-complete. At the same time, the reasoning is soundness-guaranteed, and demonstrated to give high recall in practice.

Although both of these two approaches are theoretically incomplete, for many application scenarios incompleteness is not the end of the world, where it may enable many practical advantages, in particular relating to scalability and performance. Instead of fixating on what we might lose by not fully supporting a complete approach, we should instead focus on the more important issue with respect to what we can *gain* through an incomplete approach—especially in scenarios such as Linked Data, where messiness and scale are major challenges.

Similar sentiments—playing down the importance of completeness for reasoning engines operating over Web data—have been presented in the literature. Firstly, we have Fensel et al. [21] who state that:

> *"The Web is open, with no defined boundaries. Therefore, completeness is a rather strange requirement for an inference procedure in this context. Given that collecting all relevant information on the Web is neither possible nor often even desirable (usually, you want to read the first 10 Google hits but don't have the time for the remaining two million), requesting complete reasoning on top of such already heavily incomplete facts seems meaningless."*

—[21]

Again, the emphasis here is on incremental improvements of query-answering results through incomplete reasoning, as opposed to *requiring* completeness over Web data, which is in any case published under an Open World Assumption and which is largely incomplete with respect to its universal domain. Similar arguments for incomplete reasoning—in scenarios such as that faced when reasoning over Linked Data—are laid out by Hitzler and van Harmelen [37] who state:

> *"[...] we would advocate* [viewing] *the formal semantics of a system (in whatever form it is specified) as a "gold standard", that need not necessarily be obtained in a system (or even be obtainable). What is required from systems is not a proof that they satisfy this gold standard, but rather a precise description of the extent to which they satisfy this gold standard."*

—[37]

In [37], the authors generally discuss the benefits of using Information Retrieval inspired precision and recall measures (as used for our OWL 2 RL reasoning approach) for reasoning systems, as opposed to forcing completeness on them as an immutable requirement.

According to the evaluation results presented herein, our two incomplete approaches can perform reasoning at a level of scale and performance which would likely not be feasible for a complete reasoner. In general, this makes them useful compensations for complete reasoners, and even promising alternatives in scenarios where scalability, noise, time pressure and incomplete knowledge are pressing concerns—i.e., scenarios such as Linked Data.

# References

1. Allemang, D., Hendler, J.A.: Semantic Web for the Working Ontologist: Effective Modeling in RDFS and OWL. Morgan Kaufmann, San Francisco (2008)
2. Auer, S., Ngomo, A.-C.N., Lehmann, J.: Introduction to Linked Data. In: Polleres, A., et al. (eds.) Reasoning Web 2011. LNCS, vol. 6848, pp. 251–327. Springer, Heidelberg (2011)
3. Baader, F., Brandt, S., Lutz, C.: Pushing the $\mathcal{EL}$ Envelope. In: Proceedings IJCAI (2005)
4. Baader, F., Calvanese, D., McGuinness, D.L., Nardi, D., Patel-Schneider, P.F.: The Description Logic Handbook: Theory, Implementation and Application. Cambridge University Press, Cambridge (2002)
5. Baader, F., Calvanese, D., McGuinness, D.L., Nardi, D., Patel-Schneider, P.F. (eds.): The Description Logic Handbook: Theory, Implementation, and Applications. Cambridge University Press, Cambridge (2003)
6. Barabási, A.L., Albert, R.: Emergence of scaling in random networks. Science 286, 509–512 (1999)
7. Beckett, D.: RDFS 3.0. In: W3C Workshop on RDF Next Steps, Stanford, Palo Alto, CA, USA (June 2010)
8. Beckett, D., Berners-Lee, T.: Turtle – Terse RDF Triple Language. W3C Team Submission (January 2008), http://www.w3.org/TeamSubmission/turtle/
9. Belnap, N.: A Useful Four-Valued Logic. Modern Uses of Multiple-Valued Logic, 5–37 (1977)
10. Berners-Lee, T.: Linked Data. W3C Design Issues (July 2006), http://www.w3.org/DesignIssues/LinkedData.html (retrieved October 27, 2010)
11. Bernstein, A.: Scalable non-standard reasoning on the Semantic Web. In: Polleres, A., et al. (eds.) Reasoning Web 2011. LNCS, vol. 6848, Springer, Heidelberg (2011)
12. Birbeck, M., McCarron, S.: CURIE Syntax 1.0 – A syntax for expressing Compact URIs. W3C Recommendation (January 2009), http://www.w3.org/TR/curie/
13. Bishop, B., Kiryakov, A., Ognyanoff, D., Peikov, I., Tashev, Z., Velkov, R.: OWLIM: A family of scalable semantic repositories. Semantic Web Journal (in press, 2011), http://www.semantic-web-journal.net/sites/default/files/swj97_0.pdf
14. Bizer, C., Cyganiak, R., Heath, T.: How to Publish Linked Data on the Web, linkeddata.org Tutorial (July 2008), http://linkeddata.org/docs/how-to-publish
15. Bizer, C., Lehmann, J., Kobilarov, G., Auer, S., Becker, C., Cyganiak, R., Hellmann, S.: DBpedia - A crystallization point for the Web of Data. J. Web Sem. 7(3), 154–165 (2009)
16. Cheng, G., Ge, W., Wu, H., Qu, Y.: Searching Semantic Web Objects Based on Class Hierarchies. In: Proceedings of Linked Data on the Web Workshop (2008)
17. Cosmadakis, S.S., Gaifman, H., Kanellakis, P.C., Vardi, M.Y.: Decidable Optimization Problems for Database Logic Programs (Preliminary Report). In: STOC, pp. 477–490 (1988)

18. de Bruijn, J., Heymans, S.: Logical foundations of (e)RDF(S): Complexity and reasoning. In: Aberer, K., Choi, K.-S., Noy, N., Allemang, D., Lee, K.-I., Nixon, L.J.B., Golbeck, J., Mika, P., Maynard, D., Mizoguchi, R., Schreiber, G., Cudré-Mauroux, P. (eds.) ASWC 2007 and ISWC 2007. LNCS, vol. 4825, pp. 86–99. Springer, Heidelberg (2007)
19. Dean, J., Ghemawat, S.: MapReduce: Simplified Data Processing on Large Clusters. In: OSDI, pp. 137–150 (2004)
20. Delbru, R., Polleres, A., Tummarello, G., Decker, S.: Context Dependent Reasoning for Semantic Documents in Sindice. In: Proc. of 4th SSWS Workshop (2008)
21. Fensel, D., van Harmelen, F.: Unifying Reasoning and Search to Web Scale. IEEE Internet Computing 11(2), 94–95 (2007)
22. Fielding, R.T., Gettys, J., Mogul, J.C., Frystyk, H., Masinter, L., Leach, P.J., Berners-Lee, T.: Hypertext Transfer Protocol – HTTP/1.1. RFC 2616 (June 1999), http://www.ietf.org/rfc/rfc2616.txt
23. Ghilardi, S., Lutz, C., Wolter, F.: Did i damage my ontology? a case for conservative extensions in description logics. In: Proceedings of the Tenth International Conference on Principles of Knowledge Representation and Reasoning, pp. 187–197 (2006)
24. Glimm, B.: Using SPARQL with RDFS and OWL entailment. In: Polleres, A., et al. (eds.) Reasoning Web 2011. LNCS, vol. 6848, pp. 251–327. Springer, Heidelberg (2011)
25. Golbreich, C., Wallace, E.K.: OWL 2 Web Ontology Language: New Features and Rationale. W3C Recommendation (October 2009), http://www.w3.org/TR/owl2-new-features/
26. Grau, B.C., Horrocks, I., Parsia, B., Ruttenberg, A., Schneider, M.: OWL 2 Web Ontology Language: Mapping to RDF Graphs. W3C Recommendation (October 2009), http://www.w3.org/TR/owl2-mapping-to-rdf/
27. Grau, B.C., Motik, B., Wu, Z., Fokoue, A., Lutz, C.: OWL 2 Web Ontology Language: Profiles. W3C Recommendation (October 2009), http://www.w3.org/TR/owl2-profiles/
28. Groot, P., Stuckenschmidt, H., Wache, H.: Approximating description logic classification for semantic web reasoning. In: Gómez-Pérez, A., Euzenat, J. (eds.) ESWC 2005. LNCS, vol. 3532, pp. 318–332. Springer, Heidelberg (2005)
29. Grosof, B., Horrocks, I., Volz, R., Decker, S.: Description Logic Programs: Combining Logic Programs with Description Logic. In: 13th International Conference on World Wide Web (2004)
30. Guo, Y., Pan, Z., Heflin, J.: LUBM: A benchmark for OWL knowledge base systems. J. Web Sem. 3(2-3), 158–182 (2005)
31. Gutierrez, C.: Models for the Web of Data. In: Polleres, A., et al. (eds.) Reasoning Web 2011. LNCS, vol. 6848, pp. 251–327. Springer, Heidelberg (2011)
32. Harth, A., Kinsella, S., Decker, S.: Using Naming Authority to Rank Data and Ontologies for Web Search. In: Bernstein, A., Karger, D.R., Heath, T., Feigenbaum, L., Maynard, D., Motta, E., Thirunarayan, K. (eds.) ISWC 2009. LNCS, vol. 5823, pp. 277–292. Springer, Heidelberg (2009)
33. Hayes, P.: RDF Semantics. W3C Recommendation (February 2004), http://www.w3.org/TR/rdf-mt/
34. Heath, T., Bizer, C.: Linked Data: Evolving the Web into a Global Data Space, 1st edn. Synthesis Lectures on the Semantic Web: Theory and Technology, vol. 1. Morgan & Claypool, San Francisco (2011), http://linkeddatabook.com/editions/1.0/
35. Hepp, M.: Product Variety, Consumer Preferences, and Web Technology: Can the Web of Data Reduce Price Competition and Increase Customer Satisfaction? In: Di Noia, T., Buccafurri, F. (eds.) EC-Web 2009. LNCS, vol. 5692, pp. 144–144. Springer, Heidelberg (2009)
36. Hitzler, P.: OWL and Rules. In: Polleres, A., et al. (eds.) Reasoning Web 2011. LNCS, vol. 6848, pp. 251–327. Springer, Heidelberg (2011)
37. Hitzler, P., van Harmelen, F.: A Reasonable Semantic Web. Semantic Web Journal – Interoperability, Usability, Applicability 1(1) (2010)

38. Hitzler, P., Vrandečić, D.: Resolution-based approximate reasoning for OWL DL. In: Gil, Y., Motta, E., Benjamins, V.R., Musen, M.A. (eds.) ISWC 2005. LNCS, vol. 3729, pp. 383–397. Springer, Heidelberg (2005)

39. Hogan, A.: Exploiting RDFS and OWL for Integrating Heterogeneous, Large-Scale, Linked Data Corpora. PhD thesis, Digital Enterprise Research Institute, National University of Ireland, Galway (2011), http://aidanhogan.com/docs/thesis/

40. Hogan, A., Harth, A., Polleres, A.: Scalable Authoritative OWL Reasoning for the Web. Int. J. Semantic Web Inf. Syst. 5(2) (2009)

41. Hogan, A., Harth, A., Umbrich, J., Kinsella, S., Polleres, A., Decker, S.: Searching and Browsing Linked Data with SWSE: the Semantic Web Search Engine. Technical Report DERI-TR-2010-07-23, Digital Enterprise Research Institute, Galway (2010), http://www.deri.ie/fileadmin/documents/DERI-TR-2010-07-23.pdf

42. Hogan, A., Pan, J.Z., Polleres, A., Decker, S.: SAOR: Template Rule Optimisations for Distributed Reasoning over 1 Billion Linked Data Triples. In: Patel-Schneider, P.F., Pan, Y., Hitzler, P., Mika, P., Zhang, L., Pan, J.Z., Horrocks, I., Glimm, B. (eds.) ISWC 2010, Part I. LNCS, vol. 6496, pp. 337–353. Springer, Heidelberg (2010)

43. Hollunder, B., Nutt, W., Schmidt-Schauß, M.: Subsumption Algorithms for Concept Description Languages. In: ECA 1990, pp. 348–353. Pitman Publishing (1990)

44. Horrocks, I., Kutz, O., Sattler, U.: The Even More Irresistible SROIQ. In: KR 2006 (2006)

45. Horrocks, I., Sattler, U., Tobies, S.: Practical Reasoning for Very Expressive Description Logics. Logic Journal of the IGPL 8, 2000 (2000)

46. Huang, Z., van Harmelen, F.: Using Semantic Distances for Reasoning with Inconsistent Ontologies. In: Sheth, A.P., Staab, S., Dean, M., Paolucci, M., Maynard, D., Finin, T., Thirunarayan, K. (eds.) ISWC 2008. LNCS, vol. 5318, pp. 178–194. Springer, Heidelberg (2008)

47. Er, K.: Hudek and Grant Weddell. Binary Absorption in Tableaux-Based Reasoning for Description Logics. In: Proc. DL 2006 (2006)

48. Jiménez-Ruiz, E., Grau, B.C., Sattler, U., Schneider, T., Llavori, R.B.: Safe and economic re-use of ontologies: A logic-based methodology and tool support. In: Proceedings of the 21st International Workshop on Description Logics, DL 2008 (May 2008)

49. Jones, N.D., Gomard, C.K., Sestoft, P., Andersen, L.O., Mogensen, T.: Partial Evaluation and Automatic Program Generation. Prentice Hall International, Englewood Cliffs (1993)

50. Kazakov, Y.: SRIQ and SROIQ are Harder than SHOIQ. In: DL 2008 (2008)

51. Y. Kazakov Consequence-Driven Reasoning for Horn SHIQ Ontologies. In: IJCAI 2009 (2009)

52. Kiryakov, A., Ognyanoff, D., Velkov, R., Tashev, Z., Peikov, I.: LDSR: a Reason-able View to the Web of Linked Data. In: Semantic Web Challenge, ISWC 2009 (2009)

53. Kobilarov, G., Scott, T., Raimond, Y., Oliver, S., Sizemore, C., Smethurst, M., Bizer, C., Lee, R.: Media Meets Semantic Web – How the BBC Uses DBpedia and Linked Data to Make Connections. In: Aroyo, L., Traverso, P., Ciravegna, F., Cimiano, P., Heath, T., Hyvönen, E., Mizoguchi, R., Oren, E., Sabou, M., Simperl, E. (eds.) ESWC 2009. LNCS, vol. 5554, pp. 723–737. Springer, Heidelberg (2009)

54. Kolovski, V., Wu, Z., Eadon, G.: Optimizing Enterprise-Scale OWL 2 RL Reasoning in a Relational Database System. In: Patel-Schneider, P.F., Pan, Y., Hitzler, P., Mika, P., Zhang, L., Pan, J.Z., Horrocks, I., Glimm, B. (eds.) ISWC 2010, Part I. LNCS, vol. 6496, pp. 436–452. Springer, Heidelberg (2010)

55. Komorowski, H.J.: Partial Evaluation as a Means for Inferencing Data Structures in an Applicative Language: A Theory and Implementation in the Case of Prolog. In: POPL, pp. 255–267 (1982)

56. Lembo, D., Lenzerini, M., Rosati, R., Ruzzi, M., Savo, D.F.: Inconsistency-Tolerant Semantics for Description Logics. In: Hitzler, P., Lukasiewicz, T. (eds.) RR 2010. LNCS, vol. 6333, pp. 103–117. Springer, Heidelberg (2010)

57. Lloyd, J.W.: Foundations of Logic Programming, 2nd edn. Springer, Heidelberg (1987)

58. Lloyd, J.W., Shepherdson, J.C.: Partial Evaluation in Logic Programming. J. Log. Program. 11(3&4), 217–242 (1991)

59. Lutz, C., Walther, D., Wolter, F.: Conservative extensions in expressive description logics. In: IJCAI 2007, Proceedings of the 20th International Joint Conference on Artificial Intelligence, pp. 453–458 (2007)

60. Ma, Y., Hitzler, P.: Paraconsistent Reasoning for OWL 2. In: Polleres, A., Swift, T. (eds.) RR 2009. LNCS, vol. 5837, pp. 197–211. Springer, Heidelberg (2009)

61. Maier, F.: Extending Paraconsistent $\mathcal{SROIQ}$. In: Hitzler, P., Lukasiewicz, T. (eds.) RR 2010. LNCS, vol. 6333, pp. 118–132. Springer, Heidelberg (2010)

62. Meditskos, G., Bassiliades, N.: DLEJena: A practical forward-chaining OWL 2 RL reasoner combining Jena and Pellet. J. Web Sem. 8(1), 89–94 (2010)

63. Motik, B.: Web Ontology Reasoning with Logic Databases. PhD thesis, AIFB, Karlsruhe, Germany (2004)

64. Motik, B., Patel-Schneider, P.F., Parsia, B.: OWL 2 Web Ontology Language Structural Specification and Functional-Style Syntax. W3C Recommendation (October 2009), http://www.w3.org/TR/owl2-syntax/

65. Motik, B., Shearer, R., Horrocks, I.: Optimized reasoning in description logics using hypertableaux. In: Pfenning, F. (ed.) CADE 2007. LNCS (LNAI), vol. 4603, pp. 67–83. Springer, Heidelberg (2007)

66. Motik, B., Shearer, R., Horrocks, I.: Hypertableau Reasoning for Description Logics. Submitted to a Journal (2008)

67. Muñoz, S., Pérez, J., Gutierrez, C.: Minimal Deductive Systems for RDF. In: Franconi, E., Kifer, M., May, W. (eds.) ESWC 2007. LNCS, vol. 4519, pp. 53–67. Springer, Heidelberg (2007)

68. Muñoz, S., Pérez, J., Gutierrez, C.: Simple and Efficient Minimal RDFS. J. Web Sem. 7(3), 220–234 (2009)

69. Oren, E., Delbru, R., Catasta, M., Cyganiak, R., Stenzhorn, H., Tummarello, G.: Sindice.com: a document-oriented lookup index for open linked data. IJMSO 3(1), 37–52 (2008)

70. Oxford-Benchmark. Oxford Benchmark (2009), http://hermit-reasoner.com/2009/JAIR_benchmarks/

71. Page, L., Brin, S., Motwani, R., Winograd, T.: The PageRank Citation Ranking: Bringing Order to the Web. Technical report, Stanford Digital Library Technologies Project (1998)

72. Pan, J.Z., Thomas, E.: Approximating OWL-DL Ontologies. In: AAAI 2007, pp. 1434–1439 (2007)

73. Prud'hommeaux, E., Seaborne, A.: SPARQL Query Language for RDF. W3C Recommendation (January 2008), http://www.w3.org/TR/rdf-sparql-query/

74. Ramakrishnan, R., Srivastava, D., Sudarshan, S.: Rule Ordering in Bottom-Up Fixpoint Evaluation of Logic Programs. In: Proc. of 16th VLDB, pp. 359–371 (1990)

75. Ren, Y., Pan, J.Z., Zhao, Y.: Towards soundness preserving approximation for abox reasoning of owl2. In: Description Logics Workshop 2010, DL 2010 (2010)

76. Rudolph, S.: Foundations of Description Logics. In: Polleres, A., et al. (eds.) Reasoning Web 2011. LNCS, vol. 6848, pp. 251–327. Springer, Heidelberg (2011)

77. Schaerf, M., Cadoli, M.: Tractable Reasoning via Approximation. Artificial Intelligence 74, 249–310 (1995)

78. Spackman, K.: Managing clinical terminology hierarchies using algorithmic calculation of subsumption: Experience with SNOMED-RT. JAMIA (2000)

79. Stonebraker, M.: The Case for Shared Nothing. IEEE Database Eng. Bull. 9(1), 4–9 (1986)
80. Stuckenschmidt, H., Niepert, M.: Combining Probabilistic and Logical Reasoning for Web Data Processing. In: Polleres, A., et al. (eds.) Reasoning Web 2011. LNCS, vol. 6848, pp. 251–327. Springer, Heidelberg (2011)
81. Stuckenschmidt, H., van Harmelen, F.: Approximating Terminological Queries. In: Andreasen, T., Motro, A., Christiansen, H., Larsen, H.L. (eds.) FQAS 2002. LNCS (LNAI), vol. 2522, Springer, Heidelberg (2002)
82. ter Horst, H.J.: Combining RDF and Part of OWL with Rules: Semantics, Decidability, Complexity. In: Gil, Y., Motta, E., Benjamins, V.R., Musen, M.A. (eds.) ISWC 2005. LNCS, vol. 3729, pp. 668–684. Springer, Heidelberg (2005)
83. ter Horst, H.J.: Completeness, decidability and complexity of entailment for RDF Schema and a semantic extension involving the OWL vocabulary. Journal of Web Semantics 3, 79–115 (2005)
84. Trček, D.: Trust management methodologies for the Web. In: Polleres, A., et al. (eds.) Reasoning Web 2011. LNCS, vol. 6848, pp. 251–327. Springer, Heidelberg (2011)
85. Tsarkov, D., Horrocks, I.: Efficient Reasoning with Range and Domain Constraints. In: DL 2004 (2004)
86. Tsarkov, D., Horrocks, I., Patel-Schneider, P.F.: Optimizing Terminological Reasoning for Expressive Description Logics. J. Autom. Reason. 39(3), 277–316 (2007)
87. Ullman, J.D.: Principles of Database and Knowledge Base Systems. Computer Science Press (1989)
88. Urbani, J., Kotoulas, S., Maassen, J., van Harmelen, F., Bal, H.: OWL reasoning with webPIE: Calculating the closure of 100 billion triples. In: Aroyo, L., Antoniou, G., Hyvönen, E., ten Teije, A., Stuckenschmidt, H., Cabral, L., Tudorache, T. (eds.) ESWC 2010. LNCS, vol. 6088, pp. 213–227. Springer, Heidelberg (2010)
89. Urbani, J., Kotoulas, S., Oren, E., van Harmelen, F.: Scalable distributed reasoning using mapReduce. In: Bernstein, A., Karger, D.R., Heath, T., Feigenbaum, L., Maynard, D., Motta, E., Thirunarayan, K. (eds.) ISWC 2009. LNCS, vol. 5823, pp. 634–649. Springer, Heidelberg (2009)
90. Vanderbei, R.J.: Linear Programming: Foundations and Extensions, 3rd edn. Springer, Heidelberg (2008)
91. Vrandečíc, D., Krötzsch, M., Rudolph, S., Lösch, U.: Leveraging Non-Lexical Knowledge for the Linked Open Data Web. Review of Fool's day Transactions (RAFT) 5, 18–27 (2010)
92. Wache, H., Groot, P., Stuckenschmidt, H.: Scalable instance retrieval for the semantic web by approximation. In: Dean, M., Guo, Y., Jun, W., Kaschek, R., Krishnaswamy, S., Pan, Z., Sheng, Q.Z. (eds.) WISE 2005 Workshops. LNCS, vol. 3807, pp. 245–254. Springer, Heidelberg (2005)
93. Weaver, J., Hendler, J.A.: Parallel materialization of the finite RDFS closure for hundreds of millions of triples. In: Bernstein, A., Karger, D.R., Heath, T., Feigenbaum, L., Maynard, D., Motta, E., Thirunarayan, K. (eds.) ISWC 2009. LNCS, vol. 5823, pp. 682–697. Springer, Heidelberg (2009)
94. Yardeni, E., Shapiro, E.Y.: A Type System for Logic Programs. J. Log. Program 10(1/2/3/4), 125–153 (1991)
95. Zhang, R., Artale, A., Giunchiglia, F., Crispo, B.: Using description logics in relation based access control. In: Grau, B.C., Horrocks, I., Motik, B., Sattler, U. (eds.) Description Logics. CEUR Workshop Proceedings, vol. 477 (2009), CEUR-WS.org
96. Zhang, X., Xiao, G., Lin, Z.: A tableau algorithm for handling inconsistency in OWL. In: Aroyo, L., Traverso, P., Ciravegna, F., Cimiano, P., Heath, T., Hyvönen, E., Mizoguchi, R., Oren, E., Sabou, M., Simperl, E. (eds.) ESWC 2009. LNCS, vol. 5554, pp. 399–413. Springer, Heidelberg (2009)

# A    OWL 2 RL/RDF Rules

Herein, we list the rule tables categorising supported OWL 2 RL/RDF rules [27] according to terminological and assertional arity of atoms in the body. Note that herein we (ab)use Turtle syntax [8] and highlight authoritative variable positions (TAVars($R$)— see § 3.5) in bold.

## A.1    "A-Linear" OWL 2 RL/RDF Rules

**Table 15.** OWL 2 RL/RDF rules with empty body (axiomatic triples)

| | Body($R$) = ∅ | |
|---|---|---|
| **ID** | **Head** | **Notes** |
| prp-ap | ?p a owl:AnnotationProperty | For each built-in annotation property |
| cls-thing | owl:Thing a owl:Class . | - |
| cls-nothing | owl:Nothing a owl:Class . | - |
| dt-type1$^a$ | ?dt a rdfs:Datatype . | For each built-in datatype |
| ~~dt-type2~~$^a$ | ~~?l a ?dt .~~ | ~~For all ?l in the value-space of datatype ?dt~~ |
| ~~dt-eq~~$^a$ | ~~?l₁ owl:sameAs ?l₂ .~~ | ~~For all ?l₁ and ?l₂ with the same data value~~ |
| ~~dt-diff~~$^a$ | ~~?l₁ owl:differentFrom ?l₂ .~~ | ~~For all ?l₁ and ?l₂ with different data values~~ |

$^a$ These rules mandate (naïvely) infinite materialised inferences, and so we exclude them.

**Table 16.** OWL 2 RL/RDF rules containing only T-atoms in the body

| | TBody($R$) ≠ ∅, ABody($R$) = ∅ | |
|---|---|---|
| **ID** | **Body** *terminological* | **Head** |
| cls-oo | ?c owl:oneOf (?x₁...?xₙ) . | ?x₁...?xₙ a ?c . |
| scm-cls | ?c a owl:Class . | ?c rdfs:subClassOf ?c , owl:Thing ; owl:equivalentClass ?c . owl:Nothing rdfs:subClassOf ?c . |
| scm-sco | ?c₁ rdfs:subClassOf ?c₂ . ?c₂ rdfs:subClassOf ?c₃ . | ?c₁ rdfs:subClassOf ?c₃ . |
| scm-eqc1 | ?c₁ owl:equivalentClass ?c₂ . | ?c₁ rdfs:subClassOf ?c₂ . ?c₂ rdfs:subClassOf ?c₁ . |
| scm-eqc2 | ?c₁ rdfs:subClassOf ?c₂ . ?c₂ rdfs:subClassOf ?c₁ . | ?c₁ owl:equivalentClass ?c₂ . |
| scm-op | ?p a owl:ObjectProperty . | ?p rdfs:subPropertyOf ?p . ?p owl:equivalentProperty ?p . |
| scm-dp | ?p a owl:DatatypeProperty . | ?p rdfs:subPropertyOf ?p . ?p owl:equivalentProperty ?p . |
| scm-spo | ?p₁ rdfs:subPropertyOf ?p₂ . ?p₂ rdfs:subPropertyOf ?p₃ . | ?p₁ rdfs:subPropertyOf ?p₃ . |
| scm-eqp1 | ?p₁ owl:equivalentProperty ?p₂ . | ?p₁ rdfs:subPropertyOf ?p₂ . ?p₂ rdfs:subPropertyOf ?p₁ . |
| scm-eqp2 | ?p₁ rdfs:subPropertyOf ?p₂ . ?p₂ rdfs:subPropertyOf ?p₁ . | ?p₁ owl:equivalentProperty ?p₂ . |
| scm-dom1 | ?p rdfs:domain ?c₁ . ?c₁ rdfs:subClassOf ?c₂ . | ?p rdfs:domain ?c₂ . |
| scm-dom2 | ?p₂ rdfs:domain ?c . ?p₁ rdfs:subPropertyOf ?p₂ . | ?p₁ rdfs:domain ?c . |
| scm-rng1 | ?p rdfs:range ?c₁ . ?c₁ rdfs:subClassOf ?c₂ . | ?p rdfs:range ?c₂ . |
| scm-rng2 | ?p₂ rdfs:range ?c . ?p₁ rdfs:subPropertyOf ?p₂ . | ?p₁ rdfs:range ?c . |
| scm-hv | ?c₁ owl:hasValue ?i : owl:onProperty ?p₁ . ?c₂ owl:hasValue ?i : owl:onProperty ?p₂ . ?p₁ rdfs:subPropertyOf ?p₂ . | ?c₁ rdfs:subClassOf ?c₂ . |

**Table 16.** *(Continued)*

| | | |
|---|---|---|
| scm-svf1 | $?c_1$ owl:someValuesFrom $?y_1$ ; owl:onProperty $?p$ .<br>$?c_2$ owl:someValuesFrom $?y_2$ ; owl:onProperty $?p$ .<br>$?y_1$ rdfs:subClassOf $?y_2$ . | $?c_1$ rdfs:subClassOf $?c_2$ . |
| scm-svf2 | $?c_1$ owl:someValuesFrom $?y$ ; owl:onProperty $?p_1$ .<br>$?c_2$ owl:someValuesFrom $?y$ ; owl:onProperty $?p_2$ .<br>$?p_1$ rdfs:subPropertyOf $?p_2$ . | $?c_1$ rdfs:subClassOf $?c_2$ . |
| scm-avf1 | $?c_1$ owl:allValuesFrom $?y_1$ ; owl:onProperty $?p$ .<br>$?c_2$ owl:allValuesFrom $?y_2$ ; owl:onProperty $?p$ .<br>$?y_1$ rdfs:subClassOf $?y_2$ . | $?c_1$ rdfs:subClassOf $?c_2$ . |
| scm-avf2 | $?c_1$ owl:allValuesFrom $?y$ ; owl:onProperty $?p_1$ .<br>$?c_2$ owl:allValuesFrom $?y$ ; owl:onProperty $?p_2$ .<br>$?p_1$ rdfs:subPropertyOf $?p_2$ . | $?c_1$ rdfs:subClassOf $?c_2$ . |
| scm-int | $?c$ owl:intersectionOf $(?c_1...?c_n)$ . | $?c$ rdfs:subClassOf $?c_1...?c_n$ . |
| scm-uni | $?c$ owl:unionOf $(?c_1...?c_n)$ . | $?c_1...?c_n$ rdfs:subClassOf $?c$ . |

**Table 17.** OWL 2 RL/RDF rules containing some T-atoms and precisely one A-atom in the body

| | | |
|---|---|---|
| | ABody$(R) \neq \emptyset$, TBody$(R) = \emptyset$ | |
| **ID** | **Body**<br>*assertional* | **Head** |
| ~~eq-ref~~[a] | ~~$?s$ $?p$ $?o$ .~~ | ~~$?s$ owl::sameAs $?s$ . $?p$ owl::sameAs $?p$ . $?o$ owl::sameAs $?o$ .~~ |
| eq-sym | $?x$ owl::sameAs $?y$ . | $?y$ owl::sameAs $?x$ . |

[a] We typically omit this rule which adds unnecessary bulk to the materialised inferences, and could be more easily supported by backward-chaining.

**Table 18.** OWL 2 RL/RDF rules with no T-atoms, but one A-atom in the body

| | | | |
|---|---|---|---|
| | TBody$(R) \neq \emptyset$ and $|$ABody$(R)| = 1$ | | |
| **ID** | **Body**<br>*terminological* | *assertional* | **Head** |
| prp-dom | $?p$ rdfs:domain $?c$ . | $?x$ $?p$ $?y$ . | $?x$ a $?c$ . |
| prp-rng | $?p$ rdfs:range $?c$ . | $?x$ $?p$ $?y$ . | $?y$ a $?c$ . |
| prp-symp | $?p$ a owl:SymmetricProperty . | $?x$ $?p$ $?y$ . | $?y$ $?p$ $?x$ . |
| prp-spo1 | $?p_1$ rdfs:subPropertyOf $?p_2$ . | $?x$ $?p_1$ $?y$ . | $?x$ $?p_2$ $?y$ . |
| prp-eqp1 | $?p_1$ owl:equivalentProperty $?p_2$ . | $?x$ $?p_1$ $?y$ . | $?x$ $?p_2$ $?y$ . |
| prp-eqp2 | $?p_1$ owl:equivalentProperty $?p_2$ . | $?x$ $?p_2$ $?y$ . | $?x$ $?p_1$ $?y$ . |
| prp-inv1 | $?p_1$ owl:inverseOf $?p_2$ . | $?x$ $?p_1$ $?y$ . | $?y$ $?p_2$ $?x$ . |
| prp-inv2 | $?p_1$ owl:inverseOf $?p_2$ . | $?x$ $?p_2$ $?y$ . | $?y$ $?p_1$ $?x$ . |
| cls-int2 | $?c$ owl:intersectionOf $(?c_1 ... ?c_n)$ . | $?x$ a $?c$ . | $?x$ a $?c_1, ..., ?c_n$ . |
| cls-uni | $?c$ owl:unionOf $(?c_1...?c_n)$ . | $?x$ a $?c_i$ . | $?x$ a $?c$ . |
| cls-svf2 | $?x$ owl:someValuesFrom owl:Thing ; owl:onProperty $?p$ . | $?u$ $?p$ $?v$ . | $?u$ a $?x$ . |
| cls-hv1 | $?x$ owl:hasValue $?y$ ; owl:onProperty $?p$ . | $?u$ a $?x$ . | $?u$ $?p$ $?y$ . |
| cls-hv2 | $?x$ owl:hasValue $?y$ ; owl:onProperty $?p$ . | $?u$ $?p$ $?y$ . | $?u$ a $?x$ |
| cax-sco | $?c_1$ rdfs:subClassOf $?c_2$ . | $?x$ a $?c_1$ . | $?x$ a $?c_2$ . |
| cax-eqc1 | $?c_1$ owl:equivalentClass $?c_2$ . | $?x$ a $?c_1$ . | $?x$ a $?c_2$ . |
| cax-eqc2 | $?c_1$ owl:equivalentClass $?c_2$ . | $?x$ a $?c_2$ . | $?x$ a $?c_1$ . |

## A.2  Unsupported OWL 2 RL/RDF Rules

**Table 19.** OWL 2 RL/RDF rules containing no T-atoms, but multiple A-atoms in the body—all relate to supporting the positive semantics of `owl:sameAs`, and all give quadratic materialisation

| | $\lvert \mathsf{ABody}(R) \rvert > 1, \mathsf{TBody}(R) = \emptyset$ | |
|---|---|---|
| **ID** | **Body** | **Head** |
| | *assertional* | |
| eq-trans | *?x* owl:sameAs *?y* . *?y* owl:sameAs *?z* . | *?x* owl:sameAs *?z* . |
| eq-rep-s | *?s* owl:sameAs *?s′* . *?s ?p ?o* . | *?s′ ?p ?o* . |
| eq-rep-p | *?p* owl:sameAs *?p′* . *?s ?p ?o* . | *?s ?p′ ?o* . |
| eq-rep-o | *?o* owl:sameAs *?o′* . *?s ?p ?o* . | *?s ?p ?o′* . |

**Table 20.** OWL 2 RL/RDF rules containing some T-atoms and multiple A-atoms in the body

| | $\mathsf{TBody}(R) \neq \emptyset$ and $\lvert \mathsf{ABody}(R) \rvert > 1$ | | |
|---|---|---|---|
| **ID** | **Body** | | **Head** |
| | *terminological* | *assertional* | |
| | *linear materialisation w.r.t. assertional data* | | |
| cls-int1 | *?c* owl:intersectionOf (*?c_1 ... ?c_n*) . | *?y* a *?c_1* , ... , *?c_n* . | *?y* a *?c* . |
| cls-svf1 | *?x* owl:someValuesFrom *?y* ; owl:onProperty *?p* . | *?u ?p ?v* . *?v* a *?y* . | *?u* a *?x* . |
| cls-avf | *?x* owl:allValuesFrom *?y* ; owl:onProperty *?p* . | *?u ?p ?v* ; a *?x* . | *?v* a *?y* . |
| | *quadratic materialisation w.r.t. assertional data* | | |
| prp-fp | *?p* a owl:FunctionalProperty . | *?x ?p ?y_1* , *?y_2* . | *?y_1* owl:sameAs *?y_2* . |
| prp-ifp | *?p* a owl:InverseFunctionalProperty . | *?x_1 ?p ?y* . *?x_2 ?p ?y* . | *?x_1* owl:sameAs *?x_2* . |
| prp-key | *?c* owl:hasKey (*?p_1 ... ?p_n*) | *?x ?p_1 ?z_1* ; ... ; *?p_n ?z_n* , a *?c* . *?y ?p_1 ?z_1* ; ... ; *?p_n ?z_n* , a *?c* . | *?x* owl:sameAs *?y* . |
| cls-maxc2 | *?x* owl:maxCardinality 1 ; owl:onProperty *?p* . | *?u a ?x* ; *?p ?y_1* , *?y_2* . | *?y_1* owl:sameAs *?y_2* . |
| cls-maxqc3 | *?x* owl:maxQualifiedCardinality 1 . *?x* owl:onProperty *?p* ; owl:onClass *?c* . | *?u a ?x* ; *?p ?y_1* , *?y_2* . *?y_1* a *?c* . *?y_2* a *?c* . | *?y_1* owl:sameAs *?y_2* |
| cls-maxqc4 | *?x* owl:maxQualifiedCardinality 1 . owl:onProperty *?p* ; owl:onClass owl:Thing . | *?u a ?x* ; *?p ?y_1* , *?y_2* . | *?y_1* owl:sameAs *?y_2* |
| prp-trp | *?p* a owl:TransitiveProperty . | *?x ?p ?y* . *?y ?p ?z* . | *?x ?p ?z* |
| prp-spo2 | *?p* owl:propertyChainAxiom (*?p_1 ... ?p_n*) . | *?u_1 ?p_1 ?u_2* . *?u_2 ?p_2 ?u_3* . ... *?u_n ?p_n ?u_{n+1}* . | *?u_1 ?p ?u_{n+1}* . |

**Table 21.** OWL 2 RL/RDF "constraint" rules

| | Head($R$) $= \perp$ | |
|---|---|---|
| **ID** | **Body** | |
| | *terminological* | *assertional* |
| eq-diff1 | - | $?x$ owl:sameAs $?y$ .<br>$?x$ owl:differentFrom $?y$ . |
| *eq-diff2* | - | $?x$ a owl:AllDifferent ;<br>owl:members ($?z_1...?z_n$) .<br>$?z_i$ owl:sameAs $?z_j$ . ($i{\neq}j$) |
| *eq-diff3* | - | $?x$ a owl:AllDifferent ;<br>owl:distinctMembers ($?z_1...?z_n$) .<br>$?z_i$ owl:sameAs $?z_j$ . ($i{\neq}j$) |
| *prp-irp* | $?\boldsymbol{p}$ a owl:IrreflexiveProperty . | $?x$ $?p$ $?x$ . |
| *prp-asyp* | $?\boldsymbol{p}$ a owl:AsymmetricProperty . | $?x$ $?p$ $?y$ . $?y$ $?p$ $?x$ . |
| *prp-pdw* | $?\boldsymbol{p_1}$ owl:propertyDisjointWith $?\boldsymbol{p_2}$ . | $?x$ $?p_1$ $?y$ ; $?p_2$ $?y$ . |
| *prp-adp* | $?x$ a owl:AllDisjointProperties ; owl:members ($?\boldsymbol{p_1}...?\boldsymbol{p_n}$) . | $?u$ $?p_i$ $?y$ ; $?p_j$ $?y$ . ($i{\neq}j$) |
| *prp-npa1* | - | $?x$ owl:sourceIndividual $?i_1$ .<br>$?x$ owl:assertionProperty $?p$ .<br>$?x$ owl:targetIndividual $?i_2$ .<br>$?i_1$ $?p$ $?i_2$ . |
| *prp-npa2* | - | $?x$ owl:sourceIndividual $?i$ .<br>$?x$ owl:assertionProperty $?p$ .<br>$?x$ owl:targetValue $?lt$ .<br>$?i$ $?p$ $?lt$ . |
| cls-nothing2 | - | $?x$ a owl:Nothing . |
| cls-com | $?\boldsymbol{c_1}$ owl:complementOf $?\boldsymbol{c_2}$ . | $?x$ a $?c_1$ , $?c_2$ . |
| cls-maxc1 | $?\boldsymbol{x}$ owl:maxCardinality 0 ; owl:onProperty $?\boldsymbol{p}$ . | $?u$ a $?x$ ; $?p$ $?y$ . |
| *cls-maxqc1* | $?\boldsymbol{x}$ owl:maxQualifiedCardinality 0 ;<br>  owl:onProperty $?\boldsymbol{p}$ ; owl:onClass $?\boldsymbol{c}$ . | $?u$ a $?x$ ; $?p$ $?y$ . $?y$ a $?c$ . |
| *cls-maxqc2* | $?\boldsymbol{x}$ owl:maxQualifiedCardinality 0 ;<br>  owl:onProperty $?\boldsymbol{p}$ ; owl:onClass owl:Thing . | $?u$ a $?x$ ; $?p$ $?y$ . |
| cax-dw | $?\boldsymbol{c_1}$ owl:disjointWith $?\boldsymbol{c_2}$ . | $?x$ a $?c_1$ , $?c_2$ . |
| *cax-adc* | $?x$ a owl:AllDisjointClasses ; owl:members ($?\boldsymbol{c_1}...?\boldsymbol{c_n}$) . | $?z$ a $?c_i$ , $?c_j$ . ($i{\neq}j$) |
| dt-not-type | - | $?lt$ a $?dt$ . (s.t. $?lt$ is an ill-typed literal) |

**Table 22.** Enumeration of the coverage of inferences in case of the omission of rules in Table 16 wrt. inferencing over assertional knowledge by recursive application of rules in Table 17: underlined rules are not supported, and thus we would encounter incompleteness wrt. assertional inference (would not affect a full OWL 2 RL/RDF reasoner which includes the underlined rules).

| ID | partially covered by recursive rule(s) |
|---|---|
| scm-cls | *incomplete* for `owl:Thing` membership inferences[a] |
| scm-sco | cax-sco |
| scm-eqc1 | cax-eqc1, cax-eqc2 |
| scm-eqc2 | cax-sco |
| scm-op | *no unique assertional inferences* |
| scm-dp | *no unique assertional inferences* |
| scm-spo | prp-spo1 |
| scm-eqp1 | prp-eqp1, prp-eqp2 |
| scm-eqp2 | prp-spo1 |
| scm-dom1 | prp-dom, cax-sco |
| scm-dom2 | prp-dom, prp-spo1 |
| scm-rng1 | prp-rng, cax-sco |
| scm-rng2 | prp-rng, prp-spo1 |
| scm-hv | prp-rng, prp-spo1 |
| scm-svf1 | *incomplete:* <u>cls-svf1</u>, cax-sco |
| scm-svf2 | *incomplete:* <u>cls-svf1</u>, prp-spo1 |
| scm-avf1 | *incomplete:* <u>cls-avf</u>, cax-sco |
| scm-avf2 | *incomplete:* <u>cls-avf</u>, prp-spo1 |
| scm-int | cls-int2 |
| scm-uni | cls-uni |

[a] In our scenario, are not concerned—we filter out such statements and rules such as **cls-svf2** and **cls-maxqc2** encode direct support for `owl:Thing`.

# B  CURIE Prefixes Used

Herein, we enumerate the CURIE prefixes [12] used throughout these notes to abbreviate URIs.

**Table 23.** Prefixes used

| Prefix | URI |
|--------|-----|
| aifb: | http://www.aifb.kit.edu/id/ |
| avtimbl: | http://www.advogato.org/person/timbl/foaf.rdf# |
| bmpersons: | http://www4.wiwiss.fu-berlin.de/bookmashup/persons/ |
| b2r2008: | http://bio2rdf.org/bio2rdf-2008.owl# |
| contact: | http://www.w3.org/2000/10/swap/pim/contact# |
| dblpperson: | http://www4.wiwiss.fu-berlin.de/dblp/resource/person/ |
| dbpedia: | http://dbpedia.org/resource/ |
| dc: | http://purl.org/dc/elements/1.1/ |
| dct: | http://purl.org/dc/terms/ |
| doap: | http://usefulinc.com/ns/doap# |
| ex*: | *arbitrary example namespace* |
| fb: | http://rdf.freebase.com/ns/ |
| foaf: | http://xmlns.com/foaf/0.1/ |
| frbr: | http://purl.org/vocab/frbr/core# |
| geonames: | http://www.geonames.org/ontology# |
| identicauser: | http://identi.ca/user/ |
| mo: | http://purl.org/ontology/mo/ |
| opiumfield: | http://rdf.opiumfield.com/lastfm/spec# |
| owl: | http://www.w3.org/2002/07/owl# |
| po: | http://purl.org/ontology/po/ |
| plink: | http://buzzword.org.uk/rdf/personal-link-types# |
| pres: | http://www.w3.org/2004/08/Presentations.owl# |
| rail: | http://ontologi.es/rail/vocab# |
| rdf: | http://www.w3.org/1999/02/22-rdf-syntax-ns# |
| rdfs: | http://www.w3.org/2000/01/rdf-schema# |
| skos: | http://www.w3.org/2004/02/skos/core# |
| swid: | http://semanticweb.org/id/ |
| sworg: | http://data.semanticweb.org/organization/ |
| timblfoaf: | http://www.w3.org/People/Berners-Lee/card# |
| wgs84: | http://www.w3.org/2003/01/geo/wgs84_pos# |
| wn: | http://xmlns.com/wordnet/1.6/ |
| xfn: | http://vocab.sindice.com/xfn# |
| yagor: | http://www.mpii.de/yago/resource/ |

# Rules and Logic Programming for the Web

Adrian Paschke

Freie Universität Berlin
AG Corporate Semantic Web
Königin-Luise-Str. 24/26, 14195 Berlin, Germany
paschke@inf.fu-berlin.de
http://www.corporate-semantic-web.de/

**Abstract.** This lecture script gives an introduction to rule based knowledge representation on Web. It reviews the logical foundations of logic programming and derivation rule languages and describes existing Web rule standard languages such as RuleML, the W3C Rule Interchange Format (RIF), and the Web rule engine Prova.

## 1 Introduction to Rule Based Knowledge Representation

Knowledge representation (KR) focuses on methods for describing the world in terms of high-level, abstracted models which can be used to build intelligent applications, i.e., it provides methods to find implicit consequences of explicitly represented knowledge. Approaches can be roughly divided into *logic based formalisms*, usually a variant of first-order predicate calculus and *non-logic based formalisms* such as graphical semantic networks, object frames or (early) production rule systems. Non-logic based approaches, which are often based on ad hoc data structures and graphical representations, typically lack a precise formal semantics which makes it hard to verify the correctness of drawn consequences. On the other hand, logic based approaches use the powerful and general semantics of first-order logic (FOL) (typically a decidable subset of FOL) which allows a precise characterization of the meaning of a world by expressing it as a knowledge base (KB) of statements in a language which has a truth theory. While the syntax may differ, the semantics of FOL KBs is often given in a Tarski-style semantics.

Rule based systems have been investigated comprehensively in the realms of declarative programming and expert systems over the last decades. Using (inference) rules has several advantages: reasoning with rules is based on a semantics of formal logic, usually a variation of first order predicate logic, and it is relatively easy for the end user to write rules. The basic idea is that users employ rules to express *what* they want, the responsibility to interpret this and to decide on *how* to do it is delegated to an interpreter (e.g., an inference engine or a just in-time rule compiler). Traditionally, rule based systems have been supported by two types of inferencing algorithms: forward chaining and backward chaining.

A. Polleres et al. (Eds.): Reasoning Web 2011, LNCS 6848, pp. 326–381, 2011.

## 1.1   Forward Chaining Rule Systems

Forward chaining is one of the two main methods of reasoning when using "if-then" style inference rules in artificial intelligence. Forward chaining is data-driven. The inference engine makes inferences based on rules from given data. It starts with the available data and uses inference rules to extract more data until an optimal goal is reached. An inference engine using forward chaining searches the inference rules until it finds one where the *if clause* is known to be true. When found it can conclude, or infer, the *then clause*, resulting in the addition of new information to its KB. The most common form of forward chaining is the Rete algorithm. In a nutshell, this algorithm keeps the derivation structure in memory and propagates changes in the fact and rule base. There are many forward chaining implementations in the area of deductive databases and many well-known forward-reasoning engines for production rules ("if condition then action" rules) such as IBM ILOG's commercial rule system or popular open source solutions such as Drools, CLIPS or Jess which are based on variants of the Rete algorithm.

## 1.2   Backward Chaining Rule Systems

The other main reasoning method for if-then rules is backward chaining which is typically used in logic programming, where the rules are called derivation rules. Backward chaining starts with a list of goals (hypothesis) and works backwards to see if there are data available that will support any of these goals. Accordingly, backward chaining is goal-driven. An inference engine using backward chaining would search the inference rules until it finds one which has a *then clause* that matches a desired goal. If the *if clause* of that inference rule is not known to be true, then it is added to the list of goals. The common deductive computational model of logic programming uses backward-reasoning (goal-driven) *resolution* to instantiate the program clauses via goals and uses *unification* to determine the program clauses to be selected and the variables to be substituted by terms. The unification algorithm supports backtracking usually according to depth-first recursive backward chaining, but forward chaining bottom-up approaches are also possible.

## 1.3   Discussion Backward Chaining vs. Forward Chaining in the Web Context

Forward chaining, e.g. based on the Rete algorithm in production rules, can be very effective, e.g., if you just want to find out what new facts are true or when you have a small set of initial facts and when there tends to be lots of different rules which allow you to draw the same conclusion. However, in the context of reasoning on top of Web content backward chaining often qualifies to be the better choice:

- In forward-reasoning additional software must propagate changes to the memory based fact base which leads to a lot of redundancy and difficulties, e.g., a Web content database normally does not propagate changes and for dynamic real-time access the fact base and the Web database must be synchronized.

- In Web applications often large set of initial facts are provided which are likely to change. Using forward chaining, lots of rules would be eligible to fire in any cycle and a lot of irrelevant conclusions are drawn. In backward-reasoning the knowledge base can be temporarily populated with the needed facts from external Web systems to answer a particular goal at query time which can be discarded from the memory afterwards. Forward-reasoning on the Web works best only for closed scopes, e.g., firing rules when certain events occur.
- Open-distributed environments such as the Web are usually based on a pull-model and most implementations of push-architectures (the push model relates to active event processing) are basically pull-concepts, i.e., the push functionality is simulated by frequently issuing queries, e.g., a mail client which queries the mailbox every second for new mails. Therefore, a goal-driven backward-reasoning system perfectly fits to those architectures.
- Forward-reasoning production rules have an operational semantics but no clear logical semantics and a restricted expressiveness, e.g. no recursion, only inflationary negation etc.

The further paper is structured as follows: Section 2 describes logical foundations of logic programming, rules and reasoning. Section 3 introduces standard Web rule languages on the platform independent *interchange* level and the platform specific *execution* level. In particular, the W3C Rule Interchange Format (RIF), RuleML and the Prova rule language (ISO Prolog like syntax) are detailed in this section. Finally, the conclusion summarizes the current state-of-art and future trends.

## 2    Logic Foundations

This section reviews general background knowledge about logic and logic programming and its use for rule based knowledge representation and reasoning.

### 2.1    First-Order Logic

This subsection recalls the definition of a first order logic (FOL) language and classical FOL models (structures) under Tarski semantics adopted from [62, 63, 46]. Both are interrelated concepts and play a central role in logic and form a general basis that allows to cover a wide range of logical formalisms for rule based reasoning and knowledge representation.

**Syntax.** This subsection defines the syntax of a first order language according to [62, 63, 46].

**Definition 1.** *(Signature) $S$ is a signature if $S$ is a four-tuple $\langle \overline{P}, \overline{F}, arity, \overline{c} \rangle$ where:*

1. $\overline{P}$ is a finite sequence of predicate symbols $\langle P_1, .., P_n \rangle$.
2. $\overline{F}$ is a finite sequence of function symbols $\langle F_1, .., F_m \rangle$
3. For each $P_i$ respectively each $F_j$, $arity(P_i)$ resp. $arity(F_j)$ is a non-zero natural number denoting the arity of $P_i$ resp. $F_i$.
4. $\overline{c} = \langle c1, .., c_o \rangle$ is a finite or infinite sequence of constant symbols.

A signature is called function-free if $\overline{F} = \emptyset$.

**Definition 2. (Alphabet)** An alphabet $\Sigma$ consists of the following class of symbols:

1. A signature $S = \langle \overline{P}, \overline{F}, arity, \overline{c} \rangle$.
2. A collection of variables $V$ which will be denoted by identifiers starting with a capital letter like $U, V, X$
3. Logical connectives / operators: $\neg$. (negation), $\wedge$ (conjunction), $\vee$ (disjunction), $\rightarrow$ (implication), $\equiv$ (syntactical equivalent), $=$ (equivalence), $\perp$ (bottom), $\top$ (top).
4. Quantifier: $\forall$ (forall), $\exists$ (exists).
5. Parentheses and punctation symbols: $($, $)$ and $,$.

**Definition 3. (Terms)** A term is defined inductively as follows:

1. A variable is a term.
2. A constant in $\overline{c}$ is a term.
3. If $f$ is a function symbol with arity $n$ and $t_1, .., t_n$ are terms, then $f(t_1, .., t_n)$ is a (complex) term.

Function symbols are written in prefix notation whereby a function always precedes its terms. However, usually terms are also composed of both prefix and infixed symbols, e.g., $(f(2) - f(1)/f(1))$. There are standard ways of dealing with these issues.

**Definition 4. (Atom)** Let $p$ be a predicate symbol with arity $n \in \aleph$. Let $t_1, .., t_n$ be terms, then $p(t_1, .., t_n)$ is an atomic formula of terms. A ground atom is an atomic formula without variables.

**Definition 5. (Well-formed Formula)** A (well-formed) formula is defined as follows:

1. An atom is a formula.
2. If $H$ and $G$ are formulas then
   - $\neg H$ is a formula (negation)
   - $(H \wedge G)$ is a formula (conjunction)
   - $(H \vee G)$ is a formula (disjunction)
   - $(H \rightarrow G)$ is a formula (implication)
   - $(H \equiv G)$ is a formula (equivalence)
3. If $H$ is a formula and $X$ is a variable, then $(\forall X H)$ and $(\exists X H)$ are formulas.

The following precedences are defined:

1. ¬, ∀, ∃
2. ∧, ∨
3. →, ≡

**Definition 6.** *(**First-Order Language**) A FOL language is defined over an alphabet Σ where the signature S may vary from language to language. It consists of the set of all formulas that can be constructed according to the definitions of well-founded formulas using the symbols of Σ. A FOL language is called function-free if the signature is function-free.*

Thus, a language in addition to a signature also contains the logical symbols and a list of variables. The notion "first-order" refers to the fact that quantification is over individuals rather than classes (or functions).

**Definition 7.** *(**Scope of Variables**) Let X be a variable and H be a formula. The scope of ∀X in ∀XH and of ∃X in ∃XH is H. Combinations of ∀X and ∃X bind every occurrence of X in their scope. Any occurrences of variables that are not bound are called free.*

**Definition 8.** *(**Open and Closed Formula**) A formula is open if it has free variables. A formula is closed if it has no free variables.*

**Definition 9.** *(**Literal**) A literal L is an atom or the negation of an atom.*

**Definition 10.** *(**Complement**) Let L be a literal. The complement −L of L is defined as follows:*

$$-L := \begin{cases} \neg A \text{ if } L \equiv A \\ A \text{ if } L \equiv \neg A \end{cases}$$

*where A is an atom.*

**Definition 11.** *(**Theory**) A FOL theory Φ or FOL knowledge base is a set of formulas in a FOL language Σ: Φ ⊆ Σ. The signature S of Φ is obtained from all the constant, function and predicate symbols which occur in Φ.*

Every finite FOL knowledge base (FOL KB) is equivalent to the conjunction of its elements, i.e., it might be equivalently written as a conjunction of formulas.

**Interpretations and Models.** This subsection is concerned with attributing meaning (or truth values) to sentences (well-formed formulas) in a FOL language. The definitions follow [62, 63, 46]. Informally, the sentences are mapped to some statements about a chosen domain through a process known as interpretation. An interpretation which gives the value true to a sentence is said to satisfy the sentence. Such an interpretation is called a model for the sentence and an interpretation which does not satisfy a sentence is called a counter-model.

**Definition 12.** *(**Interpretation / Structure**) Let $S = \langle \overline{P}, \overline{F}, arity, \overline{c} \rangle$ be a signature. I is called an interpretation (or a structure) for S if $I = \langle |M|, \overline{P}^I, \overline{F}^I, \overline{c}^I \rangle$ consists of:*

1. *a non-empty set $|M|$ called the universe of $I$ or the domain of the interpretation. The members of $|M|$ are called individuals of $I$.*
2. $\overline{P}^I = \langle P_1^I, .., P_k^I \rangle$ *associates with each predicate $P_i$ in $S$ of arity $n = arity(P_i)$ an $n$-ary relation $P_i^I$ on $|M|$, i.e., $P_i^I \subseteq |M|^n$, where $|M|^n$ denotes the collection of all $n$-tuples from $|M|$.*
3. $\overline{F}^I = \langle F_1^I, .., F_l^I \rangle$ *is an interpretation for each function symbol $F_j$ of arity $m$, where $F_j^I$ is an $m$-place function $\overline{F}_j^I : |M|^m \rightarrow |M|$, i.e., $F_j^I$ is defined on the set of $m$-tuples of individuals $|M|^m$ with values in $|M|$.*
4. $\overline{c}^I = \langle c^I | c = constant \rangle$ *is an interpretation for the constants of $S$: $c \in S$, where $c^I$ is an individual of $M$: $c^I \in |M|$.*

**Definition 13.** *(Assignment)*

1. Variable Assignment*: Let $\Sigma$ be a FOL language with $\overline{X}$ its set of variables, and $I$ an interpretation for $\Sigma$. An assignment is a function $\sigma$ from $\overline{X}$ into the universe of $\Sigma$.*
2. Term Assignment*: Let $I$ be an interpretation of a FOL language $\Sigma$ with domain $|M|$ and variable assignment $\sigma$. The term assignment wrt $\sigma$ of the term in $\Sigma$ is defined as:*
   *- Each variable is given its assignment according to $\sigma$.*
   *- Each constant is given its assignment according to $I$.*
   *- If $t_1', .., t_n'$ are the term assignments of $t_1, .., t_n$ and $f'$ is the assignment of the function symbols $f$ with arity $n$, then $f'(t_1', .., t_n') \in |M|$ is the term assignment of $f(t_1, .., t_n)$.*

That is, given an assignment $\sigma$, any variable term of the language that is in the domain of $\sigma$ is given a constant value in $|M|$.

**Definition 14.** *(Truth Values)* Let $I$ be an interpretation of a FOL language $\Sigma$ with domain $|M|$ and $\sigma$ be a variable assignment. A formula $F \in \Sigma$ can be given a truth value "false" or "true" as follows:

1. *If the formula is an atom $p(t_1, .., t_n)$ then the truth value is obtained by calculating the value of $p'(t_1', .., t_n')$ where $p'$ is the mapping assigned to $p$ by $I$ and $t_1', .., t_n'$ are the term assignments of $t_1, ..t_n$ wrt $I$ and $\overline{X}$.*
2. *The truth values of the following formulas is given by the following table:*

| $F$ | $G$ | $\neg F$ | $F \wedge G$ | $F \vee G$ | $F \rightarrow G$ | $F = G$ |
|-------|-------|-------|-------|-------|-------|-------|
| true | true | false | true | true | true | true |
| true | false | false | false | true | false | false |
| false | true | true | false | true | true | false |
| false | false | true | false | false | true | true |

3. *If $\exists X F$, then the truth value of the formula is true if there exists $c \in |M|$ such that the formula $F$ has truth value "true" wrt $I$ and $\sigma(X/c)$; otherwise it is false.*
4. *If the formula has the form $\forall X F$, then the truth value of the formula is true if, for all $c \in |M|$ $F$ is "true" wrt $I$ and $\sigma(X/c)$; otherwise, its truth value is false.*

The satisfaction relation $\models$ goes back to A. Tarski and is a major achievement in logic.

**Definition 15. (*Satisfaction*)** *If $F$ is a formula and $\sigma$ is an assignment to the interpretation $I$ of a FOL language $\Sigma$, then the relation $I \models F[\sigma]$ means that $F$ is true in $I$ when there is a substitute for each free variable $X$ of $F$ with the value of $\sigma(X)$. The inductive requirements of "$\models$" are:*

1. *For any atomic formula of the form $p(t_1, ..t_n)[\sigma]$ iff $\langle t_1^\sigma, .., t_n^\sigma \rangle \in p^I$.*
2. *$I \models \neg F[\sigma]$ iff it is not the case that $I \models F[\sigma]$*
3. *$I \models (F \wedge G)[\sigma]$ iff both $I \models F[\sigma]$ and $I \models G[\sigma]$. Similarly, for the other statements.*
4. *$I \models \exists X F[\sigma]$ iff there exists some assignment $\sigma'$ such that*
   - *for every variable $Y$ different from $X$ $\sigma'(Y) = \sigma(Y)$*
   - *$\sigma'(X)$ is defined and $I \models F[\sigma']$*
5. *$I \models \forall X F[\sigma]$ iff for any assignment $\sigma'$, if $\sigma'(X)$ is defined and $\sigma'$ is equal to $\sigma$ on each variable different from $X$, then $I \models F[\sigma']$*

*Accordingly, a formula $F$ is satisfied by an interpretation $I$ ($F$ is true in $I$: $I \models F$ ) iff $I \models_\sigma F$ for all variable assignments $\sigma$. $F$ is valid iff $I \models F$ for every interpretation $I$.*

**Definition 16. (*Model*)** *Let $I$ be an interpretation of a FOL language $\Sigma$. Then $I$ is a model of a closed formula $F$, if $F$ is true wrt $I$. Further, $I$ is a model of a set $\overline{F}$ of closed formulas, if $I$ is a model of each formula of $\overline{F}$. $I$ is a model of an FOL KB $\Phi$ iff $I \models F$ for every formula $F \in \Phi$: $I \models \Phi$.*

**Definition 17. (*Logical Consequence, Entailment, Logical Implication*)** *A formula $F \in \Sigma$ is a logical consequence of a FOL KB $\Phi$ written as $\Phi \models F$, i.e., $\Phi$ entails $F$ iff for all models $I \in \Sigma$ for which $I \models \Phi$ also $I \models F$. For a fixed FOL language (and signature) $\Sigma$ let $\Phi$ and $\Psi$ be two sets of sentences (two KBs), then $\Phi \rightarrow \Psi$ means that for every interpretation $I$ of $\Sigma$, if $I$ is a model for $\Phi$ then it is also a model for $\Psi$.*

*$\Phi \rightarrow \Psi$ is also meaningful when $\Phi$ and $\Psi$ are sets of formulas with variables, i.e., for every interpretation $I$ of $\Sigma$ and every assignment $\sigma$ in $I$, if $I[\sigma]$ satisfies every formula in $\Phi$ then it also satisfies every formula in $\Psi$.*

## 2.2   Logic Programming

Full first-order logic is not suitable as a declarative programming language, e.g. due to the following reasons:

- unrestricted FOL is in general undecidable
- the results are not always unique
- finding (most general) unifier and solving formula is highly complex
- large search domains, which must be restricted using complex control structures
- danger of implementation incompleteness

Hence, logic programming is based on a subset of FOL which deals with a specific class of well-formed formulas, so called statement clauses which consist of an antecedent part and a consequent. The declarative meaning for such clauses is that the consequent part is true, if the antecedents are true. The procedural meaning is, that the consequent is proven by reducing it to a set of sub-goals given by the antecedent part. The most common form of logic programming is based on Horn Logic where clauses in normal form only have one positive literal which is the consequent. Such programs are called definite LPs or Horn LPs. The semantics of definite Logic Programs (LPs) is based on minimal Herbrand models. Although definite LPs are expressive enough to model many problems the formulation is often neither easy nor elegant. Hence, extensions to definite LPs like different forms of negations have been proposed. This subsection introduces relevant terms, concepts, syntax and semantics of different classes of logic programs (LPs) derived from [62, 63, 46].

### Syntax of Logic Programs

**Definition 18.** *(Clause)* *A clause is a formula such as* $\forall \overline{X}(L_1 \vee .. \vee L_m)$ *where each* $L_i$ *is a literal and* $\overline{X} = \{X_1, .., X_n\}$ *are all the variables occurring in* $L_1 \vee .. \vee L_m$.

Different classes of clauses are distinguished: *propositional clauses, Datalog clauses, definite Horn clauses, normal clauses, extended clauses, positive clauses, positive-disjunctive clauses, disjunctive clauses,* and *extended disjunctive clauses.* Associated with each type of clause is a class of logic programs: *propositional LP, Datalog LP, definite LP, stratified LP, normal LP* (aka general LP), *extended LP, disjunctive LP* and combinations of classes, with an increasing expressiveness as illustrated in figure 1 for several classes of LPs. Each class can be propositional (without terms), Datalog (without functions) or with terms and variables.
    These LPs are defined as follows:

**Definition 19.** *(Logic Programs and Rules)* *Given a FOL language* $\Sigma$, *a (disjunctive extended) logic program P consists of logical rules (or program clause) of the form*
    $A_1, .., A_k \leftarrow B_1, .., B_m, not\, C_1, .., not\, C_n$

*or equivalently*
    $\forall \overline{X}(A_1 \vee .. \vee A_k \leftarrow B_1 \wedge .. \wedge B_m \wedge not\, C_1 \wedge .. \wedge not\, C_n)$

*which is a convenient notation for a FOL clause where all variables* $X_i \in \overline{X}$ *occurring in the literals* $A_i$, $B_j$, $C_k$ *are universally quantified* $\forall X_1..\forall X_s$, *the commas in the antecedent denote conjunction and the commas in the consequent denote disjunction, and not denotes negation by default, rather than classical negation. For short a rule is denoted in set notation as:*
    $A \leftarrow B \wedge not\, C$
    *where* $A = A_1 \vee .. \vee A_k$, $B = B_1 \wedge .. \wedge B_m$, $C = C_1 \vee .. \vee C_n$. *Note that* $C$ *is a disjunction and according to De Morgan's law not* $C$ *is taken to be a conjunction.*

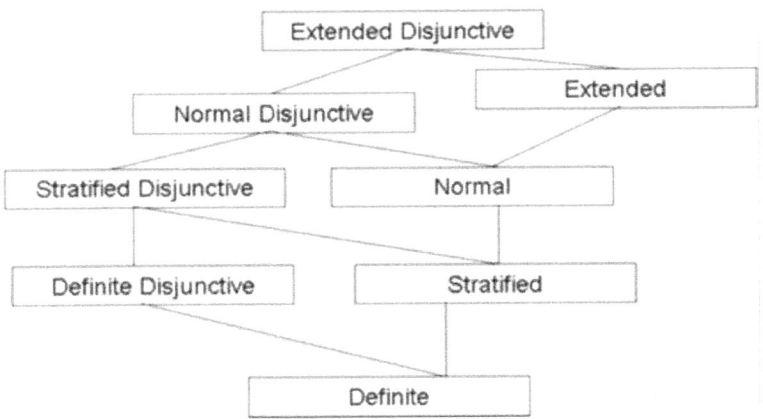

**Fig. 1.** Classes of LPs

*The A is called the* rule head *which consists of the set of head literals and B and C is called the* rule body *which consists of the set of body literals. Note that this set notation is legitime because the conjunction is commutative.*
   *A* clause *is called:*

- a fact *if $m = n = 0$, i.e., $A \leftarrow \emptyset$*
- a query *(or goal) if $k = 0$, i.e., $\leftarrow B \wedge C$. A query or goal is called atomic if it consists of a single literal $B_1$, i.e., $m = 1$ and $n = 0$.*
- a propositional rule *if the arity of all predicates is 0, i.e., all literals are propositional ones. If all rules in a program $P$ are propositional the $P$ is called a* propositional LP.
- a Datalog rule *if it contains no functions, i.e., is function-free and no predicate symbol of the input schema appears in the rule head. A* Datalog LP *(aka deductive database) is a function-free LP.*
- a definite or positive rule *(or Horn clause) if all literals are atoms, $n = 0$ and $k = 1$, i.e., it neither contains negation nor disjunction. The corresponding LP is called* positive or definite LP *(or Horn Program).*
- positive-disjunctive rule *if all literals are atoms and $n = 0$, i.e., it does not contain negation. The corresponding LP is called* positive-disjunctive LP.
- normal rule *if all literals are atoms and $k = 1$, i.e., it does not contain disjunction. The corresponding program is called a* normal LP.
- extended rule *if $A_i$, $B_i$ and $C_i$ are literals, i.e., are atoms or explicitly negated atoms . The corresponding programm is called an* extended LP.
- disjunctive rule *if $k > 1$, i.e., it does contain a disjunction. The corresponding program is called a* disjunctive LP.
- range-restricted *if all variable symbols occurring in the head also occur in the positive body.*
- ground *if no variables occur in it.*

**Semantics of Logic Programs.** Proof-theoretically the semantics of a logic program $P$ is defined as a set of literals that is (syntactically) derivable from $P$ using a particular derivation mechanism such as SLDNF resolution. Model-theoretically, a semantics for a logic program $P$ is concerned with attributing meaning (truth values) to clauses (rules). The properties of soundness and completeness establish a relation between the notions of syntactic ($\vdash$) and semantic ($\models$) entailment in logic programming. This subsection reviews several approaches to define proof-theoretic and model-theoretic semantics for different types of logic programs.

*Substitution and Unification.* At first, the concepts of substitution and unification from [62, 52] are introduced which are at the heart of *proof-theoretic semantics* of non-ground LPs.

**Definition 20. (Substitution)** *A substitution $\theta$ in a language $\Sigma$ is a finite set of the form $\{X_1/t_1, .., X_n/t_n\}$, where each $X_i$ is a variable in $\Sigma$, each $t_i$ is a term in $\Sigma$ distinct from $X_i$ and the variables $X_1, .., X_n$ are pairwise distinct. Each element $X_i/t_i$ is called a binding for $X_i$. $\theta$ is called a ground substitution if the $t_i$ are all ground terms. $\theta$ is called a variable-pure substitution if the $t_i$ are all variables.*

**Definition 21. (Expression)** *An expression $E$ is either a term, a literal or a conjunction or disjunction of literals.*

**Definition 22. (Instance)** *Let $\theta = \{X_1/t_1, .., X_n/t_n\}$ be a substitution and $E$ be an expression then $E\theta$ is the instance of $E$ by $\theta$ is the expression obtained from $E$ by simultaneously replacing each occurrence of the variable $X_i$ in $E$ by the term $t_i$ for $i = 1, .., n$. if $E\theta$ is ground then $E\theta$ is called a ground instance of $E$.*

**Definition 23. (Variant)** *Let $E$ and $D$ be expressions. $E$ and $D$ are variants if there exists substitutions $\theta$ and $\sigma$ such that $E = D\theta$ and $D = E\sigma$.*

**Definition 24. (Renaming Substitution)** *Let $E$ be an expression and $\overline{X}$ be the set of variables occurring in $E$. A renaming substitution for $E$ is a variable-pure substitution $\{X_1/Y_1, .., X_n/Y_n\}$ such that $\{X_1, .., X_n\} \subseteq \overline{X}$, the $Y_i$ are pairwise distinct and $(\overline{X} \setminus \{X_1, .., X_n\}) \cap \{Y_1, .., Y_n\} = \emptyset$.*

**Definition 25. (Composition)** *Let $\theta = \{X_1/s_1, .., X_m/s_m\}$ and $\sigma = \{Y_1/t_1, .., Y_n/t_n\}$ be substitutions. The composition $\theta\sigma$ of $\theta$ and $\sigma$ is the substitution obtained from the set*
$\{X_1/s_1\sigma, .., X_m/s_m\sigma, Y_1/t_1, .., Y_n/t_n\}$
*by deleting any binding $X_i/s_i\sigma$ for which $X_i = s_i\sigma$ and deleting any binding $Y_j/t_j$ for which $Y_j \in \{X_1, .., X_m\}$.*

**Definition 26. (Most General Unifier (MGU))** *Let $\overline{E}$ be a finite set of expressions. A substitution $\theta$ is called a unifier for $\overline{E}$ if $\overline{E}\theta$ is a singleton. An unifier for $\overline{E}$ is called most general unifier (MGU) for $\overline{E}$ if for each unifier $\sigma$ of $\overline{E}$ there exists a substitution $\gamma$ such that $\sigma = \theta\gamma$. $\overline{E}$ is called unifiable if there exists a unifier for $\overline{E}$.*

Note that a MGU for a set of expressions is unique modulo renaming if there exists a MGU at all.

*Minimal Herbrand Model.* For the *model-theoretic semantics* first the minimal or least Herbrand model semantics is introduced which is considered as the natural interpretation of a definite LP. Then the minimal Herbrand semantics is extended for other more expressive subclasses of LPs and further (declarative) semantics are introduced together with their proof-theoretic counterparts for logic programming.

**Definition 27. (*Herbrand Universe*)** *The Herbrand universe of a program $P$ defined over the alphabet $\Sigma$, denoted $U_P$, is the set of all ground terms which can be formed out of the constants and function symbols of the signature $S$ of $\Sigma$.*

**Definition 28. (*Herbrand Base*)** *The Herbrand base of a program $P$, denoted $B_P$, is the set of all ground atomic literals which can be formed by using the predicate symbols in the signature $S$ of $\Sigma$ with the ground terms in $U_P$ as arguments.*

**Definition 29. (*Herbrand Instantiation aka Grounding*)** *The Herbrand instantiation ground$(P)$ of $P$ consists of all ground instances of all rules in $P$ wrt the Herbrand universe $U_P$ which can be obtained as follows: The ground instantiation of a rule $r$ is the collection of all formulas $r[X_1/t_1, .., X_n/t_n]$ with $X_1, .., X_n$ denoting the variables which occur in $r$ and $t_1, .., t_n$ ranging over all terms in $U_P$.*

**Definition 30. (*Herbrand Interpretation*)** *The Herbrand interpretation $I^{Herb}$ of $P$ is a consistent subset of $B_P$. The interpretation is given as follows:*

1. *The domain of the interpretation is the Herbrand universe $U_P$.*
2. *Constants are assigned themselves in $U_P$.*
3. *If $f$ is a function in $P$ with arity $n$ then the mapping $f' : U_P^n \mapsto U_P$ assigned to $f$ is defined by $f'(t_1, ..t_n) := f(t_1, .., t_n)$.*

Note that since the assignment to constant and function symbols is fixed for Herbrand interpretations, it is possible to identify a Herbrand interpretation with a subset of the Herbrand base. For any Herbrand interpretation, the corresponding subset of the Herbrand base is the set of all ground atoms which are true wrt the interpretation.

**Definition 31. (*Herbrand Model*)** *Let $P$ be a positive / definite program. A Herbrand interpretation $I^{Herb}$ of $P$ is a model of $P$, denoted as $M^{Herb}$, iff for every rule $H \leftarrow B_1, .., B_n \in ground(P)$ the following holds: If $B_1, .., B_n \in I^{Herb}$ then $H \in I^{Herb}$.*

The Herbrand model $M^{Herb}$ satisfies the *unique name assumption*, i.e., for any two distinct ground terms in $B_P$, their interpretations are distinct as well.

**Definition 32. (*Unique Name Assumption and Domain Closure Assumption*)** *Let $\Sigma$ be a given language. The unique name assumption (UNA)*

restricts the model $M^{Herb}$, where syntactically different ground terms $t_1, t_2$ are interpreted as non-identical elements: $t_1^{M^{Herb}} \neq t_2^{M^{Herb}}$.
The domain closure assumption (DCA) is a restriction to those models $M^{Herb}$ where for any element $a$ in $M^{Herb}$ there is a term $t$ that represents this element: $a = t^{M^{Herb}}$.

Model-theoretically the intended meaning of a LP is that a formula should be true if it is a logical consequence of the program, i.e., it is true in all models of the program. For definite LPs this intention leads to a semantics that coincides with the intuition because of the model intersection property.

**Definition 33. (Model Intersection Property)** Let $\overline{M}^{Herb}$ be the set of all Herbrand models of a program $P$. The intersection of all Herbrand models $\bigcap \overline{M}^{Herb}(P)$ of a definite LP $P$ is also a Herbrand model of $P$.

Note that since every definite LP $P$ has $B_P$ as an Herbrand model, the set of all Herbrand models for $P$ is always non-empty: $\bigcap \overline{M}^{Herb}(P) \neq \emptyset$.

**Definition 34. (Minimal Herbrand Model)** Let $P$ be a definite LP then the minimal or least Herbrand model $M_P^{Herb}$ of $P$ is the intersection of all Herbrand models for $P$.

The constructive computational characterization of the minimal Herbrand model of a definite LP $P$ is based on the least fixpoint of the immediate consequence operator of $P$. A detailed description of the theory of lattices and fixpoints can be found in [62,52]. Here the relevant definitions are recalled.

**Definition 35. (Immediate Consequence Operator)** Let $P$ be a definite LP. Let $I^{Herb} \subseteq B_P$ be a set of atoms. The set of immediate consequences of $I^{Herb}$ wrt $P$ is defined as follows:
$$T_P(I^{Herb}) := \{A \ — \ there \ is \ A \leftarrow B \in ground(P) \ with \ B \subseteq I^{Herb}\}.$$

**Definition 36. (Monotonic Mapping)** Let $T : P(U) \to P(U)$ be a mapping then $T$ is monotonic if $T(X) \subseteq T(Y)$, whenever $X \subseteq Y$.

**Definition 37. (Ordinal Power of T)** Let $T : P(U) \to P(U)$ be a monotonic mapping then:
$T \uparrow 0 = \emptyset$
$T \uparrow a = T(T \uparrow (a - 1))$ if $a$ is a successor ordinal
$T \uparrow a = \bigcup(T \uparrow b | b < a)$ if $a$ is a limit ordinal

**Definition 38. (Fixpoint of operators)** An operator $T$ is a function $T : P(U) \to P(U)$, where $P(U)$ denotes the powerset of a countable set $U$. A set $X \subseteq U$ is called a fixpoint of the operator $T : P(U) \to P(U)$ iff $T(X) = X$

**Definition 39. (Least Fixpoint)** Let $T : P(U) \to P(U)$ be a mapping. An element $e \in P(U)$ is called a least fixpoint $lfp(T)$ iff $e$ is a fixpoint of $T$ and for all fixpoints $f$ of $T$ it is that $e \subseteq f$.

According to the Fixpoint Theorem of Knaster and Tarski (see [56] for more details) each monotonic operator $T$ has a least fixpoint $lfp(T)$, which is the least upper bound of the sequence $T^0 = \emptyset$, $T^{i+1} = T(T^i)$ for $i \geq 0$. It appears that for each set $P$ of clauses $lfp(T)$ coincides with the unique least Herbrand model of $P$, where a model $M^{Herb}$ is smaller than a model $N^{Herb}$, if $M^{Herb} \subset N^{Herb}$ [43].

**Definition 40.** *(Fixpoints of Monotonic Mappings) Let $T$ be a monotonic mapping. Then $T$ has a least fixpoint $lfp(T)$. For every ordinal $a$, $T \uparrow a \subseteq lfp(T)$. Moreover, there exists an ordinal $b$ such that $c \geq b$ implies $T \uparrow c = lfp(T)$.*

If the operator $T_P$ is not only monotonic but also continuous[1], then a least fixpoint of $T_P$ is always reached not later than at the first upper ordinal (see [62]). By Kleene's theorem (see [37]) $lfp = T \uparrow \omega$.

**Theorem 1.** *(Fixpoint Characterization of the Minimal Herbrand Model) Let $P$ be a definite LP then $M_P^{Herb} = lfp(T_P) = T_P \uparrow \omega$.*

In summary, the semantics of LPs is now defined as follows:

**Definition 41.** *(Herbrand Semantics of Logic Programs) Let the grounding of a clause $r$ in a language $\Sigma$ be denoted as $ground(r, \Sigma)$ where $ground(r, \Sigma)$ is the set of all clauses obtained from $r$ by all possible substitutions of elements of $U_\Sigma$ for the variables in $r$. For any definite LP $P$*
$ground(P, \Sigma) = \bigcup_{r \in P} ground(r, \Sigma)$
*The operator $T_P : 2^{B_P} \to 2^{B_P}$ associates with $P$ is defined by $T_P = T_{ground(P)}$, where $ground(P)$ denotes $ground(P, \Sigma(P))$, and accordingly:*
$SEM_{Herb}(P) = M_{ground(P)}^{Herb}.$

Generating $ground(P)$ is often a very complex task, since, even in case of function-free languages, it is in general exponential in the size of $P$. Moreover, it is not always necessary to compute $M_{ground(P)}^{Herb}$ in order to determine whether $P \models A$ for some particular atom $A$. In practice, various proof-theoretic strategies of deriving atoms from a LP have been proposed. These strategies are based on variants of Robinson's famous *Resolution Principle* [87]. The major variant is SLD-resolution [57].

*SLD Resolution.* In a nutshell, in SLD a goal is a conjunction of atoms. A substitution $\theta$ is a function that maps variables to terms. Asking a query $Q?$, where $Q?$ may contain variables, to a program $P$ means asking for all possible substitutions $\theta$ of the variables in $Q?$ such that $Q\theta$ follows from $P$, i.e., $\theta$ is the answer to $Q$. In other words, SLD resolution repeatedly transforms the initial goal by applying the resolution rule to an atom $Q_i$ from the query/goal and a rule from $P$, unifying $Q_i$ with the head H of the rule, i.e., it tries to find a substitution $\theta$ such that $H\theta = Q_i\theta$. The typical selection rule is to choose always the first atom in the query. This step is repeated until all goals are resolved and the empty goal is obtained.

---

[1] In the sense of a Scott-continuous function, which is one that preserves all directed suprema.

*Example 1.* (**Linear resolution computation step**)

$$\frac{\neg Q_1, .., \neg Q_n \qquad\qquad \neg A_1, .., \neg A_m, H}{\theta = unify(Q_1, \neg H)}$$

Remark: The common deductive computational model of logic programming uses backward-reasoning (goal-driven) resolution to instantiate the program clauses via goals and uses unification to determine the program clauses to be selected and the variables to be substituted by terms. In logic programming unification is used to derive specific information out of general rules which assert general information about a problem. The rules are instantiated via goals, leading to specific instances of these rules. A goal $G$? initiates a refutation attempt unifying the goal $G$? with the head of an appropriate rule $H \leftarrow B$ leading to an instance of the rule $(H \leftarrow B)'$ if there exists a substitution $\theta = \{V_1/t_1, .., V_n/t_n\}$ which assigns terms $t_i$ to variables $V_i$ such that $(H \leftarrow B)' = (H \leftarrow B)\theta$. Applying a substitution $\theta$ to a term, atom or rule (program clause) yields the instantiated term, atom or clause. For example, the rule $son(X, Y) : -parent(Y, X), male(X).$ and a goal $son(adrian, Y)$? leads to the more specialized instance $son(adrian, Y) : -parent(Y, adrian), male(adrian).$ The instance body $B\theta$ is the goal reduction (sub goal) for further derivation leading to more specific instances. Repeating this process leads to an instance order $(H \leftarrow B) \geq (H \leftarrow B)' \geq$ whereas $\geq$ denotes the relation "more general as". The unification algorithm finds the greatest lower bounds (glb) of terms under this instance order $\geq$, i.e. if $\theta$ is a most general unifier (MGU) for a set of terms $T$ then $T\theta$ is the glb of $T$.

For a more precise account see [5,62] and [59] for resolution on normal clauses. The task to find substitutions $\theta$ such that $Q\theta$ is derivable from the program $P$ as well as $M_P^{Herb}$ is closely related to SLD. The following properties are equivalent:

**Theorem 2.** *(Soundness and Completeness of SLD)*

- $P \models \forall Q\theta$, *i.e* $\forall Q\theta$ *is true in all models of* $P$,
- $M_P^{Herb} \models \forall Q\theta$,
- *SLD computes an answer* $\tau$ *that subsumes* $\theta$ *wrt* $Q$, *i.e.,* $\exists \sigma : Q\tau\sigma = Q\theta$.

Since SLD resolution is a top-down approach which starts with the query, the main feature of it is, that it automatically ensures, that it only considers those rules that are relevant for the query to be answered (see also section 1.3 for a discussion of backward vs. forward reasoning). Rules that are not at all related are simply not considered in the course of the proof. Note that there are also several bottom up approaches for computing the least Herbrand model $M_P^{Herb}$ from below. However, the bottom-up approach has two serious shortcomings:

1. The "goal-orientedness" from top-down approaches is lost, i.e the whole $M_P^{Herb}$ has to be computed even for those facts that have nothing to do with the query.
2. In any step facts that are already computed before are recomputed again.

Partial solutions have been proposed, e.g., semi-naive bottom-up evaluation [98, 26] or Magic Sets techniques [13]. However, as discussed in section 1 top-down semantics are more appropriate in Web knowledge representation and the focus is on backward-reasoning logic programming techniques.

**Theory of Logic Programming with Negation.** Definite LPs are typically not expressive enough for general knowledge representation on the Web which is used to represent e.g. decision logics and situational logics. They e.g., exclude negative information and (non-monotonic) default statements such as *normally a implies c, unless something abnormal holds*. Such statements and the computation of default negation where the main motivation for alternative formulations of non-monotonic reasoning by circumscription [66], default reasoning [84] or autoepistemic reasoning [65]. Independently of these work in non-monotonic reasoning the proof-theory for *negation-as-finite-failure* (NAF), the well-known *SLDNF resolution* (SLD+NAF), originated from SLD resolution. In short, negation-as-finite-failure can be characterized as: A (default) negated literal $\sim L$ succeeds, if $L$ finitely fails. See [62, 5] for the formal definition of SLDNF resolution and NAF. The implementation is often given as a cut-fail test[2]:

```
not([P|Args]) :-
    derive([P|Args]), % derive P(Args)
    !, % cut
    fail(). % fail
not([P|Args]). % positive answer
```

The corresponding model-theoretic semantics is defined by *Clark's completion* (COMP) [33] whose idea was to interpret "$\leftarrow$" in rules as "$\leftrightarrow$" in the classical sense.

**Definition 42.** *(Clark's Completion COMP) Clark's completion semantics COMP for a program P is given by the set of all classical models $\overline{M}(comp(P))$ of the completion theory $comp(P)$.*

See Clark's Equational Theory for more details [33]. COMP gives two rules for inferring negative information:

- Infer $\neg A$ iff $B_P \setminus \overline{M}(comp(P)) \models \neg A$
- Infer $\neg A$ iff $\overline{M}(comp(P)) \models \neg A$

But, (two-valued) COMP is incomplete and does not characterize the transitive closure correctly. In [80] various problems with loops in COMP were discussed. Therefore, Fitting [44] introduced a three-valued formulation $comp_3(P)$ of the

---

[2] The basic idea behind this implementation is to make a closed world assumption (i.e. all knowledge is completely known to the inference interpreter) and positively proof the existence of the negated goal literal, which would refute the negation.

two-valued COMP. It was shown by Kunen [58] that SLDNF is sound and complete wrt $COMP_3$ for propositional LPs and correct but not complete in the predicate logic case [93].

SLDNF resolution suffers from problems with loops and floundering and its implementation is only a simple test, i.e., no variable bindings are produced. See [94] for a discussion of unsolvable problems related to SLDNF. Much work has been done to define restriction properties (on the dependency graph whose vertices are the predicate symbols from a program $P$) for which SLDNF is complete. The important ones are briefly reviewed here:

- stratified: no predicate depends negatively on itself
- strict: there are no dependencies that are both even and odd
- allowedness: at least every variable occurring in a clause must occur in at least one positive literal of the body
- call-consistent: no predicate depends oddly on itself.
- hierarchical: no form of recursion is allowed

*Stratified LPs* for which the rules do not have recursion through negation have been defined by [7]. The predicates of stratified LPs can be placed into strata so that one can compute over the strata. The model-theoretic semantics, the *supported Herbrand model* $M_P^{supp}$, is defined by declaring $M_P^{supp}$ as the intended model among all minimal Herbrand models of $comp(P)$ which could be obtained by iterating over the strata. Przymusinski [77] showed that the selected model was the so-called *perfect model*. The semantics of definite and stratified LPs lead to the unique minimal model semantics which is generally accepted to be the semantics for these classes of LPs.

However, this is not the case for more expressive LPs. Here are several possible ways to determine the semantics and various approaches based on extensions of the 2-valued classical logic to three-valued logics have been proposed, e.g., Fitting [44] or Kunen [58] semantics which are based on Kleene's strong three-valued logics, or the *well founded semantics* (WFS) [101] which is an extension of the perfect model semantics. Another approach is based on the tradition of non-monotonic reasoning in which the definition of entailment is based on the notion of beliefs. The *stable model semantics* (STABLE) [47] is based on this approach. For a discussion of the relationships between non-monotonic theories and logic programming see [67]. In the following, the (declarative) semantics and theory of more expressive types of LPs will be reviewed. Different semantics have been defined in the past. Table 1 gives an incomplete overview.

In the following, the prominent semantics will be described, namely well-founded semantics (WFS) and stable model semantics (STABLE) for normal LPs with its extension answer set semantics (ASS) for extended LPs.

*Stable Model Semantics.* The Gelfond-Lifschitz transformation $P^M$ [47] of a normal LP $P$ wrt to its interpretation $I$ is obtained from the ground instance $ground(P)$ of $P$ as follows:

**Table 1.** Semantics for LP Classes (adapted from [36])

| Class | Semantics | Ref. |
|---|---|---|
| Definite LPs | Least Herbrand model: $M_p$ | [7] |
| Stratified LPs | Supported Herbrand model: $M_p^{supp}$ | [7] |
| Normal LPs | Clark's Completion: $COMP$ | [33] |
| | 3-valued Completion: $COMP_3$ | [58, 44] |
| | Well-founded Semantics: $WFS$ | [101] |
| | $WFS^+$ and $WFS'$ | [34] |
| | $WFS_C$ | [91] |
| | Strong Well-founded Semantics: $WFS_E$ | [27] |
| | Stable Model Semantics: $STABLE$ | [47] |
| | Generalized WFS: $GWFS$ | [10] |
| | $STABLE^+$ | [35] |
| | $STABLE_C$ | [91] |
| | $STABLE^{rel}$ | [34] |
| | Pereira's $O - SEM$ | [74] |
| | Partial Model Semantics: $PARTIAL$ | [90] |
| | Regular Semantics: $REG - SEM$ | [102] |
| | Preferred Semantics: $PREFERRED$ | [39] |
| Extended LPs | Extended Well-founded Semantics: $WFS_S$ | [54] |
| | Answer Set Semantics: $ASS$ | [48, 49] |
| | Extended Well-founded Semantics: $WFSX$ | [73] |
| General Disjunctive | Disjunctive WFS: $DWFS$ | [21] |
| | Generalized Disjunctive WFS: $GDWFS$ | [11] |
| | Disjunctive Stable: $DSTABLE$ | [82] |
| Stratified Disjunctive | Perfect model $PERFECT$ | [77] |
| | Weakly Perfect: $WPERFECT$ | [75] |
| | Generalized Closed World Assumption: $GCWA$ | |
| Positive Disjunctive | Weak generalized closed world assumption: $WGCWA$ | [83] |

**Definition 43.** (*Gelfond-Lifschitz transform*) *Let $P$ be a program and $M \subseteq B_P$. The Gelfond-Lifschitz transform $P^M$ of $P$ (aka reduct of $P$) wrt $M$ is defined by $P^M = r^M |r \in ground(P)$. It is obtained from $ground(P)$ by:*

1. *Replace in every ground rule $A \leftarrow B \wedge notC \in ground(P)$ the negative body by its truth value wrt $M$.*
2. *Deleting each rule $r$ in $P$ with $B^-(r) \cap M \neq \emptyset$ where $B^-$ denotes the set of negated atoms in the body of the rule $r$.*

Based on $P^M$ the concepts of stable models [47] and partial stable models [82] have been defined:

**Definition 44.** (*Stable Model*) *An interpretation $I$ of a normal LP $P$ is a stable model $M^{Stable}$ of $P$ if $I$ is a minimal model of $P^M$:*
$$SEM_{Stable}(P) = \bigcap\nolimits_{M^{Stable} \in SEM_{Stable}(P)} (M^{Stable} \cup neg(B_P \setminus M^{Stable}))$$

**Definition 45.** *(**Partial Stable Model**) A partial Herbrand interpretation is called a partial stable model of $P$ if it is a partial minimal model of $P^M$.*

It can be shown that stable models are always partial models and that every stratified LP $P$ has a unique stable model where stratified and stable semantics coincide.

*Answer Set Semantics.* Gelfond and Lifschitz [48,49] have extended the concept of stable models to extended and disjunctive LPs based on the notion of answer sets. The proposed answer-set semantics is defined as follows:

**Definition 46.** *(**Answer Set Semantics**) Let $P$ be an extended (disjunctive) LP. $P$ is transformed to a (explicit) negation-free program $P'$ by replacing all negative literals $\neg A$ by positive literals $A'$ over new predicate symbols. Every stable model $M^{Stable}$ of $P'$ defines an answer set of $P$, which is a set of literals:*
$\overline{L} = A \in B_P | M^{Stable}(A) = t \cup \neg A \in \neg B_P | M^{Stable}(A') = t$
*If $\overline{L}$ does not contain complementary pairs $A, \neg A$ of literals, then the answer set is $\overline{L}$ else it is $B_P \cup \neg B_P$ is the set of all ground literals.*

Associated with $SEM_{Stable}$ are two entailment relations:

**Definition 47.** *(**Cautious Entailment**) An extended LP $P$ cautiously entails a ground atomic formula $a$ iff $a \in I$ for every answer set $M^{Stable}$ of $P$.*

**Definition 48.** *(**Brave Entailment**) An extended program $P$ bravely entails a ground atomic formula $a$ iff $a \in I$ for some answer set $M^{Stable}$ of $P$.*

*Well-founded Semantics.* There exists several definitions to well-founded semantics (WFS), e.g., [101,45,12,81]. Van Gelder, Ross and Schilpf [101] were the first to extend the work of Apt et al. [7] to the class of normal logic programs. The well-founded semantics (WFS) of Gelder et al. is a three-valued logic: *true*, *false* and *unknown*. WFS is an extension of the perfect model semantics, in contrast to Fitting and Jacob's semantics which is based on Kleene's strong three valued logic. For instance, WFS (as well as perfect model semantics) assigns the truth value "false" to a clause $p \leftarrow p$ while Fitting and Jacob assign "unknown". Following the definition from [101] WFS is defined as follows.

**Definition 49.** *(**Partial Interpretation**) Let $P$ be a normal LP. A partial interpretation $I$ is a set of ground literals such that for no atom $A$ both $A$ and not $A$ are contained in $I$, i.e., $pos(I) \cap neg(I) = \emptyset$ and whose atoms are contained in $B_P$ of $P$, i.e., $pos(I) \cup neg(I) \subseteq B_P$. $I$ is a total interpretation, if $I$ is a partial interpretation and for every atom $A \in B_P$ it contains $A$ or not $A$, i.e., $pos(I) \cup neg(I) = B_P$.*

**Definition 50.** *(**Unfounded Set**) Let $P$ be a normal LP. Let $I$ be a partial interpretation. Let $\alpha \subseteq B_P$ be a set of ground atoms. $\alpha$ is an unfounded set of $P$ wrt $I$, if for every atom $A \in \alpha$ and every ground rule instance $A \leftarrow \beta \in ground(P)$ at least one of the following conditions holds:*

1. *at least one body literal $L \in \beta$ is false in $I$.*
2. *at least one positive body literal $B \in \beta$ is contained in $\alpha$.*

**Definition 51.** *(**Greatest Unfounded Set**) Let $P$ be a normal LP. Let $I$ be a partial interpretation. The greatest unfounded set of $P$ wrt $I$ is the union of all unfounded sets of $P$ wrt $I$.*

**Definition 52.** *(**Pos. and Neg. Immediate Consequences**) For a ground normal LP $P$ and a partial interpretation $I \subseteq B_P$ the following monotonic transformation operators are defined:*

- $T_P(I) := A \in B_P | \exists (A \leftarrow \beta) \in ground(P) : \beta \subseteq I$
- $U_P(I) :=$ *the greatest unfounded set of $P$ wrt $I$*
- $W_P(I) := T_P(I) \cup \sim U_P(I)$

**Lemma 1.** *$T_P$, $U_P$ and $W_P$ are monotonic operators.*

**Theorem 3.** *Let $P$ be a normal LP. For every countable ordinal $\alpha$, $W_P \uparrow \alpha$ is a partial model of $P$.*

**Definition 53.** *(**Well-founded Model**) The least fixpoint of $W_P$ is the well-founded (partial) model of $P$ denoted $W_P^*$. The least fixpoint can be computed as follows, $lfp(W_P) = W_P^\infty(\emptyset)^3$. If $lfp(W_P) \subseteq B_P$ is a total interpretation of $P$ then $lfp(W_P)$ is a well-founded model. An atom $A \in B_P$ is well-founded (resp. unfounded) wrt $P$ iff $A$ (resp. $\neg A$) is in $lfp(W_P)$.*

WFS is defined for the grounding of an arbitrary normal LP: $ground(P)$, i.e., it defines a mapping $SEM_{WFS}$, which assigns to every normal LP $P$ a set $SEM_{WFS}(P)$ of (partial) models of $P$ such that $SEM_{WFS}(P) = SEM_{WFS}(ground(P))$ (i.e., $SEM_{WFS}$ is instantiation invariant).

**Definition 54.** *(**Well-founded Semantics**) The Well-founded semantics (WFS) assigns to every normal LP $P$ the well-founded partial model $W_P^*$ of $P$:*

$SEM_{WFS}(P) := \{W_P^*\}.$

Remark: In the (van Gelder)-Definition of the well-founded semantics, $W_P$ is not a function on the set of all three-valued interpretations, i.e. it is not well-defined. Indeed, there are three-valued interpretations $I$ such that $W_P(I)$ is not three-valued (it becomes four-valued). However, this is not a serious problem because the iterates $W_P \uparrow \alpha$ are provably still always all three-valued. [52]

**Definition 55.** *(**Entailment**) A normal LP $P$ entails a ground atom $a$ under WFS, denoted by $P \models a$, if it is true in $SEM_{WFS}(P)$.*

WFS can be considered an approximation of stable models, i.e., if a program has stable models, then if an atom is true resp. false wrt the WFS then it is true resp. false wrt STABLE. [81] Moreover, for weakly stratified LPs [76] WFS coincides with STABLE. However, there are three important distinction between STABLE and WFS:

1. WFS is a three-valued semantics, whereas STABLE is two-valued.
2. every normal LP has exactly one WFS model, whereas every normal LP has zero or more stable models.
3. Irrelevant clauses (tautologies) lead to the non-existence of stable models, e.g., $p \leftarrow \neg p$ has no stable model.

While the alternating fixpoint on normal logic programs only captures the negation of positive existential closure such as e.g. transitive closure, it does not capture the negation of positive universal closure. As shown by van Gelder [101] the constructive characterization of the well-founded semantics for normal logic programs in terms of alternating fixpoint partial models can be further extended towards an alternating fixpoint semantics for general logic programs. There have been also several proposals for extending WFS by classical negation leading to a well-founded semantics for extended LPs - see e.g., [39,40,9,61,73,24]. Decidable and semi-decidable fragments of the WFS have been discussed in [32].

*Procedural Semantics for Normal and Extended LPs.* Existing procedural semantics for the computation of the well-founded model can be divided into two groups: (1) *bottom-up approaches* such as the alternating fixpoint approach [99,100,64], the magic set approach [89,55,68,95] and transformation based (aka residual program) approaches [25,41,22,23] and (2) *top down approaches* such as *non-tabling based approaches* such as Global SLS resolution [78,88] or tabling based approaches such as extensions to OLDT resolution [96], e.g., WELL [14], XOLDTNF [29] or the approach of Bol and Degerstedt [16], SLT resolution [92] or the well-known SLG resolution [28] (another prominent extension of OLDT). There are also some proof procedures for well-founded semantics for extended logic programs (WFSX) such as [97] or SLX resolution [3].

The well-known 2-valued top-down SLDNF (classical LP Prolog) resolution [33], a resolution based method derived from SLD resolution [57,8], as a procedural semantics for LPs has many advantages. Due to its linear derivations it can be implemented using efficient stack based memory structures, it supports very useful sequential operators such as cut, denoted by !, or *assert/retract* and the negation-as-finite failure test is computationally quite efficient. Nevertheless, it is a too weak procedural semantics for unrestricted LPs with negations. It does not support goal memoization and suffers from well-known problems such as redundant computations of identical calls, non-terminating loops or floundering. It is not complete for LPs with negation or infinite functions. Moreover, it can not answer free variables in negative subgoals since the negation as finite failure rules is only a simple test. For more information on SLDNF resolution I refer to [62,6]. For typical unsolvable problems related to SLDNF see e.g. [94].

SLG resolution [30,28] is the most prominent tabling based top-down method for computing the well-founded semantics for normal LPs and it has been show in [31] how SLG can be used for query evaluation of general logic programs under WFS alternating fixpoint semantics. SLG resolution overcomes infinite loops and redundant computations by tabling. The basic idea of tabling, as implemented e.g., in ODLT resolution [96], is to answer calls (goals) with the memorized answers from earlier identical goals which are stored in a table. However, SLG resolution is a non-linear approach. SLG is based on program transformations using six basic transformation rules, instead of the tree based approach of SLDNF. It distinguishes between solution nodes, which derive child nodes using the clauses from the program and look-up nodes, which produce child nodes using the memorized answers in the tables. Since all variant subgoals derive answers from the same solution node, SLG resolution essentially generates a search graph instead of a search tree and jumps back and forth between lookup and solution nodes, i.e., it is non-linear. Special delaying literals are used for temporarily undefined negative literals and a dependency graph is maintained to identify negative loops. Calls to look-up nodes will be suspended until all answers are collected in the table, in contrast to the linear SLD style where a new goal is always generated by linearly extending the latest goal. It is up to this non-linearity of SLG that tabled calls are not allowed to occur in the scope of sequential operators such as cut.

Global-SLS resolution [78, 79, 88] for WFS is a procedural semantics which directly extends SLDNF resolution and hence preserves the linearity property of SLDNF. In contrast to SLDNF-trees, SLS-trees treat infinite derivations as failed and recursions through negation as undefined. However, it assumes a positivistic computation rule that selects all positive literals before negative ones and inherits the problem of redundant computations from SLDNF. Moreover, a query fails if the SLS-trees for the goal either end at a failure leave or are infinite, which makes Global-SLS computationally ineffective [88]. To avoid redundant computations in SLS, a tabling approach called tabulated SLS resolution [15] was proposed. But the approach, like SLG, is based on non-linear tabling.

SLX resolution [3] is a procedural semantics for extended LPs which is sound and complete wrt WFSX semantics. As in SLS resolution it uses a failure rule to solve the problems of infinite positive recursions and distinguishes two kinds of derivations for proving verity (SLX-T tree) and proving non-falsity (SLX-TU tree) in the well-founded model in order to fail or succeed literals involved in recursion through negation. Thus, SLX does not consider a temporal undefined status as the other top-down approaches for WFS do, but implements the following derivations: if a goal $L$ is to be undefined wrt WFS it must be failed, if it occurs in a SLX-T derivation and refuted if it occurs in a SLX-TU derivation. To fulfill the coherence requirement of WFSX a default negated literal $\sim L$ is removed from a goal if there is no SLX-TU refutation for $L$ or if there is one SLX-T refutation for $\sim L$. In short, SLX is very close to SLDNF resolution. As already pointed out by the authors [3,4] it is only theoretically complete, does not guarantee termination since it lacks loop detection mechanisms, is in general not efficient and makes redundant computations since tabling is not supported. Its implementation is given as a meta program in Prolog.

SLE resolution (Linear resolution with Selection function for Extended WFS) extends linear SLDNF with goal memoization based on linear tabling and loop cutting. In short, it resolves infinite loops and redundant computations by tabling without violating the linearity property of SLD style resolutions. SLE resolution is based on four truth values: $t$ (true), $f$ (false), $u$ (undefined) and $u'$ (temporarily undefined) with $t = \neg f$, $\neg f = t$, $\neg u = u$ and $\neg u' = u$ and a truth ordering

$\neg f > t > u > u'$. $u'$ will be used if the truth value of a subgoal is temporarily undecided. SLE resolution follows SLDNF, where derivation trees are constructed by resolution. For more information on the notion of trees for describing the search space of top-down proof procedures see e.g. [62]. In SLE a node in a tree is defined by $N_i : G_i$, where $N_i$ is the node name and $G_i$ is the first goal labelling the node. Tables are used to store intermediate results. In contrast to SLG resolution, there is no distinction between lookup and solution nodes in SLE. The algorithm, always, first tries to answer the call (goal) with the memorized answers in the tables. If there are no answers available in a table the call is resolved against program clauses which are selected in the same top-down order as in SLDNF. This avoids redundant computations. To preserve the order the answers stored in a table are used in a FIFO (first-in-first-out) style, i.e., the first memorized answer is first used to answer the call. In case of loops the two main issues in top-down procedural semantics for WFS are solutions to infinite positive recursions (positive loops) and infinite recursion through negation by default (negative loops).

## 3   Web Rule Languages

Web rule languages have been developed for the declarative representation of, e.g., privacy policies, business rules, and Semantic Web rules. Rules are central to knowledge representation for the Semantic Web and are often considered as being side by side with ontologies, e.g. in W3C's hierarchical Semantic Web architecture (2007 version shown in Figure 2).

There are different types of rules which can be used on the Web such as

- *Derivation rules* are sentences of knowledge that are derived from other knowledge by an inference or mathematical calculation.
- *Reaction rules* are behavioral rules which react on occurred events or changed conditions by executing actions.
- *Integrity rules* (or constraints) are assertions which express conditions that must be always satisfied.
- *Deontic rules* describe rights and obligations of roles in the context of evolving states (situations triggered by events/actions) and state transitions.
- *Transformation rules* - specify term rewriting, which can be considered as derivation rules of logics with (oriented) equality
- *Facts* might describe various kinds of information such as events (event/action messages, event occurences), (object-oriented) object instances, class individuals (of ontology classes), norms, constraints, states (fluents), conditions of various forms, actions, data (e.g., relational, XML), etc., which might be qualified, e.g., by priorities, temporally, etc.

Rules can influence the operational and decision processes of Web systems.

- Derivation rules (deduction rules): establish / derive new information from existing Web data that is used, e.g. in a decision process.
- Reaction rules that establish when certain activities should take place:

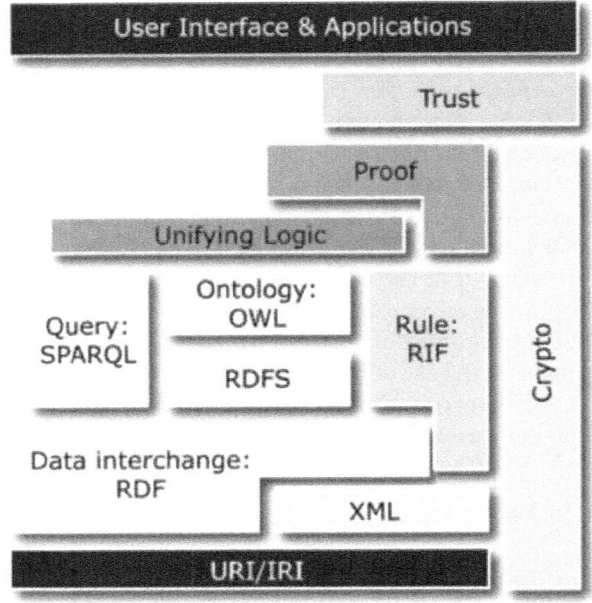

**Fig. 2.** Semantic Web Layer Cake [adapted from (W3C, 2007)]

- Condition-Action rules (production rules)
- Event-Condition-Action (ECA) rules + variants (e.g. ECAP)
- Messaging reaction rules (event message reaction rules)

Rules can also act as constraints on the Web systems structure, behavior and information.

- Structural constraints (e.g. deontic assignments).
- Integrity constraints and state constraints
- Process and flow constraints.

Web rule markup languages provide the required expressiveness enabling machine-interpretation, automated processing and translation into other such Web languages, some of which also being the execution syntaxes of rule engines. One of these languages may act as a lingua franca to interchange rules and integrate with other markup languages, in particular with Web languages based on XML and with Semantic Web languages (e.g. RDF Schema, OWL and OWL 2) for ontologies based on RDF or directly on XML. Web rule languages may also be used for publication purposes on the Web and for the serialization of external data sources, e.g. of native online XML databases or RDF stores. Recently, there have been several efforts aiming at rule interchange and building a general, practical, and deployable rule markup standard for the (Semantic) Web. These include several important general standardization or standards-proposing efforts including RuleML (www.

rulɘml.org), the W3C member submission SWRL (www.w3.org/Submission/SWRL/), the W3C recommendation RIF (www.w3.org/2005/rules/), and others.

A complete specification of Web rule languages consists of a formalization of their *syntax, semantics* and, often left implicit, *pragmatics*. The syntax of Web rule markup languages always includes the concrete syntax of (XML) markup, perhaps indirectly through other languages such as via RDF/XML. Often, there is another more or less concrete syntax such as a compact shorthand or presentation syntax, which may be parsed into the XML markup. While a presentation syntax can already disregard certain details, an abstract syntax systematically replaces character sequences with abstract constructors, often in a (UML) diagram form or as an abstract syntax tree (AST). Together with different token dictionaries, it can be used to generate corresponding concrete syntaxes. The semantics is formalized in a model-theoretic, proof-theoretic, or procedural manner, sometimes in more than one. When rules and speech-act-like performatives, such as queries and answers, are transmitted between different systems, their pragmatic interpretation, including their pragmatic context, becomes relevant, e.g. in order to explain the effects of performatives - such as the assertion or retraction of facts - on the internal knowledge base [72].

A general distinction of three rule modeling layers can be adopted from OMG's model driven architecture (MDA) engineering approach (http://www.omg.org/mda/):

- A platform specific model (PSM) which encodes the rule statements in the language of a specific execution environment
- A platform independent model (PIM) which represents the rules in a common (standardized) interchange format, a rule markup language
- A computational independent model (CIM) with rules represented in a natural or visual language

The CIM level comprises visual and verbal rendering and rule modeling, e.g. via graphical representation or a controlled natural language syntax for rules, mainly intended for human consumption. Graphical representations such as UML diagrams or template-driven/controlled languages can also be used as presentation languages.

The PIM level should enable platform-independent machine interpretation, processing, interchange and translation into multiple PSM execution syntaxes of concrete rule engines. Hence, the concrete XML (or RDF/XML based) syntax of a Web rule language such as RuleML, SWRL or RIF resides on this level, whereas the abstract syntax is on the borderline between the PIM and CIM levels.

The PSM level is the result of translating/mapping PIM rule (interchange) languages into execution syntaxes, such as ISO Prolog, POSL, Prova (http://prova.ws/), which can be directly used in a specific execution environment such as a rule engine. A general distinction can be made between a compiled language approach, where the rules are statically translated into byte code (at compile time) versus interpreted scripting languages, which are dynamically interpreted (at run-time). While the compiled approach has obvious efficiency benefits, the interpreted approach is more dynamic and facilitates, e.g., updates at run-time. Often, Semantic Web Rule Languages are directly executable by their respective rule engines; hence reside on the PSM level. As an intermediate step between the concrete PSM level and the PIM level an abstract representation is often introduced, such as N3, which provides an abstract rule syntax based on the RDF syntax.

The correct execution of an interchanged PIM-level rule set serialized in a rule markup language depends on the semantics of both the rule program and the platform-specific rule inference engine (IE). To address this issue, the IE and the interchanged rule set must reveal their intended/implemented semantics. This may be solved via explicit annotations based on a common vocabulary, e.g. an (Semantic Web) ontology which classifies the semantics. Annotations describing the semantics of an interchanged rule set could even be used to find appropriate IEs on the Web to correctly and efficiently interpret and execute the rule program; for example, (1) by configuring the rule engine for a particular semantics in case it supports different ones, (2) by executing an applicable variant of several interchanged semantic alternatives of the rule program, or (3) by automatic transformation approaches which transform the interchanged rule program into a rule program with an applicable semantics.

In the following two subsections languages on the PIM and PSM level will be described.

### 3.1   Platform Independent Web Rule Languages

In the following, three prominent platform independent Web Rule languages are introduced.

**RuleML**
The Rule Markup Language (RuleML, www.ruleml.org) is a markup language developed to express a family of Web rules in XML for deduction, rewriting, and reaction, as well as further inferential, transformational, and behavioral tasks. It is defined by the Rule Markup Initiative (www.ruleml.org), an open network of individuals and groups from both industry and academia that was formed to develop a canonical Web language for rules using XML markup and transformations from and to other rule standards/systems. It develops a modular, hierarchical specification for different types of rules comprising facts, queries, derivation rules, integrity constraints (consistency-maintenance rules), production rules, and reaction rules (Reaction RuleML, http://reaction.ruleml.org), as well as tools and transformations from and to other rule standards/systems. Datalog RuleML is defined over both data constants and individual constants with an optional attribute for IRI (URI) webizing. Atomic formulas have n arguments, which can be positional terms or, in Object-Oriented Datalog, slots (F-logic-like key→term pairs); OO Datalog also adds optional types and RDF-like oids/anchors, via IRIs (Boley, 2003). Inheriting all of these Datalog features, Hornlog RuleML adds positional or slotted functional expressions as terms. In Hornlog with equality, such uninterpreted (constructor-like) functions are complemented by interpreted (equation-defined) functions. This derivation rule branch is extended upward

towards First Order Logic, has subbranches with Negation-As-Failure, strong-Negation, or combined languages, and is parameterized by 'pluggable' built-ins.

## SWRL

The Semantic Web Rule Language (SWRL, www.w3.org/Submission/SWRL/) is defined as a language combining sublanguages of the OWL Web Ontology Language (OWL DL and Lite) with those of the Rule Markup Language (Unary/Binary Datalog). The specification was submitted to W3C in May 2004 by the National Research Council of Canada, Network Inference (since acquired by web-Methods), and Stanford University in association with the Joint US/EU ad hoc Agent Markup Language Committee. Compared to Description Logic Programs (DLP) [50], a slightly earlier proposal for integrating description logic and Horn rule formalisms by an overlapping authoring team, SWRL takes the opposite integration approach: DLP can be seen as the intersection of description logic and Horn logic; SWRL, as roughly their union. For DLP, the resulting rather inexpressive language corresponds to a peculiar looking description logic imitating special rules. It is hard to see the DLP restrictions, which stem from Lloyd-Topor transformations, being either natural or satisfying. On the other hand, SWRL retains the full power of OWL DL, but adds rules at the price of undecidability and a lack of complete implementations, although the SWRL Tab of Protege has become quite popular (http://protege.cim3.net/cgi-bin/wiki.pl?SWRLTab). Rules in SWRL are of the form of an implication between an antecedent (body) conjunction and a consequent (head) conjunction, where description logic expressions can occur on both sides. The intended interpretation is as in classical first-order logic: whenever the conditions specified in the antecedent hold, then the conditions specified in the consequent must also hold.

SWRL [53] is a homogeneous approach combining rules with ontoligies with relatively high complexity bounds for the ontology reasoning part (due to the fact that standard rule engines are not optimized for DL reasoning). In general, the works on combining rules and ontologies can be basically classified into two basic approaches: *homogeneous and heterogeneous integrations*. Starting from the early Krypthon language [20] among the heterogeneous approaches, which hybridly use DL reasoning techniques and tools in combination with rule languages and rule engines are e.g., CARIN [60], Life [2], Al-log [38], non-monotonic dl-programs [42] and r-hybrid KBs [85]. Among the homogeneous approaches which combine the rule component and the DL component in one homogeneous framework sharing the combined language symbols are e.g., DLP [50], KAON2 [69] or SWRL [53]. Both integration approaches have pros and cons and different integration strategies such as reductions or fixpoint iterations are applied with different restrictions to ensure decidability. These restrictions reach from the intersection of DLs and Horn rules [50] to leaving full syntactic freedom for the DL component, but restricting the rules to *DL-safe rules* [69], where DL variables must also occur in a non DL-atom in the rule body, or *role-safe rules* [60], where at least one variable in a binary DL-query in the body of a hybrid rule must also appear in a non-DL atom in the body of the rule which never appears in the consequent of any rule in the program or to tree-shaped rules [51]. Furthermore, they can be distinguished according to their information flow which might be *uni-directional* or *bi-directional*. For instance, in homogeneous approaches bi-directional information flows between the rules and the ontology part are naturally supported and new DL constructs introduced in the rule heads can be directly used in the integrated ontology inferences, e.g., with the restriction that the variables also appear in the rule body (safeness condition). However, in these approaches the DL reasoning is typically solved completely by the rule engine and benefits of existing optimized DL reasoners using, e.g. variants of tableau based algorithms, are lost. On the other hand, heterogenous approaches, benefit from the hybrid use of both reasoning concepts exploiting the advantages of both (using LP reasoning and tableaux based DL reasoning), but bi-directional information flow and fresh DL constructs in rule heads are much more difficult to implement. A more complete survey and discussion of the combination of rules and ontologies is given in chapter "OWL and Rules" of this lecture book. [1]

**W3C RIF**

The W3C Rule Interchange Format (RIF) Working Group [86] is an effort, influenced by RuleML, to define a standard Rule Interchange Format for facilitating the exchange of rule sets among different systems and to facilitate the development of intelligent rule based applications for the Semantic Web. For these purposes, RIF Use Cases and Requirements (RIF-UCR) have been developed. The RIF architecture is conceived as a family of languages, called dialects. A RIF dialect is a rule based language with an XML syntax and a well-defined semantics.

The W3C RIF recommendation defines the Basic Logic Dialect (RIF-BLD), which corresponds to a definite Horn rule language with equality. RIF-BLD has a number of syntactic extensions with respect to 'regular' Horn rules, including internationalized resource identifiers (IRIs) as identifiers for concepts, F-logic-like frames and slots, and a standard system of built-ins drawn from Datatypes and Built-Ins (RIF-DTB). RIF Core (RIF-Core) in the intersection of RIF-BLD and the Production Rule Dialect (RIF-PRD) influenced by OMG's PRR, which can then be further extended or supplemented by reaction rules. The connection to other W3C Semantic Web languages is established via RDF and OWL Compatibility (RIF-SWC). Moreover, RIF-BLD is a general Web language in that it supports the use of IRIs (Internationalized Resource Identifiers) and XML Schema data types. The RIF Working Group has also defined the Framework for Logic Dialects (RIF-FLD). RIF-FLD uses a uniform notion of terms for both expressions and atoms in a higher order logic (HiLog)-like manner.

In the following, the syntax and semantics of the basic logic dialect of RIF will be summarized.

**Definition 56.** *(**Alphabet**): The alphabet of the non-normative presentation language of RIF-BLD, which maps to the normative XML syntax of RIF, consists of*

- *a countably infinite set of **constant symbols** Const*
- *a countably infinite set of **variable symbols** Var*
- *a countably infinite set of **argument/slot names**, Arg*
- ***connective symbols** And, Or, and : −*

- **quantifiers** *Forall and Exists*
- *the symbols* $\rightarrow$, *External,* =, #, ##, *Import, Prefix, and Base*
- *the symbols Group and Document*
- *the auxiliary symbols* (, ), [, ], <, >, *and* ^^.

Constants in RIF are written as $literal^{\wedge\wedge}symbolspace$, where literal is a sequence of Unicode characters and symbolspace is an identifier for a symbol space consisting of an identifier and a lexical space. Symbol spaces supported in RIF are

- **identifiers of Web entities**, where the lexical space consists of strings that syntactically are internationalized resource identifiers (IRIs), e.g.,
  `http://www.w3.org/2007/rif#iri`
- **datatypes** supported by RIF, e.g.
  `http://www.w3.org/2001/XMLSchema#integer`
- **rif:local** which is used for function and predicate symbols that are local to a rule document.

Slot/argument names $Arg$ and variables $Var$ are unicode strings. Variables start with the symbol ?, e.g. $?x$. The symbol $\rightarrow$ is used in terms that have named arguments and in frame formulas. Equality in RIF is denoted by =. The symbols #, and ## are used in formulas that define class membership and subclass relationships. The symbol $External$ defines an external atomic formula or a function term defined by a RIF built-in. The symbol $Document$ is used to specify RIF-BLD documents. The symbol $Import$ is used for importing documents, and the symbol $Group$ is used to organize RIF-BLD formulas into rule sets.

Using the above alphabet the language of RIF-BLD is constructed as a set of formulas. The main building blocks that are used to construct formulas are **terms**. RIF-BLD defines several kinds of terms: **constants** and **variables**, **positional** (as in normal logic programs) and **named-argument** (slotted) terms (as in F-Logic), and additionally **equality, membership, subclass, frame**, and **external** terms.

Positional terms of the form $p(v_1...v_n)$ and unpositional named arguments $p(s_1 \rightarrow v_1...s_n \rightarrow v_n)$, where $p$ is a (webized) predicate symbol, are atomic formula. Equality, subclass, membership, and frame terms are atomic formulas, too. $External(\varphi)$, where $\varphi$ is an atomic formula, is an externally defined atomic formula.

The *condition language* of RIF-BLD constructs condition formula from atomic formula using **Conjunction**: $And(\varphi_1...\varphi_n)$ to build a conjunctive formula, **Disjunction**: $Or(\varphi_1...\varphi_n)$, to build disjunctive formula, and **Existentials**: $Exists?V_1...?V_n(\varphi)$ to build existential formula.

Condition formulas are used inside the premises of rules in the RIF *rule language* dialects (RIF Core, RIF BLD and RIF PRD). In RIF-BLD definite horn rules are defined as **rule implications**: $\psi : -\varphi$ which are universally quantified $Forall ?V_1...?V_n(\psi : -\varphi)$. A set of rules is grouped in a **group formula**: $Group(\varphi_1...\varphi_n)$, where $\varphi_i$ is either a universal fact, variable-free rule implication, variable-free atomic formula, or another group formula. Finally, RIF-BLD **document formula** are expressions of the form: $Document(directive_1...directive_n \Pi)$, where $\Pi$ is an optional group formula (the knowledge base of the RIF document) and the optional directives are

- import directive of the form $Import(iri)$ or $Import(iriprofile)$, where $iri$ indicates the location of another RIF document to be imported plus and optional $profile$ for import.
- base directives of the form $Base(iri)$ defining syntactic shortcuts for expanding relative IRIs into full IRIs
- prefix directive of the form $Prefix(pv)$ defining a syntactic shortcut to enable a compact URI representation for $rif : iri$ constants.

Additionally, RIF-BLD allows every term and formula to be optionally preceded by an annotation of the form $(* \ id \ \varphi \ *)$, where $id$ is a $rif : iri$ constant and $\varphi$ is a formula.

The non-normative presentation syntax corresponds to the normative XML syntax of RIF-BLD which uses the element and attribute names listed below:

- **And**: conjunction
- **Or**: disjunction
- **Exists**: quantified formula for existentials, containing declare and formula roles
- **declare**: declare role, containing a Var
- **formula**: formula role, containing a FORMULA
- **Atom**: atom formula, positional or with named arguments
- **External**: external call, containing a content role
- **content**: content role, containing an Atom, for predicates, or Expr, for functions
- **Member**: prefix version of member formula #
- **Subclass**: prefix version of subclass formula ##
- **Frame**: Frame formula
- **object**: Member/Frame role, containing a TERM or an object description
- **op**: Atom/Expr role for predicates/functions as operations
- **args**: Atom/Expr positional arguments role, with fixed 'ordered' attribute, containing n TERMs
- **instance**: Member instance role
- **class**: Member class role
- **super**: Subclass super-class role
- **sub**: Subclass sub-class role
- **slot**: prefix version of Name/TERM → TERM pair as an Atom/Expr or Frame slot role, with fixed 'ordered' attribute
- **Equal**: prefix version of term equation '='
- **Expr**: expression formula, positional or with named arguments
- **left**: Equal left-hand side role
- **right**: Equal right-hand side role
- **Const**: individual, function, or predicate symbol, with optional 'type' attribute
- **Name**: name of named argument
- **Var**: serialized version of logic '?' variable

- **id**: identifier role, containing IRICONST
- **meta**: meta role, containing metadata as a Frame or Frame conjunction
- **Document**: document, containing optional directive and payload roles
- **directive**: directive role, containing Import
- **payload**: payload role, containing Group
- **Import**: importation, containing location and optional profile
- **location**: location role, containing IRICONST
- **profile**: profile role, containing PROFILE
- **Group**: nested collection of sentences
- **sentence**: sentence role, containing RULE or Group
- **Forall**: quantified formula for 'Forall', containing declare and formula roles
- **Implies**: prefix version of logic ':-' implication, containing if and then roles
- **if**: antecedent role, containing FORMULA
- **then**: consequent role, containing ATOMIC or conjunction of ATOMICs

Like RuleML, the XML syntax of RIF divides all XML tags into class descriptors starting with upper case letters, called *type tags*, and property descriptors starting with lower case letters, called *role tags*. [17]

The semantics of RIF-BLD is an adaptation of the standard semantics for Horn clauses. It is specified using general models.

**Definition 57.** *(**Semantic Structure**) A semantic structure, $I$, is a tuple of the form ¡TV, DTS, D, $D_{ind}$, $D_{func}$, $I_C$, $I_V$, $I_F$, $I_{frame}$, $I_{NF}$, $I_{sub}$, $I_{isa}$, $I_=$, $I_{external}$, $I_{truth}$¿, where $D$ is a non-empty set of elements called the domain of $I$, and $D_{ind}$, $D_{func}$ are nonempty subsets of $D$. $D_{ind}$ is used to interpret the elements of Const that play the role of individuals and $D_{func}$ is used to interpret the constants that play the role of function symbols. DTS denotes a set of identifiers for primitive datatypes as defined in RIF-DTB. $I_C$ maps Const to $D$. $I_V$ maps Var to $D_{ind}$. $I_F$ maps $D$ to functions $D*_{ind} \rightarrow D$ with $D*_{ind}$ being a set of all finite sequences over the domain $D_{ind}$. $I_{NF}$ maps $D$ to the set of total functions $SetOfFiniteSets(ArgNames \times D_{ind}) \rightarrow D$, where ArgNames are named arguments. $I_{frame}$ maps $D_{ind}$ to total functions of the form $SetOfFiniteBags(Dind \times Dind) \rightarrow D$. $I_{sub}$ is a mapping of the form $D_{ind} \times D_{ind} \rightarrow D$. $I_{isa}$ is a mapping of the form $D_{ind} \times D_{ind} \rightarrow D$. $I_=$ is a mapping of the form $D_{ind} \times D_{ind} \rightarrow D$. $I_{truth}$ is a mapping of the form $D \rightarrow TV$. Finally, $I_{external}$ is a mapping of symbols into Const described as external to fixed n-ary functions.*

RIF-BLD also defines a generic mapping from terms to $D$ as follows:

- $I(k) = I_C(k)$, if $k$ is a symbol in *Const*
- $I(?v) = I_V(?v)$, if $?v$ is a variable in *Var*
- $I(f(t_1...t_n)) = I_F(I(f))(I(t_1), ..., I(t_n))$
- $I(f(s_1 \rightarrow v_1...s_n \rightarrow v_n)) = I_{NF}(I(f))(< s_1, I(v_1) >, ..., < s_n, I(v_n) >)$
- $I(o[a_1 \rightarrow v_1...a_k \rightarrow v_k]) = I_{frame}(I(o))(< I(a_1), I(v_1) >, ..., < I(a_n), I(v_n) >)$
  Note, that in RIF $I(o[a \rightarrow b \, a \rightarrow b]) = I(o[a \rightarrow b])$.
- $I(c_1 \#\# c_2) = I_{sub}(I(c_1), I(c_2))$
- $I(o\#c) = I_{isa}(I(o), I(c))$

– $I(x = y) = I_{=}(I(x), I(y))$
– $I(External(p(s_1...s_n))) = I_{external}(p)(I(s_1), ..., I(s_n))$.

The truth value of (non-document) formulas in RIF BLD is determined from the semantic structures by the following truth valuation.

**Definition 58. (Truth Valuation)** *The truth valuation $TVal_I$ is defined as follows:*

– *Positional atomic formulas: $TVal_I(r(t_1...t_n)) = I_{truth}(I(r(t_1...t_n)))$*
– *Atomic formulas with named arguments: $TVal_I(p(s_1 \rightarrow v_1...s_k \rightarrow v_k)) = I_{truth}(I(p(s_1 \rightarrow v_1...s_k \rightarrow v_k)))$*
– *Equality: $TVal_I(x = y) = I_{truth}(I(x = y))$ with $I_{truth}(I(x = y)) = t$ if $I(x) = I(y)$ and that $I_{truth}(I(x = y)) = f$ otherwise*
– *Subclass: $TVal_I(sc\#\#cl) = I_{truth}(I(sc\#\#cl))$*
– *Membership: $TVal_I(o\#cl) = I_{truth}(I(o\#cl))$*
– *Frame: $TVal_I(o[a_1 \rightarrow v_1...a_k \rightarrow v_k]) = I_{truth}(I(o[a_1 \rightarrow v_1...a_k \rightarrow v_k]))$*
– *Externally defined atomic formula: $TVal_I(External(t)) = I_{truth}(I_{external}(t))$*
– *Conjunction: $TVal_I(And(c_1...c_n)) = t$ if and only if $TVal_I(c_1) = ... = TVal_I(c_n) = t$. Otherwise, $TVal_I(And(c_1...c_n)) = f$.*
– *Disjunction: $TVal_I(Or(c_1...c_n)) = f$ if and only if $TVal_I(c_1) = ... = TVal_I(c_n) = f$. Otherwise, $TVal_I(Or(c_1...c_n)) = t$.*
– *Quantification:*
  • *$TVal_I(Exists?v_1...?v_n(\varphi)) = t$ if and only if for some $TVal_{I*}(\varphi) = t$*
  • *$TVal_I(Forall?v_1...?v_n(\varphi)) = t$ if and only if $TVal_{I*}(\varphi) = t$, where $I*$ is a semantic structure with the special mapping $I*_V$ which coincides with $I_V$ on all variables except, possibly, on $?v_1, ..., ?v_n$.*
– *Rule implication:*
  • *$TVal_I(conclusion:-condition) = t$, if either $TVal_I(conclusion) = t$ or $TVal_I(condition) = f$.*
  • *$TVal_I(conclusion:-condition) = f$ otherwise.*
– *Groups of rules: If $\Pi$ is a group formula of the form $Group(\varphi_1...\varphi_n)$ then*
  • *$TVal_I(\Pi) = t$ if and only if $TVal_I(\varphi_1) = t, ..., TVal_I(\varphi_n) = t$.*
  • *$TVal_I(\Pi) = f$ otherwise.*

Since RIF allows to import other documents which can have $rif : local$ constants, semantic multi-structures are introduced for the interpretation of documents. Semantic multi-structures are essentially similar to regular semantic structures, as defined above, but, in addition, they allow to interpret $rif : local$ symbols that belong to different documents differently.

The following logical entailment defines what it means for a set of RIF-BLD rules to entail another RIF-BLD formula, in particular entailment of RIF condition formulas.

**Definition 59. (Models)** *A multi-structure $I$ is a model of a formula, $\varphi$, written as $I \models \varphi$, iff $TVal_I(\varphi) = t$.*

**Definition 60.** *(Logical Entailment) Let $\varphi$ and $\psi$ be formulas, then $\varphi$ entails $\psi$, written as $\varphi \models \psi$, if and only if for every multi-structure $I$ for which both $TVal_I(\varphi)$ and $TVal_I(\psi)$ are defined, $I \models \varphi$ implies $I \models \psi$.*

For a more detailed account of RIF and RIF BLD we refer to the W3C recommendation [86] and [19].

The following subsection 3.3 exemplifies how platform independent rule languages such as RuleML and RIF are mapped into platform specific rule languages such as Prova.

## 3.2   Prova - A Platform Specific Web Rule Language

Prova (http://www.prova.ws/) is both a (Semantic) Web rule language and a highly expressive distributed (Semantic) Web rule engine which supports complex reaction rule based workflows, rule based complex event processing, distributed inference services, rule interchange, rule based decision logic, dynamic access to external data sources, Web Services, and Java APIs. Prova follows the spirit and design principles of the W3C Semantic Web initiative and combines declarative rules, ontologies and inference with dynamic object-oriented programming and access to external data sources via query languages such as SQL, SPARQL, and XQuery. One of the key advantages of Prova is its separation of logic, data access, and computation as well as its tight integration of Java, Semantic Web technologies and enterprise service-oriented computing and complex event processing technologies.

Semantically Prova provides the expressiveness of serial Horn logic with a linear resolution for extended logic programs (SLE resolution) and with several extra logical features which will be described in the following subsequent subsections. Syntactically Prova builds on top of the ISO Prolog syntax (ISO Prolog ISO/IEC 13211-1:1995), but it extends it syntactically and semantically. The following diagram 3 gives an overview on the Prova 3 language structure and its main language elements.

The basic syntactic structures of the Prova language are rules (head :- body), facts (rule heads with no body), and goals (rules with no head). Prova supports atomic terms (constants and variables) and complex terms (functions internally represented as lists). Constants in Prova can be simple strings starting with lower case letters (e.g. const) or text in single or double quotes (e.g. "Constant 1"), numeric data (e.g. 12, -300L), as well as fully qualified static or instance fields in Java objects (e.g. java.lang.Double(1.3)) or (Description Logic) individuals of ontology concepts (e.g. $10^{\wedge\wedge}$math:Percentage). Variables start with upper case letters (e.g. $X$). They can be typed (e.g. Integer.$X$) and assume the type of the assigned constant (e.g. $X = 1$). Like in Prolog anonymous variables begin with underscore (_). Special global variables and constants have names starting with '\$' (e.g. \$Counter). Complex terms in Prova are functions which can be equally represented as generic lists, where the first head element is the function operator and the list tail are its arguments, e.g. $f(X, Y)$ can be equally represented as $[f, X, Y]$ or $[f|R]$ (which binds the list tail to the variable $R$). Prova supports positional literals as in Prolog, e.g. $p(arg_1, ..., arg_n)$ as well as unpositional

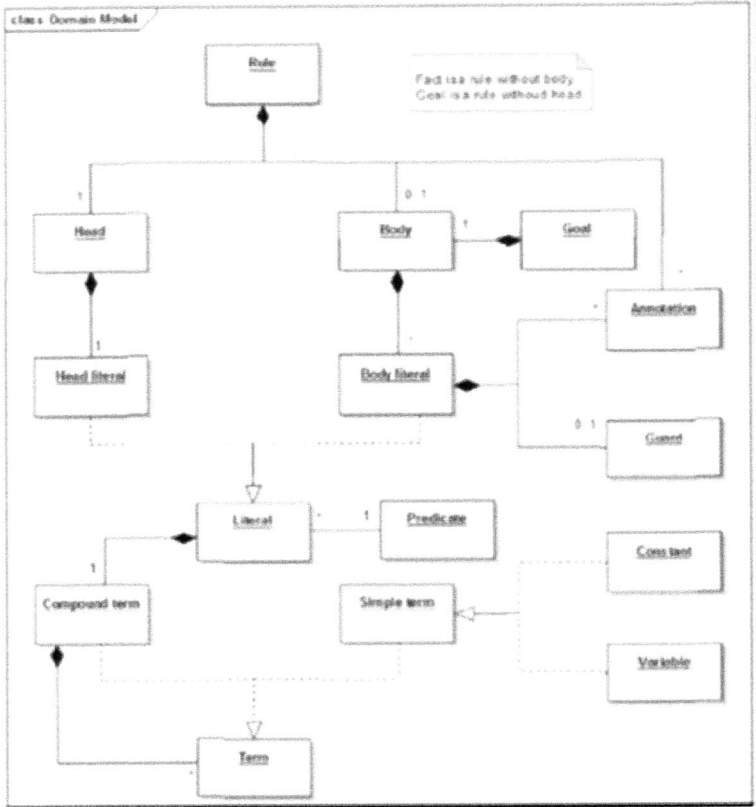

**Fig. 3.** Main Prova 3 Language Elements

slotted literals, as in slotted and object-oriented logics such as F-Logic, e.g. $p(slot_1 -> arg_1, ..., slot_n -> arg_n)$. In the following subsection we show how RIF and RuleML syntactically maps to Prova. Prova distinguishes between for all quantified *solve* goals and existential *eval* goals. For solve goals, for all successful inference all assignments for the variables in the goal predicate satisfying the query are handed back. For eval goals, the engine executes an exhaustive existential search of the rules and facts until no more backtracking is possible. Additionally, the built-in meta-predicate *derive* allows to define (sub) goals dynamically with the predicate symbol unknown until run-time, e.g. $p(F) : -derive([F|Args])$. where the variable $F$ is assigned the function name at runtime.

### 3.3   Mapping from RIF to RuleML and Prova

This section by means of examples shows how RIF can be mapped into RuleML and Prova (Prolog). These examples largely correspond to the partial mappings defined for Datalog RuleML and the RIF-Core subset of RIF BLD [18]. While RIF only supports neutral constants (Const), RuleML supports specialized constant

| RIF | RuleML | Prova |
|---|---|---|
| `<Const type="&xs;string">`<br>`ABC`<br>`</Const>` | `<Data xsi:type="xs:string">`<br>`ABC`<br>`</Data>` | `"ABC"` |
| `<Var>x</Var>` | `<Var>x</Var>` | `X` |
| `<Expr>`<br>`  <op>`<br>`    <Const type="&rif;iri">`<br>`      &func;f`<br>`    </Const>`<br>`  </op>`<br>`  <args ordered="yes">`<br>`    <Var>X</Var>`<br>`  </args>`<br>`</Expr>` | `<Expr>`<br>`  <Fun iri="func:f"`<br>`       per="value"/>`<br>`  <Var>X</Var>`<br>`</Expr>` | `func:f(X)` |
| `<Atom>`<br>`  <op>`<br>`    <Const type="&rif;iri">`<br>`      &cpt;discount`<br>`    </Const>`<br>`  </op>`<br>`  <args ordered="yes">`<br>`    <Var>cust</Var>`<br>`    <Var>prod</Var>`<br>`    <Var>val</Var>`<br>`  </args>`<br>`</Atom>` | `<Atom>`<br>`  <Rel iri="cpt:discount"/>`<br>`  <Var>cust</Var>`<br>`  <Var>prod</Var>`<br>`  <Var>val</Var>`<br>`</Atom>` | `cpt:discount(Cust,Prod,Val)` |
| `<Equal>`<br>`  <left>`<br>`    <Var>X</Var>`<br>`  </left>`<br>`  <right>`<br>`    <Var>Y</Var>`<br>`  </right>`<br>`</Equal>` | `<Equal oriented="yes">`<br>`  <Var>X</Var>`<br>`  <Var>Y</Var>`<br>`</Equal>` | `X=Y` |
| `<Member>`<br>`  <instance>`<br>`  <Const type="&rif;iri">`<br>`    &ppl;Adrian`<br>`  </Const>`<br>`  </instance>`<br>`  <class>`<br>`  <Const type="&rif;iri">`<br>`    &ppl;Person`<br>`  </Const>`<br>`  </class>`<br>`</Member>` | `<Ind iri="ppl:Adrian"`<br>`     type="ppl:Person"/>` | `ppl:Adrian^^ppl:Person` |
| `<External>`<br>`  <content>`<br>`  <Expr>`<br>`  <op>`<br>`  <Const type="&rif;iri">`<br>`    &rifb;numeric-add`<br>`  </Const>`<br>`  </op>`<br>`  <args ordered="yes">`<br>`  <Const type="&xs;integer">`<br>`    1`<br>`  </Const>`<br>`  <Const type="&xs;integer">`<br>`    1`<br>`  </Const>`<br>`  </args>`<br>`  </Expr>`<br>`  </content>`<br>`</External>` | `<Expr>`<br>`<Fun iri="rifb:numeric-add"`<br>`     per="value"/>`<br>`<Data xsi:type="xs:integer">`<br>`  1`<br>`</Data>`<br>`<Data xsi:type="xs:integer">`<br>`  1`<br>`</Data>`<br>`</Expr>` | `1+1` |
| `<Atom>`<br>`  <op>`<br>`  <Const type="&rif;iri">`<br>`    &ex;gold`<br>`  </Const>`<br>`  </op>`<br>`  <slot>`<br>`  <Name>customer</Name>`<br>`  <Var>Customer</Var>`<br>`  </slot>`<br>`</Atom>` | `<Atom>`<br>`  <Rel iri="ex:gold"/>`<br>`  <slot>`<br>`   <Ind>customer</Ind>`<br>`   <Var>Customer</Var>`<br>`  </slot>`<br>`</Atom>` | `ex:gold({customer->Customer})` |

terms which distinguish data constants (Data) from individuals/instances object constants (Ind). RIF does not support a multi-sorted logic with type definitions as in RuleML (@type attribute). The special member built-in in RIF (Member) can be used to define instances of classes which can be interpreted as an explicit type definition. External functions (built-ins) in RIF are restricted to the predefined RIF datatypes and built-ins (DTB) library which can be reused in RuleML together with other external built-in libraries (e.g. from SWRL, XPath etc.). Unpositional named arguments as well as positional arguments are supported by both RIF and RuleML and can be mapped into positional terms in Prova like in Prolog standard logic programs or into unpositional slotted terms.

In the following some of the extra logical extensions of Prova will be introduced.

### 3.4    Access to External Data, Type Systems and Procedural Attachments

Prova follows the spirit and design of the W3C Semantic Web initiative and combines declarative rules, ontologies and inference with dynamic object-oriented programming and access to external data sources and type systems. Therefore, Prova assumes not just a single universe of discourse, but several domains, so called sorts (types) which are interpreted in a multi-sorted logic. The extension of the signature and the typed variables of the language alphabet with sorts (aka types) is defined as follows.

**Definition 61. (Multi-sorted Signature)** *The multi-sorted signature $S$ of Prova is defined as a tuple $\langle \overline{T}, \overline{P}, \overline{F}, arity, \overline{c}, sort \rangle$ where:*

1. *$\overline{P}$ is a finite sequence of predicate symbols $\langle P_1, .., P_n \rangle$.*
2. *$\overline{F}$ is a finite sequence of function symbols $\langle F_1, .., F_m \rangle$*
3. *For each $P_i$ respectively each $F_j$, $arity(P_i)$ resp. $arity(F_j)$ is a non-zero natural number denoting the arity of $P_i$ resp. $F_i$.*
4. *$\overline{c} = \langle c1, .., c_o \rangle$ is a finite or infinite sequence of constant symbols,*
5. *and, $\overline{T} = \{T_1, .., T_n\}$ is a set of sort/type symbols called sorts.*

*The function sort associates with each predicate, function or constant its sorts:*

- *if $c$ is a constant, then $sort(c)$ returns the type $T$ of $c$.*
- *if $p$ is a predicate of arity $k$, then $sort(p)$ is a $k$-tuple of sorts $sort(p) = (T_1, .., T_k)$ where each term $t_i$ of $p$ is of some type $T_j$, i.e., $t_i : T_j$.*
- *if $f$ is a function of arity $k$, then $sort(f)$ is a $k + 1$-tuple of sorts $sort(f) = (T_1, .., T_k, T_{k+1})$ where $(T_1, .., T_k)$ defines the sorts of the domain of $f$ and $T_{k+1}$ defines the sorts of the range of $f$*

*Prova supports the following three basic types of sorts*

1. *primitive sorts are given as a fixed set of primitive data types such as integer, string, etc.*

2. *function sorts are complex sorts constructed from primitive sorts $T_1 \times ... \times T_n \rightarrow T_{n+1}$ and other complex sorts defined in the external type alphabet*
3. *Boolean sorts are a (predicate) statement of the form $T_1 \times ... \times T_n$*

**Definition 62.** *(**Multi-sorted Logic**) Prova's multi-sorted logic associates which each term, predicate and function a particular sort:*

1. *Any constant or variable t is a term and its sort $T$ is given by $sort(t)$*
2. *Let $f(t_1, .., t_n)$ be a function then it is a term of sort $T_{n+1}$ if $sort(f) = \langle T_1, .., T_n, T_{n+1} \rangle$, i.e., f takes argument of sort $T_1, .., T_n$ and returns arguments in sort $T_{n+1}$.*

The intuitive meaning is that a predicate or function holds only if each of its terms is of the respective sort given by *sort*.

The alphabet of the Prova language builds on top of the standard ISO Prolog syntax standard, but further extends it. For typing each variable $X_j$ in the multi-sorted alphabet of the Prova language is associated with a specific sort $sort(X_j) = T_i$, written as $X_j : T_i$, where $X_j$ is a variable and $T_i$ is a type sort associated with the variable. That is, the extended Prova language considers external sort/type alphabets. The combined signatures of the Prova rule language and the external type languages form the basis for combined hybrid knowledge bases and the integration of external type systems into the rule system.

**Definition 63.** *(**Type alphabet**) An external type alphabet $\overline{T}$ is a finite set of monomorphic sort/type symbols built over the distinct set of terminological class concepts of a (external type) language.*

**Definition 64.** *(**Combined Signature**) A combined signature $\overline{S}$ is the union of all its constituent finite signatures: $\overline{S} = \langle S_1 \cup .. \cup S_n \rangle$*

The type systems considered in Prova are order-sorted (i.e., with sub-type relations):

**Definition 65.** *(**Order-sorted Type System**) A finite order-sorted type system $TS$ comes with a partial order $\leq$, i.e., $TS$ under $\leq$ has a greatest lower bound $glb(T_1, T_2)$ for any two types $T_1$ and $T_2$ having a lower bound at all. Since $TS$ is finite also a least upper bound $lub(T_1, T_2)$ exists for any two types $T_1$ and $T_2$ having an upper bound at all.*

**Definition 66.** *(**Combined Knowledge Base**) The combined knowledge base of a typed Prova $\overline{KB} = \langle \Phi, \Psi \rangle$ consists of a finite set of (order-sorted) type systems / type knowledge bases $\Psi = \{\Psi_1 \cap .. \cap \Psi_n\}$ and a typed Prova KB $\Phi$.*

The combined signature is the union of all constituent signatures, i.e., each interpretation of a Prova rule program has the set of ground terms of the combined signature as its fixed universe.

**Definition 67.** (***Extended Herbrand Base***) *Let $\overline{KB} = \langle \Phi, \Psi \rangle$ be a typed combined Prova KB P. The extended Herbrand base of P, denoted $\overline{B}(P)$, is the set of all ground literals which can be formed by using the predicate/function symbols in the combined signature with the ground typed terms in the combined universe $\overline{U}(P)$, which is the set of all ground typed terms which can be formed out of the constants, type and function symbols of the combined signature.*

The grounding of the combined KB is computed wrt the composite signature.

**Definition 68.** (***Grounding***) *Let P be a typed (combined) Prova KB and $\overline{c}$ its set of constant symbols in the combined signature. The grounding ground(P) consists of all ground instances of all rules in P w.r.t to the combined multi-sorted signature which can be obtained as follows:*

- *The ground instantiation of a rule r is the collection of all formulas $r[X_1 : T_1/t_1, .., X_n : T_n/t_n]$ with $X_1, .., X_n$ denoting the variables and $T_1, .., T_n$ the types of the variables (which must not necessarily be disjoint) which occur in r and $t_1, .., t_n$ ranging over all constants in $\overline{c}$ wrt to their types.*
- *For every explicit query/goal $Q[X_1 : T_1, .., X_m : T_m]$ to the type system, being either a fact with one or more free typed variables $X_1 : T_1, .., X_m : T_m$ or a special built-in Prova query literal rdf(...) with variables as arguments in the triple-like query, the grounding ground(Q) is an instantiation of all variables with constants (individuals) in $\overline{c}$ according to their types.*

Using equalities Prova assumes a notion of default inequality for the combined set of individuals/constants which leads to a default unique name assumption:

**Definition 69.** (***Default Unique Name Assumption***) *Two ground terms are assumed to be unequal, unless equality between the terms can be derived.*

The interpretation $I$ of a typed Prova program $P$ then is a subset of the extended Herbrand base $\overline{B}(P)$.

**Definition 70.** (***Multi-sorted Interpretation***) *Let $\overline{KB} = \langle \Phi, \Psi \rangle$ be a combined KB and $\overline{c}$ its set of constant symbols. An interpretation I for a multi-sorted combined signature $\overline{S}$ consists of*

1. *a universe $|M| = T_1^I \cup T_2^I \cup .. \cup T_n^I$, which is the union of the types (sorts), and*
2. *the predicates, function symbols and constansts/individuals $\overline{c}$ in the combined signature, which are interpreted in accordance with their types.*

*The assignment function $\sigma$ from the set of variable $\overline{X}$ of P into the combined universe $\overline{U}(P)$ must respect the sorts/types of the variables (in order-sorted type systems also subtypes). That is, if $X_i$ is a variable of type T, then $\sigma(X) \in T^I$. In general, if $\phi$ is a typed predicate or function in $\Phi$ and $\sigma$ an assignment to the interpretation I, then $I \models \phi[\sigma]$, i.e., $\phi$ is true in I when each variable X of $\phi$ is substituted by the values $\sigma(X)$ wrt to its type. Since the assignment to constant and function symbols is fixed and the domain of discourse corresponds one-to-one with the constants $\overline{c}$ in the combined signature $\overline{U}(P)$, it is possible to identify an interpretation I with a subset of the extended Herbrand base: $I \subseteq \overline{B}(P)$.*

The assignment function in Prova is given as a query from the rule component to the type system, so that there is a separation between the inferences in a type system and the rule component. Moreover, explicit queries to a type system (Java or Semantic Web) defined in the body of a rule, e.g., procedural attachments, built-ins or ontology queries (special *rdf* query or free DL-typed facts) are based on this hybrid query mechanism.

**Definition 71. (*Semantic Multi-Structure Model*)** *Let* $\overline{KB} = \langle \Phi, \Psi \rangle$ *be a combined KB of a typed Prova program P.*
*An interpretation I is a model of an untyped ground atom $A \in \overline{KB}$ or I satisfies A, denoted $I \models A$ iff $A \in I$.*
*I is a model for a ground typed atom $A : T \in \overline{KB}$, or I satisfies $A : T$, denoted $I \models A : T$, iff $A : T \in I$ and for every typed term $t_i : T_j$ in A the type query $T_j = sort(t_i)$, denoting the type check "is $t_i$ of type $T_i$", is entailed in $\overline{KB}$, i.e., $\overline{KB} \models T_i = sort(t_i)$ (note, in an order sorted type system subtypes are considered, i.e., $t_i$ is of the same or a subtype of $T_j$).*
*I is an interpretation of an ground explicit query/goal Q to the type system $\Psi$ if $\Psi \models Q$.*
*I is a model of a ground rule $r : H \leftarrow B$ iff $I \models H(r)$ whenever $I \models B(r)$. I is a model a typed program P (resp. a combined knowledge base $\overline{KB}$), denoted by $I \models P$, if $I \models r$ for all $r \in ground(P)$.*

Informally, a typed Prova knowledge base consists of rules with logic programming literals which have typed terms and a set of external (order-sorted) type systems in which the types (sorts) are defined over their type alphabets. An external type system might possibly define a complete knowledge base with types/sorts (Java classes or T-Box in DL) and individuals associated with these types (Java object instances of the classes or A-box in DL). Restricted built-in predicates and procedural attachment predicates or functions which construct or return individuals of a certain type (boolean or object-valued) are also considered to be part of the external type system(s), i.e., part of the external signature. The combined signature is then the union of the two (or more) signatures, i.e., the combination of the signature of the rule component and the signatures of the external type systems / knowledge bases combining their type alphabets, their functions and predicates and their individuals.

The operational semantics of typed Prova is implemented as hybrid polymorphic order-sorted unification. [71] In contrast to other hybrid (DL-typing) approaches which apply additional constraint literals as type guards in the rule body and leave the usual machinery of resolution and unification unchanged, the operational semantics for prescriptive types in Prova's typed logic is implemented by an order-sorted unification. Here the specific computations that are performed in the typed language are intimately related to the types attached to the atomic term symbols. The order-sorted unification yields the term of the two sorts (types) in the given sort hierarchy. This ensures that type checks apply directly during typed unification of terms at runtime enabling ad-hoc polymorphism of variables leading e.g.. to different optimized rule variants and early constrained search trees. Thus, the order-sorted mechanism provides higher level of abstraction, providing more compact and complete solutions and avoiding possibly expensive backtracking.

Prova provides support for two external order-sorted type systems, namely *Java* class hierarchies and ontological type systems (e.g. OWL or RDFS ontologies) respectively *Description Logic* knowledge bases.

**Description Logic Type Systems / Ontologies.** An external type systems supported by Prova are Semantic Web ontologies (Description Logic KBs) represented e.g. in RDFS or OWL. That is, the combined signature $\overline{S}_{DL}$ consisting of the finite signature $S$ of the rule component and the finite signature(s) $S_i$ of the ontology language(s).

The type alphabet $TS$ is a finite set of monomorphic type symbols built over the distinct set of terminological atomic concepts $\overline{T}$ in a Semantic Web ontology language $\Sigma^{DL}$, i.e., defined by the atomic classes in the T-Box model.

Note, that restricting types to atomic concepts is not a real restriction, because for any complex concept such as $(T_1 \sqcap T_2)$ or $(T_1 \sqcup T_2)$ one may introduce an atomic concept $T_3$ in the T-Box and use $T_3$ as atomic type instead of the complex concept. This approach is also reasonable from a practical point of view since dynamic type checking must be computationally efficient in order to be usable in an order-sorted typed logic with possible very large rule derivation trees and many typed unification steps, i.e., fast type checks are crucial during typed term unification. We assume that the type alphabet is fixed (but arbitrary), i.e., no new terminological concepts can be introduced in the T-Box by the rules at runtime. This ensure completeness of the domain and enables static type checking on the used DL-types in Prova programs at compile time (during parsing the Prova script).

The set of constants/individuals $\overline{c}$ is built over the set of individual names in $\Sigma^{DL}$, but Prova do not fix the constant names and allow arbitrary fresh constants (individuals) (under default UNA) to be introduced in the head of rules and facts of the rule base. However, new individuals which are introduced in rules or facts apply locally within the scope of the rules in which they are defined, i.e., within a local reasoning chain; in contrast to the individuals defined in the A-box model of the type system which apply globally as individuals of a class. DL-typed terms in Prova are defined as follows:

**Definition 72.** *(**DL-typed Terms**) A DL-type is a terminological concept/class defined in the DL-type system (T-Box model). A typed DL-typed Prova term is denoted by the relation $t^{\wedge\wedge}T$ where $t$ is the term and $T$ is the DL-type of term.*

The type ontologies are typically provided as Web ontologies (RDFS or OWL) where types and individuals are represented as resources having an webized URI. Namespaces can be used to avoid name conflicts and namespace abbreviations facilitate are more readable language.

```
% A customer gets 10 percent discount, if the customer is a gold customer

discount(X^^business:Customer, 10^^math:Percentage) :-
   gold(X^^business:Customer).

% fact with free typed variable acts as instance query on the ontology A-box
gold(X^^business:Customer).
```

Free DL-typed variables are allowed in facts. They act as free instance queries on the ontology layer, i.e., they query all individuals of the given type and bind them to the typed variable.

**Java Type System, Procedural Attachments and Built-Ins.** For external Java type systems, the combined multi-sorted signature $\overline{S}_{Java}$ uses the fully qualified order-sorted Java class hierarchy as type symbols. In order to type a variable with a Java type the fully qualified name of the Java class to which the variable should belong must be specified as a prefix separated from the variable by a dot ".".

```
java.lang.Integer.X       variable X is of type Integer
java.util.Calendar.T      variable T is of type Calendar
java.sql.Types.STRUCT.S    variable S is of SQL type Struct
```

To sense the environment and trigger actions, query data from external sources such as databases, call external procedural code such as Enterprise Java Beans, and receive / send messages from / to other agents or external services, Prova provides a set of built-in functions and additionally can dynamically instantiate any Java object and call its API methods at runtime.

Java objects, as instances of Java classes, can be dynamically constructed by calling their constructors or static methods using extra logical procedural attachments. The returned objects, might then be used as individuals/constants that are bound by an equality relation (denoting typed unification equality) to appropriate variables, i.e., the variables must be of the same type or of a super type of the Java object.

A procedural attachment is a function that is implemented by an external procedure (i.e., a Java method). They are used in Prova to dynamically call external procedural methods during runtime, i.e., they enable the (re)use of procedural code and allow dynamic access to external data sources and tools using their programming interfaces (APIs). They are a crucial extension to traditional logic programming, combining the benefits of object-oriented languages (Java) with declarative rule based programming, e.g., in order to externalize mathematical computations such as aggregations to highly optimized procedural code in Java or use query languages such as SQL by JDBC to select and aggregate facts from external data sources.

**Definition 73.** *(Procedural Attachments) A procedural attachment is a function or predicate whose implementation is given by an external procedure. Two types of procedural attachments are distinguished:*

- *Boolean-valued attachments (or predicate attachments) which call methods which return a Boolean value, i.e., which are of Boolean sort (type).*
- *Object-valued attachments (or functional attachments) which are treated as functions that take arguments and return one or more objects, i.e., which are of a function sort.*

Functional Java attachments have a left-hand side with which the results (the returned object(s)) of the call are unified by a unification equality relation =, e.g., $C = java.util.Calendar.getInstance()$. If the left-hand side is a free (unassigned) variable the latter stores the result of the invocation. If the left-hand side is a bound variable or a list pattern the unification can succeed or fail according to the typed unification and consequently the call itself can succeed or fail. List structures are used on the left-hand side to allow matching of sets of constructed/returned objects to specified list patterns. A predicate attachment is assumed to be a test in such a way that the call succeeds only if a true Boolean variable is returned. Static, instance and constructor calls are supported in both predicate and functional attachments depending on their return type. Constructor calls follow the Java syntax with the fully qualified name of the class and the constructor arguments, e.g., $X = java.lang.Long(123)$. Static method calls require fully qualified class names to appear before the name of the static method followed by arguments, e.g., $Z = java.lang.Math.min(X, Y)$. Instance methods are mapped to concrete classes dynamically based on the type of the variable, i.e., the method of a previously bound Java object is called. They require a variable before the name of an instance method followed by the arguments, e.g., $S = X.toString()$.

```
add(java.lang.Integer.In1,java.lang.Integer.In2,Result):-
   Result = java.lang.Integer.In1 + java.lang.Integer.In2.
```

The first rule takes two Integer variables $In1$ and $In2$ as input and returns the result which is bound to the untyped variable $Result$. Accordingly, a query $add(1, 1, Result)$? succeeds with an Integer object 2 bound to the $Result$ variable, while a query $add("abc", "def", Result)$? will fail.

It is important to note, that Java objects can be bound to variables and their methods can be dynamically used as procedural attachment functions anywhere during the reasoning process, i.e., in other rules. This enables a tight and highly expressive integration of external object oriented functions into declarative agent's rules' execution.

**Definition 74.** (*Built-in Predicates or Functions*) *Built-in predicates or functions (built-ins) are special restricted procedural attachment predicate resp. function symbols in the Prova language for concrete domains, e.g., integers or strings, that may occur in the body of a rules.*

Examples are $+$, $=$, *assert*, *bound*, *free* etc. For instance, Prova provides a rich library of built-ins for query languages such as SQL, SPARQL, and XQuery:

```
File Input / Output
   ..., fopen(File,Reader), ...
XML (DOM)
   document(DomTree,DocumentReader) :-  XML(DocumenReader),...
SQL
   ... ,sql_select(DB,cla,[pdb_id,"1alx"],[px,Domain]).
RDF
   ...,rdf(http://...,"rdfs",Subject,"rdf_type","gene1_Gene"),...
XQuery
 ..., XQuery = 'for $name in StatisticsURL//Author[0]/@name/text()
   return $name', xquery_select(XQuery,name(ExpertName)),...
SPARQL
   ...,sparql_select(SparqlQuery,...
```

The following rule uses a SPARQL query built-in to access an RDF Friend-of-a-Friend (FOAF) profile published on the Web. The selected data is assigned to variables which can be used within an agent's rule logic, e.g. to expose the agent's contact data.

```
exampleSPARQLQuery(URL,Type) :-
 QueryString = ' PREFIX foaf:
              PREFIX rdf:
              SELECT ?contributor ?url ?type
              FROM
              WHERE {
                    ?contributor foaf:name "Bob DuCharme" .
                    ?contributor foaf:weblog ?url .
                    ?contributor rdf:type ?type . } ',
 sparql_select(QueryString,url(URL),type(Type)).
```

Note, that the structures in Java type systems are usually not considered as interpretations in the strict model-theoretic definition, but are composite structures involving several different structures whose elements have a certain inner composition. However, transformations of composite structures into their flat model theoretic presentations is in the majority of cases possible. From a practical point of view, it is convenient to neglect the inner composition of the elements of the universe of a structure. These elements are just considered as "abstract" points devoid of any inherent meaning. This structural mapping between objects from their interpretations in the Java universe to their interpretation in the rule system ignoring finer-grained differences that might arise from the respective definitions is given by the following isomorphism.

**Definition 75.** *(Isomorphism) Let $I_1$, $I_2$ be two interpretations of the combined signature $\overline{S} = \{T_1, .., T_n\}$, then $f_\cong : |M_1| \to |M_2|$ is an isomorphism of $I_1$ and $I_2$ if $f_\cong$ is a one-to-one mapping from the universe $|M_1|$ of $I_1$ onto the universe $|M_2|$ of $I_2$ such that:*

1. *For every type $T_i$, $t \in T_i^{I_1}$, iff $f_\cong(t) \in T_i^{I_2}$*
2. *For every constant $c$, $f_\cong(c^{I_1}) \cong c^{I_2}$*
3. *For every n-ary predicate symbol $p$ with n-tuple $t_1, .., t_n \in |M_1|$, $\langle t_1, .., t_n \rangle \in p^{I_1}$ iff $\langle f_\cong(t_1), .., f_\cong(t_n) \rangle \in p^{I_2}$*
4. *For every n-ary function symbol $f$ with n-tuple $t_1, .., t_n, \in |M_1|$, $f_\cong(f^{I_1}(t_1, .., t_n)) \cong f^{I_2}(f_\cong(t_1), .., f_\cong(t_n))$*

For instance, in Prova an isomorphism between Boolean Java objects and their model-theoretic truth value is defined, which makes it possible to treat boolean-valued procedural attachments as conditional body literals in rules and establish a model-theoretic interpreation as defined above between the Java type system and the model-theoretic semantics of the typed logic of the rule component. Other examples are String objects which are treated as standard constants in rules, i.e., the Java String object maps with the untyped theory of logic programming. Primitive datatype values, from the ontology resp. XML domain (XSD datatypes) can be mapped similarly.

## 3.5   Modularization, Scopes and Guards

To capture the distributed, open structure of Web based rule bases and enable scoped queries on explicitly closed parts of open and distributed knowledge, Prova supports principles of information hiding and modularization, which makes it easier to maintain and manage (distributed) rule sets.

**Metadata Based Modularization and Module Imports/Updates.** Prova has a flexible approach towards modularization of the knowledge base which allows constructing metadata based views on the knowledge base, so called *scopes*. Therefore, Prova extends the rule language to a labelled logic programming rule language (LLP) with metadata annotations such as rule labels, module (rule sets in rule bases) labels and arbitrary other (Semantic Web) annotations (e.g., Dublin Core author, date etc). These metadata annotations are used to manage the rules and facts in the knowledge base.

In analogy to the multi-sorted extension for types, the meta-data extension of the Prova language is defined over a combined signature $\overline{S}$ which is the union of the signature of the rule language and the signatures of the used metadata vocabularies (e.g. Dublin Core).

**Definition 76.** *(**Combined Signature with Metadata Annotations**) The combined metadata annotated signature $\overline{S}$ is defined as a tuple $\langle \overline{T}, \overline{P}, \overline{F}, arity, \overline{c}, sort, meta \rangle$ where $\overline{P}$ is the union of the predicate symbols define in the signature of the core Prova rule language and the metadata predicate symbols (denoting metadata key properties) defined in the signature(s) of the metadata vocabularie(s) and $\overline{c}$ is the union of constant symbols defined in the rule signature and in the metadata signature(s) (denoting metadata values). meta is a special unary function which returns the assigned metadata.*

To explicitly annotate clauses in a Prova program $P$ with an additional set of metadata labels a general 1-ary built-in function @ is introduced in the Prova language.

**Definition 77.** *(**Metadata Annotation Labels**) The special 1-ary built-in function @ is a partial injective labelling function that assigns a set of metadata annotations m (property-value pairs) to a clause cl in P, e.g.*
$@(L_1, .., L_n) \ H : -B$
*where $L_i$ are a finite set of unary positive literals (positive metadata literals) which denote an arbitrary metadata property(value) pair, e.g., @label(rule1).*

The implicit form $@(L_1), .., @(L_n) \ H : -B$ of the metadata function expresses that $@(H : -B) = L_1, .., L_n$. The explicit @() annotation is optional, i.e., a Prova program $P$ without metadata annotated clauses coincides with a standard unlabelled logic program.

Clauses in Prova are treated as objects in KB having an unique *object id* (*oid*) which might be user-defined, i.e., explicitly defined by a metadata annotation $@label(oid) \ H : -B$ or system-defined i.e., all rules are automatically "labelled" with an auto-incremented *oid* (an increasing natural number) provided by the system at compile time. Rules and facts might be bundled to clause sets, so called *modules*, which also have an object id, the module oid. By default the module oid is the URI or full document name of the Prova script which contains the module. But the module oid might also be user-defined @*src*(*moduleoid*). All clauses (rules and facts) defined in a module are automatically annotated with the module oid @*src*(*moduleoid*) $H : -B$. The oids are used to manage the knowledge in the (distributed) knowledge base, e.g., to import a rule set from an URI which is then used as the module oid or remove a module from the KB by its oid. Beside oids arbitrary other semantic annotations such as Dublin Core data might be specified in the @ annotation function.

```
@label(r1) @dc:author("Adrian") @dc:date(2006-11-12)
    p(X):-q(X).
@label(f1)
    q(1).
```

The example shows a rule with rule label $r1$ and two additional Dublin Core annotations $dc : author("Adrian")$ and $dc : date(2006 - 11 - 12)$ and a fact with fact label $f1$. Since there is no explicitly user-defined module oid in the meta-data labels, the default module oid for both clauses is the URI or document name of the Prova script in which they are defined, e.g. @*src*("*http* : //*prova.ws*/*example*1. *prova*").

In Prova it is possible to consult (import/load) distributed rulebases from local files, a Web address, or from incoming messages transporting a rulebase. Furthermore, Prova supports update built-ins such as *assert* and *retract*.

```
%load from a local file
:- eval(consult("organization2009.prova")).
% import from a Web address
:- eval(consult("http://ruleml.org/organization2010.prova")).
```

The imported rulebases are managed as modules in the knowledge base, which are uniquely identified by their source object id $src(moduleOID)$. Since multiple nested imports are possible, modules might be nested, i.e. a module denoting a rule base (e.g. a Prova script) might consist of several nested submodules (e.g. sets of rules and facts).

Similar to imports of external type systems and built-ins (procedural attachments) which query and compute external data, the semantics for modules in Prova is defined over the combined knowledge base of the modules, an extended state based Herbrand Base and semantic multi-structures.

**Definition 78. (*Combined Knowledge Base*)** *The combined knowledge base of a modular Prova* $\overline{KB} = \langle \Phi, \Psi \rangle$ *consists of a finite set of modules* $\Psi = \{\Psi_1 \cap .. \cap \Psi_n\}$ *and an initial primary Prova KB* $\Phi$.

Prova supports knowledge updates which import modules (*consult*) and add or remove clauses (*assert, retract*). Each update leads to a new knowledge state of the combined KB.

**Definition 79.** *(**Knowledge State**) A knowledge state represents the combined knowledge base* $KB_k$ *at this particular state, where* $k \in \aleph$.

Note that according to the modularized logic in Prova a state, i.e., a combined knowledge base $KB_k$, might consist of nested submodules, each having an unique ID (the module oid). Intuitively, a state represents the union of all clauses stored in all modules in the combined knowledge base.

An update is then a transition which adds or removes facts and/or rules and changes the knowledge base. That is, the KB transits from the initial state $KB_1$ to a new state $KB_2$. We define the following notion of positive (assert) and negative(retract) transition:

**Definition 80.** *(**Positive Update Transition**) A positive update transition, or simply positive update, to a knowledge state* $KB_k$ *is defined as a finite set* $U^{pos}_{oid} :=$ $\{r_N : H : -B, fact_M : A\}$ *with A an atom denoting a fact,* $H : -B$ *a rule,* $N = 0, .., n$ *and* $M = 0, ..m$ *and oid being the update oid which is also used as module oid to manage the knowledge as a new module in the KB. Applying* $U^{pos}_{oid}$ *to* $KB_k$ *leads to the extended state* $KB_{k+1} = \{KB_k \cup U^{pos}_{oid}\}$. *Applying several positive updates as an increasing finite sequence* $U^{pos}_{oid_j}$ *with* $j = 0, .., k$ *and* $U^{pos}_{oid_0} := \emptyset$ *to* $KB_0$ *leads to a state* $KB_k = \{KB_0 \cup U^{pos}_{oid_0} \cup U^{pos}_{oid_1} \cup ... \cup U^{pos}_{oid_k}\}$.

That is a state $KB_k$ is decomposable in the previous knowledge state $k - 1$ plus the update: $KB_k = \{KB_{k-1} \cup U^{pos}_k\}$. We define $KB_0 = \{\emptyset \cup U^{pos}_{oid_0}\}$ and $U^{pos}_{oid_0} = \{KB :$ the set of rules and facts defined in the program $P\}$, i.e., importing the initial Prova program $P$ from a Prova script document is the first update leading to the knowledge state $KB_1$.

Likewise, We define a *negative update transition* as follows:

**Definition 81.** *(**Negative Update Transition**) A negative update transition, or for short a negative update, to a knowledge state* $KB_k$ *is a finite set* $U^{neg}_{oid} :=$ $\{r_N : H : -B, fact_M : A\}$ *with* $A \in KB_k$, $H : -B \in P$, $N = 0, .., n$ *and* $M = 0, ..m$, *which is removed from* $KB_k$, *leading to the reduced program* $KB_{k+1} = \{KB_k \setminus U^{neg}_{oid}\}$.

Applying arbitrary sequences of positive and negative updates leads to a sequence of KB states $KB_0, .., KB_k$ where each state $KB_i$ is defined by either $KB_i = KB_{i-1} \cup U^{pos}_{oid_i}$ or $KB_i = KB_{i-1} \setminus U^{neg}_{oid_i}$. In other words, $KB_i$, i.e., the set of all clauses in the KB at a particular knowledge state $i$, is decomposable in the previous knowledge state plus/minus an update, whereas the previous state consists of the state $i - 2$ plus/minus an update and so on. Hence, each particular knowledge state can be decomposed in the initial state $KB_0$ and a sequence of updates. Although an update might insert more than one rule or fact, i.e., insert or remove a complete module, it nevertheless is treated as an elementary update, a so called bulk update, which transits the current knowledge state to the next state in an elementary transition: $\langle KB_i, U^{pos/neg}_{oid}, KB_{k+1} \rangle$. Intuitively, one might think of it as a complex atomic update action which performs all knowledge inserts resp. removes simultaneously.

Elementary updates have both a truth value, i.e. they may succeed or fail, and a side effect on the knowledge base leading to the transition of the knowledge state. The extended Herbrand Base is defined on the notion of knowledge states and transitions from one state to another.

**Definition 82.** *(**Extended State based Herbrand Base**) Let $P$ be the combined KB at a particular knowledge state $KB_k$. The extended Herbrand base of $P$, denoted $\overline{B}(P)$, is the set of all ground literals which can be formed by using the predicate/function symbols in the combined signature with the ground typed terms in the combined universe $\overline{U}(P)$, which is the set of all ground typed terms which can be formed out of the constants, type and function symbols of the combined signature of $KB_k$.*

**Definition 83.** *(**Modular semantic multi-structure**) A modular multi-structure $I$ is model of a modular program $P$ (resp. the knowledge state $KB_k$ of the combined knowledge base $\overline{KB}$), denoted by $I \models P$, if $I \models c$ for all clauses $c \in ground(P)$, where $I \models c$ is a usual multi-sorted model for providing the interpretation of Prova clauses.*

Accordingly, all queries to a Prova program apply on the extended resp. reduced transition knowledge state of the program, i.e., the truth valuation of a goal $G$ depends on its model at the current knowledge state $KB_k$, denoted by $TVal_{KB_k \models G}(G)$.

Based on this modular knowledge state transition semantics and the metadata based control of the knowledge state updates which are treated as modules in the combined KB, Prova provides supports for transactional updates, where failing sequences of knowledge updates can be rolled back by removing the associated modules from the combined Prova KB. In the non-transactional style updates in (serial) Prova rules are not rolled-back to the original state if the derivation fails and the system backtracks. Typically this "weak" non-transactional semantics is intended when external Prova script are imported (consult) or new rule sets are added (assert) as modules. That is, independently, of whether the particular derivation in which the update is performed fails from some reason the update transition to the next knowledge state subsists and is not rolled back in case of failures.

**Scoped Reasoning.** The metadata annotation of rules/facts and rule sets (modules) enables scoped (meta) reasoning with the semantic annotations. The metadata can act as an explicit scope for constructive queries (creating a view) on the knowledge base. For instance, the metadata annotations might be used to constrain the level of generality of a scoped goal literal to a particular module, i.e., to consider only the set of rules and facts which belong to the specified module.

**Definition 84.** *(**Scoped Literal**) A scoped literal is of the form @$\overline{C}$ $L$ where $L$ is a positive or negative literal and @$\overline{C}$ is the scope definition which is a set of one or more metadata constraints. Scoped literals are only allowed in the body of a rule.*

Informally, the semantics of scoped literals allows to explicitly close the domain of discourse to certain parts of the KB.

**Definition 85.** *(**Metadata based Scope**) Let $\overline{KB}$ be a combined KB consisting of a set of submodules $\overline{KB} = \{KB_1 \cup .. \cup KB_k\}$. The scope $KB'$ of a scoped literal @$\overline{C}$ $L$ is the set of clauses $KB' = \{m'_1 cl_1, .., m'_n cl_n\} \in \overline{KB}$, where for all clauses $cl_i(m'_i) \in \overline{KB'}$ its set of metadata annotations $m'_i$ satisfy the scope constraints $\overline{C}$ of the scoped literal $L$, i.e., $m'_i \models \overline{C}$.*

Accordingly, a scope (aka constructiv view) is constructed by one or more metadata constraints, e.g., the module oid @$src(URI/Filename)$ or Dublin Core values @$dc : author(...)$.

**Definition 86.** *(**Closure**) Let $\overline{KB}$ be a combined KB. The closure of $\overline{KB}$, denoted $Cl(\overline{KB})$, is defined by $\overline{KB}$ plus all modules $KB_k$ which are in the scope of any scoped literal in $\overline{KB}$.*

   *A scoped literal @$\overline{C}$ $L$ is closed if each rule in $\overline{KB}$ which unifies with the literal $L$ is also closed, i.e., its body literals are closed in $Cl(\overline{KB})$.*

Intuitively, this means that the closure of a Prova program depends on the scopes of the literals in the bodies of its rules. Obviously, if one of the subsequently used goal literals in a proof attempt is open, i.e., without a scope, the closure expands to the open KB.

**Definition 87.** *(**Scoped Semantics**) Given a scoped $\overline{KB}'$, where all literals are scoped with closure $Cl(\overline{KB}')$, the truth value of a scoped literal @$\overline{C}$ $L$ depends on the partial model of the clauses of $\overline{KB}'$ wrt the scope definition $\overline{C}$, i.e., $I_{partial^{\overline{C}}}(\overline{KB}') \models L$.*

Syntactically the scope definitions use the syntax of Prova metadata annotations.

```
@label(rule1) r1(X):-q(X).
@label(rule2) r2(X):-q(X).
@label(rule3) p1(X):-
            @label(rule1) r1(X). % scoped goal literal
q(1).

:-solve(p1(Y)).
```

The example shows three metadata annotated rules. They query $p1(Y)$ will return only one solution with $Y = 1$, since the subgoal $r1(X)$ of $rule3$ applies only in the scope of the rule with label $rule1$, but not on $rule1$ and $rule2$, which would be the case if there would be no scope constraint defined for the subgoal.

   Prova allows variables in the scope definitions which are bound to the annotated metadata values. The following example shows the definition of a scope, that constraints the application of the subgoal $r2(X)$ on the rule with label $rule3$ and on the module with source name $AgentRole1.prova$.

```
% get module label
r1(X,Y):-
   @src(Y) @label(rule3)
   r2(X).
   :-solve(r1(X,"AgentRole1.prova")).
```

**Guards.** In addition to scopes Prova supports literal *guards* which act as additional pre-condition constraints.

Guards in Prova are syntactically specified in the Prova rule language using brackets after the goal literal. The model-theoretic semantics of guards is like for goal literals, however in the proof-theoretic semantics guards act like pre-conditions before the proofs of the standard goal literals starts.

For instance, the following rule makes decisions on the basis of rules which haven been authored by different persons and only applies those rules from trusted authors.

```
%simplified decision rules of an agent
@author(dev22) r2(X):-q(X).
@author(dev32) r2(X):-s(X).
q(2).
s(-2).

% for simplicity this is a fact, but could be also a complex rule
% which computes the trust value from the reputation value of dev22
trusted(dev22).

% Author dev22 is trusted but dev32 is not, so one solution is found: X=2
p1(X):-
 @author(A)
 r2(X) [trusted(A)].

% for all query
:-solve(p1(X1)).
```

This example uses metadata annotations on rules for the head literals $r2/1$ and a scopes on the literal $r2(X)$ in the body of the rule for $p1(X)$. Since variable $A$ in $@author(A)$ is initially free, it gets instantiated from the matching target rule(s). Once $A$ is instantiated to the target rule's $@author$ annotation's value ($dev22$, for the first $r2$ rule), the body of the target rule is dynamically non-destructively modified to include all the literals in the additional guard $trusted(A)$ before the body start, after which the processing continues. Since $trusted(dev22)$ is true but $trusted(dev32)$ is not, only the first rule for predicate r2 is used and so one solution $X1 = 2$ is returned by $solve(p1(X1))$.

### 3.6 Prova Serial Horn Rules for Messaging

For communication between distributed rule agents Prova supports special built-ins for asynchronously sending and receiving event messages within serial Horn rules. The main language constructs of messaging reaction rules are: *sendMsg* predicates to send messages, reaction *rcvMsg* rules which react to inbound messages, and *rcvMsg* or *rcvMult* inline reactions in the body of messaging reaction rules to receive one or more context-dependent multiple inbound event messages:

```
sendMsg(XID,Protocol,Agent,Performative,Payload |Context)
rcvMsg(XID,Protocol,From,Performative,Paylod|Context)
rcvMult(XID,Protocol,From,Performative,Paylod|Context)
```

Here, *XID* is the conversation identifier (conversation-id) of the conversation to which the message will belong. *Protocol* defines the communication protocol.

*Agent* denotes the target party of the message. *Performative* describes the pragmatic envelope for the message content. A standard nomenclature of performatives is, e.g., the FIPA Agents Communication Language (ACL). *Payload* represents the message content sent in the message envelope. It can be a specific query or answer or a complex interchanged rule base (set of rules and facts). For instance, the following rule snippet shows how a query is sent to an agent via an Enterprise Service Bus (esb) and then an answer is received from this agent.

```
...
sendMsg(Sub_CID,esb,Agent,acl:query-ref, Query),
rcvMsg(Sub_CID,esb,Agent,acl:inform-ref, Answer),
...
```

Interchanged messages besides the conversation's metadata and payload also carry the pragmatic context of the conversation such as communicative situations / acts, mentalistic notions, organizational and individual norms, purposes or individual goals and values. The payload of incoming event messages is interpreted with respect to the local conversation state, which is denoted by the conversation id, and the pragmatic context, which is given by a pragmatic performative. For instance, a standard nomenclature of pragmatic performatives, which can be integrated as external (semantic) vocabulary/ontology, is e.g., defined by the Knowledge Query Manipulation Language (KQML) (Finin et al. 1993), by the FIPA Agent Communication Language (ACL), which gives several speech act theory based communicative acts, or by the Standard Deontic Logic (SDL) with its normative concepts for obligations, permissions, and prohibitions. Depending on the pragmatic context, the message payload is used, e.g. to update the internal knowledge of the agent (e.g., add new facts or rulebases), add new tasks (goals), or detect a complex event pattern (from the internal event instance sequence). For instance, the following example shows a reaction rule that sends a complete rule base, which is loaded from a local *File* to an agent service *Remote* using JMS as transport protocol.

*Example 2*

```
% Upload a rule base read from File to the host
% at address Remote via JMS
upload_mobile_code(Remote,File) :-
    % Opening a file returns an instance
    % of java.io.BufferedReader in Reader
    fopen(File,Reader),
    Writer = java.io.StringWriter(),
    copy(Reader,Writer),
    Text = Writer.toString(),
    % variable SB will encapsulate the whole content of File
    SB = StringBuffer(Text),
    % send the complete rule base to the receiver agent "Remote"
    sendMsg(XID,jms,Remote,acl:inform,consult(SB)).
```

The corresponding receiving reaction rule of the remote agent is:

```
% wait for incoming messages with pragmatic context $acl:inform$
rcvMsg(XID,jms,Sender,acl:inform,[Predicate|Args]):-
    % derive the message payload, i.e. consult the received rule set to the internal KB
    derive([Predicate|Args]).
```

This rule receives incoming JMS based messages with the pragmatic context $acl : inform$ and derives the message content, i.e. consults the received rule base to the local knowledge base of the remote agent. It is important to note that via the conversation id several reaction rule reasoning processes might run in parallel, local to their conversation flows. Inactive reactions (conversation partitions) are removed from the system, e.g. by timeouts. Self-activations by sending a message to the receiver "self" are possible. With the pragmatic performatives it is possible to implement different coordination and negotiation protocols. For instance, if an agent does not understand the semantics of the interchanged message payload, it can inform the sender about this, using, e.g., the $acl : not - understood$ performative, so that the sender can additionally send the semantic information, e.g. a pointer to the ontology that defines the concepts of the payload, and the receiving agent can import this ontology to its internal knowledge base.

By using messaging reaction rules a Prova rule engine can be deployed as a distributed rule inference service, e.g. in the Rule Responder agent architecture [72], or e.g. as an OSGI component enabling massive parallelization of Prova agent nodes in grid/cloud environments and (smart) devices (e.g. RFID networks) which communicate via event messages.

# 4    Conclusion

Rule based systems have been investigated comprehensively in the realms of declarative logic programming and expert systems in the past decades. Logic programming has been a very popular paradigm and one of the most successful representatives of declarative programming in general. It is based on solid and well-understood theoretical concepts and has been proven to be very useful for rapid prototyping and describing problems on a high abstraction level. In recent years rule based technologies have experienced a remarkable come back namely in two areas: business rules processing, and reasoning in the context of the (Semantic) Web. The first trend is caused by the need to accelerate the slow and expensive software development life cycle. The vision of treating application logic as declarative business rules is particularly interesting for businesses with rapidly changing business logic. The second trend is related to the Semantic Web initiative of the W3C. The vision is that intelligent Semantic Web agents with their rule-based decision and reaction logic are capable of processing the cross referenced, machine processable knowledge on the Web in a platform independent manner. They are able to infer new knowledge and make intelligent, possibly pro-active and self-autonomous decisions and reactions. Emerging standards for rules operating in the context of the Semantic Web include RuleML (and SWRL) and the new W3C RIF recommendation.

A general rule markup language such as RuleML or RIF covers many different rule types and rule families. Their syntax builds on well establish Web data representation standards such as XML, RDF, URIs/IRIs etc. Some of the language families such as classical production rules historically only define an operational semantics, while other rule families such as logical rules are based on a model-theoretic and/or proof-theoretic semantics. An open research question is whether there exists a unifying semantic framework for all different rule types. Work in this direction is pursued, e.g. in the RIF Framework for Logic Dialects (RIF FLD) and in Reaction RuleML for reaction rule and complex event processing semantics. However, since there is no general consensus on one particular semantics for all expressive rule languages, an exclusive commitment to one particular semantics for a Web rule language should be avoided (even in well-researched fields such as logic programming several semantics such as well-founded semantics and answer set semantics are competing). Nevertheless, for certain subfamilies a preferred semantics can still be given and semantic mappings between rule families be defined.

Another crucial extension to the classical theory of rule-based logic programming in modern Web rule engines such as Prova is that they include practical language constructs which might not (yet) have a standard formal semantics based on classical model-theoretic logic. For instance, procedural calls to external (object) functions, operational systems, data sources and terminological descriptions, are often vital to deal with practical real-world settings of distributed Web applications. Recent research, e.g. in Prova, is done on adopting such practical language constructs without a standard formal semantics but with a non-standard extra logical one which allows for a hybrid knowledge representation. Further examples of useful practical constructs are the annotation of rules and rule sets with additional metadata such as rule qualifications, rule names, module names, Dublin Core annotations, etc., which eases, e.g., the modularization of rules into rule sets (bundling of rules), the creation of constructive views over internal and external knowledge (scoped reasoning), as well as the publication and interchange of rules / rule sets on the Web (rule messaging). Advanced rule qualifications such as validity periods or rule priorities might for example safeguard dynamic updates (e.g. the incorporation of interchanged rules into the existing rule base), where conflicts are resolved by rule prioritizations. Although these extra logical features have no direct formalization in first order logic, the benefits for a practical rule-based Web system, which needs to cope with large problem sizes and which needs to efficiently interoperate with existing systems and data sources on the Web, prevail. The hybrid KR design which allows the integration of external vocabulary types, methods and data into rule execution combines the benefits of declarative and imperativ (object-oriented) programming and helps to overcome typical problems of declarative programming, e.g., wrt to computational efficiency of certain tasks. While there is a risk that these concessions to non-standard semantics might endanger the benefits of formal semantics for the overall rule language, they turn out to be a crucial means to avoid limitations of standard rule representations. The rule component will rarely run in isolation, but interact with various external components, hence call for functionalities such as efficient object-oriented, relational/SQL-style, and RDF data retrieval and aggregation methods that are common in modern Web information systems.

Another domain of research is the engineering and maintenance of large rule-based applications, where the rules are serialized and managed in a distributed manner, and are interchanged across domain boundaries. This calls for support of verification, validation and integrity testing (V&V&I), e.g., by test cases that are written in the same rule markup language and are stored and interchanged together with the rule program [70].

**Acknowledgements.** This work has been partially supported by the "Inno Profile-Corporate Semantic Web" project funded by the German Federal Ministry of Education and Research (BMBF) and the BMBF Innovation Initiative for the New German Länder - Entrepreneurial Regions.

# References

1. Krisnadhi, F.M.A.A., Hitzler, P.: Owl and rules. In: 7th International Summer School 2011 - Tutorial Lectures. LNCS, Springer, Heidelberg (2011)
2. Ait-Kaci, H., Podelski, A.: Towards the meaning of life. In: Małuszyński, J., Wirsing, M. (eds.) PLILP 1991. LNCS, vol. 528, pp. 255–274. Springer, Heidelberg (1991)
3. Alferes, J., Damasio, C., Pereira, L.M.: Slx: a top-down derivation procedure for programs with explicit negation. In: Bruynooghe, M. (ed.) International Logic Programming Symp., pp. 424–439 (1994)
4. Alferes, J.J., Damasio, C., Pereira, L.M.: A logic programming system for non-monotonic reasoning. J. of Automated Reasoning 14(1), 93–147 (1995)
5. Apt, K.: Logic programming. In: Leeuwen, J.v. (ed.) Handbook of Theoretical Computer Science, vol. B, ch. 10, pp. 493–574. Elsevier, Amsterdam (1990)
6. Apt, K., Blair, H.: Logic Programming and Negation: A Survey. J. of Logic Programming 19(20), 9–71 (1994)
7. Apt, K., Blair, H., Walker, A.: Towards a theory of declarative knowledge. In: Minker, J. (ed.) Foundations of Deductive Databases, pp. 89–148. Morgan Kaufmann, San Francisco (1988)
8. Apt, K., Emden, M.H.: Contributions to the theory of logic programming. J. of ACM 29(3), 841–862 (1982)
9. Baral, C., Gelfond, M.: Logic programming and knowledge representation. J. of Logic Programming 19, 20, 73–148 (1994)
10. Baral, C., Lobo, J., Minker, J.: Generalized well-founded semantics for logic programs. In: Stickel, M.E. (ed.) International Conference on Automated Deduction. Springer, Heidelberg (1990)
11. Baral, C., Lobo, J., Minker, J.: Generalized disjunctive well-founded semantics for logic programs. Annals of Math and Artificial Intelligence 11(5), 89–132 (1992)
12. Baral, C., Subrahmanian, V.S.: Dualities between alternative semantics for logic programming and non-monotonic reasoning. In: Int. Workshop of Logic Programming and Non-Monotonic Reasoning, pp. 69–86. MIT Press, Cambridge (1991)
13. Beeri, C., Ramakrishnan, R.: On the power of magic. The Journal of Logic Programming 10, 255–299 (1991)
14. Bidoit, N., Legay, P.: Well!: An evaluation procedure for all logic programs. In: Int. Conf. on Database Theory, pp. 335–348 (1990)
15. Bol, R.: Tabulated resolution for the well-founded semantics. Journal of Logic Programming 34(2), 67–109 (1998)
16. Bol, R., Degerstedt, L.: Tabulated resolution for well founded semantics. In: Intl. Logic Programming Symposium (1993)

17. Boley, H.: Object-oriented ruleML: User-level roles, URI-grounded clauses, and order-sorted terms. In: Schröder, M., Wagner, G. (eds.) RuleML 2003. LNCS, vol. 2876, pp. 1–16. Springer, Heidelberg (2003)
18. Boley, H.: RIF RuleML Rosetta Ring: Round-Tripping the Dlex Subset of Datalog RuleML and RIF-Core. In: Governatori, G., Hall, J., Paschke, A. (eds.) RuleML 2009. LNCS, vol. 5858, pp. 29–42. Springer, Heidelberg (2009), http://dx.doi.org/10.1007/978-3-642-04985-9
19. Boley, H., Kifer, M.: A guide to the basic logic dialect for rule interchange on the web. IEEE Trans. on Knowl. and Data Eng. 22, 1593–1608 (2010)
20. Brachman, R.J., Gilbert, P.V., Levesque, H.J.: An essential hybrid reasoning system: Knowledge and symbol level accounts for krypton. In: Int. Conf. on Artificial Inelligence (1985)
21. Brass, S., Dix, J.: Characterizations of the disjunctive wellfounded semantics: Confluent calculi and iterated gcwa. Journal of Automated Reasoning (1997)
22. Brass, S., Dix, J.: Characterizations of the disjunctive well-founded semantics. Journal of Logic Programming 34(2), 67–109 (1998)
23. Brass, S., Dix, J., Zukowski, U.: Transformation based bottom-up computation of the well-founded model. Theory and Practice of Logic Programming 1(5), 497–538 (2001)
24. Brewka, G.: Well-founded semantics for extended logic programs with dynamic preferences. Journal of Artificial Intelligence Research 4, 19–36 (1996)
25. Bry, F.: Negation in logic programming: A formalization in constructive logic. In: Karagiannis, D. (ed.) IS/KI 1990 and KI-WS 1990. LNCS, vol. 474, pp. 30–46. Springer, Heidelberg (1991)
26. Bry, F.: Query evaluation in recursive databases: bottom-up and top-down reconciled. Data and Knowlege Engineering 5, 289–312 (1990)
27. Chen, J., Kundu, S.: The strong semantics for logic programs. In: Proceedings of the 6th Int. Symp. on Methodologies for Intelligent Systems, Charlotte, NC (1991)
28. Chen, W., Swift, T., Warren, D.S.: Efficient top-down computation of queries under the well-founded semantics. J. of Logic Programming 24(3), 161–199 (1995)
29. Chen, W., Warren, D.S.: A goal-oriented approach to computing well-founded semantics. In: Intl. Conf. and Symposium on Logic Programming (1992)
30. Chen, W., Warren, D.S.: Query evaluation under the well-founded semantics. In: Proceedings of Symp. on the Principles of Database Systems (1993)
31. Chen, W.: Query evaluation in deductive databases with alternating fixpoint semantics. ACM Transactions on Database Systems 20, 239–287 (1995)
32. Cherchago, N., Hitzler, P., Hölldobler, S.: Decidability under the well-founded semantics. In: Marchiori, M., Pan, J.Z., Marie, C.d.S. (eds.) RR 2007. LNCS, vol. 4524, pp. 269–278. Springer, Heidelberg (2007)
33. Clark, K.L.: Negation as failure. In: Gallaire, H., Minker, J. (eds.) Logic and Data-Bases, New York, pp. 293–322 (1978)
34. Dix, J.: A framework for representing and characterizing semantics of logic programs. In: Nebel, B., Rich, C., Swartout, W. (eds.) Principles of Knowledge Representation and Reasoning: Proceedings of the Third International Conference (KR 1992), pp. 591–602. Morgan Kaufmann, San Mateo (1992)
35. Dix, J.: A classification-theory of semantics of normal logic programs: Ii. weak properties. Fundamenta Informaticae XXII(3), 257–288 (1995)
36. Dix, J.: Semantics of logic programs: Their intuitions and formal properties. an overview. In: Fuhrmann, A., Rott, H. (eds.) Essays on Logic in Philosophy and Artificial Intelligence, pp. 241–327. DeGruyter, Berlag-New York (1995)

37. Doets, K.: From Logic to Logic Programming. MIT Press, Camebridge (1994)
38. Donini, F.M., Lenzerini, M., Nardi, D., Schaerf, A.: A hybrid system with data-log and concept languages. In: Ardizzone, E., Sorbello, F., Gaglio, S. (eds.) AI*IA 1991. LNCS (LNAI), vol. 549, pp. 88–97. Springer, Heidelberg (1991)
39. Dung, P.M.: Negation as hypotheses: An abductive foundation for logic programming. In: 8th Int. Conf. on Logic Programming, MIT Press, Cambridge (1991)
40. Dung, P.M.: An argumentation semantics for logic programming with explicit negation. In: 10th Logic Programming Conf., MIT Press, Cambridge (1993)
41. Dung, P.M., Kanchansut, K.: A natural semantics of logic programs with negation. In: 9th Conf. on Foundations of Software Technology and Theoretical Computer Science, pp. 70–80 (1989)
42. Eiter, T., Lukasiewicz, T., Schindlauer, R., Tompits, H.: Combining answer set programming with description logics for the semantic web. In: KR 2004 (2004)
43. Emden, M.H., Kowalski, R.: The semantics of predicate logic as a programming language. JACM 23, 733–742 (1976)
44. Fitting, M.: A kripke-kleene semantics of logic programs. Journal of Logic Programming 4, 295–312 (1985)
45. Fitting, M.: Well-founded semantics, generalized. In: Int. Symposium of Logic Programming, pp. 71–84. MIT Press, San Diego (1990)
46. Fitting, M.: First-Order Logic and Automated Theorem Proving, 2nd edn. Springer, Heidelberg (1996)
47. Gelfond, M., Lifschitz, V.: The stable model semantics for logic programming. In: Kowalski, R., Bowen, K. (eds.) 5th Conference on Logic Programming, pp. 1070–1080 (1988)
48. Gelfond, M., Lifschitz, V.: Logic programs with classical negation. In: ICLP 1990, pp. 579–597. MIT Press, Cambridge (1990)
49. Gelfond, M., Lifschitz, V.: Classical negation in logic programs and disjunctive databases. New Generation Computing 9, 365–385 (1991)
50. Grosof, B.N., Horrocks, I., Volz, R., Decker, S.: Description logic programs: Combining logic programs with description logic. In: International World Wide Web Conference, ACM, New York (2003)
51. Heymans, S., Van Nieuwenborgh, D., Hadavandi, E.: Nonmonotonic ontological and rule-based reasoning with extended conceptual logic programs. In: Gómez-Pérez, A., Euzenat, J. (eds.) ESWC 2005. LNCS, vol. 3532, pp. 392–407. Springer, Heidelberg (2005)
52. Hitzler, P., Seda, A.K.: Mathematical Aspects of Logic Programming Semantics. Studies in Informatics. Chapman and Hall/CRC Press (2010)
53. Horrocks, I., Patel-Schneider, P.F., Boley, H., Tabet, S., Grosof, B., Dean, M.: Swrl: A semantic web rule language combining owl and ruleml (2004), http://www.w3.org/submission/swrl/ (accessed January 2006)
54. Hu, Y., Yuan, L.Y.: Extended well-founded model semantics for general logic programs. in koichi furukawa, editor, In: Int. Conf. on Logic Programming, Paris, pp. 412–425 (1991)
55. Kemp, D.B., Srivastava, D., Stuckey, P.J.: Bottom-up evaluation and query optimization of well-founded models. Theor. Comput. Sci. 146, 145–184 (1995)
56. Khamsi, M.A., Misane, D.: Fixed point theorems in logic programming. Ann. Math. Artif. Intell. 21(2-4), 231–243 (1997)
57. Kowalski, R., Kuehner, D.: Linear resolution with selection function. Artifical Intelligence 2, 227–260 (1971)
58. Kunen, K.: Negation in logic programming. Journal of Logic Programming 4, 289–308 (1987)

59. Leitsch, A.: The Resolution Calculus. Springer, Heidelberg (1997)
60. Levy, A., Rousset, M.-C.: A representation language combining horn rules and description logics. In: European Conference on Artificial Intelligence, ECAI 1996 (1996)
61. Lifschitz, V.: Foundations of declarative logic programming. Principles of Knowledge Representation. CSLI publishers (1996)
62. Lloyd, J.W.: Foundations of logic programming, 2nd extended edn. Springer, New York (1987)
63. Lobo, J., Minker, J., Rajasekar, A.: Foundations of disjunctive logic programming. MIT Press, Cambridge (1992)
64. Lonc, Z., Truszcynski, M.: On the problem of computing the well-founded semantics. Theory and Practice of Logic Programming 1(5), 591–609 (2001)
65. Marek, V.W.: Autoepistemic logic. Journal of the ACM 38(3), 588–619 (1991)
66. McCarthy, J.: Circumscription - a form of non-monotonic reasoning. Journal of Artificial Intelligence 13(1-2), 27–39 (1980)
67. Minker, J.: An overview of nonmonotonic reasoning and logic programming. Journal of Logic Programming 17(2-4), 95–126 (1993)
68. Morishita, S.: An extension of van gelder's alternating fixpoint to magic programs. Journal of Computer and System Sciences 52, 506–521 (1996)
69. Motik, B., Sattler, U., Studer, R.: Query answering for owl-dl with rules. Journal of Web Semantics 3(1), 41–60 (2005)
70. Paschke, A.: Verification, validation, integrity of rule based policies and contracts in the semantic web. In: 2nd International Semantic Web Policy Workshop (SWPW 2006), Athens, GA, USA, November 5-9 (2006)
71. Paschke, A.: A typed hybrid description logic programming language with polymorphic order-sorted dl-typed unification for semantic web type systems. CoRR, abs/cs/0610006 (2006)
72. Paschke, A., Boley, H., Kozlenkov, A., Craig, B.L.: Rule responder: Ruleml-based agents for distributed collaboration on the pragmatic web. In: ICPW, pp. 17–28 (2007)
73. Pereira, L.M., Alferes, J.J.: Well founded semantics for logic programs with explicit negation. Proceedings of ECAI 1992 (1992)
74. Pereira, L.M., Alferes, J.J., Aparicio, J.N.: Adding closed world assumptions to well founded semantics. In: Fifth Generation Computer Systems, pp. 562–569 (1992)
75. Przymusinska, H., Przymusinski, T.C.: Weakly perfect semantics for logic programs. In: 5th International Conference and Symposium on Logic Programming, pp. 1106–1121 (1988)
76. Przymusinska, H., Przymusinski, T.C.: Weakly stratified logic programs. Fundamenta Informaticae 13, 51–65 (1990)
77. Przymusinski, T.C.: Perfect model semantics. In: 5th Int. Conf. and Symp. on Logic Pro- gramming, pp. 1081–1096. MIT Press, Cambridge (1988)
78. Przymusinski, T.C.: Every logic program has a natural stratification and an iterated fixed point model. Proceedings of ACM Symp. on Principles of Database Systems, 11–21 (1989)
79. Przymusinski, T.C.: On the declarative and procedural semantics of logic programs. Journal of Automated Reasonig 5, 167–205 (1989)
80. Przymusinski, T.C.: Non-monotonic reasoning vs. logic programming: A new perspective. In: Partridge, D., Wilks, Y. (eds.) The Foundations of Artifical Intelligence - A Sourcebook, Cambridge University Press, London (1990)

81. Przymusinski, T.C.: The well-founded semantics coincides with the three-valued stable semantics. Fundamenta Informaticae 13, 445–463 (1990)
82. Przymusinski, T.C.: Stable semantics for disjunctive programs. New Generation Computing 9, 401–424 (1991)
83. Rajasekar, A., Lobo, J., Minker, J.: Weak generalized closed world assumption. Journal of Automated Reasonig 5(3), 293–307 (1989)
84. Reiter, R.: A logic for default reasoning. Journal of Artificial Intelligence 13, 81–132 (1980)
85. Riccardo, R.: On the decidability and complexity of integrating ontologies and rules. Journal of Web Semantics 3(1) (2005)
86. RIF. W3c rif: Rule interchange formant (2010), http://www.w3.org/2005/rules/ (accessed october 2010)
87. Robinson, J.: A machine-oriented logic based on the resolution-principle. JACM 12(1), 23–41 (1965)
88. Ross, K.: A procedural semantics for well-founded negation in logic programs. Journal of Logic Programming 13(1), 1–22 (1992)
89. Ross, K.: Modular stratification and magic sets for datalog programs with negation. Journal of the ACM 41(6), 1216–1266 (1994)
90. Sacca, D., Zaniolo, C.: Partial models and three-valued models in logic programs with negation. In: Workshop of Logic Programming and Non-Monotonic Reasoning, Washington D.C, pp. 87–104. MIT Press, Cambridge (1991)
91. Schlipf, J.: Formalizing a logic for logic programming. Annals of Mathematics and Artificial Intelligence, 5, 279–302 (1992)
92. Shen, Y.-D., Yuan, L.-Y., You, J.-H.: Slt-resolution for the well-founded semantics. Journal of Automated Reasoning 28(1), 53–97 (2002)
93. Shepherdson, J.C.: Negation in logic programming. In: Minker, J. (ed.) Foundations of Deductive Databases, pp. 19–88. Morgan Kaufmann, San Francisco (1988)
94. Shepherdson, J.C.: Unsolvable problems for sldnf resolution. J. of Logic Programming, 19–22 (1991)
95. Stuckey, P.J., Sudarsham, S.: Well-founded ordered search: Goal-directed bottom-up evaluation of well-founded models. The Journal of Logic Programming 32(3), 171–205 (1997)
96. Tamaki, H., Sato, T.: Old resolution with tabulation. In: 3rd Int. Conf. on Logic Programming, London, pp. 84–98 (1986)
97. Teusink, F.: A proof procedure for extended logic programs. In: ILPS 1993. MIT Press, Cambridge (1993)
98. Ullman, J.D.: Principles of Database and Knowlegebase Systems, vol. 2. Computer Science Press, Rockville (1989)
99. Van Gelder, A.: The alternating fixpoint of logic programs with negation. In: 8th ACM SIGACT-SIGMOND-SIGART Symposium on Principles of Database Systems, pp. 1–10 (1989)
100. Van Gelder, A.: The alternating fixpoint of logic programs with negation. Journal of Computer and System Sciences 47(1), 185–221 (1993)
101. Van Gelder, A., Ross, K., Schlipf, J.: The well-founded semantics for general logic programs. JACM 38(3), 620–650 (1991)
102. You, L.H., Yuan, L.Y.: Three-valued formalization of logic programming: is it needed. In: Proceedings of 9th ACM SIGACT-SIGMOD-SIGART Symposium on Principles of Database Systems, pp. 172–182. ACM Press, New York (1990)

# OWL and Rules

Adila Krisnadhi, Frederick Maier, and Pascal Hitzler

Kno.e.sis Center, Wright State University, Dayton, Ohio

**Abstract.** The relationship between the Web Ontology Language OWL and rule-based formalisms has been the subject of many discussions and research investigations, some of them controversial. From the many attempts to reconcile the two paradigms, we present some of the newest developments. More precisely, we show which kind of rules can be modeled in the current version of OWL, and we show how OWL can be extended to incorporate rules. We finally give references to a large body of work on rules and OWL.

## 1 Introduction

Since research into the Semantic Web began, there have been different paradigms for modeling ontologies. Two prominent approaches discussed at the very beginning are description logics [4] and rules, the latter in the wider sense of logic programming (e.g., in the form of F-Logic [41]). While both of these approaches are based on classical logic, they are sufficiently different that naive attempts to combine them were unsuccessful.

The Web Ontology Language OWL [33,61], which is now a W3C standard, was the primary DL-based formalism that resulted from these discussions [34,84]. Nevertheless, rule-based formalisms [76] proved successful, including in commercial applications, and they continued to be pursued after the development of OWL. This eventually led to the development of the W3C Recommendation RIF (Rule Interchange Format) [5,6].

The modeling split between description logics and rules has naturally led to a considerable number of efforts to understand the relationships between the two paradigms and to establish workable combinations of them. Some of the resulting formalisms and systems have proved to be successful (we give a partial list in Section 5). However, a formalism that successfully combines the two paradigms into a single ontology language—while at the same time remaining conceptually true to both of them and remaining computationally viable—has not been developed.

In this paper, we focus on new results in developing such a language. Specifically, we discuss adding what are called *nominal schemas* (first described in [51]) to description logics. The resulting language is entirely in the spirit of description logics (a point which we discuss in more detail in Section 4), and yet it allows basic rule patterns to be captured. This paper can be understood as a continuation of [30], in the sense that it discusses (in the same spirit) recent work on combining rules and ontologies.

A. Polleres et al. (Eds.): Reasoning Web 2011, LNCS 6848, pp. 382–415, 2011.

After providing necessary terminology and technical preliminaries in Section 2, Sections 3 and 4 present material first described in [53] and [51], respectively. Specifically, Section 3 investigates the kinds of rules that are already expressible in the current OWL standard, and Section 4 shows how OWL can be extended to incorporate a significantly wider class of rules. In Section 5, we give pointers to other work combining rules and OWL. Section 6 concludes with some open issues for future research.

## 2  Preliminaries

For notation and terminology, and in particular for the definition of $\mathcal{SROIQ}$, we follow the chapter by Sebastian Rudolph contained in this volume [84]. For a textbook introduction, see [34]; whereas for a comprehensive treatment of description logics, see [4]. We use description logic notation throughout. Recall that the description logic $\mathcal{SROIQ}$ corresponds roughly to the OWL 2 DL profile of the Web Ontology Language [33,75]. Henceforth, by *OWL* we will understand *OWL 2 DL*. Some of the results discussed in this paper will also be closely related to the three *tractable profiles* of OWL 2 DL, namely *OWL 2 EL*, *OWL 2 RL*, and *OWL 2 QL* [64].

The description logic $\mathcal{SREL}$, also known as $\mathcal{EL}^{+}$, encompasses[1] the following concept (class) and role (property) constructs:

- concept conjunction
- existential quantification
- Self
- role chains
- the universal role

The description logic $\mathcal{SROEL}$ furthermore allows nominals. It essentially corresponds to OWL 2 EL [64]. The logic $\mathcal{SROIEL}$ further allows the use of inverse roles.

Given a first-order logic signature, a *Horn clause* is a formula of the form $(\forall x_1)\ldots(\forall x_n)(B_1 \wedge \cdots \wedge B_k \to A)$, where each $x_i$ is a variable occurring in the formula and $A$ and each $B_i$ are atomic formulas, also called *atoms*. It is usual to omit the quantifiers and abbreviate the formula as

$$B_1 \wedge \cdots \wedge B_k \to A,$$

commonly known as a *rule*. Given such a rule, $A$ is called the *head* of the rule, while $B_1 \wedge \cdots \wedge B_k$ is called the *body*, and each $B_i$ is referred to as a *body atom*.

A *function-free* Horn clause is called a *Datalog rule*, and we will see many examples below. The Rule Interchange Format of the W3C [42] encompasses the RIF Core Dialect [5], which is essentially Datalog. In our discussion on

---

[1] Some additional role characteristics are usually also included, but this is not important for our discussion.

integrating OWL and rules, we will mainly be concerned with Datalog using only unary and binary predicate symbols.

Semantically, we understand Datalog to be interpreted under the standard first-order predicate logic semantics. In some cases, we will refer to the Herbrand semantics, and will do so explicitly in each case.

# 3    Rules in OWL

In this section, we explore the question of which rules can be expressed in the current version of OWL. Results are adapted mainly from [53].

## 3.1    DLP and OWL 2 RL

It is rather obvious that certain DL axioms can be translated naively into rules:

$$A \sqsubseteq B \text{ becomes } A(x) \to B(x)$$
$$R \sqsubseteq S \text{ becomes } R(x,y) \to S(x,y)$$

DL axioms which involve only existential quantification and conjunction, and do so only on the *left hand side* of the concept inclusion, can also be translated easily:

$$A \sqcap \exists R.\exists S.B \sqsubseteq C \text{ becomes } A(x) \wedge R(x,y) \wedge S(y,z) \wedge B(z) \to C(x)$$

However, for existential quantifiers on the right hand side of concept inclusion, there is no such translation.[2]

Things become a bit trickier if we look at other DL concept constructors. Universal quantification occurring on the right hand side can be translated, but only when it is not on the left hand side.

$$A \sqsubseteq \forall R.B \text{ becomes } A(x) \wedge R(x,y) \to B(y)$$

This is so because the axiom $A \sqsubseteq \forall R.B$ is equivalent to $\exists R^-.A \sqsubseteq B$. Note, however, that the latter axiom requires an inverse role, whereas the former doesn't. Similarly, concept negation can be dealt with when occurring on the right hand side if it occurs together with disjunction, because an axiom like $A \sqsubseteq \neg B \sqcup C$ can be rewritten to $A \sqcap B \sqsubseteq C$, i.e.

$$A \sqsubseteq \neg B \sqcup C \text{ becomes } A(x) \wedge B(x) \to C(x).$$

Cardinality restrictions can be translated as long as they can be rewritten, e.g., expressions such as $\geq 1R.A$ would become $\exists R.A$, which can be handled if occurring on a left hand side. If we are allowed to use an equality symbol with the rules, then we can also express, e.g., functionality:

$$\top \sqsubseteq \leq 1R.\top \text{ becomes } R(x,y) \wedge R(x,z) \to y = z.$$

---

[2] Unless we allow Skolemization which, however, does not result in a semantically equivalent expression, only in an equisatisfiable one.

Nominals can also be dealt with. They usually translate into the use of constants, and in some cases we also need equality:

$$A \sqcap \exists R.\{b\} \sqsubseteq C \text{ becomes } A(x) \wedge R(x,b) \rightarrow C(x).$$

$$\{a\} \equiv \{b\} \text{ becomes } \rightarrow a = b.$$

If we allow truth value predicates $t$ and $f$ on the rules side, then we can also express some axioms involving $\top$ and $\bot$:

$$A \sqcap B \sqsubseteq \bot \text{ becomes } A(x) \wedge B(x) \rightarrow f.$$

Rules like the latter are usually called *integrity constraints*.

In some cases, DL axioms can be translated but result in more than one rule. This occurs, e.g., with disjunction on the left and with conjunction on the right hand side:

$$A \sqsubseteq B \wedge C \text{ becomes } A(x) \rightarrow B(x) \text{ and } A(x) \rightarrow C(x)$$

$$A \sqcup B \rightarrow C \text{ becomes } A(x) \rightarrow C(x) \text{ and } B(x) \rightarrow C(x)$$

If we look at this purely on the DL side, then the reason for this is that the first axiom indeed can be expressed as the two axioms $A \sqsubseteq B$ and $A \sqsubseteq C$, and likewise the second axiom can be expressed as the two axioms $A \sqsubseteq C$ and $B \sqsubseteq C$.

Armed with these observations, one is tempted to define a DL consisting only of axioms which can be translated into rules, e.g. as follows: *A DL axiom $\alpha$ can be translated into rules if, after translating $\alpha$ into a first-order predicate logic expression $\alpha'$, and after normalizing this expression into a set of clauses $M$, each formula in $M$ is a Horn clause (i.e., a rule).* It needs to be noted, though, that this definition is dependent on the exact translation and normalization algorithm used: Is it allowed to use Skolemization? Is it allowed to use *sophisticated* algorithms which may, for example, eliminate tautological axioms which are not directly expressible as rules?[3]

If we stick to a *naive* translation and normalization,[4] then the above observations are in fact the key idea behind the early language *DLP* [28], where the authors define a fragment of the DL $\mathcal{SHOIQ}$ (and thus for the 2004 version of OWL [61]) in this vein. DLP is discussed more in Section 5.3.

A naively adapted version of DLP, in fact, resulted in the OWL 2 profile OWL 2 RL [64].[5] In particular, in OWL 2 RL we can also deal with role chain axioms,

---

[3] We could also consider the whole DL knowledge base as input to this process, and algorithms which do a sophisticated *compilation* of the knowledge base. Indeed, such investigations have been carried out in a rather successful way, see e.g. [62], and also the notion of *Horn DLs* resulting from this [52].

[4] It is difficult to exactly define "naive"—but essentially we mean a kind of direct translation of each axiom into equivalent rules, in the spirit of the examples we have given. How exactly the notion "naive" is understood, in fact, does not matter much for our discussion. See [55] for a more conceptually inspired approach to defining rule fragments of DLs.

[5] OWL 2 RL extends a naive adaptation of the DLP language by some additional features, such as keys, which are not relevant to our discussion.

which were not present in the 2004 version of OWL, and thus not part of the
original DLP language:

$$R \circ S \sqsubseteq T \text{ becomes } R(x, y) \wedge S(y, z) \to T(x, z)$$

However, the *Self* construct from OWL 2 DL did not make it into OWL 2
RL, although it in fact mediates another rather strong relationship to rules. We
explore this in the following.

## 3.2   Rolification

Consider the sentence *"All elephants are bigger than all mice."* [85], which is
easily expressed by the rule

$$\text{Elephant}(x) \wedge \text{Mouse}(y) \to \text{biggerThan}(x, y). \tag{1}$$

It is indeed possible to translate this rule into OWL 2—however this involves a
transformation which we call *rolification*:[6] The rolification of a concept $A$ is a
(new) role $R_A$ defined by the axiom $A \equiv \exists R_A.\text{Self}$. Armed with rolification, we
can now express rule (1) by the axiom

$$R_{\text{Elephant}} \circ U \circ R_{\text{Mouse}} \sqsubseteq \text{biggerThan},$$

where $U$ is the universal role, together with the two axioms for the rolifications
of the concepts Elephant and Mouse,

$$\text{Elephant} \equiv \exists R_{\text{Elephant}}.\text{Self} \qquad \text{and} \qquad \text{Mouse} \equiv \exists R_{\text{Mouse}}.\text{Self}.$$

Note that this transformation is not exactly an equivalence transformation, since
we introduce new role names. However, it is very akin to the technique of *folding*
in logic programming, and the models of the rule stand in direct correspondence
with the models of the resulting set of DL axioms, in the sense of a conservative
extension[7].

The rolification technique now makes it possible to translate further rules into
DL syntax, in particular such rules where the rule head is a binary predicate:

$$A(x) \wedge R(x, y) \to S(x, y) \text{ becomes } R_A \circ R \sqsubseteq S$$
$$A(y) \wedge R(x, y) \to S(x, y) \text{ becomes } R \circ R_A \sqsubseteq S$$
$$A(x) \wedge B(y) \wedge R(x, y) \to S(x, y) \text{ becomes } R_A \circ R \circ R_B \sqsubseteq S$$

A natural use of this form of axiom would be in specifying when a role restricts
to a subrole, e.g., to state something like

$$\text{Woman}(x) \wedge \text{marriedTo}(x, y) \wedge \text{Man}(y) \to \text{hasHusband}(x, y),$$

---

[6] It is also called *man-man-ification*, because one of the early examples involved a
concept called *Man* [87].

[7] That is, for every model $\mathcal{I}$ of the rule, there exists a model of the DL axioms which
can be obtained from $\mathcal{I}$ by modifying the interpretation of the predicate symbols
not appearing in the rule; in this case, the new roles $R_{\text{Elephant}}$ and $R_{\text{Mouse}}$. See [60]
for further discussion about this definition.

which translates to

$$R_{\text{Woman}} \circ \text{marriedTo} \circ R_{\text{Man}} \sqsubseteq \text{hasHusband}$$

However this has to be done with caution, because it would be natural for an axiom like

$$\text{hasHusband} \sqsubseteq \text{marriedTo}$$

to appear in the same knowledge base. This, however, is not allowed since it would violate regularity conditions on the RBox (see [84]).

To give another example for the rolification technique, consider the rule

$$\text{worksAt}(x, y) \wedge \text{University}(y) \wedge \text{supervises}(x, z) \wedge \text{PhDStudent}(z)$$
$$\rightarrow \text{professorOf}(x, z),$$

which can be expressed as

$$R_{\exists \text{worksAt}.\text{University}} \circ \text{supervises} \circ R_{\text{PhDStudent}} \sqsubseteq \text{professorOf}.$$

## 3.3 Description Logic Rules

Given the previous examples, it becomes natural to ask about sufficient conditions on rules for a possible translation into DL expressions using the rolification technique. Such conditions gave rise to the notion of *Description Logic Rules* (DL Rules) as introduced in [53]. The key intuition behind DL Rules is that bodies of such rules must be *tree-shaped* in a sense which we will now formally define. An example for a body which is *not* tree-shaped is $R(x, y) \wedge S(y, z) \wedge T(x, z)$—just consider each pair of variables connected by a role as an edge in a directed graph with the variables as vertices: for this example, the graph is not a tree, hence the body is not tree-shaped.

To formally define DL Rules, we have to fix the description logic. From our examples above we can see that the following expressive features are desirable: conjunction, existential quantification, role chains, Self, and the universal role. These are available in the polynomial-time DL $\mathcal{SREL}$ (a.k.a. $\mathcal{EL}^+$). To also deal with constants, we require nominals, which are available in the polynomial DL $\mathcal{SROEL}$ (a.k.a. $\mathcal{EL}^{++}$) which contains $\mathcal{SREL}$ and is contained in OWL 2 EL. We have also seen above that inverse roles can be helpful, however they are not available in OWL 2 EL. They are available in $\mathcal{SROIEL}$ which is contained in OWL 2 DL.

Given a rule with body $B$, we construct a directed graph as follows: First rename individuals (i.e., constants) such that each individual occurs only once—a body such as $R(a, x) \wedge S(x, a)$ becomes $R(a_1, x) \wedge S(x, a_2)$. Denote the resulting new body by $B'$. The vertices of the graph are then the variables and individuals occurring in $B'$, and there is a directed edge between $t$ and $u$ if and only if there is an atom $R(t, u)$ in $B'$.

To illustrate this, consider the rule

$$C(x) \wedge R(x, a) \wedge S(x, y) \wedge D(y) \wedge T(y, a) \rightarrow P(x, y).$$

The resulting graph is $a_1 \longleftarrow x \longrightarrow y \longrightarrow a_2$ .

**Definition 1.** *We call a rule with head H tree-shaped (respectively, acyclic), if the following conditions hold.*

- *Each of the maximally connected components of the corresponding graph is in fact a tree (respectively, an acyclic graph)—or in other words, if it is a forest, i.e., a set of trees (respectively, a set of acyclic graphs).*
- *If H consists of an atom $A(t)$ or $R(t,u)$, then t is a root in the tree (respectively, in the acyclic graph).*

To give some examples, the rule $R(x,a) \wedge S(y,a) \to C(x)$ is tree-shaped, while the rule $R(x,z) \wedge S(y,z) \to T(x,y)$ is acyclic but not tree-shaped. The first rule translates to $R_{\exists R.\{a\}} \circ U \circ R_{\exists S.\{a\}} \sqsubseteq R_C$ while the second translates to $R \circ S^- \sqsubseteq T$. Note the use of the inverse role in the second example, which cannot be avoided—this is typically the case for rules which are acyclic but not tree-shaped.

We now have the following results, which are slight adaptations from results in [53].

**Theorem 1.** *The following hold.*

- *Every tree-shaped rule can be expressed in $\mathcal{SROEL}$.*
- *Every acyclic rule can be expressed in $\mathcal{SROIEL}$.*

*Description Logic Rules* as defined in [50,53] now generalize Definition 1 by allowing unary predicates in rule atoms which are in fact concept expressions from the underlying DL. It is shown that, if this is done for $\mathcal{SROIQ}$ (resulting in $\mathcal{SROIQ}$ *Rules*), then there is a polynomial transformation of such rules back into $\mathcal{SROIQ}$. If it is done for $\mathcal{SROEL}$ or for OWL 2 RL, then the resulting language is polynomial. It is furthermore shown that $\mathcal{SROEL}$ can be captured completely by tree-shaped rules with the extension that rule heads may be of the form $\exists R.A$, for a role $R$ and an atomic concept $A$.

A word of caution: Not every set of acyclic rules results in a set of axioms constituting a $\mathcal{SROIQ}$ knowledge base. This is due to the fact that not every set of $\mathcal{SROIQ}$ axioms is a $\mathcal{SROIQ}$ knowledge base: Restrictions on the use of non-simple roles must be adhered to, and the set of role chain axioms must be regular (see [84]).

We close this part with a rule that is not acyclic:

$$\text{hasReviewAssignment}(v,x) \wedge \text{hasAuthor}(x,y) \wedge \text{atVenue}(x,z)$$
$$\wedge \text{hasSubmittedPaper}(v,u) \wedge \text{hasAuthor}(u,y) \wedge \text{atVenue}(u,z) \tag{2}$$
$$\to \text{hasConflictingAssignedPaper}(v,x)$$

The corresponding graph is the following.

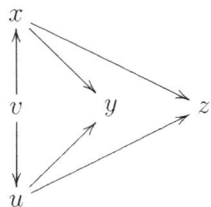

Note, however, that if $y$ and $z$ were constants, then the rule would be tree-shaped and could be expressed in $\mathcal{SROEL}$ as

$$R_{\exists\text{hasSubmittedPaper}.(\exists\text{hasAuthor}.\{y\}\sqcap\exists\text{atVenue}.\{z\})} \circ \text{hasReviewAssignment}$$
$$\circ\, R_{\exists\text{hasAuthor}.\{y\}\sqcap\exists\text{atVenue}.\{z\}}$$
$$\sqsubseteq \text{hasConflictingAssignedPaper}.$$

## 4   Rules Plus OWL

Theorem 1 allows us to identify rules expressible as DL axioms in a rather natural way. This, however, is only one step towards reconciling the rule-based and DL-based paradigms, as there are clearly additional (and desirable) things that are expressible in rules but which do not fit the format of Theorem 1. In this section, we discuss using *nominal schemas* [51] to significantly widen the class of rules expressible in a DL language. We believe nominal schemas provide one of the more seamless methods of integrating rule-based and DL-based ontology languages to date. But before we arrive at that, we will provide some relevant historical background.

### 4.1   DL-safe Rules, DL-safe Variables and ELP

Although DLs and rule languages are decidable fragments of first-order logic, it is well known that an unrestricted combination of both leads to undecidability. Intuitively, this is because many DLs rely on the so-called *tree model property* to retain decidability, and this property is lost when rules come into play [74].[8] Another related source of problems, which may similarly lead to undecidability or complexity blow-up, is the fact that DL knowledge bases typically entail the existence of anonymous individuals within a possibly infinite domain. This makes things difficult in the presence of rules, which generally apply to all individuals in the domain [54]. Therefore, a crucial step when one wants to combine the rule-based paradigm and the DL-based paradigm in one ontology is to come up with some safety criterion to ensure decidability or certain complexity bounds for reasoning over the combined language.

A prominent example of such a safety criterion is the notion of *DL-safe rules* [74] (see Sections 5.1 and 5.2). These restrict the applicability of rules in the combined knowledge base to named individuals, i.e., to individuals explicitly mentioned in the knowledge base. This guarantees decidability because there can only be a finite number of named individuals in the knowledge base.

More relevant to the current discussion is that DL-safe rules can be added to $\mathcal{SROEL}$ without losing tractability, under the restriction that there is a global

---

[8] A DL is said to have the *tree model property* when every satisfiable formula in it has a model which is of a tree-shape, where tree-shapedness is understood in a similar way as discussed in Section 3.3. Note that there are decidable DLs in which this property is not satisfied. In such DLs, decidability can be recovered by applying sophisticated strategies in the reasoning algorithm, e.g., blocking, see [4].

bound on the number of variables which can occur in each rule. The resulting language, called ELP, is a tractable ontology language based on the DL rules framework (discussed in Section 3.3) that generalizes DL-safe rules by building this safety criteria directly into the semantics of variables [54]. Syntactically, an ELP rule base is a set of rules with function-free, unary and binary atoms whose predicate symbols are formed from $\mathcal{SROEL}$ concept and role expressions.

We assume, in the signature of ELP, that the set of individuals is finite and contains only those named individuals occurring in the knowledge base. In addition, the DL-safety criteria is built into the semantics of variables as follows: from the set of variables that is a part of ELP's signature, we specify a fixed subset that contains precisely those variables which can only be assigned to named individuals. Let us revisit the following example (2) from page 388:

$$\text{hasReviewAssignment}(v, x) \wedge \text{hasAuthor}(x, y) \wedge \text{atVenue}(x, z)$$
$$\wedge \text{hasSubmittedPaper}(v, u) \wedge \text{hasAuthor}(u, y) \wedge \text{atVenue}(u, z) \tag{3}$$
$$\to \text{hasConflictingAssignedPaper}(v, x)$$

This rule is in ELP if the variables $y$ and $z$ are DL-safe variables. The intuition behind DL-safe variables is so that we can regain a tree-shape for the rule when these safe variables are replaced with named individuals from the knowledge base.

The tree-shapedness notion for ELP rules is based on Definition 1 with the following exceptions:

- there can be more than one tree edge (must be of the same direction) between two vertices; this corresponds to role conjunctions; if there is more than one tree edge between two vertices, those edges must correspond to simple roles only;
- atoms of the form $R(x, x)$ are ignored when defining a path in the tree, i.e., local reflexivity is allowed; ($R$ must be simple).

A rule base in ELP contains those rules whose atoms use $\mathcal{SROEL}$ concepts and role expressions and satisfy the tree-shapedness notion above, and which may in addition contain rules of the form $R(x, y) \to C(y)$ that satisfy: for each such rule, if the rule base contains a rule $B \to H$ with $R(t, z) \in H$, then $C(z) \in B$.

The following theorem from [54] gives the tractability result for ELP.

**Theorem 2.** *Satisfiability of any* ELP *rule base can be decided in time polynomial in the size of the rule base.*

The above result from ELP is an important milestone in the effort to reconcile DL-based and rule-based paradigms in ontology languages. Not only because of the tractability of reasoning, but also because of the fact that it subsumes both $\mathcal{SROEL}$ (i.e., OWL 2 EL) and DLP (i.e., most of OWL 2 RL) in the following sense [54].

**Theorem 3.** *Given any ground atom $\alpha$ of the form $C(a)$ or $R(a, b)$, a DLP rule base $\mathcal{R}$, and a $\mathcal{SROEL}$ knowledge base $\mathcal{K}$, there exists an ELP rule base $\mathcal{R}'$ such* *that if* $\mathcal{R} \models \alpha$ *or* $\mathcal{K} \models \alpha$ *then* $\mathcal{R}' \models \alpha$, *and if* $\mathcal{R}' \models \alpha$ *then* $\mathcal{R} \cup \mathcal{K} \models \alpha$, *and* $\mathcal{R}'$ *can be computed in linear time.*

In fact, the expressivity of ELP exceeds that of $\mathcal{SROEL}$ because it admits conjunctions of simple roles and limited range restrictions (expressed using rules). Note however, that ELP is clearly still a hybrid language because it uses both rule-based and DL-based syntax. This hybrid nature of ELP makes it rather complicated to integrate with OWL 2 DL standard which is roughly based on the DL paradigm. This becomes one of the motivations for the development of *nominal schemas* which is discussed in the sequel.

## 4.2   Nominal Schemas: Intuitive Idea

The notion of DL-safe variables in the previous section gives an insight on how to integrate rule-based and DL-based paradigms in a DL framework and how such integration can then be adapted quite easily into the current OWL syntax. The key observation is obtained from the fact that a DL-safe variable essentially represents all possible groundings to named individuals in the knowledge base. What we need is a way to specify this explicitly within DL syntax. This was realized in a new DL construct called *nominal schemas*, which syntactically resemble nominals [51]. In this paper, we consider the following DL languages: $\mathcal{SROIQV}(\mathcal{B}_s, \times)$ that is an extension of $\mathcal{SROIQ}$ (which roughly corresponds to OWL 2 DL) with Boolean operators on roles, concept products, and nominal schemas; and $\mathcal{SROELV}(\sqcap, \times)$ that is an extension of $\mathcal{SROEL}$ (which roughly corresponds to OWL 2 EL) with role conjunction, concept products and nominal schemas. For the latter, we will mainly speak about the tractable fragments $\mathcal{SROELV}(\sqcap, \times), n \geq 0$, which can be obtained from $\mathcal{SROELV}(\sqcap, \times)$ by restricting the number of occurrences of certain nominal schemas that will be introduced later.

To understand why nominal schemas allow a seamless integration of rules within DL-based syntax, note that in ELP, variables can essentially be categorized into two types: DL-safe variables which must be bound only to named individuals, and non-DL-safe variables which may represent anonymous individuals in the domain of the knowledge base. Thus, if we want to use a DL-based syntax, we can just hide the anonymous individuals inside the concept and role expressions and then deal with DL-safe variables separately. This is where nominal schemas are used.

One characteristic feature of rules that is brought into DL axioms by nominal schemas is *variable bindings*. Consider the following rule

$$\text{hasChild}(x, y) \wedge \text{hasChild}(x, z) \wedge \text{classmate}(y, z) \rightarrow C(x)$$

which defines a concept $C$ of parents with at least children which are classmates (consider the role classmate to be irreflexive). This rule is not tree-shaped as it induces two paths from $x$ to $z$. Moreover, the variable $z$ which occurs in different atoms must be bound to the same individual. This cannot be simulated in DLs unless we are equipped with nominal schemas as follows:

$$\exists \text{hasChild}.\{z\} \sqcap \exists \text{hasChild}.\exists \text{classmate}.\{z\} \sqsubseteq C$$

The following example—see (2) on page 388 and (3) on page 390— is expressed in $\mathcal{SROELV}_n(\sqcap, \times)$. It states that somebody has a conflicting review assignment (paper $x$) if this person has a paper submitted at the same event which is co-authored by one of the authors of paper $x$.

$$\exists \text{hasReviewAssignment}.(( \{x\} \sqcap \exists \text{hasAuthor}.\{y\}) \sqcap (\{x\} \sqcap \exists \text{atVenue}.\{z\}))$$
$$\sqcap \exists \text{hasSubmittedPaper}.(\exists \text{hasAuthor}.\{y\} \sqcap \exists \text{atVenue}.\{z\}) \tag{4}$$
$$\sqsubseteq \exists \text{hasConflictingAssignedPaper}.\{x\}$$

The last example does not induce tree-shaped structures, the fact of which is quite clear if we rewrite it as a rule. There, the tree-shaped structure can be recovered when $x$ is ground as a named individual. This particular insight is exploited to show the tractability of reasoning for $\mathcal{SROELV}_n(\sqcap, \times)$.

Formally, this is done by introducing the notion of *safe environment*.[9]

**Definition 2.** *An occurrence of nominal schema $\{x\}$ in a concept $C$ is* safe *if $C$ contains a sub-concept of the form $\{v\} \sqcap \exists R.D$ for some nominal schema or nominal $\{v\}$ such that $\{x\}$ is the only nominal schema that occurs (possibly more than once) in $D$. In this case, $\{v\} \sqcap \exists R.D$ is a* safe environment *for this occurrence of $\{x\}$, sometimes written as $S(v, x)$.*

The virtue of safe environments lies in the fact that, algorithmically, safe occurrences of nominal schemas can essentially be handled separately from the axiom in which they occur, thus avoiding a combinatorial explosion through grounding, provided that there is a global bound on the number of occurrences of those safe nominal schemas in each axiom [51]—we will return to this issue in the proof sketch, and subsequent examples, of Theorem 5 below. The following definition captures this idea, and it will be explained in more detail further below.

**Definition 3.** *Let $n \geq 0$ be an integer. A $\mathcal{SROELV}(\sqcap, \times)$ knowledge base $KB$ is a $\mathcal{SROELV}_n(\sqcap, \times)$ knowledge base if in each of its axioms $C \sqsubseteq D$, there are at most $n$ nominal schemas appearing more than once in non-safe form, and all remaining nominal schemas appear only in $C$.*

Note the dependency of the definition on the positive integer $n$, which is a global bound on the number of nominal schemas which can occur (more than once in non-safe form) in any axiom. Without this global bound we would not be able to retain tractability of reasoning.

Returning to our example axiom (4) above, we see that it indeed lies in $\mathcal{SROELV}_1(\sqcap, \times)$.

### 4.3   Nominal Schemas: Formal Definitions and Results

We now formally introduce syntax and semantics of nominal schema. As indicated in section 4.2, we introduce two new languages: $\mathcal{SROIQV}(\mathcal{B}_s, \times)$ and

---

[9] Definition 2 is slightly more general than the one presented in [51], leading to a slightly more general polynomial language.

$\mathcal{SROELV}_n(\sqcap, \times)$. We will start with the former and then introduce the latter as its sublanguage. Let the set of individual names $\mathsf{N}_I$, the set of concept names $\mathsf{N}_C$, and the set of role names $\mathsf{N}_R$ form the signature of the DL $\mathcal{SROIQ}$ as defined in [84]. The signature of $\mathcal{SROIQV}(\mathcal{B}_s, \times)$ is then formed from $\mathsf{N}_I, \mathsf{N}_C, \mathsf{N}_R$, and additionally the set of *variables* $\mathsf{N}_V$. We also assume that these sets are finite and pairwise disjoint. As already seen from the earlier examples, we use lower case letters $x, y, z, \ldots$ to denote variables. Furthermore, the set of role names $\mathsf{N}_R$ is partitioned into disjoint sets $N_R^s$ of *simple role names* and $N_R^n$ of *non-simple role names*. Note that this partition is fixed from the signature, i.e., is not defined based on syntactic properties, e.g., how it occurs in the TBox or ABox, etc. This simplifies the presentation.

The set of $\mathcal{SROIQV}(\mathcal{B}_s, \times)$ *roles* $\mathbf{R}$ is the union of two (non-disjoint) sets: the set of *simple roles* $\mathbf{R}^s$ and the set of *non-simple roles* $\mathbf{R}^n$ where $\mathbf{R}^s$ consists of (defined inductively):

- all simple role names;
- inverses of simple role names, i.e., $R^-$ for every simple role name $R$;
- the universal role $U$;
- $\neg R$, $R \sqcap S$ and $R \sqcup S$ where $R, S$ are simple roles in $\mathbf{R}^s$;
- the concept products $A \times B$ where $A, B$ are concept names;

and $\mathbf{R}^n$ consists of (defined inductively):

- all non-simple role names;
- inverses of non-simple role names, i.e., $R^-$ for every non-simple role name $R$;
- the universal role $U$;
- the concept products $A \times B$ where $A, B$ are concept names.

The set of $\mathcal{SROIQV}(\mathcal{B}_s, \times)$ *concepts* $\mathbf{C}$ consists of (defined inductively):

- the top concept $\top$ and the bottom concept $\bot$;
- every concept name $A \in \mathsf{N}_C$;
- $\{a\}$ for every individual name $a \in \mathsf{N}_I$;
- $\{v\}$ for every variable $v \in \mathsf{N}_V$;
- $\neg C$, $C \sqcap D$ and $C \sqcup D$ where $C, D$ are concepts;
- $\exists R.C$ and $\forall R.C$ where $R$ is a role;
- $\exists R.\mathsf{Self}$, $\leq kR.C$ and $\geq kR.C$ where $R$ is a *simple* role, $k$ any non-negative integer and $C$ concept.

Concepts $\{a\}$ with $a \in \mathsf{N}_I$ are called *nominals* and concepts $\{v\}$ with $v \in \mathsf{N}_V$ are called *nominal schemas*. Essentially, concepts and roles for $\mathcal{SROIQV}(\mathcal{B}_s, \times)$ are $\mathcal{SROIQ}$ concepts and roles extended with concept product (indicated with $\times$), nominal schema (indicated with the letter $\mathcal{V}$) and Boolean role constructors (indicated with the letter $\mathcal{B}_s$).

A $\mathcal{SROIQV}(\mathcal{B}_s, \times)$ knowledge base consist of RBox, TBox and ABox axioms with syntax defined as usual. The regularity condition for $\mathcal{SROIQ}$ knowledge bases also applies for $\mathcal{SROIQV}(\mathcal{B}_s, \times)$ knowledge bases.

The semantics of $\mathcal{SROIQV}(\mathcal{B}_s, \times)$, like that of $\mathcal{SROIQ}$, is based on interpretations $\mathcal{I} = (\Delta^{\mathcal{I}}, \cdot^{\mathcal{I}})$ with $\Delta^{\mathcal{I}}$ the domain of $\mathcal{I}$ and $\cdot^{\mathcal{I}}$ the interpretation mapping. But we need an additional component for interpretation of variables. This is realized by associating a *variable assignment* $\mathcal{Z} : \mathsf{N}_V \to \Delta^{\mathcal{I}}$ for the interpretation $\mathcal{I}$. The assignment $\mathcal{Z}$ is such that for each $v \in \mathsf{N}_V$, $\mathcal{Z}(v) = a^{\mathcal{I}}$ for some $a \in \mathsf{N}_I$. Another interpretation mapping $\cdot^{\mathcal{I},\mathcal{Z}}$ is then defined that reflects both $\mathcal{I}$ and $\mathcal{Z}$. The base definition of $\cdot^{\mathcal{I},\mathcal{Z}}$ starts from concept names, role names, individual names and variables as follows:

$$A^{\mathcal{I},\mathcal{Z}} = A^{\mathcal{I}} \subseteq \Delta^{\mathcal{I}} \qquad\qquad R^{\mathcal{I},\mathcal{Z}} = R^{\mathcal{I}} \subseteq \Delta^{\mathcal{I}} \times \Delta^{\mathcal{I}}$$
$$a^{\mathcal{I},\mathcal{Z}} = a^{\mathcal{I}} \in \Delta^{\mathcal{I}} \qquad\qquad x^{\mathcal{I},\mathcal{Z}} = \mathcal{Z}(x) \in \Delta^{\mathcal{I}}$$

Extending $\cdot^{\mathcal{I},\mathcal{Z}}$ for complex concepts and roles is straightforward and very similar to the way $\cdot^{\mathcal{I}}$ is extended to them in $\mathcal{SROIQ}$. The following are for complex concepts:

$$\top^{\mathcal{I},\mathcal{Z}} = \Delta^{\mathcal{I}} \qquad \bot^{\mathcal{I},\mathcal{Z}} = \emptyset \qquad \{t\}^{\mathcal{I},\mathcal{Z}} = \{t^{\mathcal{I},\mathcal{Z}}\} \text{ for } t \in \mathsf{N}_I \cup \mathsf{N}_V$$
$$(\exists R.C)^{\mathcal{I},\mathcal{Z}} = \{\delta \mid \text{ there is } \epsilon \text{ with } \langle \delta, \epsilon \rangle \in R^{\mathcal{I},\mathcal{Z}} \text{ and } \epsilon \in C^{\mathcal{I},\mathcal{Z}}\}$$
$$(\forall R.C)^{\mathcal{I},\mathcal{Z}} = \{\delta \mid \text{ for all } \epsilon \text{ with } \langle \delta, \epsilon \rangle \in R^{\mathcal{I},\mathcal{Z}}, \text{ we have } \epsilon \in C^{\mathcal{I},\mathcal{Z}}\}$$
$$(\exists R.\mathsf{Self})^{\mathcal{I},\mathcal{Z}} = \{\delta \mid \langle \delta, \delta \rangle \in R^{\mathcal{I},\mathcal{Z}}\}$$
$$(\neg C)^{\mathcal{I},\mathcal{Z}} = \Delta^{\mathcal{I}} \setminus C^{\mathcal{I},\mathcal{Z}}$$
$$(C \sqcap D)^{\mathcal{I},\mathcal{Z}} = C^{\mathcal{I},\mathcal{Z}} \cap D^{\mathcal{I},\mathcal{Z}} \qquad (C \sqcup D)^{\mathcal{I},\mathcal{Z}} = C^{\mathcal{I},\mathcal{Z}} \cup D^{\mathcal{I},\mathcal{Z}}$$
$$(\leq k R.C)^{\mathcal{I},\mathcal{Z}} = \{\delta \mid \#\{\langle \delta, \epsilon \rangle \in R^{\mathcal{I},\mathcal{Z}} \mid \epsilon \in C^{\mathcal{I},\mathcal{Z}}\} \leq k\}$$
$$(\geq k R.C)^{\mathcal{I},\mathcal{Z}} = \{\delta \mid \#\{\langle \delta, \epsilon \rangle \in R^{\mathcal{I},\mathcal{Z}} \mid \epsilon \in C^{\mathcal{I},\mathcal{Z}}\} \geq k\}$$

For roles, the following holds:

$$U^{\mathcal{I},\mathcal{Z}} = \Delta^{\mathcal{I}} \times \Delta^{\mathcal{I}}$$
$$(R^-)^{\mathcal{I},\mathcal{Z}} = \{\langle \delta, \epsilon \rangle \mid \langle \epsilon, \delta \rangle \in R^{\mathcal{I},\mathcal{Z}}\}$$
$$(A \times B)^{\mathcal{I},\mathcal{Z}} = \{\langle \delta, \epsilon \rangle \mid \delta \in A^{\mathcal{I},\mathcal{Z}} \text{ and } \epsilon \in B^{\mathcal{I},\mathcal{Z}}\}$$
$$(\neg R)^{\mathcal{I},\mathcal{Z}} = (\Delta^{\mathcal{I}} \times \Delta^{\mathcal{I}}) \setminus R^{\mathcal{I},\mathcal{Z}}$$
$$(R \sqcap S)^{\mathcal{I},\mathcal{Z}} = R^{\mathcal{I},\mathcal{Z}} \cap S^{\mathcal{I},\mathcal{Z}} \qquad (R \sqcup S)^{\mathcal{I},\mathcal{Z}} = R^{\mathcal{I},\mathcal{Z}} \cup S^{\mathcal{I},\mathcal{Z}}$$

Let $\mathcal{I}$ be an interpretation and $\mathcal{Z}$ a variable assignment for $\mathcal{I}$. For a $\mathcal{SROIQV}(\mathcal{B}_s, \times)$ axiom $\alpha$, we say, $\mathcal{I}$ and $\mathcal{Z}$ *satisfy* $\alpha$ (written $\mathcal{I}, \mathcal{Z} \models \alpha$) if the following holds for the corresponding form of $\alpha$:

$$\mathcal{I}, \mathcal{Z} \models A(t) \text{ iff } t^{\mathcal{I},\mathcal{Z}} \in A^{\mathcal{I},\mathcal{Z}}$$
$$\mathcal{I}, \mathcal{Z} \models R(t, u) \text{ iff } (t^{\mathcal{I},\mathcal{Z}}, u^{\mathcal{I},\mathcal{Z}}) \in R^{\mathcal{I},\mathcal{Z}}$$
$$\mathcal{I}, \mathcal{Z} \models C \sqsubseteq D \text{ iff } C^{\mathcal{I},\mathcal{Z}} \subseteq D^{\mathcal{I},\mathcal{Z}}$$
$$\mathcal{I}, \mathcal{Z} \models R \sqsubseteq S \text{ iff } R^{\mathcal{I},\mathcal{Z}} \subseteq S^{\mathcal{I},\mathcal{Z}}$$
$$\mathcal{I}, \mathcal{Z} \models R_1 \circ \ldots \circ R_n \sqsubseteq S \text{ iff } R_1^{\mathcal{I},\mathcal{Z}} \circ \ldots \circ R_n^{\mathcal{I},\mathcal{Z}} \subseteq S^{\mathcal{I},\mathcal{Z}}$$

where '$\circ$' denotes the usual composition of binary relations

$\mathcal{I}$ *satisfies* $\alpha$, written $\mathcal{I} \models \alpha$, if $\mathcal{I}, \mathcal{Z} \models \alpha$ for every variable assignment $\mathcal{Z}$ for $\mathcal{I}$. $\mathcal{I}$ *satisfies* a $\mathcal{SROIQV}(\mathcal{B}_s, \times)$ knowledge base $KB$, written $\mathcal{I} \models KB$, if $\mathcal{I} \models \alpha$ for every $\alpha \in KB$. In this case, we say $KB$ is *satisfiable (has a model)*. $KB$ *entails* an axiom $\alpha$, written $KB \models \alpha$, if all models of $KB$ are also models of $\alpha$.

It is known that reasoning in $\mathcal{SROIQ}(\mathcal{B}_s)$ is N2ExpTime-complete — thus, of the same complexity as $\mathcal{SROIQ}$ — where this logic is an extension of $\mathcal{SROIQ}$ with Boolean role operators (and concept products too, since concept products can be simulated using role negations) [83]. Reasoning in $\mathcal{SROIQV}(\mathcal{B}_s, \times)$ can thus be done by grounding the nominal schemas first, i.e., substituting each nominal schema with finitely many named individuals it may represent, resulting in a knowledge base in $\mathcal{SROIQ}(\mathcal{B}_s)$, and then proceeded with the reasoning algorithm for $\mathcal{SROIQ}(\mathcal{B}_s)$. If each axiom contains $m$ different nominal schemas, and there are a total of $n$ axioms in the knowledge base, then this naive grounding will generate $n \cdot |N_I|^m$ new axioms, i.e., a number exponential in the size of the input knowledge base if there is no global bound on $m$. However, as stated in the following theorem, adding nominal schema does not actually increase the complexity [51].

**Theorem 4.** *The problem of deciding satisfiability of a $\mathcal{SROIQV}(\mathcal{B}_s, \times)$ knowledge base is* N2ExpTime-*complete.*

Another problem of obvious interest is to identify a fragment of the language $\mathcal{SROIQV}(\mathcal{B}_s, \times)$ that admits nominal schemas as one of its constructors, but is still tractable in reasoning.

As mentioned in Section 4.2, the idea of nominal schemas is inspired from the use of DL-safe variables in ELP which is a tractable extension of $\mathcal{SROEL}$. So, obvious candidates to look at are extensions of $\mathcal{SROEL}$ with nominal schemas. In [51], the DLs $\mathcal{SROELV}_n(\sqcap, \times)$ were presented as such candidates. These DLs are extensions of $\mathcal{SROEL}(\sqcap, \times)$ which are defined for each integer $n \geq 0$. The number $n$ that is a part of the language definition provides a global bound that restricts the number of "unsafe" occurrences of nominal schemas in an axiom.

Recall that occurrences of nominal schemas in an axiom provides variable bindings which are a characteristic feature of rules, but not of DL axioms. In general, such bindings may represent complex dependencies that are difficult to simplify. The naive way to process nominal schemas is by grounding them all to every possible replacement with named individuals in the knowledge base. This obviously leads to intractability as this naive grounding introduces exponential blow-up in the size of the knowledge base.

To achieve tractability, a better reduction on the number of nominal schemas is needed. Fortunately, by borrowing insight from ELP, we understood that there are special cases in which nominal schemas on the left-hand side of TBox axioms can be eliminated or separated using independent axioms. The idea from ELP is that when the dependencies expressed in a rule body are *tree-shaped*, the rule can be reduced to a small set of normalized rules, each of which contains a limited number of variables. This idea was then exploited to obtain the tractability results of ELP [54].

Elevating this idea to $\mathcal{SROELV}_n(\sqcap, \times)$, we view variables in rules as either "hidden" in the concept expression or as occurring explicitly as nominal schemas. Note that in [54], tree-shapedness only refers to variables and not constants which correspond to nominals in our case here. Thus, nominals can be used to disconnect a dependency structure in a concept. For example, consider the concept

$$A \sqcap \exists R.\{z\} \sqcap \exists S.(B \sqcap \exists T.\{z\})$$

which corresponds to the rule body

$$A(x) \wedge R(x, z) \wedge S(x, y) \wedge B(y) \wedge T(y, z).$$

The tree-shapedness of the rule is recovered when $y$ is actually a constant. In the corresponding concept, this means a nominal in the place of the concept $B$. When this is the case, the nominal schema $\{z\}$ within the last conjunct of the example concept occurs in a *safe environment*, which is the safety criteria that we need. The formal Definition 2 generalizes this to the case where $y$, as in the example above, is a nominal schema instead of a nominal.

We now give a formal definition of the DL $\mathcal{SROELV}(\sqcap, \times)$ — and thus, of $\mathcal{SROELV}_n(\sqcap, \times)$ for every $n \geq 0$. We define a $\mathcal{SROELV}(\sqcap, \times)$ *concept* as a $\mathcal{SROIQV}(\mathcal{B}_s, \times)$ concept that may contain $\top$, $\bot$, conjunctions, existential restrictions, self restrictions, nominals and nominal schemas, but that does not contain disjunctions, negations, universal restrictions, and number restrictions. A $\mathcal{SROELV}(\sqcap, \times)$ *role* is a $\mathcal{SROIQV}(\mathcal{B}_s, \times)$ role (simple or non-simple) which may contain role conjunction (for simple roles) and the universal role, but no inverse roles, role disjunction or role negation. TBox, RBox and ABox axioms for $\mathcal{SROELV}(\sqcap, \times)$ are TBox, RBox, and ABox axioms in $\mathcal{SROIQV}(\mathcal{B}_s, \times)$ that use only $\mathcal{SROELV}(\sqcap, \times)$ concepts and roles. Furthermore, every $\mathcal{SROELV}(\sqcap, \times)$ knowledge base satisfies the following restriction.

**Definition 4.** *Let $KB$ be a knowledge base and $R$ a role name. Let $\mathsf{ran}(R)$ be the set of all concept names $B$ for which there is a set $\{R \sqsubseteq R_1, R_1 \sqsubseteq R_2, \ldots, R_{n-1} \sqsubseteq R_n, R_n \sqsubseteq A \times B\} \subseteq KB$ with $n > 0$ and $R_0 = R$. We impose that every $\mathcal{SROELV}(\sqcap, \times)$ knowledge base must satisfy* admissibility range restrictions *for every role inclusion axiom in it as follows: $R_1 \circ \ldots \circ R_n \sqsubseteq S$ implies $\mathsf{ran}(S) \subseteq \mathsf{ran}(R_n)$ and $R_1 \sqcap R_2 \sqsubseteq S$ implies $\mathsf{ran}(S) \subseteq \mathsf{ran}(R_1) \cup \mathsf{ran}(R_2)$.*

This admissibility criteria is from $\mathcal{SROEL}(\sqcap, \times)$, as defined in [49].

Finally, $\mathcal{SROELV}_n(\sqcap, \times)$ concepts and roles are $\mathcal{SROELV}(\sqcap, \times)$ concepts and roles. Also, $\mathcal{SROELV}_n(\sqcap, \times)$ knowledge bases are $\mathcal{SROELV}(\sqcap, \times)$ knowledge bases that satisfies Definition 3. For $\mathcal{SROELV}_n(\sqcap, \times)$, we have obtain the following result for every integer $n \geq 0$.

**Theorem 5.** *If $KB$ is a $\mathcal{SROELV}_n(\sqcap, \times)$ knowledge base of size $s$, satisfiability of $KB$ can be decided in time proportional to $s^n$. If $n$ is constant, then the problem is P-complete.*

A full proof of this theorem can be found in [51]. We explain the key idea of the proof by means of our running example (4). Note that a naive grounding, as explained above, would result in $|\mathsf{N}_I|^3$ new axioms (without nominal schemas, but with nominals). To decrease this figure without loss of completeness or soundness, we take advantage of safe environments—the rationale behind this being that safe environments can be handled separately from the rest of the axiom, as follows.[10]

We first replace, in the axiom, the safe environments by a single nominal, and we do this replacement for every nominal in the knowledge base. That is, we obtain $|\mathsf{N}_I|$ new axioms as follows, where $a_i$ ranges over all elements of $\mathsf{N}_I$. Note the we also replaced the remaining occurrence of the nominal schema $\{x\}$ accordingly.[11]

$$\exists \text{hasReviewAssignment}.(\{a_i\} \sqcap \{a_i\})$$
$$\sqcap \exists \text{hasSubmittedPaper}.(\exists \text{hasAuthor}.\{y\} \sqcap \exists \text{atVenue}.\{z\})$$
$$\sqsubseteq \exists \text{hasConflictingAssignedPaper}.\{a_i\}$$

Next, we replace the remaining occurrences of $\{y\}$ and $\{z\}$ (note that there can be at most one for each of these nominal schemas, per definition of the language $\mathcal{SROELV}_n(\sqcap, \times)$) by new concept names $O_y$ and $O_z$ (when subsequently converting other axioms, new concept names need to be used).

$$\exists \text{hasReviewAssignment}.(\{a_i\} \sqcap \{a_i\})$$
$$\sqcap \exists \text{hasSubmittedPaper}.(\exists \text{hasAuthor}.O_y \sqcap \exists \text{atVenue}.O_z)$$
$$\sqsubseteq \exists \text{hasConflictingAssignedPaper}.\{a_i\}$$

We furthermore conjoin the expressions $\exists U.O_y$ and $\exists U.O_z$ to the left-hand side of the axiom, where $U$ is the universal role.

$$(\exists U.O_y) \sqcap (\exists U.O_z) \sqcap \exists \text{hasReviewAssignment}.(\{a_i\} \sqcap \{a_i\})$$
$$\sqcap \exists \text{hasSubmittedPaper}.(\exists \text{hasAuthor}.O_y \sqcap \exists \text{atVenue}.O_z)$$
$$\sqsubseteq \exists \text{hasConflictingAssignedPaper}.\{a_i\}$$

Note that this results in $\mathsf{N}_I$ new axioms. Finally, add to the knowledge base the following axioms, which are constructed from the safe environments and from the elements $a_i$ of $\mathsf{N}_I$ already used:

$$\exists U.(\{a_i\} \sqcap \exists \text{hasAuthor}.\{a_j\}) \sqsubseteq \exists U.(\{a_j\} \sqcap O_y) \tag{5}$$
$$\exists U.(\{a_i\} \sqcap \exists \text{atVenue}.\{a_j\}) \sqsubseteq \exists U.(\{a_j\} \sqcap O_z) \tag{6}$$

Note that this results in $2 \cdot |\mathsf{N}_I|^2$ new axioms, for a total of $|\mathsf{N}_I| + 2 \cdot |\mathsf{N}_I|^2$ new axioms, which for large $|\mathsf{N}|_I$ is considerably smaller than the number $|\mathsf{N}_I|^3$ of new

---

[10] This obviously needs a proof, see [51].
[11] In this specific case, we could also simplify $((\{a_i\}) \sqcap (\{a_i\}))$ to $\{a_i\}$, but this is coincidental in our example.

axioms obtained from the naive grounding—and the effect is more drastic for axioms with more nominal schemas. Note, in particular, that the number of new axioms is of the order of magnitude of $|\mathsf{N}_I|^{\max\{2,n\}}$, where $n$ is the global bound from the definition of $\mathcal{SROELV}_n(\sqcap, \times)$—in particular the number is polynomially bounded for fixed $n$.

The key idea behind the transformation just described is, that the axioms (5) and (6) *constrain* the possible values for $O_y$ and $O_z$, and that this suffices for the reasoning process, since the concrete values obtained as elements of these concepts are not required for further processing.

## 4.4   Embedding Datalog under Nominal Schemas

An important feature of nominal schemas is that they can express arbitrary Datalog rules with unary and binary predicates which are interpreted as DL-safe, i.e., the predicates (and their variables) only apply to named individuals. Here, the DL-safe (Datalog) rules use a first-order logic semantics adapted using DL-safe variables—which as such is akin to a Herbrand semantics reading—which is compatible with the semantics of $\mathcal{SROIQV}(B_s, \times)$. Moreover, there is an easy syntactic transformation from DL-safe rules into $\mathcal{SROIQV}(B_s, \times)$ axioms which are semantically equivalent to the original DL-safe rules. The transformation can be done as follows:

- Each unary atom $A(x)$ is translated into $\exists U.(\{x\} \sqcap A)$.
- Each binary atom $R(x, y)$ is translated into $\exists U.(\{x\} \sqcap \exists R.\{y\})$.
- Let $B \to H$ be a DL-safe rule, $\mathsf{dl}(H)$ be the translation of the head atom $H$, and $\mathsf{dl}(B_i)$ be the translation of the atom $B_i$ for each atom $B_i$ in the body $B$. Then $B \to H$ is translated into $\bigsqcap\{\mathsf{dl}(B_i) \mid B_i \text{ in } B\} \sqsubseteq \mathsf{dl}(H)$
- Finally, the translation of a set of DL-safe rules RB is the set of axioms, each of which is the translation of an original rule from RB.

This translation clearly yields a set of axioms the size of which is linear in the size of the original rule base. Each such axiom, however, when naively grounded, results in $|\mathsf{N}_I|^n$ new axioms without nominal schemas, where $n$ is the number of variables occurring in the originating rule. This number is exponential in $n$, however with a global bound on $n$ (as we have for $\mathcal{SROELV}_n(\sqcap, \times)$), it is still polynomial in the size of the knowledge base.

By way of an example, consider the rule

$$R(x, y) \wedge A(y) \wedge S(z, y) \wedge T(x, z) \to P(z, x),$$

which after the transformation defined above becomes the axiom

$$\exists U.(\{x\} \sqcap \exists R.\{y\})$$
$$\sqcap \exists U.(\{y\} \sqcap A)$$
$$\sqcap \exists U.(\{z\} \sqcap \exists S.\{y\})$$
$$\sqcap \exists U.(\{x\} \sqcap \exists T.\{z\})$$
$$\sqsubseteq \exists U.(\{z\} \sqcap \exists P.\{x\}).$$

## 4.5   Relation to OWL Profiles

Recall that OWL 2 standards have three *tractable profiles* for which reasoning is possible in (sub)polynomial time: OWL 2 EL, OWL 2 RL and OWL 2 QL [64]. All of them include support for datatypes and concrete data values that we omit from discussion. No technical problem will occur due to this omission as datatype literals can be treated in a similar way as individuals.

First, OWL 2 EL is contained in $\mathcal{SROEL}(\sqcap, \times)$ [49]. Since $\mathcal{SROEL}(\sqcap, \times)$ is a sublanguage of $\mathcal{SROELV}_n(\sqcap, \times)$ for each $n$, our approach here then subsumes the OWL 2 EL profile without datatypes.

Next, OWL 2 RL is an extension of DLP [28] and essentially based on a *Horn Description Logic* (see section 5.3 for discussion about DLP and Horn DL). It does neither permit disjunctive information nor existential quantification, *It supports a very limited form of existential quantification, namely in such a way that it can be rewritten into a formula without existential quantification.* but it includes inverse roles and unrestricted range restrictions which are disallowed in OWL 2 EL. In general, axioms of OWL 2 RL can be reduced to normal forms given below.

$$
\begin{array}{lll}
A \sqsubseteq C & A \sqcap B \sqsubseteq C & R \sqsubseteq T \\
A \sqsubseteq \forall R.C & A \sqsubseteq\, \leq 1R.C & R \circ S \sqsubseteq T \\
A \sqsubseteq \{a\} & \{a\} \sqsubseteq C & R^- \sqsubseteq T
\end{array}
$$

All normal forms of axioms above are clearly expressible in $\mathcal{SROELV}_n(\sqcap, \times)$, save for three: $A \sqsubseteq \forall R.C$, $A \sqsubseteq\, \leq 1R.C$ and $R^- \sqsubseteq S$. But this is also not a problem because these three normal forms of axiom can be encoded using DL-safe rules which can then be translated into legal $\mathcal{SROELV}_n(\sqcap, \times)$ axioms in the sequel.

The normal form $A \sqsubseteq \forall R.C$ can be encoded as the rule $A(x) \wedge R(x, y) \to C(y)$ which, in $\mathcal{SROELV}_n(\sqcap, \times)$, becomes

$$\exists U.(\{x\} \sqcap A) \sqcap \exists U.(\{x\} \sqcap \exists R.\{y\}) \sqsubseteq \exists U.(\{y\} \sqcap C) \tag{7}$$

Meanwhile, $R^- \sqsubseteq S$ can be encoded as the rule $R(x, y) \to S(y, x)$ which can be translated into $\mathcal{SROELV}_n(\sqcap, \times)$ as

$$\exists U.(\{x\} \sqcap \exists R.\{y\}) \sqsubseteq \exists U.(\{y\} \sqcap \exists S.\{x\}) \tag{8}$$

For $A \sqsubseteq\, \leq 1R.C$, we need an auxiliary "DL-safe equality" role $R_\approx$ which is encoded using the axiom

$$\{x\} \sqcap \exists R_\approx\{y\} \sqsubseteq \exists U.(\{x\} \sqcap \{y\})$$

We can thus encode $A \sqsubseteq\, \leq 1R.C$ by the rule $A(x) \wedge R(x, y_1) \wedge C(y_1) \wedge R(x, y_2) \wedge C(y_2) \to R_\approx(y_1, y_2)$ which can be translated into $\mathcal{SROELV}_3(\sqcap, \times)$ as

$$
\begin{aligned}
\exists U.(\{x\} \sqcap A) \sqcap \exists U.(\{x\} \sqcap \exists R.\{y_1\}) \sqcap \exists U.(\{y_1\} \sqcap C) \\
\sqcap\, \exists U.(\{x\} \sqcap \exists R.\{y_2\}) \sqcap \exists U.(\{y_2\} \sqcap C) \\
\sqsubseteq \exists U.(\{y_1\} \sqcap \exists R_\approx\{y_2\})
\end{aligned}
\tag{9}
$$

Note that Equations (7), (8) and (9) are all legal axioms in $\mathcal{SROELV}_3(\sqcap, \times)$. Thus, OWL 2 RL is subsumed by $\mathcal{SROELV}_n(\sqcap, \times)$. Note however, that the translation of OWL 2 RL into $\mathcal{SROELV}_3(\sqcap, \times)$ is done under DL-safe restriction. This implies that some TBox entailments are lost because the translated axioms are not semantically equivalent to the original ontology. On the other hand, if we were to allow unrestricted combination of OWL 2 EL and OWL 2 RL, we would lose tractability as reasoning becomes 2ExpTime-complete. ABox entailments, the main inference task for OWL 2 RL, are still preserved, however.

Finally, OWL 2 QL is based on DL-Lite$_R$ [8] in which inverse roles and limited forms of existential quantification are allowed, but complex RIAs are not allowed. Similar to OWL 2 RL, OWL 2 QL can be approximated using DL-safe rules, and hence by $\mathcal{SROELV}_n(\sqcap, \times)$. In particular, inverse roles $R^-$ can be approximated by DL-safe rules $R_{\mathrm{inv}}(x, y) \rightarrow R(y, x)$ and $R(x, y) \rightarrow R_{\mathrm{inv}}(y, x)$; and axioms of the form $T \sqsubseteq \exists R^-.C$ can be expressed as $R \sqsubseteq \top \times C$. However, due to the use of DL-safe rules in the translation, some conclusions are lost as in the case of OWL 2 RL. Note, that the common usage of OWL 2 QL is for ontology-based querying large-scale datasets and this is possible since OWL 2 QL has a low data complexity which enables efficient query rewriting. This is obviously not supported in $\mathcal{SROELV}_n(\sqcap, \times)$, although, on the other hand, it provides some features not available in OWL 2 QL, e.g., role transitivity.

## 5   Pointers to Further Literature

Below we discuss several other formalisms which integrate, in some fashion or other, description logics and rules. We note that there are a great many ways to achieve integration, and there are indeed multiple ways to view integration itself. Particularly, one may distinguish between *syntactic integration*—e.g., whether a common vocabulary is used to create rules and other sorts of assertions, and to what extent rules are syntactically isolated from other components or otherwise restricted—and *semantic integration*, that is whether a common semantics is used for rules and other components or whether multiple, distinct semantics are used (and then combined in some fashion). For instance, in SWRL, rules are syntactically distinct from DL axioms—there's an ontology, and there's also a rule base—but a uniform model theoretic semantics is used for each. In contrast, in $\mathcal{AL}$-log, a knowledge base consists of a DL ontology and a separate Datalog program, but additionally, the semantics for each is distinct—an interpretation of a knowledge base consists of two interpretations, one for the DL ontology and another for the program. There are also formalisms where no syntactic distinction is made. That is, a common language is used (and expressions are interpreted according to a common semantics). DLP and the nominal schema formalism described in Section 4.2 fall into this category.

Along both the syntactic and semantic dimensions, there are degrees of integration—or at least considerable variation in how integration is achieved. In some cases, the syntactic and semantic separation between the sub-systems is extreme. For example, in dl-programs, a logic program is extended with atoms for interacting with an external description logic ontology, and an answer set semantics is provided for the program. But this method of interacting with a logic program is easily generalizable to other sorts of systems (i.e., non-DL systems). This is what is done in HEX-programs (which extend dl-programs).

The below list is not exhaustive, but it does describe several formalisms that are significant, either because they have been historically significant and influenced the field, or else because they indicate current research trends.

## 5.1   SWRL

One of the earliest formalisms combining OWL and rules is the *Semantic Web Rule Language* SWRL [36,37,38] (called ORL in [36]). Syntactically, SWRL extends the syntax of OWL DL and OWL Lite (circa 2004) with additional constructs to form Horn-style rule axioms. A SWRL knowledge base consists of a set of rules and OWL axioms. Semantically, the model theoretic semantics of OWL is extended to cover rules—the notable addition being the specification of variable bindings associated with interpretations.

Using an informal human readable syntax, each SWRL rule has the form $B \rightarrow H$ (as in Section 2), where $B$ and $H$ are possibly empty conjunctions of atoms. The atoms have one of the forms $C(x)$, $P(x,y)$, sameAs$(x,y)$, or differentFrom$(x,y)$, where $x$ and $y$ are variables or individuals, $P$ is an OWL property (role), and $C$ is a possibly complex OWL class (concept) description. Atoms involving datatypes and data values are also allowed, as are "built-in" atoms (for, e.g., arithmetic). We don't discuss them here, however.

Complex class descriptions in rules can be replaced with a new class name $A$, and the two class descriptions can be declared equivalent in the OWL ontology. Similarly, sameAs and differentFrom (when it appears in the consequent of rules) can be eliminated [37].

Variables in SWRL are typed: those ranging over individuals are distinct from those ranging over data-values. Variables must also be *safe*, in the sense that every variable in the consequent of a rule must also appear in the antecedent. Even with this restriction, however, the satisfiability problem for SWRL knowledge bases is known to be undecidable [37].

## 5.2   DL-Safe Rules

The composition of rules and OWL DL[12] axioms can be made decidable by forcing each rule to be *DL-safe* [66,73,74]. As noted above, the atoms appearing in rules may be restricted to simple unary and binary predicates (complex class descriptions can be eliminated from rules). DL-safety separates the predicates into two classes: 1) those that are names of atomic classes and roles and which are used in non-rule axioms; and 2) predicates that are not so used. Atoms making use of class and role names are called *DL-atoms*. A rule is DL-safe if every

---

[12] The papers [73,74] deal specifically with the description logic $\mathcal{SHOIN}(D)$, on which OWL DL was based; in [66] the logic used is $\mathcal{SHIQ}(D)$.

variable of the rule appears in a non-DL atom in the rule body. The combined knowledge base is DL-safe if every rule is. DL-safety ensures that each variable of the rule can be bound to only individuals explicitly named in the ontology.

A rule can be made DL-safe by adding, for each variable $x$ appearing in the rule, a special non-DL atom $O(x)$ to the body, and by simultaneously adding an assertion $O(a)$, for each individual name $a$, to the knowledge base. DL-safety can also be enforced by requiring each variable assignment to bind every variable to named elements in the universe of discourse. We followed the latter perspective in Section 4.1.

### 5.3   DLP

SWRL and DL-Safe rules do not restrict the syntax of the underlying formalisms, and DL-safety is used to ensure the decidability of the combination of rules and DL axioms. In contrast, *description logic programs* (DLP) [28,88] ensure decidability by restricting the formalisms to the fragment that can be expressed in *def-Horn* (equality- and function-free definite Horn logic) [28]. In [28], *def-LP*, the logic programming analog of def-Horn is also specified. The two differ in that the consequences of a def-LP program are restricted to ground atoms; no such restriction is applied to def-Horn. The atomic consequences of the program are precisely those found in the program's least Herbrand model (which is guaranteed to exist).

*Description Horn Logic* is defined via a set of transformation rules to def-Horn. Specifically, the rules transform a set of DL axioms into a set of logically equivalent def-Horn rules (see Section 3.1). However, since many DL axioms yield non-Horn expressions upon transformation, certain restrictions must be made. For example, neither existential restrictions nor concept unions are permitted on the right-hand side of an inclusion axiom; universal restrictions are not allowed on the left-hand side. A Description Horn Logic ontology is simply a DL ontology whose transformation is in def-Horn. A *DLP* ontology is the same ontology interpreted according to the least Herbrand model semantics.

### 5.4   $\mathcal{AL}$-log

In SWRL, the DL axioms and rules are syntactically distinct. Nevertheless, a uniform model theoretic semantics is provided for the combination. Similarly, a single semantics is used for DLP. In other approaches, rules and DL systems are allowed to interact, but they are kept as distinct components (both syntactically and semantically).

In $\mathcal{AL}$-log [11,12], a knowledge base $\langle \mathcal{O}, \mathcal{P} \rangle$ is composed of an $\mathcal{ALC}$ ontology $\mathcal{O}$ (the *structural subsystem*, itself composed of an ABox and Tbox) and a Datalog program $\mathcal{P}$ (the *relational subsystem*). The Datalog program consists of *constrained* clauses: each clause $\gamma$ is accompanied by zero or more constraints $C_1(t_1), \ldots, C_n(t_n)$, where each $C_i$ is an $\mathcal{ALC}$ concept description and each $t_i$ is constant or variable. The constraints are intended to restrict the values of variables to instances of concepts. In a valid knowledge base, the following conditions must also be met: 1) the Datalog predicates of $\mathcal{P}$ are disjoint from the set of concept and role names in $\mathcal{O}$; 2) the constants of $\mathcal{P}$ coincide with the individual names of $\mathcal{O}$, and each constant of $\mathcal{P}$ appears in $\mathcal{O}$; and 3) for each constrained clause $\gamma \& \, C_1(t_1), \ldots, C_n(t_n)$, if $t_i$ is a variable, then $t_i$ appears in $\gamma$.

The semantics of $\langle \mathcal{O}, \mathcal{P} \rangle$ is given by providing interpretations for both $\mathcal{O}$ and $\mathcal{P}$. Let $\mathcal{I}$ be an interpretation of $\mathcal{O}$ and $\mathcal{H}$ a Herbrand interpretation of $\mathcal{P}$ (the constraints are ignored). $\langle \mathcal{I}, \mathcal{H} \rangle$ is a model of $\langle \mathcal{O}, \mathcal{P} \rangle$ if and only if $\mathcal{I}$ is a model of $\mathcal{O}$, and for each ground instantiation of $\gamma \& \, C_1(t_1), \ldots, C_n(t_n)$, either there is a $C_i(t_i)$ that is not satisfied by $\mathcal{I}$ or else $\gamma$ is satisfied by $\mathcal{H}$. Entailment is defined in the usual fashion, save that if $a_1, \ldots, a_n$ is a set of ground atoms and $C_1(t_1), \ldots, C_m(t_m)$ a set of ground constraints, $\langle \mathcal{O}, \mathcal{P} \rangle \models a_1, \ldots, a_n \& \, C_1(t_1), \ldots, C_m(t_m)$ if and only if every model of $\langle \mathcal{O}, \mathcal{P} \rangle$ is a model of each $a_i$ and $C_i(t_i)$. These constitute the possible answers to queries, the latter themselves being a set of atoms together with a set of constraints. In [11,12], it is shown that query-answering for $\mathcal{AL}$-log is decidable. A query answering procedure—based on resolution—is also provided.

## 5.5   CARIN

CARIN [56,57], a family of combined DL-rule languages, is similar to $\mathcal{AL}$-log in the sense that it couples a description logic ontology to a function-free Horn-logic rule base. Unlike $\mathcal{AL}$-log, however, concept and role names are allowed to appear as predicates in rule bodies.

In [56,57], $\mathcal{ALCNR}$ is the underlying description logic used, and the problem dealt with is *existential entailment*. Two sorts of programs are examined—those with recursive rules, and those without. Without recursion, reasoning is decidable, and a sound and complete inference procedure exists. For programs with recursive rules, however, reasoning problems in CARIN-$\mathcal{ALCNR}$ are generally undecidable. Certain restrictions restore decidability, e.g. if the system employs *role safe* rules (where at least one variable of every role atom appears in a predicate that is neither in the consequent of a rule nor a concept or role name).

CARIN makes use of a classical semantics (with the unique name assumption). A single interpretation is given for both the DL ontology and rule base, and it constitutes a model of the combined knowledge base if it simultaneously satisfies both components.

## 5.6   $\mathcal{DL}+log$

$\mathcal{DL}+log$ [78,79,80,81,82] integrates description logic ontologies with disjunctive logic programs. A $\mathcal{DL}+log$ knowledge base is a tuple $\langle \mathcal{O}, \mathcal{P} \rangle$, where $\mathcal{O}$ is a DL ontology and $\mathcal{P}$ is a logic program with rules of the form

$$p_1(\overline{X_1}) \vee \ldots \vee p_n(\overline{X_n}) \leftarrow r_1(\overline{Y_1}) \wedge \ldots \wedge r_m(\overline{Y_m}) \wedge$$
$$s_1(\overline{Z_1}) \wedge \ldots \wedge s_k(\overline{Z_k}) \wedge$$
$$\text{not } u_1(\overline{W_1}) \wedge \ldots \wedge \text{not } u_h(\overline{W_h})$$

where $\overline{X_i}$, $\overline{Y_i}$, etc., are tuples of variables and constants. Each $s_i(\overline{Z_i})$ is a DL-atom (as in DL Safe rules), and every $r_i(\overline{Y_i})$ and $u_j(\overline{W_j})$ is a non-DL atom. The rules must be safe (every rule variable must appear in a positive literal of the body). Furthermore, every variable of the head must appear in one of the $r_i$ atoms. This latter condition is called *weak safeness*. A further condition of $\mathcal{P}$ is that it contains all constants of $\mathcal{O}$.

$\mathcal{DL}+log$ specifies two semantics. In the first-order semantics, the DL ontology is translated into FOL, and rules are interpreted as material implications. Negation is interpreted as classical negation. The *standard names assumption* is made: each interpretation is over a single countably infinite universe, each constant names the same element in each interpretation, and two distinct constants name distinct elements of the universe. In the nonmonotonic semantics, rules are interpreted according to a stable model semantics. Without negation, the two semantics yield the same results for the satisfiability problem: a knowledge base is satisfiable in one if and only if it is satisfiable in the other. In general, satisfiability for $\mathcal{DL}+log$ KBs is decidable, provided the problem of query containment for Boolean conjunctive queries and Boolean unions of conjunctive queries is decidable in the DL used.

### 5.7   Horn-$\mathcal{SHIQ}$

Horn-$\mathcal{SHIQ}$ [39,52,66] is a fragment of $\mathcal{SHIQ}$ in which the ability to express disjunction has been eliminated. The definition is somewhat complicated, but Horn-$\mathcal{SHIQ}$ knowledge bases can in general be translated into first-order Horn clauses, and every general concept inclusion axiom can be normalized into one of the below forms, where each $A_i$ is a concept name, $R$ and $S$ are roles (with $S$ simple), and $m \geq 1$ [15].

$$A_i \sqcap A_j \sqsubseteq A_k \qquad A_i \sqsubseteq \forall R.A_j \qquad A_i \sqsubseteq \geq mS.A_j$$
$$\exists R.A_i \sqsubseteq A_j \qquad A_i \sqsubseteq \exists R.A_j \qquad A_i \sqsubseteq \leq 1S.A_j$$

The loss of disjunction brings with it lower data-complexity. For instance, while checking satisfiability of $\mathcal{SHIQ}$ knowledge bases (where the ABox assertions $C(a)$ and $\neg C(a)$ are allowed only if $C$ is atomic) is NP-complete relative to the size of the ABox, the problem is P-complete for similarly restricted Horn-$\mathcal{SHIQ}$ knowledge bases [39].

In [15], an algorithm for conjunctive query answering in Horn-$\mathcal{SHIQ}$ is provided. It is shown there that the entailment problem for conjunctive queries is EXPTIME-complete (combined complexity). P-completeness holds for data complexity. In [40], an EXPTIME algorithm for classifying Horn-$\mathcal{SHIQ}$ ontologies, similar in spirit to the completion based algorithm for $\mathcal{EL}^{++}$, is given.

### 5.8   Hybrid MKNF

Hybrid MKNF knowledge bases [65,70,71,72] combine description logics with disjunctive logic programs interpreted according to Lifschitz's logic of *minimal knowledge and negation as failure* (MKNF) [58]. Formally, a Hybrid MKNF knowledge base $\mathcal{K} = \langle \mathcal{O}, \mathcal{P} \rangle$ consists of a DL ontology $\mathcal{O}$ together with a disjunctive logic program $\mathcal{P}$, where $\mathcal{P}$ is composed of DL-safe rules of the form

$$\mathbf{K}H_1 \vee \ldots \vee \mathbf{K}H_n \leftarrow \mathbf{K}B_1, \ldots, \mathbf{K}B_m, \mathbf{not}\ C_1, \ldots, \mathbf{not}\ C_l.$$

Each $H_i$, $B_j$, and $C_k$ is a function free atomic formula or else a binary formula using predicate $\approx$. The symbols $\mathbf{K}$ and $\mathbf{not}$ are modal operators. Roughly, $\mathbf{K}A$ is read as "$A$ is known to hold," and $\mathbf{not}A$ as "$A$ can be false" [65].

The semantics of a Hybrid MKNF knowledge base $\mathcal{K}$ is given by translating it to a formula $\pi(\mathcal{P}) \wedge \mathbf{K}\pi(\mathcal{O})$ of MKNF. $\pi(\mathcal{P})$ is just the conjunction of rules of $\mathcal{P}$, each rule read as a material implication. $\pi(\mathcal{O})$ is the formula obtained by translating $\mathcal{O}$ into function-free first order logic with equality. The underlying DL must be one where such a translation is possible. The result is interpreted according to MKNF, though interpretations are restricted to Herbrand interpretations, and the standard names assumption is made.

It is noted in [70] that Hybrid MKNF generalizes several of the formalisms already discussed here, including CARIN, $\mathcal{AL}$-log, SWRL, and DL-Safe rules. Its semantics also extends both classical DL semantics and the MKNF semantics of the rules. That is, if $\mathcal{P}$ is empty, then $\mathcal{K}$'s consequences are the same as $\mathcal{O}$'s classical consequences. Similarly, if $\mathcal{O}$ is empty, then the consequences reduce to those of $\mathcal{P}$ specified by MKNF (which, as noted in [58], correspond to those determined by the stable model semantics [23,24]).

In [70,71], an algorithm for entailment checking is given, and data complexity analyses are given for knowledge bases using programs of various kinds. Without the DL-safety requirement, the satisfiability problem for Hybrid MKNF becomes undecidable.

In a separate series of papers [1,2,25,43,45,46,47], a well-founded semantics (WFS) for Hybrid MKNF knowledge bases is discussed (the rules must be *normal*, meaning $\neg$ does not appear). The advantage here over the semantics defined above is that it is sound relative to the original semantics but of a strictly lower complexity. Interpretations are again restricted to Herbrand interpretations, but a third truth value $u$ is added (with the ordering $f < u < t$), applicable to formulas involving modal atoms only. As above, the semantics extends both the classical DL semantics and the traditional WFS of the rules. An alternating fixpoint procedure is defined in [44] for non-disjunctive Hybrid MKNF knowledge bases, yielding what they call the *well-founded partition*.

The semantics is modified in [43,47] to ensure *coherence*: i.e., if $\neg P$ holds, then so does $\mathbf{not}\ P$. This arguably yields more intuitively correct results and allows one to pinpoint inconsistencies. A fixpoint procedure is again defined, and the data complexity of computing the well-founded partition is given as $P^{\mathcal{C}}$, where $\mathcal{C}$ is the data-complexity of solving the ground atom entailment problem for the underlying description logic.

A top-down method for querying Hybrid MKNF under the WFS, avoiding the computation of the full well-founded partition, is described in [1,2]. The method—$SLG(\mathcal{O})$ resolution—alters SLG resolution [9] so that queries to an ontology reasoner can be made. That is, the ontology reasoner is used as an oracle. If certain restrictions are met by the oracle, then the $SLG(\mathcal{O})$ method remains tractable. A prototype reasoner (CDF-Rules), based on $SLG(\mathcal{O})$ and constructed in part using XSB Prolog, is described in [25].

### 5.9    dl-programs

Hybrid MKNF, like MKNF, is nonmonotonic. Another such formalism is *dl-programs* [14,16,17,18,21], which again combines description logic ontologies with extended logic programs (i.e., programs using both ¬ and **not**, the latter being default negation). The essential idea of a dl-program is that logic program rules can contain *queries* to a description logic ontology. Information flow is bidirectional—data is provided as input to the query, and answers to the queries affect what may be inferred using the rules (which are interpreted according to the answer-set semantics [24]). The two components are thus distinct in the framework and yet interact in a complex way. The DLs discussed in [14] are $\mathcal{SHIF}(D)$ and $\mathcal{SHOIN}(D)$, but the framework could be used with other DLs.

A *dl-query* is either a concept inclusion axiom or its negation, or else a positive or negative concept or role assertion—e.g., $C(t)$, $\neg R(t_1, t_2)$, where $C$ is a concept description, $R$ a role, and $t_i$ a term. A *dl-atom*, which can appear in the body of a rule but not the head, is a structure of the form $DL[S_1 \ op_1 \ p_1, \ldots, S_m \ op_m \ p_m; Q](t)$, where each $S_i$ is a role or concept, each $op_i$ is in the set $\{⊎, ⊍\}$,[13] and each $p_i$ is a predicate from the program. Each expression $S_i \ op_i \ p_i$ is interpreted relative to a Herbrand interpretation $\mathcal{I}$. $S_i ⊎ p_i$ indicates that when answering the query, atoms in the extension of predicate $p_i$—as specified by $\mathcal{I}$—should be included in the ontology as instances of $S_i$. $S_i ⊍ p_i$ indicates that such atoms should be included as instances of $\neg S_i$.

The usual notion of satisfaction by a Herbrand interpretation is extended to apply to dl-atoms, and given this, Herbrand models for positive dl-programs (those lacking **not**) are defined. Positive programs, provided they have any models at all, have unique minimal Herbrand models which can be computed via a fixpoint procedure. Canonical models for stratified programs are also defined.

The minimal models of positive programs are used to define the answer sets of arbitrary dl-programs. Given a combined knowledge base $\mathcal{K} = \langle \mathcal{O}, \mathcal{P} \rangle$, the *strong reduct* of program $\mathcal{P}$ relative to $\mathcal{I}$ and ontology $\mathcal{O}$, written $s\mathcal{P}_{\mathcal{O}}^{\mathcal{I}}$, is the set of ground rules obtained by 1) deleting from the grounding of $\mathcal{P}$ all rules with an atom **not** $A$ in the body such that $A$ is satisfied by $\mathcal{I}$; and 2) deleting all remaining such atoms. The reduct is a positive dl-program. If its minimal model exists, then it is a *strong answer set* of $\mathcal{K}$. Without dl-atoms, every strong answer set of $\mathcal{K}$ is just an answer set of $\mathcal{P}$. Weak answer-sets, in which the reduct eliminates all dl-atoms and default negation atoms from programs, are also defined. Each weak answer set is a model of the dl-program.

If $\mathcal{SHIF}(D)$ ontologies are used, the problem of deciding whether an unrestricted dl-program has an answer set (strong or weak) is NExpTime-complete.

---

[13] A further operator, ⋒, is also discussed, but it introduces another source of nonmonotonicity even in programs without default negation. In [18], it is not discussed.

It is ExpTime-complete for positive and stratified programs. For $\mathcal{SHOIN}(D)$, the problem of deciding whether a positive dl-program has a strong or weak answer set is NExpTime-complete. For stratified programs, it's $\mathrm{NP}^{\mathrm{NExpTime}}$-complete for weak answer sets and $\mathrm{P}^{\mathrm{NExpTime}}$-complete for strong answer sets. For unrestricted programs, it's $\mathrm{NP}^{\mathrm{NExpTime}}$-complete for both.

In [16], well-founded semantics for dl-programs are defined.[14] The definition proceeds by first defining unfounded sets and then the operators $T_{KB}$, $U_{KB}$, and $W_{KB}$, similar to the original account of the WFS for normal logic programs [22]. An alternating fixpoint procedure for computing the well-founded model is also given, and it is shown that the semantics for dl-programs extends the WFS for normal logic programs, and also that it approximates the strong answer set semantics: every well-founded atom in the well-founded model is true in every strong answer set, and every unfounded atom is false in every strong answer set. For dl-programs based on $\mathcal{SHIF}(D)$, determining whether a literal $l$ is in the well-founded model is ExpTime-complete. For $\mathcal{SHOIN}(D)$, the corresponding problem is $\mathrm{P}^{\mathrm{ExpTime}}$-complete.

Observe that dl-queries essentially provide an interface between a logic program and a distinct DL ontology. This basic framework permits the use of external data sources other than DLs. This is the basic idea behind *HEX-programs* (*higher order logic programs with external atoms*) [19,20,86]. Disjunctions are allowed in the heads of rules, and instead of dl-atoms, programs make use of *external* atoms of the form $\&[Y_0(Y_1, \ldots, Y_n)](X_1, \ldots, X_m)$, where $g$ is an external predicate (not used save in such atoms) and $[Y_0(Y_1, \ldots, Y_n)]$ and $(X_1, \ldots, X_m)$ are *input* and *output* lists of terms, respectively. A solver for HEX-programs, dlvhex, has been implemented (by extending the answer-set solver dlv[15]).

## 5.10  Disjunctive dl-programs

Another formalism [59] also goes by the name "dl-programs", but it is unrelated to the formalism described above. In [59], a knowledge base $\langle \mathcal{O}, \mathcal{P} \rangle$ is again formed by combining a (disjunctive) logic program $\mathcal{P}$ with a DL ontology $\mathcal{O}$, but in this case the logic program is a more typical disjunctive logic program (i.e., there are no dl-atoms). Only one form of negation, default negation, is allowed. Constants of the program are a subset of the individuals in the DL ontology, but no other special restrictions are made on the vocabulary used.

A uniform semantics is used. The basic idea is to interpret $\mathcal{P}$ using Herbrand interpretations that also satisfy $\mathcal{O}$. That is, a Herbrand interpretation $\mathcal{I}$ of a program $\mathcal{P}$ is any subset of the Herbrand base $HB$ of the program. $\mathcal{I}$ is a model of $\mathcal{O}$ if and only if $\mathcal{O} \cup \mathcal{I} \cup \{\neg a | HB - \mathcal{I}\}$ is satisfiable. $\mathcal{I}$ is a model of $\langle \mathcal{O}, \mathcal{P} \rangle$ if $\mathcal{I}$ models both $\mathcal{P}$ and $\mathcal{O}$. $\mathcal{I}$ is an answer set of $\langle \mathcal{O}, \mathcal{P} \rangle$ if it is a minimal model of $\langle \mathcal{O}, \mathcal{P}^{\mathcal{I}} \rangle$, where $\mathcal{P}^{\mathcal{I}}$ is the reduct of $\mathcal{P}$ with respect to $\mathcal{I}$.

The semantics described above extends the answer set semantics for disjunctive logic programs: If $\mathcal{O}$ is empty, then the answer sets for $\langle \mathcal{O}, \mathcal{P} \rangle$ are the answer

---

[14] The programs are normal in the sense that negative literals $\neg a$ are not allowed. Furthermore, the semantics is only defined for dl-programs not involving $\cap$.

[15] http://www.dbai.tuwien.ac.at/proj/dlv/

sets for $\mathcal{P}$. If instead $\mathcal{P}$ is empty, a ground atom $a$ is true in every answer set of $\langle \mathcal{O}, \mathcal{P} \rangle$ if and only if it is true in all first-order models of $\mathcal{O}$. It is shown in [59] that, if $\mathcal{O}$ is in $\mathcal{SHIF}(D)$ or $\mathcal{SHOIN}(D)$, then deciding whether the combined knowledge base has an answer-set is $\text{NEXP}^{\text{NP}}$-complete. Determining whether a ground atom $a$ is true in all (some) answer-sets of the knowledge base is $\text{co} - \text{NEXP}^{\text{NP}}$-complete ($\text{NEXP}^{\text{NP}}$-complete). Reasoning algorithms for deciding the existence of answer-sets are also identified, as is a class of stratified knowledge bases (based on DL-Lite). For such knowledge bases, the problems of deciding whether an answer set exists (which must be unique, if it exists), and whether a given ground atom is true in it, have polynomial data-complexity.

## 5.11   Quantified Equilibrium Logic for Hybrid Knowledge Bases

In [10], it is shown how a variation of the Quantified Equilibrium Logic (QEL) [77] can be used as a semantics for hybrid knowledge bases, one which encompasses other semantics proposed in the literature. Here, a hybrid knowledge base is defined to be a combination $\langle \mathcal{O}, \mathcal{P} \rangle$ of first order theory $\mathcal{O}$ and a disjunctive logic program $\mathcal{P}$. $\mathcal{P}$ may contain first order literals $a$ and $\neg a$. Both components are function-free and are defined using the same constants. $\mathcal{P}$'s predicates are a superset of $\mathcal{O}$'s. The *stable closure* of a hybrid knowledge base is defined (essentially by taking the union of $\mathcal{O}$ and $\mathcal{P}$ and adding $(\forall \overline{X})(p(\overline{X}) \vee \neg p(\overline{X})$ for each predicate of $\mathcal{O}$), and equilibrium models are then defined for the stable closure. It is shown that by varying restrictions on the domain of discourse, these models correspond to models of the hybrid knowledge base according to frameworks proposed by Rosati, including $\mathcal{DL}+log$ (discussed above), and according to *guarded-hybrid* (g-hybrid) knowledge bases [29].

## 5.12   Description Graphs

*Description graphs* [63,67,68,69] extend DLs with first-order rules and graphs allowing the representation of structured objects (such as the bones of a hand) not otherwise expressible in a DL. The graphs can be arranged into a hierarchy (which may be used to describe an object at differing levels of granularity).

In the framework, an *n-ary description graph* $G$ is a directed graph of $n$ vertices, with each vertex labeled with a set of atomic concepts or their negations, and each edge labeled with a set of atomic roles or their negations. Some subset of the atomic concepts is selected as constituting the *main* concepts of the graph (roughly, they indicate what the graph is about). A *graph specialization axiom* $G \lhd G'$ indicates that each vertex of $G$ is one of $G'$. A *graph alignment axiom* $G_1[v_1, \ldots, v_n] \leftrightarrow G_2[u_1, \ldots, u_n]$ is a 1-1 mapping of some subset of vertices of two graphs. A *graph box* (GBox) $\mathcal{G}$ is a finite collection of description graphs, specialization axioms, and alignment axioms. A *graph assertion* is an expression of the form $G(a_1, \ldots, a_n)$, where $G$ is an n-ary description graph and each $a_i$ is an individual.

The bodies of rules consist of conjunctions of atomic concept atoms $C(t)$, atomic role atoms $R(t_1, t_2)$, but also graph atoms $G(t_1, \ldots, t_k)$, where each $t_i$

is an individual or a variable and $G$ is a description graph. Rule heads are disjunctions of such atoms (the head may also contain equality atoms $t_1 \approx t_2$). Each rule must be *connected*: for any variables $x$ and $y$ in the rule, there is a sequence $x_1, \ldots, x_n$ of variables such that $x_1 = x$ and $x_n = y$ and for each $i < n$, $x_i$ and $x_{i+1}$ appear in the same body atom.

A *graph extended knowledge base* is a tuple $\mathcal{K} = (\mathcal{T}, \mathcal{P}, \mathcal{G}, \mathcal{A})$, where $\mathcal{T}$ is a TBox, $\mathcal{P}$ is a finite set of connected graph rules, $\mathcal{G}$ is a GBox, and $\mathcal{A}$ is an ABox possibly containing graph assertions. In an interpretation $\mathcal{I}$, each $n$-ary graph $G$ is read as an $n$-ary relation over $\Delta^{\mathcal{I}}$. An assertion $G(a_1, \ldots, a_n)$ is satisfied by $\mathcal{I}$ if and only if $(a_1^{\mathcal{I}}, \ldots, a_n^{\mathcal{I}}) \in G^{\mathcal{I}}$. The semantics is such that in any model of $\mathcal{K}$, no two distinct instances of a description graph share vertices, and the vertices are ensured to participate in the concepts and role relations indicated in the graph. $G \lhd G'$ holds if each instance of $G'$ is an instance of $G$, and $G_1[v_1, \ldots, v_n] \leftrightarrow G_2[u_1, \ldots, u_n]$ holds if, whenever instances of $G_1$ and $G_2$ share vertices $u_i$ and $v_i$, then they share all other vertex pairs in the axiom.

Under many circumstances, the satisfiability problem for graph extended knowledge bases is undecidable—for example, if $\mathcal{T}$ is empty, $\mathcal{P}$ is Horn, and no specialization or alignment axioms are used. Decidability can be regained in this example by requiring the hierarchy of graph descriptions to be "acyclic" (see [63]). In other cases, however, additional restrictions are required. In [63], it is shown that the satisfiability problem for an acyclic $\mathcal{K}$ is NExpTime-complete, provided $\mathcal{K}$ is *weakly separated* and $\mathcal{T}$ is in $\mathcal{SHOQ}^+$. Alternatively, it is NExpTime-complete if $\mathcal{K}$ is *strongly separated* and $\mathcal{T}$ is in $\mathcal{SHIQ}^+$. Here, weak separation means that the roles of $\mathcal{T}$ and $\mathcal{P}$ are disjoint. Strong separation additionally requires the roles of $\mathcal{T}$ to be disjoint with those of $\mathcal{G}$.

## 6  Conclusions

We have reported on the considerable body of work on OWL and Rules, describing integration proposals that sometimes differ substantially in terms of their underlying approach and rationale. Some approaches have been more popular than others. In some cases, it appears to be a matter of subjective judgement regarding which provide the best underpinnings for a "unified logic" in the sense of the W3C Semantic Web Stack.[16] And it's likely additional alternatives will be proposed in the future. that we will see a few more alternative proposals in the near future.

Further theoretical investigations will certainly shed more light on the issue. Concerning the proposed formalism in Section 4, for example, it would be helpful to investigate possibilities for incorporating nonmonotonic negation or other closed world features [3,7,13,26,27,47,48,65,71] which commonly occur in logic-programming-based rule approaches.[17] For example, we have recently proposed an intuitively appealing approach for extending description logics with local closed world features which retains decidability if added to the description logics

---

[16] http://www.w3.org/2007/03/layerCake.png
[17] See, e.g., [35,76].

with nominal schemas discussed herein [48]. Even more importantly, however, efficient algorithms and implementations need to be developed.

In the end, usability aspects will also play a decisive role, and it is here where the development of Semantic Web applications involving deep reasoning are often found to be lacking [31,32]. The Semantic Web requires usable tools, interfaces, design patterns, and best-practice guidelines which would allow developers to use ontologies and underlying reasoning paradigms without having to become expert logicians. We're still a long way away from that goal.

**Acknowledgements.** This work was supported by the National Science Foundation under award 1017225 *III: Small: TROn—Tractable Reasoning with Ontologies*. Adila Krisnadhi acknowledges support by a Fulbright Indonesia Presidential Scholarship PhD Grant 2010.

# References

1. Alferes, J.J., Knorr, M., Swift, T.: Queries to Hybrid MKNF Knowledge Bases through Oracular Tabling. In: Bernstein, A., Karger, D.R., Heath, T., Feigenbaum, L., Maynard, D., Motta, E., Thirunarayan, K. (eds.) ISWC 2009. LNCS, vol. 5823, pp. 1–16. Springer, Heidelberg (2009)
2. Alferes, J.J., Knorr, M., Swift, T.: Query-driven Procedures for Hybrid MKNF Knowledge Bases. CoRR abs/1007.3515 (2010), http://arxiv.org/abs/1007.3515
3. Baader, F., Hollunder, B.: Embedding defaults into terminological representation systems. J. Automated Reasoning 14, 149–180 (1995)
4. Baader, F., Calvanese, D., McGuinness, D., Nardi, D., Patel-Schneider, P. (eds.): The Description Logic Handbook: Theory, Implementation, and Applications. Cambridge University Press, Cambridge (2007)
5. Boley, H., Hallmark, G., Kifer, M., Paschke, A., Polleres, A., Reynolds, D. (eds.): RIF Core Dialect. W3C Recommendation (June 22, 2010), http://www.w3.org/TR/rif-core/
6. Boley, H., Kifer, M. (eds.): RIF Basic Logic Dialect. W3C Recommendation (June 22, 2010), http://www.w3.org/TR/rif-bld/
7. Bonatti, P., Lutz, C., Wolter, F.: Expressive Non-Monotonic Description Logics Based on Circumscription. In: Proc. of 10th Intern. Conf. on Principles of Knowledge Representation and Reasoning (KR 2006), pp. 400–410. AAAI Press, Menlo Park (2006)
8. Calvanese, D., Giacomo, G.D., Lembo, D., Lenzerini, M., Rosati, R.: Tractable reasoning and efficient query answering in description logics: The DL-Lite family. J. of Automated Reasoning 39(3), 385–429 (2007)
9. Chen, W., Warren, D.S.: Tabled evaluation with delaying for general logic programs. J. ACM 43, 20–74 (1996)
10. De Bruijn, J., Pearce, D., Polleres, A., Valverde, A.: Quantified equilibrium logic and hybrid rules. In: Marchiori, M., Pan, J.Z., de Marie, C.S. (eds.) RR 2007. LNCS, vol. 4524, pp. 58–72. Springer, Heidelberg (2007)
11. Donini, F., Lenzerini, M., Nardi, D., Schaerf, A.: A hybrid system with datalog and concept languages. In: Ardizzone, E., Sorbello, F., Gaglio, S. (eds.) AI*IA 1991. LNCS, vol. 549, pp. 88–97. Springer, Heidelberg (1991)

12. Donini, F.M., Lenzerini, M., Nardi, D., Schaerf, A.: $\mathcal{AL}$-log: Integrating datalog and description logics. J. Intell. Inf. Syst. 10, 227–252 (1998), doi:10.1023/A:1008687430626
13. Donini, F.M., Nardi, D., Rosati, R.: Description logics of minimal knowledge and negation as failure. ACM Trans. Comput. Logic 3(2), 177–225 (2002)
14. Eiter, T., Lukasiewicz, T., Schindlauer, R., Tompits, H.: Combining Answer Set Programming with Description Logics for the Semantic Web. In: Proc. of the 9th Int. Conf. on the Principles of Knowledge Representation and Reasoning (KR 2004). AAAI Press, Menlo Park (2004)
15. Eiter, T., Gottlob, G., Ortiz, M., Šimkus, M.: Query answering in the description logic horn-$\mathcal{SHIQ}$. In: Hölldobler, S., Lutz, C., Wansing, H. (eds.) JELIA 2008. LNCS (LNAI), vol. 5293, pp. 166–179. Springer, Heidelberg (2008)
16. Eiter, T., Ianni, G., Lukasiewicz, T., Schindlauer, R.: Well-founded semantics for description logic programs in the semantic web. ACM Trans. Comput. Log. 12(2), article 11 (2011)
17. Eiter, T., Ianni, G., Lukasiewicz, T., Schindlauer, R., Tompits, H.: Combining answer set programming with description logics for the semantic web. Artif. Intell. 172, 1495–1539 (2008)
18. Eiter, T., Ianni, G., Polleres, A., Schindlauer, R., Tompits, H.: Reasoning with rules and ontologies. In: Barahona, P., Bry, F., Franconi, E., Henze, N., Sattler, U. (eds.) Reasoning Web 2006. LNCS, vol. 4126, pp. 93–127. Springer, Heidelberg (2006)
19. Eiter, T., Ianni, G., Schindlauer, R., Tompits, H.: dlvhex: A prover for semantic-web reasoning under the answer-set semantics. In: 2006 IEEE / WIC / ACM International Conference on Web Intelligence (WI 2006), Hong Kong, China, December 18-22, pp. 1073–1074. IEEE Computer Society, Los Alamitos (2006)
20. Eiter, T., Ianni, G., Schindlauer, R., Tompits, H.: Effective integration of declarative rules with external evaluations for semantic-web reasoning. In: Sure, Y., Domingue, J. (eds.) ESWC 2006. LNCS, vol. 4011, pp. 273–287. Springer, Heidelberg (2006)
21. Eiter, T., Lukasiewicz, T., Schindlauer, R., Tompits, H.: Well-founded semantics for description logic programs in the semantic web. In: Antoniou, G., Boley, H. (eds.) RuleML 2004. LNCS, vol. 3323, pp. 81–97. Springer, Heidelberg (2004)
22. Gelder, A.V., Ross, K., Schlipf, J.S.: Unfounded sets and well-founded semantics for general logic programs. In: PODS 1988: Proceedings of the seventh ACM SIGACT-SIGMOD-SIGART symposium on Principles of database systems, pp. 221–230. ACM Press, New York (1988)
23. Gelfond, M., Lifschitz, V.: The stable model semantics for logic programming. In: Kowalski, R.A., Bowen, K. (eds.) Proceedings of the Fifth International Conference on Logic Programming, pp. 1070–1080. MIT Press, Cambridge (1988)
24. Gelfond, M., Lifschitz, V.: Classical negation in logic programs and disjunctive databases. New Generation Computing 9(3/4), 365–386 (1991)
25. Gomes, A.S., Alferes, J.J., Swift, T.: Implementing Query Answering for Hybrid MKNF Knowledge Bases. In: Carro, M., Peña, R. (eds.) PADL 2010. LNCS, vol. 5937, pp. 25–39. Springer, Heidelberg (2010)
26. Grimm, S., Hitzler, P.: Semantic Matchmaking of Web Resources with Local Closed-World Reasoning. International Journal of Electronic Commerce 12(2) 89–126 (2008)
27. Grimm, S., Hitzler, P.: A preferential tableaux calculus for circumscriptive $\mathcal{ALCO}$. In: Polleres, A., Swift, T. (eds.) RR 2009. LNCS, vol. 5837, pp. 40–54. Springer, Heidelberg (2009)

28. Grosof, B., Horrocks, I., Volz, R., Decker, S.: Description Logic Programs: Combining Logic Programs with Description Logic. In: Proceedings of WWW 2003, Budapest, Hungary, pp. 48–57 (May 2003)
29. Heymans, S., Predoiu, L., Feier, C., de Bruijn, J., Nieuwenborgh, D.V.: G-hybrid knowledge bases. In: Proc. of ICLP 2006 Workshop on Applications of Logic Programming in the Semantic Web and Semantic Web Services (ALPSWS 2006) (2006)
30. Hitzler, P., Parsia, B.: Ontologies and rules. In: Staab, S., Studer, R. (eds.) Handbook on Ontologies, 2nd edn., pp. 111–132. Springer, Heidelberg (2009)
31. Hitzler, P.: Towards reasoning pragmatics. In: Janowicz, K., Raubal, M., Levashkin, S. (eds.) GeoS 2009. LNCS, vol. 5892, pp. 9–25. Springer, Heidelberg (2009)
32. Hitzler, P., van Harmelen, F.: A reasonable semantic web. Semantic Web 1(1–2), 39–44 (2010)
33. Hitzler, P., Krötzsch, M., Parsia, B., Patel-Schneider, P.F., Rudolph, S. (eds.): OWL 2 Web Ontology Language: Primer. W3C Recommendation (October 27, 2009), http://www.w3.org/TR/owl2-primer/
34. Hitzler, P., Krötzsch, M., Rudolph, S.: Foundations of Semantic Web Technologies. Chapman & Hall/CRC (2009)
35. Hitzler, P., Seda, A.K.: Mathematical Aspects of Logic Programming Semantics. CRC Press, Boca Raton (2010)
36. Horrocks, I., Patel-Schneider, P.F.: A proposal for an OWL rules language. In: Proceedings of the 13th international conference on World Wide Web, WWW 2004, pp. 723–731. ACM, New York (2004)
37. Horrocks, I., Patel-Schneider, P.F., Bechhofer, S., Tsarkov, D.: OWL rules: A proposal and prototype implementation. J. of Web Semant. 3, 23–40 (2005)
38. Horrocks, I., Patel-Schneider, P.F., Boley, H., Tabet, S., Grosof, B., Dean, M.: SWRL: A Semantic Web Rule Language Combining OWL and RuleML. W3C Member Submission (May 21, 2004), http://www.w3.org/Submission/SWRL/
39. Hustadt, U., Motik, B., Sattler, U.: Data complexity of reasoning in very expressive description logics. In: Proceedings of the 19th international joint conference on Artificial intelligence, pp. 466–471. Morgan Kaufmann Publishers Inc., San Francisco (2005)
40. Kazakov, Y.: Consequence-driven reasoning for Horn $\mathcal{SHIQ}$ ontologies. In: Proceedings of the 21st international jont conference on Artifical intelligence, pp. 2040–2045. Morgan Kaufmann Publishers Inc., San Francisco (2009)
41. Kifer, M., Lausen, G., Wu, J.: Logical foundations of object-oriented and frame-based languages. Journal of the ACM 42(4), 741–843 (1995)
42. Kifer, M.: Rule interchange format: The framework. In: Calvanese, D., Lausen, G. (eds.) RR 2008. LNCS, vol. 5341, pp. 1–11. Springer, Heidelberg (2008)
43. Knorr, M., Alferes, J., Hitzler, P.: Local closed-world reasoning with description logics under the well-founded semantics. Artificial Intelligence 175(9-10), 1528–1554 (2011)
44. Knorr, M., Alferes, J., Hitzler, P.: A well-founded semantics for hybrid MKNF knowledge bases. In: Calvanese, D., Franconi, E., Haarslev, V., Lembo, D., Motik, B., Turhan, A.-Y., Tessaris, S. (eds.) Proceedings of the 2007 International Workshop on Description Logics (DL 2007), Brixen-Bressanone, Italy. CEUR Workshop Proceedings, vol. 250 (June 2007)
45. Knorr, M., Alferes, J.J.: Querying in $\mathcal{EL}^+$ with nonmonotonic rules. In: Proceeding of the 2010 conference on ECAI 2010: 19th European Conference on Artificial Intelligence, pp. 1079–1080. IOS Press, Amsterdam (2010)

46. Knorr, M., Alferes, J.J., Hitzler, P.: A well-founded semantics for hybrid MKNF knowledge bases. In: Description Logics. CEUR Workshop Proceedings, vol. 250, Description Logics. CEUR Workshop Proceedings (2007), CEUR-WS.org
47. Knorr, M., Alferes, J.J., Hitzler, P.: A coherent well-founded model for hybrid MKNF knowledge bases. In: Proceeding of the 2008 conference on ECAI 2008: 18th European Conference on Artificial Intelligence, pp. 99–103. IOS Press, Amsterdam (2008)
48. Krisnadhi, A., Sengupta, K., Hitzler, P.: Local closed world semantics: Keep it simple, stupid! Tech. rep., Kno.e.sis Center, Wright State University, Dayton, Ohio (2011), http://www.pascal-hitzler.de/
49. Krötzsch, M.: Efficient inferencing for OWL EL. In: Janhunen, T., Niemelä, I. (eds.) JELIA 2010. LNCS, vol. 6341, pp. 234–246. Springer, Heidelberg (2010)
50. Krötzsch, M.: Description Logic Rules, Studies on the Semantic Web, vol. 008. IOS Press/AKA (2010)
51. Krötzsch, M., Maier, F., Krisnadhi, A.A., Hitzler, P.: A better uncle for OWL: Nominal schemas for integrating rules and ontologies. In: Sadagopan, S., Ramamritham, K., Kumar, A., Ravindra, M., Bertino, E., Kumar, R. (eds.) Proceedings of the 20th International World Wide Web Conference, WWW 2011, Hyderabad, India, pp. 645–654. ACM, New York (2011)
52. Krötzsch, M., Rudolph, S., Hitzler, P.: Complexity boundaries for Horn description logics. In: Proceedings of the Twenty-Second AAAI Conference on Artificial Intelligence, Vancouver, British Columbia, Canada, July 22-26, pp. 452–457. AAAI Press, Menlo Park (2007)
53. Krötzsch, M., Rudolph, S., Hitzler, P.: Description Logic Rules. In: Ghallab, M., Spyropoulos, C.D., Fakotakis, N., Avouris, N.M. (eds.) Proceeding of the 18th European Conference on Artificial Intelligence, Patras, Greece, July 21-25, vol. 178, pp. 80–84. IOS Press, Amsterdam (2008)
54. Krötzsch, M., Rudolph, S., Hitzler, P.: ELP: Tractable rules for OWL 2. In: Sheth, A.P., Staab, S., Dean, M., Paolucci, M., Maynard, D., Finin, T., Thirunarayan, K. (eds.) ISWC 2008. LNCS, vol. 5318, pp. 649–664. Springer, Heidelberg (2008)
55. Krötzsch, M., Rudolph, S., Schmitt, P.H.: On the semantic relationship between datalog and description logics. In: Hitzler, P., Lukasiewicz, T. (eds.) RR 2010. LNCS, vol. 6333, pp. 88–102. Springer, Heidelberg (2010)
56. Levy, A.Y., Rousset, M.C.: CARIN: A representation language combining Horn rules and description logics. In: Wahlster, W. (ed.) Proceedings of 12th European Conference on Artificial Intelligence, Budapest, Hungary, August 11-16, pp. 323–327. John Wiley and Sons, Chichester (1996)
57. Levy, A.Y., Rousset, M.-C.: Combining Horn rules and description logics in CARIN. Artif. Intell. 104, 165–209 (1998)
58. Lifschitz, V.: Nonmonotonic databases and epistemic queries. In: Proceedings of the 12th International Joint Conference on Artificial Intelligence, vol. 1, pp. 381–386. Morgan Kaufmann Publishers Inc., San Francisco (1991)
59. Lukasiewicz, T.: A novel combination of answer set programming with description logics for the semantic web. In: Franconi, E., Kifer, M., May, W. (eds.) ESWC 2007. LNCS, vol. 4519, pp. 384–398. Springer, Heidelberg (2007)
60. Lutz, C., Walther, D., Wolter, F.: Conservative extensions in expressive description logics. In: Proc. of IJCAI 2007, pp. 453–459. AAAI Press, Menlo Park (2007)
61. McGuinness, D., van Harmelen, F. (eds.): OWL Web Ontology Language Overview. W3C Recommendation (10 February 2004), http://www.w3.org/TR/owl-features/

62. Motik, B.: Reasoning in Description Logics using Resolution and Deductive Databases. Ph.D. thesis, Universität Karlsruhe (TH), Germany (2006)
63. Motik, B., Cuenca Grau, B., Horrocks, I., Sattler, U.: Representing ontologies using description logics, description graphs, and rules. Artificial Intelligence 173(14), 1275–1309 (2009)
64. Motik, B., Cuenca Grau, B., Horrocks, I., Wu, Z., Fokoue, A., Lutz, C. (eds.): OWL 2 Web Ontology Language: Profiles. W3C Recommendation (October 27, 2009), http://www.w3.org/TR/owl2-profiles/
65. Motik, B., Horrocks, I., Rosati, R., Sattler, U.: Can OWL and Logic Programming Live Together Happily Ever After? In: Cruz, I., Decker, S., Allemang, D., Preist, C., Schwabe, D., Mika, P., Uschold, M., Aroyo, L.M. (eds.) ISWC 2006. LNCS, vol. 4273, pp. 501–514. Springer, Heidelberg (2006)
66. Motik, B.: Reasoning in Description Logics using Resolution and Deductive Databases. Ph.D. thesis, Universität Karlsruhe (TH), Germany (2006)
67. Motik, B., Grau, B.C., Horrocks, I., Sattler, U.: Representing Structured Objects using Description Graphs. In: Brewka, G., Lang, J. (eds.) Proc. of the 11th Int. Joint Conf. on Principles of Knowledge Representation and Reasoning (KR 2008), August 16–19, pp. 296–306. AAAI Press, Sydney (2008)
68. Motik, B., Grau, B.C., Horrocks, I., Sattler, U.: Modeling Ontologies Using OWL, Description Graphs, and Rules. In: Ruttenberg, A., Sattler, U., Dolbear, C. (eds.) Proc. of the 5th Int. Workshop on OWL: Experiences and Directions (OWLED 2008 EU), Karlsruhe, Germany, October 26–27 (2008)
69. Motik, B., Grau, B.C., Sattler, U.: Structured Objects in OWL: Representation and Reasoning. In: Huai, J., Chen, R., Hon, H.W., Liu, Y., Ma, W.Y., Tomkins, A., Zhang, X. (eds.) Proc. of the 17th Int. World Wide Web Conference (WWW 2008), April 21–25, pp. 555–564. ACM Press, Beijing (2008)
70. Motik, B., Rosati, R.: Closing semantic web ontologies. Tech. rep., University of Manchester, UK (2006)
71. Motik, B., Rosati, R.: A faithful integration of description logics with logic programming. In: Veloso, M.M. (ed.) IJCAI 2007: Proceedings of the 20th International Joint Conference on Artificial Intelligence, Hyderabad, India, January 6-12, pp. 477–482 (2007)
72. Motik, B., Rosati, R.: Reconciling Description Logics and Rules. Journal of the ACM 57(5), 1–62 (2010)
73. Motik, B., Sattler, U., Studer, R.: Query Answering for OWL-DL with Rules. In: McIlraith, S.A., Plexousakis, D., van Harmelen, F. (eds.) ISWC 2004. LNCS, vol. 3298, pp. 549–563. Springer, Heidelberg (2004)
74. Motik, B., Sattler, U., Studer, R.: Query answering for OWL-DL with rules. Journal of Web Semantics: Science, Services and Agents on the World Wide Web 3(1), 41–60 (2005)
75. OWL Working Group, W.: OWL 2 Web Ontology Language: Document Overview. W3C Recommendation (October 27, 2009), http://www.w3.org/TR/owl2-overview/
76. Paschke, A.: Rules and Logic Programming for the Web. In: Polleres, A., et al. (eds.) Reasoning Web 2011. LNCS, vol. 6848, pp. 384–417. Springer, Heidelberg (2011)
77. Pearce, D., Valverde, A.: Quantified equilibrium logic and the first order logic of here-and-there. Tech. rep., Univ. Rey Juan Carlos (2006)

78. Rosati, R.: Towards expressive KR systems integrating datalog and description logics: preliminary report. In: Lambrix, P., Borgida, A., Lenzerini, M., Möller, R., Patel-Schneider, P.F. (eds.) Description Logics. CEUR Workshop Proceedings, vol. 22 (1999), CEUR-WS.org
79. Rosati, R.: On the decidability and complexity of integrating ontologies and rules. J. of Web Semant. 3, 61–73 (2005)
80. Rosati, R.: Semantic and computational advantages of the safe integration of ontologies and rules. In: Fages, F., Soliman, S. (eds.) PPSWR 2005. LNCS, vol. 3703, pp. 50–64. Springer, Heidelberg (2005)
81. Rosati, R.: DL+log: Tight integration of description logics and disjunctive datalog. In: Doherty, P., Mylopoulos, J., Welty, C.A. (eds.) Proceedings, Tenth International Conference on Principles of Knowledge Representation and Reasoning, Lake District of the United Kingdom, June 2-5, pp. 68–78. AAAI Press, Menlo Park (2006)
82. Rosati, R.: Integrating ontologies and rules: Semantic and computational issues. In: Barahona, P., Bry, F., Franconi, E., Henze, N., Sattler, U. (eds.) Reasoning Web 2006. LNCS, vol. 4126, pp. 128–151. Springer, Heidelberg (2006)
83. Rudolph, S., Krötzsch, M., Hitzler, P.: Cheap boolean role constructors for description logics. In: Hölldobler, S., Lutz, C., Wansing, H. (eds.) JELIA 2008. LNCS (LNAI), vol. 5293, pp. 362–374. Springer, Heidelberg (2008)
84. Rudolph, S.: Foundations of description logics. In: Polleres, A., et al. (eds.) Reasoning Web 2011. LNCS, vol. 6848, pp. 384–417. Springer, Heidelberg (2011)
85. Rudolph, S., Krötzsch, M., Hitzler, P.: All elephants are bigger than all mice. In: Baader, F., Lutz, C., Motik, B. (eds.) Proceedings of the 21st International Workshop on Description Logics (DL 2008) CEUR Workshop Proceedings, Dresden, Germany, May 13-16, vol. 353 (2008)
86. Schindlauer, R.: Answer-Set Programming for the Semantic Web. Ph.D. thesis, Vienna University of Technology, Austria (2006)
87. Tsarkov, D., Sattler, U., Stevens, R.: A solution for the Man-Man problem in the Family History Knowledge Base. In: Hoekstra, R., Patel-Schneider, P.F. (eds.) Proceedings of the 5th International Workshop on OWL: Experiences and Directions (OWLED 2009) CEUR Workshop Proceedings, Chantilly, VA, United States, October 23-24, vol. 529 (2009)
88. Volz, R.: Web Ontology Reasoning With Logic Databases. Ph.D. thesis, Universität Fridericiana zu Karlsruhe (TH), Germany (2004)

# Modeling the Web of Data
# (Introductory Overview)

Claudio Gutierrez

Department of Computer Science,
Universidad de Chile, Chile
cgutierr@dcc.uchile.cl

**Abstract, scope and disclaimer.** These notes are meant as a companion to
a lecture on the topic at the Reasoning Web Summer School 2011. The goal of
this work is to present diverse and known material on modeling the Web from
a data perspective, to help students to get a first overview of the subject.

Methodologically, the objective is to give pointers to the relevant topics
and literature, and to present the main trends and development of a new area.
The idea is to organize the existing material without claiming completeness.
In many parts the notes have a speculative character, oriented more towards
suggesting links and generating discussion on different points of view, rather
than establishing a consolidated view of the subject.

The historical accounts and references are given with the sole objective of
aiding in the contextualization of some milestones, and should not be considered as signaling intellectual priorities.

## Introduction

From the point of view of information, the most naive –and probably also the
most understandable– model of the Web is that of an infinite library. The idea
is not new: in 1939 Jorge Luis Borges published the story *The Total Library*[1],
where he writes:

> *"Everything would be in its blind volumes. Everything: the detailed his-
> tory of the future, Aeschylus' The Egyptians, the exact number of times
> that the waters of the Ganges have reflected the flight of a falcon, the
> secret and true nature of Rome, the encyclopedia Novalis would have
> constructed, my dreams and half-dreams at dawn on August 14, 1934,
> the proof of Pierre Fermat's theorem, the unwritten chapters of Edwin
> Drood, those same chapters translated into the language spoken by the
> Garamantes, the paradoxes Berkeley invented concerning Time but didn't
> publish, Urizen's books of iron, the premature epiphanies of Stephen
> Daedalus, which would be meaningless before a cycle of a thousand years,
> the Gnostic Gospel of Basilides, the song the sirens sang, the complete
> catalog of the Library, the proof of the inaccuracy of that catalog. Every-
> thing: but for every sensible line or accurate fact there would be millions*

---

[1] J. L. Borges, La Biblioteca Total, Sur No. 59, August 1939. Trans. by Eliot Wein-
berger. In Selected Non-Fictions (Penguin: 1999).

A. Polleres et al. (Eds.): Reasoning Web 2011, LNCS 6848, pp. 416–444, 2011.

*of meaningless cacophonies, verbal farragoes, and babblings. Everything:*
*but all the generations of mankind could pass before the dizzying shelves-*
*shelves that obliterate the day and on which chaos lies-ever reward them*
*with a tolerable page."*

The view of a universal space of information as the (infinite) generalization
of a library is an extremely useful one. It includes almost all facets we would
like to incorporate when abstracting and modeling such an artifact. There is
one crucial slant, though: the library is composed of books, let us say in Web
terms, of documents. Documents (books) are artifacts produced by humans to be
consumed by humans. If one replaces data in the place of books, we essentially
have an abstract model of the "Web of Data". But this is not a minor change,
bringing with it complex challenges.

Modeling the Web of data is a relevant goal. The big excitement about current
levels of production, availability and use of data indicates that we are witnessing
a fundamental change in information practices. The tide of data was observed
a few years ago by cutting-edge technology analysts. In his widely read 2005
article that sparked the notion of Web 2.0 [66], O'Reilly wrote that "data is the
next Intel Inside." On a more academic level, the Claremont Report on Database
Research [6] centered its analysis on the challenges that this phenomena is posing,
stating that ubiquity of "Big Data" will shake up the field [of databases]. Szalay
and Gray pointing to the fact in 2006, that the amount of scientific data is
doubling every year, spoke of an "exponential world" [30] and Bell et al. [18]
called it "Data Deluge". They state that, compared to volumes generated only a
decade ago, some areas of science are facing hundred- to thousandfold increases
in data volumes from satellites, telescopes, high-throughput instruments, sensor
networks, accelerators, and supercomputers.

The phenomena is not exclusive of the scientific fields. A similar trend can be
found in almost all areas. Social networks are generating not only high volumes
of data, but complex networks of data which call for a new stage in data management. New technologies have also impacted government policies. Transparency
laws and wide-range archiving and publishing initiatives are posing similar challenges to the public sector [25]. Managing, curating and archiving of digital data
is becoming a discipline per se. Today some people are even talking about "data
science" [46].

It is no surprise that this phenomena has put data at the center of computing discipline itself, both, at the level of systems, architecture and communications (see "petascale computational systems" [17]), new database architectures
at web-scale [65,60], and at the programming and modeling levels. In these new
developments the Web, as the natural common platform for handling such data,
plays a central role.

*Data management at Web scale.* With the advance of computing power in the
last decade, the perspective on the Web is gradually shifting from a document
centric-view to a data centric-view. Originally conceived as a global hypertext
model, today the granularity of the information on the Web has reached the level

of atomic data. For example, the project *Linked Data* [44,20] views the Web as a huge collection of data and links.

How to manage data at Web scale? Since the very origins of the Web, the database community has addressed this challenge. In the late nineties the efforts to integrate the new Web phenomena and database technology provoked a heated discussion. Is the Web a database? Could the classical database techniques be of any use in this new environment?

Two main lines of thought were developed. The first one conceived the Web as a collection of documents plus hyperlinks, and extended the ideas and technologies of hypertext and followed the lines of semi-structured data and information retrieval techniques [21,2,5]. This was consistent with the view that "sites" and Web pages were the central objects of interest. This, combined with the need to model documents and the exchange and integration of information, made this conception dominant. The research centered on semi-structured data and query languages, which with the advent of XML, dominated the scene for the decade of 2000 [5].

A different perspective called for modeling the Web as a database and developed the so called Web-query languages [55]. The systematic exploration of the idea of modeling the Web as a huge repository of structured data using database techniques did not succeed, likely because the amount of structured data on the Web did not yet reach a critical level. Such ideas were too futuristic for the time, though recent developments as the one mentioned at the beginning, show that the need has reemerged.

In the meantime, several areas of research have addressed, with variable emphasis and focus, the problems of data on the Web. Among them, projects like Semantic Web, Linked data, Open data, put the the topic in the main discussion forums. From a database point of view, areas such as distributed, semi-structured and graph databases, and particular topics like incomplete information, cost models, etc., have addressed similar problems on a smaller scale. There are also other areas such as information system, multimedia, etc., that touch on problems of data on the Web, but their exhaustive enumeration would be too long to fit here.

*Notes Outline.* These notes present an overview of the work done in modeling data on the Web and discusses requirements needed to convert the current Web of documents in a Web of Data. The organization is as follows: in section 1 we study the principles of the Web as devised by their founders and the evolution of the Web. In section 2, we present basic tools and projects that have helped build the Web of Data. In section 3, we review data representations on the Web and data models of the Web. In section 4 we bring forward a group of requirements and themes that should be addressed in a model of the Web of Data. In section 5, we briefly review the work in related areas which touch on the problems, concerns and techniques faced in our "field". Finally in section 6, we round up our trip through this new area.

# 1    The Web

Tim Berners-Lee (TBL from now on), the creator of the Web, states that its "major goal was to be a shared information space through which people and machines could communicate" [11]. Let us read between the lines. He meant a "global" information space, a kind of gigantic, infinite, blackboard to write and read: "The most important thing about the Web is that it is universal" [12]. But this is not enough: another key consideration is that it should be "shared". By whom? Not by a company, not by a government, not by a particular organization: shared by all people around the world.

The problem he was addressing what that of people working at CERN, located around the world, in different research labs and academic places. This was a heterogeneous group, managing and exchanging heterogeneous type of information (addresses and phone lists, research notes, official documentation, etc.), via a heterogeneous infrastructure of terminals, servers, supercomputers, etc., with diverse operating systems, software and file formats. As Roy Fielding [27] sated, the challenge was to build a system that would provide a universally consistent interface to this structured information, available on as many platforms as possible, and incrementally deployable as new people and organizations joined the project.

## 1.1    The Classical Web

In 2001, in his Japan Lecture [12], TBL defined the Web as follows:

> *"The concept of the Web integrated many disparate information systems, by forming an abstract imaginary space in which the differences between them did not exist. The Web had to include all information of any sort on any system. The only common idea needed to tie it all together was the Universal Resource Identifier(URI) identifying a document. From that cascaded a series of designs of protocols (such as HTTP) and data formats (such as HTML) which allowed computers to exchange information, mapping their own local formats into standards which provided global interoperability."*

The architecture of the Web is based on three basic pillars:

1. URI (*Universal Resource Identifiers*), a set of global identifiers which can be created and managed in a distributed form.
2. HTTP (*Hyper Text Transfer Protocol*): a protocol for exchanging data on the Web whose basic functions are putting data in, and getting data from, this abstract space.
3. HTML (*Hyper Text Markup Language*): a language for representing information and displaying (visualizing) it to humans.

**Fig. 1.** The first proposal of the Web by TBL. Note the underlying ideas: heterogenous data, heterogeneous users, lack of hierarchies, networking, mainly documents. (Picture taken from TBL, Information Management: A Proposal).

Of these three, the global identifiers are the keystone. TBL highlights this point saying that "the Web still was designed to only fundamentally rely on one specification: the Universal Resource Identifier." The particular form of the transfer protocol and of the language, are temporal solutions with the technology and knowledge available at the time.

If one would like to generalize, the Web can be thought of as supported by three basic specifications:

1. Global Identifiers.
2. A protocol to exchange data.
3. A language to represent data.

*TBL's general requirements.* In the Japan lecture, TBL stated the following principles/requirements that should guide the development of this architecture (the text closely follows his wording):

1. *Device independence.* The same information should be accessible from many devices. The size of the screens, the means of input and output information should be independent of the hardware.

2. *Software Independence.* The Web should support diverse programs and software. The decentralization of software development was and always will be crucial to the unimpeded growth of the Web. It also prevents the Web itself from coming under the control of a given company or government through control of the software.

3. *Internationalization.* The Web should not depend on one country or culture. Internationalization should take into account not only the language, but also the direction in which text moves across the page, hyphenation conventions, and even cultural assumptions about the way people work and address each other, and the forms of organization they make.

4. *Multimedia.* Multimedia is at the heart of modern digital objects. Images, music, video have to be essential part of the design of the Web.

5. *Accessibility.* Just as people differ in the language, characters and cultures to which they belong, so they differ in terms of their capacities, regarding vision, hearing, motor or cognition. The universality includes making the Web a place which people can use irrespective of disabilities.

6. *Rhyme and Reason.* There is another axis along which information varies: its purpose and usage. At one end of the axis is the poem, at the other, the database table. Most information on the Web now contains both elements. The Web technology must allow information intended for a human to be effectively presented, and also allow machine processable data to be conveyed.

7. *Quality.* Quality notions are very subjective, and change with time, and all of them should be allowed in the Web. To support this, the technology must allow powerful filtering tools which, combining opinions and information about information from many sources, are completely under the control of the user.

8. *Independence of Scale.* Although the Web is a global phenomenon, personal, family and group information systems are part of it too. The Web must support all of those, allowing privacy of personal information to be negotiated, and groups to feel safe in controlling access to their spaces. Only in such a balanced environment can we develop a sufficiently complex and many-layered fractal structure which will respect the rights of every human being.

*Requirements for the Protocols.* Roy Fielding, one of the authors of the HTTP protocol,

In his doctoral thesis, Roy Fielding, one of the authors of the HTTP protocol, explored in depth the architecture of the Web [27]. He identified the following requirements (the text follows his wording):

*1. Low Entry-barrier.* Since participation in the creation and structuring of information was voluntary, a low entry-barrier was necessary to enable sufficient adoption. This applied to all users of the Web architecture: readers, authors, and application developers.

*2. Extensibility.* While simplicity makes it possible to deploy an initial implementation of a distributed system, extensibility allows us to avoid getting stuck forever with the limitations of what was deployed. Even if it were possible to build a software system that perfectly matches the requirements of its users, those requirements will change over time just as society changes over time. A system intending to be as long-lived as the Web must be prepared for change.

*3. Distributed Hypermedia.* Hypermedia is defined by the presence of application control information embedded within, or as a layer above, the presentation of information. Distributed hypermedia allows the presentation and control information to be stored at remote locations.

The usability of hypermedia interaction is highly sensitive to user-perceived latency: the time between selecting a link and the rendering of a usable result. Since the Web's information sources are distributed across the global Internet, the architecture needs to minimize network interactions (round-trips within the data transfer protocols).

*4. Internet-scale.* The Web is intended to be an Internet-scale distributed hypermedia system, which means considerably more than just geographical dispersion. The Internet is about interconnecting information networks across multiple organizational boundaries. Suppliers of information services must be able to cope with the demands of anarchic scalability and the independent deployment of software components.

a) *Anarchic Scalability.* Most software systems are created with the implicit assumption that the entire system is under the control of one entity, or at least that all entities participating within a system are acting towards a common goal and not at cross-purposes. Such an assumption cannot be safely made when the system runs openly on the Internet. Anarchic scalability refers to the need for architectural elements to continue operating when they are subjected to an unanticipated load, or when given malformed or maliciously constructed data, since they may be communicating with elements outside their organizational control.

b) *Independent Deployment.* Existing architectural elements need to be designed with the expectation that later architectural features will be added. Likewise, older implementations need to be easily identified so that legacy behavior can be encapsulated without adversely impacting newer architectural elements. The architecture as a whole must be designed to ease the deployment of architectural elements in a partial, iterative fashion, since it is not possible to force deployment in an orderly manner.

Based on these principles, he proposed a style of software architecture for distributed hypermedia systems called *REST* (Representational State Transfer). These are the constraints defined for it:

1. *Client-server.* Separation of concerns is the principle behind the client-server constraints. Clients are separated from servers by a uniform interface. This separation of concerns means that, for example, clients are not concerned with data storage, which remains internal to each server, so that the portability of client code is improved. Servers are not concerned with the user interface or user state, so that servers can be simpler and more scalable.

Servers and clients may also be replaced and developed independently, as long as the interface is not altered.

2. *Stateless.* Each request from client to server must contain all of the information necessary to understand the request, and cannot take advantage of any stored context on the server. Session state is therefore kept entirely on the client. This constraint induces the properties of visibility, reliability, and scalability.

3. *Cacheable.* Cache constraints require that the data within a response to a request be implicitly or explicitly labeled as cacheable or non-cacheable. If a response is cacheable, then a client cache is given the right to reuse that response data for later, equivalent requests.

4. *Uniform interface.* The central feature that distinguishes the REST architectural style from other network-based styles is its emphasis on a uniform interface between components.

5. *Layered system.* The layered system style allows an architecture to be composed of hierarchical layers by constraining component behavior such that each component cannot "see" beyond the immediate layer with which they are interacting.

*What about the language requirements?* Paradoxically, one of the reasons for the success of the Web was the weaknesses of its language HTML: loose structure (allowing the display badly-formed pages) and only oriented to visualization (by humans). The next generation language, XML, improved on both aspects: (1) strict enforcement of structure and constraints (allowing semi-structured querying); and (2) flexible to code different objects languages (for visualization, exchange, domain specific, etc.) Nevertheless, one fundamental bias remained: it was designed with a document-style organization in mind.

Today we know that, although documents are important part of our global data, there is plenty of data that has no document-style organization: table data, raw data, sensor data, streams, images, etc. What is a "good" language for a global exchange of data? We would like to advance some basic general requirements for it:

1. Include codification for data, metadata and knowledge.
2. Be flexible enough to describe most types of data.
3. Be minimalist and efficient (regarding user needs and evaluation complexity).
4. Scale in a non-centralized form.

We will come back to the language theme repeatedly, because it is one of the cornerstones of the Web of Data.

## 1.2   The Semantic Web

A review of the Web would be incomplete without covering the Semantic Web. Indeed, the original project included as an ideal target a Web where all contents shared global semantics.

There are two driving forces behind the development of the Semantic Web: first, the fact that if data and information scale to meta-human levels, the only possibility to access, organize and manage such data is via machines; Second, the problem of meaning of information: what is the meaning of each piece of information on the Web? This has to do fundamentally with the semantics and meaning of concepts (even in the same language).

The first problem is an old one and is at the root of the discipline of databases on one hand, and of information retrieval on the other. One deals with structured data and the task of the organization of data –via logic– to allow semi-automatic querying and management of it. The other deals with unstructured data and documents, and relies on statistical methods to approximate the user needs.

The basic assumptions of classic database models (closed world, known goals, well defined users, etc.) do not scale at planetary level. The statistical approach has shown to be more suited to scale, but at the cost of trading logical precision by approximate results.

The Semantic Web aims to partially solve this problem based on the simple idea of organizing information at planetary level. The Semantic Web is *the Web of machine-processable data*, writes TBL, and this amounts to standardize meanings. Is this program viable? Naive approaches in this direction, like the Esperanto language, have failed miserably. Optimistically one could think that there were basic design failures in that project: centralized approach, lack (or high cost) of extensibility, not machine processable, complex semantics, little participation of (prospective) users in their enrichment.

The Semantic Web program devised two humble goals in order to overcome these problems:

1. *Develop languages* for describing metadata, sufficiently flexible, distributively extensible, machine-processable. (Note how this fits smoothly with the requirements for a global language for the Web discussed in a previous section). Two families of languages have been developed:
   (a) *Resource Description Framework, RDF* [42]. A basic language, in the style of semantic networks and graph data specifications, based on universal identifiers. Basic tools for interconnecting (linking) data, plus a lightweight machinery for coding basic meanings.
   (b) *The Web Ontology Language, OWL* [50]. A version of logic languages adapted to cope with the Web requirements. Composed of basic logic operators plus a mechanism for defining meaning in a distributed fashion.
2. *Develop an infrastructure* for it. Among the most important building blocks for the Semantic Web are protocols, query languages, specifications and applications for accessing, consulting, publishing and exchanging data.

Goal (1) has been a successful program. As time went on, two more or less defined communities have been developing this area (see Figure 2; Information Retrieval will not be discussed here):

*The logic and knowledge representation community.* It is oriented towards developing high level and expressive languages for describing information on

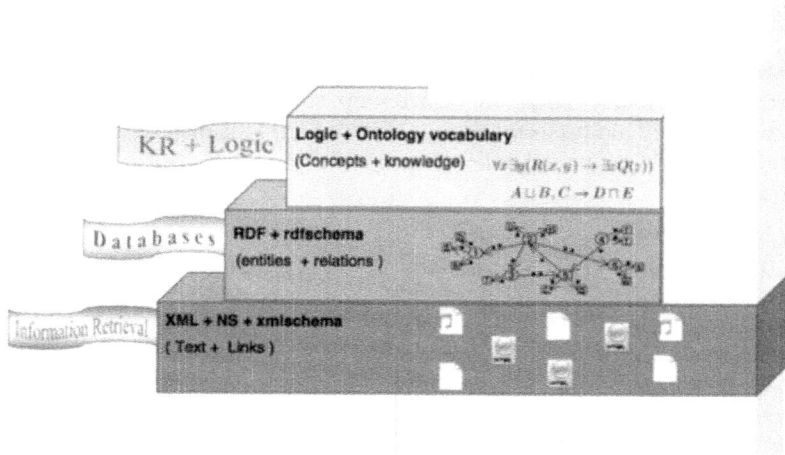

**Fig. 2.** The technical fields involved in data aspects of the Semantic Web Tower: Information Retrieval, Databases and Knowledge Representation

the Web. One could summarize its accomplishments saying that it has achieved the internationalization and global extensibility of logic languages, particularly via OWL. At the same time, it is important to understand its limitations: this approach does succeed in describing data at a massive scale. In fact, the most basic tools have a computational complexity far exceeding the needs of big-scale data management.

*The database community.* It is centered on the development of RDF and its query language SPARQL. In the next section we will expand on these languages.

Goal (2) has been partially successful. On the positive aspects, there is a solid community behind the specifications and increasing interest by different stakeholders (e.g. governments, scientific communities). On the other hand, the area is still looking for applications that show the full potentialities of these approaches (an issue that deserves a careful analysis, beyond the scope of these notes).

## 2   Towards the Web of Data

Roughly speaking, the Web of data can be defined as follows:

> *The Web of data is the global collection of data produced by the*
> *systematic and decentralized exposure and publication of (raw)*
> *data using Web protocols.*

From this point of view, one of the main question arising is how to identify the changes produced to data management when incorporating (raw) data to the classical Web model reviewed .

In this section we will address the most visible initiatives aimed either, at enhancing and overcoming the Web of documents, or at addressing new challenges to information management posed by the new developments of data at Web scale. First, we will summarize the challenges posed by the "data deluge" on data management. Then, we will review the data level of the role and perspectives of RDF in this new setting. Finally we will discuss the contributions of two important projects, Linked Data and Open data to the goals of the Web of Data.

## 2.1   The Data Deluge Structure

The data deluge described in the introduction consists of different types of data and data sources. Today there is a widespread feeling that this is beginning of a chaotic new era. I think it is important to realize that this tsunami of data is going to stabilize; that we should not act like the people shaken by the first big waves, but try to get a comprehensive picture of the process that is opening.

Any modeling of data on the Web should begin with a clear picture of the sources of such data. First of all, one has to consider the traditional publishing sources (editorials, writers, in general: sources that surely will remain, although in different formats) and scientific data that is gradually changing because of the increasing capacity and will to record and store. An important additional source of data are sensors, either capturing data directly from non-human natural processes (metereology, radioastronomy, animal behavior, etc.) or directed at humans (surveillance, logs of computer applications, medical, etc.)

Determining what types of data are more relevant is of paramount importance. The resources are finite, and hence naturally the most relevant data is the one that will constitute the main source of the sea of data.[48]

A characterization of the data itself (independent of its source) is another challenge. Classical methodologies and results about it (e.g. those of librarians) dealt essentially with human-produced data in natural language. Photography and video are still datasets that for most of us have hold little meaning outside of the human annotations (in natural language) attached to them. Clearly this is going to change.

## 2.2   RDF as Infrastructure

It should not be a surprise that the notion of Web of data has a close relationship with the Semantic Web. The most influential semantic technologies, the RDF model and the SPARQL query language, have given new impetus to the development of the idea of Web of data. Here we will briefly explain the strengths of RDF and the coming challenges. (For RDF and SPARQL, consult [10].)

RDF was designed to facilitate automatic processing of information on the Web via metadata. The 1999 Recommendation stated it clearly: *"RDF is intended for situations in which this information needs to be processed by applications, rather than only displayed to people"*. Thus, it is at the core of its goal the incorporation of machine readable information to the Web. But the design of RDF had another rather unexpected outcome: its graph nature (due to its triple structure) allows for representation of any type of data, and hence opens the door for converting the Web of documents into a Web of Data.

The power of RDF resides in the combination of two ideas: (1) a flexible model able to represent plain data as well as metadata in a uniform manner, pushing the idea of objects of information where data and metadata (schema) have the same status; (2) a graph structure that represents naturally, interconnections and relationships between data. In fact this latter feature is the one that led to the development of the Linked Data initiative.

These two ideas crystallize at the structural level of RDF in two main blocks of the RDF language: its (graph) data structure and its vocabulary.

*Data structure.* RDF triples can be considered from a logical point of view as statements. But at the same time, they naturally represent a graph structure. Hence its expressive power: the structure really represents a linked network of statements.

This graph can be considered as relational data (a set of triples is a table with three columns). This viewpoint has the advantage of dealing with a well understood object; thus, allows the reuse of well studied and proven relational technology to manage such data.

It is important to understand the implications of the fact that RDF is a graph model: we face an object of study still not well understood, but with enormous potential to represent and model information [9].

*Vocabulary.* RDF was designed to be flexible and extensive regarding vocabulary, allowing to give meanings to the relationships indicated by its graph structure. It has a few pre-defined (built-in) keywords with a light semantics (see [35]). The compromise here is the usual: the computational complexity of processing such data increases with the expressive power of its vocabulary semantics. Today one can roughly separate the vocabulary in three groups: (a) Having light (or no) semantics (essentially `type`, `subClassOf`, `subPropertyOf`, etc.); (b) RDF Schema plus some light extensions; (c) OWL, the Web ontology language. For linking and describing raw data, (a) seems to be enough.

*A Remark concerning Blank Nodes.* Blank nodes allow flexibility in structure data and representation of incomplete information. For a global model of information seems that these features are unavoidable. The problem, nevertheless, is that data with such features increases the computational complexity of processing and its semantics of querying is not simple [51].

## 2.3   Linked Data

Among the most successful world-wide projects addressing the problem of ubiquitous data on the Web, *Linked Data* stands out [44,20]. This project originated in the practice of linking data and TBL's ideas on Web architecture [14], and has become one of the main driving forces pushing the idea of exposing data on the Web. As the authors of the project state [44]:

> *Linked Data is about using the Web to connect related data that*
> *wasn't previously linked, or using the Web to lower the barriers*
> *to linking data currently linked using other methods. More specif-*
> *ically, Wikipedia defines Linked Data as "a term used to describe*
> *a recommended best practice for exposing, sharing, and connect-*
> *ing pieces of data, information, and knowledge on the Semantic*
> *Web using URIs and RDF".*

The idea is simple: thanks to the Web technologies, the possibility to produce, publish and consume data (not only documents in the form of Web pages) has become universally available. These processes are being done by different stake-holders, with different goals, in different forms and formats, in different places. Taking full advantage of this new scenario is a challenge. One of the main problems –the one addressed by the Linked Data project– is that this universe of data is not interlinked meaningfully.

The relevance of the Linked Data project has been eloquently expressed TBL [15] as follows:

*Linked Data allows different things in different datasets of all kinds to be connected.* The added value of putting data on the Web is given by the way it can be queried in combination with other data you might not even be aware of. People will be connecting scientific data, community data, social web data, enterprise data, and government data from other agencies and organizations, and other countries, to ask all kinds of interesting questions not asked before.

*Linked data is decentralized.* Each agency can source its own data without a big cumbersome centralized system. The data can be stitched together at the edges, more as one builds a quilt than the way one builds a nuclear power station.

*The Linked Open Data movement uses open royalty-free standards from W3C.* These do not bind governments nor agencies to any specific supplier.

*A virtuous circle.* There are many organizations and companies who will be motivated by the presence of the data to provide all kinds of human access to this data, for specific communities, to answer specific questions, often in connection with other data from different sites.

The TBL's "five-stars" test to measure the level of implementation of these ideas demonstrates the strategic goal of the Linked Data project:

1. Make your stuff available on the web (whatever format).
2. Make it available as structured data (e.g. excel instead of image scan of a table).
3. Use non-proprietary format (e.g. csv instead of excel).
4. Use URLs to identify things, so that people can point at your stuff.
5. Link your data to other people's data to provide context.

The Linked Data project has rapidly earned solid support among developers and governments (e.g. [25]), and is slightly gaining space in Academia [45]. The applications of database techniques to it, particularly the development of an infrastructure for querying and navigating such network, are just taking off [37,38]. There is a recent book by Heath and Bizer [39] that covers the area systematically.

## 2.4   Open Data

Open data is a movement towards facilitating both, the *production* and *dissemination* of data and information at global scale.[2] In this regard, it is closely related with the original goals of the Web project. Because of its relationship with the issues arising in the "public versus private" sphere, it has become influential in management of information in Government and big organizations. On the other hand, regarding it as the data version of similar movements for software, we can define it as follows:

> *Open data is a movement whose goal is to develop and spread open standards for data.*

The big question here is what does openness mean for data. We will follow here the methodological approach of Jon Hoem in his study of openness in communication [41], and adapt the discussion to data. There are several possible dimensions from where one can consider openness. Three important ones are: the content, the logical and the physical levels. For data, this means respectively: semantics; datatypes and formats; and applications to access the data. For communication Hoem isolates two other crucial parameters: control of production and re-use; and control of distribution and consumption.

People working with public (government) data are among the ones that have elaborated more on this subject. Early in 2007, eight principles for openness in data were proposed [62]. Although they refer to "public data", the principles offer good insights into the requirements for open data: (1) *Be Complete:* All data is made available. (Restrictions: valid privacy, security or privilege limitations); (2) *Be Unprocessed:* Data is as collected at the source, with the highest possible level of granularity, not in aggregate or modified forms; (3) *Be Timely:* Data is made available as quickly as necessary to preserve the value of the data; (4) *Be Accessible:* Data is available to the widest range of users for the widest range of purposes; (5) *Be Machine processable:* Data is reasonably structured to allow automated processing; (6) *Be Non-discriminatory:* Data is available to anyone, with no requirement of registration; (7) *Be Non-proprietary:* Data is available in a format over which no entity has exclusive control; (8) *Be License-free:* Data is not subject to any copyright, patent, trademark or trade secret regulation. Reasonable privacy, security and privilege restrictions may be allowed.

A more systematic set of parameters characterizing data can be obtained from an analysis of the cycle of (digital) data. For our purposes, the following four basic processes in that cycle give a good first approximation:

1. *Production:* producing data from the physical world; production of bits (writing, sensors, music, images, etc.)
2. *Access:* possibility of of getting (copying, locally storing) digital data.

---

[2] Usually Open data refers to open "information", understanding information as data apt for direct human consumption. For the discussion in this section, the distinction is not relevant.

**Table 1.** Examples of data openness for images. We show openness allowed by current socio-economic model and in brackets intrinsic openness of the application. (C=closed; O=open; "Any"= both models are possible).

|        | satellite | medical | surveillance | historical | leisure |
|--------|-----------|---------|--------------|------------|---------|
| Use    | Closed [O] | Closed [C] | Closed [C] | Any | Any |
| Re-use | Closed [O] | Closed [C] | Closed [C] | Any | Any |
| Access | Closed [O] | Closed [C] | Closed [C] | Any | Any |
| Prodn. | Closed [C] | Closed [C] | Open [O] | Closed [C] | Any |

3. *Use:* final (terminal) consumption of the data (can be thought of as "returning" the bits to the physical world).
4. *Re-use:* producing data using other data (already produced).

Each of these processes can have restrictions (be closed) or be available to everyone (be open). In Table 1 we show examples of the behavior of these parameters for some types of images. Note that they permit us to discriminate between some basic types of images. Note also how external factors like the socio-economic impact the openness criteria in some cases (e.g. although the satellite images could have open access, use and re-use in a future world, it is unrealistic to imagine that everybody could produce them).

Table 2, on the other hand, indicates current policies on openness for their data of some paradigmatic repositories and applications. Note how databases are closed in all four criteria. For the new "data enterprises", whose essential driving force – and business model– is to get and process data of other people like Google, Yahoo!, Facebook, Twitter, etc.; it is crucial to enforce and open model of production of data while keeping a closed model for access, re-use and use.

These perspectives on data production and consumption necessitate new requirements and pose new challenges to a Web model. The impulse to develop "open" data models has disclosed a number of activities that were, either considered as "given", or did not gain the prominence they have today. Among them: open digital windows to existing data; availability of digital data; linkage of data; building of an infrastructure for data.

Several new requirements emerge at this point: preparing data; cleaning data; pre-processing (for publication) data; logical design of the internationalization (vocabularies, models, etc.); and at the physical dimension, availability, service, formats, etc.

## 3   Modeling Data on the Web

The Web can be viewed from multiple points of view. In this section we examine ideas and abstract conceptualization of the Web. First we review the notion of data model. Then we briefly present the ideas and viewpoints that people have elaborated on regarding the "object" called Web. Then we study the most

**Table 2.** Classical repositories and applications and their current policies on openness criteria for the data they hold. (C=closed; O=open; "Any"= both models are possible.) Web2 application signifies Web applications based on data-intensive processing: search engines, social networks, etc.

|        | library | broadcasting | database | Web2 appl. | Web page |
|--------|---------|--------------|----------|------------|----------|
| Use    | Open    | Closed       | Closed   | Closed     | Open     |
| Re-use | Closed  | Closed       | Closed   | Closed     | Any      |
| Access | Closed  | Closed       | Closed   | Closed     | Open     |
| Prodn. | Open    | Closed       | Closed   | Open       | Open     |

widespread concept of the Web, that is, a collection of documents. Next, we look at models and representations of data beyond documents. Finally, we describe the most comprehensive attempts to model the Web as a whole.

### 3.1 Data Models and their Role

A data model is a set of concepts that can be used to describe the structure and operations of a database for a given domain [57], where database is defined as a collection of data with the following properties:

1. Represent some aspect of the real (or an imagined) world, the "domain" of application.
2. Be logically coherent, i.e., the data has to have some common domain and must have some purpose.
3. Directed at an intended group of users (known in advance), and usually refined to preconceived applications.

The two last conditions define a clear difference of data found in classical data management applications and data on the Web.

### 3.2 The Web as Information Artifact

One could generalize the notions of database given above, by defining a data model (in general) as a set of concepts (a conceptual framework) that describes at an abstract level an information system or artifact. Examples of information systems are libraries, databases, tables, photo albums, etc. The Web also can be viewed as an information artifact, and hence devising models of it is pertinent.

In fact, many researchers have described, characterized, and even modeled, the Web for different purposes. The following list shows the most typical characterizations of it:

1. The Web is an abstract (imaginary) space of information. (Berners-Lee, [13].)
2. The Web is not a database. (Mendelzon, [53].) The Web is a large, heterogeneous, distributed collection of documents connected by hypertext links. (Mendelzon, Mihaila, Milo, 1996 [56].)

3. The Web is one huge database. (Asilomar Report, 1998, [19].)
4. The Web is a vast collection of completely uncontrolled heterogeneous documents. (Brin and Page, 1998, [23].)
5. The Web is a huge heterogenous distributed database. (Konopnicki and Shmueli, 1999 [59].)
6. The Web provides a simple and universal standard for the exchange of information. (Abiteboul, Buneman, Suciu, 2000, [5].)
7. The pages and hyperlinks of the Web may be viewed as nodes and edges in a directed graph. (Kumar et al, 2000, [43].)

### 3.3   The Web of Documents

Perhaps the most clear expression of the most consolidated conception of the Web is the one that the creators of Google, Sergey Brin and Lawrence Page, gave in their well-known paper on Search Engines [23]: "The web is a vast collection of completely uncontrolled heterogeneous documents". Here the Web is defined by contrasting it with the world of "well controlled documents". This contrast nicely parallels the one found in databases between the Web and the world of closed and structured information.

In that paper, Brin and Page identify a core set of challenges to be addressed when dealing with information at Web scale:

1. Documents have extreme internal variation, in language, vocabulary, format, form of generation (human, machine).
2. External meta information. External meta-information was defined as information that can be inferred about a document but is not contained within it.
3. Things that are measured vary by many orders of magnitude.
4. Virtually no control over what people can put on the web; flexibility to publish anything.
5. The contrasting interests between "the enormous influence of search engines" and companies "deliberately manipulating search engines", with the user interests.
6. Metadata efforts have largely failed with search engines.

Let us extract the underlying view and characteristic that this influential design had: *heterogeneity in format and usage (items 1,4); the key idea that relationships between documents (networked data) is of fundamental importance (item 2); the understanding that scalability is a breaking point with the previous world of information management (item 3); and finally, the implicit assumption that search engines are the basic data access tools at Web scale (items 5,6).*

It is worth noting that multimedia and raw data do not play a special role in this model. On the other hand, their solution –which will be the solution implemented by an ample set of successful companies– is a centralized one. The user plays the passive role of consulting information. Brin and Page are essentially addressing the challenges of heterogeneity in content and massive access produced by the new scale. But overall, they are anchored in a Web of documents and centralized services oriented directly to a human user.

## 3.4    Models of Data on the Web

Documents are at the heart of the classical Web. Its original language for specifying data was HTML, that although has facilities to represent data (via tables), has as primarily goal the representation and visualization of documents.

As the Web became popular, the need for better formats to represent more structured data was raised. Such a language had two basic requirements: (1) to be able to represent documents, the most popular information object on the Web (and in daily practice), and (2) to have some level of structuring, so to be able to be queried much like the well known and successful relational technology (SQL).

At the abstract level the answer was the notion of *semistructured data*. The guiding motivations of semi-structured data were the paradigm shift in data management produced by the advent of the Web and the new type of data [5,21]. The main characteristic of this new type of data was that it was neither raw nor strictly typed: its structure is irregular, implicit, partial, with large, evolving and sometimes ignored schemas, and the distinction between schema and data is blurred [2]. Probably the most representative, abstract and minimalist model is OEM [61]. Grahne and Lakshmanan [31] slightly extend the OEM model to better capture the notion of data independence in these models.

As "real world" version of semistructured data emerged XML, that rapidly became a standard for exchanging data (more precisely: documents) on the Web. XML has another important feature: it unifies in one information object the data and the metadata (traditionally split in classical databases). Despite its success, XML is a verbose format not designed to codify raw data.

The data format, JSON, considered by its followers as being a "fat-free alternative to XML", is a lightweight data-interchange format, with the goal of being "easy for humans to read and write and easy for machines to parse and generate". It has become popular to code data at Web scale by its flexibility and minimality. It resembles the OEM model.

In parallel to XML, the Web Consortium developed a standard for representing metadata. It is the RDF universal model of triples for representing data, metadata and knowledge on the Web. Its structure and flexibility to represent any kind of data, and moreover, to *link* (to establish relationships between) different datasets, the feature that has made RDF a prime candidate for representing data on the Web, and a candidate for base data format for the Web of Data.

In summary, we have today a universal syntax, on the lines of a minimal semistructured model (XML, JSON, etc.) plus a model for describing and linking data (RDF).

## 3.5    Data Models of the Web

The philosophy of the formalizations we have seen so far is to develop good data models applicable at Web scale. Something more elaborated is to model the Web itself as a whole. Indeed, jointly with the explosion of the Web of documents, researchers have been trying to model the Web as a huge data system. We would like to call the reader's attention to two of the most interesting such attempts.

*Abiteboul & Vianu's model.*    Abiteboul and Vianu [3,4] presented a model of the Web which is more sophisticated than a graph. They assume that the characteristics of the Web –departing from traditional notions of databases– are its global nature and the loosely structured information it holds.

They model the Web as an infinite set of semistructured objects over the relational schema { *Obj(oid), Ref(source, label, destination), Val(oid, value)* }, where *oid* is an identifier of objects (URIs), *Ref* specifies a finite set of labeled arcs, and *Val* specifies the value of an object. (The reader can see how the triple model emerges again here.) Intuitively objects are Web pages, the value is the content of a page, and references are links. The model –departing from the traditional database notions– enrich the notion of computable query. Considerer the following simple query: list all links that point to my page. This query is not computable because we do not have global information on links. The formalization given is based in a slight generalization of the classical notion of computability according to the new scenario. They introduce the notion of *Web machine*, that essentially is a Turing machine dealing with possibly infinite inputs and outputs. Based on this machine, the notion of query on the Web is formalized.

The model explores only basic aspects of querying and computing on the Web, leaving out, among others: communication costs; the notion of locality; the essentially distributed nature of the Web; the fact that queries on the Web are intrinsically concurrent processes; updates; and the fact that users often seem to be satisfied with incomplete and imprecise answers.

*Mendelzon & Milos's model.*    Almost concurrently with Abiteboul and Vianu's, Mendelzon and Milo [53,54,55] introduced another model, assuming that the Web is not a database (mainly due to the lack of concurrency control and limited data access capabilities). The central difference with Abiteboul and Vianu's model is the infinite character of the Web. The Web is huge, but finite at any given moment, state Mendelzon and Milo. Second, the infinity assumption blurs the distinction between intractable and impossible. For example, the query "List all pages reachable from my page", in an infinite model is not computable, but in Mendelzon and Milo's is in principle computable, although intractable. The formalization is done via a Turing machine with an oracle, which simulates the navigational access from a set of URIs to the Web graph it spans. They present results on computability of queries on the Web, and introduce a Web query language, which is a generalization of the seminal WebSQL query language that integrates data retrieval based in contents, structure and their topology [56].

Mendelzon and Milo's model does not address heterogeneity of data, degrees of autonomy among users, and lack of structure. Also it is restricted to the static case, that is, it ignores updates.

The importance of these works is that they introduce the notion of a data space with peculiarities that are just emerging: the practical impossibility of accessing all data; the intrinsically distributed nature of updating and querying data; heterogeneity of data; etc. This and other open issues reaffirm the need of a model including all or most of these features.

# 4   Requirements for the Web of Data

In this section we will explore several parameters that play relevant roles in the data system underlying the Web. The general philosophical principles of the Web as declared by TBL continue to be the basis of this new artifact. There is no claim of completeness, nor theory behind them. They are presented with the goal of sparking ideas and motivating the identification of relationships between different views.

## 4.1   Architectural Views

The most comprehensive discussion of Web architectural principles is Fielding's thesis [27]. Departing from the classic Web, several ideas have been developed, in different directions making them, strict sensu, incomparable.

In Table 3 we put together three influential such models, just to let the reader grasp similarities and differences.

**Table 3.** A rough comparison of architectural styles. The Web of Data uses the existing background of the Web, and should enhance it to support massive exchange querying of data. The language for logical specification of data and metadata is the RDF model (syntax is not relevant here). The access is semi-automatized via the SPARQL query language.

|  | Classic Web | RESTful Web | Web of Data |
|---|---|---|---|
| Access Tool | Navig/Search Eng. | Web Service | SPARQL, endpoints |
| Language | HTML | XML, JSON | RDF |
| Access Protocol | HTTP | HTTP 1.1 | HTTP "++" |
| Data primitive | URI | URI | URI |

The definitive architecture of the Web of Data is yet to be designed, but should include several facets besides the ones shown in Table 3. In particular, enhancements of the current access/put protocol for data on the Web.

## 4.2   Static versus Dynamics

The classic models assume that the Web is an essentially static object. The models of Abiteboul & Vianu and Mendelzon & Milo speak of a non-dynamic Web. Also the models of the Web as a graph share the same implicit assumption. By dynamics we mean not only the addition of new data, or deletion of old data, but also modifications of it. It is useful to exemplify this difference: a library is static, in the sense that it incorporates books, disposes books, but does not create nor change them. The same happens today with image collections. On the contrary, a database management system is essentially a dataset that is constantly being modified by applications. Its essence is the volatile data that

**Table 4.** Dynamics versus openness. A rough classification of some information artifacts and projects based on these dimensions. Where should projects like Linked Data
and Open Data be classified?

|  | Static | Dynamic |
|---|---|---|
| open world | Data Govs | Classic Web |
|  |  | Web of Data |
| closed world | Libraries | Dataspaces |
|  | Archives | Databases |
|  |  | Desktops |

is created and modified constantly. Table 4 shows a classification of information
artifacts when crossing the dynamics and the openness parameters.

A closely related issue, *transactionality*, is a notion inseparable from classical
data management and its dynamics. Gray defines a transaction as "a transformation of state which has the properties of atomicity (all or nothing), durability
(effects survive failures) and consistency (a correct transformation). The transaction concept is key to the structuring of data management applications." [29]
Does this notion make sense at Web scale? Is it consistent with Web principles?

### 4.3 Data Access Methods

For the common user, access to data on the Web is accomplished either by
navigation or by filling in forms. These methods do not scale. For Web volumes
of data, semi-automatic and automatic methods are necessary. Figure 5 shows
the menu of most common access methods available today.

**Table 5.** The most popular current methods to access data on the Web. The Web of
Data currently points to structured and automatic retrieval of data.

|  | human | semi-automatic | automatic |
|---|---|---|---|
| non-structured | Navigation | Search engine | Statistical techniques |
| structured | Forms | Query language | API, Web serv., Endpoints |

As for query and transformation languages (prime methods when working
with massive data), Table 6 shows the most "popular" access languages dealing
with data on the Web.

### 4.4 Cost Models

Any model for the Web of Data has to include a corresponding cost model for
accessing and exchanging data, and even for data itself [48]. There is need for
common *cost models* to evaluate information on the Web [26]. Many of the models
proposed include cost models for accessing and exchanging data; nevertheless

**Table 6.** Most popular semi-automatic data access approaches

| | Keyword | SQL | XQuery | SPARQL |
|---|---|---|---|---|
| application | text | spreadsheets | documents | statements |
| abstract data | strings | tables | Trees | Graphs |
| data format | nat. lang. | SQL table | XML | RDF |
| technique | statistics | algebra/logic | autom./logic | algebra/patterns |

there is yet not a common approach to compare them. The need to explore and incorporate ideas from other areas (e.g. classical cost models, economic ones, response time models, communication complexity models, etc.) As large data companies have advanced in this area (see e.g. [52]), it is now important to devise models for the open world.

### 4.5 Incomplete and Partial Information

The ability to deal with incomplete or partial information is one of the basic requirements for a model of the Web of Data.

There are several database developments that partially address this issue. The most natural one is the theory of incomplete information. A whole area of research has spun off from this subject since the seminal work by Lipski [49]. This research has been partially subsumed in the area of *Probabilistic Databases* [32,63]. Theories of incomplete information deal with unknown and uncertain information, whereas probabilistic databases can be considered as numerical quantification of the uncertainty. The models mainly follow the ideas of the possible-worlds approach. Even though the Web has other facets that escape these models, they are valuable starting points.

The theories above deal essentially with the problem of how to code partial information and query it. A different perspective presents itself when trying to model the user of the Web, who can only get partial information from the network of data that constitutes the Web. This approach overlaps with the problem of the behavior of agents having bounded capacity and bounded information. Theories like *bounded rationality* [58] are worth exploring here.

### 4.6 Organizing Data

As we learned, the RDF graph model is a good candidate for a universal format for representing data and their relationships on the Web. Applications like social networks, or projects like Linked data, are increasingly showing success in this task. With these we are seeing the distributed construction of a huge network of data. Although it is possible to find partial classifications of such data, they resemble that of the imperial encyclopedia imagined by Borges where the animals were divided into categories as follows: (a) belonging to the emperor, (b) embalmed, (c) tame, (d) sucking pigs, (e) sirens, (f) fabulous, (g) stray dogs,

(h) included in the present classification, (i) frenzied, (j) innumerable, (k) drawn with a very fine camelhair brush, (l) etcetera, (m) having just broken the water pitcher, (n) that from a long way off look like flies.[3]

A natural question arises: does the task of organizing the network of data on the Web make sense? Note that the task is not impossible in principle and that librarians succeeded in organizing data coded with human languages. The experience of tagging and folksonomies is valuable, but sheds little light on the problem of organization of structured data at Web scale. Most of the challenges in this regard are still open, even at small scale (cf. the experience of graph data models [9]).

There is more. The challenges posed by the growing amount of data known as *multimedia* (images, videos, scans, etc.) is something that any model of the Web of Data should address. Until today they have been treated as collections of black boxes whose descriptions are done by tagging, with little and poor additional metadata, and no relationships among their "contents". Although this is not the place to discuss this topic in depth, it is important to call attention to the crucial role it will play in the Web of Data.

## 5    Other Relevant Related Areas

A discussion of models for the Web of Data would not be complete without mentioning other areas of research which are closely related to this goal. In this section we briefly address the most relevant of them.

### 5.1    Distributed Data Management

A *distributed database* is one that has a central control, but whose storage devices are not all attached to a common central server, that is, they are stored in multiple computers, in the same physical location or over a network of computers.

The notion of distribution (of tasks, of people, of data, etc.) is intrinsic to the Web, hence there being several characteristics from this model that are common to Web phenomena. The commonalities among distributed databases, P2P systems and the Web can be established as shown in Table 7 [16]. Without doubt, the P2P approach is the most interesting and fruitful source of ideas in this regard.

*Peer to Peer.* P2P systems became popular with Napster, Gnutella and Bit Torrent. The model of P2P has two characteristics which made it one of the closest in spirit with the Web principles: (1) The sharing of computer resources by direct exchange, rather than requiring the intermediation of a centralized server; and (2) The ability to treat instability and variable connectivity as the norm, automatically adapting to failures in both network connections of computers, as well as to a transient population of nodes [8].

---

[3] J. L. Borges, *The Analytical Language of John Wilkins*, Translation of Lilia Graciela Vázquez.

**Table 7.** A rough classification of some data systems according to their distributed nature and the intrinsic quality of their services (cf. Bernstein et al. [16])

|  | poor services | good services |
|---|---|---|
| no central control | Web | Peer to Peer |
| central control |  | Distributed DB |

Gribble et al. [33] enumerate the following principles as general characteristics of the P2P model:

1. No client/service necessary: each peer is a provider or a consumer. Everybody more or less has the same role, with the same duties and rights.
2. No central control. In particular each agent decides to enter/be part of or leave/abandon the network at his/her convenience.
3. Exchange of large, opaque and atomic objects, whose content is well described by their name. Large-granularity requests for objects by identifier.

Valdurriez and Pacitti [70], studying data management in large-scale P2P systems, indicate the main requirements for such systems:

1. *Autonomy.* Peers should be able to join or leave the system at any time, and control the data it stores.
2. *Query Expressiveness.* Allow users to describe data at the appropriate level of detail.
3. *Efficiency.* Efficient use of system resources: bandwidth, computer power, storage.
4. *Quality of Service.* Completeness of query results, data consistency, data availability, query response time, etc.
5. *Fault-tolerance.* Efficiency and quality of services should be provided despite the occurrence of peer's failures. Given the nature of peers, the only solution seems to rely on data replication.
6. *Security.* The main issue is access control (including enforcing intellectual property rights of data contents).

*Wide Distributed Systems.* The project Mariposa [67] is an influential proposal for developing architectures for distributed systems at large, that is, working over wide networks. The main goal is to overcome the main underlying assumptions on the area, that in their opinion, do not apply to wide-area networks (and less to the Web): Static data allocation; Single administrative structure; and Uniformity. The guiding principles of the new design are the following: Scalability to a large number of cooperating sites; Data mobility; No global synchronization; Total local autonomy; and Easily configurable policies.

*Dataspaces. Dataspaces* [36] is another abstraction for information management that attempts to address the "data everywhere" problem. It focuses on supporting basic functionalities of data management, such a keyword searching for loosely integrated data sources and relational-style querying for more integrated ones. Currently the project has not included the publishing-of-data agenda.

## 5.2  Logic Approaches

Under the logical framework there have been some works that model aspects of the Web. Let us show a few examples just to give a flavor of the possibilities and scope of this approach.

Himmeröder et al. [47] propose to use F-logic to model knowledge on the Web, particularly Web queries. The model, though, is just a graph of documents with arcs representing hyperlinks, where they concentrate on a language to explore the Web. From another perspective, Terzi et al. [68] present a constraint-based logic approach to modeling Web data, introducing order and path constraints, and proposing a declarative language over this model. The basic assumptions, though, are the same as those of the semi-structured model.

More recently, Datalog, the classic logic query language, has been the object of attention by the people working in distributed systems, and suggested as model for the Web. We would like to call the attention to two interesting ongoing projects in this direction. One is being developed by Joseph Hellerstein and his group [40], and centers on "data-centric computation" motivated by the urgency of parallelism at micro and macro scale. They develop an extension of Datalog that, relying on the time parameter, addresses the fundamental issues of distribution. The other project is headed by Serge Abiteboul [71], which also uses Datalog to specify the problems of distribution. As stated in their project, "the goal is to develop a universally accepted formal framework for describing complex and flexible interacting Web applications featuring notably data exchange, sharing, integration, querying and updating".

## 6    Concluding Remarks

The massive production and availability of data at world scale due to the technological advances of sensors, communication devices and processing capabilities, is a phenomena that is challenging the classic views on data management.

The Web has become the premium infrastructure to support such data deluge. Designed originally as a worldwide interrelated collection of documents, and oriented primarily to direct human visualization, today the Web is rapidly incorporating the data dimension and evolving towards automatic handling of such volumes of data.

The challenges for computer scientists are immense, as long as the new scenario involves data management, knowledge management, information systems, Web protocols, user interfaces, Web engineering, and several other disciplines and techniques. To have a unified and consistent view of the data dimension at Web scale one necessarily must have a model of such Web of Data. All indications point to the fact that such a model should follow the original Web principles of decentralization, distribution and collaborative development, and depart from small-scale and closed views of data and knowledge management that have been deployed until today.

In these notes we tried to present an introductory overview of the themes and techniques arising in such program for the development of a Web of Data.

**Acknowledgments.** Materials of this lecture have been taught to students at Universidad de Chile, Chile; Universidad de la República, Uruguay; Biblioteca del Congreso, Chile; to whom I thank for feedback and suggestions. Also thanks to R. Angles, J. Fernández, D. Hernández, J. E. Muñoz plus anonymous referees that helped with detailed comments to improve previous versions. Of course, the responsibility for what is finally said here is mine.

# References

1. Abiteboul, S., Quass, D., McHugh, J., Widom, J., Wiener, J.L.: The Lorel query language for semistructured data. International Journal on Digital Libraries 1(1) (1997)
2. Abiteboul, S.: Querying semi-structured data. In: Afrati, F.N., Kolaitis, P.G. (eds.) ICDT 1997. LNCS, vol. 1186, Springer, Heidelberg (1996)
3. Abiteboul, S., Vianu, V.: Queries and Computation on the Web. In: Afrati, F.N., Kolaitis, P.G. (eds.) ICDT 1997. LNCS, vol. 1186, Springer, Heidelberg (1996)
4. Abiteboul, S., Vianu, V.: Queries and Computation on the Web. Theor. Comput. Sci. 239(2) (2000)
5. Abiteboul, S., Buneman, P., Suciu, D.: Data on the Web. From Relations to Semistructured Data and XML. Morgan Kaufmann, San Francisco (2000)
6. Agrawal, R., et al.: The Claremont Report on Database Research (2008), http://db.cs.berkeley.edu/claremont/
7. Alex Sung, L.G., Ahmed, N., Blanco, R., Li, H., Ali Soliman, M., Hadaller, D.: A Survey of Data Management in Peer-to-Peer Systems. In: Web Data Management (2005)
8. Androutsellis-theotokis, S., Spinelis, D.: A Survey of Peer-to-Peer Content Distribution Technologies. ACM Surveys 36(4) (December 2004)
9. Angles, R., Gutierrez, C.: Survey of Graph Database Models. ACM Computing Surveys 40(1) (2008)
10. Arenas, M., Gutierrez, C., Perez, J.: Foundations of RDF Databases (Tutorial). Reasoning Web Summer School (2009)
11. Berners-Lee, T.: WWW: Past, present, and future. IEEE Computer 29(10) (October 1996)
12. Berners-Lee, T.: Commemorative Lecture The World Wide Web - Past Present and Future. Exploring Universality. Japan Prize Commemorative Lecture (2002), http://www.w3.org/2002/04/Japan/Lecture.html
13. Berners-Lee, T.: Frequently asked questions, http://www.w3.org/People/Berners-Lee/FAQ.html
14. Berners-Lee, T.: Design Issues/Linked Data, http://www.w3.org/DesignIssues/LinkedData.html
15. Berners-Lee, T.: Linked Open Data. What is the idea?, http://www.thenationaldialogue.org/ideas/linked-open-data

16. Bernstein, P.A., Giunchiglia, F., Kementsietsidis, A., Mylopoulos, J., Serafini, L., Zaihrayeu, I.: Data Management for Peer-to-Peer Computing: A Vision. In: WebDB, Workshop on Databases and the Web (2002)
17. Bell, G., Gray, J., Szalay, A.: Petascale Computational Systems: Balanced Cyber-Infrastructure in a Data-Centric World. Computer 39(1) (January 2006)
18. Bell, G., Hey, T., Szalay, A.: Beyond the Data Deluge. Science 323 (March 2009)
19. Ph. Bernstein, M., Brodie, S., Ceri, D., DeWitt, M., Franklin, H., Garcia-Molina, J., Gray, J., Held, J., Hellerstein, H.V., Jagadish, M., Lesk, D., Maier, J., Naughton, H., Pirahesh, M., Stonebraker, J.: The Asilomar report on database research. ACM SIGMOD Record 27(4) (December 1998)
20. Ch. Bizer, T., Heath, T.: Linked Data - The Story So Far. International Journal on Semantic Web and Information Systems 3 (2009)
21. Buneman, P.: Semistructured data. In: ACM PODS (1997)
22. Bray, T., Paoli, J., Sperberg-McQueen, C.M.: Extensible Markup Language (XML) 1.0, W3C Recommendation 10 (February 1998), http://www.w3.org/TR/1998/REC-xml-19980210
23. Brin, S., Page, L.: The anatomy of a large-scale hypertextual Web search engine. In: Computer Networks and ISDN Systems (1998)
24. Cai, M., Frank, M.: RDFPeers: a scalable distributed RDF repository based on a structured peer-to-peer network. In: Proc. WWW 2004 (2004)
25. DATA.gov project, http://www.data.gov/
26. Erling, O., Mikhailov, I.: Towards Web Scale RDF.In: 4th International Workshop on Scalable Semantic Web Knowledge Base Systems, SSWS 2008 (2008)
27. Fielding, R.T.: Architectural Styles and the Design of Network-based Software Architectures. Doctoral dissertation, University of California, Irvine (2000), http://www.ics.uci.edu/~fielding/pubs/dissertation/top.htm
28. Fielding, R.T., Taylor, R.N.: Principled design of the modern Web architecture. ACM Trans. Internet Technol. 2(2) (May 2002)
29. Gray, J.: The Transaction Concept, Virtues And Limitations. In: Proceedings of 7th VLDB, Cannes, France (1981)
30. Szalay, A., Gray, J.: Science in an Exponential World. Nature 440 (March 2006)
31. Grahne, G., Lakshmanan, L.V.S.: On the difference between navigating semistructured data and querying it. In: Connor, R.C.H., Mendelzon, A.O. (eds.) DBPL 1999. LNCS, vol. 1949, p. 271. Springer, Heidelberg (2000)
32. Green, T., Tannen, V.: Models for Incomplete and Probabilistic Information. In: EDBT Workshops, Munich, Germany (March 2006)
33. Gribble, S., Halevy, A., Ives, Z., Rodrig, M., Suciu, D.: What Can Databases Do for Peer-to-Peer? In: WebDB, Workshop on Databases and the Web (2001)
34. Guan, T., Saxton, L.: A complexity model for web queries. In: Fundamentals of Information Systems. ch. 1. Kluwer, Dordrecht (1999)
35. Muoz, S., Prez, J., Gutierrez, C.: Simple and Efficient Minimal RDFS. J. Web Sem. 7(3) (2009)
36. Halevy, A.Y., Franklin, M.J., Maier, D.: Principles of dataspace systems. In: PODS (2006)
37. Hartig, O., Bizer, C., Freytag, J.-C.: Executing SPARQL queries over the web of linked data. In: Bernstein, A., Karger, D.R., Heath, T., Feigenbaum, L., Maynard, D., Motta, E., Thirunarayan, K. (eds.) ISWC 2009. LNCS, vol. 5823, pp. 293–309. Springer, Heidelberg (2009)
38. Hausenblas, M., Karnstedt, M.: Understanding Linked Open Data as a Web-Scale Database. In: 1st Internat. Conf. on Advances in Databases (2010)

39. Heath, T., Bizer, C.: Linked Data: Evolving the Web into a Global Data Space. In: Synthesis Lectures on the Semantic Web: Theory and Technology. Morgan & Claypool, San Francisco (2011), http://linkeddatabook.com/editions/1.0/
40. Hellerstein, J.M.: The Declarative Imperative. Experiences and Conjectures in Distributed Logic. SIGMOD Record 39(1) (March 2010)
41. Hoem, J.: Openness in Communicaton. First Monday 11(7) (July 3, 2006), http://firstmonday.org/htbin/cgiwrap/bin/ojs/index.php/fm/article/viewArticle/1367/1286
42. Klyne, G., Carroll, J.: Resource Description Framework (RDF) Concepts and Abstract Syntax. W3C Recommendation (2004), http://www.w3.org/TR/2004/REC-rdf-concepts-20040210/
43. Kumar, R., Raghavan, P., Rajagopalan, S., Sivakumar, D., Tompkins, A., Upfal, E.: The Web as a Graph. In: Proc. PODS 2000 (2000)
44. LinkedData Project, http://www.linkeddata.org
45. Workshops and academics events on Linked Data, http://linkeddata.org/calls-for-papers
46. Loukides, M.: What is data science?, http://radar.oreilly.com/2010/06/what-is-data-science.html
47. Himmeröder, R., Lausen, G., Ludäscher, B., Schlepphorst, C.: On a Declarative Semantics for Web Queries. In: Bry, F. (ed.) DOOD 1997. LNCS, vol. 1341, Springer, Heidelberg (1997)
48. Lesk, M.: Encouraging Scientific Data Use. Posted on The Fourth Paradigm on (February 7, 2011), http://blogs.nature.com/fourthparadigm/2011/02/07/encouraging-scientific-data-use-michael
49. Lipski Jr., W.: On Databases with incomplete information. Journal of the ACM (JACM) JACM Homepage archive 28(1) (January 1981)
50. McGuinness, D.L., van Harmelen, F.: OWL Web Ontology Language Overview. W3C Recommendation 10 February (2004), http://www.w3.org/TR/owl-features/
51. Arenas, M., Consens, M., Mallea, A.: Revisiting Blank Nodes in RDF to Avoid the Semantic Mismatch with SPARQL. In: W3C Workshop: RDF Next Steps, Palo Alto, CA (2010)
52. Madhavan, J., Jeffery, S.R., Cohen, S., Dong, X., Ko, D., Yu, C., Halevy, A.: Web-scale Data Integration: You Can Only Afford to Pay As You Go. In: Proc. of Third Conference on Innovative Data System Research, CIDR 2007 (2007)
53. Mendelzon, A.O.: The Web is not a Database. In: Workshop on Web Information and Data Management (1998)
54. Mendelzon, A.O., Milo, T.: Formal Models of Web Queries.In: Proc. PODS (1997)
55. Mendelzon, A.O., Milo, T.: Formal Models of Web Queries. Inf. Syst. 23(8) (1998)
56. Mendelzon, A.O., Mihaila, G.A., Milo, T.: Querying the World Wide Web. Intl. Journal of Digit. Libr. 1 (1997)
57. Navathe, S.B.: Evolution of data modeling for databases. Communications of the ACM 35(9) (September 1992)
58. Rubinstein, A.: Modeling Bounded Rationality. MIT Press, Cambridge (1998)
59. Brin, S.: Extracting patterns and relations from the world wide web. In: Atzeni, P., Mendelzon, A.O., Mecca, G. (eds.) WebDB 1998. LNCS, vol. 1590, pp. 172–183. Springer, Heidelberg (1999)
60. No SQL, http://nosql-database.org/
61. Papakonstantinou, Y., Garcia-Molina, H., Widom, J.: Object exchange across heterogeneous information sources. In: 11th International Conference on Data Engineering, ICDE (1995)

62. Seminar on Open Government data (Open Government Working Group), December 7-8 (2007), http://resource.org/8_principles.html
63. Suciu, D.: Probabilistic Databases, Database Theory Column. SIGMOD Record (2008)
64. Spielmann, M., Tyszkiewicz, J., Van den Bussche, J.: Distributed computation of web queries using automata. In: Proc. PODS (2002)
65. Stonebraker, M., Madden, S., Abadi, D.J., Harizopoulos, S., Hachem, N., Helland, P.: The end of an architectural era (it's time for a complete rewrite). In: Proc. VLDB 2007 (2007)
66. O'Reilly, T.: What Is Web 2.0,
    http://oreilly.com/web2/archive/what-is-web-20.html
67. Stonebraker, M., Aoki, P.M., Litwin, W., Pfeffer, A., Sah, A., Sidell, J., Staelin, C., Yu, A.: Mariposa: a wide-area distributed database system. The VLDB Journal 5(1) (January 1996)
68. Terzi, E., Hacid, M.-S., Vakali, A., Hacid, S.: Modeling and Querying Web Data: A Constraint-Based Logic Approach. Information Modeling for Internet Applications book Contents (2003)
69. Horng, J.T., Tai, Y.Y.: Pattern-based approach to structural queries on the World Wide Web. Proc. Natl. Sci, Counc. ROC(A) 24(1) (2000)
70. Valduriez, P., Pacitti, E.: Data management in large-scale P2P systems. In: Daydé, M., Dongarra, J., Hernández, V., Palma, J.M.L.M. (eds.) VECPAR 2004. LNCS, vol. 3402, pp. 104–118. Springer, Heidelberg (2005)
71. Webdam Project. Foundations of Web Data Management,
    http://webdam.inria.fr

# Trust Management Methodologies for the Web

Denis Trček

Faculty of Computer and Information Science, University of Ljubljana,
Tržaška c. 25, 1000 Ljubljana, Slovenia, EU
denis.trcek@fri.uni-lj.si

**Abstract.** Trust and its support with appropriate trust management methodologies and technologies is becoming one crucial element for wider acceptance of web services. In the computing society trust and related issues were addressed already in the nineties of the former century, but the approaches from that period were about security, more precisely security services and security mechanisms. These approaches were followed by more advanced ones, where the first branch was based on Bayesian statistics, the second branch was based on Dempster-Shafer theory of evidence and its successors, most notably subjective logic, and the third branch originated from game theory. It is, however, important to note that at the core of trust there are cognition, assessment processes, and they are governed by various factors. Consequently, trust management methodologies should take these factors, which may ne rational, irrational, contextual, etc., into account. This research contribution will therefore provide an extensive overview of existing methodologies in the computer sciences field, followed by their evaluation in terms of their advantages and disadvantages. Further, some latest experimental results will be given that identify and evaluate some of those most important factors mentioned above. Finally, we will present a new trust management methodology called Qualitative Assessment Dynamics, QAD (aka Qualitative Algebra) that complements existing methodologies mentioned above, and that is aligned with the results of the latest experimental findings.

**Keywords:** web technologies, ergonomic methodologies, trust, trust management, qualitative assessment dynamics, and simulation.

## 1 Introduction

Before focusing on current situation in the area of computerized trust management research, it is very instructive to have a look at trust through the main epochs of development of distributed computing (web computing is certainly a kind of distributed computing).

In the mid-nineties e-business emerged and it changed the landscape of business processes significantly. In this context trusting distributed computing was (and still is) mainly about security of businesses (prevention of financial loss).

Nowadays, the Internet (as the most prominent implementation of distributed computing paradigm) is mainly entering our private domain, so trusting distributed web computing is largely related to users' personal integrity and privacy.

A. Polleres et al. (Eds.): Reasoning Web 2011, LNCS 6848, pp. 445–459, 2011.

In the near future, web computing will intensively integrate sensor networks with the Internet, and trusting such sensor networks extended internet will add mainly questions of safety.

Now focusing on trust, the following early methodologies and solutions should be mentioned that are in fact about security (security services) and not about the core of trust. Certainly, trust can be influenced by security, but there exist important distinctions between these two terms. So the early trust related solutions are the following ones:

- Trusted Computer System Evaluation Criteria, known as the Orange Book - this standard was published by the US Department of Defense in 1985. Although it was originally intended for military systems, it became accepted for security classifications in the computer industry. And, as stated, although it was said to be about trusted computer systems, it was actually about their security.
- Platform for Internet Content Selection (PICS), which was another standard that was about access control, more precisely, web-sites filtering [1].
- PolicyMaker, which was a solution aimed at addressing trust management problems in distributed services environments. By deploying digital certificates, PolicyMaker bounded access rights to the owner of a public key. In turn, owner's identity was tied to this key by means of a certificate. Clearly, this was a PKI based trust enabling implementation [2].
- Trust Establishment Module, which was based on a dedicated language and implemented in Java. It was similar to PolicyMaker and enabled trusting relationships between unknown entities by deploying public key certificates (so this was another PKI deploying implementation) [3].

Many other early approaches are described in detail in a survey by Grandison and Sloman [4], and the reader is referred to it for additional details.

As to more contemporary and current methodologies that are used for trust management in information systems (ISs), and web environments in general, the following solutions should be considered. The simplest ones, like eBay's system, should be mentioned first. eBay's (reputation) system sums positive scores about an entity as well as negative ones, and the difference of these two results presents reputation of a particular entity. Similar approach, but a slightly more sophisticated one, and also deployed by Amazon, uses averaging, so the final score is the average of all ratings.

These two basic trust management related approaches are now being upgraded by various new, more sophisticated methodologies. This is not to say that they useless – on the contrary. They certainly have their merit, but they are essentially reputation systems. This is an important distinction, and the reasons will become clear through the presentation of current trust management methodologies in the next section.

## 2  Some Most Important Methodologies

One of the basic research streams for trust management in web environments is based on Bayesian statistics. This stream starts with the Bayes theorem, which states that the posterior probability of a hypothesis $H$ after observing datum $D$ is given by

$$p(H|D) = p(D|H) * \frac{p(H)}{p(D)}, \tag{1}$$

where $p(H)$ is the prior probability of $H$ before $D$ is observed, $p(D|H)$ is the probability that $D$ will be observed when $H$ is true, and $p(D)$ is the unconditional probability of $D$. Similarly, for more data, Bayes theorem can be extended (and step by step generalized) as follows:

$$p\big(H\big|(D_1, D_2)\big) = \frac{p(H, D_1, D_2)}{p(D_1, D_2)} = \cdots = p\big(D_1\big|(H, D_2)\big) * \frac{p(H|D_2)}{p(D_1|D_2)}. \qquad (2)$$

## 2.1 Naïve Trust Management

Being the basis, Bayes theorem has served for so called naïve trust management implementations [5]. This methodology goes as follows. Suppose one is concerned with trust in competence in file providers on the web, where competences include files types, files quality, and files download speed. This problem can be represented by a Bayesian network (see Fig. 1).

With this approach every agent develops a Bayesian network for each file provider that it has interacted with. Each leaf under the root presents provider's capability in certain aspect through associated conditional probability.

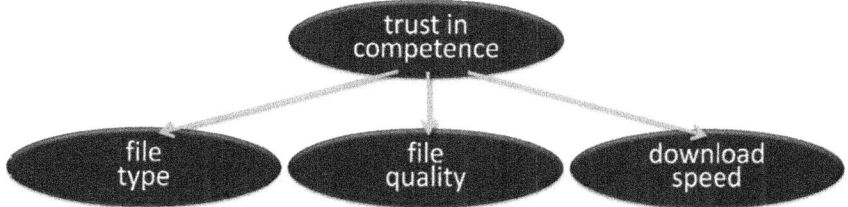

**Fig. 1.** Bayesian network for trust calculations related to file provider's competences

The root of the network is assigned 1 for "satisfying", and 0 for "unsatisfying" interaction. Thus $p(T=1) = m/n$ is the percentage of satisfying, and $p(T=0) = (n-m)/n$ the percentage of unsatisfying interactions, where $m$ stands for number of satisfying, and $n$ for number of all interactions. The leaf nodes represent various aspects of file provider capability, so each leaf node has associated conditional probability table:

**Table 1.** The conditional probability table for a certain file provider

|  | T = 1 | T = 0 |
|---|---|---|
| music | $p(FT = \text{music} \mid T = 1)$ | $p(FT = \text{music} \mid T = 0)$ |
| movie | $p(FT = \text{movie} \mid T = 1)$ | $p(FT = \text{movie} \mid T = 0)$ |
| document | $p(FT = \text{document} \mid T = 1)$ | $p(FT = \text{document} \mid T = 0)$ |
| image | $p(FT = \text{image} \mid T = 1)$ | $p(FT = \text{image} \mid T = 0)$ |
| software | $p(FT = \text{software} \mid T = 1)$ | $p(FT = \text{software} \mid T = 0)$ |

In the above table $p(FT = music \mid T = 1)$ means the probability that the involved interaction is exchange of a music file given the interaction is satisfying. According to definition, it can be obtained as $p(FT = music, T = 1)/p(T = 1)$. In this equation, $p(FT = music, T = 1)$ is the probability that interactions are satisfying and that files involved are music files, while $p(T = 1)$ is the probability of satisfying interaction. $p(FT = music \mid T = 1)$ is computed as the number of satisfying interactions $m_1$ when files involved are music files, divided by the total number of interactions, i.e. $m_1/n$. Similarly, values are computed for file quality ($FQ$), where this quality can be "high", "medium" or "low", and for download speed ($DS$), which can be "fast", "medium", or "slow". This way conditional probability tables for $DS$ and $FQ$ are obtained.

Having these conditional probability values for nodes in Bayesian networks, an agent can calculate the probabilities about trustworthiness of a certain file provider in various aspects by using Bayes rules. For example, an agent can obtain probability that the file provider is trustworthy in providing music, $p(T = 1 \mid FT = music)$, or the file provider is trustworthy in providing music files with high quality, $p(T = 1 \mid FT = music, FQ = high)$. Agents update their corresponding Bayesian nets after each interaction and in case of satisfaction $m$ and $n$ are increased, otherwise only $n$ is increased.

## 2.2  Theory of Evidence and Josang's Logic / Algebra

Generalization of Bayes theorem leads to the Dempster – Shaffer Theory of evidence, or ToE [6]. Its starting point is a set of possible (atomic) states, called a *frame of discernment* $\Theta$. Within $\Theta$, exactly one state is assumed to be true at any time.

Based on $\Theta$, *basic probability assignment*, or *BPA* (also called *belief mass*) function is defined as

$$m: 2^{\Theta} \rightarrow [0,1], \tag{3}$$

where $m\{ \ \} = 0$, and $\sum_{A \subseteq \Theta} m(A) = 1$. A belief mass $m_{\Theta}(X)$ expresses the belief assigned to the set $X$ as a whole, and does not express any belief in subsets of $X$. Now for a subset $A \subseteq \Theta$, the belief function $bel(A)$ is defined as the sum of the beliefs committed to the possibilities in $A$.

To illustrate ToE in a simple trust related scenario, let the frame of discernment be given by $\Theta = \{T, \neg T\}$, where "$T$" *means* that the target is trustworthy, while "$\neg T$" means that the target is untrustworthy. A *basic probability assignment* for the above $\Theta$ has to be such that $m(\{T\}) + m(\{\neg T\}) + m(\{T, \neg T\}) = 1$. Now for a subset $A = \{T, \neg T\} \subseteq \Theta$, the belief function $bel(A)$ presents the sum of the beliefs committed to the possibilities in $A$: $bel(A) = m(\{T\}) + m(\{\neg T\}) + m(\{T, \neg T\})$. For example, let $m(\{T\}) = 0.7$, $m(\{\neg T\}) = 0$, and $m(\{T, \neg T\}) = 0.3$. Now if $A=\{T\}$ then $bel(\{T\}) = m(\{T\}) = 0.7$, and if $A=\{\neg T\}$ then $bel(\{\neg T\}) = m(\{\neg T\}) = 0$.

ToE serves as a basis for subjective logic and algebra, developed by Jøsang that is also often used in computational trust management solutions [7]. This algebra introduces many new operators for modeling trust like consensus and recommendation. Trust $\omega$ is represented by a triplet $(b, d, u)$, where $b$ stands for belief (belief function in ToE), $d$ for disbelief and $u$ for uncertainty, and where values $b, d, u$ are obtained from the closed interval [0, 1] as follows:

$$b(x) = \sum_{y \subseteq x} m(y), \qquad d(x) = \sum_{x \cap y = \emptyset} m(y),$$

$$u(x) = 1 - \big(b(x) + d(x)\big), \qquad x, y \in 2^{\Theta}. \tag{4}$$

Belief in a state has to be interpreted as an observer's total belief that a particular state is true. Similarly, an observer's disbelief has to be interpreted as the total belief that a state is not true. Let $\Theta = \{x_1, x_2, x_3, x_4\}$ be a frame of discernment, and let the part of power set of $\Theta$ with certain assigned values $m$ be as given in Fig. 2 (this means that all other subsets that are not presented in Fig. 2 are assigned $m = 0$):

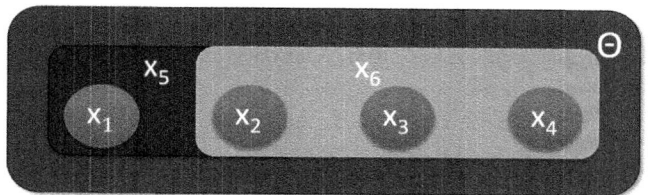

**Fig. 2.** An example scenario for derivation of $b$, $d$, and $u$

Belief in, for example, $x_5$ in Fig.2, i.e. $b(x_5)$, is the sum of belief masses assigned to $x_1$ and $x_2$. Disbelief in state $x_5$ is the sum of the belief masses on the states $x_3$ and $x_4$ (i.e. all those that have an empty intersection with $x_5$). Finally, the uncertainty about $x_5$, i.e. $u(x_5)$, is the sum of belief masses on set $x_6$ and on set $\Theta$.

As mentioned, one main contribution of subjective algebra are various trust modeling operators that preserve sound mathematical basis of ToE. Examples for conjunction, recommendation and consensus are defined as follows.

**Definition 1.** Let $\omega_p^A = \{b_p^A, d_p^A, u_p^A\}$ and $\omega_q^A = \{b_q^A, d_q^A, u_q^A\}$ be agent A's opinion about two distinct binary statements $p$ and $q$. Then the conjunction of and representing A's opinion about both $p$ and $q$ being true is defined by

$$\omega_{p \wedge q}^A = \omega_p^A \wedge \omega_q^A = \{b_{p \wedge q}^A, d_{p \wedge q}^A, u_{p \wedge q}^A\}, \tag{5}$$

where

$$b_{p \wedge q}^A = b_p^A b_q^A,$$
$$d_{p \wedge q}^A = d_p^A + d_q^A - d_p^A d_q^A, \text{ and}$$
$$u_{p \wedge q}^A = b_p^A u_q^A + u_p^A b_q^A + u_p^A u_q^A.$$

**Definition 2.** Let A and B be two agents where $\omega_B^A = \{b_B^A, d_B^A, u_B^A\}$ is A's opinion about B's recommendations, and let $p$ be a binary statement where $\omega_p^B = \{b_p^B, d_p^B, u_p^B\}$ is B's opinion about $p$ expressed in a recommendation to A. Then A's opinion about $p$ as the result of the recommendation from B is defined by

$$\omega_p^{AB} = \omega_B^A \otimes \omega_p^B = \{b_p^{AB}, d_p^{AB}, u_p^{AB}\}, \tag{6}$$

where

$$b_p^{AB} = b_B^A b_p^B,$$

$$d_p^{AB} = d_B^A + d_p^B, \text{ and}$$
$$u_p^{AB} = d_B^A + u_B^A + b_B^A u_p^B.$$

**Definition 3.** *Let* $\omega_p^A = \{b_p^A, d_p^A, u_p^A\}$ *and* $\omega_p^B = \{b_p^B, d_p^B, u_p^B\}$ *be opinions held by agents A and B about a binary statement p. Then the consensus opinion is defined by*

$$\omega_p^{AB} = \omega_B^A \oplus \omega_p^B = \{b_p^{AB}, d_p^{AB}, u_p^{AB}\}, \tag{7}$$

*where*

$$b_p^{AB} = (b_p^A u_p^B + b_p^B u_p^A)/(u_p^A + u_p^B - u_p^A u_p^B),$$
$$d_p^{AB} = (d_p^A u_p^B + d_p^B u_p^A)/(u_p^A + u_p^B - u_p^A u_p^B), \text{ and}$$
$$b_p^{AB} = (u_p^A u_p^B)/(u_p^A + u_p^B - u_p^A u_p^B).$$

An application case of subjective algebra follows. Assume an authentication scenario where agents $B$ and $C$ pass the recommendation received from $D$ about $E$ to $A$ (see Fig. 3).

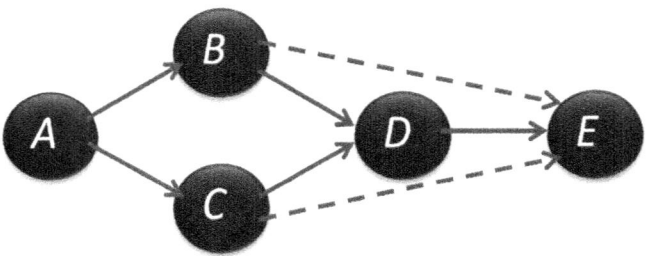

**Fig. 3.** PKI authentication structure example

This situation requires the following calculations [8] (solid arrows denote first hand evidence, while dashed ones denote second hand evidence):

$$\omega_{KA(k_E)}^{(AB,AC)D} = (((\omega_{RT(B)}^A \wedge \omega_{KA(k_B)}^A) \otimes (\omega_{RT(D)}^B \wedge \omega_{KA(k_D)}^B)) \oplus$$
$$((\omega_{RT(C)}^A \wedge \omega_{KA(k_C)}^A) \otimes (\omega_{RT(D)}^C \wedge \omega_{KA(k_D)}^C))) \otimes \omega_{KA(k_E)}^D.$$

In the above equation $k_A$ denotes $A$'s public key, $\omega_{KA(k_B)}^A$ denotes $A$'s opinion about authenticity of $k_B$, and $\omega_{RT(B)}^A$ denotes $A$'s opinion (trust) about recommendation trustworthiness of $B$.

### 2.3  Yu's and Singh's ToE Based Methodology

An approach that is similar and also based on ToE is given in [9] and [10], where $\Theta = \{T, \neg T\}$ and where evidence mass function $m$ is obtained as follows:

$$m(\emptyset) = 0, \quad m(\{T\}) = \frac{N^+}{N}, \quad m(\{\neg T\}) = N^-/N, \quad m(\Theta) = N^\circ/N \tag{8}$$

In the above equations $N$ stands for total interactions, of which $N^+$ denotes positive interactions, $N^-$ negative interactions, and $N^\circ$ "inappreciable" interactions. This enables derivation of *bel* function (mappings) as $bel(\{T\}$ and $bel(\{\neg T\})$. An entity decides to trust another entity iff $bel(\{T\}) - bel(\{\neg T\}) \geq \rho$, where $\rho$ is its referred to as cautiousness level.

## 2.4 Game Theoretic Methodologies for Trust Management

Another stream of approaches is based on game theory [11], [12]. In this theory a game consists of a set of players, a set of actions that are realizations of certain strategies available to the players, and a set of payoffs for each strategy.

One key concept in game theory is Naish equilibrium, NE. NE is important, because it represents action(s) that no other agent would prefer to deviate from, assuming that other agents also stick to it. For example, Alice and Bob are in NE if Alice makes her best decision she can while taking into account Bob's decision, and Bob makes his best decision while taking into account Alice's decision. To illustrate the central idea of game theory, the well-known prisoner's dilemma is presented in Fig. 4.

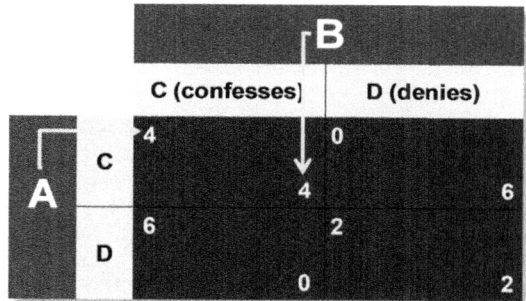

**Fig. 4.** The prisoners' dilemma game (C stands for confesses, D for denies)

In prisoner's dilemma two suspects are caught and offered the following choices: If they both confess the crime, they both get sentenced for four years (see the upper left dark quadrant in Fig. 4). If they deny it, both are sentenced for two years (see the lower right dark quadrant in Fig. 4). However, if one confesses while the other one not, the latter suspect gets six years of prison, while the confessing one is free (see the upper right dark quadrant and lower left quadrant in Fig. 4). This scenario clearly shows that if suspects rationally follow maximization of their self-interests they are worse off as if they were in case of cooperation (i.e., not confessing the crime and therefore not betraying the other suspect).

The formal definition of NE is as follows [11], [12]: Let $N$ be the set of players in the game, $A$ the set of strategy profiles, $A_i$ the set of strategies available to player $i$, and $u_i$ player $i$'s utility function. Then a profile $a \in A$ is a Nash equilibrium (NE) if

$$\forall i \in N: a_i \epsilon\, br_i(a_{-i}), \tag{9}$$

where $br_i(a_{-i})$ for $i \in N, a_{-i} \epsilon\, A_{-i}$ denotes the set of best responses of $i$ to $a_{-i}$:

$$\arg\max_{a_i \in A_i} \{u_i(a_i, a_{-i})\}.$$

Now looking at the above game theoretic approach through trust perspective, one can observe that it is actually addressing recommendations and not trust. Therefore such approaches are about recommendation, and not trust management systems. In [13] authors present such solution, called axiomatic approach, where they study properties (termed axioms) that characterize particular aggregation rules, and analyze whether particular desired properties can be simultaneously satisfied. This axiomatic approach is tailored to personalized ranking systems with the following four basic axioms:

1. An agent would be ranked at the top of his own personalized rank.
2. An agent preferred by more highly trusted agents, should be ranked higher than an agent preferred by less trusted agents.
3. Under the perspective of any agent, the relative ranking of two other agents would depend only on the pair wise comparisons between the rank of the agents that prefer them.
4. An agent cannot gain trust by any agent's perspective by manipulating its reported trust preference.

The settings, within which the personalized ranking systems are searched for, are the domains of graphs and linear orderings. More precisely, these two definitions are the central ones for personalized ranking systems:

**Definition 4.** *Let A be some set. A relation $R \subseteq A \times A$ is called an ordering on A if it is reflexive, transitive, and complete. Let $L(A)$ denote the set of orderings on A.*

**Definition 5.** *Let $\mathbb{G}_V^s$ be the set of all directed graphs $G = (V, E)$ such that for every vertex $v \in V$, there exists a directed path in E from s to v. A personalized ranking system F is a function that for every finite vertex set V and for every source $s \in V$ maps every graph $G \in \mathbb{G}_V^s$ to an ordering $\preccurlyeq_{G,s}^F \in L(V)$.*

In the above definitions, let $\preccurlyeq$ be an ordering, then $\simeq$ is the equality predicate of $\preccurlyeq$, and $\prec$ is the strict order induced by $\preccurlyeq$; formally, $a \simeq b$ if and only if $a \preccurlyeq b$ and $b \preccurlyeq a$; and $a \prec b$ if and only if $a \preccurlyeq b$ but not $b \preccurlyeq a$.

Last but not least, not all game-theory based approaches are (almost) a synonym for reputation systems. There exist game-theory based approaches that do address the core of trust, and they are typically applied in multi-agent systems or MAS. Many such examples can be found in [14].

To round up this section, let us mention that the above methodological approaches are not the only ones that are used for trust management. For more extensive overview of existing approaches with focus on web environments the reader is referred to [15] and [16].

# 3   An Analysis of Existing Approaches

Having provided a rather extensive overview of existing trust management methodologies that have domicile in computer sciences field, we should now analyze them through trust perspective. The common shortcomings of Bayesian statistics based approaches (naïve trust management), Theory of evidence and subjective algebra are the following ones. First, agents are not (always) rational. Second, assuming that agents are rational they may still have problems with the basic notion of probability. Third, even if they do not have problems with the basic issues related to probability, they will likely not understand sophisticated mathematics that is required for ToE and subjective algebra. And finally, is trust really perceived by agents as something that can be described with $\omega = (b, d, u)$?

As to game theoretic approaches, game theory assumes rational agents, too. Further, the second tenet of game theoretic approaches is that there exists some preference. The third tenet is transitivity of preferences. But agents are not necessarily rational, or may be rational in certain contexts, but not in other contexts, e.g. the problem of irrationality in economic contexts has already been described in some outstanding research [17]. Further, experiments that will be presented in this paper indicate that for many people (or in many contexts) trust is not transitive. Moreover, agents (people) may even not have preferences when it comes to trust. This all limits application of the game theoretic approaches. There are other interesting cognitive specifics when it comes to trust – on the basis of some preliminary tests (that are yet to be experimentally proved on a wider scale) we anticipate that in certain contexts transitive preferences may become circular.

Clearly, the above trust management methodologies have certain merit. However, there exists a need for complementary methodologies, and that is where qualitative assessment dynamics, QAD, comes in. But before proceeding further, it has probably become clear to the reader that the so far presented issues are very much related to the core question, which is: *What, actually, is trust?*

We therefore need appropriate definition of trust, which will be tight enough to enable formal treatment and consequently, appropriate support with trust management applications in computing environments. In the literature there exist many definitions of trust, but one of the most authoritative ones is in the Merriam-Webster dictionary, which states that trust is assured reliance on the character, ability, strength, or truth of someone or something. This definition, although consistent, is not appropriate for our purposes, and needs further refinement and focus on web environments. One of the best candidates for our purpose is the definition given in the first half of the nineties by D.J. Dennig: Trust is an assessment that is driven by experience, shared through a network of people interactions and continually remade each time the system is used [18]. Now we can formally define trust for supporting trust management in web environments:

**Definition 6.** *Trust is an assessment relation between agents A and B that can be totally trusted, partially trusted, undecided, partially distrusted, and totally distrusted; it is denoted by $\alpha_{A,B}$, which means agent's A assessment of agent B.*

## 4   Qualitative Assessments Dynamics - QAD

Now that we have presented trust management methodologies, which are the most important and widely cited ones in the computer sciences domain, the next basic scientific question is: "How well do these methodologies reflect reality?" Or restated:

"How well are existing trust management methodologies aligned with users' behavior and mental processes when it comes to trust in computerized environments?"

To find answers to above questions we started a development of a questionnaire battery in line with methodological principles for survey research in IT area in 2005 (our earlier research results can be found in [19] and [20]). The literature about these principles of research in this field is extensive, and one good example is described in [21]. It is specifically concerned with surveys for computerized applications; it is the basis for our research methodology, which is intended to get the basic knowledge about trust phenomenon for its management in computing (web) environments. These are the related fundamental questions:

- What are the main demographics of our population / sample?
- What kind of metrics is preferred when it comes to trust – quantitative, probability related assessments, or qualitative assessments?
- What is the most appropriate number of qualitative (ordinal) descriptions for users' trust assessments?
- Do agents perceive trust as a reflexive, or symmetric, or transitive relation?
- What is the influence of a society on particular agent's trust decisions?
- How is a certain trust assessment, when set for the first time, formed?
- How frequently is trust assessment changed because of no apparent reason?
- Would users allow computers to decide on their behalf when it comes to trust, or do they want to be directly involved?

Our goal with the QAD is the following. Suppose that a complementary methodology that we are aiming at should meet the requirements and expectations of such a number of users that it would be the second player on the market. Now what does it mean to be the second player? To determine this figure, one can look at market shares of most commonly used IT solutions like operating systems, web browsers and search engines. The market shares of the first three most important players in these areas are given below (see http://marketshare.hitslink.com, data as of May 2010):

- Operating systems: 91.3% Windows, 5.26% MacOS, 1.1% Linux.
- Search engines: 84.8% Google, 6.19% Yahoo, 3.24% Bing.
- Web browsers: 59.75% Internet Explorer, 24.32% Firefox, 7.04% Chrome.

It follows that to become the second player in the field, the threshold can be set as low as approx. 6% in case of operating systems, while in case of web browsers it has to be set above approx. 25%. We will set it high enough to exceed all above thresholds - to 30%.

It is now possible to state the relevant hypotheses ($H_1 - H_{11}$) for computationally supported trust management methodologies and solutions (these hypotheses serve to find out if our assumptions about trust management formalism properties, operators and operands, are aligned with reality or not):

- More than 30% of users would prefer direct trust management.
- More than 30% of users would prefer qualitative assessment of trust.
- More than 30% of users have problems with conforming to the basic definition of probability when it comes to trust.

- More than 30% of users would choose five levels ordinal scale for trust assessments.
- To more than 30% of users trust is not reflexive.
- To more than 30% of users trust is not symmetric.
- To more than 30% of users trust is not transitive.
- To more than 30% of users that belong to a certain group their trust assessment may generally differ from the (aggregated) assessment of the group.
- More than 30% of users may occasionally change trust assessment on a non-identifiable basis.
- To more than 30% of users that assess a certain group as a whole equals to their assessment about the majority of the members of this group.
- In more than 30% of users trust may be initialized on a non-identifiable factors basis.

Our research aimed at confirming / refuting the above hypotheses has an extensive history of almost six years. In order to make a long story short, the latest results will be briefly given. The last experiment took place in May 2010 over the web to a sample of B.Sc. students' population of computer and information sciences at FAMNIT, University of Primorska. Invitation e-mails were sent through e-mail to all 109 B.Sc. students, and the response rate was 24.1 %. Due to the conditions (anonymous participation, no benefits of whatever kind were offered, etc.) we can assume negligible response and non-response bias. Therefore we treated respondents as a random sample of the above population.

After getting and analyzing the results, we were able to confirm all hypotheses for this population, except hypothesis H3 that had to be refuted. The results are as follows (confidence interval is set to 95%, i.e. $Z = 1.96$): $H1 = 0.77 \pm 0.16$, $H2 = 0.81 \pm 0.15$, $H3 = 0.42 \pm 0.19$, $H4 = 0.62 \pm 0.19$, $H5 = 0.69 \pm 0.18$, $H6 = 0.54 \pm 0.19$, $H7 = 0.69 \pm 0.18$, $H8 = 0.73 \pm 0.17$, $H9 = 0.62 \pm 0.19$, $H10 = 0.54 \pm 0.19$, $H11 = 0.58 \pm 0.19$.

Taking the experimental results into account, we have developed QAD. It is aligned with THE observation that significant number of users prefers direct interaction with trust management system. Further, users prefer support of qualitative assessments on an ordinal scale, where this ordinal scale has five (descriptive, qualitative) levels. In addition, users do not perceive trust as being reflexive, symmetric, or transitive relation. Also the existence of preferences should not be assumed at all. Further, trust is driven by the community assessments, and agents choose initial trust assessments randomly (the same often holds true for already assigned trust values).

**Definition 7.** *Propagated trust in agents comunities is given by a trust matrix* **A**, *where elements* $\alpha_{i,j}$ *denote assessment (trust relations) of i-th agent towards j-th agent, and where their values are taken from the set* $\Lambda = \{2, 1, 0, -1, -2, -\}$. *These values denote trusted, partially trusted, undecided, partially distrusted and distrusted relationships. The last symbol, "−", denotes an undefined relation, meaning that an agent is either not aware of existence of another agent, or does not want to disclose its trust.*

A general form of trust matrix **A** of a certain society with $n$ agents is defined as follows:

$$A = \begin{bmatrix} \alpha_{1,1} & \alpha_{1,2} & \cdots & \alpha_{1,n} \\ \alpha_{2,1} & \alpha_{2,2} & \cdots & \alpha_{2,n} \\ \vdots & \vdots & \ddots & \vdots \\ \alpha_{n,1} & \alpha_{n,2} & \cdots & \alpha_{n,n} \end{bmatrix}$$

**Definition 8.** *In a trust matrix* **A**, *columns represent society trust vector, which states society assessments about particular agent k, i.e.* $\mathbf{A}_{n,k} = (\alpha_{1,k}, \alpha_{2,k}, ..., \alpha_{n,k})$, *while rows represent agent's k trust vector, i.e.* $\mathbf{A}_{k,n} = (\alpha_{k,1}, \alpha_{k,2}, ..., \alpha_{k,n})$, *where k = 1, 2,..., n. Further, excluding undefined relations from trust vector results in a society assessment sub-vector, denoted by* $\underline{\mathbf{A}}_{n1,k} = (\alpha_{1,k}, \alpha_{2,k}, ..., \alpha_{n1,k})$, *where index "$n_1$" denotes number of non-undefined values in a society trust vector.*

Based on the above definitions, it is possible to present an example society with trust relations, qualitative weights and corresponding matrix:

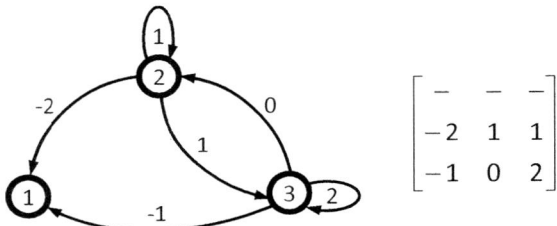

**Fig. 5.** An example society graph and corresponding matrix

**Definition 9.** *QAD operators belong to the set* $\Psi = \{\Uparrow, \Downarrow, \uparrow, \downarrow, \rightsquigarrow, \leftrightarrow, \odot, \updownarrow\}$, *where the symbols denote extreme optimistic assessment, extreme pessimistic assessment, moderate optimistic assessment, moderate pessimistic assessment, centralistic consensus seeker assessment, non-centralistic consensus-seeker assessment, self-confident assessment and assessment-hoping. These operators are functions* $f_j \in \Psi$, *such that* $f_j : \mathbf{A}_{n,i} = (\alpha_{1,i}^-, \alpha_{2,i}^-, \alpha_{3,i}^-, ..., \alpha_{j,i}^-, ..., \alpha_{n,i}^-) \rightarrow \alpha_{j,i}^+$, $j = 1, 2,..., n$, *where "j" denotes the j-th agent, superscript "-" denotes pre-operation value, superscript "+" post-operation value, and where mappings for particular operators are defined as follows:*

a) $\alpha_{j,i}^- \neq -:$

- $\Uparrow_j:$ 
$$max(\alpha_{1,i}^-, \alpha_{2,i}^-, \alpha_{3,i}^-, ..., \alpha_{j,i}^-, ..., \alpha_{n,i}^-) \rightarrow \alpha_{j,i}^+ \qquad i = 1,2,...,n$$

- $\Downarrow_j:$ 
$$min(\alpha_{1,i}^-, \alpha_{2,i}^-, \alpha_{3,i}^-, ..., \alpha_{j,i}^-, ..., \alpha_{n,i}^-) \rightarrow \alpha_{j,i}^+ \qquad i = 1,2,...,n$$

- $\uparrow_j:$ 
$$\begin{cases} \alpha_{j,i}^- \rightarrow \alpha_{j,i}^+ \\ \lfloor \alpha_{j,i}^- + 1 \rfloor \rightarrow \alpha_{j,i}^+ \end{cases} \qquad \begin{array}{l} if \ \frac{1}{n_1}\sum_{i=1}^{n_1} \alpha_{i,k}^- \leq \alpha_{j,i}^- \\ otherwise \end{array}$$

$$\bullet \quad \downarrow_j: \quad \begin{cases} \alpha_{j,i}^- \to \alpha_{j,i}^+ & \text{if } \dfrac{1}{n_1}\sum_{i=1}^{n_1} \alpha_{i,k}^- \geq \alpha_{j,i}^- \\[2ex] \lceil \alpha_{j,i}^- - 1 \rceil \to \alpha_{j,i}^+ & otherwise \end{cases}$$

$$\bullet \quad \leadsto_j: \quad \begin{cases} \left\lceil \dfrac{1}{n_1}\sum_{i=1}^{n_1} \alpha_{i,k}^- \right\rceil \to \alpha_{j,i}^+ & \text{if } \dfrac{1}{n_1}\sum_{i=1}^{n_1} \alpha_{i,k}^- < 0 \\[2ex] \left\lfloor \dfrac{1}{n_1}\sum_{i=1}^{n_1} \alpha_{i,k}^- \right\rfloor \to \alpha_{j,i}^+ & otherwise \end{cases}$$

$$\bullet \quad \leftrightarrow_j: \quad \begin{cases} \left\lceil \dfrac{1}{n_1}\sum_{i=1}^{n_1} \alpha_{i,k}^- \right\rceil \to \alpha_{j,i}^+ & \text{if } \dfrac{1}{n_1}\sum_{i=1}^{n_1} \alpha_{i,k}^- > 0 \\[2ex] \left\lfloor \dfrac{1}{n_1}\sum_{i=1}^{n_1} \alpha_{i,k}^- \right\rfloor \to \alpha_{j,i}^+ & otherwise \end{cases}$$

$$\bullet \quad \odot_j: \quad \alpha_{j,i}^- \to \alpha_{j,i}^+ \qquad\qquad i = 1,2,\dots,n$$

$$\bullet \quad \updownarrow_j: \quad rand(-2,-1,0,1,2) \to \alpha_{j,i}^+ \qquad i = 1,2,\dots,n$$

$$\textbf{b)} \quad \alpha_{j,i}^- = -: \\ \quad - \to \alpha_{j,i}^+ \qquad\qquad i = 1,2,\dots,n$$

## 5   Analyzing Agents Behavior with QAD

A demonstration of the presented apparatus follows. Suppose we want to analyze behavior of a society that can be considered as one typical example society. It consists of 100 agents, where all agents are initially undecided about one another. Further, 90% of them are initially governed by extreme optimistic operator, while 10% are governed by extreme pessimistic operator. Now in each step 10% of population randomly changes its operator (all possible values for newly assigned random assessments and operators are equally likely). Running 30 simulation runs on this society, each of them taking 45 steps, the following histogram, presented in Fig. 7, has been obtained.

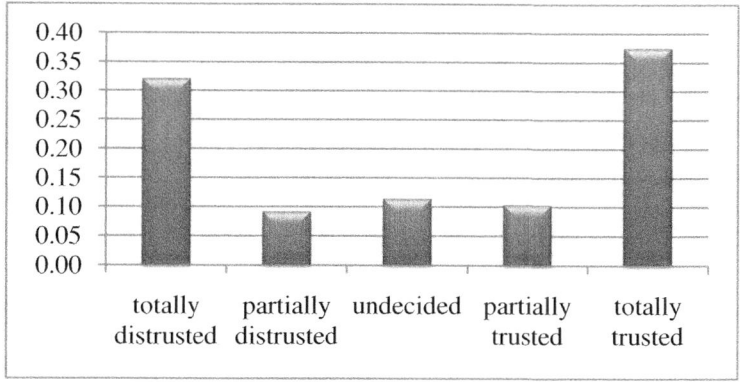

**Fig. 6.** Simulation results – histogram of trust values

It follows that in such community of agents, under given conditions, the resulting distribution is (almost) bimodal. More precisely, 32% of agents become totally distrusted, and 37% of them become totally trusted. Further, 9% becomes partially distrusted, and 10% partially trusted, while for 11% of agents' population other agents remain undecided.

This is a very interesting result. Despite the fact that initially everyone was undecided about others (assuming the distribution of operators and random changes during the simulation), clear assessment patterns emerge. Further, these patterns seem to tend towards extreme assessments, so a notable polarization within the society becomes visible.

## 6  Conclusions

Trust has been an important topic for quite a long time in social sciences area. However, with web expansion, and e-media solutions in general, our lives have become more and more dependent on IT. This has triggered computer and information sciences researchers to start investigating intensively this area as well. This research has been further stimulated even by high ranking politicians like EU Commissioner V. Reding claiming that lack of trust is critical for wider acceptance of e-solutions, which in turn are critical for economic prosperity of the EU [22].

Certainly, the research in this area is important, not only for the web environments, because it has visible wider implications. So we anticipate that in the near future this research will be even more intensive, while at the same time the trust management infrastructure will have to be developed [23] in order to support those trust management methodologies that users will accept as the most appropriate one(s). Based on our research and experimental findings we believe that one of them may be, at least partially, Qualitative Assessment Dynamics.

Last but not least, it is worth to mention that QAD nicely complements other research efforts in this area [24]. It seems that there will not exist "one-size fits all" trust management solution when it comes to trust in web environments. Therefore some of them will be most appropriate for one kind of uses, some of them for another kind of uses. Moreover, it may even be the case that certain combinations of these methodologies may turn out to be most useful for certain practical uses and applications.

**Acknowledgement.** The Slovene Research Agency has financed this research through research program P2-0359. Special thanks go also to both reviewers for their helpful comments and suggestions.

## References

[1] Miller, J., Resnick, P., Singer, D.: PICS Rating Services and Rating Systems. W3C (1996), http://www.w3c.org/TR/REC-PICS-services
[2] Blaze, M., Feigenbaum, J., Lacy, J.: Decentralized Trust Management. In: Proceedings of the 1996 IEEE Symposium on Security and Privacy, Oakland, pp. 164–173 (1996)
[3] Herzberg, A., et al.: Access Control Meets Public Key Infrastructure. In: Proc. of the IEEE Conf. on Security and Privacy, Oakland, pp. 2–14 (2000)

[4] Grandison, T., Sloman, M.: A survey of trust in internet applications. IEEE Communications Surveys 3(4), 2–13 (2000)

[5] Wang, Y., Vassileva, J.: Trust and Reputation Model in Peer-to-Peer Networks. In: Proc. Of the 3rd Int. Conference on Peer-to-Peer Computing (P2P 2003), p. 150, Linkoping (2003)

[6] Shafer, G.: A Mathematical Theory of Evidence. Princeton University Press, Princeton (1976)

[7] Jøsang, A.: A logic for uncertain probabilities. Int. Journal of Uncertainty, Fuzziness and Knowledge-Based Systems 9(3), 279–311 (2001)

[8] Jøsang, A.: An Algebra for Assessing Trust in Certification Chanis. In: Proceedings of the Network and Distributed Systems Security Symposium, NDSS 1999, pp. 618–644. The Interent Society, San Diego (1999)

[9] Yu, B., Singh, M.P.: Distributed Reputation Management for e-Commerce. In: Proc. of the 1st AA-MAS Conference, Bologna (2002)

[10] Paul-Amaury, M., Morge, M., Toni, F.: Combining statistics and arguments to compute trust. In: Proc. of the 9th Int. Conf. on Autonomous Agents and Multiagent Systems (AA-MAS), Toronto, pp. 209–216 (2010)

[11] Tennenholtz, M.: Game-Theoretic Recommendations: Some Progress in an Uphill Battle.In: Proc. of AAMAS 2008, Estoril , pp. 10 –16, (2008)

[12] Harish, M., Anandavelu, N., Anbalagan, N., Mahalakshmi, G.S., Geetha, T.V.: Design and analysis of a game theoretic model for P2P trust management. In: Janowski, T., Mohanty, H. (eds.) ICDCIT 2007. LNCS, vol. 4882, pp. 110–115. Springer, Heidelberg (2007)

[13] Altman, A., Tennenholtz, M.: An axiomatic approach to personalized ranking systems. In: Proceedings of the 20th International Joint Conference on Artificial Intelligence (IJCAI 2007), pp. 1187–1192. Morgan & Kaufmann, San Francisco (2007)

[14] Sabater, J., Sierra, C.: Review on Computational Trust and Reputation Models. Artificial Intelligence Review 24(1), 33–60 (2005)

[15] Golbeck, J.: Trust on the World Wide Web: A survey. Foundation and Trends in Web Science 1(2), 131–197 (2006)

[16] Artz, D., Gil, Y.: A survey of trust in computer science and the Semantic Web. Software Engineering and the Semantic Web 5(2), 58–71 (2007)

[17] Kahneman, D., Slovic, P., Tversky, A. (eds.): Judgment Under Uncertainty, 22nd reprint. Cambridge University Press, Cambridge (2006)

[18] Denning, D.: A new Paradigm for trusted systems. In: Proc. of ACM SIGSAC New Security Paradigms Workshop, pp. 36–41. ACM, New York (1993)

[19] Trček, D.: A formal apparatus for modeling trust in computing environments. Mathematical and Computer Modeling 49(1-2), 226–233 (2009)

[20] Trček, D.: Ergonomic trust management in pervasive computing environments - qualitative assessment dynamics. In: Proceedings of the ICPCA 2010, pp. 1–7. IEEE Press, Maribor (2010)

[21] Pfleeger, S.L., Kitchenham, B.A.: Principles of Survey Research, Parts 1-6. ACM Software Engineering Notes 26-28 (2001-2003)

[22] Reding, V.: The need for a new impetus to the European ICT R & I Agenda. In: Int. High Level Research Seminar on "Trust in the Net", Vienna (2006)

[23] Kovač, D., Trček, D.: Qualitative trust modeling in SOA. Journal of Systems Architecture 55(4), 255–263 (2009)

[24] Grabner-Kraeuter, S., Kaluscha, E.A.: Empirical Research in on-line trust: a review and critical assessment. International Journal of Human Computer Studies 2003(58), 783–812 (2003)

# Application and Evaluation of Inductive Reasoning Methods for the Semantic Web and Software Analysis*

Christoph Kiefer and Abraham Bernstein

Dynamic and Distributed Information Systems Group,
Department of Informatics, University of Zurich
Binzmuehlestrasse 14, CH-8050 Zurich, Switzerland
lastname@ifi.uzh.ch
http://www.ifi.uzh.ch/ddis

**Abstract.** Exploiting the complex structure of relational data enables to build better models by taking into account the additional information provided by the links between objects. We extend this idea to the Semantic Web by introducing our novel SPARQL-ML approach to perform data mining for Semantic Web data. Our approach is based on traditional SPARQL and statistical relational learning methods, such as Relational Probability Trees and Relational Bayesian Classifiers. We analyze our approach thoroughly conducting four sets of experiments on synthetic as well as real-world data sets. Our analytical results show that our approach can be used for almost any Semantic Web data set to perform instance-based learning and classification. A comparison to kernel methods used in Support Vector Machines even shows that our approach is superior in terms of classification accuracy.

**Keywords:** Inductive Reasoning, Semantic Web, Machine Learning, SPARQL, Evaluation.

**Please Note:** This paper represents part of the summer school lecture. It contains one critical, previously unpublished element: the description of inductive reasoning as an important component for non-traditonal reasoning on the Semantic Web. The lecture will also cover analogical reasoning [28,27,25], Markov Logic Networks [38], and the use of modern distributed techniques to run graph algorithms such as SIGNAL/COLLECT [43], Pregel [31], or MapReduce [12] with the Hadoop infrastructure (http://hadoop.apache.org/).
*A. B.*

## 1    Introduction

The vision of the Semantic Web is to interlink data from divers heterogeneous sources using a semantic layer as "glue" technology. The result of this combination

---

* This paper is a significant extension and complete rewrite of [26], which won the best paper award at ESWC2008.

A. Polleres et al. (Eds.): Reasoning Web 2011, LNCS 6848, pp. 460–503, 2011.

process constitutes the often cited *Web of data* that makes data accessible on the traditional Web such that other applications can understand and reuse it more easily [5,1].

The above-mentioned semantic glue basically comprises a *rule-based meta-data layer* to *expose the meaning of data* in a machine-readable format. The term *rule-based* refers to the logic-based foundations of the Semantic Web that uses a number of description logic (DL) languages to represent the terminological knowledge of a domain (*i.e.*, a data source) in a structured and theoretically sound way. *Meta-data* means *self-describing*, that is, the raw data is tagged with additional information to express its meaning in the format of these DL languages.

The most universal DL languages in the Semantic Web are the *Resource Description Framework (RDF)*[1] and the *Web Ontology Language (OWL)*.[2] These languages/formats enable (i) to combine heterogeneous data under a common representation scheme by the use of *ontologies* and (ii) to give the data some well-defined, logic-based semantics, turning the otherwise meaningless data into information typically stored in a *knowledgebase (KB)*. Hence, ontologies serve as a formal specification of the conceptualization of this knowledge in terms of *classes* and *relations* among them [18].

## 1.1   Description Logic Reasoning

At this point, we are able to transfer the data that comes, for instance, from traditional relational databases to Semantic Web knowledgebases by using ontologies to specify the structure of the knowledge and a set of description logic languages to define the (logical) relations between these structure elements.

Typically, the information in a knowledgebase is stored as *asserted* (*i.e.*, *atomic*) facts. Such a piece of information could, for example, be the proposition "The type of service A is tourism", or in triples notation [ serviceA type tourism ].

Now suppose the knowledgebase additionally includes the information [serviceB type serviceA ] to express that service B is a specification of service A (B might, for instance, deliver information about hotels in a given city). One of the underpinnings of the Semantic Web and, therefore, a strength of any such semantic architecture is the ability to reason from the data, that is, to derive new knowledge (new facts) from base facts. In other words, the information that is already known and stored in the knowledgebase is extended with the information that can be *logically deduced from the ground truth*.

This situation is also depicted in Figure 1 that shows schematically by the leftmost arrow the typical description logic reasoning process to infer additional, derived triples from a set of asserted triples in a knowledgebase. To summarize, the above service example is a simple application of *classical deductive logic* where the rule of inference over the type (subclass) hierarchy makes the proposition of B being of type tourism a valid conclusion.

---

[1]  http://www.w3.org/RDF/
[2]  http://www.w3.org/TR/owl-features/

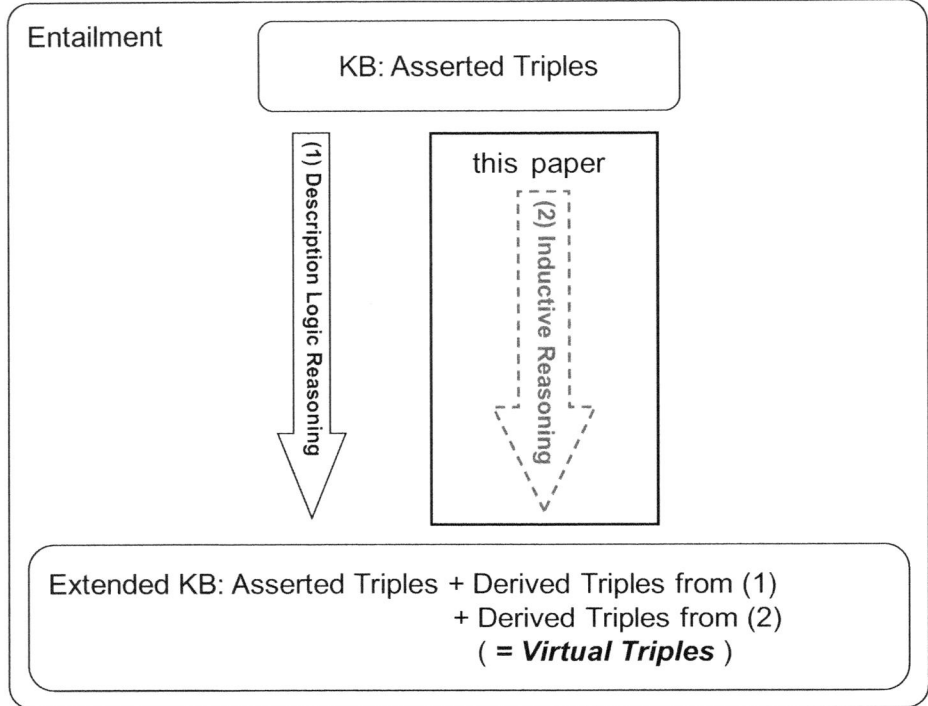

**Fig. 1.** The traditional Semantic Web infrastructure supports a logic-based access to the Semantic Web. It offers a retrieval (or reasoning) approach of data based on facts and classical deductive description logic reasoning (left arrow). Our novel reasoning extension presented and evaluated in this paper, on the other hand, extends the traditional Semantic Web infrastructure with *inductive reasoning* that is realized by *virtual triple patterns* and *statistical induction techniques* (right arrow).

## 1.2   What Is This Paper All About?

Metaphorically speaking, if this world would only be black and white, this is all that we could expect from a classical deductive reasoning system as supported by the current Semantic Web infrastructure. All the conclusions that could be drawn given some well-defined semantics such as the ones that come with the RDF/OWL languages will always be true if and only if the premises (*i.e.*, asserted knowledge, ground truth) are true. Otherwise they will be false, without any exception.

But the world is (fortunately!) not only black and white. The truth is, the world does generally not fit into a fixed, predetermined logic system of zeros and ones. Everyday life demonstrates again and again that we are performing some kind of *reasoning under uncertainty*, which does not follow the strict rules of formal logic.

Consider, for example, a doctor having to provide a medical diagnosis for one of his patients. Although he knows from his experiences and similar courses of disease that this special therapy seems to be best, there is, however, some risk involved, as such an inference is defeasible (*i.e.*, can be called into question)— medical advances may invalidate old conclusions. In other words, our actions are almost always driven by our heart and spirit (*i.e.*, by belief, experience, vague assumptions) rather than by formal logical implication.

To account for this, especially to deal with uncertainty inherent in the physical world, different models of human reasoning are required. Philosophers and logicians (among others) have, therefore, established new science fields in which they investigate and discuss such new types of human reasoning [32]. One prominent way to model human reasoning to some extend is *inductive reasoning* that denotes the process of *reasoning from sample-to-population* (*i.e.*, evidence-based reasoning). In inductive reasoning, the premises are only believed to support the conclusions but they cannot be (logically) entailed.

*This paper transfers the idea of inductive reasoning to the Semantic Web and Software Analysis. To this end, it extends the well-known RDF query language SPARQL with our novel, non-deductive reasoning extension in order to enable inductive reasoning.*

Traditional RDF query languages such as SPARQL [37] or SeRQL [9] support a logic-based access to the Semantic Web. They offer a retrieval approach of data based on facts and classical deductive description logic reasoning. The extension presented and evaluated in this paper, on the other hand, extends traditional Semantic Web query answering with inductive reasoning facilities.

Inductive reasoning is realized by *statistical induction techniques* which are applied to draw conclusions about an individual given some statistical quantities such as probabilities, averages, or deviations from a previous examined population. In other words, by the use of statistical induction techniques, additional triples are derived based on some (precomputed) statistics about these data.

**Example 1 (Statistical Induction).** *Suppose that from a set of 5 services, 3 are related to the tourism sector (i.e., have type* tourism*) and 2 to the medical sector (see Figure 2). Given only this information, we could conclude that for a new, not yet examined service F (one that is outside the original sample of five services), there is a probability estimate of 0.6 (i.e., $\frac{3}{5}$) that the service is of type* tourism*. Because the probability estimate for type* medical *is only 0.4 (i.e., $\frac{2}{5}$), we infer that F must also be located in the tourism sector. Such inferences are also called* quantitative probabilistic reasoning *[32].*

The inductive reasoning approach presented in this paper works similarly: it involves a prediction/classification step performed by the SPARQL query engine to predict, for instance, the membership of a data sample (individual/instance) to a particular class with some prediction accuracy. For the classification task, this approach employs algorithms from machine learning such as decision trees, support vector machines (SVMs), and regression models [45].

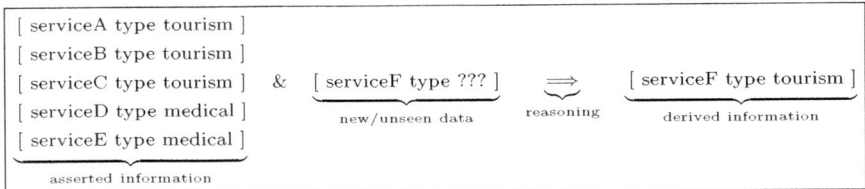

**Fig. 2.** Novel inductive reasoning process using statistical inference methods

### 1.3   Our Approach

To address these issues, specifically to implement our novel reasoning variant by using SPARQL, this paper introduces the concept of *virtual triple patterns (VTPs)*. Figure 1 shows the relation between asserted, 'ordinary' derived (1), and extraordinary derived triples (2). Ordinary triples are inferred using the traditional description logic reasoning system of the Semantic Web by applying the fundamental RDF/OWL inference rules. The extraordinary triples are the result of applying our novel inductive reasoning methods to the Semantic Web.

Typically, a Semantic Web dataset is made of a large number of RDF triples which model the relations among all data instances in terms of a so called `subject` and `object`, and a `predicate` to link them up. As an example, consider the triple pattern [ `serviceA hasName name` ] that relates service A to its name by the `hasName` predicate. An RDF dataset can then be thought of as a graph which is spanned by these triples. Query evaluation, can thus, essentially be reduced to the task of matching a number of triple patterns (called graph patterns) to an RDF graph.

VTPs, on the other hand, are triple patterns which are *not* matched against an RDF graph. Instead, they perform pattern matching as the result of calling some user-defined piece of code. VTPs can conceptually be thought of as ordinary function calls which consist of the function name followed by a list of arguments in parentheses, and which have a return value. VTPs are presented in details in Section 3.

### 1.4   Importance to the Semantic Web and Software Analysis

Regarding inductive reasoning, a number of past researches have highlighted the crucial element of statistics for the Semantic Web (*e.g.*, [17] or [21]). Two prominent tasks that can benefit from the use of statistics are *Semantic Web service classification* and *(semi-) automatic semantic data annotation*. Therefore, the support from tools that are able to work autonomously is needed to add the required semantic annotations. Consequently, a big challenge for Semantic Web research is not if, but *how* to extend the existing Semantic Web infrastructure with statistical inferencing capabilities.

In Software Analysis, researchers heavily deal with the analysis of software source code and abstract software models. Software Analysis and its subdisciplines have grown tremendously, which can also be observed from the increasing number of diverse papers submitted to the largest Software Analysis/Engineering conferences and workshops such $ICSE^3$ and $MSR^4$ in the past years. In order to show the advantages of inductive reasoning for Software Analysis via virtual triple patterns and Statistical Relational Learning methods, we have decided to perform a *bug prediction experiment* where the goal is to predict whether or not a piece of code is likely to have bugs or not. Roughly speaking, software bug prediction (aka defect prediction) is about finding locations in source code that are likely to be error-prone. We, thus, argue that the development and testing of tools that are able to detect such defect locations are crucial to (i) increase software quality and (ii) to reduce software development cost (among others).

To summarize, as we will show in this paper, the Semantic Web and Software Analysis can substantially benefit from our novel inductive reasoning extension to SPARQL. Our proposed, unified, SPARQL-based framework not only helps to solve these important research tasks, but also helps to establish the semantic glue mentioned at the very beginning of this work by (semi-) automatic semantic annotation (through classification).

Specifically, the contributions can be summarized as follows: For our inductive reasoning extension, we first present our *SPARQL-ML* approach to create and work with statistical induction/data mining models in traditional SPARQL (see Section 3). The major contribution of our proposed SPARQL-ML framework is, therefore, the ability *to support data mining tasks for knowledge discovery in the Semantic Web*.

Second, our presented SPARQL-ML framework is validated using not less than four case studies ranging over three heavily researched Semantic Web tasks and one Software Analysis task. For the Semantic Web, we perform two general data classification tasks (Sections 4.1 and 4.2) and one specific semantic service classification task (*i.e.*, service annotation; see Section 4.3). For Software Analysis, we perform a bug prediction task using semantically annotated software source code (Section 4.4).

By applying our approaches to these different tasks, we hope to show the approach's generality, ease-of-use, extendability, and high degree of flexibility in terms of customization to the actual task. Finally, we close the paper with a discussion of the results in Section 5, and our conclusions and some insights into future work in Section 6.

## 2    Related Work

This chapter briefly reviews the most important related work. We start with a short summary of some important Semantic Web publications to set this work

---

[3] International Conference on Software Engineering, http://www.icse-conferences.org/

[4] International Working Conference on Mining Software Repositories, http://msr.uwaterloo.ca/

into perspective in Section 2.1. Specifically, we review a couple of studies that influenced the history and development of SPARQL. Section 2.2 proceeds with some related approaches to inductive reasoning. Section 2.3 proceeds with some of the most important related works regarding the tasks we use to validate/evaluate our SPARQL-ML framework.

## 2.1  Semantic Web

In 1989, Alexander Borgida [8] presented his work about the CLASSIC language that can be regarded as an early approach to the Semantic Web. CLASSIC is a language for structural, partial descriptions of objects in a relational database management system. It is worth to mention this work for several reasons: first, CLASSIC allows the user to describe both the intensional structure of objects as well as their extensional relations to other objects (which in RDF terminology is achieved through data and object type properties); second, using CLASSIC it is possible to describe objects only partially and to add more information about it over time; third, CLASSIC can be used both as a data description as well as data query language; and fourth, the CLASSIC system is able to infer new knowledge about objects (*i.e.*, it performs an early kind of reasoning by applying a limited form of forward-chaining rules [39]).

12 years later, in 2001, Tim Berners-Lee [2] published his famous article about his vision of a true Semantic Web as an extension of the current Web in which data is given well-defined meaning through *ontologies*. This is an important improvement to, for instance, XML that allows the user to structure the data but does not say what the data in fact means. Such semantically enriched data can then be meaningfully manipulated by autonomous computer programs also referred to as *agents*.

Furthermore, one of the most important building blocks of the Semantic Web are, as argued in [2], automated reasoning facilities, which denote the process of deriving new information from existing, asserted information through classical deductive description logic (DL) reasoning rules. Pure deductive DL reasoning is, however, not sufficient for some tasks. On the contrary, as we will show in this work, tasks such as semantic service classification can substantially benefit from our novel, inductive reasoning facility.

Five years later, Shadbolt, Hall, and Berners-Lee [41] critically revisited some of the statements made in [2]. Specifically, they emphasized on the need for shared semantics which is badly needed for data integration—a task that is of particular importance in the life sciences [30]. As explained in [41], most of the motivation for a Semantic Web came from the tremendous amount of valuable information stored in traditional relational databases. This information must be exported into a system of URIs and, hence, given well-defined meaning. "The data exposure revolution has, however, not yet happened", which should increase the amount of available RDF data to push the Semantic Web even further.

**RDF Query Language SPARQL.** In recent years, the RDF query language SPARQL has gained increasing popularity in the Semantic Web. SPARQL stands for *SPARQL Protocol and RDF Query Language* and offers well-known constructs from database technology, such as SELECT, FILTER, and ORDER BY. Furthermore, the SPARQL specifications define a protocol for the communication between a query issuer and a query processor. The SPARQL language has currently the status of a W3C Recommendation and is extensively described in [37].

As the language was used more and more over time by different parties for different applications, it became clear that it needed a more mathematical basis in terms of an algebra, similar to relational algebra for relational databases [10]. This was especially important as the need for optimization of SPARQL queries also arose as people wanted to use ever growing RDF datasets for their experiments. Among those who dealt with the development of an algebra for SPARQL, it was Cyganiak [11] who described as one of the first how to transform (a subset of) SPARQL into relational algebra that is, as argued by Cyganiak, the language of choice when analyzing queries in terms of query planning and optimization. Furthermore, he defined the semantics of the relational algebra operators and discussed a translation into SQL, which is important to execute the queries against traditional relational databases storing the RDF data.

One year after Cyganiak's work was published, Pérez [35] conducted an extensive analysis of the semantics and complexity of SPARQL, focusing on the algebraic operators JOIN, UNION, OPTIONAL, and FILTER. The semantics and complexity of these operators are studied in great detail and insights into query optimization possibilities are presented. In particular, they introduced well-defined graph patterns that can be transformed to patterns in normal form, which when matched against the underlying RDF dataset results in improved query execution time. The presented theoretical framework in [35] is build around sets of solution mappings which are created in the process of matching the query's basic graph patterns (BGP) to the underlying RDF graph.

It is important to say, that the study of Pérez *et al.* highly influenced the work presented in this paper. Our proposed inductive reasoning extension to SPARQL is based on *virtual triple patterns* (see Section 3.2) that are theoretically defined in the algebraic notation of [35]. *ARQ property functions*[5]—the implementational foundations of virtual triple patterns—are, however, not addressed in [35]. It is, therefore, one of the contributions of this work to reflect on the semantics of such property functions, as our SPARQL-ML framework heavily relies on them.

## 2.2   Inductive Reasoning

Our proposed inductive reasoning extension relies on statistics (*i.e.*, machine learning techniques) and elements from probability theory to reason from data. In this section, we will briefly review some of the inductive reasoning (machine learning) approaches from the Semantic Web literature which are relevant in the context of this work. Specifically, as our novel reasoning extension heavily relies

---

[5] http://jena.sourceforge.net/ARQ/library-propfunc.html

on *Statistical Relational Learning (SRL)* algorithms, we shortly summarizes the two SRL methods we use in this paper. The section closes with an overview of some related works regarding the Semantic Web and Software Analysis tasks we chose to evaluate our inductive reasoning extension.

Little work has been done so far on seamlessly integrating knowledge discovery capabilities into SPARQL. Recently, Kochut and Janik [29] presented *SPARQL-eR*, an extension of SPARQL to perform semantic association discovery in RDF (*i.e.*, finding complex relations between resources). One of the main benefits of our inductive reasoning approach through SPARQL-ML is that we are able to use a multitude of different, pluggable machine learning techniques to not only perform semantic association discovery, but also prediction/classification and clustering.

Getoor and Licamele [16] highlighted the importance of link mining for the Semantic Web. They state that the links between resources form graphical patterns which are helpful for many data mining task, but usually hard to capture with traditional statistical learning approaches. With our SPARQL-ML framework we, therefore, apply SRL algorithms that are able to exploit these patterns to improve the performance of the pure statistical approaches (see Section 3.2).

Similarily, Gilardoni [17] argued that machine learning techniques are needed to build a semantic layer on top of the traditional Web. Therefore, the support from tools that are able to work autonomously is needed to add the required semantic annotations. We show that our inductive reasoning extension to SPARQL offers this support, and thus, facilitates the process of (semi-) automatic semantic annotation (through classification).

We are aware of two other independent studies that focus on data mining techniques for Semantic Web data using Progol—an Inductive Logic Programming (ILP) system.[6] In the first study, Edwards [14] conducted an empirical investigation of the quality of various machine learning methods for RDF data classification, whereas in the second study, Hartmann [19] proposed the *ARTEMIS* system that provides data mining techniques to discover common patterns or properties in a given RDF dataset. Our work extends their suggestions in extending the Semantic Web infrastructure in general with machine learning approaches, enabling the exploration of the suitability of a large range of machine learning techniques (as opposed to few ILP methods) to Semantic Web tasks without the tedious rewriting of RDF datasets into logic programming formalisms.

Last but not least, Bloehdorn and Sure [6] explored an approach to classify ontological instances and properties using SVMs (*i.e.*, kernel methods). They presented a framework for designing such kernels that exploit the knowledge represented by the underlying ontologies. Inspired by their results, we conducted the same experiments using our proposed SPARQL-ML approach (see Section 4.3). Initial results show that we can outperform their results by a factor of about **10%**.

---

[6] http://www.doc.ic.ac.uk/~shm/progol.html

**Statistical Relational Learning Methods.** Our SPARQL-ML framework employs machine learning-based, statistical relational reasoning techniques to create and work with data mining models in SPARQL (see Section 3). These techniques are *Relational Probability Trees (RPTs)* and *Relational Bayesian Classifiers (RBCs)* that model not only the intrinsic attributes of objects, but also the extrinsic relations to other objects and, thus, should perform at least as accurate as traditional, propositional learning techniques. Both algorithms enable to perform inductive reasoning for the Semantic Web, in other words, they enable to induce statistical models *without prior propositionalization of the data* (*i.e.*, translation to a single table) [13], which is a cumbersome and error-prone task.

RPTs [33] extend standard probability estimation trees (also called *decision trees*) to a relational setting, in which data instances are heterogeneous and interdependent. This procedure is explained in more details in Section 3.2.

The RBCs used to perform inductive reasoning through SPARQL-ML were also proposed by Neville in [34]. An RBC is a modification of the traditional Simple Bayesian Classifier (SBC) for relational data [45]. Please refer to Section 3.2 for more details about RBCs.

## 2.3  SPARQL-ML Evaluation/Validation Tasks

Sabou [40] stated that the Semantic Web can facilitate the discovery and integration of web services. The addition of ontologies, containing knowledge in the domain of the service such as the types of input/output parameters, offers new background information, which can be exploited by machine learning algorithms. We evaluate this assumption in this work in the context of our semantic web service classification experiment by comparing the results of data mining with and without the enhancement of ontologies (see Section 4.2).

Furthermore related is the study of Heß [21], in which a machine learning approach for semi-automatic classification of web services is described. Their proposed application is able to determine the category of a WSDL web service and to recommend it to the user for further annotation. They treated the determination of a web service's category as a text classification problem and applied traditional data mining algorithms, such as Naïve Bayes and Support Vector Machines [45]. Our conducted experiment is similar in that it employs OWL-S service descriptions instead of WSDL descriptions. In contrast to [21], we employ SRL algorithms such as RPTs and RBCs and additional background information provided by ontologies to perform semantic service classification. Regarding bug/defect prediction in source code, many approaches have been proposed in the past to accomplish this task. In Fenton [15], an extensive survey and critical review of the most promising learning algorithms for bug prediction from the literature is presented. [15] proposed to use Bayesian Belief Networks (BBNs) to overcome some of the many limitations of the reviewed bug prediction algorithms. BBNs are based on applying Bayes' rule that assumes that all attributes of training and testing examples are independent of each other given the value of the class variable (which is called conditional independence). It is important to note that the RBCs validated in this case study is an extension of the simple Bayesian classifier (that applies Bayes's rule for classification) to a relational data setting (see Section 3.2).

Bernstein [3] proposed an approach based on a non-linear model on temporal features for predicting the number and location of bugs in source code. In their experiments, six different models were trained using Weka's J48 decision tree learner. The data they used to evaluate their prediction models were collected from six plug-ins of the Eclipse open source project.[7]

These data were then enhanced with temporal information extracted from Eclipse's concurrent versions system (CVS) and information from Bugzilla.[8] Using this approach, they successfully showed that the use of a non-linear model in combination with a set of temporal features is able to predict the number and location of bugs with a very high accuracy.

In order to demonstrate the usefulness and applicability of inductive reasoning on semantically annotated software source code, we perform the same experiment using our proposed SPARQL-ML framework (see Section 4.4). As we will show in the remainder of this paper, inductive reasoning techniques for this kind of task and dataset provide a powerful means to quickly analyze source code.

## 3    Inductive Reasoning with SPARQL-ML

This chapter presents our novel inductive reasoning approach that intends to complement the classical deductive description logic reasoning facilities of the traditional Semantic Web. In a nutshell, inductive reasoning enables to draw conclusions about an unseen object (not included in the original set of observed samples) based on statistical induction/inferencing techniques. Basically, this comprises (1) the learning of a statistical model mirroring the characteristics of the observed samples and (2) the application of the model to the population. In Semantic Web terminology, inductive reasoning denotes the process of deriving new triples from the set of asserted triples based on the statistical observations of a sufficiently large, representative set of resources.

To add inductive reasoning support to the current Semantic Web infrastructure, specifically to integrate it with SPARQL, we focus on a special class of statistical induction techniques called *statistical relational learning (SRL)* methods. As we will show in our experiments, the large and continuously growing amount of interlinked Semantic Web data is a perfect match for SRL methods due to their focus on *relations between objects* in addition to features/attributes of objects of traditional, propositional learning techniques.

Our inductive reasoning extension to SPARQL is called *SPARQL-ML (SPARQL Machine Learning)*. SPARQL-ML supports the integration of traditional Semantic Web techniques and machine learning-based, statistical inferencing to create and work with data mining models in SPARQL. To that end, SPARQL-ML introduces new keywords to the official SPARQL syntax to facilitate the induction of models.

---

[7] http://www.eclipse.org/
[8] http://www.bugzilla.org/

```
D = {
 (SP1 profile:name ''CityLuxuryHotelInfoService''),
 (SP1 profile:desc ''Often used service to get
                         information about luxury hotels.''),
 (SP1 profile:hasInput _CITY),
 (SP1 profile:hasInput _COUNTRY),
 (SP1 profile:hasOutput _LUXURYHOTEL),
 (SP1 profile:hasCategory ''travel''),
 (SP2 profile:name ''CityCountryHotelInfoService''),
 (SP2 profile:desc ''Accommodation and restaurant
                         information service.''),
 (SP2 profile:hasInput _CITY),
 (SP2 profile:hasOutput _HOTEL),
 (SP2 profile:hasCategory ''travel''),
 (SP3 profile:name ''CityCountryInfoService''),
 (SP3 profile:desc ''Hotels and sports facilities
                         information service.''),
 (SP3 profile:hasInput _SPORT),
 (SP3 profile:hasOutput _CAPITAL),
 (SP3 profile:hasCategory ''education'') }
```

**Fig. 3.** Example dataset $D$ that lists services A, B, and C in triple notation

For the prediction/classification of unseen objects in a dataset, SPARQL-ML makes use of our proposed *virtual triple pattern* approach [27] to call customized, external prediction functions implemented as ARQ property functions (Section 3.2).

The two SRL methods used in SPARQL-ML are *Relational Probability Trees (RPTs)* and *Relational Bayesian Classifiers (RBCs)* proposed in [33] and [34], respectively. The use of these methods enables to induce statistical models without prior propositionalization (*i.e.*, translation to a single table) [13]—a cumbersome and error-prone task.

To ensure the extensibility of our inductive reasoning approach with other learning methods, the *SPARQL Mining Ontology (SMO)* is proposed to enable the seamless integration of additional machine learning techniques (see Section 3.3).

### 3.1 Preliminaries

In this chapter, the dataset $D$ shown in Figure 3 will be used for all examples. $D$ describes three semantic services A, B, and C in triple notation (with profile names SP1, SP2, and SP3 respectively). In triple notation, each characteristic of the services is written as a simple triple of *subject*, *predicate*, and *object*, in that order. Note that all the queries in the remainder of this chapter use the prefixes shown in Listing 1.1.

### 3.2 Theoretical Foundations

The theory introduced in this chapter heavily relies on our virtual triple pattern approach presented in [27] and Statistical Relational Learning learning methods. This section, therefore, (i) briefly reviews the most important elements of the semantics of SPARQL and virtual triples, and (ii), shortly summarizes Relational Bayesian Classifiers (RBCs) and Relational Probability Trees (RPTs).

```
PREFIX pf: <java:ch.uzh.ifi.ddis.pf>
PREFIX grounding:
  <http://www.daml.org/services/owl-s/1.1/Grounding.owl#>
PREFIX owl: <http://www.w3.org/2002/07/owl#>
PREFIX process:
  <http://www.daml.org/services/owl-s/1.1/Process.owl#>
PREFIX profile:
  <http://www.daml.org/services/owl-s/1.1/Profile.owl#>
PREFIX rdf:
  <http://www.w3.org/1999/02/22-rdf-syntax-ns#>
PREFIX rdfs: <http://www.w3.org/2000/01/rdf-schema#>
PREFIX service:
  <http://www.daml.org/services/owl-s/1.1/Service.owl#>
PREFIX sml: <java:ch.ifi.ddis.pf.sml>
PREFIX smo: <http://www.ifi.uzh.ch/ddis/sparql-ml/>
PREFIX xsd: <http://www.w3.org/2001/XMLSchema#>
```

**Listing 1.1.** Query prefixes used in this paper

**Semantics of SPARQL.** To explain our virtual triple pattern approach, the concept of SPARQL solution mappings is central. According to [37], a solution mapping is defined as follows:

**Definition 1 (Solution Mapping).** *A solution mapping $\mu(?v \mapsto t)$ maps a query variable $?v \in V$ to an RDF term $t$ where $V$ is the infinite set of query variables and $t$ a member of the set union of literals, IRIs, and blank nodes called RDF-T. The domain of $\mu$, $dom(\mu)$, is the subset of $V$ where $\mu$ is defined.*

**Example 2 (Solution Mappings).** *Matching the basic graph pattern {* `SP1 profile:name ?name` *} against dataset D will result in a simple solution mapping, i.e.,*

$$\mu(?name \mapsto \text{``CityLuxuryHotelInfoService''}).$$

*The domain of $\mu$ is $dom(\mu) = \{$ ?name $\}$ (i.e., $\mu$ is defined for precisely one variable). Matching the graph pattern {* `SP1 ?predicate ?name` *} against D will additionally find a mapping for variable* `?predicate`*, i.e.,*

$$\mu(?predicate \mapsto profile{:}name,$$
$$?name \mapsto \text{``CityLuxuryHotelInfoService''}).$$

*In this case, the domain of $\mu$ is $dom(\mu) = \{$ ?predicate, ?name $\}$.*

In [35], it is stated that the evaluation of a graph pattern over a dataset results in a *(multi-) set of solution mappings $\Omega$.*

**Example 3 (Set of Solution Mappings).** *The basic graph pattern {* `?profile profile:name ?name` *} specifies both the subject and the object of the triple pattern as variable. The graph matching algorithm will return a set of solution mappings $\Omega$ including precisely three solution mappings when matching the pattern against dataset D, i.e.,*

$$\Omega = \{ \ \mu_1(?profile \mapsto SP1,$$
$$?name \mapsto \text{``CityLuxuryHotelInfoService''}),$$

```
1  SELECT ?descLower WHERE
2    { SP1         profile:desc     ?desc .
3      ?descLower  pf:lower-case  ( ?desc ) .
4    }
```

**Listing 1.2.** SPARQL query with a single virtual triple pattern expression including property function **lower-case** to convert the text argument to lower case.

$$\mu_2(?profile \mapsto SP2,$$
$$?name \mapsto \text{``}CityCountryHotelInfoService\text{''}),$$
$$\mu_3(?profile \mapsto SP3,$$
$$?name \mapsto \text{``}CityCountryInfoService\text{''}) \}.$$

**Virtual Triple Pattern Approach.** Our proposed approach to enable inductive reasoning via SPARQL exploits ARQ property functions (aka *magic properties*).[9] The concept behind property functions is simple: whenever the predicate of a triple pattern is prefixed with a special name, a call to a customized, external prediction function (CPF) is made and arguments are passed to the function (in this case by the object of the triple pattern). The passed object may be an arbitrary list of query variables for which solution mappings were already found during query execution. The property function determined by the property URI computes a value and returns it to the subject variable of the triple pattern.

We call this the *virtual triple pattern approach* as such triple pattern expressions including property functions are not matched against the underlying ontology graph, but against the only virtually existing class membership of the resource specified in the pattern expression. More formally, a virtual triple pattern expression $vt$ is defined as a triple employing a particular kind of property function reference by a property URI:

**Definition 2 (Virtual Triple Pattern).** *A virtual triple pattern $vt$ is a triple of the form { **?v pf:funct ArgList** } where **pf:funct** is a property function and **ArgList** a list of solution mapping arguments $\mu(?x_1 \mapsto t_1), \mu(?x_2 \mapsto t_2), \ldots, \mu(?x_n \mapsto t_n)$. The value computed by **pf:funct** is bound to the subject variable **?v**.*

Similarly to the definition of solution mappings, *virtual solution mappings* can now be defined.

**Definition 3 (Virtual Solution Mapping).** *A virtual solution mapping $\mu_v(?v \mapsto t)$ maps a query variable $?v \in V$ to an RDF term $t$ where $V$ is the infinite set of query variables and $t$ an RDF literal **not included** in the queried RDF graph. The domain of $\mu_v$, $dom(\mu_v)$, is the subset of $V$ where $\mu_v$ is defined.*

The sets of virtual solution mappings $\mu_v$ are defined as $\Omega_{VGP}$ and the sets of solution mappings found by basic graph pattern matching as $\Omega_{BGP}$. Furthermore, based on the description of basic graph patterns in [37], *virtual graph patterns* $VP$ are defined as sets of virtual triple patterns $vt$.

---

[9] http://jena.sourceforge.net/ARQ/extension.html#propertyFunctions

**Example 4 (Virtual Solution Mapping).** *Consider the query shown in Listing 1.2. Matching the first triple pattern on line 2 against dataset D results in the following set of solution mappings:*

$$\Omega_{BGP} = \{ \ \mu(?desc \mapsto \text{``Often used service to get}$$
$$\text{information about luxury hotels.''}) \ \}.$$

*The evaluation of the virtual triple pattern on line 3 results in a call to the property function* lower-case, *which results in the set $\Omega_{VGP}$ of a single virtual solution mapping, i.e.,*

$$\Omega_{VGP} = \{ \ \mu_v(?descLower \mapsto \text{``often used service to}$$
$$\text{get information about luxury hotels.''}) \ \}.$$

**Statistical Relational Learning (SRL) Methods.** SRL methods have been shown to be very powerful as they model not only the *intrinsic attributes* of objects, but also the *extrinsic relations* to other objects, thus, should perform at least as accurate as traditional, propositional learning techniques (cf. [13], [33], and [34]).

Note that in accordance with [33], we refer to such objects with links to intrinsic and extrinsic attributes as *subgraphs*: "The SRL algorithms take a collection of subgraphs as input. Each subgraph contains a single target object to be classified; The objects and links in the subgraph form its relational neighborhood."

**Example 5 (Intrinsic vs. Extrinsic Attributes).** *Consider Figure 4 that shows service A represented as relational subgraph. The subgraph on the left contains the service profile of A, the links to its name, description, and category, as well as the links to its in- and output concepts. These objects and links in the subgraph are called* intrinsic *as they are directly associated with A.*

*The subgraph on the right basically models the same information about A but is extended with* extrinsic *relations to other objects. In this example, these relations are the* subClassOf *links to the super concepts of A's asserted (i.e., direct) I/O concepts. Of course, these relations could again have other relations to other objects resulting in an even larger relational neighborhood of A.*

**Relational Bayesian Classifiers (RBCs).** An RBC is a modification of the traditional Simple Bayesian Classifier (SBC) for relational data [34] (also called *Naïve Bayes Classifier*). SBCs assume that the attributes of an instance $C$ are conditionally independent of each other given the class of the instance. Hence, the probability of the class given an example instance can be computed as the product of the probabilities of the example's attributes $A_1, \ldots, A_n$ given the class, *i.e.,*

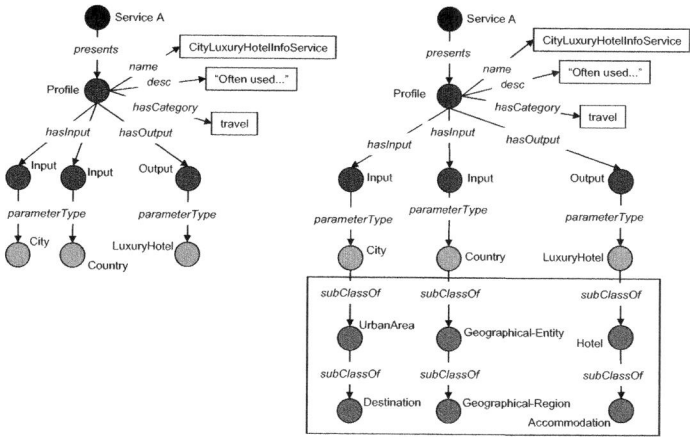

**Fig. 4.** Intrinsic vs. extrinsic attributes of the semantic service A. The subgraph on the left contains A's intrinsic relations to its attributes, whereas on the right the extrinsic relations are shown. These extrinsic relations are the `subClassOf` links to the super concepts of A's asserted (*i.e.*, direct) I/O concepts.

$$
\begin{aligned}
Pr(C = c_i \mid A_1, \ldots, A_n) \\
= \alpha Pr(A_1, \ldots, A_n \mid C = c_i) Pr(C = c_i) \\
= \alpha Pr(C = c_i) \times \prod_{i=1}^{n} Pr(A_i \mid C = c_i).
\end{aligned}
\tag{1}
$$

Equation 1 is exactly Bayes' rule of conditional probability where $\alpha$ is a scaling factor dependent only on the attributes $A_1, \ldots, A_n$.

RBCs apply this independence assumption to relational data. The RBC algorithm transforms the heterogeneous subgraphs in Figure 4 to homogenous sets of attributes as shown in Tables 1 and 2. Each row in the tables stands for a subgraph (*i.e.*, semantic service), each column represents one of its attributes, and the cells contain the multisets (or distributions) of values of attributes. These attributes include the service category as well as the asserted and inferred I/O concept distributions of the semantic services.

Learning an RBC model then basically consists of estimating probabilities for each attribute and/or attribute-value distribution. Such probability estimation techniques include, but are not limited to, *average-value* and *random-value* estimations (cf. [34]).

**Relational Probability Trees (RPTs).** RPTs extend standard probability estimation trees (also called *decision trees*) to a relational setting, in which data instances are heterogeneous and interdependent [33].[10] Similar to RBCs, RPTs

---

[10] Actually, when it comes to predicting numeric values, decision trees with averaged numeric values at the leaf nodes are called *regression trees*.

**Table 1.** The relational subgraphs of the semantic services A, B, and C are decomposed by attributes. The table lies the focus on the input concepts. Each column represents one of the service's attributes and the cells contain the multisets (or distributions) of values of these attributes.

| Service | Category | Inputs | Input Super Concepts (ISCs) |
|---------|----------|--------|------------------------------|
| A | travel | { travel.owl#City, portal.owl#Country } | { Destination, Generic-Agent, Geographical-Entity, Geographical-Region, Location, support.owl#Thing, Tangible-Thing, Temporal-Thing, UrbanArea } |
| B | travel | { portal.owl#City, portal.owl#Country } | { Generic-Agent, Geographical-Entity, Geographical-Region, Location, Municipal-Unit, support.owl#Thing, Tangible-Thing, Temporal-Thing } |
| C | education | { Sports } | { Activity } |

**Table 2.** The relational subgraphs of the semantic services A, B, and C are decomposed by attributes (focus on output concepts)

| Service | Category | Outputs | Output Super Concepts (OSCs) |
|---------|----------|---------|-------------------------------|
| A | travel | { LuxuryHotel } | { Accommodation, Hotel } |
| B | travel | { Hotel } | { Accommodation } |
| C | education | { Capital } | { Destination, travel.owl#City, UrbanArea } |

look beyond the intrinsic attributes of objects, for which a prediction should be made; it also considers the effects of adjacent objects (extrinsic relations) on the prediction task.

As is the case for RBCs, the RPT algorithm first transforms the relational data (the semantic services represented as subgraphs) to multisets of attributes. It then attempts to construct an RPT by searching over the space of possible binary splits of the data based on the relational features, until further processing no longer changes the class distributions significantly. The features for splitting these (training) data are created by mapping the multisets of values into single-value summaries with the help of aggregation functions. These functions are for instance *count, mode/average, degree, proportion, minimum, maximum,* and *exists* (see [33]).

**Example 6 (RPT Classification).** *As an example, consider the RPT shown in Figure 5 that predicts the value for a semantic service's* hasCategory *attribute. The value of this attribute should be one out of* communication, economy, education, food, medical, travel, *and* weapon. *The root node in the RPT starts by examining the super concepts of the service's direct (i.e., asserted) output concepts in the current subgraph. If the proportion of all super concepts being* concept.owl#UntangibleObjects *is greater than or equal to 0.0345, the left edge in the RPT is traversed. Assume no, the next test looks at how many super concepts have type* my_ontology.owl#Liquid *in the subgraph, represented by* count(link_subClassOf.outputSuper = my_ontology.owl#-Liquid). *Specifically, we test whether the subgraph contains at least one such*

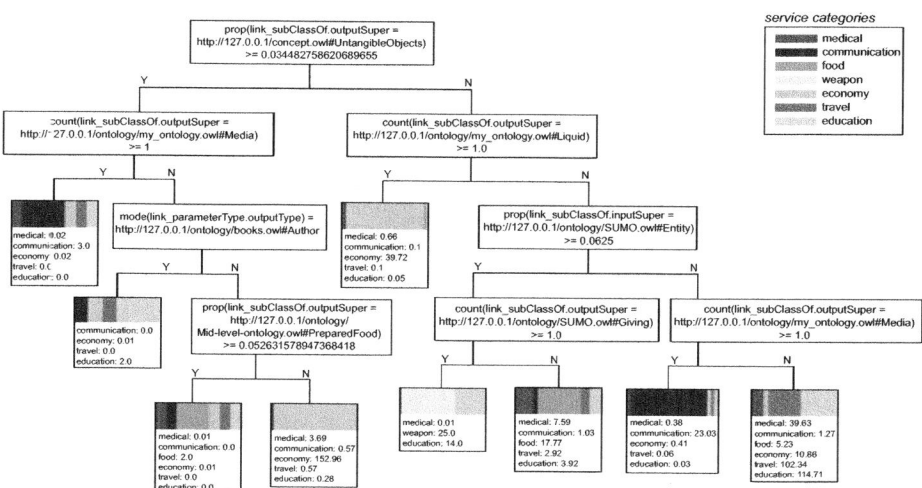

**Fig. 5.** Example RPT to predict the value for a semantic service's `hasCategory` attribute

*super concept. If this is the case, we pass this test and traverse the left edge to the leaf node.*

*The leaf nodes show the distribution of the training examples (that "reached the leaf") and the resulting class probabilities of the `hasCategory` target attribute. In other words, the leaf nodes hold the probabilistic counts (out of all services from the training set that reach this leaf node) for each potential classification of this service. We can observe that services that reach this leaf node have much more likely category `economy` than any other category. Therefore, this model would predict that this service (subgraph) has category `economy`.*

### 3.3  Adding Inductive Reasoning Support to SPARQL via SRL Methods

SPARQL-ML is an extension of SPARQL that extends the Semantic Web query language with knowledge discovery capabilities. Our inductive reasoning extensions add new syntax elements and semantics to the official SPARQL grammar described in [37]. In a nutshell, SPARQL-ML facilitates the following two tasks on any Semantic Web dataset: (1) induce a model based on training data using the new CREATE MINING MODEL statement (Section 3.3); and (2), apply a model to make predictions via two new ARQ property functions (Section 3.3). The model created in the CREATE MINING MODEL step follows the definitions in our *SPARQL Mining Ontology (SMO)* presented in Section 3.3.

**Table 3.** Extended SPARQL grammar for the `CREATE MINING MODEL` statement

| [1] | *Query* | ::= | Prologue( SelectQuery \| ConstructQuery \| DescribeQuery \| AskQuery \| CreateQuery ) |
|---|---|---|---|
| [100] | *CreateQuery* | ::= | CREATE MINING MODEL' SourceSelector '{' Var 'RESOURCE' 'TARGET' ( Var ( 'RESOURCE' \| 'DISCRETE' \| 'CONTINUOUS' ) 'PREDICT'? )+ '}' DatasetClause* WhereClause SolutionModifier UsingClause |
| [102] | *UsingClause* | ::= | 'USING' SourceSelector BrackettedExpression |

SPARQL-ML is implemented as an extension to ARQ—the SPARQL query engine for Jena.[11] The current version of SPARQL-ML supports, but is not limited to *Proximity*[12] and *Weka*[13] as data mining modules.

**Step 1: Learning a Model Syntax and Grammar.** SPARQL-ML enables to induce a classifier (model) on any Semantic Web training data using the new `CREATE MINING MODEL` statement. The chosen syntax was inspired by the Microsoft Data Mining Extension (DMX) that is an extension of SQL to create and work with data mining models in Microsoft SQL Server Analysis Services (SSAS) 2005.[14] The extended SPARQL grammar is tabulated in Table 3. Listing 1.3 shows a particular example query to induce an RPT model for the prediction of the category of a semantic service.

Our approach adds the `CreateQuery` symbol to the official SPARQL grammar rule of *Query* [37]. The structure of `CreateQuery` resembles the one of `SelectQuery`, but has complete different semantics: the `CreateQuery` expands to Rule 100 adding the new keywords `CREATE MINING MODEL` to the grammar followed by a `SourceSelector` to define the name of the trained model. In the body of `CreateQuery`, the variables (attributes) to train the model are listed. Each variable is specified with its content type, which is currently one of the following: `RESOURCE`—variable holds an RDF resource (IRI or blank node), `DISCRETE`—variable holds a discrete/nominal literal value, `CONTINUOUS`—variable holds a continuous literal value, and `PREDICT`—tells the learning algorithm that this feature should be predicted. The first attribute is additionally specified with the `TARGET` keyword to denote the resource for which a feature should be predicted (also see [33]).

After the usual `DatasetClause`, `WhereClause`, and `SolutionModifier`, we introduced a new `UsingClause`. The `UsingClause` expands to Rule 102 that adds the new keyword `USING` followed by a `SourceSelector` to define the name and parameters of the learning algorithm.

**Semantics.** According to [35], a SPARQL query consists of three parts: the *pattern matching part*, the *solution modifiers*, and the *output*. In that sense, the semantics of the `CREATE MINING MODEL` queries is the construction of new triples

---

[11] http://jena.sourceforge.net/

[12] http://kdl.cs.umass.edu/proximity/index.html

[13] http://www.cs.waikato.ac.nz/ml/weka/

[14] http://technet.microsoft.com/en-us/library/ms132058.aspx

```
1  CREATE MINING MODEL <http://www.ifi.uzh.ch/services>
2  { ?service     RESOURCE TARGET
3    ?category    DISCRETE PREDICT
4                 { 'communication', 'economy',
5                   'education', 'food', 'medical',
6                   'travel', 'weapon' }
7    ?profile     RESOURCE
8    ?output      RESOURCE
9    ?outputType  RESOURCE
10   ?outputSuper RESOURCE
11   ?input       RESOURCE
12   ?inputType   RESOURCE
13   ?inputSuper  RESOURCE
14 }
15 WHERE
16 { ?service service:presents    ?profile ;
17           service:hasCategory ?category .
18
19   OPTIONAL
20   { ?profile profile:hasOutput    ?output .
21     ?output  process:parameterType ?outputType .
22
23     OPTIONAL
24     { ?outputType rdfs:subClassOf ?outputSuper . }
25   }
26
27   OPTIONAL
28   { ?profile profile:hasInput     ?input .
29     ?input   process:parameterType ?inputType .
30
31     OPTIONAL
32     { ?inputType rdfs:subClassOf ?inputSuper . }
33   }
34 }
35 USING <http://kdl.cs.umass.edu/proximity/rpt>
```

**Listing 1.3.** SPARQL-ML CREATE MINING MODEL query for semantic service classification. The goal of this query is to induce an RPT model that predicts the value for a service's hasCategory attribute that should be one out of communication, economy, education, food, medical, travel, and weapon as defined by the DISCRETE PREDICT keywords on line 4–5.

describing the metadata of the trained model (*i.e.*, SPARQL-ML introduces a new output type). An example of such metadata for the model induced in Listing 1.3 is shown in Listing 1.4, which follows the definitions of our *SPARQL Mining Ontology (SMO)* in Figure 6. The ontology enables to permanently save the parameters of a learned model, which is needed by the predict queries (see next section).

The ontology includes the model name, the used learning algorithm, all variables/features being used to train the classifier, as well as additional information, such as where to find the generated model file. In Listing 1.4, lines 1–11 show the constructed triples of a model with name *services*, while lines 13–28 show the metadata for two particular features of the model.

**Step 2: Making Predictions Via Virtual Triple Patterns.** The second step to perform inductive reasoning with SPARQL-ML is to apply the previously induced model to draw conclusions about new samples from the population. After the induction of the model with the CREATE MINING MODEL statement, SPARQL-ML allows the user to make predictions via two new ARQ property functions. In the following, these functions are called sml:predict and sml:mappedPredict.

Property functions are called whenever the predicate of a triple pattern is prefixed with a special name (*e.g.*, sml). In that case, a call to an external function is made and arguments are passed to the function (by the object of

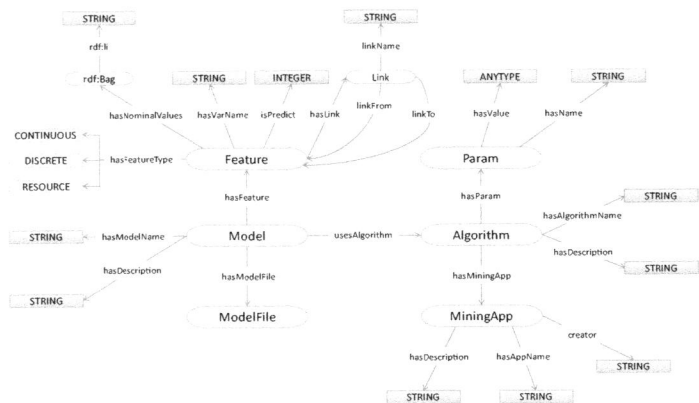

**Fig. 6.** SPARQL-ML Mining Ontology (SMO)

the triple pattern). For inductive reasoning, we are particularly interested in the following form of virtual triple pattern expressions:

$$\underbrace{(pred\ prob)}_{\text{subject}}\ \underbrace{predictionFunction}_{\text{predicate}}\ \underbrace{(arg1\ \ldots\ argN)}_{\text{object}}$$

In a nutshell, such pattern expressions define a list of arguments that are passed to a customized prediction function (CPF) by the object of the pattern expression. In our case, the first argument in this list (`arg1`) is a URI reference to the previously induced model that will be applied for making predictions. The rest of the arguments describe the new resource for which a prediction should be made.

**Example 7 (SPARQL-ML Prediction Query).** *Consider the SPARQL-ML query shown in Listing 1.5 that includes a single virtual triple pattern expression on lines 22–27. The goal of the query is to predict the value of a semantic service's* `hasCategory` *attribute by applying the previously induced model in Listing 1.3. The model is referenced by the model URI* `http://www.ifi.uzh.ch/ddis/services` *and passed as the first argument to the prediction function. The rest of the arguments define the attributes/features of the service that should be used for predicting its category. The result of the prediction (one out of* `communication`, `economy`, `education`, `food`, `medical`, `travel`, *or* `weapon`*), and its probability are finally bound on line 22 to the variables* `?prediction` *and* `?probability` *respectively.*

**Syntax and Grammar.** The extended SPARQL-ML grammar for the prediction queries is shown in Table 4. To implement the virtual triple approach in SPARQL-ML, a new symbol called *PredictionBlockPattern* is added to the official SPARQL grammar rule of *GraphPatternNotTriples* [37]. The structure

```
 1  <http://www.ifi.uzh.ch/ddis/services>
 2   a smo:Model ;
 3     smo:hasFeature
 4       <http://www.ifi.uzh.ch/ddis/services#output> ,
 5       <http://www.ifi.uzh.ch/ddis/services#category> ,
 6       <http://www.ifi.uzh.ch/ddis/services#input> ;
 7     smo:hasModelFile
 8       <http://www.ifi.uzh.ch/ddis/models/services.xml> ;
 9     smo:hasModelName   "services" ;
10     smo:usesAlgorithm
11       <http://kdl.cs.umass.edu/proximity/rpt> .
12
13  <http://www.ifi.uzh.ch/ddis/services#category>
14   a smo:Feature ;
15     smo:hasFeatureType   "DISCRETE" ;
16     smo:hasNominalValues
17     [ a rdf:Bag ;
18         rdf:_1   "education" ;
19         rdf:_2   "travel"
20     ] ;
21     smo:hasVarName   "category" ;
22     smo:isPredict   "1" .
23
24  <http://www.ifi.uzh.ch/ddis/services#service>
25   a smo:Feature ;
26     smo:hasFeatureType   "RESOURCE" ;
27     smo:hasVarName   "service" ;
28     smo:isRootVar   "YES" .
```

**Listing 1.4.** Part of the metadata generated from inducing the RPT model as shown in Listing 1.3.

**Table 4.** SPARQL-ML grammar rules for the `PREDICTION` statement

| [22] | *GraphPatternNotTriples* ::= | OptionalGraphPattern \| GroupOrUnionGraphPattern \| |
| | | GraphGraphPattern \| PredictionBlockPattern |
| [22.1] | *PredictionBlockPattern* ::= | 'PREDICTION' '{' ( ( Var1 FunctionCall )+ Filter? )+ '}' |
| [28] | FunctionCall ::= | IRIref ArgList |

of *PredictionBlockPattern* resembles the one of *OptionalGraphPattern* but has completely different semantics: instead of matching patterns in the RDF graph, the triples in a *PredictionBlockPattern* act as virtual triple patterns that are interpreted by the query processor. A *PredictionBlockPattern* expands to Rule [22.1] that adds the new keyword `PREDICTION` to the grammar, which is followed by a number of virtual triples and optional `FILTER` statements.

**Semantics.** The semantics of a *PredictionBlockPattern* is basically that of a *prediction join*:[15] (1) the CPF maps the variables in the basic graph patterns of the query to the features in the specified model; (2) the CPF creates instances out of the mappings according to the induced model; (3) the model is used to classify an instance as defined in the `CREATE MINING MODEL` query; and (4), the values of the prediction and its probability are bound to variables in the predict query.

More formally, in the notation of Pérez [35], the semantics of a *PredictionBlockPattern* can be defined as follows. In [35], Pérez discussed four different SPARQL query types: join queries, union queries, optional queries, and filter queries. In accordance to [35], prediction joins are, thus, introduced as a new type of SPARQL queries for which the semantics is subsequently investigated in the remainder of this section. The new type is specified as follows (displayed in original SPARQL syntax on the left and algebraic syntax on the right):

---

[15] http://msdn2.microsoft.com/en-us/library/ms132031.aspx

```
1   SELECT DISTINCT ?service ?prediction ?probability
2   WHERE
3   { ?service service:presents ?profile .
4
5     OPTIONAL
6     { ?profile profile:hasOutput      ?output .
7       ?output   process:parameterType ?outputType .
8
9       OPTIONAL
10        { ?outputType rdfs:subClassOf ?outputSuper . }
11    }
12
13    OPTIONAL
14    { ?profile profile:hasInput       ?input .
15      ?input    process:parameterType ?inputType .
16
17      OPTIONAL
18        { ?inputType rdfs:subClassOf ?inputSuper . }
19    }
20
21    PREDICTION
22    { ( ?prediction ?probability )
23        sml:predict
24        ( <http://www.ifi.uzh.ch/ddis/services>
25          ?service ?profile ?output ?outputType
26          ?outputSuper ?input ?inputType
27          ?inputSuper ) .
28    }
29 }
```

**Listing 1.5.** SPARQL-ML query to predict the the value of a service's hasCategory attribute

**Definition 4 (Prediction Join Query).** *Prediction join queries involve basic graph patterns $P$ and virtual graph patterns $VP$ which trigger a call to a customized prediction function, i.e.,*

$$\{ P \; PREDICTION \; \{ \; VP \; \} \; \} \Longleftrightarrow ( P \; PREDJOIN \; VP ).$$

Similarly to the definition of the join of ordinary sets of solution mappings, the prediction join of sets $\Omega_{BGP}$ and $\Omega_{VGP}$ can now be defined:

**Definition 5 (Prediction Join Operation).** *A prediction join $\bowtie_p$ of basic graph pattern expressions $P$ and virtual graph pattern expressions $VP$ extends the sets $\Omega_{BGP}$ from basic graph pattern matching with the sets of virtual solution mappings $\Omega_{VGP}$ from virtual graph pattern matching. The prediction join of $\Omega_{BGP}$ and $\Omega_{VGP}$ is defined as:*

$$\Omega_{BGP} \bowtie_p \Omega_{VGP} = \{ \; \mu_1 + \mu_2 \; |$$
$$\mu_1 \in \Omega_{BGP}, \; \mu_2 \in \Omega_{VGP}, \; \mu_1, \; \mu_2 \; are$$
$$compatible, \; and \; 1 \leq card[\Omega_{VGP}](\mu_2) \leq 2 \; \}$$

**Example 8 (Prediction Join Operation).** *Consider the query shown in Listing 1.6 for the prediction of the value of the hasCategory attribute of a semantic service (assume an appropriate induction model was induced in an earlier step). Focusing only on service A, the evaluation of the basic triple patterns results in the set of solution mappings $\Omega_1$, i.e.,*

$$\Omega_1 = \{ \; \mu_{11}(?profile \mapsto SP1, ?input \mapsto \_CITY,$$
$$?ouput \mapsto \_LUXURYHOTEL\}.$$

```
1  SELECT ?prediction ?probability
2  WHERE
3    { ?profile   profile:hasInput    ?input ;
4                 profile:hasOutput   ?ouput .
5
6      ( ?prediction ?probability )
7          sml:predict ( <modelURI> ?profile
8                                    ?input ?output ) .
9    }
```

**Listing 1.6.** SPARQL-ML query exemplifying a prediction join operation

*The evaluation of the virtual triple pattern that specifies the property function for making predictions returns a set of virtual solution mappings $\Omega_2$ that contains the values of the prediction and its probability. Assume the prediction model returns the following values, i.e.,*

$$\Omega_2 = \{\ \mu_{21}(?prediction \mapsto travel,$$
$$?probability \mapsto 0.99)\ \}.$$

*Finally, the prediction join operation merges $\Omega_1$ and $\Omega_2$ into the set of solution mappings $\Omega_3$:*

$$\Omega_3 = \{\ \mu_{31}(?profile \mapsto SP1, ?input \mapsto \_CITY,$$
$$?ouput \mapsto \_LUXURYHOTEL,$$
$$?prediction \mapsto travel, ?probability \mapsto 0.99)\ \}.$$

In [27], the semantics of virtual graph patterns were defined as an evaluation function $[[vt]]$ that takes a virtual triple pattern $vt$ and returns a virtual solution mapping $\mu_v$. Adapting this equation to the inductive reasoning scenario in this paper, the evaluation of a SPARQL-ML predict query over a dataset $D$ can be defined recursively as follows:

$$[[vt]] = \{\ \mu_v(?v_1 \mapsto pre, ?v_2 \mapsto pro)\ |\ (pre, pro)$$
$$= \text{pf:funct}\ (\ \mu(?x_1 \mapsto t_1), \dots, \mu(?x_n \mapsto t_n)\ )\ \} \qquad (2)$$
$$[[(P\ \text{PREDJOIN}\ VP)]]_D = [[P]]_D \bowtie_p [[VP]]$$

Again, the first part of Equation 2 takes a virtual triple pattern expression and returns a set of virtual solution mappings $\Omega_{VGP}$. New solution mappings are generated that assign the value of a prediction and its probability to query variables (*i.e.*, $?v1$ and $?v2$) (note that Equation 2 only shows the case were both values are returned).

**Pros and Cons.** The following list summarizes the pros and cons of the virtual triple pattern approach to perform inductive reasoning with our SPARQL-ML framework.

+ A number of different prediction models can be used in the same query (which is useful to compare their performance).
+ The integration of inductive reasoning support into SPARQL provides an easy-to-use and flexible approach to quickly create and work with data mining models in SPARQL.
+ The values of the predictions and its probabilities are assigned to query variables, thus, can be reused in the query for filtering and ranking, or can be returned for arbitrary further processing.
+ Solution modifiers such as `ORDER BY` and `LIMIT` are applicable to the calculated prediction (probability) values.
+ A very simple adaption of `sml:predict` allows us to also apply the induced model on a dataset with a different ontology structure (*i.e.*, `sml:mappedPredict`).
− The virtual triple pattern expressions we use for prediction are somehow 'overloaded' (*i.e.*, the property functions potentially have a long parameter list). Furthermore, the functions may return a list of prediction-probability values.
− The SPARQL grammar needs to be extended to account for the `PREDICTION` statements (which requires an adaptation of the query engines).
− Queries using property functions depend on a query engine extension currently only implemented in Jena ARQ and, hence, have limited interoperability.

## 4    Evaluation/Validation of SPARQL-ML

Our inductive reasoning method presented in Section 3 relies on statistical induction to reason over Semantic Web data. We have implemented inductive reasoning as an extension to the RDF query language SPARQL. More specifically, we use virtual triple patterns as key technology to integrate inductive reasoning with the traditional Semantic Web infrastructure.

This section is devoted to the application and evaluation of this novel reasoning method for three Semantic Web and one Software Analysis task. These tasks along with the datasets we used to evaluate them are listed in Table 5. In the following, we will briefly give an overview of each of these tasks.

**Business Project Success Experiment.** In order to show the ease-of-use and predictive capability of our inductive reasoning framework SPARQL-ML, we put together a proof of concept setting with a small, artificially created dataset. To that end, in our first experiment in Section 4.1, we show that using a synthetic dataset, the combination of statistical inference with logical deduction produces superior performance over statistical inference only.

**Semantic Web Service Classification Experiment.** The goal of our semantic service classification experiment in Section 4.2 is to evaluate our novel inductive reasoning extension to the task of performing automatic service classification. To that end, we perform a Semantic Web service category prediction

**Table 5.** The four tasks and datasets we considered to evaluate/validate our novel inductive reasoning extension

| Evaluation/Validation Task | Dataset(s) |
|---|---|
| Business Project Success Experiment | synthetic business project dataset |
| Semantic Web Service Classification Experiment | OWL-S TC v2.1 |
| SVM-Benchmark Experiment | SWRC/AIFB dataset |
| Bug Prediction Experiment | Eclipse `updateui`, `updatecore`, `search`, `pdeui`, `pdebuild`, and `compare` plug-ins |

experiment (*i.e.*, automatically generate semantic annotation/metadata for semantic services). As benchmarking dataset, we use a large OWL-S semantic service retrieval test collection.

**SVM-Benchmark Experiment.** In our third experiment—the SVM-benchmark experiment—we compare the prediction performance of our SPARQL-ML approach to another state-of-the-art kernel-based Support Vector Machine (SVM) [6] using a real-world data set.

**Software Bug Prediction Experiment.** Finally, in our bug prediction experiment in Section 4.4, we aim to show some of the advantages of inductive reasoning for *Software Analysis*. Specifically, we will use SPARQL-ML in combination with the EvoOnt software model to perform bug prediction. To that end, the defect location experiment presented in [3] is repeated.

### 4.1   Business Project Success Experiment

**Evaluation Methodology and Dataset.** The synthetic *business project dataset* consists of different business projects and the employees of an imaginary company. The company has 40 employees each of which having one out of 8 different occupations. Figure 7 shows part of the created ontology in more detail. In our dataset, 13 employees belong to the superclass `Manager`, whereas 27 employees belong to the superclass `Non-Manager`.

We then created business projects and randomly assigned up to 6 employees to each project. The resulting teams consist of 4 to 6 members. Finally, we randomly defined each project to be successful or not, with a bias for projects being more successful, if more than three team members are of type `Manager`. The resulting dataset contains 400 projects with different teams. The prior probability of a project being successful is 35%. We did a 50:50 split of the data and followed a single holdout procedure, swapping the roles of the testing and training set and averaged the results.

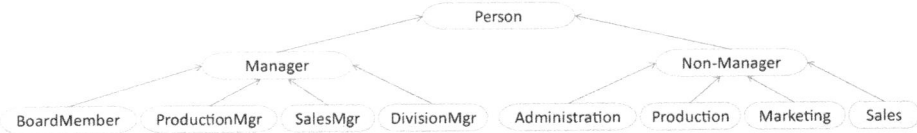

**Fig. 7.** Example business ontology

```
1  CREATE MINING MODEL <http://www.example.org/projects>
2  { ?project RESOURCE TARGET
3    ?success DISCRETE PREDICT {'Yes','No'}
4    ?member  RESOURCE
5    ?class   RESOURCE
6  }
7  WHERE
8  { ?project ex:isSuccess ?success .
9    ?project ex:hasTeam    ?member .
10   ?member  rdf:type      ?class .
11 }
12 USING <http://kdl.cs.umass.edu/proximity/rpt>
```

**Listing 1.7.** SPARQL-ML CREATE MINING MODEL query. The goal of this query is to induce an RPT model that predicts the value for a project's isSuccess attribute that should be either Yes or No as defined by the DISCRETE PREDICT keywords on line 3.

**Experimental Results.** Listing 1.7 shows the CREATE MINING MODEL query that we used in the model learning process. We tested different learning algorithms with and without the support of inferencing. With the reasoner disabled, the last triple pattern in the WHERE clause (line 10) matches only the direct type of the received employee instance (*i.e.*, if an employee is a 'direct' instance of class Manager). This is the typical situation in relational databases without the support of inheritance. With inferencing enabled, the last triple pattern also matches all inferred types, indicating if an employee is a Manager or not.

Given the bias in the artificial dataset, it is to be expected that the ability to infer if a team member is a Manager or not is central to the success of the induction procedure. Consequently, we would expect that models induced on the inferred model should exhibit a superior performance. The results shown in Figure 8 confirm our expectations. The Figure shows the results in terms of prediction accuracy (ACC; in legend), Receiver Operating Characteristics (ROC; graphed), and the area under the ROC-curve (AUC; also in legend). The ROC-curve graphs the true positive rate (y-axis) against the false positive rate (x-axis), where an ideal curve would go from the origin to the top left (0,1) corner, before proceeding to the top right (1,1) one [36]. It has the advantage to show the prediction quality of a classifier independent of the distribution (and, hence, prior) of the underlying dataset. The area under the ROC-curve is, typically, used as a summary number for the curve. Note that a random assignment whether a project is successful or not is also shown as a line form the origin (0,0) to (1,1). The learning algorithms shown are a Relational Probability Tree (RPT), a Relational Bayes Classifier (RBC), both with and without inferencing, and, as a baseline, a $k$-nearest neighbor learning algorithm ($k$-NN) with inferencing and $k = 9$ using a maximum common subgraph isomorphism metric [44] to compute the closeness to neighbors.

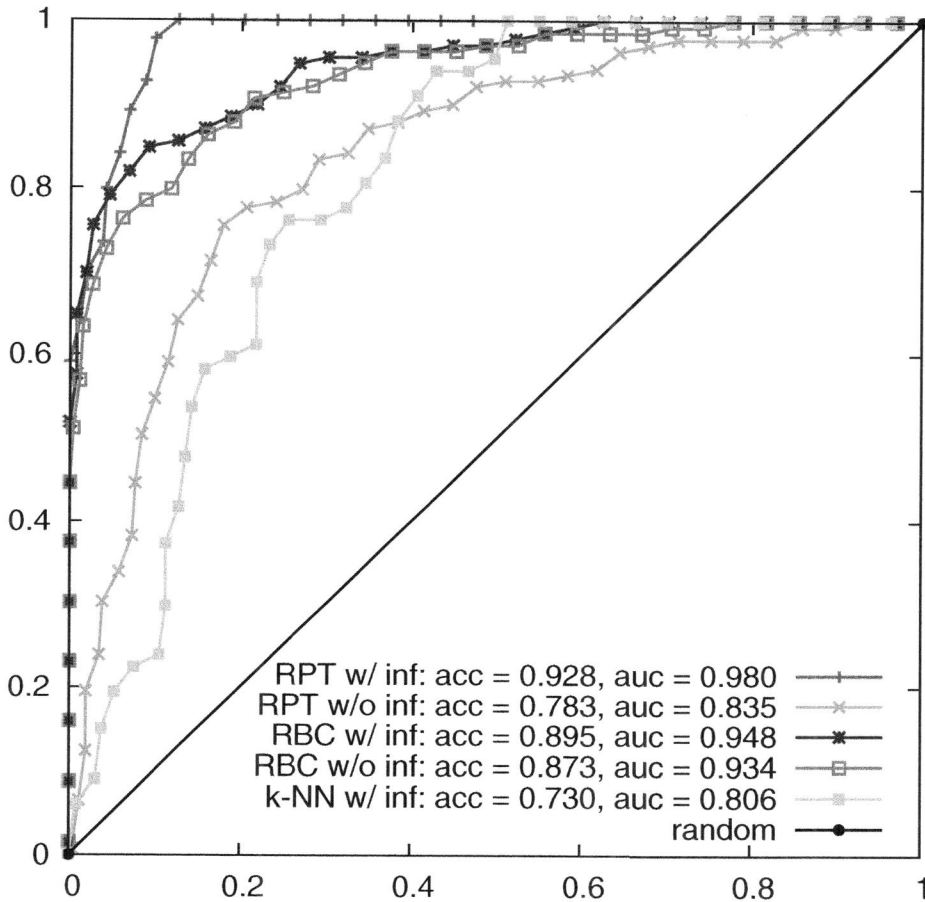

**Fig. 8.** ROC-Curves of business project success prediction

As the Figure shows, the relational methods clearly dominate the baseline
$k$-NN approach. As expected, both RPT and RBC with inferencing outperform
the respective models without inferencing. It is interesting to note, however, that
RPTs seem to degrade more with the loss of inferencing than RBCs. Actually,
the lift of an RBC with inferencing over an RBC without inferencing is only
small. These results support our assumption that the combination of induction
and deduction should outperform pure induction. The major limitation of this
finding is the artificial nature of the dataset. We, therefore, decided to conduct
further experiments with the same goals using real-world datasets, which we
present in the following sections.

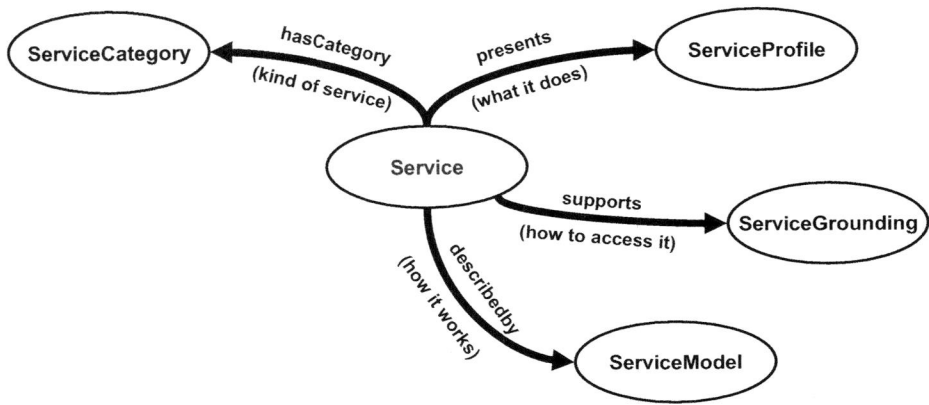

**Fig. 9.** Extended OWL-S upper ontology model. In addition to the service profile, grounding, and model, an extra relation to the service category is added to the description of a service.

## 4.2   Semantic Web Service Classification Experiment

In this section, we proceed with the evaluation of SPARQL-ML on a real-world dataset. Specifically, we show how SPARQL-ML can be used to automatically classify Semantic Web services into their most appropriate service *category*. According to [22], a web service category describes the general kind of service that is offered, such as "travel services" or "educational services".

In a nutshell, our SPARQL-ML framework is used to classify/predict the category of a semantic service, which is usually a string value, say, `travel` or `education`. This value can then be used to *tag* (annotate) the semantic service.[16]

**Evaluation Methodology and Dataset.** For all our service classification experiments we use the OWLS-TC v2.1 Semantic Web service retrieval test collection.[17] OWLS-TC contains 578 semantic service descriptions of seven different categories. These categories are `economy`, `education`, `travel`, `medical`, `communication`, `food`, and `weapon`. The prior distribution of the services is `economy` = 35.63%, `education` = 23.36%, `travel` = 18.34%, `medical` = 8.99%, `communication` = 5.02%, `food` = 4.33%, and `weapon` = 4.33% (*i.e.*, economy is the category with the most services).

In order to predict a semantic service's `hasCategory` attribute, we first had to assert this information in the dataset (as it is originally not). In other words, we had to extend the OWL-S service ontology model with an additional relation to the service category. The extended OWL-S ontology is shown in Figure 9.

---

[16] *E.g.*, in Semantic Web terminology add a new triple to the service description holding the value of the classification step. Note, however, that our focus clearly lies on service classification rather than service annotation.

[17] http://projects.semwebcentral.org/projects/owls-tc/

**Table 6.** Detailed results for the Semantic Web service classification experiments. As can be observed, the models induced on the (logically) inferred I/O concepts (*w/ inf*) perform considerably better than the ones induced on only the asserted information (*w/o inf*) across almost all measures and categories.

| Category | FP Rate | | Precision | | Recall | | F-measure | |
|---|---|---|---|---|---|---|---|---|
| | w/o inf | w/ inf | w/o inf | w/ inf | w/o inf | w/ inf | w/o inf | w/ inf |
| communication | 0.007 | 0.004 | 0.819 | 0.900 | 0.600 | 0.600 | 0.693 | 0.720 |
| economy | 0.081 | 0.018 | 0.810 | 0.964 | 0.644 | 0.889 | 0.718 | 0.925 |
| education | 0.538 | 0.090 | 0.311 | 0.716 | 0.904 | 0.869 | 0.463 | 0.786 |
| food | 0 | 0.002 | 0 | 0.960 | 0 | 0.800 | 0 | 0.873 |
| medical | 0.006 | 0.030 | 0 | 0.688 | 0 | 0.550 | 0 | 0.611 |
| travel | 0 | 0.069 | 1 | 0.744 | 0.245 | 0.873 | 0.394 | 0.803 |
| weapon | 0.002 | 0.002 | 0.917 | 0.964 | 0.367 | 0.900 | 0.524 | 0.931 |
| average | **0.091** | **0.031** | **0.551** | **0.848** | **0.394** | **0.783** | **0.399** | **0.807** |
| t-test (paired, one-tailed) | **p=0.201** | | **p=0.0534** | | **p=0.00945** | | **p=0.0038** | |

Using these extended service descriptions, we are able to write CREATE MINING MODEL queries that (i) define the instances to be use for model induction and (ii) specify the learning algorithm and its parameters. Note that in all our experiments we limited our investigations to the I/O concepts of services as we believe that they are most informative for this task (cf. [20]).

**Experimental Results.** Listing 1.3 shows the CREATE MINING MODEL query that we used in the model learning step. By using OPTIONAL patterns, we enable the inclusion of services with no outputs or inputs. The additional OPTIONAL pattern for the rdfs:subClassOf triple enables us to run the same query on the asserted and the inferred data.

We ran the experiment once on the asserted and once on the (logically) inferred model using the predict query shown in Listing 1.5. Furthermore, we performed a 10-fold cross validation where 90% of the data was used to learn a classification model and the remaining 10% to test the effectiveness of the learned model, which is standard practice in machine learning (see [45]). For our experiments, we induced a RPT to predict the service category of a service based on its input and output concepts. We chose an RPT because in all our experiments it turned out to perform superior than RBCs.

The averaged classification accuracy of the results of the 10 runs is 0.5102 on the asserted and 0.8288 on the inferred model. Hence, the combination of logical deduction with induction improves the accuracy by 0.3186 over pure induction. The detailed results of our experiments are shown in Table 6 that further confirm this result for all seven categories by listing the typical data mining measures false positive rate (FP rate), precision, recall, and F-measure for all categories. As the results of the t-test show, the differences for recall and F-measure are (highly) significant. The results for precision just barely misses significance at the 95% level.

When investigating the structure of the RPTs, the trees induced on the inferred model clearly exploit inheritance relations using the transitive *rdfs:subClassOf* property, indicating that the access to the newly derived triples improves the determination of a service's category. The SRL algorithms are

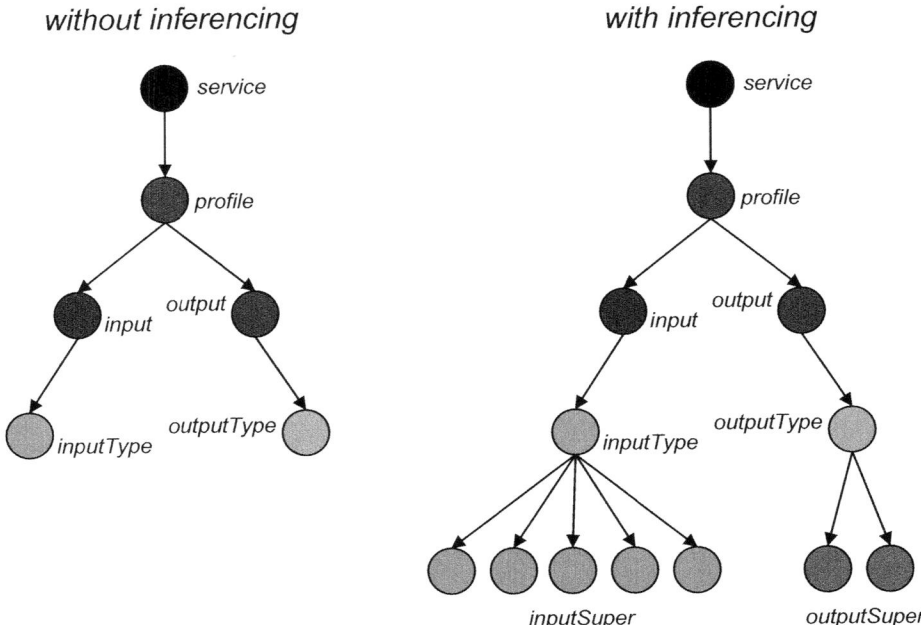

**Fig. 10.** Service subgraphs built for the Semantic Web service classification task, on the left without inferencing and on the right with inferencing

able to exploit the richer relational neighborhood to improve their performance. These observations further support our finding that a combination of deduction and induction is useful for Semantic Web tasks and can be easily achieved with SPARQL-ML.

### 4.3   SVM-Benchmark Experiment

**Evaluation Methodology and Dataset.** With our third set of experiments, we aimed to show possible advantages of SPARQL-ML over another state-of-the-art method. Specifically, we compared the off-the-shelf performance of a simple **xx**-lines SPARQL-ML statement (see Listing 1.8) with a Support Vector Machine (SVM) based approach proposed by Bloehdorn and Sure [7] following exactly their evaluation procedure.[18] In their work, they introduced a framework for the design and evaluation of kernel methods that are used in Support Vector Machines, such as $SVM^{light}$ [24]. The framework provides various kernels for the comparison of classes as well as datatype and object properties of instances. Moreover, it is possible to build customized, weighted combinations of such kernels. Their evaluations include two tasks: (1) prediction of the affiliation a person belongs to (`person2affiliation`), and (2) prediction of the affiliation

---

[18] We would like to thank them for sharing the exact dataset used in their paper.

**Table 7.** LOOCV results for the *person2affiliation* and *publication2affiliation* tasks

| person2affiliation | | | | | publication2affiliation | | | | |
|---|---|---|---|---|---|---|---|---|---|
| *algorithm* | *err* | *prec* | *rec* | *F-measure* | *algorithm* | *err* | *prec* | *rec* | *F-measure* |
| sim-ctpp-pc, c=1 | 4.49 | 95.83 | 58.13 | 72.37 | sim-cta-p, c=10 | 0.63 | 99.74 | 95.22 | 97.43 |
| RBC w/o inf | 9.43 | 79.41 | 77.94 | 78.51 | RBC w/o inf | 1.035 | 97.36 | 94.21 | 95.68 |
| RBC w/ inf | 9.39 | 80.90 | 75.46 | 77.73 | RBC w/ inf | 0.73 | 95.53 | 97.52 | 96.46 |

a publication is related to (`publication2affiliation`). As a dataset they used the SWRC ontology—a collection of OWL annotations for persons, publications, and projects, and their relations from the University of Karlsruhe.[19]

In order to understand the course of our experiment, we think a few words about the experimental procedure described in [7] are necessary. For each of the two tasks, Bloehdorn and Sure performed exactly four binary classification experiments and averaged the results for each task. More precisely, consider the `person2affiliation` task: for each of the four distinct research groups and 177 persons in the SWRC dataset, the authors conducted a two-class classification experiment to predict whether a person belongs to a research group or not. The same approach was chosen for the `publication2affiliation` task: for each of the four research groups and 1232 publication instances in the dataset, a binary classification experiment was performed in order to predict whether one of the authors of the publication is affiliated with the group.

In order to perform the identical experiment as described in [7], we first had to add the information about a person's affiliation to the dataset via a couple of `belongsToGroupX` ($X = 1 \dots 4$) datatype properties. This was necessary because we wanted to predict the value of this property (either 'Yes' or 'No') using our proposed SPARQL-ML SRL methods. An example `CREATE MINING MODEL` query is shown in Listing 1.8, where the goal is to predict whether a person belongs to the research group `Group1`. We ran this query exactly four times with different `belongsToGroupX` properties, recorded the results, and averaged them.

**Experimental Results.** Table 7 summarizes the macro-averaged results that were estimated via Leave-One-Out Cross-Validation (LOOCV). We applied both, an RBC and an RPT learning algorithm to both tasks. The table also reports the best-performing SVM results from Bloehdorn and Sure's experiments. The RBC clearly outperformed the RPT in both predictions, hence, we report only on the results given by the RBC. For both tasks the performance of the inferred model is not very different from the one induced on the asserted model. When consulting Listing 1.8 (for `person2affiliation`) it is plausible to conclude that the only inferred properties (types of persons and publications) do not help to classify a person's or a publication's affiliation with an organizational unit.

For the `person2affiliation` task, Table 7 shows that our method clearly outperforms the kernel-based approach in terms of recall, but has only marginally better F-Measure improvement. This is because our method is clearly inferior in terms of prediction error and precision. For the `publication2affiliation` task,

---

[19] http://ontoware.org/projects/swrc/

```
 1  CREATE MINING MODEL <http://example.org/svm>
 2  { ?person           RESOURCE   TARGET
 3    ?value            DISCRETE   PREDICT  {'Yes','No'}
 4    ?personType       RESOURCE
 5    ?project          RESOURCE
 6    ?topic            RESOURCE
 7    ?publication      RESOURCE
 8    ?publicationType  RESOURCE
 9  }
10  WHERE
11  { ?person  swrc:affiliation     ?affiliation ;
12             rdf:type             ?personType ;
13             uzh:belongsToGroup1  ?value .
14
15    OPTIONAL
16      { ?person  swrc:worksAtProject  ?project . }
17    OPTIONAL
18      { ?topic  swrc:isWorkedOnBy  ?person . }
19    OPTIONAL
20      { ?person       swrc:publication  ?publication .
21        ?publication  rdf:type          ?publicationType .
22      }
23  }
24  USING <http://kdl.cs.umass.edu/proximity/rbc>
```

**Listing 1.8.** CREATE MINING MODEL query for the person2affiliation task

**Table 8.** LOOCV results for the *person2affiliation* and *publication2affiliation* tasks

| person2affiliation | | | | publication2affiliation | | | | |
|---|---|---|---|---|---|---|---|---|
| *algorithm* | *err* | *prec* | *rec* | *F-measure* | *algorithm* | *err* | *prec* | *rec* | *F-measure* |
| sim-ctpp-pc, c=1 | 4.49 | 95.83 | 58.13 | 72.37 | sim-cta-p, c=10 | 0.63 | 99.74 | 95.22 | 97.43 |
| RBC w/o inf | 3.53 | 87.09 | 80.52 | 83.68 | RBC w/o inf | 0.09 | 98.83 | 99.61 | 99.22 |
| RBC w/ inf | 3.67 | 85.72 | 80.18 | 82.86 | RBC w/ inf | 0.15 | 97.90 | 99.25 | 98.57 |

the results are even worse: turning the reasoner on improves, at least, the results compared to no reasoner used, however, the results are still inferior compared to the kernel-based approach by Bloehdorn and Sure across all performance measures.

Because these results were not very promising, we asked ourselves how we could achieve better prediction performance. We thought, why not perform a real multi-class prediction experiment instead of four rather tedious individual experiments and averaging the results. Luckily, with our SPARQL-ML approach we are able to perform exactly this kind of prediction experiment. The corresponding example query is shown in Listing 1.9 and the results in Table 8. Note that this query can use the ontology *as is*, *i.e.*, the dataset does not have to be extended with additional relations (as was the case in Listing 1.8).

As Table 8 clearly shows, our multi-class prediction method outperforms the kernel-based approach in terms of prediction error, recall, and F-Measure, while having an only slightly lower precision. The slightly lower precision could be a result of the limitation to just a few properties used by an off-the-shelf approach without a single parameter setting, whereas the SVM approach is the result of extensive testing and tuning of the kernel method's properties and parameters.

We conclude from this experiment, that writing a SPARQL-ML query is a simple task for everyone familiar with the data and the SPARQL-ML syntax. Kernels, on the other hand, have the major disadvantage that the user has to choose from various kernels, kernel modifiers, and parameters. This constitutes a major problem for users not familiar with kernels and SVM algorithms.

```
1   CREATE MINING MODEL <http://example.org/svm>
2   { ?person           RESOURCE   TARGET
3     ?affiliation      DISCRETE   PREDICT   {'ID1','ID2','ID3','ID4'}
4     ?personType       RESOURCE
5     ?project          RESOURCE
6     ?topic            RESOURCE
7     ?publication      RESOURCE
8     ?publicationType  RESOURCE
9   }
10  WHERE
11  { ?person   swrc:affiliation     ?affiliation ;
12              rdf:type             ?personType .
13
14    OPTIONAL
15    { ?person  swrc:worksAtProject  ?project . }
16    OPTIONAL
17    { ?topic  swrc:isWorkedOnBy  ?person . }
18    OPTIONAL
19    { ?person        swrc:publication  ?publication .
20      ?publication  rdf:type           ?publicationType .
21    }
22  }
23  USING <http://kdl.cs.umass.edu/proximity/rbc>
```

**Listing 1.9.** CREATE MINING MODEL query for the person2affiliation task

## 4.4   Bug Prediction Experiment

In our last experiment, we evaluate the applicability and usefulness of our novel inductive reasoning framework SPARQL-ML for *bug prediction*. We, therefore, evaluated the predictive power of our SPARQL-ML approach on several real-world software projects modeled in the EvoOnt format (see [28]). To that end, we will compare the off-the-shelf performance of SPARQL-ML with a traditional, propositional data mining approach proposed in [3] following exactly their evaluation procedure. To achieve this goal, we use *historical/evolutionary information* about the software projects in all our experiments. This information is provided by a concurrent versions system (CVS) and a bug-tracking system (*i.e.*, Bugzilla).[20]

**Evaluation Methodology and Datasets.** The data used in this case study was collected from six plug-ins of the Eclipse open source project in the overall time span from January 3, 2001 to January 31, 2007.[21] The plug-ins are compare, pdebuild, pdeui, search, updatecore, and updateui, which are all available at the CVS repository at dev.eclipse.org.

In a nutshell, the experimental procedure can be summarized as follows: first, along with the data from CVS and Bugzilla, we exported each of the plug-ins into our EvoOnt format; second, we created a small extension to EvoOnt to take into account the 22 extra features from [3] that are used for model induction and making predictions; and third, we wrote SPARQL-ML queries for the induction of a mining model on the training set as well as for the prediction of bugs on the test set. The queries in Listings 1.10 and 1.11 show an example of the CREATE MINING MODEL and PREDICT statements we used for the model induction and prediction tasks respectively.

---

[20] http://www.bugzilla.org/

[21] http://www.eclipse.org/

Addressing the first step, exporting the information from CVS and Bugzilla into our EvoOnt format, the information from the first releases up to the last one released in January 2007 was considered. For the second step, the extension of the EvoOnt model with the additional features for learning and predicting, we exploited the fact that EvoOnt (and more generally, the OWL data format) is easily extendable with additional classes and properties. We had to extend EvoOnt with a total number of 22 additional features, which were all computed in a preprocessing step and added to the OWL class `File` in EvoOnt's Version Ontology Model (VOM) via a set of new OWL datatype properties (*e.g.*, `vom:-loc`, `vom:lineAddedIRLAdd`, etc.). Furthermore, for each ontologized file of the plug-ins, an additional `vom:hasError` property is added. The value of the property is either 'Yes' or 'No' depending on wether the file was mentioned in a bug report from Bugzilla.

In the experiments in [3], six different models were trained using Weka's J48 decision tree learner. The first model does not take into account any temporal features whilst the second to fifth model all use a variation of different temporal and non-temporal features for model induction. Finally, the sixth model is a *summary model* that uses only those features that turned out to be most significant/discriminant in the other models.

For each set of discriminating features, we created a `CREATE MINING MODEL` query to induce a model using either a Relational Probability Tree or a Relational Bayesian Classifier as prediction algorithm. Listing 1.10 shows the corresponding SPARQL-ML query for inducing a model using only the most significant features from [3]. For model induction, all the files of the plug-ins that were released before January 31, 2007 are considered (lines 16–17). Variable `?file` is the target variable that is linked to variable `?error` for which a prediction should be made (either 'Yes' or 'No') expressing if the file is likely to be error-prone or not (lines 2 and 3). Finally, the induced model is available for predictions via its model URI `<http://www.example.org/bugssignificant>`.

To test the model, we applied the predict query shown in Listing 1.11. The query first selects the source code files for which a revision was made before January 31, 2007 (line 6), and second, applies the previously induced model to classify a file as either buggy or non-buggy (lines 24–32).[22] The result of the prediction and its probability are finally bound on line 25 to the variables `?prediction` and `?probability`. [23]

**Experimental Results.** The results of the bug prediction experiments are summarize in Figures 11 and 12 that illustrate the performance of the temporal

---

[22] Note that every file we considered has at least one revision (*i.e.*, for when it was created/checked into CVS).

[23] Furthermore note that the prediction query in Listing 1.11 is only shown for illustration purposes. This kind of query is useful to predict if a *new, unseen* file is likely to be buggy or not. However, as we use the same set of files for training and testing, we currently run a variation of the scripts proposed in [23] (pages 102–108) to perform cross-validation.

```
1   CREATE MINING MODEL <http://www.example.org/bugs>
2   { ?file                     RESOURCE    TARGET
3     ?error                    DISCRETE    PREDICT
4                                           {'YES','NO'}
5     ?lineAddedIRLAdd          CONTINUOUS
6     ?lineDeletedIRLDel        CONTINUOUS
7     ?revision1Month           CONTINUOUS
8     ?defectAppearance1Month   CONTINUOUS
9     ?revision2Months          CONTINUOUS
10    ?reportedIssues3Months    CONTINUOUS
11    ?reportedIssues5Months    CONTINUOUS
12  }
13  WHERE
14  { ?file       vom:hasRevision ?revision .
15    ?revision   vom:creationTime ?creation .
16    FILTER ( xsd:dateTime(?creation)
17               <= "2007-01-31T00:00:00"^^xsd:dateTime )
18
19    ?file       vom:hasError ?error .
20
21    OPTIONAL { ?file vom:lineAddedIRLAdd
22                     ?lineAddedIRLAdd . }
23    OPTIONAL { ?file vom:lineDeletedIRLDel
24                     ?lineDeletedIRLDel . }
25    OPTIONAL { ?file vom:revision1Month
26                     ?revision1Month . }
27    OPTIONAL { ?file vom:defectAppearance1Month
28                     ?defectAppearance1Month . }
29    OPTIONAL { ?file vom:revision2Months
30                     ?revision2Months . }
31    OPTIONAL { ?file vom:reportedIssues3Months
32                     ?reportedIssues3Months . }
33    OPTIONAL { ?file vom:reportedIssues5Months
34                     ?reportedIssues5Months . }
35  }
36  USING <http://kdl.cs.umass.edu/proximity/rpt>
```

**Listing 1.10.** SPARQL-ML CREATE MINING MODEL query to induce a model using the most significant code features from [3]

and non-temporal feature models using RPTs and RBCs. The results are again presented in terms of prediction accuracy (acc; in legend), Receiver Operating Characteristics (ROC; graphed), and the area under the ROC curve (auc; also in legend).

Figures 11 and 12 show the performance of the best model from [3] as a baseline (the black line with bullet points; acc = 0.992, auc = 0.925). This is the model that was trained with only the most significant/discriminating features. As can be seen, the best SRL model is the RBC model induced on the 3-months features (auc = 0.977), closely followed by the RPT model on only the most significant features from [3] (auc = 0.972). It can be observed, that with the exception of the RPT model for the most significant features, all the RBC models slightly outperform the RPT models in terms of area under the curve. Examining accuracy, the RPT models, on the other hand, outperform the RBC models.

Furthermore, is is interesting to observe that all but the models trained on the 1-month features outperform the traditional, propositional learning approach of [3] in terms of area under the curve. For both the RPT and RBC algorithm, the 1-month model shows the worst performance compared with the baseline as well as with the rest of the temporal/non-temporal feature models. This is contrary to the findings of [3] where the 1-month model was second best in terms of accuracy and at third position for auc.

```
1   SELECT DISTINCT ?file ?prediction ?probability
2   WHERE
3   { ?file      vom:hasRevision  ?revision .
4     ?revision vom:creationTime ?creation .
5
6     FILTER ( xsd:dateTime ( ?creation )
7                <= "2007-01-31T00:00:00"^^xsd:dateTime )
8
9     OPTIONAL { ?file vom:lineAddedIRLAdd
10                        ?lineAddedIRLAdd . }
11    OPTIONAL { ?file vom:lineDeletedIRLDel
12                        ?lineDeletedIRLDel . }
13    OPTIONAL { ?file vom:revision1Month
14                        ?revision1Month . }
15    OPTIONAL { ?file vom:defectAppearance1Month
16                        ?defectAppearance1Month . }
17    OPTIONAL { ?file vom:revision2Months
18                        ?revision2Months . }
19    OPTIONAL { ?file vom:reportedIssues3Months
20                        ?reportedIssues3Months . }
21    OPTIONAL { ?file vom:reportedIssues5Months
22                        ?reportedIssues5Months . }
23
24    PREDICT
25    { ( ?prediction ?probability )
26         sml:predict
27         ( <http://www.example.org/bugs>
28           ?file ?lineAddedIRLAdd ?lineDeletedIRLDel
29           ?revision1Month ?defectAppearance1Month
30           ?revision2Months ?reportedIssues3Months
31           ?reportedIssues5Months ) .
32    }
33  }
```

**Listing 1.11.** SPARQL-ML predict query to classify a source code file as either buggy or non-buggy

The traditional model is, however, better in terms of prediction/classification accuracy (acc = 0.992). Note that the use of accuracy as a measure for the quality of the prediction is, however, misleading as it does not relate the prediction to the prior probability of the classes (*i.e.*, 'Yes'/'No' for the value of vom:hasError). As pointed out in [3], this is especially problematic in datasets which are heavily skewed (*i.e.*, that have a distribution of values far from being normal). As shown by the authors, the bug prediction dataset is indeed heavily skewed with a total number of 3691 non-buggy and 14 buggy classes. Hence, as mentioned earlier, the ROC curves and the area under the curve are more meaningful measures as they provide a prior-independent approach for comparing the quality of predictors.

Last but not least, note that the best performing RPT/RBC models (significant features for RPT, 3-months features for RBC) also have the highest prediction/-classification accuracy among the SRL models (acc = 0.985 and acc = 0.977).

## 5  Discussion and Limitations

We briefly discuss some of the limitations of our novel inductive reasoning approach. SPARQL-ML's major drawback is the use of virtual triple patterns that some might deem as conceptually problematic. However, in this work we regard virtual triple patterns simply as part of the inferred knowledgebase; in other words, the specification of a prediction function is akin to the specification of an additional inferencing rule. Another limitation of the virtual triple pattern approach lies, of course, in the need for extending existing SPARQL query engines with the necessary language statements.

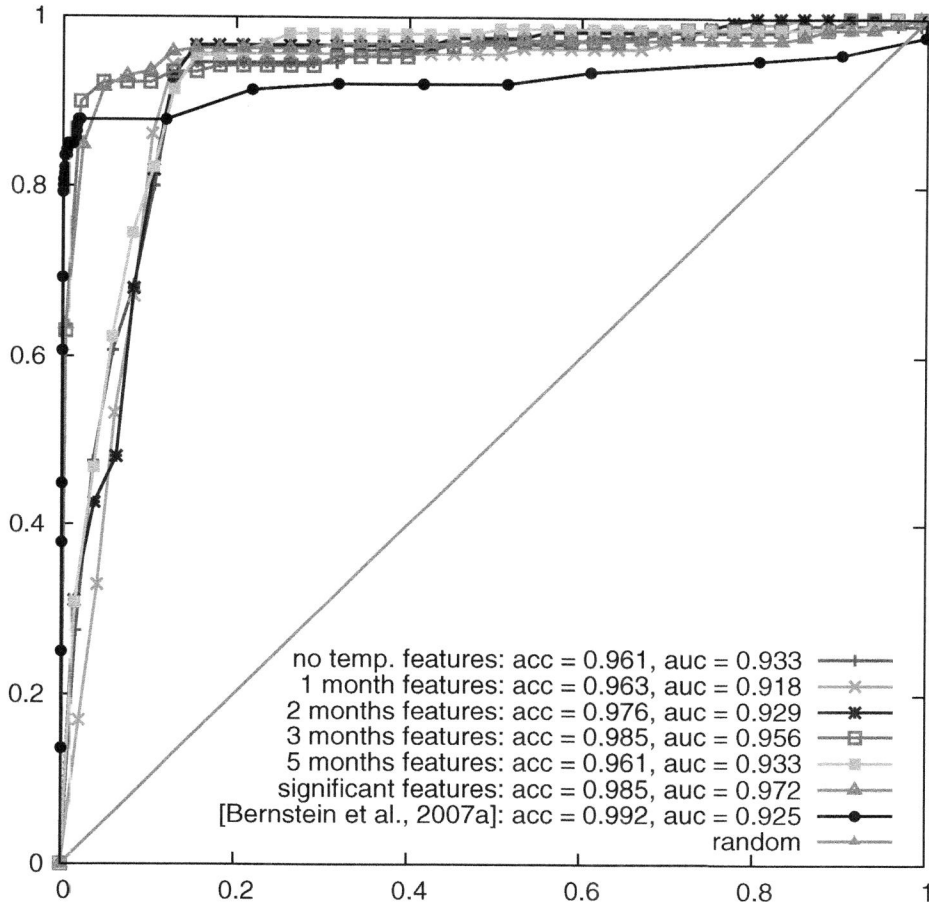

**Fig. 11.** ROC curves for all of the temporal and non-temporal models of the bug prediction experiments using RPTs. The model induced on the most significant features reported in [3] outperforms the baseline (black line) as well as all the other RPT models in terms of area under the curve.

Regarding semantic service classification, the performance of the prediction/-classification task might heavily depend on the expressiveness of the used ontologies. The Semantic Web services used in our experiments define their I/O concepts using extensive (*i.e.*, deep) ontologies (*e.g.*, the *portal.owl* and *travel.owl* ontologies), which enables to derive extensive, additional knowledge about the I/Os. Using ontologies with flatter inheritance structures will, therefore, likely result in inferior results. We note, however, that this performance loss is a limitation of the used ontologies and not of the SRL algorithms themselves. Therefore, we speculate that the loss could be eliminated by using more comprehensive ontologies.

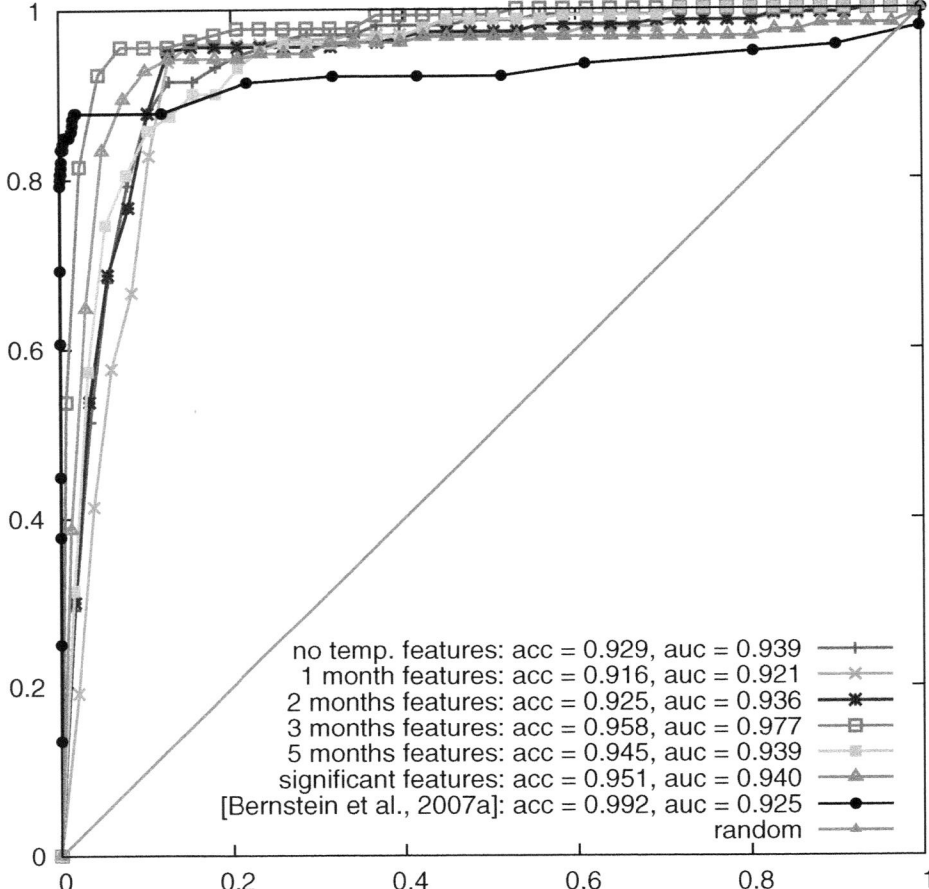

**Fig. 12.** ROC curves for all of the temporal and non-temporal models of the bug prediction experiments using RBCs. The 3-months feature model outperforms all the other models (including the baseline) in terms of area under curve.

Regarding our bug prediction experiments, we note as a technical limitation that we are currently not able to perform cross-validation through the query engine. Thus, if we want to use the same dataset for training and testing, we currently have to use specialized scripts for making predictions and calculating the performance measures.

## 6    Conclusions and Perspectives

In this paper, we have introduced our novel inductive reasoning extension to the Semantic Web. This extension aims at complementing the classical deductive description logic reasoning facilities of the traditional Semantic Web infrastructure (*i.e.*, it allows us to draw conclusions from the asserted facts in a knowledgebase which are otherwise *not* deducible by the classical approaches). Our extension is tightly integrated with the RDF query language SPARQL, providing access to the newly derived knowledge through the query engine. To that end, our extension exploits SPARQL virtual triple patterns that perform pattern matching by calling a customized, external piece of code, rather than matching triple patterns against an RDF graph.

To evaluate/validate our novel extension, we performed four sets of experiments using synthetic and real-world datasets. In our first case study, we fully analyzed SPARQL-ML on a synthetic dataset to show its excellent prediction/-classification quality in a proof-of-concept setting. Secondly, we have shown the benefits of Statistical Relational Learning (SRL) algorithms (particularly Relational Probability Trees) to perform Semantic Web service classification using a well-known Semantic Web benchmarking dataset.

By enabling/disabling ontological inference support in our experiments, we came to the conclusion that the combination of statistical inference with logical deduction produces superior performance over statistical inference only. These findings support our assumption that the interlinked Semantic Web data is a perfect match for SRL methods due to their focus on relations between objects (extrinsic attributes) in addition to features/attributes of objects of traditional, propositional learning techniques (intrinsic attributes).

In our third set of experiments, we have shown SPARQL-ML's superiority to another related, kernel-based approach used in Support Vector Machines. Finally, in the bug prediction case study, we have demonstrated, that inductive reasoning enabled by our SPARQL-ML framework allows us to easily perform bug prediction on semantically annotated software source code. Our empirical findings suggest that SPARQL-ML is indeed able to predict bugs with a very good accuracy, which, ultimately, makes SPARQL-ML a suitable tool to help improve the quality of software systems.

## 6.1 Future Work

**Reasoning.** The focus of this paper is clearly on the application and evaluation of our inductive reasoning extension to complement the classical deductive reasoning approaches of the current Semantic Web infrastructure (see Figure 13). There exist, however, yet different types of (human) reasoning as described in [32], which were not addressed in this paper. These types are, for instance, *non-monotonic reasoning* and *temporal reasoning*. Generally speaking, in a non-monotonic reasoning system, additional/new information not considered when drawing the original conclusions can *change the reasonableness of these conclusions* [32]. In other words, the original correct conclusions are probably no longer valid and have to be revised. On the other hand, in a temporal reasoning system, the goal is to draw conclusions about the resources in the knowledgebase depending on some *notion of time*.

Without going into the details of either concepts, we think that it would make perfect sense to allow for non-monotonic and temporal reasoning facilities

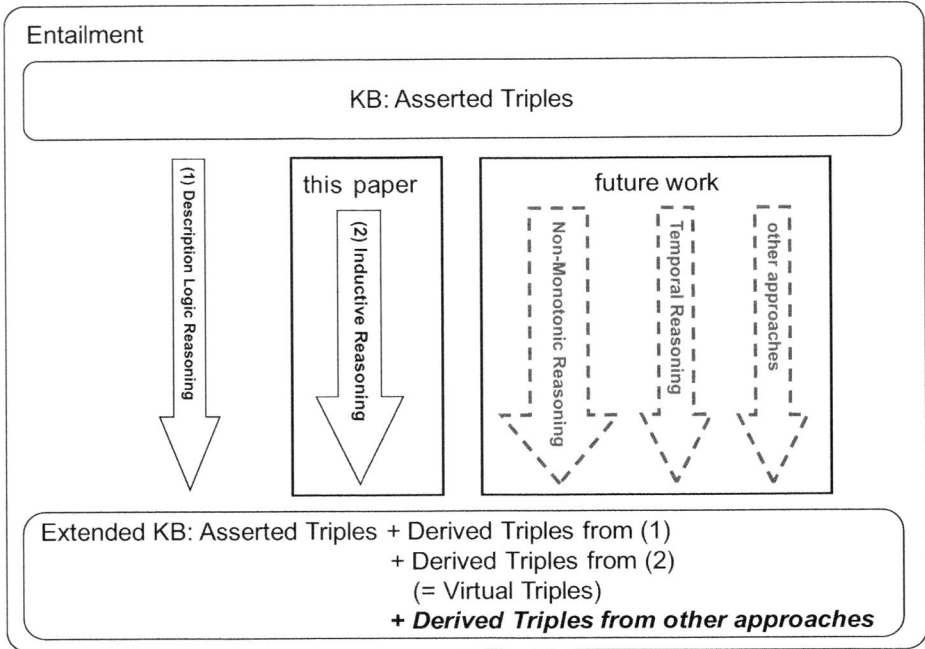

**Fig. 13.** Possible future reasoning extensions to the Semantic Web—*non-monotonic reasoning* and *temporal reasoning*

through the SPARQL query engine. This would allow us to derive even more additional knowledge from the asserted facts in a knowledgebase which can neither be derived by the classical deductive nor our presented inductive reasoning facilities.

**Optimization.** Another possible path for future work is optimization. Besides the work achieved for SPARQL basic graph pattern optimization through selectivity estimation (which we presented in [42] and [4]), we did not yet consider SPARQL-ML optimization techniques.

Generally speaking, we suggest having a closer look at virtual triple pattern optimization. Optimization in this direction will probably be twofold: first, the externally called functions need to be improved. For inductive reasoning, this implies faster algorithms to make predictions. Second, and probably more important, the query engine might need some modifications to perform query evaluation including virtual triple patterns more efficiently. This is especially important if our novel reasoning approach should be scalable and applicable to datasets which are much larger than the ones used in this work (*i.e.*, if it should *scale to the Web*).

**Algorithms, Datasets, and Tasks.** We think that our approach's applicability to different validation tasks should be systematically investigated. An example of such a task that could substantially benefit from inductive reasoning is the classification of semantically annotated, scientific publications (as presented in the SwetoDBLP dataset).[24]

Moreover, future work should definitely evaluate the pros and cons of other relational learning methods such as the ones proposed by NetKit[25] or Alchemy.[26] This would help to underline the usefulness of this kind of learning methods for the Semantic Web.

# References

1. Auer, S., Bizer, C., Kobilarov, G., Lehmann, J., Cyganiak, R., Ives, Z.: DBpedia: A Nucleus for a Web of Open Data. In: Aberer, K., Choi, K.-S., Noy, N., Allemang, D., Lee, K.-I., Nixon, L.J.B., Golbeck, J., Mika, P., Maynard, D., Mizoguchi, R., Schreiber, G., Cudré-Mauroux, P. (eds.) ASWC 2007 and ISWC 2007. LNCS, vol. 4825, Springer, Heidelberg (2007)
2. Berners-Lee, T., Hendler, J., Lassila, O.: The Semantic Web. Scientific American 284(5) (May 2001)
3. Bernstein, A., Ekanayake, J., Pinzger, M.: Improving Defect Prediction Using Temporal Features and Non-Linear Models. In: Proceedings of the 9th International Workshop on Principles of Software Evolution (IWPSE), pp. 11–18. ACM Press, New York (2007)
4. Bernstein, A., Kiefer, C., Stocker, M.: OptARQ: A SPARQL Optimization Approach based on Triple Pattern Selectivity Estimation. Tech. Rep. IFI-2007.02, Department of Informatics, University of Zurich (2007)
5. Bizer, C., Heath, T., Ayers, D., Raimond, Y.: Interlinking Open Data on the Web. In: Proceedings of the Demonstrations Track of the 4th European Semantic Web Conference, ESWC (2007)
6. Bloehdorn, S., Sure, Y.: Kernel Methods for Mining Instance Data in Ontologies. In: Aberer, K., Choi, K.-S., Noy, N., Allemang, D., Lee, K.-I., Nixon, L.J.B., Golbeck, J., Mika, P., Maynard, D., Mizoguchi, R., Schreiber, G., Cudré-Mauroux, P. (eds.) ASWC 2007 and ISWC 2007. LNCS, vol. 4825, pp. 58–71. Springer, Heidelberg (2007)
7. Bloehdorn, S., Sure, Y.: Kernel Methods for Mining Instance Data in Ontologies. In: Aberer, K., Choi, K.-S., Noy, N., Allemang, D., Lee, K.-I., Nixon, L.J.B., Golbeck, J., Mika, P., Maynard, D., Mizoguchi, R., Schreiber, G., Cudré-Mauroux, P. (eds.) ASWC 2007 and ISWC 2007. LNCS, vol. 4825, pp. 58–71. Springer, Heidelberg (2007)
8. Borgida, A., Brachman, R.J., McGuinness, D.L., Resnick, L.A.: CLASSIC: A Structural Data Model for Objects. In: Proceedings of the ACM SIGMOD International Conference on Management of Data, pp. 58–67. ACM, New York (1989)
9. Broekstra, J., Kampman, A.: SeRQL: A Second Generation RDF Query Language. In: Proceedings of the SWAD-Europe Workshop on Semantic Web Storage and Retrieval (2003)
10. Codd, E.F.: A Relational Model of Data for Large Shared Data Banks. Communications of the ACM 13(6), 377–387 (1970)

---

[24] http://lsdis.cs.uga.edu/projects/semdis/swetodblp/
[25] http://www.research.rutgers.edu/~sofmac/NetKit.html
[26] http://alchemy.cs.washington.edu/

11. Cyganiak, R.: A relational algebra for SPARQL. Tech. Rep. HPL-2005-170, Hewlett-Packard Laboratories, Bristol (2005)
12. Dean, J., Ghemawat, S.: Mapreduce: simplified data processing on large clusters. Commun. ACM 51, 107–113 (2008), http://doi.acm.org/10.1145/1327452.1327492
13. Džeroski, S.: Multi-Relational Data Mining: An Introduction. ACM SIGKDD Explorations Newsletter 5(1), 1–16 (2003)
14. Edwards, P., Grimnes, G.A., Preece, A.: An Empirical Investigation of Learning from the Semantic Web. In: Proceedings of the Semantic Web Mining Workshop (SWM) co-located with 13th European Conference on Machine Learning (ECML) and the 6th European Conference on Principles and Practice of Knowledge Discovery in Databases (PKDD), pp. 71–89 (2002)
15. Fenton, N.E., Neil, M.: A Critique of Software Defect Prediction Models. IEEE Transactions on Software Engineering 25(5), 675–689 (1999)
16. Getoor, L., Licamele, L.: Link Mining for the Semantic Web. In: Dagstuhl Seminar (2005)
17. Gilardoni, L., Biasuzzi, C., Ferraro, M., Fonti, R., Slavazza, P.: Machine Learning for the Semantic Web: Putting the user into the cycle. In: Dagstuhl Seminar (2005)
18. Gruber, T.R.: Toward Principles for the Design of Ontologies Used for Knowledge Sharing. International Journal Human-Computer Studies 43(5-6), 907–928 (1995)
19. Hartmann, J., Sure, Y.: A Knowledge Discovery Workbench for the Semantic Web. In: International Workshop on Mining for and from the Semantic Web (MSW), pp. 62–67 (2004)
20. Hau, J., Lee, W., Darlington, J.: A Semantic Similarity Measure for Semantic Web Services. In: Proceedings of the Workshop Towards Dynamic Business Integration co-located with the 14th International World Wide Web Conference, WWW (2005)
21. Heß, A., Johnston, E., Kushmerick, N.: Machine Learning for Annotating Semantic Web Services. In: Semantic Web Services: Papers from the 2004 AAAI Spring Symposium Series. AAAI Press, Menlo Park (2004)
22. Heß, A., Kushmerick, N.: Learning to Attach Semantic Metadata to Web Services. In: Fensel, D., Sycara, K., Mylopoulos, J. (eds.) ISWC 2003. LNCS, vol. 2870, pp. 258–273. Springer, Heidelberg (2003)
23. Jensen, D.: Proximity 4.3 Tutorial. Knowledge Discovery Laboratory, University of Massachusetts Amherst (2007), tutorial, available at http://kdl.cs.umass.edu/proximity/documentation.html
24. Joachims, T.: SVM light—Support Vector Machine (2004), software, available at http://svmlight.joachims.org/
25. Kiefer, C., Bernstein, A., Lee, H.J., Klein, M., Stocker, M.: Semantic Process Retrieval with iSPARQL. In: Franconi, E., Kifer, M., May, W. (eds.) ESWC 2007. LNCS, vol. 4519, pp. 609–623. Springer, Heidelberg (2007)
26. Kiefer, C., Bernstein, A., Locher, A.: Adding Data Mining Support to SPARQL Via Statistical Relational Learning Methods (Best paper award!). In: Bechhofer, S., Hauswirth, M., Hoffmann, J., Koubarakis, M. (eds.) ESWC 2008. LNCS, vol. 5021, pp. 478–492. Springer, Heidelberg (2008)
27. Kiefer, C., Bernstein, A., Stocker, M.: The Fundamentals of iSPARQL: A Virtual Triple Approach for Similarity-Based Semantic Web Tasks. In: Aberer, K., Choi, K.-S., Noy, N., Allemang, D., Lee, K.-I., Nixon, L.J.B., Golbeck, J., Mika, P., Maynard, D., Mizoguchi, R., Schreiber, G., Cudré-Mauroux, P. (eds.) ASWC 2007 and ISWC 2007. LNCS, vol. 4825, pp. 295–309. Springer, Heidelberg (2007)
28. Kiefer, C., Bernstein, A., Tappolet, J.: Analyzing Software with iSPARQL. In: Proceedings of the 3rd International Workshop on Semantic Web Enabled Software Engineering, SWESE (2007)

29. Kochut, K.J., Janik, M.: SPARQLeR: Extended Sparql for Semantic Association Discovery. In: Franconi, E., Kifer, M., May, W. (eds.) ESWC 2007. LNCS, vol. 4519, pp. 145–159. Springer, Heidelberg (2007)

30. Lam, H.Y.K., Marenco, L., Clark, T., Gao, Y., Kinoshita, J., Shepherd, G., Miller, P., Wu, E., Wong, G., Liu, N., Crasto, C., Morse, T., Stephens, S., Cheung, K.-H.: AlzPharm: integration of neurodegeneration data using RDF. BMC Bioinformatics 8(3) (2007)

31. Malewicz, G., Austern, M.H., Bik, A.J.C., Dehnert, J.C., Horn, I., Leiser, N., Czajkowski, G.: Pregel: a system for large-scale graph processing. In: Proceedings of the 2010 International Conference on Management of Data, SIGMOD 2010, pp. 135–146. ACM Press, New York (2010), http://doi.acm.org/10.1145/1807167.1807184

32. Mohanan, K.P.: Types of Reasoning: Relativizing the Rational Force of Conclusions. Academic Knowledge and Inquiry (2008), http://courses.nus.edu.sg/course/ellkpmoh/critical/reason.pdf

33. Neville, J., Jensen, D., Friedland, L., Hay, M.: Learning Relational Probability Trees. In: Proceedings of the 9th ACM SIGKDD International Conference on Knowledge Discovery and Data Mining (KDD), pp. 625–630. ACM, New York (2003)

34. Neville, J., Jensen, D., Gallagher, B.: Simple Estimators for Relational Bayesian Classifiers. In: Proceedings of the 3rd IEEE International Conference on Data Mining (ICDM), pp. 609–612. IEEE Computer Society Press, Washington, DC (2003)

35. Pérez, J., Arenas, M., Gutierrez, C.: Semantics and Complexity of SPARQL. In: Cruz, I., Decker, S., Allemang, D., Preist, C., Schwabe, D., Mika, P., Uschold, M., Aroyo, L.M. (eds.) ISWC 2006. LNCS, vol. 4273, pp. 30–43. Springer, Heidelberg (2006)

36. Provost, F., Fawcett, T.: Robust Classification for Imprecise Environments. Machine Learning 42(3), 203–231 (2001)

37. Prud'hommeaux, E., Seaborne, A.: SPARQL Query Language for RDF. Tech. rep., W3C Recommendation, January 15 (2008), http://www.w3.org/TR/rdf-sparql-query/

38. Richardson, M., Domingos, P.: Markov logic networks. Mach. Learn. 62, 107–136 (2006), http://portal.acm.org/citation.cfm?id=1113907.1113910

39. Russell, S.J., Norvig, P.: Artificial Intelligence: A Modern Approach. Prentice-Hall, Englewood Cliffs (2003)

40. Sabou, M.: Learning Web Service Ontologies: Challenges, Achievements and Opportunities. In: Dagstuhl Seminar (2005)

41. Shadbolt, N., Berners-Lee, T., Hall, W.: The Semantic Web Revisited. IEEE Intelligent Systems 21(3), 96–101 (2006)

42. Stocker, M., Seaborne, A., Bernstein, A., Kiefer, C., Reynolds, D.: SPARQL Basic Graph Pattern Optimization Using Selectivity Estimation. In: Proceedings of the 17th International World Wide Web Conference (WWW), pp. 595–604. ACM Press, New York (2008)

43. Stutz, P., Bernstein, A., Cohen, W.: Signal/Collect: Graph Algorithms for the (Semantic) Web. In: Patel-Schneider, P.F., Pan, Y., Hitzler, P., Mika, P., Zhang, L., Pan, J.Z., Horrocks, I., Glimm, B. (eds.) ISWC 2010, Part I. LNCS, vol. 6496, pp. 764–780. Springer, Heidelberg (2010)

44. Valiente, G.: Algorithms on Trees and Graphs. Springer, Heidelberg (2002)

45. Witten, I.H., Frank, E.: Data Mining: Practical Machine Learning Tools and Techniques. Morgan Kaufmann, San Francisco (2005)

# Probabilistic-Logical Web Data Integration

Mathias Niepert, Jan Noessner, Christian Meilicke, and Heiner Stuckenschmidt

KR & KM Research Group,
University of Mannheim, B6 26, 68159 Mannheim, Germany
{mathias,jan,christian,heiner}@informatik.uni-mannheim.de

**Abstract.** The integration of both distributed schemas and data repositories is a major challenge in data and knowledge management applications. Instances of this problem range from mapping database schemas to object reconciliation in the linked open data cloud. We present a novel approach to several important data integration problems that combines logical and probabilistic reasoning. We first provide a brief overview of some of the basic formalisms such as description logics and Markov logic that are used in the framework. We then describe the representation of the different integration problems in the probabilistic-logical framework and discuss efficient inference algorithms. For each of the applications, we conducted extensive experiments on standard data integration and matching benchmarks to evaluate the efficiency and performance of the approach. The positive results of the evaluation are quite promising and the flexibility of the framework makes it easily adaptable to other real-world data integration problems.

## 1 Introduction

The growing number of heterogeneous knowledge bases on the web has made data integration systems a key technology for sharing and accumulating distributed data and knowledge repositories. In this paper, we focus on (a) the problem of aligning description logic ontologies and (b) the problem of object reconciliation in open linked datasets[1].

Ontology matching, or ontology alignment, is the problem of determining correspondences between concepts, properties, and individuals of two or more different formal ontologies [12]. The alignment of ontologies allows semantic applications to exchange and enrich the data expressed in the respective ontologies. An important results of the yearly ontology alignment evaluation initiative (OAEI) [11,13] is that there is no single best approach to all existing matching problems. The factors influencing the quality of alignments range from differences in lexical similarity measures to variations in alignment extraction approaches. This insight provides justification not only for the OAEI itself but also for the

---

[1] The present chapter provides a more didactical exposition of the principles and methods presented in a series of papers of the same authors published in several conferences such as AAAI, UAI, and ESWC.

A. Polleres et al. (Eds.): Reasoning Web 2011, LNCS 6848, pp. 504–533, 2011.

development of a framework that facilitates the comparison of different strategies with a flexible and declarative formalism. We argue that Markov logic [39] provides and excellent framework for ontology matching. Markov logic (ML) offers several advantages over existing matching approaches. Its main strength is rooted in the ability to combine *soft* and *hard* first-order formulas. This allows the inclusion of both *known* logical and *uncertain* statements modeling potential correspondences and structural properties of the ontologies. For instance, hard formulas can help to reduce incoherence during the alignment process while soft formulas can factor in lexical similarity values computed for each correspondence. An additional advantage of ML is joint inference, that is, the inference of two or more interdependent hidden predicates. Several results show that joint inference is superior in accuracy when applied to a wide range of problems such as ontology refinement [53] and multilingual semantic role labeling [32].

Identifying different representations of the same data item is called object reconciliation. The problem of object reconciliation has been a topic of research for more than 50 years. It is also known as record linkage [14], entity resolution [3], and instance matching [15]. While the majority of the existing methods were developed for the task of matching database records, modern approaches focus mostly on graph-based data representations such as the resource description framework (RDF). Using the proposed Markov logic based framework for data integration, we leverage schema information to exclude logically inconsistent correspondences between objects improving the overall accuracy of instance alignments. In particular, we use logical reasoning and linear optimization techniques to compute the overlap of derivable types of objects. This information is combined with the classical similarity-based approach, resulting in a novel approach to object reconciliation that is more accurate than state-of-the-art alignment systems.

We demonstrate how description logic axioms are modeled within the framework and show that alignment problems can be posed as linear optimization problems. These problems can be efficiently solved with integer linear programming methods also leveraging recent meta-algorithms such as cutting plane inference and delayed column generation first proposed in the context of Markov logic.

The chapter is organized as follows. First, we briefly introduce some basic formalism such as description logics and Markov logic. Second, we define ontology matching and object reconciliation and introduce detailed running examples that we use throughout the chapter to facilitate a deeper understanding of the ideas and methods. We also introduce the syntax and semantics of the ML framework and show that it can represent numerous different matching scenarios. We describe probabilistic reasoning in the framework of Markov logic and show that a solution to a given matching problem can be obtained by solving the maximum a-posteriori (MAP) problem of a ground Markov logic network using integer linear programming. We then report the results of an empirical evaluation of our method using some of the OAEI benchmark datasets.

# 2    Data Integration on the Web

The integration of distributed information sources is a key challenge in data and knowledge management applications. Instances of this problem range from mapping schemas of heterogeneous databases to object reconciliation in linked open data repositories. In the following, we discuss two instances of the data integration problem: ontology matching and object reconciliation. Both problems have been in the focus of the semantic web community in recent years. We investigate and assess the applicability and performance of our probabilistic-logical approach to data integration using these two prominent problems. In order to make the article comprehensive, however, we first briefly cover description logics and ontologies as these logical concepts are needed in later parts of the document.

## 2.1    Ontologies and Description Logics

An Ontology usually groups objects of the world that have certain properties in common (e.g. cities or countries) into concepts. A specification of the shared properties that characterize a set of objects is called a concept definition. Concepts can be arranged into a subclass–superclass relation in order to further discriminate objects into subgroups (e.g. capitals or European countries). Concepts can be defined in two ways, by enumeration of its members or by a concept expression. The specific logical operators that can be used to formulate concept expressions can vary between ontology languages.

Description logics are decidable fragments of first order logic that are designed to describe concepts in terms of complex logical expressions[2] The basic modeling elements in description logics are concepts (classes of objects), roles (binary relations between objects) and individuals (named objects). Based on these modeling elements, description logics contain operators for specifying so-called concept expressions that can be used to specify necessary and sufficient conditions for membership in the concept they describe. These modeling elements are provided with a formal semantics in terms of an abstract domain interpretation mapping $\mathcal{I}$ mapping each instance onto an element of an abstract domain $\Delta^{\mathcal{I}}$. Instances can be connected by binary relations defined as subsets of $\Delta^{\mathcal{I}} \times \Delta^{\mathcal{I}}$. Concepts are interpreted as a subset of the abstract domain $\Delta$. Intuitively, a concept is a set of instances that share certain properties. These properties are defined in terms of concept expressions. Typical operators are the Boolean operators as well as universal and existential quantification over relations to instances in other concepts.

A description logic knowledge base consists of two parts. The A-box contains information about objects, their type and relations between them, the so-called T-Box consists of a set of axioms about concepts (potentially defined in terms of complex concept expressions and relations. The first type of axioms can be used to describe instances. In particular, axioms can be used to state that an instance

---

[2] Details about the relation between description logics and first-order logic can be found in [4] and [51].

**Table 1.** Axiom patterns for representing description logic ontologies

| DL Axiom | Semantics | Intuition |
|---|---|---|
| | A-Box | |
| $C(x)$ | $x^{\mathcal{I}} \in C^{\mathcal{I}}$ | x is of type C |
| $r(x,y)$ | $(x^{\mathcal{I}}, y^{\mathcal{I}}) \in r^{\mathcal{I}}$ | x is related to y by r |
| | T-Box | |
| $C \sqsubseteq D$ | $C^{\mathcal{I}} \subseteq D^{\mathcal{I}}$ | C is more specific than D |
| $C \sqcap D \sqsubseteq \bot$ | $C^{\mathcal{I}} \cap D^{\mathcal{I}} = \emptyset$ | C and D are disjoint |
| $r \sqsubseteq s$ | $r^{\mathcal{I}} \subseteq s^{\mathcal{I}}$ | r is more specific than s |
| $r \equiv s^{-}$ | $r^{\mathcal{I}} = \{(x,y)\|(y,x) \in s^{\mathcal{I}}\}$ | r is the inverse of s |
| $\exists r.\top \sqsubseteq C$ | $(x^{\mathcal{I}}, y^{\mathcal{I}}) \in r^{\mathcal{I}} \Rightarrow x^{\mathcal{I}} \in C^{\mathcal{I}}$ | the domain of r is restricted to C |
| $\exists r^{-}.\top \sqsubseteq C$ | $(x^{\mathcal{I}}, y^{\mathcal{I}}) \in r^{\mathcal{I}} \Rightarrow y^{\mathcal{I}} \in C^{\mathcal{I}}$ | the range of r is restricted to C |

belongs to a concept or that two instances are in a certain relation. It is easy to see, that these axioms can be used to capture case descriptions as labeled graphs. The other types of axioms describe relations between concepts and instances. It can be stated that one concept is a subconcept of the other (all its instances are also instances of this other concept). Further, we can define a relation to be a subrelation or the inverse of another relation. The formal semantics of concepts and relations as defined by the interpretation into the abstract domain $\Delta^{\mathcal{I}}$ can be used to automatically infer new axioms from existing definitions. Table 1 lists a few examples of DL axioms, their semantics, and the intuition behind them.

Encoding ontologies in description logics is beneficial, because it enables inference engines to reason about ontological definitions. In this context, deciding subsumption between two concept expressions, i.e. deciding whether one expression is more general than the other one is one of the most important reasoning tasks as it has been used to support various tasks including information integration [47], product and service matching [27] and query answering over ontologies [2].

### 2.2   Ontology Matching

Ontology matching is the process of detecting links between entities in heterogeneous ontologies. Based on a definition by Euzenat and Shvaiko [12], we formally introduce the notion of *correspondence* and *alignment* to refer to these links.

**Definition 1 (Correspondence and Alignment).** *Given ontologies $\mathcal{O}_1$ and $\mathcal{O}_2$, let q be a function that defines sets of matchable entities $q(\mathcal{O}_1)$ and $q(\mathcal{O}_2)$. A correspondence between $\mathcal{O}_1$ and $\mathcal{O}_2$ is a triple $\langle 3, e_1, e_2 \rangle\, r$ such that $e_1 \in q(\mathcal{O}_1)$, $e_2 \in q(\mathcal{O}_2)$, and r is a semantic relation. An alignment between $\mathcal{O}_1$ and $\mathcal{O}_2$ is a set of correspondences between $\mathcal{O}_1$ and $\mathcal{O}_2$.*

The generic form of Definition 1 captures a wide range of correspondences by varying what is admissible as matchable element and semantic relation. In the

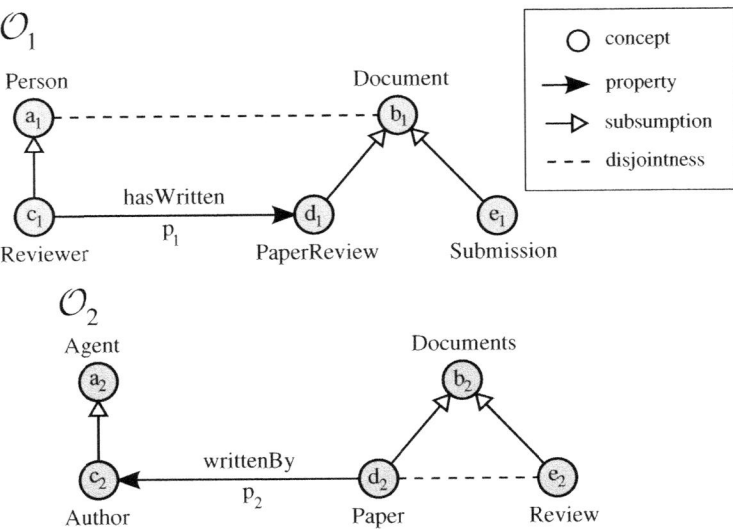

**Fig. 1.** Example ontology fragments

context of ontology matching, we are only interested in equivalence correspondences between concepts and properties. In the first step of the alignment process most matching systems compute a-priori similarities between matching candidates. These values are typically refined in later phases of the matching process. The underlying assumption is that the degree of similarity is indicative of the likelihood that two entities are equivalent. Given two matchable entities $e_1$ and $e_2$ we write $\sigma(e_1, e_2)$ to refer to this kind of a-priori similarity. Before presenting the formal matching framework, we motivate the approach by a simple instance of an ontology matching problem which we use as a running example.

*Example 1.* Figure 1 depicts fragments of two ontologies describing the domain of scientific conferences. The following axioms are part of ontology $\mathcal{O}_1$ and $\mathcal{O}_2$, respectively. If we apply a similarity measure $\sigma$ based on the Levenshtein distance [26] there are four pairs of entities such that $\sigma(e_1, e_2) > 0.5$.

$$\sigma(Document, Documents) = 0.88 \tag{1}$$
$$\sigma(Reviewer, Review) = 0.75 \tag{2}$$
$$\sigma(hasWritten, writtenBy) = 0.7 \tag{3}$$
$$\sigma(PaperReview, Review) = 0.54 \tag{4}$$

The alignment consisting of these four correspondences contains two correct (1 & 4) and two incorrect (2 & 3) correspondences resulting in a precision of 50%.

**Table 2.** Discription logics axioms in the ontology of Figure 1

| Ontology $\mathcal{O}_1$ | | Ontology $\mathcal{O}_2$ | |
|---|---|---|---|
| $\exists hasWritten \sqsubseteq Reviewer$ | | $\exists writtenBy \sqsubseteq Paper$ | |
| $PaperReview \sqsubseteq Document$ | | $Review \sqsubseteq Documents$ | |
| $Reviewer \sqsubseteq Person$ | | $Paper \sqsubseteq Documents$ | |
| $Submission \sqsubseteq Document$ | | $Author \sqsubseteq Agent$ | |
| $Document \sqsubseteq \neg Person$ | | $Paper \sqsubseteq \neg Review$ | |

## 2.3 Object Reconciliation

The problem of object reconciliation has been a topic of research for more than 50 years. It is also known as the problem of record linkage [14], entity resolution [3], and instance matching [15]. While the majority of the existing methods were developed for the task of matching database records, modern approaches focus mostly on graph-based data representations extended by additional schema information. We discuss the problem of object reconciliation using the notion of instance matching. This allows us to describe it within the well-established ontology matching framework [12]. Ontology matching is the process of detecting links between entities in different ontologies. These links are annotated by a confidence value and a label describing the type of link. Such a link is referred to as a *correspondence* and a set of such correspondences is referred to as an *alignment*.

In the following we refer to an alignment that contains correspondences between concepts and properties as *terminological alignment* and to an alignment that contains correspondences between individuals as *instance alignment*. Since instance matching is the task of detecting pairs of instances that refer to the same real world object [15], the semantic relation expressed by an instance correspondence is that of identity. The confidence value of a correspondence quantifies the degree of trust in the correctness of the statement. If a correspondence is automatically generated by a matching system this value will be computed by aggregating scores from multiple sources of evidence.

*Example 2.* An A-box is a set of membership statements of the following form: $C(a), P(a,b)$ where a,b are constants, C is a concept name and P is a property name. Further, we extend the notion of an A-Box by also allowing membership statements of the form $\neg C(a)$ and $\neg P(a,b)$ stating that object a is not a member of Concept C and that the objects a and b are not in relation R, respectively. We illustrate the problem of object reconciliation using the following example A-Boxes and their corresponding graphs.

A-Boxes can be regarded as labeled directed multi-graphs, where object constants are represented by nodes and binary relations between objects are represented by links labeled with the name of the corresponding relation. Object reconciliation is the task of finding the 'right' mapping between the nodes in different A-Box graphs. The basis for finding the right mapping between different

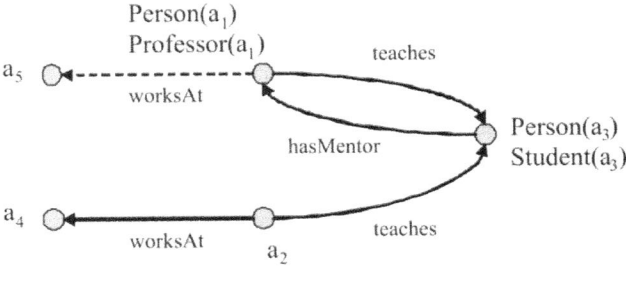

(a) Graph for A-Box $\mathcal{A}_1$

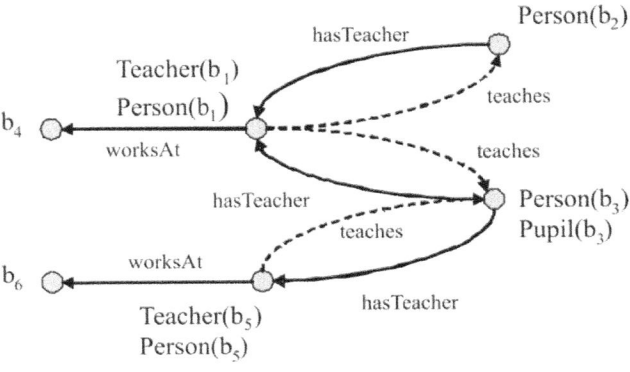

(b) Graph for A-Box $\mathcal{A}_2$

**Fig. 2.** Examples of A-Boxes

objects is typically based on a measure of similarity between the nodes that is determined on the local or global structures in the corresponding graph. Typical features for determining the similarity of two objects are:

- the similarity of their labels
- the similarity of the classes the objects belong to
- the similarity of relations and related objects

Based on these features, we would generate a priori similarities. For the example we would receive high values for $\sigma(a_5, b_4)$, $\sigma(a_1, b_1)$, $\sigma(a_3, b_3)$, $\sigma(a_3, b_2)$, $\sigma(a_2, b_5)$ and $\sigma(a_4, b_6)$. Besides the similarity between objects, in the case where the A-Box is based on an ontology, the logical constraints from the ontologies should be taken into account in the matching process. In particular, objects should not be maps on each other if they have incompatible types. In the example this means that assuming the underlying ontology contains a statement $student \perp pupil$ declaring the classes 'student' and 'pupil' as disjoint, the objects $a_3$ and $b_3$ should not be mapped on each other, despite the high a priori similarity.

# 3   Probabilistic-Logical Languages and Ontologies

Data integration for heterogeneous knowledge bases typically involves both purely logical and uncertain data. For instance, the description logic axioms of the ontologies are known to be true and, therefore, should be modeled as logical rules – the alignment system should not alter the logical structure of the input ontologies. Conversely, matching systems usually rely on degrees of confidence that have been derived through the application of lexical similarity, data mining, and machine learning algorithms. The presence of both known logical rules and degrees of uncertainty requires formalism that allow the representation of both deterministic and uncertain aspects of the problem. In the following, we introduce such a probabilistic-logical framework based on Markov logic and show how description logic ontologies are represented in the language. Moreover, we describe the application of an efficient probabilistic inference algorithm that uses integer linear programming.

## 3.1   Markov Logic

Markov logic combines first-order logic and undirected probabilistic graphical models [39]. A Markov logic network (MLN) is a set of first-order formulas with weights. Intuitively, the more evidence we have that a formula is true the higher the weight of this formula. To simplify the presentation of the technical parts we do *not* include functions. In addition, we assume that all (ground) formulas of a Markov logic network are in clausal form and use the terms *formula* and *clause* interchangeably.

**Syntax.** A signature is a triple $S = (O, H, C)$ with $O$ a finite set of observable predicate symbols, $H$ a finite set of hidden predicate symbols, and $C$ a finite set of constants. A Markov logic network (MLN) is a set of pairs $\{(F_i, w_i)\}$ with each $F_i$ being a function-free first-order formula built using predicates from $O \cup H$ and each $w_i \in \mathbb{R}$ a real-valued weight associated with formula $F_i$. We can represent hard constraints using large weights.

**Semantics.** Let $M = (F_i, w_i)$ be a Markov logic network with signature $S = (O, H, C)$. A *grounding* of a first-order formula $F$ is generated by substituting each occurrence of every variable in $F$ with constants in $C$. Existentially quantified formulas are substituted by the disjunctions of their groundings over the finite set of constants. A formula that does not contain any variables is *ground*. A formula that consists of a single predicate is an *atom*. Note that Markov logic makes several assumptions such as (a) different constants refer to different objects and (b) the only objects in the domain are those representable using the constants [39]. A set of ground atoms is a *possible world*. We say that a possible world $W$ *satisfies* a formula $F$, and write $W \models F$, if $F$ is true in $W$. Let $\mathcal{G}_F^C$ be the set of all possible groundings of formula $F$ with respect to $C$. We say that $W$ satisfies $\mathcal{G}_F^C$, and write $W \models \mathcal{G}_F^C$, if $F$ satisfies every formula in $\mathcal{G}_F^C$. Let $\mathcal{W}$

be the set of all possible worlds with respect to $S$. Then, the probability of a possible world $W$ is given by

$$p(W) = \frac{1}{Z} \exp \left( \sum_{(F_i, w_i)} \sum_{G \in \mathcal{G}^C_{F_i} : \; W \models G} w_i \right).$$

Here, $Z$ is a normalization constant. The score $s_W$ of a possible world $W$ is the sum of the weights of the ground formulas implied by $W$

$$s_W = \sum_{(F_i, w_i)} \sum_{G \in \mathcal{G}^C_{F_i} : \; W \models G} w_i. \tag{5}$$

We will see later that, in the data integration context, possible worlds correspond to possible alignments. Hence, the problem of deriving the most probably alignment given the evidence can be interpreted as finding the possible world $W$ with highest score.

### 3.2   Representing Ontologies and Alignments in Markov Logic

Our approach for data integration based on logics and probability is now based on the idea of representing description logic ontologies as Markov logic networks and utilizing the weights to incorporate similarity scores into the integration process [34]. The most obvious way to represent a description logic ontology in Markov logic would be to directly use the first-order translation of the ontology. For instance, the axiom $C \sqsubseteq D$ would be written as $\forall x \; C(x) \Rightarrow D(x)$. In other words, the representation would simply map between concepts and unary predicates and roles and binary predicates. However, we take a *different* approach by mapping axioms to predicates and use constants to represent the classes and relations in the ontology. Some typical axioms with their respective predicates are the following:

$$\begin{aligned}
C \sqsubseteq D & \mapsto sub(c, d) \\
C \sqcap D \sqsubseteq \bot & \mapsto dis(c, d) \\
\exists r.T \sqsubseteq C & \mapsto dom(r, c) \\
\exists r^{-1}.T \sqsubseteq C & \mapsto range(r, c)
\end{aligned}$$

This way of representing description logic ontologies has the advantage that we can model some basic inference rules and directly use them in the probabilistic reasoning process. For example, we can model the transitivity of the subsumption relation as

$$sub(x, y) \wedge sub(y, z) \Rightarrow sub(x, z)$$

and the fact that two classes that subsume each other cannot be disjoint at the same time

$$\neg sub(x, y) \vee \neg dis(x, y)$$

While the use of such axioms in a Markov logic network does not guarantee consistency and coherence of the results, they often cover the vast majority of

**Table 3.** The description logic $\mathcal{EL}^{++}$ without nominals and concrete domains

| Name | Syntax | Semantics |
|---|---|---|
| top | $\top$ | $\Delta^{\mathcal{I}}$ |
| bottom | $\bot$ | $\emptyset$ |
| conjunction | $C \sqcap D$ | $C^{\mathcal{I}} \cap D^{\mathcal{I}}$ |
| existential restriction | $\exists r.C$ | $\{x \in \Delta^{\mathcal{I}} \mid \exists y \in \Delta^{\mathcal{I}} : (x,y) \in r^{\mathcal{I}} \wedge y \in C^{\mathcal{I}}\}$ |
| GCI | $C \sqsubseteq D$ | $C^{\mathcal{I}} \subseteq D^{\mathcal{I}}$ |
| RI | $r_1 \circ ... \circ r_k \sqsubseteq r$ | $r_1^{\mathcal{I}} \circ ... \circ r_k^{\mathcal{I}} \subseteq r^{\mathcal{I}}$ |

conflicts that can exist in an ontology, especially in cases where the ontology is rather simple and does not contain a complex axiomatization.

For certain description logics, it is possible to completely capture the model using the kind of translation described above. In particular, if an ontology can be reduced to a normal form with a limited number of axiom types, we can provide a complete translation based on this normal form. An example for such a description logic is $\mathcal{EL}^{++}$, a light weight description logic that supports polynomial time reasoning. Table 3 shows the types of axioms an $\mathcal{EL}^{++}$ Model can be reduced to.

We can completely translation any $\mathcal{EL}^{++}$ model into a Markov Logic representation using the following translation rules:

$$
\begin{aligned}
C_1 \sqsubseteq D &\mapsto sub(c_1, d) \\
C_1 \sqcap C_2 \sqsubseteq D &\mapsto int(c_1, c_2, d) \\
C_1 \sqsubseteq \exists r.C_2 &\mapsto rsup(c_1, r, c_2) \\
\exists r.C_1 \sqsubseteq D &\mapsto rsub(c_1, r, d) \\
r \sqsubseteq s &\mapsto psub(r, s) \\
r_1 \circ r_2 \sqsubseteq r_3 &\mapsto pcom(r_1, r_2, r_3)
\end{aligned}
$$

In principle, such a complete translation is possible whenever there is a normal form representation of a description logic that reduces the original model to a finite number of axiom types that can be captured by a respective predicate in the Markov logic network.

Finally, being interested in data integration, we often treat correspondences between elements from different models separately although in principle they could be represented by ordinary DL axioms. In particular, we often use the following translation of correspondences to weighted ground predicates of the Markov logic network

$$(e_1, e_2, R, c) \mapsto \langle map_R(e_1, e_2), c \rangle$$

where c is a a-priori confidence values.

### 3.3    MAP Inference and Integer Linear Programming

If we want to determine the most probable state of a MLN, we need to compute the set of ground atoms of the hidden predicates that maximizes the probability given both the ground atoms of observable predicates and all ground formulas. This is an instance of MAP (maximum a-posteriori) inference in the ground Markov logic network. Let $\mathbf{O}$ be the set of all ground atoms of observable predicates and $\mathbf{H}$ be the set of all ground atoms of hidden predicates both with respect to $C$. We make the closed world assumption with respect to the observable predicates. Assume that we are given a set $\mathbf{O}' \subseteq \mathbf{O}$ of ground atoms of observable predicates. In order to find the most probable state of the MLN we have to compute

$$\operatorname*{argmax}_{\mathbf{H}' \subseteq \mathbf{H}} \sum_{(F_i, w_i)} \sum_{G \in \mathcal{G}_{F_i}^C : \ \mathbf{O}' \cup \mathbf{H}' \models G} w_i.$$

Every $\mathbf{H}' \subseteq \mathbf{H}$ is called a *state*. It is the set of *active* ground atoms of hidden predicates. Markov logic is by definition a declarative language, separating the formulation of a problem instance from the algorithm used for probabilistic inference. MAP inference in Markov logic networks is essentially equivalent to the weighted MAX-SAT problem and, therefore, NP-hard. Integer linear programming (ILP) is an effective method for solving exact MAP inference in undirected graphical models [41,50] and specifically in Markov logic networks [40]. ILP is concerned with optimizing a linear objective function over a finite number of integer variables, subject to a set of linear constraints over these variables [43]. We omit the formal details of the ILP representation of a MAP problem and refer the reader to [40].

*Example 3.* Consider a small instance of the ontology alignment problem which involves both soft and hard formulas. ML was successfully applied to ontology matching problems in earlier work [34]. Let $\mathcal{O}_1$ and $\mathcal{O}_2$ be the two ontologies in Figure 3 with the (a-priori computed) string similarities between the concept labels given in Table 4. Let $S = (O, H, C)$ be the signature of a MLN $M$ with $O = \{sub_1, sub_2, dis_1, dis_2\}$, $H = \{map\}$, and $C = \{a_1, b_1, c_1, a_2, b_2\}$. Here, the observable predicates model the subsumption and disjointness relationships between concepts $C$ in the two ontologies and $map$ is the hidden predicate modeling the sought-after matching correspondences. We also assume that the predicates are typed meaning that, for instance, valid groundings of $map(x, y)$ are those with $x \in \{a_1, b_1, c_1\}$ and $y \in \{a_2, b_2\}$. Furthermore, let us assume that the MLN $M$ includes the following formula with weight $w = 10.0$:

$$\forall x, x', y, y' : dis_1(x, x') \wedge sub_2(y, y') \Rightarrow (\neg map(x, y) \vee \neg map(x', y'))$$

The formula makes those alignments less likely that match concepts $x$ with $y$ and $x'$ with $y'$ if $x$ is disjoint with $x'$ in the first ontology and $y'$ subsumes $y$ in the second. We also include cardinality formulas with weight 10.0 forcing alignments to be one-to-one and functional:

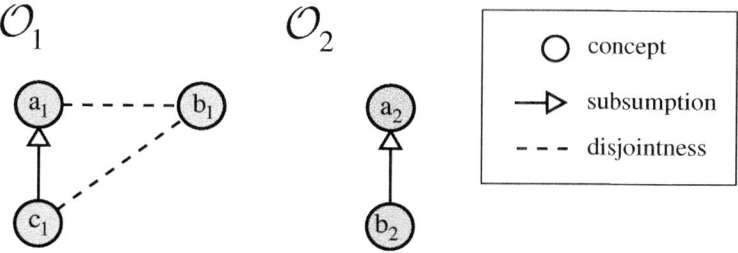

**Fig. 3.** Small fragments of two ontologies

**Table 4.** A-priori similarities between concept labels

|       | $a_1$ | $b_1$ | $c_1$ |
|-------|-------|-------|-------|
| $a_2$ | 0.95  | 0.25  | 0.12  |
| $b_2$ | 0.55  | 0.91  | 0.64  |

$$\forall x, y, z : map(x, y) \wedge map(x, z) \Rightarrow y = z$$
$$\forall x, y, z : map(x, y) \wedge map(z, y) \Rightarrow x = z$$

In addition, we add the formulas $map(x, y)$ with weight $\sigma(x, y)$ for all $x \in \{a_1, b_1, c_1, d_1\}$ and $y \in \{a_2, b_2\}$ where $\sigma(x, y)$ is the label similarity from Table 4. The observed ground atoms are $sub_1(c_1, a_1), dis_1(a_1, b_1), dis_1(b_1, a_1), dis_1(b_1, c_1), dis_1(c_1, b_1)$ for ontology $\mathcal{O}_1$ and $sub_2(b_2, a_2)$ for ontology $\mathcal{O}_2$. This results in the following relevant ground formulas for the coherence reducing constraint where each observable predicates has been substituted with its observed value:

$$\neg map(a_1, b_2) \vee \neg map(b_1, a_2) \tag{6}$$
$$\neg map(b_1, b_2) \vee \neg map(a_1, a_2) \tag{7}$$
$$\neg map(b_1, b_2) \vee \neg map(c_1, a_2) \tag{8}$$
$$\neg map(c_1, b_2) \vee \neg map(b_1, a_2) \tag{9}$$

For instance, the ground formulas (2) is encoded in an ILP by introducing a new binary variable $y$ which is added to the objective function with coefficient 10.0 and, in addition, by introducing the following linear constraints enforcing the value of $y$ to be equivalent to the truth value of the formula:

$$-x_{a,b} - y \leq -1$$
$$-x_{b,a} - y \leq -1$$
$$x_{a,b} + x_{b,a} + y \leq 2$$

The binary ILP variables $x_{a,b}$ and $x_{b,a}$ correspond to ground atoms $map(a_1, b_2)$ and $map(b_1, a_2)$, respectively. The ILP for our small example includes 19 variables (columns) and 39 linear constraints (12 from the coherence and 27 from the cardinality formulas) which we omit due to space considerations. The preprocessing step of grounding only those clauses that can evaluate to *false* given the current state of observable variables is similar to the approach presented in [44]. The ILP optimizations used for the inference procedures are not the focus of this article and we refer the reader to [40] and [33] for the details. However, in the following section we will show a typical matching formalization in Markov logic, the resulting ground formulas, and the corresponding integer linear program.

# 4   Markov Logic and Ontology Matching

We provide a formalization of the ontology matching problem within the probabilistic-logical framework. The presented approach has several advantages over existing methods such as ease of experimentation, incoherence mitigation during the alignment process, and the incorporation of a-priori confidence values. We show empirically that the approach is efficient and more accurate than existing matchers on an established ontology alignment benchmark dataset.

## 4.1   Problem Representation

Given two ontologies $\mathcal{O}_1$ and $\mathcal{O}_2$ and an initial a-priori similarity $\sigma$ we apply the following formalization. First, we introduce observable predicates $O$ to model the structure of $\mathcal{O}_1$ and $\mathcal{O}_2$ with respect to both concepts and properties. For the sake of simplicity we use uppercase letters $D, E, R$ to refer to individual concepts and properties in the ontologies and lowercase letters $d, e, r$ to refer to the corresponding constants in $C$. In particular, we add ground atoms of observable predicates to $\mathcal{F}^h$ for $i \in \{1, 2\}$ according to the following rules:

$$
\begin{aligned}
\mathcal{O}_i \models D \sqsubseteq E & \quad\longmapsto\quad & sub_i(d, e) \\
\mathcal{O}_i \models D \sqsubseteq \neg E & \quad\longmapsto\quad & dis_i(d, e) \\
\mathcal{O}_i \models \exists R.\top \sqsubseteq D & \quad\longmapsto\quad & sub_i^d(r, d) \\
\mathcal{O}_i \models \exists R^{-1}.\top \sqsubseteq D & \quad\longmapsto\quad & sub_i^r(r, d) \\
\mathcal{O}_i \models \exists R.\top \sqsupseteq D & \quad\longmapsto\quad & sup_i^d(r, d) \\
\mathcal{O}_i \models \exists R^{-1}.\top \sqsupseteq D & \quad\longmapsto\quad & sup_i^r(r, d) \\
\mathcal{O}_i \models \exists R.\top \sqsubseteq \neg D & \quad\longmapsto\quad & dis_i^d(r, d) \\
\mathcal{O}_i \models \exists R^{-1}.\top \sqsubseteq \neg D & \quad\longmapsto\quad & dis_i^r(r, d)
\end{aligned}
$$

The knowledge encoded in the ontologies is assumed to be true. Hence, the ground atoms of observable predicates are added to the set of hard constraints $\mathcal{F}^h$, making them hold in every computed alignment. The hidden predicates $map_c$ and $map_p$, on the other hand, model the sought-after concept and property correspondences, respectively. Given the state of the observable predicates, we are interested in determining the state of the hidden predicates that maximize the a-posteriori probability of the corresponding possible world. The ground atoms of these hidden predicates are assigned the weights specified by the a-priori similarity $\sigma$. The higher this value for a correspondence the more likely the correspondence is correct *a-priori*. Hence, the following ground formulas are added to $\mathcal{F}^s$, the set of soft formulas:

$$(map_c(c,d),\quad \sigma(C,D)) \qquad \text{if C and D are concepts}$$
$$(map_p(p,r),\quad \sigma(P,R)) \qquad \text{if P and R are properties}$$

Notice that the distinction between $m_c$ and $m_p$ is required since we use typed predicates and distinguish between the *concept* and *property* type.

**Cardinality Constraints.** A method often applied in real-world scenarios is the selection of a functional one-to-one alignment [7]. Within the ML framework, we can include a set of hard cardinality constraints, restricting the alignment to be functional and one-to-one. In the following we write $x, y, z$ to refer to variables ranging over the appropriately typed constants and omit the universal quantifiers.

$$map_c(x,y) \wedge map_c(x,z) \Rightarrow y = z$$
$$map_c(x,y) \wedge map_c(z,y) \Rightarrow x = z$$

Analogously, the same formulas can be included with hidden predicates $map_p$, restricting the property alignment to be one-to-one and functional.

**Coherence Constraints.** Incoherence occurs when axioms in ontologies lead to logical contradictions. Clearly, it is desirable to avoid incoherence during the alignment process. Some methods of incoherence removal for ontology alignments were introduced in [30]. All existing approaches, however, remove correspondences *after* the computation of the alignment. Within the ML framework we can incorporate incoherence reducing constraints *during* the alignment process for the first time. This is accomplished by adding formulas of the following type to $\mathcal{F}^h$, the set of hard formulas.

$$dis_1(x,x') \wedge sub_2(x,x') \Rightarrow \neg(map_c(x,y) \wedge map_c(x',y'))$$
$$dis_1^d(x,x') \wedge sub_2^d(y,y') \Rightarrow \neg(map_p(x,y) \wedge map_c(x',y'))$$

The second formula, for example, has the following purpose. Given properties $X, Y$ and concepts $X', Y'$. Suppose that $\mathcal{O}_1 \models \exists X.\top \sqsubseteq \neg X'$ and $\mathcal{O}_2 \models \exists Y.\top \sqsubseteq Y'$. Now, if $\langle X, Y, \equiv \rangle$ and $\langle X', Y', \equiv \rangle$ were both part of an alignment the merged ontology would entail both $\exists X.\top \sqsubseteq X'$ and $\exists X.\top \sqsubseteq \neg X'$ and, therefore, $\exists X.\top \sqsubseteq \bot$. The specified formula prevents this type of incoherence. It is known that such constraints, if carefully chosen, can avoid a majority of possible incoherences [29].

**Stability Constraints.** Several existing approaches to schema and ontology matching propagate alignment evidence derived from structural relationships between concepts and properties. These methods leverage the fact that existing evidence for the equivalence of concepts $C$ and $D$ also makes it more likely that, for example, child concepts of $C$ and child concepts of $D$ are equivalent. One such approach to evidence propagation is *similarity flooding* [31]. As a reciprocal idea, the general notion of stability was introduced, expressing that an alignment should not introduce new structural knowledge [28]. The *soft* formula below, for instance, decreases the probability of alignments that map concepts $X$ to $Y$ and $X'$ to $Y'$ if $X'$ subsumes $X$ but $Y'$ does *not* subsume $Y$.

$$\langle sub_1(x,x') \wedge \neg sub_2(y,y') \Rightarrow map_c(x,y) \wedge map_c(x',y'), \ w_1 \rangle$$
$$\langle sub_1^d(x,x') \wedge \neg sub_2^d(y,y') \Rightarrow map_p(x,y) \wedge map_c(x',y'), \ w_2 \rangle$$

Here, $w_1$ and $w_2$ are *negative* real-valued weights, rendering alignments that satisfy the formulas possible but less likely.

The presented list of cardinality, coherence, and stability constraints is by no means meant to be exhaustive. Other constraints could, for example, model known correct correspondences or generalize the one-to-one alignment to m-to-n alignments. Moreover, a novel hidden predicate could be added modeling correspondences between instances of the ontologies. To keep the discussion of the approach simple, however, we leave these considerations to future research.

*Example 4.* We apply the previous formalization to Example 1. To keep it simple, we only use a-priori values, cardinality, and coherence constraints. Given the two ontologies $\mathcal{O}_1$ and $\mathcal{O}_2$ in Figure 1, and the matching hypotheses (1) to (4) from Example 1, the ground MLN would include the following relevant ground formulas. We use the concept and property labels from Figure 1 and omit ground atoms of observable predicates.

**A-priori similarity**

$$\langle map_c(b_1,b_2), 0.88 \rangle, \langle map_c(c_1,e_2), 0.75 \rangle, \langle map_p(p_1,p_2), 0.7 \rangle, \langle map_c(d_1,e_2), 0.54 \rangle$$

**Cardinality constraints**

$$map_c(c_1,e_2) \wedge map_c(d_1,e_2) \Rightarrow c_1 = d_1 \tag{10}$$

**Coherence constraints**

$$dis_1^d(p_1,b_1) \wedge sub_2^d(p_2,b_2) \Rightarrow \neg(map_p(p_1,p_2) \wedge map_c(b_1,b_2)) \tag{11}$$
$$dis_1(b_1,c_1) \wedge sub_2(b_2,e_2) \Rightarrow \neg(map_c(b_1,b_2) \wedge map_c(c_1,e_2)) \tag{12}$$
$$sub_1^d(p_1,c_1) \wedge dis_2^d(p_2,e_2) \Rightarrow \neg(map_p(p_1,p_2) \wedge map_c(c_1,e_2)) \tag{13}$$

Let the binary ILP variables $x_1, x_2, x_3$, and $x_4$ model the ground atoms $map_c(b_1,b_2), map_c(c_1,e_2), map_p(p_1,p_2)$, and $map_c(d_1,e_2)$, respectively. The set of ground formulas is then encoded in the following integer linear program:

**Maximize:**   $0.88x_1 + 0.75x_2 + 0.7x_3 + 0.54x_4$

**Subject to**

$$x_2 + x_4 \leq 1 \tag{14}$$
$$x_1 + x_3 \leq 1 \tag{15}$$
$$x_1 + x_2 \leq 1 \tag{16}$$
$$x_2 + x_3 \leq 1 \tag{17}$$

The a-priori confidence values of the potential correspondences are factored in as coefficients of the objective function. Here, the ILP constraint (9) corresponds to ground formula (5), and ILP constraints (10),(11), and (12) correspond to the coherence ground formulas (6), (7), and (8), respectively. An optimal solution to the ILP consists of the variables $x_1$ and $x_4$ corresponding to the correct alignment $\{m_c(b_1, b_2), m_c(d_1, e_2)\}$. Compare this with the alignment $\{map_c(b_1, b_2), map_c(c_1, e_2), map_p(p_1, p_2)\}$ which would be the outcome without coherence constraints.

### 4.2   Experiments

We use the Ontofarm dataset [49] as basis for our experiments. It is the evaluation dataset for the OAEI conference track which consists of several ontologies modeling the domain of scientific conferences [11]. The ontologies were designed by different groups and, therefore, reflect different conceptualizations of the same domain. Reference alignments for seven of these ontologies are made available by the organizers. These 21 alignments contain correspondences between concepts and properties including a reasonable number of non-trivial cases. For the a-priori similarity $\sigma$ we decided to use a standard lexical similarity measure. After converting the concept and object property names to lowercase and removing delimiters and stop-words, we applied a string similarity measure based on the Levensthein distance. More sophisticated a-priori similarity measures could be used but since we want to evaluate the benefits of the ML framework we strive to avoid any bias related to custom-tailored similarity measures. We applied the reasoner Pellet [45] to create the ground MLN formulation and used The-Beast[3] [40] to convert the MLN formulations to the corresponding ILP instances. Finally, we applied the mixed integer programming solver SCIP[4] to solve the ILP. All experiments were conducted on a desktop PC with AMD Athlon Dual Core Processor 5400B with 2.6GHz and 1GB RAM. The software as well as additional experimental results are available at http://code.google.com/p/ml-match/.

The application of a threshold $\tau$ is a standard technique in ontology matching. Correspondences that match entities with high similarity are accepted while correspondences with a similarity less than $\tau$ are deemed incorrect. We evaluated our approach with thresholds on the a-priori similarity measure $\sigma$ ranging from 0.45 to 0.95. After applying the threshold $\tau$ we normalized the values to the range $[0.1, 1.0]$. For each pair of ontologies we computed the $F_1$-value, which is the harmonic mean of precision and recall, and computed the mean of this value over all 21 pairs of ontologies. We evaluated four different settings:

---

[3] http://code.google.com/p/thebeast/
[4] http://scip.zib.de/

- **ca**: The formulation includes only cardinality constraints.
- **ca+co**: The formulation includes only cardinality and coherence constraints.
- **ca+co+sm**: The formulation includes cardinality, coherence, and stability constraint, and the weights of the stability constraints are determined manually. Being able to set *qualitative* weights manually is crucial as training data is often unavailable. The employed stability constraints consist of (1) constraints that aim to guarantee the stability of the concept hierarchy, and (2) constraints that deal with the relation between concepts and property domain/range restrictions. We set the weights for the first group to $-0.5$ and the weights for the second group to $-0.25$. This is based on the consideration that subsumption axioms between concepts are specified by ontology engineers more often than domain and range restriction of properties [10]. Thus, a pair of two correct correspondences will less often violate constraints of the first type than constraints of the second type.
- **ca+co+sl**: The formulation also includes cardinality, coherence, and stability constraint, but the weights of the stability constraints are learned with a simple online learner using the perceptron rule. During learning we fixed the a-priori weights and learned only the weights for the stability formulas. We took 5 of the 7 ontologies and learned the weights on the 10 resulting pairs. With these weights we computed the alignment and its $F_1$-value for the remaining pair of ontologies. This was repeated for each of the 21 possible combinations to determine the mean of the $F_1$-values.

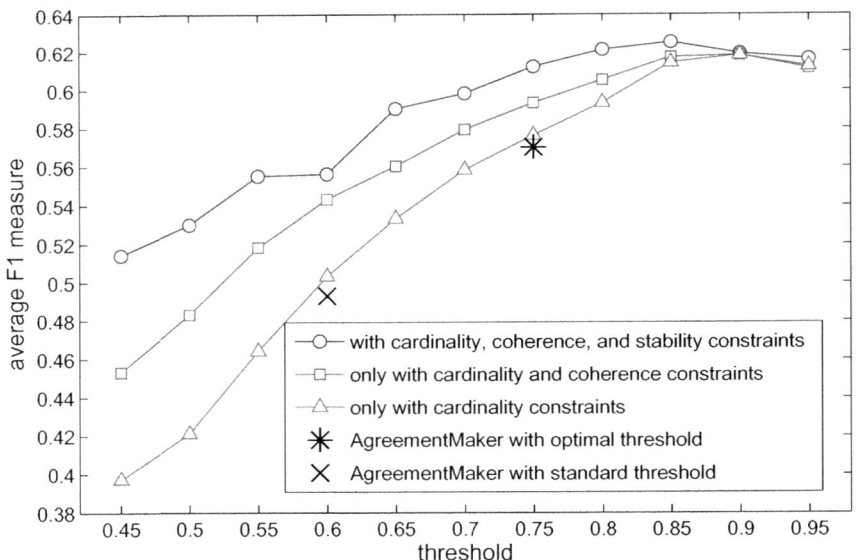

**Fig. 4.** $F_1$-values for **ca**, **ca+co**, and **ca+co+sm** averaged over the 21 OAEI reference alignments for thresholds ranging from 0.45 to 0.95. AgreementMaker was the best performing system on the conference dataset of the latest ontology evaluation initiative in 2009.

The lower the threshold the more complex the resulting ground MLN and the more time is needed to solve the corresponding ILP. The average time needed to compute one alignment was 61 seconds for $\tau = 0.45$ and 0.5 seconds for $\tau = 0.85$. Figure 4 depicts the average $F_1$-values for **ca**, **ca+co**, and **ca+co+sm** compared to the average $F_1$-values achieved by AgreementMaker [7], the best-performing system in the OAEI conference track of 2009. These average $F_1$-values of AgreementMaker were obtained using two different thresholds. The first is the default threshold of AgreementMaker and the second is the threshold at which the average $F_1$-value attains its maximum.

The inclusion of coherence constraints (**ca+co**) improves the average $F_1$-value of the alignments for low to moderate thresholds by up to 6% compared to the **ca** setting. With increasing thresholds this effect becomes weaker and is negligible for $\tau \geq 0.9$. This is the case because alignments generated with **ca** for thresholds $\geq 0.9$ contain only a small number of incorrect correspondences. The addition of stability constraints (**ca+co+sm**) increases the quality of the alignments again by up to 6% for low to moderate thresholds. In the optimal configuration (**ca+co+sl** with $\tau = 0.85$) we measured an average $F_1$-value of 0.63 which is a 7% improvement compared to AgreementMaker's 0.56. What is more important to understand, however, is that our approach generates more accurate results over a wide range of thresholds and is therefore more robust to threshold estimation. This is advantageous since in most real-world matching scenarios the estimation of appropriate thresholds is not possible. While the **ca** setting generates $F_1$-values $> 0.57$ for $\tau \geq 0.75$ the **ca+co+sm** setting generates $F_1$-values $> 0.59$ for $\tau \geq 0.65$. Even for $\tau = 0.45$, usually considered an inappropriate threshold choice, we measured an average $F_1$-value of 0.51 and average precision and recall values of 0.48 and 0.60, respectively. Table 5 compares the average $F_1$-values of the ML formulation (a) with manually set weights for the stability constraints, (b) with learned weights for the stability constraints, and (c) without any stability constraints. The values indicate that using stability constraints improves alignment quality with both learned and manually set weights.

**Table 5.** Average $F_1$-values over the 21 OAEI reference alignments for manual weights (ca+co+sm) vs. learned weights (ca+co+sl) vs. formulation without stability constraints (ca+co); thresholds range from 0.6 to 0.95.

| threshold | 0.6 | 0.65 | 0.7 | 0.75 | 0.8 | 0.85 | 0.9 | 0.95 |
|---|---|---|---|---|---|---|---|---|
| ca+co+sm | 0.56 | **0.59** | **0.60** | **0.61** | **0.62** | **0.63** | 0.62 | **0.62** |
| ca+co+sl | **0.57** | 0.58 | 0.58 | **0.61** | 0.61 | 0.61 | **0.63** | **0.62** |
| ca+co | 0.54 | 0.56 | 0.58 | 0.59 | 0.61 | 0.62 | 0.62 | 0.61 |

# 5   Markov Logic and Object Reconciliation

We are primarily concerned with the scenario where both A-Boxes are described in terms of the same T-Box. The presented approach does not rely on specific types of axioms or a set of predefined rules but on a well defined semantic similarity measure. In particular, our approach is based on the measure proposed by Stuckenschmidt [48]. This measure has originally been designed to quantify the similarity between two ontologies that describe the same set of objects. We apply a modified variant of this measure to evaluate the similarity of two A-Boxes described in terms of the same T-Box. Furthermore, our method factors in a-priori confidence values that quantify the degree of trust one has in the correctness of the object correspondences based on lexical properties. The resulting similarity measure is used to determine an instance alignment that induces the highest agreement of object assertions in $\mathcal{A}_1$ and $\mathcal{A}_2$ with respect to $\mathcal{T}$.

### 5.1    Problem Representation

The current instance matching configuration leverages terminological structure and combines it with lexical similarity measures. The approach is presented in more detail in [37]. The alignment system uses one T-Box $\mathcal{T}$ but two different A-Boxes $\mathcal{A}_1 \in \mathcal{O}_1$ and $\mathcal{A}_2 \in \mathcal{O}_2$. In cases with two different T-Boxes the T-Box matching approach is applied as a preprocessing step to merge the two aligned T-Boxes first. The approach offers complete conflict elimination meaning that the resulting alignment is always consistent for OWL DL ontologies. To enforce consistency, we need to add constraints to model conflicts, that is, we have to prevent an equivalence correspondence between two individuals if there exists a positive class assertion for the first individual and a negative for the second for the same class. These constraints are incorporated for both property and concept assertions. Analogous to the concept and property alignment before, we introduce the hidden predicate $map_i$ representing instance correspondences. Let $C$ be a concept and $P$ be a property of T-Box $\mathcal{T}$. Further, let $A \in \mathcal{A}_1$ and $B \in \mathcal{A}_2$ be individuals in the respective A-Boxes. Then, using a reasoner such as Pellet, ground atoms are added to the set of *hard* constraints $\mathcal{F}^h$ according to the following rules:

$$\mathcal{T} \cup \mathcal{A}_1 \models C(A) \qquad \wedge \; \mathcal{T} \cup \mathcal{A}_2 \models \neg C(B) \qquad \mapsto \quad \neg map_i(a,b)$$
$$\mathcal{T} \cup \mathcal{A}_1 \models \neg C(A) \qquad \wedge \; \mathcal{T} \cup \mathcal{A}_2 \models C(B) \qquad \mapsto \quad \neg map_i(a,b)$$
$$\mathcal{T} \cup \mathcal{A}_1 \models P(A,A') \qquad \wedge \; \mathcal{T} \cup \mathcal{A}_2 \models \neg P(B,B') \qquad \mapsto \quad \neg map_i(a,b) \vee \neg map_i(a',b')$$
$$\mathcal{T} \cup \mathcal{A}_1 \models \neg P(A,A') \qquad \wedge \; \mathcal{T} \cup \mathcal{A}_2 \models P(B,B') \qquad \mapsto \quad \neg map_i(a,b) \vee \neg map_i(a',b')$$

In addition to these formulas we included cardinality constraints analogous to those used in the previous concept and property alignment problem. In the instance matching formulation, the a-priori similarity $\sigma_\mathbf{c}$ and $\sigma_\mathbf{p}$ measures the *normalized overlap* of concept and property assertions, respectively. For more details on these measures, we refer the reader to [37]. The following formulas are added to the set of soft formulas $\mathcal{F}^s$:

$$\langle map_i(a,b), \; \sigma_\mathbf{c}(A,B) \rangle \qquad \qquad \text{if A and B are instances}$$
$$\langle map_i(a,b) \wedge map_i(c,d), \; \sigma_\mathbf{p}(A,B,C,D) \rangle \qquad \text{if A, B, C, and D are instances}$$

**Algorithm 1.** $\sigma(entity_1, entity_2)$

---

if $entity_1$ and $entity_2$ are either concepts or properties **then**
    $value \leftarrow 0$
    **for all** Values $s_1$ of URI, labels, and OBOtoOWL constructs in $entity_1$ **do**
        **for all** Values $s_2$ of URI, labels, and OBOtoOWL constructs in $entity_1$ **do**
            $value \leftarrow Max(value, sim(s_1, s_2))$
        **end for**
    **end for**
    **return** $value$
**end if**
if $entity_1$ and $entity_2$ are individuals **then**
    $Map\langle URI, double \rangle \; similarities \leftarrow null$
    **for all** dataproperties $dp_1$ of $entity_1$ **do**
        $uri_1 \leftarrow$ URI of $dp_1$
        **for all** dataproperties $dp_2$ of $entity_2$ **do**
            **if** $uri_1$ equals URI of $dp_2$ **then**
                $value \leftarrow sim(value of dp_1, value of dp_2)$
                **if** $uri_1$ is entailed in $similarities$ **then**
                    update entry $\langle uri_1, old\_value \rangle$ to $\langle uri_1, $ Minimum $(old\_value + value, 1) \rangle$
                    in $similarities$
                **else**
                    add new entry pair $\langle uri1, value \rangle$ in $similarities$
                **end if**
            **end if**
        **end for**
    **end for**
    **return** (sum of all values in $similarities$)/(length of $similarities$)
**end if**

---

## 5.2 Similarity Computation

Algorithm 1 was used for computing the a-priori similarity $\sigma(entity_1, entity_2)$. In the case of concept and property alignments, the a-priori similarity is computed by taking the maximal similarity between the URIs, labels and *OBO to OWL* constructs. In case of instance matching the algorithm goes through all data properties and takes the average of the similarity scores.

## 5.3 Experiments

The IIMB benchmark is a semi-automatically generated benchmark for instance matching. IIMB 2010 is created by extracting individuals from Freebase[5], an open knowledge base that contains information about 11 million real objects including movies, books, TV shows, celebrities, locations, companies and more. Data extraction has been performed using the query language JSON together with the Freebase JAVA API[6]. From this large dataset, 29 concepts, 20 object

---

[5] http://www.freebase.com/
[6] http://code.google.com/p/freebase-java/

properties, 12 data properties and a fraction of their underlying data have been chosen for the benchmark. The benchmark has been generated in a small version consisting of 363 individuals and in a large version containing 1416 individuals, respectively. Furthermore, the dataset consists of 80 different test cases divided into 4 sets of 20 test cases each. These sets have been designed according to the Semantic Web INstance Generation (SWING) approach presented in [16]. In the following, we will explain the SWING approach and its different transformation techniques resulting in the 80 different test cases in more detail.

*Data acquisition techniques.* SWING provides a set of techniques for the acquisition of data from the repositories of linked data and their representation as a reference OWL ABox. In SWING, we work on open repositories by addressing two main problems featuring this kind of data sources. First, we support the evaluation designer in defining a subset of data by choosing both the data categories of interest and the desired size of the benchmark. Second, in the data enrichment activity, we add semantics to the data acquired. In particular, we adopt specific ontology design patterns that drive the evaluation designer in defining a data description scheme capable of supporting the simulation of a wide spectrum of data heterogeneities. These techniques include

- adding super classes and super properties,
- converting attributes to class assertions,
- determining and adding new disjointness restrictions,
- enriching the ontology with additional inverse properties, and
- specifying additional domain and range restrictions.

*Data transformation techniques.* In the subsequent *data transformation* process the TBox is unchanged, while the ABox is modified in several ways by generating a set of new ABoxes, called *test cases*. Each test case is produced by transforming the individual descriptions in the reference ABox in new individual descriptions that are inserted in the test case at hand. The goal of transforming the original individuals is twofold: on one hand, we provide a simulated situation where data referred to the same objects are provided in different data sources; on the other hand, we generate a number of datasets with a variable degree of data quality and complexity.

The applied transformation techniques are categorized as followed:

- *Data value transformation* operations work on the concrete values of data properties and their datatypes when available. The output is a new concrete value. This category has been applied to the test cases 1-20 of the IIMB 2010 benchmark.
- *Data structure transformation* operations change the way data values are connected to individuals in the original ontology graph and change the type and number of properties associated with a given individual. They are implemented in the transformations 21-40 of the IIMB 2010 benchmark.
- *Data semantic transformation* operations are based on the idea of changing the way individuals are classified and described in the original ontology. This category was utilized in test cases 41-60.

**Table 6.** Results for the OAEI IIMB track for the small (large) dataset

| Transformations | 0-20 | 21-40 | 41-60 | 61-80 | overall |
|---|---|---|---|---|---|
| Precision | 0.99 (0.98) | 0.95 (0.94) | 0.96 (0.99) | 0.86 (0.86) | 0.94 (0.95) |
| Recall | 0.93 (0.87) | 0.83 (0.79) | 0.97 (0.99) | 0.54 (0.53) | 0.83 (0.80) |
| $F_1$-value | 0.96 (0.91) | 0.88 (0.85) | 0.97 (0.99) | 0.65 (0.63) | 0.87 (0.85) |

– *Combination* This fourth set is obtained by combining together the three kinds of transformations and constitute the last test cases 61-80 in IIMB.

*Data evaluation techniques.* Finally, in the *data evaluation* activity, we automatically create a ground-truth in form of a reference alignment for each test case. A reference alignment contains the correct correspondences (in some contexts called "links") between the individuals in the reference ABox and the corresponding transformed individuals in the test case. These mappings are what an instance matching application is expected to find between the original ABox and the test case.

**Results.** The results of our approach on the IIMB 2010 benchmark are summarized in Table 6. The first numbers are the results of the small IIMB dataset containing 363 individuals, while the numbers in brackets represent our results for the large IIMB benchmark consisting of 1416 individuals. When examining the differences between the small and the large dataset, we notice that the values are slightly better for the small dataset. The $F_1$-values for the first category of the large dataset decrease by 0.05 compared to the small one, for the second category the disparity is 0.03, respectively. The third and forth category both have 0.02 lower $F_1$-values for the large dataset compared to the small one.

Since the large dataset is slightly more challenging, we report the results compared to other matching systems over the large version. Figures 5 and 6 illustrate the results for all of the participating matching systems at OAEI. Our object reconciliation approach has been implemented in the combinatorial optimization for data integration (CODI) system [36]. Besides our CODI matching application, the systems ASMOV [22] and RiMOM [52] participated in this particular track of the OAEI. ASMOV uses a weighted average of measurements of similarity along different features of ontologies, and obtains a pre-alignment based on these measurements. It then applies a process of semantic verification to reduce the amount of semantic inconsistencies. RiMOM implements several different matching strategies which are defined based on different ontological information. For each individual matching task, RiMOM can automatically and dynamically combine multiple strategies to generate a composed matching result.

Figure 5 compares the matching results with respect to precision, recall, and $F_1$-value. In the first category (data transformation) the ASMOV and the RiMOM system having $F_1$-values of 0.98 and 1.00 outperformed CODI's $F_1$-value of 0.91. The reason for CODI's worse performance in this category is due to the naïve lexical similarity measures CODI applies as shown in Algorithm 1.

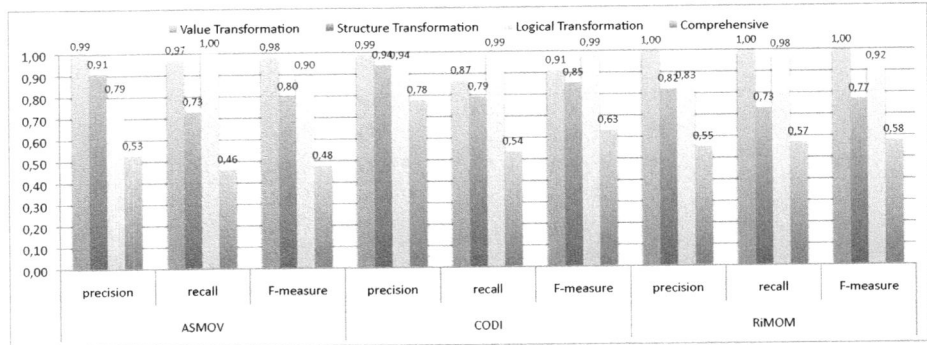

**Fig. 5.** Results for the large IIMB subtrack of the OAEI 2010

However, leveraging terminological structure for instance matching with Markov logic, like described in Section 5, leads to a significant improvement of CODI in the structure transformation category and the semantic transformation category. Our results compared to the ones of the ASMOV system are 5 per-cent higher in $F_1$-value for the structure transformation category and 9 per-cent in the semantic transformation category, respectively. The RiMOM system has 7 per-cent lower $F_1$-values in both the structure and the transformation category. In the last and most challenging category where all three transformation categories are combined, CODI achieved a $F_1$-value of 0.63 outperforming RiMOM (0.58) and ASMOV (0.48).

The precision and recall diagram in Figure 6 shows the aggregated values for recall on the x-axis and precision on the y-axis. For recall values ranging from 0.0 up to 0.6 the CODI system has the highest precision values compared to the ASMOV and RiMOM system. Only for recall values of 0.7 and higher, first the precision values of RiMOM (for recall values between 0.7 and 0.9) and then the precision values of ASMOV (for recall value 1.0) are higher.

Aggregated over all 80 test cases CODI reaches an $F_1$-value of 0.87 which is 5 per-cent higher than the result of ASMOV ($F_1$-value of 0.82) and 3 per-cent higher than RiMOM ($F_1$-value of 0.84)[7]. In summary, it is evident that utilizing the probabilistic-logical framework based on Markov logic for object reconciliation outperforms state-of-the-art instance matching systems.

## 6   Related Work

There have been a number of approaches for extending description logics with probabilistic information in the earlier days of description logics. Heinsohn [18] was one of the first to propose a probabilistic notion of subsumption for the logic ALC. Jaeger [21] investigated some general problems connected with the

---

[7] We refer the reader to http://www.instancematching.org/oaei/imei2010/iimbl.html for detailed results of every single test case and their aggregation.

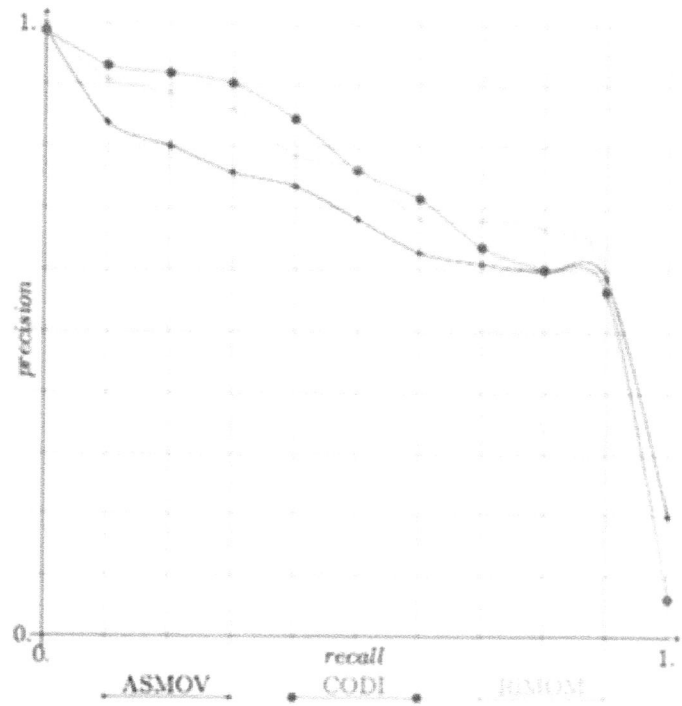

**Fig. 6.** Precision/recall of tools participating in the IIMB subtrack

extension of T-Boxes and ABoxes with objective and subjective probabilities and proposed a general method for reasoning with probabilistic information in terms of probability intervals attached to description logic axioms. Recently, Giugno and Lukasiewicz proposed a probabilistic extension of the logic SHOQ along the lines sketched by Jaeger [17]. A major advantage of this approach is the integrated treatment of probabilistic information about Conceptual and Instance knowledge based on the use of nominals in terminological axioms that can be used to model uncertain information about instances and relations. An alternative way of combining description logics with probabilistic information has been proposed by Koller et al. [24]. In contrast to the approaches mentioned above, the P-CLASSIC approach is not based on probability intervals. Instead it uses a complete specification of the probability distribution in terms of a Bayesian network which nodes correspond to concept expressions in the CLASSIC description logic. Bayesian networks have also been used in connection with less expressive logics such as TDL [55]. The approaches for encoding probabilities in concept hierarchies using Bayesian networks described in the section preliminaries and background can be seen as a simple special case of these approaches.

More recently proposals for combining the web ontology language OWL with probabilistic information have been proposed. The first kind of approach implements a loose coupling of the underlying semantics of OWL and probabilistic models. In particular these approaches use OWL as a language for talking about probabilistic models. An example of this approach is the work of Yang and Calmet that propose a minimal OWL ontology for representing random variables and dependencies between random variables with the corresponding conditional probabilities [54]. This allows the user to write down probabilistic models that correspond to Bayesian networks as instances of the OntoBayes Ontology. The encoding of the model in OWL makes it possible to explicitly link random variables to elements of an OWL ontology, a tighter integration on the formal level, however, is missing. A similar approach is proposed by Costa and Laskey. They propose the PR-OWL model which is an OWL ontology for describing first order probabilistic models [5]. More specifically, the corresponding ontology models Multi-Entity Bayesian networks [25] that define probability distributions over first-order theories in a modular way. Similar to OntoBayes, there is no formal integration of the two representation paradigms as OWL is used for encoding the general structure of Multi-entity Bayesian networks on the meta-level. The second kind of approaches actually aims at enriching OWL ontologies with probabilistic information to support uncertain reasoning inside OWL ontologies. These approaches are comparable with the work on probabilistic extensions of description logics also presented in this section. A survey of the existing work reveals, however, that approaches that directly address OWL as an ontology language are less ambitious with respect to combining logical and probabilistic semantics that the work in the DL area. An example is the work of Holi and Hyvonnen [19] that describe a framework for representing uncertainty in simple classification hierarchies using Bayesian networks. A slightly more expressive approach called BayesOWL is proposed by Ding and others [9]. They also consider Boolean operators as well as disjointness and equivalence of OWL classes and present an approach for constructing a Bayesian network from class expressions over these constructs. An interesting feature of BayesOWL is some existing work on learning and representing uncertain alignments between different BayesOWL ontologies reported in [38]. An additional family of probabilistic logics are log-linear description logics [35] which integrate lightweight description logics and probabilistic log-linear models.

Probabilistic approaches to ontology matching based on undirected probabilistic graphical models have recently produced competitive matching results [1]. There are numerous other non-probabilistic approaches to ontology matching and to mention all of them would be beyond the scope of this article. We refer the reader to the systems participating in the OAEI [13] which are described in the respective papers. More prominent systems with a long history of OAEI participation are Falcon [20], Aroma [8], ASMOV [23], and AgreementMaker [6].

The commonly applied methods for object reconciliation include structure -based strategies as well as strategies to compute and aggregate value similarities. Under the notion of instance matching, similarities between instance labels and datatype properties are mostly used to compute confidence values for instance correspondences. Examples of this are realized in the systems RiMOM [56] and OKKAM [46]. Both systems particpated in the instance matching track of the Ontology Alignment Evaluation in 2009. Additional refinements are related to a distinction between different types of properties. The developers of RiMOM manually distinguish between *necessary* and *sufficient* datatype properties. The FBEM algorithm of the OKKAM project assigns higher weights to certain properties like names and IDs. In both cases, the employed methods focus on appropriate techniques to interpret and aggregate similarity scores based on a comparison of datatype property values. Another important source of evidence is the knowledge encoded in the T-Box. RiMOM, for example, first generates a terminological alignment between the T-Boxes $\mathcal{T}_1$ and $\mathcal{T}_2$ describing the A-Boxes $\mathcal{A}_1$ and $\mathcal{A}_2$, respectively. This alignment is then used as a filter and only correspondences that link instances of equivalent concepts are considered valid [56]. An object reconciliation method applicable to our setting was proposed in [42] where the authors combine logical with numerical methods. For logical reasons it is in some cases possible to preclude that two instances refer to the same object while in other cases the acceptance of one correspondence directly entails the acceptance of another. The authors extend this approach by modeling some of these dependencies into a similarity propagation framework. However, their approach requires a rich schema and assumes that properties are defined to be functional and/or inverse functional. Hence, the approach cannot be used effectively to exploit type information based on a concept hierarchy and is therefore not applicable in many web of data scenarios.

# 7    Conclusion

We introduced a declarative framework for web data integration based on Markov logic capturing a wide range of matching strategies. Since these strategies are expressed with a unified syntax and semantics we can isolate variations and empirically evaluate their impact. While we focused only on a small subset of possible alignment strategies the results are already quite promising. We have also successfully learned weights for soft formulas within the framework. In cases where training data is not available, weights set manually by experts still result in improved alignment quality.

We have demonstrated that both ontology matching and object reconciliation problems can be expressed in the framework. Due to the declarative nature of the approach numerous algorithms can be applied to compute the final alignments. Based on our experience, however, integer linear programming in combination with cutting plane inference and delayed column generation strategies are especially suitable since they guarantee that the hard formulas are not violated. The framework allows one to combine lexical a-priori similarities between matchable entities with the terminological knowledge encoded in the ontology. We argued that most state-of-the-art approaches for ontology and instance matching focus solely on ways to compute lexical similarities. These approaches are sometimes extended by a structural validation technique where class membership is used as a matching filter. However, even though useful in some scenarios, these methods are neither based on a well-defined theoretical framework nor generally applicable without adjustment. Contrary to this, our approach is grounded in a coherent theory and incorporates terminological knowledge during the matching process. Our experiments show that the resulting method is flexible enough to cope with difficult matching problems for which lexical similarity alone is not sufficient to ensure high-quality alignments.

**Acknowledgement.** We thank Alfino Ferrara for providing us the IIMB benchmark and for the initiative at `http://www.instancematching.org/`.

# References

1. Albagli, S., Ben-Eliyahu-Zohary, R., Shimony, S.E.: Markov network based ontology matching. In: Proceedings of the International Joint Conference on Artificial Intelligence, pp. 1884–1889 (2009)
2. Bechhofer, S., Horrocks, I., Turi, D.: The OWL instance store: System description. In: Nieuwenhuis, R. (ed.) CADE 2005. LNCS (LNAI), vol. 3632, pp. 177–181. Springer, Heidelberg (2005)
3. Bhattacharya, I., Getoor, L.: Entity resolution in graphs. In: Mining Graph Data, Wiley, Chichester (2006)
4. Borgida, A.: On the relative expressiveness of description logics and predicate logics. Artificial Intelligence 82(1-2), 353–367 (1996)
5. Costa, P.C.G., Laskey, K.B.: Pr-owl: A framework for probabilistic ontologies. In: Bennett, B., Fellbaum, C. (eds.) Proceedings of the International Conference on Formal Ontology in Information Systems (FOIS). Frontiers in Artificial Intelligence and Applications, pp. 237–249. IOS Press, Amsterdam (2006)
6. Cruz, I.F., Stroe, C., Caci, M., Caimi, F., Palmonari, M., Antonelli, F.P., Keles, U.C.: Using AgreementMaker to Align Ontologies for OAEI 2010. In: Proceedings of the 5th Workshop on Ontology Matching (2010)
7. Cruz, I., Palandri, F., Antonelli, Stroe, C.: Efficient selection of mappings and automatic quality-driven combination of matching methods. In: Proceedings of the ISWC 2009 Workshop on Ontology Matching (2009)
8. David, J., Guillet, F., Briand, H.: Matching directories and OWL ontologies with AROMA. In: Proceedings of the 15th Conference on Information and knowledge management (2006)
9. Ding, L., Kolari, P., Ding, Z., Avancha, S.: Bayesowl: Uncertainty modeling in semantic web ontologies. In: Ma, Z. (ed.) Soft Computing in Ontologies and Semantic Web, Springer, Heidelberg (2006)
10. Ding, L., Finin, T.W.: Characterizing the semantic web on the web. In: Cruz, I., Decker, S., Allemang, D., Preist, C., Schwabe, D., Mika, P., Uschold, M., Aroyo, L.M. (eds.) ISWC 2006. LNCS, vol. 4273, pp. 242–257. Springer, Heidelberg (2006)
11. Euzenat, J., Hollink, A.F.L., Joslyn, C., Malaisé, V., Meilicke, C., Pane, A.N.J., Scharffe, F., Shvaiko, P., Spiliopoulos, V., Stuckenschmidt, H., Sváb-Zamazal, O., Svátek, V., dos Santos, C.T., Vouros, G.: Results of the ontology alignment evaluation initiative 2009. In: Proceedings of the ISWC 2009 workshop on Ontology Matching (2009)

12. Euzenat, J., Shvaiko, P.: Ontology matching. Springer, Heidelberg (2007)
13. Euzenat, J., et al.: First Results of the Ontology Alignment Evaluation Initiative 2010. In: Proceedings of the 5th Workshop on Ontology Matching (2010)
14. Fellegi, I., Sunter, A.: A theory for record linkage. Journal of the American Statistical Association 64(328), 1183–1210 (1969)
15. Ferrara, A., Lorusso, D., Montanelli, S., Varese, G.: Towards a Benchmark for Instance Matching. In: Sheth, A.P., Staab, S., Dean, M., Paolucci, M., Maynard, D., Finin, T., Thirunarayan, K. (eds.) ISWC 2008. LNCS, vol. 5318, Springer, Heidelberg (2008)
16. Ferrara, A., Montanelli, S., Noessner, J., Stuckenschmidt, H.: Benchmarking Matching Applications on the Semantic Web. In: The Semantic Web: Research and Applications (2011)
17. Giugno, R., Lukasiewicz, T.: P-$\mathcal{SHOQ}(\mathbf{D})$: A probabilistic extension of $\mathcal{SHOQ}(\mathbf{D})$ for probabilistic ontologies in the semantic web. In: Flesca, S., Greco, S., Leone, N., Ianni, G. (eds.) JELIA 2002. LNCS (LNAI), vol. 2424, p. 86. Springer, Heidelberg (2002)
18. Heinsohn, J.: A hybrid approach for modeling uncertainty in terminological logics. In: Kruse, R., Siegel, P. (eds.) ECSQAU 1991 and ECSQARU 1991. LNCS, vol. 548, pp. 198–205. Springer, Heidelberg (1991)
19. Holi, M., Hyvönen, E.: Modeling uncertainty in semantic web taxonomies. In: Ma, Z. (ed.) Soft Computing in Ontologies and Semantic Web, Springer, Heidelberg (2006)
20. Hu, W., Chen, J., Cheng, G., Qu, Y.: ObjectCoref & Falcon-AO: Results for OAEI 2010. In: Proceedings of the 5th International Ontology Matching Workshop (2010)
21. Jaeger, M.: Probabilistic reasoning in terminological logics. In: Doyle, J., Sandewall, E., Torasso, P. (eds.) Proceedings of the 4th International Conference on Principles of Knowledge Representation and Reasoning, pp. 305–316. Morgan Kaufmann, San Francisco (1994)
22. Jean-Mary, Y.R., Shironoshita, E.P., Kabuka, M.R.: ASMOV: Results for OAEI 2010. Ontology Matching, 126 (2010)
23. Jean-Marya, Y.R., Patrick Shironoshitaa, E., Kabuka, M.R.: Ontology matching with semantic verification. Web Semantics 7(3) (2009)
24. Koller, D., Levy, A., Pfeffer, A.: P-classic: A tractable probabilistic description logic. In: Proceedings of the 14th AAAI Conference on Artificial Intelligence (AAAI 1997), pp. 390–397 (1997)
25. Laskey, K.B., Costa, P.C.G.: Of klingons and starships: Bayesian logic for the 23rd century. In: Proceedings of the 21st Conference in Uncertainty in Artificial Intelligence, pp. 346–353. AUAI Press (2005)
26. Levenshtein, V.I.: Binary codes capable of correcting deletions and insertions and reversals. In: Doklady Akademii Nauk SSSR, pp. 845–848 (1965)
27. Li, L., Horrocks, I.: A software framework for matchmaking based on semantic web technology. International Journal of Electronic Commerce 8(4), 39 (2004)
28. Meilicke, C., Stuckenschmidt, H.: Analyzing mapping extraction approaches. In: Proceedings of the Workshop on Ontology Matching, Busan, Korea (2007)
29. Meilicke, C., Stuckenschmidt, H.: An efficient method for computing alignment diagnoses. In: Proceedings of the International Conference on Web Reasoning and Rule Systems, Chantilly, Virginia, USA, pp. 182–196 (2009)
30. Meilicke, C., Tamilin, A., Stuckenschmidt, H.: Repairing ontology mappings. In: Proceedings of the Conference on Artificial Intelligence, Vancouver, Canada, pp. 1408–1413 (2007)

31. Melnik, S., Garcia-Molina, H., Rahm., E.: Similarity flooding: A versatile graph matching algorithm and its application to schema matching. In: Proceedings of ICDE, pp. 117–128 (2002)

32. Meza-Ruiz, I., Riedel, S.: Multilingual semantic role labelling with markov logic. In: Proceedings of the Conference on Computational Natural Language Learning, pp. 85–90 (2009)

33. Niepert, M.: A Delayed Column Generation Strategy for Exact k-Bounded MAP Inference in Markov Logic Networks. In: Proceedings of the 25th Conference on Uncertainty in Artificial Intelligence (2010)

34. Niepert, M., Meilicke, C., Stuckenschmidt, H.: A Probabilistic-Logical Framework for Ontology Matching. In: Proceedings of the 24th AAAI Conference on Artificial Intelligence (2010)

35. Niepert, M., Noessner, J., Stuckenschmidt, H.: Log-Linear Description Logics. In: Proceedings of the International Joint Conference on Artificial Intelligence (2011)

36. Noessner, J., Niepert, M.: CODI: Combinatorial Optimization for Data Integration–Results for OAEI 2010. In: Proceedings of the 5th Workshop on Ontology Matching (2010)

37. Noessner, J., Niepert, M., Meilicke, C., Stuckenschmidt, H.: Leveraging Terminological Structure for Object Reconciliation. In: The Semantic Web: Research and Applications, pp. 334–348 (2010)

38. Pan, R., Ding, Z., Yu, Y., Peng, Y.: A bayesian network approach to ontology mapping. In: Gil, Y., Motta, E., Benjamins, V.R., Musen, M.A. (eds.) ISWC 2005. LNCS, vol. 3729, pp. 563–577. Springer, Heidelberg (2005)

39. Richardson, M., Domingos, P.: Markov logic networks. Machine Learning 62(1-2) (2006)

40. Riedel, S.: Improving the accuracy and efficiency of map inference for markov logic. In: Proceedings of the Conference on Uncertainty in Artificial Intelligence (2008)

41. Roth, D., Yih, W.-t.: Integer linear programming inference for conditional random fields. In: Proceedings of ICML, pp. 736–743 (2005)

42. Saïs, F., Pernelle, N., Rousset, M.-C.: Combining a logical and a numerical method for data reconciliation. Journal on Data Semantics 12, 66–94 (2009)

43. Schrijver, A.: Theory of Linear and Integer Programming. Wiley, Chichester (1998)

44. Shavlik, J., Natarajan, S.: Speeding up inference in markov logic networks by preprocessing to reduce the size of the resulting grounded network. In: Proceedings of the 21st International Joint Conference on Artifical intelligence, pp. 1951–1956 (2009)

45. Sirin, E., Parsia, B., Grau, B.C., Kalyanpur, A., Katz, Y.: Pellet: a practical OWL-DL reasoner. Journal of Web Semantics 5(2), 51–53 (2007)

46. Stoermer, H., Rassadko, N.: Results of OKKAM feature based entity matching algorithm for instance matching contest of OAEI 2009. In: Proceedings of the ISWC 2009 Workshop on Ontology Matching (2009)

47. Stuckenschmidt, H., van Harmelen, F.: Information Sharing on the Semantic Web. Advanced Information and Knowledge Processing. Springer, Heidelberg (2005)

48. Stuckenschmidt, H.: A Semantic Similarity Measure for Ontology-Based Information. In: Andreasen, T., Yager, R.R., Bulskov, H., Christiansen, H., Larsen, H.L. (eds.) FQAS 2009. LNCS, vol. 5822, pp. 406–417. Springer, Heidelberg (2009)

49. Svab, O., Svatek, V., Berka, P., Rak, D., Tomasek, P.: Ontofarm: Towards an experimental collection of parallel ontologies. In: Poster Track of ISWC, Galway, Ireland (2005)

50. Taskar, B., Chatalbashev, V., Koller, D., Guestrin, C.: Learning structured prediction models: a large margin approach. In: Proceedings of ICML, pp. 896–903 (2005)
51. Tsarkov, D., Riazanov, A., Bechhofer, S., Horrocks, I.: Using vampire to reason with OWL. In: McIlraith, S.A., Plexousakis, D., van Harmelen, F. (eds.) ISWC 2004. LNCS, vol. 3298, pp. 471–485. Springer, Heidelberg (2004)
52. Wang, Z., Zhang, X., Hou, L., Zhao, Y., Li, J., Qi, Y., Tang, J.: RiMOM Results for OAEI 2010. Ontology Matching, 195 (2010)
53. Wu, F., Weld, D.S.: Automatically refining the wikipedia infobox ontology. In: Proceeding of the International World Wide Web Conference, pp. 635–644 (2008)
54. Yang, Y., Calmet, J.: Ontobayes: An ontology-driven uncertainty model. In: Proceedings of the International Conference on Computational Intelligence for Modelling, Control and Automation and International Conference on Intelligent Agents, Web Technologies and Internet Commerce (CIMCA-IAWTIC 2005), pp. 457–463 (2005)
55. Yelland, P.M.: An alternative combination of bayesian networks and description logics. In: Cohn, A., Giunchiglia, F., Selman, B. (eds.) Proceedings of of the 7th International Conference on Knowledge Representation (KR 2000), pp. 225–234. Morgan Kaufman, San Francisco (2002)
56. Zhang, X., Zhong, Q., Shi, F., Li, J., Tang, J.: RiMOM results for OAEI 2009. In: Proceedings of the ISWC 2009 workshop on ontology matching (2009)

# An Introduction to Constraint Programming and Combinatorial Optimisation

Barry O'Sullivan

Cork Constraint Computation Centre,
Department of Computer Science, University College Cork, Cork, Ireland
b.osullivan@4c.ucc.ie

**Abstract.** Computers play an increasingly important role in helping individuals and industries make decisions. For example they can help individuals make decisions about which products to purchase or industries make decisions about how best to manufacture these products. Constraint programming provides powerful support for decision-making; it is able to search quickly through an enormous space of choices, and infer the implications of those choices. This tutorial will teach attendees how to develop models of combinatorial problems and solve them using constraint programming, satisfiability and mixed integer programming techniques. The tutorial will make use of Numberjack, an open-source Python-based optimisation system developed at the Cork Constraint Computation Centre. The focus of the tutorial will be on various network design problems and optimisation challenges in the Web.

A. Polleres et al. (Eds.): Reasoning Web 2011, LNCS 6848, p. 534, 2011.

# Author Index